MW00514071

HISTORICAL STUDIES *in the*
SOCIETAL IMPACT
of SPACEFLIGHT

HISTORICAL STUDIES *in the* SOCIETAL IMPACT *of* SPACEFLIGHT

Steven J. Dick

Editor

National Aeronautics and Space Administration

Office of Communications
NASA History Program Office
Washington, DC
2015

NASA SP-2015-4803

Library of Congress Cataloging-in-Publication Data

Historical studies in the societal impact of spaceflight / Steven J. Dick, editor.
        p. cm. — (Societal impact series ; v. 3)
    Includes bibliographical references and index.
    "NASA SP-2015-4803."
1. Astronautics—Technology transfer—History. 2. Astronautics—Public
opinion—History. 3. Astronautics—Social aspects—History. 4. United
States. National Aeronautics and Space Administration. I. Dick, Steven J.
    TL865.H58 2010
    338.973'06--dc22
                                        2009030014

This publication is available as a free download at
http://www.nasa.gov/ebooks.

ISBN 978-1-62683-026-4

CONTENTS

INTRODUCTION vii

PART I   OPINION

CHAPTER 1. The Impact of Space Exploration on Public Opinions,
Attitudes, and Beliefs
William Sims Bainbridge. 1

PART II   SPINOFF?

CHAPTER 2. Societal Impact of NASA on Medical Technology
William Sims Bainbridge. 77

CHAPTER 3. NASA's Role in the Manufacture of Integrated Circuits
Andrew J. Butrica. 149

CHAPTER 4. NASA's Role in the Development of MEMS
(Microelectromechanical Systems)
Andrew J. Butrica. 251

PART III   THE WORLD AT LARGE

CHAPTER 5. Powering Space Exploration: U.S. Space Nuclear Power,
Public Perceptions, and Outer Planetary Probes
Roger D. Launius 331

CHAPTER 6. NASA and the Environment: An Evolving Relationship
W. Henry Lambright 383

CHAPTER 7. Societal Impacts of Applications Satellites
David J. Whalen. 427

CHAPTER 8.   Impacts of the Apollo Program on NASA, the Space
             Community, and Society
             Eligar Sadeh 491

CHAPTER 9.   An Astrosociological Perspective on the Societal Impact
             of Spaceflight
             Jim Pass 535

             ABOUT THE AUTHORS 577

             THE NASA HISTORY SERIES 583

             INDEX 599

Introduction

Following the publication in 2007 of the *Societal Impact of Spaceflight* volume in the NASA History series, the NASA History Division commissioned a series of more in-depth studies on specific subjects.[1] This volume presents those studies to scholars and the public, and represents what is hoped will be a continuing series in the effort to understand the mutual interaction of space exploration and society—part of a larger need to understand the relationship between science, technology, and society.

Emphasizing the importance of public attitudes toward space, the volume opens with sociologist William Sims Bainbridge's study of the impact of space exploration on public attitudes. Based on seven decades of questionnaire survey data, and combining historical and social science approaches, the chapter considers both changes in public opinion over time and key themes that have shaped public opinion. Because the study surveyed vast ranges and quantities of data, it uncovered a number of historical and social science questions that deserve more focused study in the future, integrating historical data and methodologies into statistical analysis of questionnaire survey data. Because NASA has entered a new era of space development, it is ever more important to understand changing public opinion in a historical context.

"Spinoff" is the first aspect that comes to mind for most people who think at all about the impact of space exploration, those technologies that are thought—wrongly or rightly—to have emanated from the space program. Part II consists of case studies of specific potential spinoffs and explicitly raises the difficult questions of what can be considered spinoff and how much of any particular claimed spinoff can be attributed to NASA—thus the interrogatory "Spinoff?" title for this section rather than the usual declarative "Spinoff." Though NASA claims many spinoffs and publishes an annual

---

1. Steven J. Dick and Roger D. Launius, eds., *Societal Impact of Spaceflight* (Washington, DC: NASA SP-2007-4801, 2007), available online at *http://history.nasa.gov/sp4801.pdf.*

Spinoff report,[2] it seldom parses its claims very finely. The three chapters in this part aim to do just that. Bainbridge's study of medical technology reinforces the judgment of social scientists who wrote 30 to 40 years ago that spinoffs are a problematic concept: they may not reflect the most important channels by which NASA contributes to scientific and technological progress, even if they do provide coherent stories to communicate with the general public about the history of space exploration.

Andrew J. Butrica tackles the oft-made claim that NASA played a major role in the early development and use of integrated circuits. In particular, he addresses a specific question: What was the role of NASA in improving the manufacture of integrated circuits during the Apollo era? Butrica finds that the answer is not so simple. In a second and related essay, he shows that another claim—that the multibillion-dollar industry known as MEMS (microelectromechanical systems) originated at NASA—was actually such a close collaboration with nearby Stanford University that this story is also much more complex than usually thought. Butrica's conclusions are also in accord with an earlier finding that even if a particular spinoff can be attributed in whole or in part to NASA, attribution to individuals is still more difficult. As James E. Tomayko found in writing his report on *Computers in Spaceflight: The NASA Experience*, "often in corporations and government agencies individual achievement is buried within the institution. NASA is no exception. It was exceedingly difficult to get people both in the [A]gency and in contractor organizations to identify who did what, or even to take personal credit where appropriate."[3] This reminds us that, for better or worse, we have come a long way from the lone figure working in a laboratory.

Part III encompasses a variety of diverse studies of NASA's impact on the world at large, ranging from the technology of radioisotope thermal generators and the public controversy over the use of these nuclear components in spacecraft (Roger D. Launius's chapter), to NASA and the environment (W. Henry Lambright's chapter), the impact of applications satellites (David J. Whalen's chapter), and the impact of the Apollo program (Eligar Sadeh's chapter). At another level, space exploration has spawned new disciplines— ranging from astrobiology and astrochemistry to astrogeology—and has enlarged the boundaries of age-old problems by contemplating such areas as

---

2.   Issues of *Spinoff* are available at *http://spinoff.nasa.gov/index.html* (accessed 20 April 2015).

3.   James E. Tomayko, *Computers in Spaceflight: The NASA Experience*, NASA CR-182505 (Washington, DC: NASA Scientific and Technical Information Division, 1988), available at *http://history.nasa.gov/computers/Compspace.html*, Preface.

astrotheology.[4] One of the least developed disciplines, but ripe for exploration, is astrosociology—the subject of the final chapter (by Jim Pass) in this volume. This section demonstrates that our entry into space has altered the intellectual landscape of the 20th and 21st centuries in ways large and small, broadening our horizons in ways we sometimes fail to recognize.

This volume is the third in the NASA History subseries on the societal impact of spaceflight and follows a book entitled *Cosmos and Culture: Cultural Evolution in a Cosmic Context.*[5] That volume makes clear, far beyond the scope of the present work, how much cosmos and culture have become intertwined in the human experience. NASA and other space agencies around the world have contributed much to our understanding of the universe, enriching cultural worldviews and revealing the potential for other cultures throughout the universe. Not a bad legacy for 50 years of activity beyond the Earth's atmosphere.

Steven J. Dick
Former NASA Chief Historian
Washington, DC
April 2015

---

4. On the history of astrobiology—the study of life in the universe—see Steven J. Dick and James E. Strick, *The Living Universe: NASA and the Development of Astrobiology* (New Brunswick, NJ: Rutgers University Press, 2004). On astrotheology, see Steven J. Dick, ed., *Many Worlds: The New Universe, Extraterrestrial Life, and the Theological Implications* (Philadelphia: Templeton Press, 2000).

5. Steven J. Dick and Mark Lupisella, eds., *Cosmos and Culture: Cultural Evolution in a Cosmic Context* (Washington, DC: NASA SP-2009-4802, 2009), online at *http://history.nasa.gov/SP-4802.pdf.*

Part I

OPINION

Chapter 1

THE IMPACT OF SPACE EXPLORATION ON PUBLIC OPINIONS, ATTITUDES, AND BELIEFS

William Sims Bainbridge

1. Introduction

Since July 1944, when a Gallup poll asked two questions indirectly related to the German V-2 rocket program, scores of major questionnaires have included items about space exploration. The end of the Space Shuttle era is a good time to survey the history of public understanding and enthusiasm, because there have already been several historical periods and episodes during which influences have differed. The aims of this project are to survey the full sweep of American questionnaire studies offering insights about the public impact of space exploration and to connect the findings solidly to concrete historical developments.

Surveys of public opinion serve at least three functions in modern society. First, they support democratic institutions by informing policymakers about the mood of the citizenry. Research by political scientist Alan Monroe showed that American public policy was largely consistent with the results of opinion polls on policy issues, although more consistent in the 1960s and 1970s than afterward.[1] Second, polls provide interesting stories for journalists, and the results of a poll are often treated as news themselves. Beginning in 1967, CBS began doing its own polls in association with the *New York Times*; NBC started polling in 1973; ABC followed suit in 1981; and CNN partnered with Gallup from 1992 until 2006.[2] Third, polls offer a wealth of data for social scientists interested in tracing trends or testing theories. This study expands the social-scientific function into a historical method for understanding the past. It is important to realize that different functions imply somewhat different questionnaire designs and analytic techniques.

---

1. Alan D. Monroe, "Public Opinion and Public Policy, 1980–1993," *Public Opinion Quarterly* 61, no. 1 (1998): 6–28.
2. Seymour Sudman, "The Network Polls: A Critical Review," *Public Opinion Quarterly* 47, no. 4 (1983): 490–496.

The 1936 U.S. presidential election was a watershed for opinion poll methodology, because a Gallup poll based on rigorous sampling procedures correctly predicted Roosevelt's victory, whereas a much more massive poll done by the *Literary Digest* following its traditional unsystematic methods incorrectly predicted a Landon victory.[3] Subsequent national polls, like many reported throughout this chapter, employed complex sampling procedures, partly based on the principle that large random samples reduce many of the biases that more convenient samples would introduce and partly based on quota sampling to make sure that groups in the population are properly represented. It is important to understand that there are two very different but equally valid traditions of questionnaire research in social science, and this project will use them both:

1. Opinion research, in political science and sociology, which attempts to use random samples of the general public and is usually limited to a small number of very simple space-related items, frequently as few as one, that can be understood by everybody.

2. Research on the clustering of beliefs, attitudes, and values, typically social-psychological (in psychology or sociology), which places less emphasis on random samples and employs a large number of questions, with many aimed at respondents who are better educated and more knowledgeable than average.

Random samples have two primary advantages. First, they are the best way of estimating population parameters—the percentage of the larger population that holds a particular attitude, or the fraction of registered voters who plan to vote for a particular candidate. Second, random sampling is required for some statistical procedures, notably estimates of statistical significance. Unfortunately, it has become increasingly difficult to achieve anything like a simple random sample in practice. Costs have forced survey researchers to use stratified or quota samples; increasing fractions of the public refuse to participate, and the changing nature of telephone service makes sampling by random digit-dialing extremely problematic. Another disadvantage is that high cost has limited the number of questions that can be included on any given topic.

Nonrandom samples were disparaged for many years in social science because they lacked the parameter-estimation and statistical-significance advantages of random samples, but the increasing problems with random sampling and the new opportunities for research over the Internet have muted this earlier criticism. The chief advantage of online polling is the cost-effectiveness

---

3. George Gallup, *The Sophisticated Poll Watcher's Guide* (Ephrata, PA: Science Press, 1976).

when the aim is to include many questions on a given topic. Multiple questions allow statistical analysis of reliability of measurement, construction of multi-item scales that measure a phenomenon more precisely than any one item could achieve, and the use of methods such as factor analysis to identify clusters of items or dimensions of meaning that reveal much about the conceptual structure of the topic. In the case of Web-based questionnaires, which may have very large numbers of respondents, the lack of a random sample can be compensated for to a great extent by exploring the impact of control variables and by conducting internal replication that compares results across subgroups among the respondents.

Different methodologies naturally connect to different kinds of theories and are best suited for addressing different types of questions. Random samples naturally fit the democratic ideals that each adult citizen should have a vote equal to every other, and that public policies should reflect the will of the citizenry. More specialized samples can be justified on the traditional anthropological basis that some individuals are especially well qualified to represent their culture or subculture. There is no need to decide between these different approaches, with their distinctive advantages and disadvantages, and here I present valuable results from opinion studies of many kinds.

Consider the two questions in the July 1944 Gallup poll: "A Swedish newspaperman says the Germans are now building robot bombs which can hit cities on our East Coast. Do you believe this is true? Do you think that in another twenty-five years such flying bombs will be able to cross the Atlantic Ocean?" At that point in the war, the V-1 "buzz bomb" cruise missile had just been introduced. The V-2 rocket had not yet made its appearance in war but was being tested over the North Sea, and Swedes had heard about it. In fact, German rocket engineers were working on early designs for a three-stage transatlantic rocket, what would have been the first ICBM if it had been completed, but the public knew nothing about it. Indeed, few respondents probably had a sound basis for answering the questions. However, their answers were not far off the mark, because only 20 percent felt the Germans were already building such a weapon, but 70 percent thought one would exist in a quarter century.[4] Thus a fundamental issue is how well informed the public was and whether it had an adequate basis for responding to a particular question.

In October 1947 a Gallup poll asked, "How long do you think it will be before man will be able to fly to the moon?" The largest group, 38 percent, said "never." Another 23 percent would not venture a guess, and 16 percent

---

4.   George H. Gallup, ed., *The Gallup Poll: Public Opinion 1935–1971* (New York: Random House, 1972), p. 456.

failed to answer. Only 21 percent mentioned a specific time, but the median guess of 20 to 29 years turned out to be on the mark.[5] A decade later, in October 1957, just days after the launch of Sputnik, Gallup asked, "How long do you think it will be before men in rockets will reach the moon?" This time 52 percent were able to answer, and the median answer, 12 years, turned out to be exactly right.[6] In December 1949 and again in January 1955 the Gallup poll asked, "In the next 50 years, do you think men in rockets will be able to reach the moon?" The percentage who said "yes" increased from 15 percent to 38 percent over this span of five years. The fact that these optimists turned out to be right does not prove their superior understanding of the technical challenges. In 1949, 63 percent believed that "trains and airplanes" would be run by atomic power in 50 years, and 88 percent believed that an "absolute cure for cancer" would be found in the second half of the 20th century.[7] Neither of these breakthroughs has in fact occurred.

In May 1961, the month after Yuri Gagarin became the first human to orbit Earth, 21 percent of Americans believed men in rockets would reach the Moon in fewer than five years, and another 13 percent said exactly five years. In contrast with these optimists, 4 percent said six to nine years, 14 percent said ten years, 6 percent said more than ten years, and 9 percent said "never." The remaining third said they did not know.[8] Thus, in the first four years of the space age, the fraction lacking an opinion dropped from 48 to 33 percent. In retrospect, questions such as these can help us to understand how poorly informed many people were at the beginning of space history and allow us to trace their developing awareness of the potential of space exploration as the years passed.

---

5.   Poll of about 1,500 American adults by Gallup, 24–29 October 1947, Roper Center USGALLUP.47-406.QKT11. Note: The websites and polling data referenced in this chapter were accessed early in 2007, and many may have changed or vanished since then. Note that some URLs default to a different page from the one where data for this study were originally found, and other pages were revised over time. In addition, researchers may find the "Wayback Machine" tool at *http://archive.org/index.php* helpful to locate previous versions of Web pages. Polling data was derived from the following sites: the Gallup Organization, *http://www.galluppoll.com/*; the Roper Center, *http://www.ropercenter. uconn.edu/*; the Odum Institute for Research in Social Science, *http://152.2.32.107/ odum/jsp/home.jsp*; the Computer-assisted Survey Methods Program, *http://sda.berkeley. edu/archive.htm*; and the Pew Research Center for the People and the Press, *http://people-press.org/*.

6.   Poll of 1,573 American adults by Gallup, 10–15 October 1957, Roper Center USGALLUP.57-590.Q005A.

7.   George Gallup, "Number Who Think Trip to Moon Possible Has Doubled in 5 Years," press release, 28 January 1955, American Institute of Public Opinion, Princeton, NJ.

8.   Poll of 1,545 American adults by Gallup, 17–22 May 1961, Roper Center USGALLUP.61-645.R003.

The way a poll frames a question can shape the answers it gets. For example, a December 2003 Gallup poll done for CNN and *USA Today* split respondents at random into two groups and gave them different questions about the space program. One group was asked, "Would you favor or oppose a new U.S. space program that would send astronauts to the moon?" A majority (53 percent) said they favored the idea. The other group was asked, "Would you favor or oppose the U.S. government spending billions of dollars to send astronauts to the moon?" With this wording, stressing the cost, a majority (67 percent) opposed the idea.[9] In understanding the results of historical polls, we need to keep this wording issue in mind, even when the differences are less striking. For example, in October 1965 a Harris poll asked, "If you had to choose, do you think it more important or less important to spend 4 billion a year on the space program than to spend it on reducing the national debt?" Given this trade-off, 54 percent said that they would prefer to reduce the national debt versus 46 percent who would continue the space program. In comparison with "another tax cut," the space program did slightly better, just 51 percent preferring the tax cut and 49 percent preferring the space program.[10] In 1982 a Yankelovich poll set the stage for a long-term trade-off question: "Some authorities have said that it is important that the country build more jails and/or increase their capacity if our prison system is to be more effective in curbing crime. This will require the use of tax dollars. Do you think it is more important to use tax money to build and expand our prisons, or more important to spend the money on the space program." Given the context of curbing crime, 61 percent would have expanded prisons, and only 26 percent defended the space program.[11]

Even the attempt to avoid setting a context can have the effect of setting one. For example, here is how a Roper poll in February 1987 introduced a group of funding questions:

> Regardless of how you feel about the overall amount in the budget, you may think we should spend more or less on certain items. Here is a list of the major items in the budget. Would you go down that list and for each one tell me whether you think we should be spending more than President Reagan has proposed in his budget, or spending less than he has proposed, or that he has proposed spending about the right amount on it?

---

9. A CNN/USA Today/Gallup poll, 5–7 December 2003, reported under "Science and Nature," *http://www.PollingReport.com.*

10. Poll of 1,250 American adults by Harris, October 1965, Roper Center USHARRIS. 110165.R3E, USHARRIS.110165.R3I.

11. Poll of 1,010 American adults by Yankelovich, Skelly, & White, 8–10 June 1982, Roper Center USYANK.828611.R34A.

The pollster says to ignore the overall amount in the budget but then mentions President Reagan, thus possibly contaminating results with attitudes toward him. With this introduction, just 17 percent said we should be spending more on the space program; 34 percent said Reagan had proposed the right amount; 42 percent felt we should be spending less; and 8 percent did not know.[12]

However, the wording of the question does not overwhelm respondents when the issue is clear and they actually have an opinion. For example, in 1987 the General Social Survey (GSS) introduced a battery of funding items thus: "We are faced with many problems in this country, none of which can be solved easily or inexpensively. I'm going to name some of these problems, and for each one I'd like you to tell me whether you think we're spending too much money on it, too little money, or about the right amount." In response, 16 percent said we were spending too little on the "space exploration program," thus wanting us to spend more. The group who felt the right amount was being spent made up 38 percent; 41 percent of respondents felt we were spending too much and thus wanted us to spend less; and 6 percent were not sure.[13] Despite the very different wordings, these Roper and GSS polls are actually very close in their results: 17 percent versus 16, 34 percent versus 38, and 42 percent versus 41.

This chapter uses data from public opinion polls and other questionnaire studies to examine the impact of the space program on the American public, considering such issues as how much the public has supported the program, what the program has meant to people, and how it has affected people's thinking. I consider public reaction to historically significant events, such as the first Moon landing and the two Shuttle catastrophes, and I assess variations in support for space exploration across subgroups in the population, including changing patterns over time. I see evidence that space exploration promotes public interest in science and technology more generally, potentially deepening popular understanding of the infinite universe we live in. This is a historical study, but the space program has always concerned the future. Therefore, I examine images of the future of the space program as held at different times by the general public, such as levels of support for the Moon and Mars programs.

This study draws not only on academic publications such as articles in journals like *Public Opinion Quarterly* and press releases from polling

---

12. Poll of 1,996 American adults by Roper, 14–28 February 1987, Roper Center USROPER.87-3.R11G.

13. Computer-assisted Survey Methods Program (CSM) at the University of California, Berkeley, *http://sda.berkeley.edu/index.htm.*

organizations; it also freshly analyzes the raw data from about two dozen historical polls. I am therefore greatly indebted to the following archives: the Gallup Organization in Princeton, New Jersey; the Roper Center at the University of Connecticut; the Odum Institute for Research in Social Science at the University of North Carolina–Chapel Hill; the Computer-assisted Survey Methods (CSM) Program at the University of California, Berkeley; and the Pew Research Center for the People and the Press in Washington, DC.[14] The marvelous data contained in these archives will continue to be of great value for researchers who wish to look even more deeply into issues of public opinion and space exploration than I am able to accomplish here.

2. From Sputnik Through Apollo

When the Soviet Union launched Sputnik into orbit on 4 October 1957, the American public was not very well prepared to understand the meaning of the event. To be sure, Americans who concluded that the Soviets possessed advanced capabilities to build long-range military rockets were correct, but those who assumed the USSR was ahead of the United States in this area were wrong. Those who concluded that American science, technology, or national education system was inferior were also wrong, although this belief strengthened those institutions by channeling funding in their direction over the subsequent years. Rather, political decisions had given higher priority to perfecting ICBMs outside public view and to preparing a low-cost scientific satellite project called Vanguard. In 1956, Wernher von Braun's team working for the U.S. Army could have launched a satellite and did so on short notice after Sputnik, but his group was politically tainted for having earlier developed the V-2 rocket for Nazi Germany.

Arguably, the United States could have launched a satellite as early as 1950 if it had possessed the will to do so. On 24 February 1949, the United States had launched an American-designed WAC Corporal rocket to an altitude of 250 miles, using a captured German V-2 as a booster, although a much larger booster would have been required to achieve orbit. Work to develop the Atlas ICBM began in 1946 but was halted a year later and not resumed until the mid-1950s because the aircraft of the Strategic Air Command were given the long-range bombing role.[15] A balanced understanding of the historical period would require considerable knowledge of technical and political

---

14. See *http://www.galluppoll.com/*, *http://www.ropercenter.uconn.edu/*, *http://152.2.32.107/ odum/jsp/home.jsp*, *http://sda.berkeley.edu/archive.htm*, *http://people-press.org/*.

15. William Sims Bainbridge, *The Spaceflight Revolution* (New York: Wiley-Interscience, 1976).

factors, including information not available to the public at the time. The euphoria after victory in World War II was short-lived, followed by the Berlin crisis, the Korean War, and a Cold War in which the very future of democracy was in question. In such a context Sputnik caused great shock.

In October 1957, Gallup interviewers asked a nationwide sample of Americans an open-ended question about why they thought the Russians were able to launch a satellite before the United States did. Four main kinds of answer, in order from most common to least common, were: 1) They worked harder; 2) They had better scientists, including Germans; 3) The U.S. program was badly organized; and 4) Russia invested more money.[16] Late in the next month, a Gallup poll asked another open-ended question: "Where, specifically, would you put the blame, if anywhere, for letting the Russians get ahead of us in developing rockets and missiles?" Only 3.8 percent of respondents rejected the assumption in the question that Russia was ahead, although at the time the United States probably had a lead in most areas of military rocketry. Other respondents cast blame in many directions, including President Eisenhower (5.4 percent), the administration more generally (3.8 percent), unnamed government leaders (3.6 percent), Congress (1.6 percent), the budget (4.7 percent), the Defense Department (2.7 percent), interservice rivalry (5.0 percent), earlier administrations (1.9 percent), all Americans (2.5 percent), "our complacency, smugness, cocksureness, neglectfulness" (4.8 percent), Russian espionage (4.3 percent), not enough good scientists (1.3 percent), the failure of the United States to get the best German scientists (1.3 percent), not enough emphasis on rockets (2.8 percent), and "inadequate educational preparation for science" (4.9 percent).[17]

A Gallup press release dated 18 December 1957 claims that 4.1 million Americans had seen either Sputnik or Sputnik II as it passed overhead, on the basis that 4 percent of respondents to a poll had done so.[18] This is a rather extreme extrapolation from limited data, because of 1,505 people polled by Gallup, just 38 claimed to have seen Sputnik, and 28 said they saw Sputnik II.

James Swinehart and Jack McLeod compared a survey on science awareness administered to 1,919 Americans six months before Sputnik with a similar poll of 1,547 done six months afterward.[19] Before Sputnik, 54 percent

---

16. George Gallup, "Russia First with Satellite by 'Harder Work'—U.S. Public," press release, 24 October 1957, American Institute of Public Opinion, Princeton, NJ.

17. Poll of 1,499 American adults by Gallup, Gallup poll #592, 25 November 1957, Gallup Organization, Princeton, NJ.

18. Press release, 18 December 1957, American Institute of Public Opinion, Princeton, NJ; I myself saw Sputnik II.

19. James W. Swinehart and Jack M. McLeod, "News About Science: Channels, Audiences, and Effects," *Public Opinion Quarterly* 24, no. 4 (1960): 583–589.

claimed to have heard nothing about Earth satellites, while afterward this level of complete ignorance had dropped to only 8 percent. However, the fraction possessing detailed scientific information had not increased, remaining at 11 or 12 percent. Instead, those having only very general information about satellites had risen from 8 percent to 16 percent; awareness that competition between the United States and Russia was involved had increased from 1 percent to 20 percent; and a sense that satellites were connected to unspecified future possibilities went from 0 percent to 17 percent. In both polls the remainder of 25–27 percent possessed misinformation or were only vaguely aware that something had happened. A number of polls found that only small fractions of the public, often as low as 10 percent, understood how the balance between gravity and "centrifugal force" kept a satellite in orbit.[20] Thus the news about Sputnik seemed to have alerted many people to the topic but not to have informed them very deeply about it.

Serena Wade and Wilbur Schramm analyzed the same poll data to compare how well informed people were who got their news from different sources. Prior to Sputnik, only 10 percent of those who got their news from radio had some kind of science information about purposes and possibilities, whether detailed or not. The percentages were larger for those who got their news from television (16 percent had some science information), newspapers (22 percent), and magazines (38 percent). A year later, the fraction with at least some science information about satellites was greater for all four groups: radio (19 percent), television (25 percent), newspapers (34 percent), and magazines (47 percent).[21]

The chief meaning of the Sputniks for public opinion was announcing that the Soviet Union was technologically more capable than many people had realized and tilting the international prestige competition in its favor. In a comprehensive review article published in 1960, Gabriel Almond reports that polls in many nations demonstrated that large majorities knew about the launch. In Norway, 97 percent had heard about a satellite, and 94 percent knew it was Russian. In descending order, here are the percentages of people in various countries knowing the satellite was Russian: France (93 percent), Austria (92 percent), Belgium (91 percent), Germany (90 percent), Italy (88 percent), Canada (83 percent), Japan (78 percent), Britain (73 percent), Mexico (67 percent), and Brazil (51 percent). Polls from four nations, carried out both a month after the Sputnik launch and again a year later, supported

---

20. Donald N. Michael, "The Beginning of the Space Age and American Public Opinion," *Public Opinion Quarterly* 24, no. 4 (1960): 573–582.
21. Serena Wade and Wilbur Schramm, "The Mass Media as Sources of Public Affairs, Science, and Health Knowledge," *Public Opinion Quarterly* 33, no. 2 (1969): 197–209.

Almond's argument that the impact on public perceptions of the United States and Russia was great.[22]

Respondents in France, Great Britain, Italy, and West Germany were asked, "All things considered, do you think the U.S. or Russia is ahead in scientific development at the present time?" They were also asked which country was ahead "in total military strength." Table 1.1 reports results, leaving out those who volunteered that the two nations were equal or who expressed no opinion. A month after Sputnik, more people in Britain, France, and Italy believed Russia was ahead in science, and only West Germans gave the United States a slight edge. After a year, the United States had pulled ahead in three nations. On 31 January 1958, the United States launched its first satellite, Explorer 1; so at the time of the October 1958 polls, both the United States and Russia had proven they could launch spacecraft.

Domestically, something like a "Sputnik panic" energized the creation of NASA and provided ammunition for an existing social movement that wanted to improve American science education.[23] On the fortieth anniversary

TABLE 1.1. Prestige Competition Between the United States and Russia

Public Perception	In Scientific Discovery		In Military Strength	
	November 1957	October 1958	November 1957	October 1958
Great Britain				
United States Leads Russia	20%	43%	19%	26%
Russia Leads United States	58%	30%	50%	41%
West Germany				
United States Leads Russia	36%	44%	38%	24%
Russia Leads United States	32%	23%	23%	23%
France				
United States Leads Russia	11%	20%	17%	19%
Russia Leads United States	49%	34%	25%	28%
Italy				
United States Leads Russia	23%	33%	34%	38%
Russia Leads United States	37%	30%	22%	23%

22. Gabriel A. Almond, "Public Opinion and the Development of Space Technology," *Public Opinion Quarterly* 24, no. 4 (1960): 553–572.

23. Thomas N. Bonner, "Sputniks and the Educational Crisis in America," Journal of Higher Education 29, no. 4 (1958): 177–184, 232; and Clarence B. Hilberry, "Sputnik and the Universities," *Journal of Higher Education* 29, no. 7 (1958): 375–380.

of Sputnik, the Center for Science, Mathematics, and Engineering Education held a symposium at the National Academy of Sciences devoted to the satellite's enduring impact on educational reform. Participants found much to criticize, notably the unrealistic expectations for "the new math" that had tried to get schoolchildren to think abstractly like professional mathematicians; but on balance they agreed that the American educational system had improved its teaching of science in the wake of Sputnik. Several of the written papers noted that leading educators had been promoting reform since the end of the Second World War, and Sputnik was a useful tool to garner public support for their efforts.[24]

The competition between the USSR and the United States came to be called the Space Race.[25] At the beginning of 1960, 44 percent of Americans responding to a Gallup poll thought Russia "will be first to send a man into outer space," versus 34 percent who thought the United States would be first and 22 percent with no opinion.[26] A year later, 40 percent thought Russia would be first; 35 percent nominated the United States; and 25 percent had no opinion. On the premise that better-educated people had more solidly grounded opinions, Gallup reported that fully 54 percent of college-educated Americans thought Russia would win this stage of the space race.[27]

A poll carried out in Britain by the United States Information Agency (USIA) in late April 1960 showed how inaccurate popular impressions can be. More than a month earlier, the United States had launched the space probe Pioneer 5, which achieved solar orbit. Britain's Jodrell Bank radio telescope had picked up the weak signal from Pioneer 5, and this fact had been publicized. The poll asked, "What would be your best guess as to how many of the space satellites still in the sky are American, and how many are Russian?" While 28 percent of respondents had no opinion, and 25 percent thought the numbers were about equal, only 17 percent thought America had more satellites still in the sky, compared with 30 percent who thought Russia had more. In fact, a total of 11 satellites were still in Earth's orbit or solar

24. "Reflecting on Sputnik: Linking the Past, Present, and Future of Educational Reform," symposium held 4 October 1997, at the National Academy of Sciences, Washington, DC; papers available at *http://www.nas.edu/sputnik/agenda.htm*.

25. Klaus Knorr, "On the International Implications of Outer Space," *World Politics* 12, no. 4 (1960): 564–584; and Lincoln P. Bloomfield, "Outer Space and International Cooperation," *International Organization* 19, no. 3 (1965): 603–621.

26. "Man into Space?," press release, 16 January 1960, Public Opinion News Service, Princeton, NJ.

27. "Public Expects Russia To Be First To Put Man into Space," press release, 22 January 1961, American Institute of Public Opinion, Princeton, NJ.

orbit, 10 American and only 1 Russian.[28] In 1961, Americans were evenly split on whether the United States or Russia "is further ahead in the field of space research"—38 percent versus 38 percent, with 24 percent holding no opinion. But by June 1965 the United States had pulled ahead to 47 percent versus 24 percent saying Russia was ahead and 29 percent with no opinion.[29] When asked, "How important do you think it is for the United States to be ahead of Russia in space exploration," 51 percent of Americans said "very important." Another 21 percent said "fairly important," while 23 percent said "not too important," and 5 percent were not sure.[30]

Well-known aerospace historian Roger Launius has noted that today many aerospace professionals and fans of the space program wrongly believe that public support was strong during the heroic days of Apollo, marshaling much evidence to prove that this was not in fact the case.[31] Back in 1969, Raymond A. Bauer wrote: "At no point have any poll data indicated strong general support for the space program."[32] Depending on how one defines "strong," this statement may be slightly too categorical, but it is certainly the case that the majority of people never demanded an aggressive program of space exploration. At the end of May 1961, a Gallup press release reported, "Kennedy Must Convince Public of Value of Moon Shot Project," because 58 percent of Americans did not want the estimated $40 billion spent on this, compared with 33 percent who did.[33] In January 1962, 22 percent of Americans believed there was a "great and urgent need for action" to "land an American astronaut on the moon." Another 30 percent saw "some need," meaning that a slim majority of 52 percent saw a need to go to the Moon. In contrast, 42 percent saw "little or no" need.[34] By March 1963 these numbers had changed only slightly, to 53 percent against 42 percent.[35] In November

---

28. Harold Leland Goodwin, *The Images of Space* (New York: Holt, Rinehart and Winston, 1965), p. 112.

29. Press release, 23 July 1965, American Institute of Public Opinion, Princeton, NJ; the poll was administered 7–12 November 1957.

30. Poll of 1,625 American adults by Gallup, 23–28 June 1961, Roper Center USGALLUP.61-647.R029.

31. Roger D. Launius, "Public Opinion Polls and Perceptions of U.S. Human Spaceflight," *Space Policy* 19 (2003): 163–175.

32. Raymond A. Bauer, *Second-Order Consequences* (Cambridge, MA: MIT Press, 1969), p. 84.

33. "Kennedy Must Convince Public of Value of Moon Shot Project," press release, 31 May 1961, American Institute of Public Opinion, Princeton, NJ, based on a national poll of 1,447 adults.

34. Poll of 1,413 American adults by Opinion Research Corporation, Roper Center USORC.62APR.R09P.

35. Poll of 1,000 American adults by Opinion Research Corporation, 15 March–15 April 1963, Roper Center USORC.63AUG.R14P.

1962, 41.6 percent agreed "with President Kennedy's objective of putting a man in space on the moon by 1970," but 42.0 percent disagreed, with 16.4 percent "not sure."[36]

Alan Shepard's suborbital Mercury flight of 5 May 1961 helped Americans feel their nation was catching up to the Soviet Union in space and even pulling ahead in the related field of missiles. Gallup included this question in four polls: "Which country—the United States or Russia—do you think is farther ahead in the field of long-range missiles and rockets?" In October 1958, 40 percent thought Russia was ahead, and this view strengthened to 47 percent in February 1960 before falling to 30 percent in February 1961 and 20 percent immediately after Shepard's flight. At the same points in time, the fractions thinking the United States was ahead were 37, 33, 49, and 54 percent.[37]

In October 1964, Americans were asked, "Do you think the U.S. should go all out to beat the Russians in a manned-flight to the moon—or don't you think this is too important?" Only 26 percent wanted an all-out effort, compared with 66 percent who felt beating the Russians was "not too important" and 8 percent who did not know.[38] By February 1967, 33 percent of Americans had come to believe it was "important to send a man to the moon before Russia does," compared with 61 percent who felt it was not important.[39]

On 18 March 1965 the cosmonaut Aleksei Leonov exited the Voskhod 2 spacecraft while in orbit, accomplishing the very first spacewalk. A Harris poll done later that month reminded respondents, "the Russians recently sent two men into space, and one of them left the space ship and floated in outer space for ten minutes," then asked which of four different feelings they experienced. The event made no impression "one way or the other" on 38.0 percent, and 28.1 percent said, "I was proud that man had taken a major step in conquering space." Another 17.6 percent were "concerned because it showed how far ahead of the United States the Russians are," and 11.4 percent "didn't really believe they did it."[40] In June 1965 a Gallup poll asked, "Would you like to see the amount of money being spent on space exploration increased,

---

36. Analysis of original data from Harris study no. 1285, a November 1963 poll of 1,283 American adults, Odom Institute.

37. "U.S. Seen Leading in Missile Race in Poll Since Space Shots," press release, 7 June 1961, American Institute of Public Opinion, Princeton, NJ.

38. Poll of 1,564 American adults by Gallup, October 1964, Roper Center USGALLUP.637POS.Q17.

39. Poll of 2,344 American adults by Gallup, 16–21 February 1967, Roper Center USGALLUP.741.Q04.

40. Poll of 1,083 American adults by Harris, March 1965, Odom Institute, Harris 1522 Q12K.

decreased, or kept about the same as it is now?" Only 16 percent of respondents wanted funding increased, 42 percent were content to see it kept the same, 33 percent wanted it cut, and 9 percent expressed no opinion. This is a standard pattern found in poll after poll. Optimists may add the 16 percent calling for an increase with the 42 percent wanting current funding to continue and conclude that a majority of 58 percent supported the space program. Pessimists would argue that twice as many wanted funding cut as increased, so the opinion balance actually favors budget reductions.[41]

That same month, Harris poll respondents were asked, "Does it matter a lot to you that the Russians have been ahead of us in our space program?" Only 10.4 percent said it did. Another 17.4 percent said it mattered to them "some but not a lot," 16.0 percent said "only a little," and a majority of 54.4 percent said "not at all."[42] Respondents were happy with "the way our man in space program is being handled," with 38.6 percent rating it "excellent" and 41.7 percent "pretty good." However, this did not reflect great enthusiasm for the Apollo program. The poll went on to tell respondents: "It could cost the United States 40 billion dollars to get a man on the moon." With this in mind, 31.7 percent wanted "the moon shot program" cut out entirely, and 3.5 percent wanted it slowed down. The largest group, 37.9 percent, wanted the program kept as is, and only 18.1 percent wanted it speeded up.

Sometimes, after asking a fixed-choice question about support for the space program, national polls add an open-ended question asking respondents to explain their answer. Unfortunately, these verbal responses are seldom included in computer archives of the study. This poll was an exception. Those who wanted the space program speeded up mentioned justifications like "scientific knowledge," "more knowledge," "prestige abroad for the US," "should beat USSR," "a defense move because we are behind now," "do it well," and "get it over with." Those who wanted the program continued at its present rate mentioned "speeding it up costs more," "should spend money elsewhere," "safety is important," the "program works fine now," and a "good job is more important" than a quick one. Among the justifications other people gave for ending the program were such responses as: it is "impossible to live on the moon"; "biblical statements" and ideas like "if God wanted man" on the Moon, he would already be there; there are "enough problems

---

41. Press release, 23 July 1965, American Institute of Public Opinion, Princeton, NJ; poll of 2,534 American adults by Gallup, 24–29 June 1965, Roper Center USGALLUP.713. Q014A.

42. Analysis of original data from Harris study no. 1531, a poll done in June 1965 of 522 American adults, Odom Institute.

here on Earth"; "see no point in it"; "money should be spent elsewhere"; and "waste of money."[43]

In April 1967, only a few weeks after the tragic Apollo 1 fire that killed astronauts Gus Grissom, Roger Chaffee, and Ed White, the Harris poll included this question: "It could cost the United States $4 billion a year for the next 10 years to finally put a man on the moon and to explore other planets and outer space. All in all, do you feel the space program is worth spending that amount of money on or do you feel it isn't worth it?" Only 32.9 percent of respondents said it was worth it, fully 54.5 percent said it was not, and the remaining 12.6 percent were not sure. Then Harris asked a follow-up question: "If the Russians were not in space, and we were the only ones exploring space, would you favor or oppose continuing our space program at the present rate?" Of 383 respondents who felt the space program was worth it, 26.4 percent would oppose continuing at the present rate if the Russians were not in space, and another 3.7 percent were not sure. Of the entire group of 1,250 respondents, only 32.9 percent favored "the space project aim of landing a man on the moon," and 54.5 percent were opposed.[44]

Harris asked these same questions again in January 1969, the month after Apollo 8 successfully circled the Moon. Now, 30.4 percent felt that the cost of the program was worth it. This level of support after the success of Apollo 8 is only 4 percentage points higher than the level reported after the failure of Apollo 1. One big difference is the solidity of support among those who felt it was worth it. In 1967, 70.0 percent of this group supported continuing at the present rate even if the Russians were not in space, but in 1969 this loyal fraction was 90.7 percent. Only six months before the first actual Moon landing, support for "the space project aim of landing a man on the moon" had strengthened to 39.2 percent, but this was still less than the 48.6 percent that opposed the plan.[45]

Harris also asked respondents to consider a list of 11 government activities, including the space program, asking, "If one program had to be reduced, which one would you cut first?" Far and away the most common first choice for the budget ax was the space program, with 40.7 percent citing it. Remarkably, this was more than twice as many as the 18.4 percent who wanted to cut the second choice, "financing the war in Vietnam." The nine

---

43. Analysis of original data from Harris study no. 1531, a poll done in June 1965 of 522 American adults, Odom Institute.

44. Analysis of original data from Harris study no. 1718, a poll done in April 1967 of 1,250 American adults, Odom Institute.

45. Analysis of original data from Harris study no. 1877, a poll done in January 1969 of American adults with 1,560 responding to these questions, Odom Institute.

other options were far less likely to be cut: "welfare and relief" (9.5 percent), "building more highways" (8.6 percent), "subsidies for farmers" (6.5 percent), "anti-poverty program" (6.1 percent), "aid to cities" (5.1 percent), "anti-air and anti-water pollution programs" (1.7 percent), "Medicaid" (1.6 percent), "anti-crime and law enforcement programs" (1.1 percent), and "aid to education" (0.6 percent).

Back in December 1968, just as NASA was preparing to send Apollo 8 around the Moon and Richard Nixon was preparing to take over the presidency, a Harris poll asked respondents to react to ideas about six "specific areas where it has been suggested the U.S. military defense be strengthened." One idea was, "Convert the space program into a system of nuclear weapon space stations." Almost exactly a quarter, 25.1 percent, felt this ought to be done. The largest group, 48.6 percent, felt it should not be done, and the remaining 26.4 percent were not sure. Another of the ideas became important just 15 years later in the Reagan administration: "Build up a system of anti-missile defenses." Fully 60.1 percent felt this should be done, and only 23.2 percent were opposed.[46]

In June 1969, on the eve of the first lunar landing, Harris repeated its questions about support for the effort. Again, only a minority felt it was worth the $4-billion price tag, just 35.4 percent compared with 56.4 percent of respondents who were convinced the Apollo program was not worth it. However, Apollo 10 had just zoomed low over the lunar surface and practiced the maneuvers necessary for a landing, so many people may have felt that taking the next "small step" actually to get down to the surface was reasonable. At this point 50.1 percent favored "the space project aim of landing a man on the moon," and only 41.6 percent were still opposed.[47]

A Gallup poll begun the day the Apollo 11 astronauts returned from the first Moon landing in 1969 found that only 39 percent of Americans favored "attempting to land a man on the planet Mars," compared with 53 percent who opposed the idea.[48] About the same time, the Harris poll asked respondents whether they favored or opposed four possible "next steps for the U.S. space program. Only 31.5 percent favored "putting a permanent U.S. space station up in space with 100 men on it," and 50.7 percent opposed this idea. Opinions were similar, 34.2 percent in favor and 51.3 percent opposed, about

---

46. Analysis of original data from Harris study no. 1900, a December 1968 poll of 1,544 American adults, Odom Institute. The factor analyses employed pairwise deletion, calling for two factors.

47. Analysis of original data from Harris study no. 1936, a June 1969 poll of 1,589 American adults, Odom Institute.

48. Poll of 1,555 American adults by Gallup, 24–29 July 1969, Roper Center USGALLUP. 785.Q02.

"putting a scientific exploration station with 100 men in it on the moon." "Sending men to land on Mars" got a more positive response, 39.2 percent in favor compared with 49.1 percent opposed. And 38.0 percent were in favor of "making a permanent station in space a United Nations space station," whereas 48.3 percent were opposed. In each case more people opposed the option than favored it.[49]

In July 1999 the Gallup poll asked, "It is now thirty years since the United States first landed men on the moon. Do you think the space program has brought enough benefits to this country to justify its costs, or don't you think so?" While just 5 percent of respondents expressed no opinion, 55 percent said the space program had brought enough benefits, compared with 40 percent who said it had not.[50] Given the persistent claims of fringe groups that Apollo was a hoax, it is worth noting that only 6 percent of respondents believed "the government staged or faked the Apollo moon landing."[51] Fully 53 percent of those who were old enough said they had watched the Apollo 11 landing on television.[52] Later that same year, exactly half of American adults felt that "landing a man on the moon in 1969" was one of the most important events of the 20th century, a further 30 percent considered it "important but not the most important," and 15 percent considered it "somewhat important." Notably, only 5 percent considered the first Moon landing "not important."[53]

After the Moon flights that ended in December 1972, Apollo technology was used to loft the Skylab space station and send three crews to it in 1973 and 1974. The last Apollo flight came in 1975, with a symbolic mission to promote international cooperation and detente by linking up in orbit with a Soviet Soyuz spacecraft. By that time work had begun on the Space Shuttle program. Because the 1973–1975 flights were limited to Earth's orbit, they marked the beginning of a new and less aggressive phase in the history of space exploration, and for many observers the heroic early era of human spaceflight ended with the safe return of Apollo 17 on 19 December 1972.

---

49. Analysis of original data from Harris study no. 1944, a July 1969 poll of 1,601 American adults, Odom Institute.
50. Poll of 1,061 American adults by Gallup, 13–14 July 1999, Roper Center USGALLUP.072099.R2.
51. Poll of 1,061 American adults by Gallup, 13–14 July 1999, Roper Center USGALLUP.072099.R9.
52. Poll of 1,061 American adults by Gallup, 13–14 July 1999, Roper Center USGALLUP.072099.R6.
53. Poll of 1,011 American adults by Gallup, 4–7 November 1999, Roper Center USGALLUP.99NOM04.R21M.

3. Trends in Support After Apollo

Beginning in 1973, interviewers from the General Social Survey (GSS) have asked whether the funding level for the "space exploration program" was too much money, too little, or "about the right amount." Figure 1.1 graphs the percentage of respondents giving each of these responses through 2004, a span of more than three decades.

The Gallup poll provides a comparable set of data, sometimes agreeing with the GSS and sometimes disagreeing to a modest extent. For both polls, the low point in support since the 1970s came in 1993, when they estimated that only 9 percent of those polled wanted funding increased. Gallup's high marks were higher than those from the GSS, hitting 26 percent in 1989 and 24 percent in 2003. A Gallup poll carried out 23–25 June 2006 found that 17 percent of Americans wanted funding increased, 48 percent wanted it kept at current levels, and 33 percent wanted funding reduced or eliminated altogether.[54] Analyzing Gallup data, Mark Gillespie has noted that confidence in NASA reacts measurably to news events, peaking during John Glenn's flight on a Space Shuttle mission in 1998 and dropping after failed Mars missions

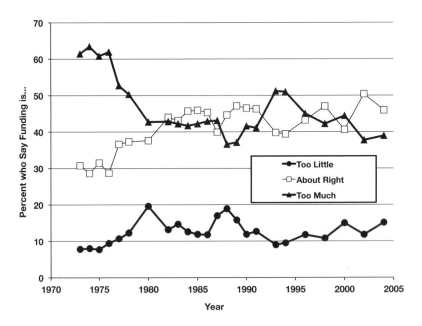

FIGURE 1.1. Attitudes Toward Space Program Funding Expressed in the General Social Survey

---

54. Joseph Carroll, "Public Divided over Money Spent on Space Shuttle Program," press release, 30 June 2006, Gallup News Service, Princeton, NJ.

in 1993 and 1999.[55] Confidence in the Agency and willingness to invest in it are related in complex ways, however, and the public might occasionally want to increase funding in order to fix a problem and fail to reward successes with increased appropriations if it is satisfied that NASA is already on the right track.[56]

The most striking feature of the GSS graph is the low level of support for the space program in the early 1970s, improving until about 1980 and then holding roughly steady since then. In 1973 only 7.8 percent felt that too little was being spent, compared with 61.4 percent who felt that that spending was too high. The year 1978 was the last time the GSS found that a majority, 50.3 percent, wanted space funding reduced, and in that year 12.3 percent wanted funding increased. The first Space Shuttle launch was on 12 April 1981, incidentally exactly 20 years after the first orbital flight by Yuri Gagarin; so public support in 1980 was clearly not a response to its success. Of course, publicity for the Shuttle had been building for years, notably during 1977, when the Enterprise repeatedly demonstrated the ability of a Shuttle to glide safely to a landing.

A number of writers who have analyzed the polls of this period have suggested that the beginning of this period was anomalous in its low public enthusiasm for the space program. For example, Roger D. Launius has noted "a significant dip in support in the early 1970s."[57] Sylvia D. Fries has observed: "The proportion of Americans opposed to more government expenditures in space from 1965 to 1975 increased from one-third to one-half of all adult Americans."[58] Sylvia K. Kraemer has argued that public support for the space program may have deteriorated because "more Americans saw the Apollo program as another effort to 'beat the Russians' than as an essential goal of U.S. space exploration."[59] Other observers would argue that the very success of Apollo, coupled with the lack of a comparably exciting post-Apollo goal, was responsible. As one team of poll analysts put the point: "Without the

55. Mark Gillespie, "Confidence in NASA Slips After Failed Mars Missions," press release, 16 December 1999, Gallup News Service, Princeton, NJ.

56. Frank Newport, "Despite Recent High Visibility, Americans Not Enthusiastic About Spending More Money on Space Program," press release, 28 July 1999, Gallup News Service, Princeton, NJ.

57. Roger D. Launius, "Public Opinion Polls and Perceptions of US Human Spaceflight," *Space Policy* 19, no. 3 (2003): 163–175, esp. p. 166.

58. Sylvia D. Fries, "Opinion Polls and the U.S. Space Program," paper given at the meetings of the American Institute for Aeronautics and Astronautics, 29 April 1992, NASA TM-109700, 1–11, esp. p. 6.

59. Sylvia K. Kraemer, "Opinion Polls and the U.S. Civil Space Program," *Journal of the British Interplanetary Society* 46, no. 11 (November 1993): 444–446.

compelling need to regain international leadership in space, the U.S. civil space program foundered in the 1970s. Decreasing public interest in the now seemingly routine Moon landings was echoed in a declining national space budget."[60]

However, there is much evidence for a very different explanation, focusing not on the specifics of space history but on the broader status of science in society. As Georgine Pion and Mark Lipsey have documented, using a variety of polls, the 1970s were marked by increased distrust in science and technology, associated with distrust in most institutions of society.[61] In 1966 and again in 1971, the National Opinion Research Center (NORC) of the University of Chicago asked a random sample of Americans how much confidence they had in "science" and "other institutional areas." The fraction having "a great deal of confidence" in science dropped from 56 percent to 32 percent over these five years. Those with great confidence in "major U.S. companies" dropped from 55 percent to 27 percent, and the drop was from 61 percent to 37 percent for education. These questions were incorporated in NORC's General Social Survey beginning in 1973, and Table 1.2 shows a close connection between confidence in science and support for the space program.

After reviewing their poll evidence, Pion and Lipsey conclude that "science did suffer from the general disillusionment experienced by all major social institutions during the late 1960s and early 1970s."[62] It is worth recalling that

TABLE 1.2. Confidence in Science and Support for the Space Program

Support for Space Program Spending	1973: Confidence in Science			2004: Confidence in Science		
	A Great Deal	Only Some	Hardly Any	A Great Deal	Only Some	Hardly Any
Too little	12.0%	6.4%	2.2%	19.7%	7.9%	0.0%
About right	37.4%	29.4%	15.4%	54.0%	47.1%	26.9%
Too much	50.7%	64.2%	82.4%	26.2%	45.0%	73.1%
Total	100%	100%	100%	100%	100%	100%
Respondents	535	673	91	169	215	22

60. Stephanie A. Roy, Elaine C. Gresham, and Carissa Bryce Christensen, "The Complex Fabric of Public Opinion on Space," *Acta Astronautica* 47, Issues 2–9 (2000): 665–675, esp. p. 668.

61. Georgine M. Pion and Mark W. Lipsey, "Public Attitudes Toward Science and Technology: What Have the Surveys Told Us?," *Public Opinion Quarterly* 45, no. 3 (1981): 303–316.

62. Pion and Lipsey, p. 313.

this was the peak period for the Vietnam War, the Watergate scandal, and the psychedelic counterculture. In 1973 the mass media ballyhooed expectations that Kohoutek would be the comet of the century, rock musicians dedicated albums to it, drug guru Timothy Leary wrote from his prison cell calling it "Starseed," and Moses David of the Children of God proclaimed it was ushering in the millennium.[63] In the words of Steven Tipton, the early 1970s were spent "getting saved from the sixties," but by 1980 that painful job had largely been accomplished and America was looking for more mundane benefits from space.[64]

In May 1981, a month after the first Space Shuttle launch, the Harris poll reminded 1,250 respondents: "There are a number of practical uses that the space shuttle may provide by taking as many as 400 flights into space and back over the next several years." Then respondents rated five of these practical uses in terms of how important they thought they were. "Doing experiments with new pharmaceutical products that can help cure disease" was rated "very important" by 81.9 percent. The other four practical uses were: "developing a military capability in space beyond what the Russians are doing" (68.0 percent); "putting new communications satellites in space at a much lower cost" (65.4 percent); "doing scientific research on metals, chemicals, and living in space" (55.6 percent); and "picking up other U.S. space satellites and repairing them in space" (47.9 percent). Analysis of correlations between these five items shows that developing pharmaceutical products was quite distinct from the other four, and the military item was also somewhat distinct. More than a quarter century after this poll, it is worth noting that new pharmaceutical products of any value have not resulted from the Space Shuttle program; the Challenger accident ended the Shuttle's mission lofting communications satellites and severely limited repair missions, and there were not 400 but only 114 launches during that period—two of them ending in total destruction of the spacecraft and crew.[65]

When Ronald Reagan became president in 1981, détente with the Soviet Union had already ended, with the Soviet invasion of Afghanistan and the U.S. withdrawal from the Moscow Olympics of 1980—followed by the Soviet

---

63. Timothy Leary, "Starseed," 1973, *http://www.lycaeum.org/books/books/starseed/starseed. shtml*; Moses David [David Berg], "The Christmas Monster!," MO Letter GP#269, 8 September 1973; and "40 DAYS!—And Nineveh Shall Be Destroyed! (Jonah 3:4)," MO Letter GP#280, 12 November 1973.

64. Steven M. Tipton, *Getting Saved from the Sixties* (Berkeley: University of California Press, 1982).

65. Analysis of original data from Harris study no. 812106, a May 1981 poll of 1,250 American adults, Odom Institute. The average correlations linking each of the items with the others, in the order cited, was 0.16, 0.24, 0.32, 0.30, and 0.31.

boycott of the Los Angeles Olympics four years later. Thus a period began during which the military potential of the space program was emphasized. In August 1981, halfway between the first two Shuttle flights, a poll done by NBC and the Associated Press asked, "Should the emphasis of the U.S. space program be primarily on national defense or on scientific exploration?" A near majority, 49 percent, answered "national defense," while 32 percent said "scientific exploration," 10 percent volunteered "both," and the remaining 9 percent were "not sure."[66]

On 23 March 1983, Reagan proposed development of a system to defend the United States against missile attack, the Strategic Defense Initiative (SDI): "I call upon the scientific community who gave us nuclear weapons to turn their great talents to the cause of mankind and world peace: to give us the means of rendering these nuclear weapons impotent and obsolete."[67] Even to those who paid close attention to public information about SDI, it was never entirely clear how prominent a role the civilian space program or orbiting military satellites would play, in contrast to ground-based defensive missiles. There was much talk about space-based lasers or particle beams that might destroy warheads at a range of thousands of miles, but such weapons have not been developed in the subsequent decades and may have been an extreme fantasy when they were widely discussed in the 1980s.[68]

Over the next several years, numerous polls examined attitudes toward SDI.[69] During this period, many Americans would have preferred to negotiate an end to the militarization of space. A January 1985 *Los Angeles Times* poll inquired, "Generally speaking, do you favor or oppose an agreement to outlaw the use of all military weapons in outer space?" By far the largest group, 43 percent, said they "favor strongly" such an agreement, and another 16 percent said they "favor somewhat," meaning that a majority of Americans backed international agreement to outlaw space weapons. On the other side of the issue, 15 percent "opposed somewhat" and 16 percent opposed strongly, with 10 percent "not sure."[70]

---

66. Poll of 1,601 American adults by NBC News and the Associated Press, 10–11 August 1981, Roper Center USNBCAP.69.R23.
67. Ronald Reagan, "Address to the Nation on National Security," 23 March 1983, *http://www.commonwealthclub.org/missiledefense/reagansp.html.*
68. Robert H. Gromoll, "SDI and the Dynamics of Strategic Uncertainty," *Political Science Quarterly* 102, no. 3 (1987): 481–500.
69. Thomas W. Graham and Bernard M. Kramer, "The Polls: ABM and Star Wars: Attitudes Toward Nuclear Defense, 1945–1985," *Public Opinion Quarterly* 50, no. 1 (1986): 125–134.
70. Poll of 1,454 American adults by the *Los Angeles Times*, 19–24 January 1985, Roper Center USLAT.93.R089.

One 1985 poll found that 31 percent of Americans had never "heard or read anything about a program called the Strategic Defense Initiative, or SDI, also known as 'Star Wars.'"[71] Thus the polls needed to explain SDI to respondents, therefore shaping public opinion even as they were trying to measure it. For example, a July 1985 Roper poll asked this massive question:

> President Reagan has proposed that the United States build a space-based defense system (sometimes called "Star Wars") against incoming missiles. Many people think that this is a good idea because it would give us an advantage over the Russians in this area, which would help deter a Soviet attack. Many others feel that a space-based defense system is a bad idea because it would escalate the arms race and increase the risk of a nuclear confrontation with Russia. How do you feel—do you think the United States should or should not build a space-based defense system?

With this explanation in mind, 43 percent felt "we should build it," 35 percent felt "we should not build it," and fully 22 percent did not know what to think.[72]

That same poll included this preface to a series of complex questions:

> Here are some arguments that have been made in favor of a space-based anti-missile defense system. For an argument to be convincing it has to be both important and true. If it isn't important, or isn't true, it isn't convincing. Would you tell me for each of those arguments whether you find it a very convincing argument for a space-based anti-missile defense system, or somewhat convincing, or not very convincing, or not at all convincing?

Just 19 percent found the following argument very convincing: "The world would be safer if the U.S. and the Soviet Union could each rely on a space-based anti-missile defense system for their security rather than relying, as they do now, on offensive missile systems to deter each other from launching an attack." Another 30 percent found it somewhat convincing, 21 percent judged it to be not very convincing, and 16 percent said it was not at all convincing.[73]

---

71. Poll of 1,008 American adults by Marttila & Kiley, September 1985, Roper Center USGALLUP.47-406.QKT11.
72. Poll of 1,997 American adults by Roper, 13–20 July 1985, Roper Center USROPER.85-7. R08.
73. Poll of 1,997 American adults by Roper, 13–20 July 1985, Roper Center USROPER.85-7. R09D.

Some polls saved their more complex questions for subsets of respondents who claimed to be especially knowledgeable. A Gallup poll in October 1985 focused on the 61 percent of respondents who claimed to have followed the "Star Wars" discussion very closely or fairly closely. Of these, 48 percent believed that developing a space-based defensive system would increase "the likelihood of reaching a nuclear arms agreement with the Soviet Union," while 36 percent believed it would decrease the chances, and 16 percent had no opinion.[74] One 1985 *Los Angeles Times* poll gave respondents many options. While only 10 percent felt SDI would "someday be a leakproof umbrella against enemy missiles," 32 percent felt it could at least "be able to reduce the number of missiles that can get through." Just 5 percent felt SDI would be "effective mainly against enemy satellites," and 6 percent felt it would "be able to protect small areas where missiles are stored." In contrast, 22 percent were convinced "a Star Wars system will never be effective at all," and 25 percent were not prepared to venture an opinion.[75] However, judgments of the feasibility of Star Wars were probably based neither on adequate knowledge nor firm feelings. A year later, a poll for *Time* magazine found that 57 percent of Americans felt "the Star Wars (space-based) defense system is likely to work," far more than the 20 percent who felt it was "not likely to work."[76]

A poll by CBS and the *New York Times* asked: "If it came down to only these choices, what should the United States do—work to develop a Star Wars system (a defense system in space to destroy incoming missiles) and give up negotiations, or work to negotiate a reduction in nuclear missiles and give up Star Wars?" A majority of 53 percent was happy to negotiate Star Wars away, while 33 percent wanted to develop the system even at the cost of forgoing negotiations, and 14 percent could not decide.[77] It is difficult to tell how closely the public connected Star Wars with the civilian space program, but one poll question suggests not very closely. Immediately after the Challenger disaster, Gallup asked: "Does the Shuttle explosion cast doubt in your mind on the ability of scientists to construct a reliable space-based 'star wars' defense against nuclear attack—or does one have very little to do

74. Poll of 1,540 American adults by Gallup, 11–14 October 1985, Roper Center USGALLUP.111785.R2.
75. Poll of 2,041 American adults by the *Los Angeles Times*, 1–7 November 1985, Roper Center USLAT.100.R77.
76. Poll of 806 American adults by Yankelovich Clancy Shulman for *Time* magazine on 15 October 1986, Roper Center USYANKCS.102086.R13.
77. Poll of 1,659 American adults by CBS News and the *New York Times*, 6–10 November 1985, Roper Center USCBSNYT.NOV85.R44.

with the other?" Only 16.2 percent felt the explosion cast doubt on SDI, and 72.9 percent asserted that the Challenger had little to do with "Star Wars."[78]

To the extent that the civilian space program did become associated with military defense in the public mind during the 1980s, it became implicated in political disagreements about defense. Immediately before the 1988 elections, in which Reagan's vice president, George Bush, defeated Michael Dukakis for the presidency, a Yankelovich poll asked registered voters how much should be spent on SDI annually. Just 14 percent said "about $5 billion as requested by President Reagan," while 38 percent said "about $4 billion as legislated by the U.S. Congress," and 31 percent said "about $1 billion as proposed by Michael Dukakis." Another 6 percent volunteered that no money should be spent at all.[79]

Rosita Thomas has suggested that the public is more likely to associate space with defense during periods of international tensions and to be more receptive to joint missions during periods of détente.[80] To the extent that a given administration's defense policies are controversial, promoting military applications for the space program can polarize attitudes toward space exploration. This works against the development of a broadly based constituency for the program and pits long-range goals, such as gaining scientific knowledge, against narrow tactics to deal with current world conditions. Although reconnaissance satellites have been a valuable aid to national defense since the beginning of the 1960s, space-based laser weapons continue to be apparently beyond our technical capabilities more than three decades after Reagan's 1983 speech. In retrospect, Star Wars was a fantasy that served, intentionally or unintentionally, to put pressure on the Soviet Union, which evolved through détente toward disintegration in the latter part of the decade. SDI did little to clarify the tension between scientific discovery and mundane applications of space technology.

The Space Shuttle Challenger was destroyed during launch on 28 January 1986, at the moment when Voyager 2's highly publicized encounter with Uranus was promoting the value of robotic missions. Before the day was over, President Reagan had spoken to the nation, saying, in part: "We'll continue our quest in space. There will be more shuttle flights and more shuttle crews and, yes, more volunteers, more civilians, more teachers in space. Nothing

---

78. Poll of 533 American adults by Gallup, 29–30 January 1986, Gallup Organization, Princeton, NJ.

79. Poll of 1,006 registered voters by the Daniel Yankelovich Group, 4–7 November 1988, Roper Center USDYG.ATS11.R13.

80. Rosita M. Thomas, *American Public Opinion and the Space Program* (Washington, DC: Congressional Research Service of the Library of Congress, 1991), p. 39.

ends here; our hopes and our journeys continue."[81] Over the following two days, Gallup carried out a telephone survey for *Newsweek*, asking: "Some people say the National Aeronautics and Space Administration should concentrate on unmanned missions like the Voyager probe that is now sending back information from the planet Uranus. Others say it is important to maintain and develop a manned space program like the Shuttle as well as unmanned missions. Which comes closer to your view?" Just 21 percent wanted to concentrate on unmanned missions, 67 percent wanted manned missions as well, and 12 percent did not know.[82]

Another telephone poll conducted at the same time by the Roper Organization for *U.S. News and World Report* offered respondents two rather complex statements, asking which they agreed with more. Just 36 percent agreed with this: "I think the space administration was pushing too hard. It was under great economic pressure to get too many flights up into space too quickly." A slim majority, 55 percent, agreed with this: "The space administration has always been very safety conscious and I don't see any indication that pressure caused it to depart from its usual commitment to safety."[83] A *USA Today* poll asked respondents to assume "an investigation of the shuttle explosions shows that a similar incident can be avoided," then asked what should be done. A large majority, 72.6 percent, said "the Shuttle Program should resume its original schedule." Much smaller groups said the program "should be cut back" (16.1 percent) or be "ended altogether" (6.3). Only 21.0 percent felt "future shuttle crews [should] be limited to military and NASA personnel," while 72.8 percent felt "civilians such as school teacher Christa McAuliffe [should] be allowed to participate."[84]

The polls for *Newsweek*, *U.S. News and World Report*, and *USA Today* were done hastily with about 500 to 800 respondents. Over the next week, Harris was able to poll 1,255 American adults. When reminded that "it is costing the U.S. government billions of dollars to develop the full potential of the space shuttle," 66.5 percent said they felt the Shuttle program was worth it. The poll also presented respondents with four statements, asking whether they felt this way or not. Fully 77.5 percent felt that "no further shuttle flights should be conducted until they find out what went wrong in the one that

81. Ronald W. Reagan, "Explosion of the Space Shuttle Challenger: Address to the Nation," 28 January 1986, *http://history.nasa.gov/reagan12886.html.*

82. Poll of 533 American adults by Gallup for *Newsweek*, 29–30 January 1986, Roper Center USGALNEW.86056.R04.

83. Poll of 502 American adults by the Roper Organization for *U.S. News and World Report*, 29–30 January 1986, Roper Center USROPER.644075.Q05.

84. Analysis of original data from *USA Today* study 9111, a poll done by Gordon S. Black Corporation in 1986 of 808 American adults, Odom Institute.

blew up." Only 37.1 percent felt that "the practice of putting civilians on board space flights should be put off until a much later date." A solid majority (63.7 percent) felt that "the practice of putting astronauts on the shuttle flights should be reviewed carefully to see if more flights can be taken which don't require risking human life." Exactly 50.0 percent felt that "they should concentrate on putting up unmanned craft like the Voyager, which can conduct important experiments and learn important facts without risking any human life." Some 44.7 percent definitely did not hold this opinion, and 5.3 percent were not sure.[85]

Nearly a month after the Challenger disaster, a *Los Angeles Times* poll told respondents: "NASA (National Aeronautics and Space Administration) says that the benefits of research performed by live astronauts in space far outweigh any possible dangers to their safety. Other people say that unmanned space probes would cost less and do the same job without risk to life." When asked their views, 27 percent of Americans felt there should be an even greater emphasis on human spaceflight, 41 percent felt the United States should "continue with the same emphasis as before," and only 24 percent felt there should be more emphasis on piloted missions.[86] When asked whether "the government should spend one and a half billion dollars for another space shuttle to replace Challenger," 52 percent said yes, compared with only 42 percent who opposed this investment.[87] Five months after the Challenger accident, only 11 percent of respondents to an NBC News and *Wall Street Journal* poll felt "the manned shuttle program should be discontinued for good," and 85 percent stated it should not be discontinued.[88]

In June 1986 the Rogers Commission Report was published, identifying the causes of the Challenger disaster and recommending improvements. NASA published its response, "Actions to Implement the Recommendations of The Presidential Commission on the Space Shuttle Challenger Accident," in July.[89] In August 1986 a poll conducted for Rockwell International, Challenger's builder, noted "that the Roger's [sic] Commission on the space

---

85. Analysis of original data from Harris study no. 861201, a February 1986 poll of 1,255 American adults, Odom Institute.

86. Poll of 2,241 American adults by the *Los Angeles Times,* 20–25 February 1986, Roper Center USLAT.103.R90.

87. Poll of 2,241 American adults by the *Los Angeles Times,* 20–25 February 1986, Roper Center USLAT.103.R86.

88. Poll of 1,599 American adults by NBC News/*Wall Street Journal*, 2–3 June 1986, Roper Center USNBCWSJ.061186.R28.

89. NASA, "Actions to Implement the Recommendations of The Presidential Commission on the Space Shuttle Challenger Accident" (also called the *Rogers Commission Report*), 14 July 1986, *http://history.nasa.gov/rogersrep/actions.pdf.*

shuttle accident has finished its work and NASA has announced its plans to follow their recommendations." With this in mind, 72 percent of respondents believed "that the space program can proceed," while 26 percent felt "there should be more investigation of the causes of the accident."[90]

In September a Harris poll referred to the report of the Rogers Commission and asked a number of questions about the Shuttle program. Some 43.8 percent of respondents felt that "the full story of what happened in the disaster of the Challenger has come out." Despite harboring doubts, 72.8 percent felt that "NASA has made, or is making, the basic changes necessary to get the space program back on track." Harris interviewers read a long paragraph explaining President Reagan's plans to replace the Challenger with a new Shuttle estimated to cost $2.8 billion, then asked respondents whether they thought several aspects of the new space policies were right. Only 20.2 percent thought that "it [would] be possible to pay for the new space shuttle out of the current NASA budget," and 39.2 percent feared "that such important other programs as expendable rockets, reusable unmanned vehicles, and space stations [would] be neglected."

Given that a tenth of respondents were unsure about most of the policies, a plurality of 48.4 percent felt "the President was right to ban the shuttle from getting paid to launch commercial satellites because doing so could prove to be too hazardous." A large majority, 72.3 percent, felt "the main emphasis in the space program now likely will be the military, especially the Air Force, which will have the money to finance shuttle trips and to undertake programs that will move toward new discoveries in space." In conclusion, 67.6 percent of respondents thought the space program was worth continuing at the current level, but only 36.8 percent were willing to provide more money to overcome the Challenger setback.[91] However, Figure 1.1 (see page 18) shows that after the Challenger accident there was a marked rise in the fraction of the public that wanted space funding increased and a reduction in the fraction who wanted funding reduced.

As the Voyager 2 space probe was approaching Neptune in 1989, a Gallup poll told respondents: "Some people feel the U.S. space program should concentrate on unmanned missions like Voyager 2, which will send back information from the planet Neptune. Others say we should concentrate on maintaining a manned space program like the space shuttle." Some

---

90. Poll of 1,200 American adults by Market Opinion Research for Rockwell International, 1–12 August 1986, Roper Center USMOR.86SPAC.R14.

91. Analysis of original data from Harris study no. 1255, a September 1986 poll of 1,255 American adults, Odom Institute. Note: question Q1A_3 in the interview schedule seems to have a typographical error in the last word (*fired* when *fined* was probably meant), and Q1B_5 rambles so much that its meaning is unclear; so they will not be analyzed here.

39.6 percent wanted more emphasis on "unmanned" missions, compared with 42.6 percent who preferred "manned" missions.[92] A year later, when the successful Neptune flyby had passed into history, Gallup asked the question again, and only 34.3 percent favored "unmanned" exploration, while 47.6 percent favored "manned" missions.[93] This comparison suggests that spectacular successes can increase enthusiasm for robot probes but that many people expect the space program to be about human exploration.

In 1991, George Gallup Jr. and Frank Newport sought to interpret the deeper meaning of the public opinion percentages the Gallup organization had been reporting over the eight previous years. They observed that most Americans felt NASA was doing a good or excellent job, and the United States was far ahead of other nations in space.

> Perhaps as a result of these feelings of "space supremacy," the race to be first on Mars has become significantly less important to Americans…. Despite the improvement in NASA's ratings, space exploration remains a low priority for most Americans' tax dollars in comparison to other government programs. Overall, more than half of all Americans (56%) think the money this country has invested in space research would have been better spent on programs such as health care and education.[94]

This was in May, after the breakup of the Soviet empire and just seven months before the USSR formally dissolved, so it may already have seemed clear to many Americans that their nation was the sole remaining "superpower" that did not any longer need to be competing for propaganda triumphs in outer space.

Although Americans respect scientists, few citizens really give very high priorities to scientific discovery unless it promises some immediate positive impact on their lives—for example, in development of valuable new medical treatments. The general public is interested in people before it is interested in ideas. Thus piloted spaceflight provides human-interest stories that can help ordinary people identify with the entire program. It is perhaps understandable, if not entirely fair, that these stories always seem to focus on the astronauts who have what Tom Wolfe called "the right stuff," rather than on the scientists and engineers who have invented the vehicles the astronauts ride

---

92. Poll of 2,051 American adults by Gallup, 6–9 July 1989, Gallup Organization, Princeton, NJ.

93. Poll of 2,020 American adults by Gallup, 19–22 July 1990, Gallup Organization, Princeton, NJ.

94. George Gallup Jr. and Frank Newport, "NASA Rating Up but Public Reluctant To Spend More Money on Space," *Gallup Poll News Service* 56, no. 2 (10 May 1991): 1–3.

and thus arguably have even "righter stuff."[95] For instance, immediately after the 29 October 1998 launch of Discovery, 77.3 percent of respondents to a Gallup poll approved "of NASA sending U.S. Senator John Glenn, a former astronaut, back into space this week on a space shuttle mission," and only 15.8 percent disapproved.[96]

On 1 February 2003, the Columbia disintegrated during reentry, and the next day Gallup did a quick-action poll of 462 people. An overwhelming majority, 81.5 percent, felt the "manned space shuttle program" should continue. When asked how much confidence they had that NASA "will be able to prevent accidents like this from happening in the future," 37.8 percent said "a great deal," 43.6 percent replied "a fair amount," 11.3 percent admitted "not very much," and 5.8 percent said "none at all." Logically, views on this issue are the product of beliefs about two issues: how difficult spaceflight objectively is and how good a job NASA is doing. Some 44.8 percent felt NASA was doing an excellent job, 37.3 percent a good job, 13.0 percent a fair job, and only 2.4 percent of respondents said NASA was doing a poor job. When asked how much money should be spent on the space program, 23.7 percent said an increased amount; the majority, 56.3 percent, said funding should be kept at current levels; and 16.0 percent wanted funding decreased (including 7.3 percent who wanted the program ended).[97] Unlike the case with Challenger, a replacement Shuttle was not built; so a substantial increase in funding was unnecessary.

Five days after the disaster, a Harris poll asked a much larger sample, "In light of what happened to the Columbia last week (February 1, 2003), do you think that continuing the space shuttle program is or is not worth the risk to human life?" Fully 71 percent considered that it is worth the risks, compared with 25 percent who judged it was not and 4 percent who were not sure.[98] In response to another poll, 77 percent felt "the manned space program is worth continuing," even "given the costs and risks involved."[99]

In August 2003 a Gallup poll challenged respondents with this difficult question: "As you may know, there have been two space shuttle crashes that have killed fourteen astronauts since the first space shuttle was launched in

95. Tom Wolfe, *The Right Stuff* (New York: Farrar, Straus, and Giroux, 1979).

96. Poll of 1,045 American adults by Gallup, 30–31 October 1998, Gallup Organization, Princeton, NJ.

97. "Space Shuttle Crash Reaction Poll," a Gallup poll of 462 adults, 2 February 2003, Gallup Organization, Princeton, NJ.

98. Poll of 1,003 American adults by *Time*/Cable News Network conducted by Harris Interactive, 6 February 2003, Roper Center USHARRIS.Y021803.R37.

99. Poll of 747 American adults by CBS News/*New York Times*, 10–12 February 2003, Roper Center USCBSNYT.200302B.Q68.

1981. Which of the following would you consider to be an acceptable price to pay for the U.S. to achieve its goals with the space program?" Just 6.2 percent felt a fatal crash every 10 missions would be an acceptable price, compared with 7.4 percent who said a crash every 20 missions, 19.3 percent who said a crash every 50 missions, and fully 42.9 percent who would not accept fatal crashes more often than 1 in every 100 missions. Another 16.9 percent said "no space shuttle crashes at all."[100] In September a University of Connecticut poll posed this question:

> Suppose an editorial in the *New York Times* said, "America should stop funding the manned space shuttle program. A recent report stated that the [1 February 2003] crash of the space shuttle Columbia resulted from mismanagement by NASA officials and lax safety standards. Space shuttle flights are too risky for the benefits they provide." Would you agree or disagree with the editorial?

Whereas 10 percent could not decide, 31 percent agreed (14 percent strongly agreed, 17 percent somewhat agreed), and 59 percent disagreed (30 percent strongly disagreed, 29 percent somewhat disagreed).[101]

In January 2004 an Associated Press survey asked: "On the whole, do you think our investment in space research is worthwhile or do you think it would be better spent on domestic programs such as health care and education?" Those who felt it would be better to spend on domestic programs outnumbered those who felt space research is worthwhile, 55 percent to 42 percent.[102] In late June and early July 2004, when Shuttles were still grounded in the wake of the Columbia disaster, the Space Foundation had Gallup ask again about the proper balance between manned and unmanned programs. Two-thirds of Americans agreed: "It is important for our nation to have a space program that uses both manned exploration with astronauts and unmanned exploration using robotics, like the recent Mars Rovers" (38 percent strongly agreed, 28 percent somewhat agreed). Some 20 percent of respondents were "neutral" about this statement, and 13 percent disagreed (7 percent strongly disagreed, 6 percent somewhat disagreed).[103]

---

100. Poll of 1,003 American adults by Gallup, 4–6 August 2003, Gallup Organization, Princeton, NJ.

101. Poll of 1,005 American adults by Center for Survey Research and Analysis, University of Connecticut, 12 September–1 October 2003, Roper Center USCSRA.03MEDIA.RSS2A.

102. Poll of 1,000 American adults by IPSOS-Public Affairs for the Associated Press, 9–11 January 2004, Roper Center USIPSOSR.011204A.R4.

103. Poll of 1,000 American adults by Gallup for the Space Foundation, 22 June–7 July 2004, Roper Center USGALLUP.04SPACE.R3.

An April 2005 Gallup poll reminded respondents: "As you may know, NASA has scheduled a space shuttle launch for July, which would be the first launch since a space shuttle was lost in an accident in 2003." A solid majority, 67.4 percent, felt NASA was moving at the right pace. In contrast, only 10.2 percent felt NASA was "moving too slowly in re-starting the space shuttle program," and 17.5 percent felt NASA was moving too quickly.[104] In June 2005, 74 percent of Americans felt "the manned space shuttle program" should continue, compared with 21 percent who felt it should end.[105]

On 26 July 2005, the Discovery was the first Shuttle launched since the loss of the Columbia, and unexpectedly a substantial piece of foam came off the external fuel tank, the same type of failure that had doomed Columbia. A Gallup poll asked, "How confident are you that the space shuttle that is currently in space will land safely?" Respondents were surprisingly positive, with 36.2 percent saying they were "very confident" and another 45.7 percent saying they were "somewhat confident." Only 14.4 percent were "not too confident," and 2.2 percent were "not at all confident."[106]

During the Shuttle era, NASA twice started development of a successor vehicle, first the National Aerospace Plane then the X-33, but both projects were eventually cancelled because of a combination of daunting technical problems and insufficient political support.[107] However, the Columbia accident made it clear that the Shuttle's days were numbered, bringing to a close the period of sustained but moderate progress that followed the revolutionary period from Sputnik through Apollo. A new period has already begun, and a deeper examination of public opinion data can help us understand how well prepared American society is for this next phase.

4. Personal Impact of Space Exploration

This section considers how the space program has affected people's feelings and their knowledge, with particular attention to the fact that it has different psychological impacts on different segments of the population that are more or less interested and attentive. Significant fractions of the population

104. Poll of 1,006 American adults by Gallup, 4 April–1 May 2005, Gallup Organization, Princeton, NJ.

105. Poll of 1,009 American adults by Gallup for Cable News Network and USA Today, 24 June–26 June 2005, Roper Center, USGALLUP.05JE024.R07.

106. Poll of 1,004 American adults by Gallup, 5–7 July 2005, Gallup Organization, Princeton, NJ.

107. John M. Logsdon, "'A Failure of National Leadership': Why No Replacement for the Space Shuttle," in Steven J. Dick and Roger D. Launius, eds., *Critical Issues in the History of Spaceflight* (Washington, DC: NASA, 2006), pp. 269–300.

have felt the emotional impact of major events in the history of spaceflight. In May 1971 a Harris poll included a series of items about television, including one about coverage of space exploration. Fully 81 percent of Americans agreed: "Nothing can equal seeing the astronauts land and walk on the moon as it happened live on TV." While 3 percent were not sure, just 16 percent disagreed with this statement.[108] In September 1997 a Harris poll asked people: "Do you remember exactly where you were, what you were doing, or who you were with when you heard [about eleven newsworthy events]?" Psychologists call vivid recollections like this *flashbulb memories*, and they reflect deep impressions the events made.[109] Of those respondents old enough to have experienced the news "the Challenger had blown up," fully 78.0 percent had a flashbulb memory. This placed the Challenger disaster just below "the news about the Oklahoma City bombing (80.5 percent) and well ahead of "President Reagan had been shot" (55.8 percent).[110] As time passes, of course, fewer and fewer people recall historical events, not just from failing memory but because many will have passed away. In 1999 a poll for the Pew Research Center for the People and the Press asked "if you happen to remember EXACTLY where you were or what you were doing the MOMENT you heard the news." Whereas 78 percent vividly recalled "the Challenger explosion," only 54 percent remembered "Armstrong walking on the moon."[111]

Many people feel an emotional bond with astronauts or sympathize with them when they face challenges and dangers. In April 1970, concerning "the Apollo 13 trip to the moon this past week," the Harris poll asked: "Did you personally feel very worried over whether the men in the spaceship would get back to earth, somewhat worried, or not very worried?" Fully 54.5 percent said they were "very worried," and another 24.4 percent were "worried." While 0.9 percent were "not sure," the remaining 20.1 percent were "not very worried."[112] Naturally, there are at least three ways a person could fail to be worried. First, they may not have known Apollo 13 had run into trouble, but the television news coverage was quite intense. Second, they could have known but been confident that NASA could handle the problem. Third, they could have anticipated anything might happen to the astronauts but simply

108. Poll of 1,600 American adults by Louis Harris and Associates, May 1971, Roper Center USHARRIS.71MAY.R32D.
109. Roger Brown and James Kulik, "Flashbulb Memories," *Cognition* 5 (1977): 73–99.
110. Poll of 1,011 American adults by Harris, September 1997, Odom Institute, H-718316.
111. "Technology Triumphs, Morality Falters," press release, 3 July 1999, Pew Research Center, *http://people-press.org/reports/display.php3?PageID=279*.
112. Analysis of original data from Harris study no. 2025, an April 1970 poll of 1,520 American adults, Odom Institute.

not cared, given that people are dying every day from accidents that do not attract public concern.

A little insight about what was going on inside people's minds can come from other items in the poll. One poll asked respondents whether getting men on the Moon had been worth the cost. Only 38.3 percent felt it was, while a majority (56.3 percent) felt the accomplishment was not worth it. Of those who felt Apollo was worth the cost, 61.3 percent were very worried, compared with only 50.4 percent of those who felt Apollo was not worth the cost. Thus part of the worry was a reflection of enthusiasm for the space program, rather than just public interest concern about human beings who were in danger. The poll also asked: "Do you expect that on one of the space shots an accident will take place and the astronauts won't get back alive, or do you think that probably won't happen?" A large majority, 71.2 percent, expected a fatal accident would happen, compared with only 17.4 percent who felt it would not and with 11.3 percent who were not sure.

In describing their personal reaction "when the space shuttle Challenger blew up" in 1986," 63.0 percent said they were "deeply upset," 28.1 percent were "somewhat upset," for a combined total of 91.1 percent who said they were upset to a significant degree.[113] Such feelings were often shared with family members. In a *USA Today* poll done right after the Challenger disaster, 49.1 percent of 599 parents said "yes" when asked: "Did you discuss the destruction of the shuttle with your children?"[114] In January 1987, 68.3 percent of respondents to a *USA Today* poll said they could remember exactly what they were doing when they heard of the Challenger disaster a year earlier. Respondents were also asked, "Are you more or less interested in the U.S. space program since the Challenger tragedy last January," or "Is your interest about the same?" The largest group, 72.7 percent, said their interest was the same. However, the fraction whose interest was greater (18.7 percent) was more than twice as large as the fraction whose interest was less (7.2 percent).[115] The day after the Columbia disaster in 2003, 57.9 percent reported being "deeply upset," and a further 36.0 were "somewhat upset."[116] Perhaps the fact that this was the second loss of a Shuttle in many people's memories diminished the shock slightly. A majority of 70.6 percent indicated

113. Analysis of original data from Harris study no. 861201, a February 1986 poll of 1,255 American adults, Odom Institute.

114. Analysis of original data from *USA Today* study 9111, a poll of 808 American adults done by Gordon S. Black Corporation in 1986, Odom Institute.

115. Analysis of original data from *USA Today* study 3002, a poll of 817 American adults done by Gordon S. Black Corporation in January 1987, Odom Institute.

116. "Space Shuttle Crash Reaction Poll," a Gallup poll of 462 adults, 2 February 2003, Gallup Organization, Princeton, NJ.

that "yesterday's tragedy was regrettable, but you thought something like this would happen again sooner or later."

Over the years, various polls have asked people whether they themselves would like to travel into space. Such questions are extremely hypothetical, but answers presumably reflect personality traits such as risk tolerance and practical considerations like good health and lack of family responsibilities. A February 1986 poll that focused on the Challenger disaster asked: "If you personally were selected as a civilian to go up on a space shuttle, would you do it or not?" Willingness to fly on a Shuttle did not differ much across groups that were more or less upset by the accident. Of those who were deeply upset, 44.5 percent said they "would do it," compared with 46.9 percent of those "somewhat" upset and 37.8 percent of those who were "not much" upset.[117] It is possible that positive and negative feelings to some extent cancel out. Perhaps people who were very enthusiastic about the space program were both more likely to want to go into space themselves and were upset by this setback to the space program. In any case, we cannot take answers to questions about personally flying in space at face value.

However, those who answer in the affirmative may include a hard-core, pro-space group in the population that is worthy of notice. In 1955 just 8 percent of respondents to a Gallup poll answered "yes" to the question "If you were asked to go along on the first rocket ship to the moon, would you want to go or not?"[118] Soon after the launch of Sputnik, Gallup asked: "Would you volunteer to be the first person to go up in an earth satellite?" Only 5 percent said they would, but Gallup pumped up the newsworthiness of this figure by extrapolating to the entire population: "5,100,000 Would-be Spacemen Volunteer for Satellite Trip!"[119] For many years we have known that three variables powerfully predict answers to questions about the space program: age, education, and gender. For example, in 1965 just 13 percent of American adults said they would like to go to the Moon, but the fraction was higher among men (18 percent), college-educated people (24 percent), and young adults aged 21–29 years (25 percent).[120] For another example, a 1969 news story based on a Gallup poll reported that 54 percent of people aged 21–29 supported an effort to land a man on the planet Mars, compared with

---

117. Analysis of original data from Harris study no. 861201, a February 1986 poll of 1,255 American adults, Odom Institute.

118. George Gallup, "Number Who Think Trip to Moon Possible Has Doubled in 5 Years," press release, 28 January 1955, American Institute of Public Opinion, Princeton, NJ.

119. George Gallup, "5,100,000 Would-be Spacemen Volunteer for Satellite Trip," press release, 18 December 1957, American Institute of Public Opinion, Princeton, NJ.

120. Poll of 2,534 American adults by Gallup, 24–29 June 1965, Roper Center USGALLUP.713. Q016; press release, 23 July 1965, American Institute of Public Opinion, Princeton, NJ.

only 40 percent of those aged 30–49 and 28 percent of those aged 50 and older. Among people with college training, 52 percent supported the idea, compared with 49 percent of those who had only attended high school and 25 percent of those with only a grade-school education.[121]

Until the 1980s public opinion researchers tended to discount gender differences in attitudes about public policies, except for noting a slight tendency of more women than men to say they did not have an opinion. Since then, some differences may have increased, even as researchers were discovering retrospectively that such differences had long existed in certain areas. Coauthors Robert Shapiro and Harpreet Mahajan have argued that men tended to be more favorable about policies involving force or violence and women to some degree favored compassionate policies or policies that protected people against dangers, but they placed the space program in an "ambiguous" category that did not fit any of these descriptions.[122] In January 1987 a poll done for *USA Today* reminded respondents of "the explosion of the space shuttle Challenger a year ago" and asked: "If you were offered the opportunity to ride on a shuttle mission, would you do it?" Some 58.2 percent of men claimed they would, compared with 26.6 percent of women.[123] Perhaps this difference merely represents greater bravado among men or an unwillingness to admit fear.

In 1966, Gallup reported the gender difference thus: "Men appear to be the more adventurous sex—at least in terms of space travel. Twenty-three per cent of men say they would like to go on the first trip to the moon, compared to 13 percent of women."[124] Even when there are high levels of agreement about a particular question, these differences exist. For example, a Gallup poll soon after the Challenger disaster found that 80 percent of Americans favored continuing the Shuttle program. Support was at 88 percent among those aged 18–29 but 72 percent among those 65 and older. Fully 89 percent of college graduates favored continuation, but just 60 percent of those who had not finished grade school did. The sex difference was similarly great: 87 percent of men wanting the Shuttle program continued, compared with 73 percent of women.[125]

---

121. "Young Adults Favor Space Push To Mars; Older Persons Opposed," press release, 7 August 1969, American Institute of Public Opinion, Princeton, NJ.

122. Robert Y. Shapiro and Harpreet Mahajan, "Gender Differences in Policy Preferences," *Public Opinion Quarterly* 50, no. 1 (1986): 42–61.

123. Analysis of original data from *USA Today* study 3002, a poll of 817 American adults done by Gordon S. Black Corporation in January 1987, Odom Institute.

124. "First Trip to Moon?—18 Million Americans Are Ready," press release, 3 August 1966, Gallup poll, Princeton, NJ.

125. "8 in 10 Favor Continuation of Manned Space Shuttle Program," press release, 23 March 1986, Gallup poll, Princeton, NJ.

TABLE 1.3. Gender Difference in Support for the Space Program

Year	Percentage Wanting Funding Increase		Correlation Between Male Respondents and Funding Increase
	Men	Women	
1974	11.6	4.8	0.34
1984	18.0	7.3	0.49
1994	14.0	6.0	0.31
2004	19.3	9.3	0.23

Table 1.3 uses data from the General Social Survey to compare the sexes at four points in time, separated by decades. In each year twice as many men as women felt that funding for the space program was too low. One way to compare the differences across years is in terms of the correlation between being male and giving more supportive answers to the question about funding, as measured by a coefficient called *gamma*. Note that the gamma is largest, indicating a greater tilt toward males, in 1984 when President Reagan's Strategic Defense Initiative was at its height. This is not just a matter of a male tendency to support the military, because in fact the sex differences are much greater for the space program than for defense budgets. Notably, in 1984, 18.3 percent of men felt military funding was too low, compared with 17.5 percent, nearly as high a rate, among women.

The trend from 1984 to 2004 in Table 1.3 hints that the gender difference may be declining, although it remains quite large. A Harris poll administered right after Pathfinder landed on Mars in 1997 asked: "If it were possible for people to travel to Mars, how interested would you be in going?" An identical question was asked about going to the Moon. Among men, 26.9 percent were "very interested" in going to Mars, compared with 30.3 percent very interested in taking the shorter and presumably less risky trip to the Moon. Among women, the percentages were 13.1 and 13.5 percent, respectively.[126]

Heather Mason Kiefer has interpreted gender-related results of a January 2004 online Gallup poll of teenagers thus:

> Since Sally Ride became the first American woman in space in 1983, more women have taken part in space missions, and one has even commanded a space shuttle. However, the fact remains that most astronauts are men. This may explain why boys seem to be more eager than girls to go to the Moon and Mars. Seventy-four percent of boys told Gallup they want to go to the Moon

---

126. Analysis of original data from Harris study no. 718288, a 9–14 July 1997 poll of 1,002 American adults, Odom Institute.

someday, compared with 43% of girls. About two-thirds (64%) of boys would like to be the first person on Mars, as would less than a third (31%) of girls.[127]

In 1998 a major online poll of children ages 13–15 found that 39.9 percent of 1,461 boys and 26.0 percent of 1,671 girls strongly agreed with the statement "If I were asked to go along on the first rocket trip to Mars, I would go." Combining both genders, 32.5 percent strongly agreed, and the percentage who merely agreed (without "strongly") was identical between the sexes, at 29.0 percent. Especially large numbers strongly agreed among children who were interested in the following activities: astronomy (47.1 percent), archaeology (42.6 percent), science (39.6 percent), rock climbing (41.0 percent), martial arts (42.3 percent), and scouts or guides (41.9 percent).[128]

Space travel is objectively risky, and males have traditionally been more ready to take physical risks than women. It also appears to be a test of physical and mental endurance. Thus perhaps people who say they want to travel in space may be males who engage in demanding physical activities here on Earth, such as outdoor sports. A poll done in 1989 for the newspaper *USA Today*, primarily devoted to outdoor activities, lets us explore this possibility. The crucial question was: "If you were asked to go along on a rocket trip to the moon, and such a trip were possible, would you want to go or not?" Altogether, 50.7 percent of the respondents claimed they wanted to go—66.1 percent of men and 35.0 percent of women. Other questions asked whether the respondent had participated during the previous year in the two dozen outdoor activities that are listed in Table 1.4.[129] The fact that spaceflight may be quite arduous and require physical stamina may be one reason older people are less interested in flying into space personally, but that factor cannot easily explain age differences in attitudes about the value of the program. Among 12,840 people who responded to the GSS from 1973 through 1983, the age group showing the greatest support for the space program was those aged 30–39, with 15.4 percent of them calling for a funding increase. People in their twenties showed a slightly lower level of enthusiasm, with 13.4 percent feeling that current funding for the space program was too little. The figure was 12.5 percent among those in their 40s. Older people showed progressively lower levels of support: ages 50–59 (10.2 percent), 60–69

127. Heather Mason Kiefer, "Would Today's Teens Take a Space Odyssey?," press release, 20 April 2004, Gallup News Service, Princeton NJ.

128. William Sims Bainbridge, "The First Martians," *Analog* 120, no. 7 (2000): 81–89.

129. Analysis of original data from *USA Today* study no. 3159, a poll of 612 American adults conducted by Gordon S. Black Corporation, July 1989, Odom Institute.

(7.5 percent), 70–79 (5.0 percent), and just 4.9 percent of the 288 elderly respondents in their 80s wanted space funding increased.

Both social scientists and historians are alert to the difference between *age effects* and *cohort effects*. An age effect relates simply to how old a person is when the data are collected—for example, the fact that more people over age 65 will report that they are retired—and this factor will hold over a wide

TABLE 1.4. Percentage of Respondents Wanting to Go to the Moon

Outdoor Activities	Men Whose Answer About the Activity Is		Women Whose Answer About the Activity Is		Average Difference Between Men and Women
	YES	NO	YES	NO	
Racquetball	92%	61%	63%	34%	30%
Windsurfing	100%	65%	57%	35%	29%
Scuba diving	83%	65%	71%	34%	28%
Martial arts	63%	66%	83%	34%	23%
Weight lifting or training	76%	60%	56%	31%	20%
Swimming laps	78%	61%	52%	29%	20%
Jogging or running	79%	55%	43%	31%	18%
Water skiing	80%	63%	51%	33%	18%
Sailing	78%	64%	55%	34%	17%
Snow skiing	76%	64%	51%	33%	15%
Tennis	78%	63%	47%	33%	14%
Some other individual sport	74%	62%	47%	31%	14%
Bicycling	74%	59%	41%	29%	13%
Hunting or shooting	68%	65%	55%	34%	12%
Canoeing or kayaking	69%	65%	51%	32%	11%
Traditional team sport	75%	59%	38%	34%	10%
Mountain or rock climbing	78%	64%	39%	35%	9%
Aerobics or dancercise	76%	65%	39%	32%	9%
Ice or figure skating	68%	66%	48%	34%	8%
Tent camping	73%	63%	38%	34%	7%
Hiking or backpacking	70%	65%	40%	33%	6%
Fishing	66%	66%	39%	33%	3%
Walking for exercise	63%	73%	37%	21%	3%
Golf	68%	66%	32%	35%	−1%

range of years when different studies are done. A cohort is a group of people born at roughly the same time, who therefore have a similar experience of historical events. For example, men born around 1920 were especially likely to serve in the military, because the massive mobilization for the Second World War came when they had entered early adulthood. A survey of the general public, asking about military service, will get different answers about the connection between age and service, depending on when the survey is administered historically in relation to the dates of past wars.

One could argue that older people are less interested in the space program because they do not expect to live long enough to benefit from its future accomplishments. Or one could argue that people tend to become more conservative as the years pass, becoming set in their ways and resisting innovation. Whatever the merits of such age-based theories, it is also possible to argue that the connection between age and support for the space program in 1973 through 1983 is really a cohort effect. People who came to adulthood before the space program existed learned to be content in a world without spaceflight, or they did not learn in school the facts about space that would help them understand the program today.

Table 1.5 looks for a cohort effect in four GSS datasets separated by decades, dividing respondents into four broadly defined birth cohorts. Each cell in the table containing a number is based on at least 100 respondents and usually far more, but two cells are empty because of the exigencies of births and deaths. Frankly, the numbers do not show very smooth patterns, and this may reflect complex responses of the cohorts to historical events as well as the random variation stemming from somewhat small numbers of respondents for some cells in the table. Without wanting to read too much into the trends, it may be worth comparing the last two rows of the table. The 1940–1959 cohort were growing up during the early, heroic years of the space program. When Sputnik was launching in 1957, they were younger than 18; when Apollo 11 landed on the Moon in 1969, they ranged in age from about 10 to 29. In contrast, the 1960–1979 cohort matured after Apollo and may not have attained the same level of excitement as the earlier cohort. For each year in the table, the 1940–1959 cohort shows more enthusiasm for space than the 1960–1979 cohort, although their figures have converged over the years.

In GSS data collected from 1994 through 2004, we see a somewhat different pattern of support by age from 1973 through 1983. Among those aged 20–29, 13.9 percent say space program funding is too low, compared with 11.9 percent among those aged 30–39, 13.6 percent of those in their 40s, and 14.9 percent of those in their 50s. The figure drops to 9.3 percent for those aged 60–69, 5.2 percent of those ages 70–79, and 5.3 percent of those aged 80–89. This pattern suggests a mixture of age effects and cohort effects. People

TABLE 1.5. Birth Cohort, Year of Survey, and Support for the Space Program

Year Respondent Was Born	Percentage Saying Too Little Is Being Spent on the Space Program			
	1974	1984	1994	2004
1900–1919	6.1	1.2	5.6	
1920–1939	10.6	9.1	4.4	10.9
1940–1959	7	19.9	12.2	15.3
1960–1979		9.3	10.3	14.2

who came of age during the exciting dawn of the space age continue to show somewhat greater enthusiasm, and elderly people show little enthusiasm.

Poll after poll has confirmed that people who are more educated respond more positively to questions about the space program, and this is probably to a very great extent the result of greater understanding of its nature and potential. In the early days of spaceflight few people had yet been informed, but over time the best-educated or technologically sophisticated individuals have developed a reasonably comprehensive understanding. A September 1969 Harris poll asked: "Do you recall or not having seen or heard about any events on TV which were transmitted over great distances by a communications satellite system?" A huge majority, 82.8 percent, said they had. Then the poll sought to learn how well informed the respondents were about communication satellites. "Now let me ask you if it is your impression that a satellite communications system, such as that which carried [a] live picture of the splash-down of the astronauts in the Pacific Ocean, works this way or not?" A majority, 65.9 percent, understood that "a TV picture is taken and then is sent to a satellite in the sky which relays the picture to the mainland." Fewer, only 46.7 percent, were confident that "the satellites in the sky go around the earth and that is why they can send pictures from Europe at only certain times." Most respondents, 69.0 percent, realized that "the satellite system does away with costly cables and wires." The final statement is especially interesting, because it alone was manifestly false: "The satellite system works on the principle of using light rays to carry pictures through space." A majority (53.5 percent) agreed that satellites use light rays rather than radio waves. Only 16.1 percent of respondents said this was not the way communication satellites work, and fully 30.4 percent were not sure.[130]

---

130. Analysis of original data from Harris study no. 1948, a poll of 2,465 American adults, September 1969, Odom Institute.

Periodically from 1979 through 2001, the Surveys of Public Understanding of Science and Technology commissioned by the National Science Foundation (NSF) asked respondents how well informed they were on "issues in the news," including "space exploration."[131] Of 21,288 people who were asked this question over the years, 12.3 percent said "very well informed," 50.6 percent said "moderately informed," and 36.9 percent admitted to being "poorly informed." Table 1.6 shows that well-informed people were also very interested, but a questionnaire survey like this does not allow us to establish cause and effect. Perhaps they were interested first and then informed themselves. But the data certainly are consistent with the view that NASA's programs to inform the public create interest, and the other rows of the table suggest they have wider beneficial effects as well. The NSF surveys contain a science and technology quiz, covering areas both close to space exploration and far from it. Especially for test items related conceptually to space or to the physical sciences, people who say they are well informed about the space program are more likely to give the correct answers.

The impact of space exploration depends in significant measure on the receptiveness of groups within the general public. It is also true that groups differ in their influence, even within a democracy. As we have seen, people with more education than the average are more supportive of the space program, and they tend to achieve more influential positions within society. Modern society is based on advanced science and technology, so the electorate needs some level of knowledge, both so voters can make their own decisions and so they can at least roughly evaluate the decisions of their leaders.[132] Michael Delli Carpini and Lee Sigelman noted in the mid-1980s that so-called yuppies (young urban professionals) were more supportive of the space program, with 21 percent calling for an increase in funding versus 12 percent among non-yuppies at the time; and these were people who certainly thought of themselves as potential leaders.[133] A study by Joan Black has shown that people are indeed influenced by informal opinion leaders among their friends and family members in deciding how to feel about the space program.[134]

---

131. Analysis of original data from the NSF's Surveys of Public Understanding of Science and Technology, 1979–2001, available from the Roper Center at the University of Connecticut, *http://www.ropercenter.uconn.edu/data_access/data/datasets/nsf.html*.

132. Jean M. Converse, "Predicting No Opinion in the Polls," *Public Opinion Quarterly* 40, no. 4 (1976–1977): 515–530.

133. Michael Delli Carpini and Lee Sigelman, "Do Yuppies Matter? Competing Explanations of Their Political Distinctiveness," *Public Opinion Quarterly* 50, no. 4 (1986): 502–518.

134. Joan S. Black, "Opinion Leaders: Is Anyone Following?," *Public Opinion Quarterly* 46, no. 2 (1982): 169–176.

TABLE 1.6. Space Awareness and Correct Answers on Science and Technology Quiz

True/False Statement	How Well Informed Is the Respondent Concerning Issues About Space Exploration?		
	Very	Moderately	Poorly
Respondent is very interested in issues about space exploration.	75.3%	31.2%	7.3%
Does the Earth go around the Sun, or does the Sun go around the Earth? (EARTH AROUND SUN)	85.2%	79.0%	68.7%
Which travels faster: light or sound? (LIGHT)	85.9%	80.6%	71.1%
The universe began with a huge explosion. (TRUE)	54.1%	42.3%	30.0%
Electrons are smaller than atoms. (TRUE)	60.5%	50.7%	39.8%
All radioactivity is man-made. (FALSE)	85.9%	82.3%	72.1%
The center of the Earth is very hot. (TRUE)	89.6%	83.7%	76.2%
The continents on which we live have been moving their location for millions of years and will continue to move in the future. (TRUE)	89.3%	83.4%	74.5%
Lasers work by focusing sound waves. (FALSE)	62.4%	46.8%	32.2%
The oxygen we breathe comes from plants. (TRUE)	87.5%	86.7%	82.5%
Cigarette smoking causes lung cancer. (TRUE)	96.2%	94.6%	93.2%
Antibiotics kill viruses as well as bacteria. (FALSE)	46.2%	46.0%	39.4%
It is the father's gene which decides whether the baby is a boy or a girl. (TRUE)	63.6%	67.7%	67.3%
Human beings, as we know them today, developed from earlier species of animals. (TRUE)	60.6%	49.7%	41.9%
The earliest humans lived at the same time as the dinosaurs. (FALSE)	58.5%	51.3%	44.5%

I conclude this section by examining a variable that is related to education and to achievement but that focuses on the person's ability to be a competent opinion leader, whether learned or trained—namely a measure of intelligence. The General Social Survey contains a 10-item word test, and the sum of the number of right answers is a rough measure of IQ. To be sure, one would never want to rely on such a short test to advise an individual about educational or career options; but averaged across large numbers of respondents, it is a fairly good proxy for a more complete IQ test.

Combining data across all years of the GSS from 1973 through 2004, just 12.4 percent of American adults felt that funding for the space exploration program was too little, but 18.7 percent of those who scored a perfect 10 on the word test did so.[135] In contrast, only 6.6 percent of those scoring zero on the word test wanted funding increased. The relationship was consistent across the full range of scores on the test, within the limits of statistical certainty. Among those who scored 7, 14.8 percent wanted funding increased, compared with 11.7 percent of those scoring 6, bracketing the average of 12.4 percent. In its 1993–1994 surveys, the GSS included an item about belief in astrology that can help us think more deeply about intelligence and support for space exploration. Respondents were asked to react to this statement: "Astrology—the study of star signs—has some scientific truth." Among 2,592 GSS respondents, 9.9 percent said this was "definitely true," and 42.6 responded "probably true." On the other side of the issue, 24.9 percent felt the pro-astrology statement was "probably not true," and 22.6 percent said it was "definitely not true." Thus more than half of the public is prepared to believe that astrology has a scientific basis, and less than a quarter shares the view of scientists that it is complete nonsense.

Figure 1.2 graphs this astrology item and the space program funding item across the range of scores on the verbal IQ test. The two lines on the graph are not perfectly straight because of random fluctuations in the data across the eleven groups of respondents, but they show a clear negative correlation between IQ and belief in astrology and a positive correlation between IQ and support for the space program. These results remind us that education about the realities and potentials of spaceflight face an uphill battle against ignorance as well as indifference. A pessimist might argue the struggle is hopeless, on the assumption that IQ is somehow biologically innate in the individual. However, scores on the word test are partly the result of education and the individual's motivation to learn.

Eric Chaisson has written compellingly that accessible results like the pictures from the Hubble Space Telescope can serve a powerful educational function for the public, communicating vividly the nature of the universe in which we dwell. He is appalled by the findings of the NSF study that so many people are ignorant of science, but he is also hopeful that vigorous space-related educational efforts can improve the situation.[136] In March 1997, Gallup asked 744 children aged 7–12 whether they considered various subjects "so exciting that you would like to study some more about it in school." Some 60.7 percent felt that "space exploration, including the planets, space travel, and special projects like the Hubble Telescope," was this exciting. They

---

135. Analysis based on original GSS data, with appropriate handling of the 1982 and 1987 oversamples and 2004 weighting procedures; data provided by the Computer-assisted Survey Methods Program at the University of California, Berkeley.
136. Eric J. Chaisson, *The Hubble Wars* (New York: HarperCollins, 1994), p. 31.

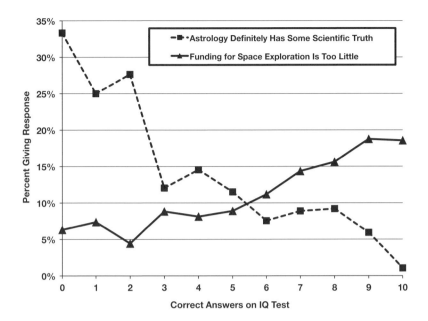

FIGURE 1.2. Correlation of IQ, Support for the Space Program, and Belief in Astrology

showed about the same level of excitement for "new advances in computer technology, such as faster processing chips and more sophisticated software" (61.8 percent) and "medical research such as cloning and hi-tech ways to study and treat human diseases" (65.8 percent).[137] Thus NASA's educational efforts are not merely public relations for the space program. Rather, they inform attentive people of all ages about the nature of the universe, and they encourage all people to exercise their minds to the fullest extent, which is beneficial in all aspects of their lives.

5. Motivations for Space Exploration

One way to understand public perceptions of the value of space exploration is to examine how people connect it to other government activities. Earlier I analyzed trends in support using an item from the GSS that asked whether the government was spending "too much money on it, too little money, or about the right amount." In 1973, 1,044 people gave valid answers for the "space exploration program" and 10 other funding areas. I performed

---

137. "Teenage Study," a Gallup poll of 744 children aged 7–12, 20–27 March 1997, Gallup Organization, Princeton, NJ.

a factor analysis of these eleven variables.[138] A factor analysis first calculates correlations linking all pairs of variables, then seeks to find a smaller number of dimensions of variation that reduce the complexity of the data without losing important information. The result was four factors, essentially four mathematical dimensions along which each of the eleven government activities could be graphed. Figure 1.3 is a map in terms of the first two factors or dimensions.

The first factor was dominated by two social problem areas, "halting the rising crime rate" and "dealing with drug addiction." The second factor emphasized "improving the conditions of Blacks" and "welfare" but also brought in "foreign aid," which was not involved in the first factor. Two other areas of government funding were ranked about equally on both of these factors, "solving the problems of the big cities" and "improving the nation's education system." Note that these factors concern social problem areas and domestic government programs about which people have significant disagreements and about which the political parties are constantly debating. At

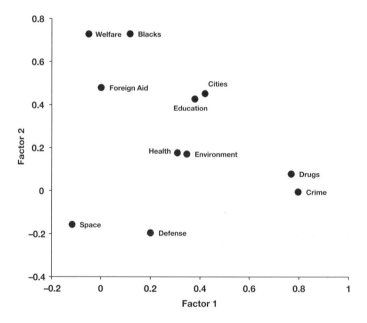

FIGURE 1.3. Factor Analysis of Funding Support in Eleven Areas

138. In my analysis I used the original data obtained from the Computer-assisted Survey Methods Program at the University of California, Berkeley, and employed listwise deletion of cases with missing data, principal components analysis, focusing on factors with eigenvalues greater than 1.00 and employing varimax rotation.

the risk of oversimplification, Factor 1 seems to be concerned with imposing control on dangerous situations, whereas Factor 2 concerns offering help to people in need.

The third factor, which is not graphed in Figure 1.3, is rather different from the other three, in that some variables are loaded positively on it while others are loaded negatively. Most strongly connected is "the military, armaments and defense," which is at the negative end of that third dimension. Also on the negative end is foreign aid, although not so far. If we could project Figure 1.3 in three dimensions, "defense" would be perhaps three inches behind the plane of the printed page, and "foreign aid" would be two inches behind. Two other items would appear in front of the plane of the printed page—"improving and protecting the environment" and "improving and protecting the nation's health"—reflecting their positive loadings on the third factor. It makes sense that these items are paired in this way because foreign aid and defense both concern international relations, and there are many connections between the health of a person and the health of the environment.

Where is the "space exploration program?" It heads the fourth factor, with the most powerful connection of any of the eleven programs to any single factor. "Improving and protecting the environment" shows a moderate connection to the factor, as well, but nothing like the powerful connection of the space program. This analysis shows that Americans in 1973 saw space exploration as a separate issue, largely unconnected to any of the domestic or international issues. Examining the individual associations between space exploration and the others, there are small but statistically significant negative correlations with "improving the conditions of Blacks" and "welfare" ($r = -0.11$, $r = -0.12$), perhaps reflecting the view that money saved by cutting the space program could be used to help poor people.

The same kind of analysis is often instructive when applied to other poll data, going back at least as early as 1962, allowing us to see how space connected to other issues in people's minds just a year after President John Kennedy had set the goal of reaching the Moon before the end of the decade. From May through September, Harris had done statewide surveys in Colorado, Illinois, Indiana, Kentucky, and New Hampshire, asking respondents "how you think President Kennedy has done" in dealing with 20 things, using a four-step response scale from "excellent" down to "poor." Such polls of presidential performance had already become a significant part of political culture.[139] Combining the five states, with no attempt to weight the sum for

---

139. Charles W. Ostrom Jr. and Dennis M. Simon, "Promise and Performance: A Dynamic Model of Presidential Popularity," *Public Opinion Quarterly* 79, no. 2 (1985): 334–358.

their populations, I carried out a factor analysis that produced three factors, based on data from 807 respondents who expressed views on all 20 issues.[140]

The first factor emphasized government economic policies, and its top five items were: "cutting down on spending" (loading = 0.74), "handling of the stock market" (0.74), "keeping the economy healthy" (0.73), "handling U.S. Steel's attempt to raise prices" (0.70), and "medical care for the aged" (0.69). "Catching up to the Russians in space development" headed the second factor, with a loading of 0.73. The next four items in the factor with space concerned the competition with the Soviet Union: "conducting tests of atomic weapons" (0.73), "handling the Berlin crisis" (0.66), "handling Khrushchev" (0.59), and "handling the crisis in Laos" (0.55). The third factor was dominated by one item on the issue of whether Kennedy's Catholic religion would influence his policies, "not giving federal aid to parochial schools" (loaded = 0.79). Reasonably enough, "federal aid to education" was also loaded on this factor but way back at a loading of 0.45, given that this item was also involved in the first factor.

This same method can be used with several other sets of data, although sometimes the results are rather more complex. For example, immediately before President Kennedy was assassinated in November 1963, the Harris poll asked a national sample: "How would you rate the job President Kennedy has done in handling this country's space program?" Of 1,264 people answering, 27.6 percent said he was doing an excellent job, 47.4 percent said a pretty good job, 14.3 said only fair, and 4.1 said poor, with the remainder not sure. The poll later asked people how good a job Kennedy was doing in 16 other areas, so I tried a factor analysis following the procedures just described for these 16 plus space. Only two factors emerged, and space was equally connected to both of them. It is possible to instruct the Statistical Package for the Social Sciences (SPSS) program I was using to look for a particular number of factors, so I experimented with expanding the data into more dimensions.[141]

With six factors the space program wound up heading the sixth factor, and no other item was strongly loaded with it. So, on some level, in 1963 people saw space exploration as a very different issue from the others, just as they would do 10 years later in the GSS. Trying an analysis with five factors, the space program was strongly loaded on the second factor, right behind

---

140. Analysis based on a dataset combining five Harris polls (#1172 in May 1962 in Colorado, #1166 in June 1962 in Kentucky, #1173 in June 1962 in New Hampshire, #1195 in July 1962 in Illinois, and #1212 in September 1962 in Indiana), Odom Institute.

141. Analysis of original data from Harris study no. 1285, a November 1963 poll of 1,283 American adults, Odom Institute. Using listwise deletion, the factor analysis was based on the 555 respondents who expressed an opinion about all 17 aspects of Kennedy's performance.

"standing firm on Berlin," "keeping the military defense of the country strong," "working for peace in the world," "handling Khrushchev," and "getting Russia to agree to an atomic test ban." In June of that year, Kennedy had given his famous "Ich bin ein Berliner" speech, defending the city which at that time was still divided by the Berlin Wall and surrounded by Communist East Germany. Kennedy's long-standing competition with Soviet leader Nikita Khrushchev was among the motivations that led Kennedy to propose the Apollo Moon program. In this five-factor solution the first factor was primarily about the quality of government, and the first six items, in descending order, were: "handling farm problems," "keeping spending under control," "getting Congress to act on [the president's] programs," "moving the country ahead," "keeping the economy healthy," and "keeping corruption out of government." Thus Americans were somewhat of two minds in their conceptualization of space exploration. In part, it was a separate issue, unrelated to many other things the government was doing, and in part it was connected to the superpower competition between the United States and the Soviet Union.

A 1965 poll of college students asked respondents to make two judgments about a list of 13 "problems." First, they judged how much progress had been made "in the last few years." With respect to "getting to the moon and other planets," 78.3 percent said "a lot" of progress had been achieved, and a further 20.6 percent said "some." Factor analysis allows us to see how the college students categorized the issues. "Getting to the moon and other planets" was most strongly loaded (0.66) on the fourth factor, right between "finding a cure for cancer" (0.71) and "curing the common cold" (0.54). Factor 1 was headed by "keeping marriages from breaking up" (10.69) and "controlling narcotics use" (0.62). Factor 2 was headed by "eliminating economic depressions" (0.64) and "eliminating racial barriers" (0.62). Factor 3 was headed by "enabling people to use their creative talents fully" (0.77) and "helping the individual to work things out" (0.64). So students had placed space exploration in the same science-related category as two medical issues, given that the other categories concerned problems facing societal institutions that were not related to the natural sciences or engineering.[142]

Students were also asked to answer this question about each problem: "Do you think that your generation will make major strides in [this area] or do you think this will continue to be a major problem?" Fully 93.7 percent felt their generation could make major strides toward "getting to the moon and other planets." Only one other problem came close—"finding a cure

---

142. Analysis of original data from Harris study no. 1431, a poll of 793 college students, February 1965, Odom Institute; the factor analyses used the criterion of eigenvalues greater than 1.00 with pairwise deletion of cases with missing data.

for cancer"—where 81.4 percent thought major strides would be made. A factor analysis for these predictions might not be expected to give the same results, in part because people may view the future very differently from the past and because there was so much agreement about space exploration that statistical variation might be too weak to measure respondents' categorization. However, "getting to the moon and other planets" (loading = 0.51) again wound up in the same factor with "finding a cure for cancer" (0.78) and "curing the common cold" (0.66).

In June 1966, Harris asked 1,130 Americans questions about President Lyndon Johnson's performance in 22 areas, including "handling the space problem." The answers were similar to those for Kennedy three years earlier, with 27.3 percent saying he was doing an "excellent" job, 43.9 percent a "pretty good" job, 10.0 percent only "fair," and 6.3 percent "poor." When I instructed the computer to give me an analysis with only two factors, the space item headed the second factor with a loading of 0.69. Other items following the space issue were "getting Congress to pass his program" (loading = 0.68), "keeping the military defense of the country strong" (0.66), "handling Russia" (0.65), "getting aid to education" (0.63), and "working for peace in the world" (0.57). The other factor was headed by "keeping federal spending under control" (0.81), "keeping the cost of living down" (0.78), and "keeping corruption out of government" (0.70).[143]

Three factors did a better job of mapping the 22 items, and I also tried four-, five-, and six-dimensional solutions. In a three-factor solution the space item headed the third factor with a loading of 0.70, followed by "getting Medicare for older people" (0.65), "getting aid to education" (0.61), and "getting Congress to pass his [President Johnson's] program" (0.60). To me, these sound like examples of discretionary spending. Factor 1 in this analysis concerned government management and economic issues: "keeping federal spending under control" (0.79), "keeping the cost of living down" (0.71), "anti-poverty program" (0.68), "handling foreign aid" (0.66), "keeping corruption out of government" (0.64), and "handling taxes" (0.61). Factor 2 was all about international tensions: "handling Red China" (0.73), "handling the situation in the Dominican Republic" (0.70), "handling the war in Vietnam" (0.68), "working for peace in the world" (0.68), "handling Russia" (0.61), "handling Castro and Cuba" (0.52), and "giving leadership to the free world" (0.52).

---

143. Analysis of original data from Harris study no. 1635, a June 1966 poll of 1,130 American adults, Odom Institute. The factor analyses employed pairwise deletion, and the three-factor solution was the one that preserved all factors with eigenvalues greater than 1.00.

Notice that "handling Russia" and "handling the space problem" wound up in different factors in these 1966 data. During the 1960s, doubts progressively grew about whether Russia was really racing the United States to the Moon, perhaps beginning with comments Khrushchev had made in a press conference in October 1963, suggesting his country was not planning Moon flights and was content to live on Earth.[144] As far as alert members of the public could tell, the United States had caught up with Russia a year before this poll, when the first American spacewalk followed the first Russian one by only three months. By the time of the poll, the U.S. Gemini program was in the dramatic early stages of experimenting with orbital rendezvous.

In July 1969, immediately after the landing of Apollo 11, a Harris poll included questions about space, the first in a set concerning President Richard Nixon's performance. A factor analysis looking for two factors from one set of 12 items included "his talking to the astronauts when they were on the moon," which led Factor 2 with a loading of 0.81. Other items near the top of this factor were: "his trip abroad" (loading = 0.76), "inspiring confidence personally in the White House" (0.69), and "his press conferences" (0.68). Clearly this factor is about Nixon's attempts to communicate. The first factor was defined by domestic economic and social issues: "keeping down the cost of living" (0.83), "his approach to taxes and spending," "keeping the economy healthy" (0.73), "his approach to crime and law and order" (0.65), and "his handling of the race and civil rights question" (0.62). One way to read these data is to infer that Nixon talking to the astronauts was simply a politician playing his role as a communicator.

Another battery of items listed eight "possible American-Russian areas of negotiation and agreement," including "agreement to joint exploration of space." The poll first asked respondents whether each was a possible area "where some agreement might be reached." Then the poll asked the respondent: "Regardless of whether you think this is an area where agreement might be reached or not, would you favor or oppose agreement between the U.S. and Russia?" For the space item, 56.0 percent thought agreement was possible, 12.2 were not sure, and 31.9 percent thought it was not possible. Setting that judgment aside, 61.8 percent favored joint exploration of space, 9.9 percent were not sure, and 28.3 percent opposed it. Table 1.7 shows the percentage favoring cooperation in each of the eight areas, the percentage thinking it was possible, and results of factor analyses for both sets of questions. Interestingly, space exploration heads Factor 1 for favoring cooperation and is tied for first place with thinking it was possible. The two parallel

---

144. Joseph G. Whelan, "The Press and Khrushchev's 'Withdrawal' from the Moon Race," *Public Opinion Quarterly* 32, no. 2 (1968): 233–250.

factor analyses divide the eight items into the same two sets of five concerning improved relations between the United States and Russia, and three items concerning joint action toward other nations. It is also interesting to see the space item taking such a leading role in definition of the factors, apparently because Americans found the idea of cooperation in space quite clear and logically connected to other kinds of cooperation.[145]

TABLE 1.7. In the Month Apollo 11 Landed: Cooperation with Russia

	Percentage In Favor	Loadings		Percentage Possible	Loadings	
		Factor Fav-1	Factor Fav-2		Factor Pos-1	Factor Pos-2
Improved Interactions with Each Other						
Agreement to joint exploration of space	61.8%	0.82	0.08	56.0%	0.74	0.13
Agreement for joint exploration of the oceans	70.6%	0.74	0.25	62.1%	0.66	0.34
Agreement to greatly expand trade between the two countries	74.1%	0.70	0.27	69.2%	0.74	0.19
Agreement to limit anti-missile (ABM) systems	71.3%	0.62	0.23	54.4%	0.60	0.29
Agreement for more exchange of scholars and cultural groups	79.2%	0.60	0.37	78.2%	0.71	0.12
Joint Action Toward Other Nations						
Agreement to act together to achieve peace in the Middle East	82.7%	0.33	0.72	58.1%	0.27	0.70
Agreement to take joint military action if another nation threatens to use nuclear weapons	66.6%	0.20	0.81	54.7%	0.22	0.78
Agreement on joint action to prevent Communist China from starting wars	73.9%	0.19	0.86	57.5%	0.14	0.82

145. Analysis of original data from Harris study no. 1944, a July 1969 poll of 1,601 American adults, Odom Institute. The two factor analyses employed pairwise deletion and varimax rotation and specified that two factors should result.

In a different July 1969 Harris poll, the interviewer handed the respondent a card, saying, "This card shows the percentage of the federal budget now spent for various purposes. Considering priorities, would you like to see more or less of the federal money go into each of these purposes?" In the case of the space program, only 14.5 percent wanted more spent, while 63.7 percent wanted less spent, and 21.8 percent were not sure. Absolute majorities wanted more spent on veterans' benefits (62.3 percent); education (77.1 percent); natural resources such as improvement of environment, pollution control, and parks (65.7 percent); and housing and community development (52.4 percent).

A factor analysis of the 10 items found four dimensions. Factor 1 was headed by natural resources (loading = 0.72) and education (0.54). The second factor highlighted health, labor, and welfare (0.82) along with housing and community development (0.71). The space program led the third factor with a loading of 0.75, followed by international affairs (0.66). The fourth factor combined veterans' benefits (0.75) with national defense (0.62). The remaining items (commerce and transportation, agriculture) were not heavily loaded on any factor. The factor analysis reflects the connections among all 10 items, but on the level of pairs of items, the space program correlated strongly not only with international affairs (0.19) but also with commerce and transportation (0.17), with its correlation with defense coming third (0.12)—keeping in mind that correlations are generally much smaller numbers than factor loadings are. Note that among this group of issues, which focus on the space program in the international context, it was perhaps seen as a mode of transportation rather than primarily in terms of its military relevance.[146] In the 1960s many scholars and historians of technology conceptualized spaceflight by analogy with earlier modes of transportation, including even railroads.[147]

An October 1969 Harris poll listed 16 "problems the United States faces at home," asking how much attention each one deserved. A factor analysis produced four factors. "Space exploration" wound up in the fourth factor, with a very high 0.86 loading. Only one other item was also strongly loaded, "missile defense" at 0.78. The first factor clearly collected social welfare problems: poverty (loading = 0.74), racial discrimination (0.71), health care (0.65), education (0.60), and housing (0.58). The second factor was civil

146. Analysis of original data from Harris study no. 1939, a July 1969 poll of 2,089 American adults (about two-thirds of whom were asked this set of questions), Odom Institute. The factor analysis employed pairwise deletion and preserved all factors with eigenvalues greater than 1.00.

147. Bruce Mazlish, ed., *The Railroad and the Space Program: An Exploration in Historical Analogy* (Cambridge, MA: MIT Press, 1965).

infrastructure: highway construction (0.70), mass transportation in and around cities (0.61), recreation (0.58), and consumer protection (0.53). The third factor concerned crime and taxes: crime on the streets (0.73), organized crime (0.65), and taxes and inflation (0.62).[148] Among respondents to a January 1971 Harris poll that asked about funding for eight government programs, only 16.8 percent wanted the space program increased, as opposed to 49.2 percent who wanted it decreased. Some 29.5 percent did not want any change in the funding level, and 4.6 percent were not sure. Space wound up in the second factor with a loading of 0.71, between "the ABM missile system" (0.75) and "the war in Vietnam" (0.54). The first factor consisted of "aid to cities" (0.77), "aid to public schools" (0.70), and "programs to control air and water pollution" (0.65). The third factor combined "support prices for farmers" (0.74) with "foreign military and economic aid" (0.63).[149]

A December 1974 Harris poll asked respondents to evaluate nine "proposals which have been made for possible agreement between the United States and the Soviet Union." A solid majority of 60.9 percent favored "undertaking joint space missions," with 27.8 opposing and 11.3 percent not sure. Given the fact that all the items concerned the narrow topic of agreement between the United States and the USSR, it is not surprising that the correlation matrix resisted factoring; but when four factors were called for, a rather clear structure appeared. "Undertaking joint space missions" loaded 0.83 on the third factor, with "exchanging scientists and other technical missions" close behind at 0.79. The top two items in the first factor were "undertaking joint efforts to solve the world energy shortage" (0.83) and "undertaking joint efforts to curb air and water pollution" (0.81). Most heavily loaded on the second factor were "giving the Soviet Union the same trade treatment that we give other countries" (0.85) and "expanding trade between the United States and the Soviet Union" (0.80). The remaining factor was dominated by "reducing the number of American and Soviet troops in Europe" (0.85) and "substantially limiting the number of nuclear missiles each country has" (0.72).[150]

---

148. Analysis of original data from Harris study no. 1970, an October 1969 poll of 1,982 American adults, Odom Institute. The factor analyses employed pairwise deletion and varimax rotation, preserving all factors with eigenvalues greater than 1.00.

149. Analysis of original data from Harris study no. 2055, a January 1971 poll of 3,092 American adults, Odom Institute. This combines the "not change" and "not sure" responses as the middle category between decrease and increase, pairwise deletion, factors with eigenvalues greater than 1.00, and varimax rotation.

150. Analysis of original data from Harris study no. 2436, a December 1974 poll of 1,843 American adults, Odom Institute.

A March 1976 poll asked respondents how serious a loss they felt it would be if the federal government abolished each of 14 programs. In the case of "space programs," only 21.3 percent felt it would be a "very serious loss," and another 29.7 percent felt cancellation would be "only a moderate loss." The largest group, 40.8 percent, said ending the space program would be "hardly a loss" at all, and 8.1 percent were not sure. In a factor analysis seeking four factors, "space programs" was loaded 0.73 on the fourth factor, alongside defense at 0.75. The first factor brought together "environmental controls" (0.70), "aid to cities" (0.67), "business regulation" (0.63), "welfare" (0.62), and "revenue sharing" (0.58). The second factor combined "education" (0.83), "health" (0.81), and "jobs for [the] unemployed" (0.62). The third factor consisted of "law enforcement" (0.75) and "social security" (0.66), two personal security issues.[151]

A July 1977 Harris poll asked about President Jimmy Carter's performance, including "his continuing the space shuttle program." On this, 12.4 percent rated him "excellent," 37.5 percent "pretty good," 16.6 percent "only fair," and 10.4 percent "poor." There were fully 60 other items in the list, and a standard analysis produced a bewildering 11 factors. Factor 7 consisted of this space item with a loading of 0.69, flanked by two nuclear technology items: "his support for production funds for the neutron bomb" (0.73) and "his wanting to speed up construction of conventional nuclear plants" (0.58). It is worth noting that of all recent presidents, Carter probably understood the technical aspects of these issues best, having received some graduate education in nuclear technology. Because of the vast list of performance areas respondents were asked to judge, this dataset is probably not comparable to the others; but the connection between space technology and nuclear technology is interesting to see.[152]

A month after President Ronald Reagan announced plans for the Strategic Defense Initiative in 1983, a Harris poll asked respondents to rate him on his performance in 15 areas, including "his proposing to construct new weapons systems in outer space that would be capable of shooting down nuclear missiles with laser and particle beams." Just 18.9 percent called his performance in this area "excellent," compared with 26.4 percent "pretty good," 27.2 percent "only fair," and 27.5 percent "poor." A factor analysis of responses to all 15 items put this space issue in the second factor with a strong loading of

---

151. Analysis of original data from Harris study no. 7681, a March 1976 poll of 1,523 American adults, Odom Institute.

152. Analysis of original data from Harris study no. P3725, a July 1977 poll of 1,515 American adults, Odom Institute. The factor analyses employed pairwise deletion and preserved all factors with eigenvalues greater than 1.00.

0.72. Only one item was more strongly loaded on this factor, at 0.74 only slightly higher: "his handling of nuclear arms reduction negotiations with the Russians" (0.74). The next two items strongly associated with this factor also stressed defense: "his saying he would not accept cuts from Congress on defense spending" (0.70) and "his proposing to the Russians to agree on cutting down but not eliminating nuclear missile installations in Europe" (0.67). The other factor concerned economic issues: "handling the economy" (loading = 0.79), "his handling of Social Security" (0.76), "getting the country out of the recession" (0.74), "helping the unemployed, small business, farmers, and others in economic trouble" (0.73), and "his handling of federal jobs programs" (0.71).[153]

The Strategic Defense Initiative is a complex topic, in many ways quite separate from the civilian space program. The SDI technologies that have come closest to maturity involved not orbiting lasers or particle beams but surface-launched rocket interceptors. As noted earlier, all the publicity about hypothetical space-based defense systems in the mid-1980s may have strengthened the connection between space and defense in the public mind. This connection may have weakened subsequently, as space-based weapons faded from the news after the mid-1980s. NSF's Surveys of Public Understanding of Science and Technology can confirm the changing popular association of space with military applications.[154] In 1983, interest in space correlated at a loading of 0.32 with interest in "military, defense policy." By 1985, after two years of SDI publicity, the correlation had strengthened to 0.44. Figure 1.4 maps the changes from 1985 to 2001, showing two superimposed factor analyses of these two items plus "new scientific discoveries," "the use of new inventions and technologies," and "new medical discoveries." For each of the five topics, the open circle plots the 1983 factor loadings, while the solid circle plots them for 2001. Naturally, there is some random wandering of the points, but notably space moves away from defense.

A Harris poll in 1993 asked respondents to say whether the country should spend more or less on eight different kinds of scientific research. Only 9.2 percent wanted to spend "a lot more" on "space research on space exploration and development." Another 20.5 percent said "a little more," while much larger fractions said "a little less" (36.1 percent) and "a lot less" (34.2 percent).

---

153. Analysis of original data from Harris study no. 832103, an April 1983 poll of 1,250 American adults, Odom Institute. The factor analyses employed pairwise deletion and preserved all factors with eigenvalues greater than 1.00.

154. Analysis of original data from the NSF's Surveys of Public Understanding of Science and Technology, 1979–2001, available from the Roper Center at the University of Connecticut, *http://www.ropercenter.uconn.edu/data_access/data/datasets/nsf.html*.

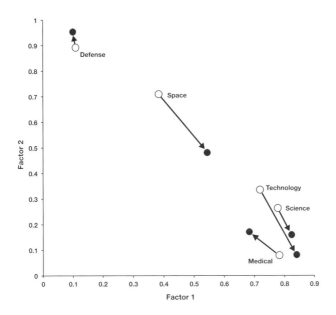

FIGURE 1.4. Two Factor Analyses Showing Changes Over 16 Years

(A related question found that only 1.4 percent of respondents thought space research was most valuable, compared with 66.2 percent for medical research and 17.5 percent for environmental research). Figure 1.5 maps the first two factors of a three-factor analysis. The third factor was dominated by "defense research to develop new weapon systems," which had a 0.84 loading. The space item was the only other one strongly loaded on the factor, with 0.62. In terms of ordinary correlations (which are generally much smaller than factor loadings), the space item was most strongly associated with "electronic research on improved television and electronic equipment" (0.28), followed by the defense item (0.24) and "computer research to improve access to information" (0.21). Thus the factor analysis shows space somewhat apart from the other seven fields of research but loosely tied to defense, and simple correlations show it connected to defense, computers, and electronics.[155]

A 1999 poll done for the Pew Center for the People and the Press asked respondents their views on "a list of some changes that have taken place over the last 100 years." "Space exploration" received good ratings, with 72 percent saying it had been a change for the better, versus only 6 percent who said it was a change for the worse and 17 percent judging it "hasn't made

155. Analysis of original data from Harris study no. 931107, an 11–15 November 1993 poll of 1,254 American adults, Odom Institute. The factor analyses employed pairwise deletion and preserved all factors with eigenvalues greater than 1.00.

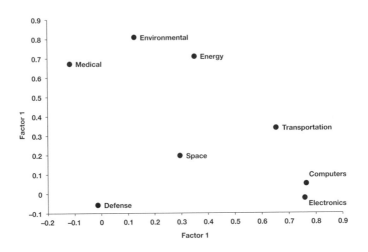

FIGURE 1.5. Factor Analysis of Eight Kinds of Scientific Research

much difference."[156] The inventions judged by more people to have been improvements were radio (96 percent), automobile (91 percent), computer (87 percent), highway system (84 percent), airline travel (77 percent), television (73 percent), and birth control pills (72 percent). Two kinds of technology rated much lower than the space program—nuclear energy (48 percent) and nuclear weapons (19 percent). Indeed a majority of 63 percent felt that nuclear weapons had made things worse. Respondents were divided into two groups, each presented with only half the items; but factor analysis can help us see how the space program connects to eight other items. Space exploration was most heavily loaded on the third factor, 0.77, and airline travel was not far behind at 0.67. The first factor combined two items related to domestic intimacy: more acceptance of divorce (0.76) and birth control pills (0.48). The second factor linked the civil rights movement (0.72) to rock and roll music (0.71). The final factor was headed by the invention of the computer (0.86) and the invention of television (0.55). "The development of nuclear energy" was not strongly loaded on any factor.[157]

How people conceptualize space exploration depends on the hotly debated issues of their historical period and on the mental connections they make between it and some of these other issues. Thus mental maps are not

---

156. "Technology Triumphs, Morality Falters," press release, 3 July 1999, Pew Research Center, *http://people-press.org/reports/display.php3?PageID=279*.

157. Analysis of original data from a poll of 1,546 adults, done by Princeton Survey Research Associates, 6 April–6 May 1999, Pew Center.

static but vary over time and across different contexts presented by pollsters. Clearly in the 1960s the space program was implicated in the prestige contest between the United States and the USSR, and in the early 1980s it took on added military connotations. Depending on the other issues presented to poll respondents, it could also be connected to transportation and to various scientific and technological activities. At the same time, space stood far apart from major social issues, such as government management of health and welfare programs, and to a very significant extent was conceptualized by Americans as a wholly unique issue.

6. Goals in Space

Some of the questionnaires most useful for assessing the value of space exploration in detail were not administered to random samples of the population but to special groups who might be expected to be more knowledgeable or thoughtful. For example, in 1960, Raymond A. Bauer asked 1,717 readers of the *Harvard Business Review* to rate five alternative justifications for the space program, listed here in Table 1.8. In 1963, Edward E. Furash administered the same items to 3,300 *Review* readers, and in 1986, I obtained responses from 1,007 Harvard students.[158] All three groups of respondents put "pure science research" in first place, but the two polls from the early 1960s placed military reasons second. Although the 1986 Harvard students are not strictly a comparable group, and their much lower interest in military goals may reflect their political liberalism and opposition to the Reagan administration, this difference also probably reflects a change over time. The initial impetus for the space program may have been international competition, but over time this motive has become less important.

On the assumption that physicists should have a better-grounded perspective on space technology than almost any other group in society, Donald Strickland surveyed 211 members of the American Physical Society late in 1963. When asked to rank the motives for the American space program, 32 percent selected "propaganda and prestige first," and 14 percent placed "military" in second place. Some 5 percent each selected "domestic political" motives, "exploration," and "basic research in natural sciences." Another

---

158. Raymond A. Bauer, "Executives Probe Space," *Harvard Business Review* (September–October 1960): 6–14, 174–200; Edward E. Furash, "Businessmen Review the Space Effort," *Harvard Business Review* (September–October 1963): 14–32, 173–190; and William Sims Bainbridge, *Goals in Space* (Albany: State University of New York Press, 1991), p. 19.

TABLE 1.8. Priorities for the Space Program

Five Possible Objectives for the Space Program	Sources of Space Program Objective Rating		
	Harvard Business Review		Harvard Students
	1960	1963	1986
Pure science research and gaining of knowledge	47%	43%	55.4%
Control of outer space for military and political reasons	31%	31%	4.1%
Tangible economic payoffs and research results for everyday life on Earth	14%	18%	30.5%
Meeting the challenge and adventure of new horizons	8%	8%	9.2%
Winning the prestige race with the Soviet Union	3%	5%	0.8%

4 percent cited "economic" goals. Apparently the remaining 35 percent were unwilling to evaluate the motives for the space program.[159]

In January 1987 respondents to a *USA Today* poll were given a list of seven things that might be "man's next great achievement in space" and asked to select the goal they would set. The largest number, 25.2 percent, selected "live in space station," the goal President Reagan had announced three years earlier. More than twice as large a group, 54.0 percent, could not select one of the proffered goals and chose the default "other" answer. Additional choices included "go to other planet" (9.9 percent), "regular space trips" (3.2 percent), "colony on moon" (2.6 percent), and "return to moon" (2.2 percent). Perhaps, mercifully, only tiny fractions selected the two goals that were effectively impossible: 1.7 percent wanting the space program to "go to other solar system" and 1.1 percent preferring "time travel."[160]

Five days after Pathfinder landed on Mars in 1997, Harris asked respondents whether they favored or opposed seven different plans for future space exploration. All the plans received favorable responses, but the most support was expressed for the least innovative plan: 86.2 percent favored "continuing to use the space shuttle for scientific research." In second place, with 79.9 percent favoring the idea, was "joint space missions involving Americans, Russians, and people from other countries." Three other plans were essentially

---

159. Donald A. Strickland, "Physicists' Views of Space Politics," *Public Opinion Quarterly* 29, no. 2 (1965): 223–235.

160. Analysis of original data from *USA Today* study 3002, a poll done by Gordon S. Black Corporation in January 1987 of 817 American adults, Odom Institute.

tied: "sending more unmanned probes to explore other planets in the solar system" (favored by 70.9 percent), "putting a manned US space station in space" (70.9 percent), and "putting a manned joint US and International space station in Earth's orbit" (69.6 percent). The two lowest-rated plans involved bold human expeditions: "sending a manned rocket to land on Mars" (63.7 percent) and "sending a manned rocket to the Moon" (62.9 percent). Without special instructions, the statistical software would have produced a factor analysis with only one factor, and Figure 1.6 shows the results when it was instructed to seek two factors. The seven plans are nearly in a straight line, from modest plans promoting international cooperation on Earth at the upper left to ambitious human missions away from Earth at the lower right.[161]

A remarkable 1980 Harris poll of adults living in Kentucky included two unusually detailed questions about the space program, which are analyzed in Table 1.9. One question asked respondents to judge how much each of eight "sectors of the economy" benefits from the space program. As Table 1.9 shows, they judged the space program to be more beneficial for the national government than for state and local governments. Not surprisingly, large majorities

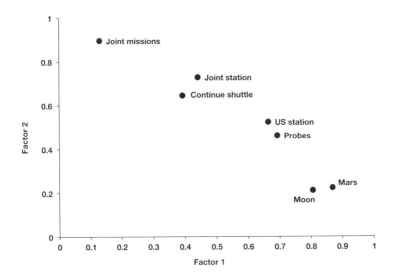

FIGURE 1.6. Factor Analysis of Plans for Future Space Exploration

161. Analysis of original data from Harris study no. 718288, a 9–14 July 1997 poll of 1,002 American adults, Odom Institute. The two factor analyses employed pairwise deletion and varimax rotation, and specified that two factors should result; the one factor analysis resulted from the criterion of selecting only factors with eigenvalues greater than 1.00.

saw great benefit for "research" and "education," but many also saw benefits for business and industry, individuals, and, to the least extent, agriculture. The second question listed 10 "products, improvements, or scientific discoveries" and asked whether each is "a result of the space program." Very large fractions correctly answered "satellite communication" and "weather prediction methods," but majorities also saw a connection to "firefighters breathing system" (perhaps like a space suit) and a number of electronic devices. Nearly half wrongly believed that Teflon, discovered in 1938 and first marketed in 1946, was a result of the space program. Less than half credited the space program with developing synthetic rubber and latex paint.[162]

A July 1989 Gallup poll asked respondents to judge which of four activities should have the highest priority for the space program. First place went to "basic research on the solar system and planets" (30 percent), followed by

TABLE 1.9. Benefits to Economic Sectors and Developments from Space

Benefits or Results of Space Program		
Governments	Federal government	63.5%
	State government	39.9%
	Local government	26.6%
Institutions	Research	85.9%
	Education	82.4%
	Business and industry	69.5%
	Individuals	55.1%
	Agriculture	45.2%
Developments	Satellite communication	88.1%
	Weather prediction methods	86.3%
	Firefighters breathing system	69.9%
	Rechargeable pacemaker	61.9%
	Hand-held calculators	57.6%
	Push-button telephones	53.7%
	Microwave ovens	53.1%
	Teflon	48.1%
	Synthetic rubber	33.4%
	Latex paint	27.0%

---

162. Analysis of original data from Harris study no. 20038, a 26 March–17 April 1980 poll of 671 Kentucky residents, Odom Institute.

"mining the natural resources on the moon and other planets" (23 percent), "developing zero-gravity manufacturing and other commercial technologies in space" (18 percent), and "developing a space-based defense shield" (14 percent).[163]

The 1986 study of Harvard students mentioned earlier involved a pair of questionnaires designed to develop and analyze a list of all the justifications for spaceflight that were current in public opinion at the time.[164] Given this goal, it was less important that the respondents were not a random sample than that they thought deeply and possessed a range of viewpoints. The first questionnaire asked 1,007 students a series of open-ended questions designed to elicit ideas about the value of the space program. From their many thousands of written utterances, statements were collated that expressed 125 separate justifications. These were then placed in the second questionnaire, and 894 students were asked to rate (on a scale of 0 to 6) how good a justification each statement was. Because adequate commercial software was not available at the time, a special cluster analysis program was written to categorize the statements into groups.

Here, I reanalyze the original data for 512 students who gave the most complete responses and for the most representative 90 items, using modern software to do a factor analysis.[165] The categorization that follows is probably an improvement over the original analysis but gives similar results. The factor analysis was done in two stages. The first stage produced 12 factors, and on the basis of the highest loadings, 89 of the 90 items could be assigned to a dozen appropriate groups. The one item left out had affinities to several factors: "From space, we could find new ways to control pollution and clean up our environment." The first factor combined fully 28 items that could be described as idealistic and emotional. For example, the two statements that led the factor with equal 0.75 loadings were: "Space exploration is a human struggle, expressing the unconquerable human spirit" and "Space exploration fulfills the human need for adventure."

Naturally, a random sample of Americans, or any other group of respondents, might rate the 125 space goals somewhat differently, but it is interesting to note that the Harvard students did not express much enthusiasm for these 28 items, rating them on average 2.4 on the 0-to-6 scale. The second

---

163. Gallup poll of 1,253 American adults, 6–9 July 1989, cited by Rosita M. Thomas, *American Public Opinion and the Space Program* (Washington, DC: Congressional Research Service of the Library of Congress, 1991), p. 43.

164. Bainbridge, *Goals in Space.*

165. William Sims Bainbridge, *Social Research Methods and Statistics: A Computer-Assisted Introduction* (Belmont, CA: Wadsworth, 1992). A computer disk included with this book holds the data plus simple software for analyzing it.

stage of the work did a second factor analysis of just the 28 idealistic and emotional items to see if they could be further categorized. This attempt was quite successful; so what follows starts with the four subfactors in Factor 1, then describes the 11 other factors. In parentheses after each statement is the loading, and for the subfactors the loading on the factor is given first, followed by the loading on the subfactor.

Subfactor 1A concerns the *spirit of exploration*, suggesting that the human need to explore is part of our fundamental nature and requires no further justification: "We should explore the unknown" (0.59/0.75); "We should boldly go where no man or woman has gone before" (0.65/0.71); "We should go into space for the same reason people climb Mt. Everest—because it's there" (0.65/0.66); "Space is the new frontier" (0.64/0.66); "Humans have an innate need to search and discover" (0.68/0.65); "We must broaden our horizons" (0.56/0.64); "Investigation of outer space satisfies human curiosity" (0.71/0.57); "Space offers new challenges, and civilization would stagnate without challenges" (0.62/0.53); "Space exploration is a human struggle, expressing the unconquerable human spirit" (0.75/0.51); and "Space exploration fulfills the human need for adventure" (0.75/0.47).

Subfactor 1B expresses emotions of pride, hope, and aspiration that can be generated by space exploration, what psychologists might justly call *ascendancy*: "The space program builds national pride" (0.43/0.83); "Space triumphs give us justified pride in our achievements" (0.62/0.77); "The exploration of space lifts morale and instills a sense of hope and optimism" (0.65/0.66); "Spaceflight reaffirms faith in man's abilities" (0.71/0.66); "Spaceflight is a noble endeavor, expressing the hopes and aspirations of humankind" (0.72/0.56); and "The space program provides a goal and a feeling of long-term purpose for humanity" (0.59/0.52).

Subfactor 1C asserts that spaceflight offers *inspiration*, giving us new perspectives: "In space, we see how small our world is and thus learn humility" (0.48/0.74); "The space program gives us new perspectives on ourselves and our world" (0.63/0.68); "New experiences and perspectives gained in space inspire art, music, and literature" (0.52/0.60); "The space program allows people to think beyond the triviality of earthbound conflicts and concerns" (0.56/0.56); "The space program encourages people to make achievements and solve problems" (0.54/0.49); "The space program inspires young people to study the sciences" (0.44/0.47); "Space stimulates the creative human imagination" (0.72/0.47); and "The exploration of space is an unselfish quest that could benefit all mankind" (0.40/0.46).

Subfactor 1D rejoices in the *excitement* that space exploration can offer individuals: "Space travel is fun" (0.45/0.80); "Space missions are exciting"

(0.64/0.77); "The beauty of space creates a sense of wonder" (0.68/0.59); and "Space gives people something to dream about" (0.74/0.45).

Factor 2 assembles 13 items with a mean rating of 2.6 on the 0-to-8 scale. They concern *colonization* of outer space: "Earth is too small for us, so we must expand off this planet" (0.80); "Space offers room for the expansion of the human species" (0.79); "Space settlements could ease the growing problem of overpopulation" (0.75); "We could find new worlds we can live on or transform a planet to make it habitable" (0.75); "We could colonize the moon, Mars, and other satellites or planets of our solar system" (0.74); "We need an alternate home planet in case Earth is destroyed by a natural catastrophe or nuclear war" (0.66); "Humans should spread life to other planets" (0.65); "We could establish manned space stations, communities in space, and space cities" (0.63); "Our future ultimately lies in space" (0.61); "Eventually, interstellar travel could be possible, taking people to distant stars" (0.58); "We could use raw materials from the moon and planets when natural resources are depleted on Earth" (0.51); "In space, we could create new cultures, lifestyles, and forms of society" (0.51); and "Farms in space and advances in terrestrial agriculture aided by the space program could increase our food supply" (0.44).

The 12 items gathered together by Factor 3 were much more popular, achieving an average rating of 3.8, and concern *technology* benefits of space development: "The space program contributes much to our technology" (0.68); "Technological spin-offs (advancements developed for the space program, then applied to other fields) improve everyday life" (0.66); "Space research provides valuable, practical information" (0.61); "The space program contributes to the advancement of science" (0.61); "The space program produces better computers, calculators, and electronics" (0.58); "The long-term, ultimate benefits of the space program could eventually be important" (0.57); "The space program has great benefits for industry" (0.55); "Space research tests our scientific theories and promises conceptual breakthroughs" (0.54); "Space has great commercial applications and many opportunities for business" (0.52); "The space program stimulates the economy and has direct economic benefits" (0.48); "Space could offer many unexpected benefits we cannot now foresee" (0.45); and "In the weightlessness and vacuum of space, we could manufacture new and better alloys, crystals, chemicals, and machine parts" (0.41).

The eight items in Factor 4 have a mean rating of 3.9 and concern applications of Earth *satellites*: "Satellites are an important component in navigation systems" (0.72); "Satellites are useful in surveying and mapping Earth" (0.71); "Satellites link all corners of the globe in a complete information and communication network" (0.68); "Meteorology satellites are great

aids for predicting the weather and understanding atmospheric patterns"
(0.67); "Communication satellites improve television transmissions" (0.59);
"Satellite photography of Earth contributes to geology, oceanography, and
archaeology" (0.59); "Observations from orbit help us find new sources of
energy and minerals on Earth" (0.55); and "An orbiting space telescope could
give astronomers a much better view of the stars" (0.42).

With a mean rating of 3.5, Factor 5 concerns the *knowledge* to be gained
by space exploration: "We could discover our origins, learning about the history of the universe and Earth" (0.70); "Through the space program, we could
learn the origin of life" (0.67); "We could gain greater understanding of the
world we live in" (0.62); "We could gain a better understanding of the universe as a whole and how it functions" (0.61); "We could gain knowledge
about ourselves" (0.51); "Space probes increase our knowledge of space, planets, comets, and the entire solar system" (0.43); and "Space research benefits
physics—in studies of the nature of matter, for example" (0.43).

Human *unity* is the theme of the five items in Factor 6, with a mean rating
of 2.6: "The space program generates national unity, encouraging cooperation between numerous sectors of society" (0.61); "The common cause
of space exploration unites the peoples of the world and could eventually
create a world community" (0.60); "The space program contributes to world
peace" (0.59); "Joint space projects between nations improve international
cooperation" (0.58); and "Competition in space is a constructive outlet for
nationalistic rivalries that otherwise would take the form of aggression and
conflict" (0.56).

Harvard students were politically liberal and skeptical of Reagan's programs, so they gave the *military* items in Factor 7 the lowest mean rating of
2.1: "A space-based antimissile system, part of the Strategic Defense Initiative,
could reduce the danger of war and nuclear annihilation" (0.85); "There are
great military applications of space" (0.84); "The space program contributes
to our defense" (0.84); and "Reconnaissance satellites help prevent war and
nuclear attack" (0.57). A different set of respondents might have rated these
items higher, but the point of the study was to categorize all the possible
motives for space development, and clearly these four objectively fit together.

The highest mean rating, 4.2, went to the *medical* items in Factor 8: "New
medicines could be manufactured in the zero gravity and vacuum of space"
(0.75); "Some medical problems could be treated more effectively in the
weightlessness of space" (0.74); and "Medical research performed in space
could benefit human health" (0.72).

Also highly rated, with 4.0, were the *resources* items in Factor 9: "New
fuels found in space or the development of fusion power in space could help
solve Earth's energy problem" (0.67); "Solar power stations in orbit could

provide clean, limitless energy to Earth" (0.51); and "We could find new mineral resources on the moon, Mars, or the asteroids" (0.49).

Two *employment* items with a mean rating of 2.6 constituted Factor 10: "The space program provides jobs for thousands of people" (0.65); and "The space program employs many engineers and scientists who otherwise would not be able to utilize their talents" (0.59).

The two items about *learning* in Factor 11 had a mean rating of 2.9: "Space travel makes us realize that Earth is a fragile, unique, unified world that deserves more respect and better care" (0.56); and "The space program is an educational tool, helping us learn from each other" (0.53).

Finally, the last pair of items in Factor 12 were rated 2.6 and concern *pollution removal*: "We could preserve Earth's environment by moving the most polluting industries into space" (0.80); and "The moon or the sun could be used for safe disposal of toxic materials and nuclear wastes" (0.69).

This study of Harvard students, dating from 1986, probably provides the most complete picture of the ideas about the value of the space program held in American culture at the time, a period when public opinion on the topic had probably matured. As we shift into a new major period in space development, fresh efforts will be needed to understand public opinion as it evolves in new directions.

7. Attitudes Toward the Future

For six decades, opinion polls have asked Americans to consider the future of spaceflight, and as the Shuttle era comes to a close, their attitudes take on a new significance. The loss of Columbia during reentry on 1 February 2003 emphasized the need to retire this complex, aging vehicle and shift to something else. On 14 January 2004, President George W. Bush announced "a new plan to explore space and extend a human presence across our solar system."[166] With a human mission to Mars as the long-range goal, the United States would return to the Moon, using new launch vehicles comparable to improved versions of the Saturn rockets used in the Apollo program, abandoning the concept of routine orbital flights with a largely reusable, winged vehicle that had been central to the Shuttle program. Would the American public support the new vision?

A poll done in February 2003 asked respondents how they would like to see money spent if NASA had a budget increase, allowing them to select as many as three choices. The leading investment was "making shuttles safer,"

---

166. See *http://www.nasa.gov/pdf/54868main_bush_trans.pdf*.

marked by 64 percent of respondents, and third place went to "replacing space shuttles" with 34 percent. In second place was "aerospace technologies to improve flight on earth" at 46 percent, recalling that the first A in "NASA" stands for "aeronautics." In fourth place, however, was "unmanned missions to other planets" at 30 percent, a goal that looks far beyond the mundane world. Other choices were "expansion of the space station" (25 percent), "manned missions to other planets" (19 percent), and "human return to the moon" (13 percent).[167]

In January 2004 a CBS News and *New York Times* poll asked: "Given the costs and risks involved in manned space exploration, do you think building a permanent space station on the moon is worth it, or not?" A majority, 58 percent, said a Moon station would not be worth it; 35 percent said it would be; and the remaining 7 percent did not know.[168] One of the biggest challenges in opinion polls is how to prepare the respondents to express an opinion when they may know little about the given topic. A standard method is to describe a topic in a sentence or two, then follow with the question, but this technique relies heavily on the exact wording used. For example, consider the word *expanding* in both the introduction and question of a poll done for the Associated Press earlier that same month:

> As you may have heard, the Bush administration is considering expanding the space program by building a permanent space station on the Moon with a plan to eventually send astronauts to Mars. Considering all the potential costs and benefits, do you favor or oppose expanding the space program this way or do you oppose it?

The actual plans did not call for increasing the NASA budget, which is what many respondents may have understood from the word "expanding." In any case, 43 percent said they favored the idea, and 52 percent were opposed.[169]

Later that month, Harris Interactive asked what some would call a leading question, suggesting how the money devoted to Mars exploration might better be spent:

> As you may know, President Bush has proposed spending billions of dollars to send a manned mission to the moon and eventually to Mars. Most estimates

---

167. Poll of 900 American adults by Techno Metrica Institute of Policy and Politics for CBS News/*New York Times*, 3–9 February 2003, Roper Center USTIPP.03FEB.R37.

168. Poll of 1,022 American adults by CBS News/*New York Times*, 12–15 January 2004, Roper Center USCBSNYT.011704.R81.

169. Poll of 1,000 American adults by PSOS–Public Affairs for the Associated Press, 9–11 January 2004, Roper Center USIPSOSR.011204A.R1B.

of the cost of these space programs put it at hundreds of billions of dollars. If you could choose, would you spend the billions of dollars proposed for the space program on: improving education, balancing the budget, cleaning up the environment, space exploration, or enlarging the military?

Only 9 percent wanted to keep the money in the space program budget, and another 4 percent of respondents were not sure. The largest group wanted to invest in improving education (40 percent), with smaller groups preferring to balance the budget (27 percent), clean up the environment (13 percent), and enlarge the military (7 percent).[170]

At the end of January 2004, a poll carried out by Gallup for CNN and *USA Today* reminded respondents that "the U.S. has landed two robotic explorers on Mars that are sending back pictures and data" and asked them to evaluate this accomplishment. Fully 70 percent called the landing of Spirit and Opportunity "a major achievement," and another 20 percent considered it "a minor achievement." Only 9 percent said it was "not an achievement at all," and 1 percent had no opinion.[171] This poll also asked: "Do you think it is worthwhile for the United States to find out whether there were ever living creatures on Mars, or not?" A majority, 54.5 percent, felt this would not be worthwhile, but a strong minority (44.0 percent) felt it would be.[172]

Polls had been asking about a trip to Mars for many years. In 1986 fully 58 percent of respondents agreed that "a joint U.S.-Soviet Union manned mission to Mars…should be a long range goal for the civilian space program," and only 40 percent disagreed.[173] The poll was conducted by Market Opinion Research for Rockwell International, an aerospace contractor; so one wonders if something about the manner in which the poll was done produced a higher positive response than other polls about Mars. As with many social science agree/disagree questionnaire items, respondents could express different levels of agreement; and favorable respondents were evenly split between "strongly agree" and "just somewhat agree." Perhaps the key is the phrase "long range goal" in the question, suggesting something that might be desirable someday but that need not have any near-term costs. Another poll by the same organizations, done two years later, found fully

---

170. Poll of 1,003 American adults by Harris Interactive for *Time* magazine and Cable News Network, 14–15 January 2004, Roper Center USHARRIS.Y011604.R22.

171. Half sample of a poll of 1,001 American adults by Gallup for Cable News Network and *USA Today*, 29 January–1 February 2004, Roper Center USGALLUP.04JAN29.R32.

172. Half sample of a poll of 1,001 American adults by Gallup for Cable News Network and *USA Today*, 29 January–1 February 2004, Princeton, NJ.

173. Poll of 1,200 American adults by Market Opinion Research for Rockwell International, 1–12 August 1986, Roper Center USMOR.86SPAC.R33.

66 percent in favor of "sending a manned mission to Mars and establishing an outpost there for scientific observation and exploration," with just 30 percent opposed.[174] Some polls have explored possible nationalistic motives for Mars exploration. A Gallup poll in July 1989, near the 20th anniversary of the first Moon landing, asked: "How important do you think it is for the U.S. to be the first country to land a person on Mars?" Opinions ranged across the given responses: "very important" (19 percent), "somewhat important" (32 percent), "not too important" (25 percent), and "not important at all" (23 percent).[175] Seven years later, 17 percent of Americans said they would be "greatly pleased" if "some other country sent the first manned mission to Mars," 36 percent would be "somewhat pleased," 25 percent "somewhat disappointed," and 9 percent "greatly disappointed."[176]

For the 25th anniversary of Apollo in 1994, a poll for *Newsweek* magazine told respondents: "Scientists estimate it would cost 50 billion dollars over ten years to put a person on Mars." The largest group, 52 percent, felt the United States should "undertake a manned mission to Mars only with other nations to share costs and expertise," whereas just 4 percent wanted the United States to "undertake a solo manned mission to Mars" and 41 percent felt the country should "not undertake a manned mission to Mars."[177] When asked in September 1996, just 13 percent of Americans expected to see "a human colony on Mars" in the next 10 years, compared with 84 percent who did not.[178]

On 4 July 1997 the robot rover Pathfinder landed successfully on Mars. A Harris poll inquired how much attention respondents had paid to the event; 23.1 percent said "a lot of attention," and 43.9 percent said "some attention," indicating that two-thirds of respondents had been at least somewhat interested. Another 24.2 percent said they had paid "not much attention," and 8.9 percent said "none at all."[179] An ABC poll reminded respondents: "As you may know, last week the United States landed a probe on Mars that has

---

174. Poll of 1,200 American adults by Market Opinion Research for Rockwell International, July 1988, Roper Center USMOR.88SPAC.R29.

175. Poll of 1,253 American adults by Gallup, 6–9 July 1989, Roper Center USGALLUP. 071989.R05.

176. Poll of 1,000 American adults by the Institute for Social Inquiry and the Roper Center for the National Science and Technology Medals Foundation, 31 May–14 June 1996, Roper Center, USISIROP.96TECH.R58.

177. Poll of 751 American adults by Princeton Survey Research Associates for *Newsweek*, 7–8 July 1994, Roper Center USPSRNEW.N0794A.R13.

178. Poll of 601 American adults by Princeton Survey Research Associates for *Newsweek*, 25–29 September 1996, Roper Center USPSRNEW.96009B.Q18D.

179. Analysis of original data from Harris study no. 718288, a 9–14 July 1997 poll of 1,002 American adults, Odom Institute.

been exploring the surface and testing the makeup of rocks." Fully 59 percent said this success made them "feel more supportive of continued US efforts to explore Mars," 25 percent said "less supportive," and the remainder had no opinion.[180] Pathfinder ceased operation late in September, but public support continued. More than two years later, 56.3 percent of respondents to a Gallup poll said the government should "continue to fund efforts by NASA to send unmanned missions to explore the planet Mars," compared with 40.2 percent who wanted funding stopped.[181]

In March 1999, 35 percent of respondents to a Gallup poll thought there was "life of some form on Mars," but 59 percent felt there was not.[182] The remarkably successful robot explorers, Spirit and Opportunity, were launched toward Mars in June and July 2003, landing in January 2004. An Associate Press poll in July 2003 found that 49 percent felt the United States should "pursue a program to send humans to Mars," compared with 42 percent who felt it should not.[183] Thus the January 2004 announcement by President Bush came after a significant amount of public interest. However, this did not mean the majority was prepared to spend the money required.

In June 2005, fully 58 percent of Americans opposed "attempting to land an astronaut on the planet Mars," compared with only 40 percent who favored the idea.[184] These numbers represent a slight deterioration of support from July 1999, when 54 percent opposed the idea and 42 percent favored it.[185] Much more positive opinions were expressed in an August 2006 Gallup poll that asked respondents to think deeply about the future:

> In January 2004, a new plan for space exploration was announced. The plan includes a stepping-stone approach to return the space shuttle to flight, complete assembly of the space station, build a replacement for the shuttle, go back to the moon, and travel to Mars and beyond. If NASA's budget did not exceed one percent of the federal budget, to what extent would you support or oppose this new plan for space exploration?

180. Poll of 505 American adults by ABC News, 8–9 July 1997, Roper Center USABC.071197. R13.

181. Poll of 501 American adults by Gallup, 9–12 December 1999, Princeton, NJ.

182. Darren K. Carlson, "Life on Mars?," press release, 27 February 2001, Gallup News Service, Princeton, NJ.

183. Poll of 1,034 American adults by International Communications Research for the Associated Press, 11–15 July 2003, Roper Center USAP.072803.R3.

184. Poll of 1,009 American adults by Gallup for Cable News Network, 24–26 June 2005, Roper Center USGALLUP.05JE024.R08.

185. Poll of 1,060 American adults by Gallup, 13–14 July 1999, Princeton, NJ.

A majority of 65.9 percent said they supported this plan, including 19.0 percent who supported it strongly, compared with 27.4 percent who opposed it. Many members of the general public lack sufficient information about possible future Mars missions, so the careful explanation of the stepping-stone approach may have created a positive impression.[186] The proviso that no more than 1 percent of the federal budget be invested undoubtedly strengthened support as well.

In response to another item from this August 2006 poll, 68.9 percent agreed that "the scientific, technical and other benefits of space exploration are worth the risks of human space flight." A third item in the questionnaire ran:

> Both China and the U.S. have announced plans to send astronauts to the moon. China has announced plans to go to the moon by 2017 and the U.S. has announced plans to send astronauts to the moon by 2018, a year later. To what extent, if any, are you concerned that China would become the new leader in space exploration or take the lead over the U.S.?

Only 28 percent were somewhat concerned or very concerned, suggesting that most Americans did not yet feel a second space race had begun.[187]

Whatever governments might do in space, at the beginning of the 21st century there was renewed hope that private enterprise might help launch humanity into the universe, notably through space tourism.[188] In 2002 the Futron Corporation published a study of the potential market for space tourism, based on a poll of 450 affluent individuals (with a net worth of over $1 million or annual incomes of $250,000) done by the Zogby organization. Among the features of a suborbital flight that respondents thought would be very important to them, 63 percent identified "viewing Earth from space," 27 percent cited "experiencing the acceleration of a rocket launch," 26 percent cited "experiencing what only astronauts and cosmonauts have experienced," and 24 percent agreed that "experiencing weightlessness" would be very important. Among important experiences for orbital flight were "eating, sleeping, and exercising in space" (49 percent), "orbiting Earth every 90 minutes" (28 percent), and the opportunity to "stay two weeks on a space station" (24 percent).

---

186. Ragnar E. Lofstedt, "Public Perceptions of the Mars Sample Return Program," *Space Policy* 19 (2003): 283–292.

187. Poll of 1,000 American adults by Gallup for the Space Foundation, 2–19 August 2006, Princeton, NJ.

188. Tom Rogers, "Space Tourism: A Response to Continuing Decay in US Civil Space Financial Support," *Space Policy* 14 (1998): 79–81.

Several questions assessed the willingness of these wealthy respondents to buy a tourist ticket on a suborbital flight, in a service projected to rise from about 500 passengers in 2006 to 15,000 in 2021, with an estimated annual revenue in 2021 of $786 million, when a flight could cost the tourist $50,000. The first demonstration of the feasibility of suborbital space tourism, the flight of SpaceShipOne that garnered the $10 million "X-Prize," took place in 2004 on the anniversary of the first Sputnik launch. A second Futron/Zogby market study, carried out in 2006, gave similar estimates for the year 2021, despite a later assumed start for commercial space tourism operations.[189] However, SpaceShipOne employed technology that is not suitable for orbital flight, and it remains unclear whether human spaceflight could become commercially viable during the next couple of decades.

NASA's plan to return to the Moon was predicated on the assumption of constant budgets, rather than a significant increase, so an increase in public enthusiasm may not be necessary. The 2004 GSS results suggest the American public would support this level of investment. When 1,403 respondents were asked to assess current funding for the "space exploration program," 14.3 percent felt too little was being spent and 43.4 percent said about the right amount, for a total of 57.7 who were willing to continue at the present level. A large minority, 36.8 percent, felt too much was being spent, and 5.5 percent did not know. If those lacking an opinion are removed from the equation, as was done in Table 1.2, those willing to support the program at current levels constitute fully 61.0 percent.[190]

Polls are not referendums, however much journalists and some politicians might want them to be. Through polls, politicians can judge how to build the constituencies they need in order to succeed. The polls do show that spaceflight has a substantial constituency within the American public, so investment in the space program makes sense in a mix of investments that serve the interests of different constituencies. Funding levels aside, the public has positive feelings about space exploration. It is not a controversial subject, such as nuclear power may be, having powerful negative as well as positive constituencies. Rather, the consensus is that space exploration is a good thing, and any debate concerns how aggressively it should be pursued.

In a representative democracy the public does not vote on every little issue. Rather, through elections and opinion polls, the public tells government what

189. "Space Tourism Market Study," October 2002, Futron Corporation, Bethesda, MD; and "Suborbital Space Tourism Demand Revisited," 24 August 2006, Futron Corporation, Bethesda, MD.

190. Analysis of the original GSS data obtained from the Computer-assisted Survey Methods Program at the University of California, Berkeley.

general values it wants government to serve. It is up to elected representatives and technical experts to decide how to achieve those goals. The interests of all citizens need to be taken into account, and that is the reason why each adult deserves one equal vote. However, better-educated citizens are in a better position to judge how to achieve the goals the society desires, and they support space exploration more than less-educated people do. Much of government's investment must serve short-term needs in areas like national defense and human services. But some investment must be made in the longer-term future, developing new opportunities and the knowledge needed to understand ever better the universe in which we live. The American public is willing to continue the voyage of discovery and achievement into outer space.

Part II

SPINOFF?

Chapter 2

SOCIETAL IMPACT OF NASA
ON MEDICAL TECHNOLOGY[1]

William Sims Bainbridge

1. Introduction

This chapter examines the history of several medically oriented innovations reported in NASA's annual *Spinoff* publication in order to better understand the processes by which the Agency's efforts benefit humanity in often unexpected ways. The literature on this topic tends to conceptualize these innovations as spinoffs from "the space program," although there is the possibility that some originated in NASA's aviation research and development as well. The aim here is not to determine a financial estimate of the annual value of spinoffs, nor to critique the reporting of innovation by NASA's dedicated analysts and writers. Rather, by tracking the historical origins and current impact of selected spinoffs, we can uncover the fundamental factors at work. The sample of spinoffs was selected to ensure that a variety of interesting phenomena would be discovered, not to characterize a hypothetical average NASA medical spinoff.

A widely used dictionary defines *spin-off* as "a collateral or derived product or effect," equivalent to a by-product.[2] As a subcategory of technology transfer, a spinoff is a distinct invention developed in the course of aerospace work and transferred more or less intact to nonaerospace uses. The website of NASA's Scientific and Technical Information Office defines the term rather more broadly: "A spinoff is a commercialized product that incorporates NASA technology or NASA 'know how' and benefits the public."[3]

---

1. Acknowledgments: Great thanks are due to Trudy E. Bell for the original suggestion that led to this study and for her early ideas about its scope and methodology.

2. *Webster's Ninth New Collegiate Dictionary* (Springfield, MA: Merriam-Webster, 1990), *s.v.* "spin-off."

3. See *http://www.sti.nasa.gov/tto/spinfaq.htm*. Note: The websites referenced in this chapter were accessed early in 2007, and many may have changed or vanished since then. Note that some URLs default to a different page from the one where data for this study were originally found, and other pages were revised over time. Researchers may find the

Spinoffs must be distinguished from *applications*, which are intended benefits of a technology. Communications satellites, meteorology satellites, and navigation satellites like the Global Positioning System are valuable applications of space technology. A 2003 book about spinoffs, *It Came from Outer Space,* by Marjolijn Bijlefeld and Robert Burke, asserts:

> Medical technology had benefited greatly from the space program. Non-invasive breast cancer screening, less traumatic breast biopsy techniques, magnetic resonance imaging (MRI), cardiac pacemakers and implantable defibrillators, kidney dialysis machines, insulin pumps, and fetal heart monitors are just some of the devices and procedures that were introduced using technologies developed in the space program.[4]

For many years scholars have debated the importance of spinoffs, in contrast with many other kinds of secondary benefits from NASA's work, and the degree to which the concept itself is problematic. In the 1960s NASA funded a number of social science studies of the potential societal implications of space exploration. Top scholars participated, and the results were often published as books by the prestigious MIT Press. For example, one collection of essays edited by Bruce Mazlish drew analogies between the building of the railroads in 19th-century America and the future building of space-related transportation systems. The book started with the working hypothesis that space development would turn out to be far more extensive than it has proven to be to this point in history, yet qualitatively its observations are still useful. A new technology may greatly benefit the economy in ways quite remote from the technical advances themselves, and any new technology needs to be weighed in comparison with other technologies that might have been developed instead.[5]

Another collection of essays, edited by William H. Gruber and Donald G. Marquis, explored the ways that new technology may diffuse from the original application area to others, a process central to understanding spinoffs.[6] Simply put, we can imagine inventions migrating away from their birthplaces by a number of means. A spinoff consists of an invention walking on its

---

"Wayback Machine" tool at *http://archive.org/index.php* helpful for locating previous versions of Web pages.

4. Marjolijn Bijlefeld and Robert Burke, *It Came from Outer Space: Everyday Products and Ideas from the Space Program* (Westport, CT: Greenwood, 2003), p. x.

5. Bruce Mazlish, ed., *The Railroad and the Space Program: An Exploration in Historical Analogy* (Cambridge, MA: MIT Press, 1965).

6. William H. Gruber and Donald G. Marquis, eds., *Factors in the Transfer of Technology* (Cambridge, MA: MIT Press, 1969).

own legs, so to speak, and maintaining its own individual integrity. People outside the original field come into contact with it, perhaps by chance, and they adopt it as a whole unit. Perhaps later they modify it, but originally it migrates intact and in the form of a distinct, probably patentable invention. One of the main other ways in which technological transfer takes place is in the minds of people who migrate. A scientist or engineer works with NASA on a particular project, gaining experience and a wealth of ideas. Then this person takes a different job, or works on the weekends in his or her basement, creating something quite different, perhaps separately patentable. This is sometimes called *embodied diffusion*. A third path involves the *social communication* of information from one person to another. This was the type most closely studied in the 1950s and 1960s, as illustrated by the influential books *Personal Influence* by Elihu Katz and Paul F. Lazarsfeld and *Diffusion of Innovations* by Everett M. Rogers.[7]

In the language of economics, embodied diffusion and social communication could both be described as mechanisms by which spillover occurs. *Spillover* is the escape of information about an innovation that allows a corporation that did not invent it to copy it or otherwise take advantage of the innovativeness of another corporation.[8] Often, companies and governments keep trade secrets or state secrets, and patent protection is a mechanism for balancing the need of an innovator to capture the benefits of the innovation, versus the benefit to society more generally if the information is widely disseminated. Spillover might include the specific design details of a particular invention, but the term is more usefully employed for the general technical expertise and scientific knowledge of which the invention is one particular instantiation. It is possible that many of the spinoffs listed by NASA are better described as spillovers.

A third book in the NASA-MIT series, *Second-Order Consequences* by Raymond A. Bauer, documented a very wide range of kinds of technology transfer that might occur, causing changes that are economic, technological, scientific, managerial, or social. *Spinoffs* were only one subcategory among many that could be distinguished from others in a number of ways. For example, spinoffs were a kind of horizontal diffusion, in which an intact

---

7. Elihu Katz and Paul F. Lazarsfeld, *Personal Influence: The Part Played by People in the Flow of Mass Communications* (Glencoe, IL: Free Press, 1955); and Everett M. Rogers, *Diffusion of Innovations* (New York: Free Press of Glencoe, 1962).

8. Timothy F. Bresnahan, "Measuring the Spillovers from Technical Advance: Mainframe Computers in Financial Services," *American Economic Review* 76, no. 4 (1986): 742–755; and Adam B. Jaffe, Manuel Trajtenberg, and Michael S. Fogarty, "Knowledge Spillovers and Patent Citations: Evidence from a Survey of Inventors," *American Economic Review* 90, no. 2 (2000): 215–218.

innovation moves from one application to another. They could be distinguished from vertical diffusion, in which general knowledge coming from scientific research performed in the space program moves down and is reshaped into a number of inventions in potentially several different fields.[9]

The general conclusion one draws from the extensive scholarship and social science done for NASA decades ago is that spinoffs are a minor part of the benefit of the space program, and by implication also of the aeronautical research and development supported by NASA. Indeed, given all the many analytical rubrics proposed by the host of authors, it becomes rather difficult to distinguish spinoffs precisely from other but similar kinds of impacts. A 1976 summary of research on impacts reported: "These findings suggest that earlier 'space spinoff' studies, which searched for commercial adoption of complete NASA-developed systems, were misdirected."[10] Instead, the primary impact of the space program has been to accelerate technological progress across a wide front in many fields. However, there are several reasons why it can be useful to take spinoffs seriously. First, the concept may fit some innovations really well. The fact that NASA may benefit society in many other ways does not negate the significance of whatever spinoffs it may have provided. In a pair of major conferences that I helped to organize, scientists, engineers, and scholars debated the societal implications of nanotechnology. All participants were aware of concerns that nanotechnology might have negative unintended consequences, such as possibly health or environmental harm from toxic nanoparticles. But a strong consensus emerged that nanotechnology may have a number of positive unintended consequences as well, most likely outweighing the negative ones. If applications are intended consequences, then spinoffs are a category of positive unintended consequences, containing very distinct innovations.

Second, spinoffs may be easier to communicate to a wide public than the more vague unintended benefits of aerospace innovation. The public often has difficulty grasping scientific and technical ideas if they do not experience them directly in their own lives. One reason may be that the human brain is wired to interpret events in terms of narratives in which an individual person faces challenges in the pursuit of a clear goal and wins through to obvious success.[11] Thus humans find it natural to conceptualize innovations in terms

---

9.   Raymond A. Bauer, *Second-Order Consequences: A Methodological Essay on the Impact of Technology* (Cambridge, MA: MIT Press, 1969).

10. Eli Ginzberg, James W. Kuhn, Jerome Schnee, and Boris Yavitz, *Economic Impact of Large Public Programs: The NASA Experience* (Salt Lake City, UT: Olympus, 1976), p. 46.

11. H. Porter Abbott, "Unnarratable Knowledge: The Difficulty of Understanding Evolution by Natural Selection," in David Herman, ed., *Narrative Theory and the Cognitive Sciences* (Stanford, CA: Center for the Study of Language and Information, 2003), pp. 143–162;

of discrete inventions achieved by individual inventors. Social scientists have long criticized the popular view of inventions. S. C. Gilfillan has argued that it is strictly impossible to draw sharp conceptual lines between one invention and another.[12] Robert K. Merton has documented that inventions tend to be "invented" simultaneously by several different individuals, which means that inventions were really made by society, not by individuals, when the level of technical culture has reached a level that makes a valuable invention possible.[13] But the popular view exists for a reason—namely, that it helps nonspecialists think about complex technical and social issues—and it squares with the truth at least part of the time.

Third, thinking in terms of spinoffs can help organize historical research, such as the current project. We can hypothesize that a particular innovation fits the model of a spinoff, then document the extent to which it actually does. We can examine issues like the distinctness of the innovation (clear versus ambiguous), the extent to which it was a unique invention versus one of many made approximately simultaneously, and the extent to which it consisted of generally applicable scientific discovery versus specific engineering design. We can weigh at least qualitatively the extent to which the innovation actually had an impact, versus being a mere technological curiosity. We can estimate how much NASA actually contributed to the innovation and whether NASA's contribution derives from the space program or aviation research. Importantly, we may find cases in which, when tracking an innovation back to its historical origins, we find that it originated outside NASA but passed through NASA on its way toward benefiting society.

Thus there is another function NASA beneficially performs, beyond primary innovation, that can be explored by tracking spinoffs. It can serve as the channel, conduit, or vector through which an innovation travels during the diffusion process. We can guess that this would most often happen when the innovation originally occurred in military technology, because military innovations are often secret, and a route for direct diffusion from the military to the consumer economy often does not exist. If an innovation needs to travel far from its origin to become valuable applications, logically we would expect it to change during the diffusion process, perhaps becoming less distinct but broader in applicability. In their 1971 book about spinoffs, *Dividends from Space*, Frederick Ordway, Carsbie Adams, and Mitchell Sharpe have argued: "Many products, materials, and new techniques developed for use in the

and H. Porter Abbott, *The Cambridge Introduction to Narrative* (New York: Cambridge University Press, 2002).

12.　S. C. Gilfillan, *The Sociology of Invention* (Cambridge, MA: MIT Press, 1963).

13.　Robert K. Merton, *The Sociology of Science* (Chicago: University of Chicago Press, 1973).

space program have been converted into practical dividends in the form of new or improved products for home and industry, but it is often difficult to pinpoint the degree of space-inspired technology in such conversions."[14] They note especially that the aerospace industry works closely with other industries and institutions, and it is often ambiguous how significant each of their contributions to a particular innovation really was.

Ordway, Adams, and Sharpe identify one important mechanism by which space technology can foster innovation without actually carrying it out, for which they use the word *catalyst*. Their prime example is the medical application of extremely cold liquefied or cryogenic gases, which are used as propellants in space rockets. From the earliest Chinese gunpowder rockets to modern intercontinental missiles, the most practical propellants for military rockets have been solid, less energetic than some liquids but easier to store and handle. However, spaceflight requires the higher energy offered by some liquids, and their thrust can also be throttled to provide the fine control needed, for example, for landing on the Moon. The very first liquid-fuel rocket, launched in 1926 by Robert Goddard, employed two liquids—gasoline at ambient temperature and liquid oxygen, which must be maintained below −183°C. The German V-2 rockets developed in the early 1940s used alcohol and liquid oxygen. When the United States began to develop long-range military rockets in the 1950s, all but one used liquid fuels: Redstone, Jupiter, Thor, Atlas, and Titan. The exception was the submarine-launched Polaris, which used solid fuel because liquid fuels would have been difficult to handle and thus dangerous at sea.

Except for a small number of Titan II ICBMs that remained in service until the late 1980s, the military shifted over to solid fuel in the 1960s, from the first deployment of Minuteman I in 1962. Thus, after the period when military applications had supported the development of technologies for producing and handling cryogenic liquids, this function was served by the space program, notably in working with liquid hydrogen, which must be kept below −253°C. The second- and third-stage engines of the Saturn V moon rocket used two cryogenic propellants, liquid oxygen and liquid hydrogen, as did the main engines of the Space Shuttles. The space program also has employed liquid nitrogen, not as a fuel but as a coolant for infrared telescopes and for testing of components that normally contact the more dangerous liquid hydrogen or liquid oxygen.

Liquid nitrogen is widely used in medicine, notably for cryosurgery, where it destroys abnormal tissues by freezing them. Dermatologists today

---

14. Frederick I. Ordway III, Carsbie C. Adams, and Mitchell R. Sharpe, *Dividends from Space* (New York: Thomas Y. Crowell, 1971), p. 5.

commonly have a flask of liquid nitrogen in their offices, simply administering tiny drops on selected small areas of their patients' skin; but there also exist very sophisticated systems for employing liquid nitrogen in brain surgery. Ordway, Adams, and Sharpe cite the example of Dr. Irving Cooper, who played a central role in developing cryosurgery in the 1950s, in particular using a cryoprobe to reach inside the brain and freeze a tiny area implicated in a patient's Parkinson's disease. To improve this device, Cooper worked with space engineers who had developed a small thermocouple device that could monitor the temperature of the probe.[15] Note that any direct influence from the space program came only after the device had been invented and used, in facilitating an improvement of it.

One way to estimate the real impact of an innovation is to examine patent records. Cooper did not patent his cryoprobes, but a 1994 patent for a similar device notes that experiments with cryosurgery began as early as 1865, and comments: "Modern applications of cryosurgery are numerous. For example, an early cryosurgical apparatus was developed by Dr. Irving Cooper, a New York surgeon, and used for treatment of Parkinson's disease."[16] A search of the patent database for "NASA" and "cryosurgery" turned up nine patents to five inventors who cited NASA publications about cold temperature science and technology, but none of these patents document a close connection between the Agency and cryosurgical inventions.[17] Thus cryosurgery may or may not be a good example of what Ordway, Adams, and Sharpe called the catalyst function of the space program. Arguably, the extensive use of cryogenic liquids in the space program supports general development of the industry, and other applications benefit indirectly in a variety of ways that may be hard to document.

The chemical metaphor that some factor serves as a "catalyst" in facilitating innovation is overused and imprecise. The fundamental idea proposed by Ordway, Adams, and Sharpe is that space technology can strengthen an entire industry, technological subsystem of the economy, or scientific community. The reverse insight is also true. NASA itself is part of a network of agencies, and it cooperates with some of the same corporations; so it can be hard to evaluate their relative importance in the development of innovations. With the focus here on medical spinoffs, we can gain some clarity by considering the different roles played by NASA and by two other science agencies of the

---

15. Ordway, Adams, and Sharpe, *Dividends from Space*, p. 33.
16. Arthur A. Fowle, U.S. Patent 5,324,286, "Entrained cryogenic droplet transfer method and cryosurgical instrument," issued 28 June 1994.
17. U.S. Patents 6,530,234; 6,475,212; 6,451,012; 6,346,668; 6,270,494; 6,193,644; 6,190,378; 6,182,666; and 6,083,166.

U.S. federal government: the National Institutes of Health (NIH) and the National Science Foundation (NSF).

As described on its own Web site, NIH is a collection of 27 institutes and centers, within the wider Department of Health and Human Services, that constitute "the primary Federal agency for conducting and supporting medical research."[18] It calls itself "the steward of medical and behavioral research for the Nation. Its mission is science in pursuit of fundamental knowledge about the nature and behavior of living systems and the application of that knowledge to extend healthy life and reduce the burdens of illness and disability." In pursuit of these goals, it supports activities both in its own laboratories and at universities:

- in the causes, diagnosis, prevention, and cure of human diseases;
- in the processes of human growth and development;
- in the biological effects of environmental contaminants;
- in the understanding of mental, addictive, and physical disorders; and
- in directing programs for the collection, dissemination, and exchange of information in medicine and health, including the development and support of medical libraries and the training of medical librarians and other health information specialists.

The NSF is an independent agency of the U.S. government, which means it does not belong to a department represented in the president's Cabinet, as NIH does. Rather, it is governed by the Science Board appointed by the president and confirmed by Congress. It describes itself as "the premier Federal agency supporting basic research at the frontiers of discovery, across all fields, and science and engineering education at all levels."[19] However, as the guide to researchers preparing research grant proposals explains: "Research with disease-related goals, including work on the etiology, diagnosis or treatment of physical or mental disease, abnormality, or malfunction in human beings or animals, is normally not supported."[20] The NSF Directorate for Biological Sciences does support much research at universities on fundamental biological phenomena that might provide a scientific basis for new medical technologies. Medically relevant research could be funded by any of the NSF directorates. For example, an NSF book describing a vast array of applications points out the importance of fundamental work in materials science for medicine:

---

18. National Institutes of Health, *http://www.nih.gov/about/*.
19. National Science Foundation, *Investing in America's Future: Strategic Plan FY 2006–2011* (Arlington, VA: NSF, 2006), p. 1.
20. National Science Foundation, *Grant Proposal Guide* (Arlington, VA: NSF, 2004), p. 10.

Researchers at Georgia Institute of Technology, California Institute of Technology, and Massachusetts Institute of Technology (MIT) are working with physicians and biological specialists to develop polymer composites for patching wounds, biocompatible casings for cell transplants, scaffolds that guide and encourage cells to form tissue, bioreactors for large-scale production of therapeutic cells, and experimental and theoretical models that predict behavior of these materials in vivo. Biomaterials have already been developed to block unwanted reactions between transplanted cells and host tissue and to help prevent scarring during healing.[21]

Thus NSF may sometimes provide the scientific basis on which NIH can build, although NIH also supports fundamental research within the broad area of health and medicine. NASA enters the picture in three main ways: 1) research on the human factors of aviation and spaceflight, 2) research and development of broadly based technologies that are intended to support the human-centered aspects of NASA's mission among others, and 3) spinoffs that may have been developed for wholly other reasons but which have some applicability to health and medicine. Aerospace medicine has been an important topic of NASA research since the very beginning, notably when Project Mercury assessed the ability of humans to live and work in space and to endure the conditions of launch, orbit, and reentry.[22] Technologies like sensors and telemetry developed in connection with the human space program relate to other space applications as well as to terrestrial medical care.[23] A renewed interest in such research is evidenced by NASA's 2005 *Bioastronautics Roadmap*, an inventory of the research and development issues that must be addressed for the future of human spaceflight, especially for very long duration missions.[24]

2. Spinoffs from 1996

The 1996 edition of *Spinoff* devoted a page each to 45 examples of "technology twice used," six examples of which were in the health and medicine field. Given that nearly two decades have passed since that annual's publication,

21. National Science Foundation, *America's Investment in the Future* (Arlington, VA: NSF, 2000), p. 29.

22. Mae Mills Link, *Space Medicine in Project Mercury* (Washington, DC: NASA SP-4003, 1965).

23. Michele Anderson, *Manned Space Flight Benefits*, NASA TM-107998 (Houston, TX: Johnson Space Center, 1987).

24. *Bioastronautics Roadmap: A Risk Reduction Strategy for Human Space Exploration* (Washington, DC: NASA SP–2004–6113, 2005).

and many of the examples dated from much earlier, there has been ample time for impacts to develop. I examine each of these medical spinoffs to determine whether it has achieved its promise and to learn how each case can illuminate the processes by which technological progress actually takes place. Three of the six cases are relatively narrowly defined and claim relatively modest impact, so I consider them first. The final trio of 1996 spinoffs involve much the same general area of technology, even having direct connections to each other, and are potentially more important. I begin with a technically cute innovation that has not yet proven to be very important for human well-being.

Case 1: Bone Density Analyzer

The Mechanical Response Tissue Analyzer (MRTA) is a machine that can measure the stiffness or density of large segments of the human body, especially the long bones in arms and legs, by briefly vibrating them and noting the response. The machine does not violently shake the human limb but instead emits a precisely calibrated buzz between 70 to 1600 cycles per second for perhaps 5 seconds. NASA was quite understandably interested in its ability to make quick, nondestructive measurements of arm and leg bones, because the crews of the 1973–1974 Skylab space stations experienced significant calcium loss from their bones. Long-duration spaceflight, whether in Earth orbit or on the way to Mars, could weaken bones so much that they become brittle and fracture. The culprit is the zero gravity of spaceflight, and creating artificial gravity would require larger and more expensive vehicles. To assess the value of countermeasures like in-orbit exercise or medication, researchers need to be able to measure the bone loss precisely. Ideally, they would want to be able to do this during a flight, using light, portable instruments; at present, however, researchers make their measurements before and after a flight but not during it.

Beginning in 1977, NASA's Ames Research Laboratory and Stanford University cooperated in research to develop the principles for a bone density device. In 1987, listing Professor Charles R. Steele as the inventor, Stanford applied for a U.S. patent, which was granted in 1991 as number 5,006,984. The patent abstract described the device in detail, including a set of computer algorithms that extracted useful measurements from the limb's resonance to the instrument's vibration at a range of frequencies. While noting that a NASA grant had helped fund the research, the patent did not talk about the applications for astronauts but suggested the device could be useful in diagnosis and treatment of osteoporosis in postmenopausal women and in assessing the healing of broken bones. Stanford's patent application cited some prior work, including a French patent dating from 1978 and a 1975 thesis by a Stanford student. The fact that the resonance frequency of metal bars or the strings in a piano depends partly on their stiffness has been known for well over a century. The challenge of the invention was developing a workable device on this principle.

While the patent application was pending, in 1989, Gait Scan, a small business in Ridgewood, New Jersey, joined the project. The name of the company refers to a computerized instrument podiatrists use to evaluate the pressures on the feet of their patients as they walk. Thus Gait Scan was in a logically related business that possessed expertise and the motivation to develop MRTA for treatment-related diagnosis of skeletal problems. According to the *Spinoff* article, "Gait Scan is pursuing applications in monitoring the effects of exercise and rehabilitation on bone stiffness and in osteoporosis, the underlying cause of some 1,300,000 bone fractures each year that involve treatment costs estimated at close to $4 billion." The article quoted Dr. Sara Arnaud of Ames Research Center's Life Sciences Division as saying, "The major attraction of the technology is the speed and simplicity with which the measurement gives a complete picture of bone strength." The *Spinoff* article says Gait Scan "invested its own funds in refining the technology," which is true but a little misleading, because the company also received Small Business Innovation Research (SBIR) grants from NASA in 1990 and 1991 to do just this.[25] SBIR projects always involve a combination of government and company money. As of 2006, the "GaitScan" trademark was held by a Canadian orthotics and footwear company, The Orthotic Group, which sells GaitScan systems but not MRTAs.[26] The abstract describing the 1990 SBIR grant predicts that "the device would be used in research centers and in health and fitness facilities as a routine means of monitoring the bone strengthening effects of exercise, diet and other factors."

In 2003, Christopher Callaghan earned his doctorate from Virginia Tech on the basis of research establishing the reliability of MRTA in measuring the ulnas and tibias of college-age women, reporting generally favorable results.[27] As the Web site of the *Journal of Clinical Densitometry* proclaims, it "publishes the latest clinical research on the uses of bone mass and density measurements in medical practice." A 2004 article compared MRTA with the two most

---

25. See "Advanced Turbomachinery CFD Design and Analysis Program," NASA 1990 SBIR Phase 1 Solicitation, *http://www.sbir.nasa.gov/SBIR/abstracts/90.html*; and *http://www.sti.nasa.gov/tto/spinoff1996/24.html*.

26. See *http://www.theorthoticgroup.com/TOG-TOGGaitScan.html*.

27. Christopher Edward Callaghan, "Reliability of Tibial Measurement with Mechanical Response Tissue Analysis," Ph.D. dissertation, Virginia Polytechnic Institute and State University, 2003.

commonly used measurement approaches, dual-energy x-ray absorptiometry (DXA) and quantitative ultrasound (QUS). DXA estimates bone density, based on the bone's absorption of two weak x-ray beams of different energies. Similarly, QUS uses a pair of ultrasound transducers plus computer analysis to accomplish the same task. The study found some differences in results between MRTA and DXA or QUS but argued that MRTA might be giving a more direct measure of the mechanical integrity of the bone.[28]

In 2005 a team including both Sara Arnaud and Charles Steele published an article showing how MRTA was used to compare the bone strength of athletes with nonathletes.[29] At the 2006 conference of the American Society of Biomechanics, Jared Ragone and John Cotton described their research using a computer technique called *finite element simulation* to model how MRTA could measure the stiffness of a human tibia. They said that MRTA "is less expensive and more portable than DXA, and is a direct measurement of mechanical response, which indicates a potential for greater accuracy."[30] However, when Alan L. Burshell, mentioned in the *Spinoff* article as a medical researcher exploring the potential of MRTA, collaborated with four other researchers on a 2005 article measuring bone density, they used DXA instead. One obvious reason was that MRTA is limited to studies of the long bones in the arm and leg, and this particular study was designed to examine bone density as a factor in scoliosis of the spine.[31]

Steele and Arnaud, with both American and Russian colleagues, had tried MRTA out on monkeys in connection with the Soyuz 2229 biosatellite that had orbited for two weeks beginning 29 December 1992.[32] They reported their findings in a 2001 issue of the *Journal of Medical Primatology*.[33] Bone

28. Christina Djokoto, George Tomlinson, Stephen Waldman, Marc Grynpas, and Angela M. Cheung, "Relationship Among MRTA, DXA, and QUS," *Journal of Clinical Densitometry* 7, no. 4 (2004): 448–456.

29. Michael T. C. Liang, Sara B. Arnaud, Charles R. Steele, Patrick Hatch, and Alexjandro Moreno, "Ulnar and Tibial Bending Stiffness as an Index of Bone Strength in Synchronized Swimmers and Gymnasts," *European Journal of Applied Physiology* 94 (2005): 400–407.

30. Jared Ragone and John Cotton, "Finite Element Simulation of the MRTA Test of a Human Tibia," paper presented at the 2006 conference of the American Society of Biomechanics, Virginia Polytechnic Institute and State University, Blacksburg, VA.

31. R. Hank Routh, Sara Rumancik, Ram D. Pathak, Alan L. Burshell, and Eric A. Nauman, "The Relationship Between Bone Mineral Density and Biomechanics in Patients with Osteoporosis and Scoliosis," *Osteoporosis International* 16 (2005): 1857–1863.

32. "Cosmos 2229 (Bion 10)," *http://lis.arc.nasa.gov/lis2/Chapter4_Programs/Cosmos2229/Cosmos2229.html*.

33. T. M. Hutchinson, A. V. Bakulin, A. S. Rakhmanov, R. B. Martin, C. R. Steele, and S. B. Arnaud, "Effects of Chair Restraint on the Strength of the Tibia in Rhesus Monkeys," *Journal of Medical Primatology* 30 (2001): 313–332.

loss in weightlessness was one of the research topics of the 8–21 March 2001 flight of the shuttle Discovery. The techniques used to measure the astronauts' bones included not only DXA and conventional ultrasound but also a three-dimensional method called Quantitative Computed Tomography, similar to a CAT scan. However, MRTA was not used.[34] The following year, NASA again expressed confidence that MRTA could become a valuable spinoff, but it is difficult to argue that it has already achieved this status.[35]

The Mechanical Response Tissue Analyzer seems to be a technically successful innovation that is still being evaluated but has not yet earned a secure place either in medical research or clinical diagnosis. The history of technology is strewn with inventions like this. They work and they have certain advantages. However, other methods can serve the same goals reasonably well and are better established. Sometimes nothing more substantial that fashion may decide when one is used rather than another. Sometimes an unwanted technology languishes in obscurity for a very long time, until conditions change and an opportunity opens up. For example, early versions of both the harpsichord and the piano were built in Europe in the 14th century, but the harpsichord drove hammer percussion keyboard instruments out of the musical marketplace until the early 18th century, when Bartolomeo Cristofori is said to have "invented" the piano in the early 1700s. Cristofori's work was excellent, and his instruments were far more sophisticated than the ones built four centuries earlier, but the piano rose to complete dominance over the harpsichord more on the basis of changes in musical taste than objective technological superiority. Then, after a century-long lapse when not a single harpsichord was built in the entire world, this elegant instrument was revived in the early 20th century.[36]

NASA's Scientific and Technical Information Program likes to distinguish spinoffs from successes: "A spinoff is a commercialized product that incorporates NASA technology or NASA 'know how' and benefits the public, while a success is a NASA technology that is not available on the market but

---

34. "Sub-regional Assessment of Bone Loss in the Axial Skeleton in Long-Term Space Flight," NASA Fact Sheet FS-2001-02-39-MSFC, March 2001, *http://www.nasa.gov/centers/ marshall/news/background/facts/bone.html*.

35. "Weak in the Knees—The Quest for a Cure for Osteoporosis," NASA feature, 6 May 2002, *http://www.nasa.gov/vision/earth/everydaylife/weak_knees.html*.

36. Frank Hubbard, *Three Centuries of Harpsichord Making* (Cambridge, MA: Harvard University Press, 1965); Stewart Pollins, *The Early Pianoforte* (New York: Cambridge University Press, 1995); and Donald Jay Grout and Claude V. Palisca, *A History of Western Music* (New York: W. W. Norton, 1996).

still yields benefits to the public."[37] In those terms, MRTA is a success but not yet a spinoff. It has attracted a good deal of interest and is being used in some research, but it has not been deployed widely in clinical applications. Alternatively, we could suggest a different categorization: a potential spinoff that is being kept in our culture's technological inventory for possible future use.

Case 2: Balance Evaluation Systems

Before humans actually went into space, it was difficult to predict how zero gravity would affect their sense of balance. Weightlessness was sometimes called "free fall," with the implication that people would feel they were falling forever, therefore gripped by terror as well as feeling totally disoriented. It turned out, however, that many astronauts found the sensations pleasurable. Perhaps two-thirds of astronauts experienced "space sickness" (nausea and related symptoms) for a while before apparently adapting, but others were not bothered at all.[38] Some part of the human sense of balance may require gravity, but the balance sense in the inner ear senses motion or rotation from inertia—which exists as usual in space—rather than from gravity. However well astronauts adapt to zero gravity, when they return to Earth after a long-duration mission, they face a second adjustment that can be quite difficult. Naturally NASA needed to be able to measure the impact of spaceflight on human balance, as a step toward preventing or mitigating the problem.

Spinoff 1996 described two balance evaluation systems, EquiTest and Balance Master, developed by NeuroCom International, with NASA support. The EquiTest system has the patient stand on a platform inside a structure similar in size to a telephone booth but with no rear wall. Wearing a safety harness, the patient faces forward. To provide some visual orientation, and perhaps a degree of psychological comfort, the front and sides of the enclosure are decorated with a simplified outdoor scene. During the test the platform on which the person is standing will move; and the patient will react to maintain balance. A technician stands outside the structure, using a personal computer that records the person's reactions. How the patient reacts to the carefully controlled movements of the platform allows assessment of the state of the individual's sense of balance. Balance Master has some similarities but

---

37. "Spinoff Frequently Asked Questions," NASA Scientific and Technical Information (STI) Program, *http://www.sti.nasa.gov/tto/spinfaq.htm*.

38. Eric Martina and Gary Riccio, "Motion Sickness and Postural Instability After Prolonged Exposure to an Altered Gravitoinertial Force Environment (+2Gz)," in S. Stavros Valenti, ed., *Studies in Perception and Action II* (Hillsdale, NJ: Lawrence Erlbaum Associates, 1993), pp. 366–367.

does not employ an enclosure. The patient stands on a platform and moves his or her body. Movements of the body control the movements of a figure on a computer screen, and the patient is supposed to move about to get the figure to targets on the computer screen. The pictures in the NASA publication and on the company's current Web site show different patients balancing on one foot on the small platform. This platform does not move but senses the changing pressures the individual exerts on it.

It is noteworthy that both devices are still being manufactured by NeuroCom after the spinoff was publicized by NASA in 1996, and their development dates from more than a decade earlier when the company was founded in 1984 in Clackamas, Oregon. NeuroCom's Web site briefly explains the company's history thus:

> Dr. Lewis Nashner, through cooperative research programs conducted with colleagues at a number of academic institutions, developed the concepts and technology that are the basis for the company's products. Computerized Dynamic Posturography (CDP) was developed initially with grant support from NASA to evaluate the effects of space flight on vestibular function and balance control in astronauts, and with later support from the National Institutes of Health to study the effects of disease on balance and mobility functions.[39]

In 1988, Nashner received US Patent 4,738,269 ("Apparatus and method for sensory integration and muscular coordination analysis"), which is an early version of EquiTest. He had originally filed an application in 1982, withdrew it as inventors often do when patent examiners raise issues or some other problem arises, and filed the successful application in 1986, two years after founding NeuroCom. From 1989 through 1996, Nashner received several other related patents: US Patent 4,830,024 ("Apparatus and method for determining the presence of vestibular perilymph fistulae and other abnormal coupling between the air-filled middle ear and the fluid-filled inner ear"); US Patent 5,052,406 ("Apparatus and method for movement coordination analysis"); US Patent 5,269,318 ("Apparatus and method for movement coordination analysis"); US Patent 5,303,715 ("Apparatus and method for determining the presence of vestibular pathology"); US Patent 5,474,087 ("Apparatus for characterizing gait"); US Patent 5,476,103 ("Apparatus and method for assessment and biofeedback training of leg coordination and strength skills"); and US Patent 5,551,445 ("Apparatus and method for movement coordination analysis").

---

39. See *http://www.onbalance.com/neurocom/about/index.aspx.*

The patent applications cite many relevant scientific articles Nashner had published, notably three in the journal *Brain Science* in the late 1970s, but say nothing about NASA support. These facts do not diminish NASA's contribution to the development of these devices, but they demonstrate that Nashner was a highly competent scientist, who holds three degrees from MIT. In 1965 and 1967 he earned B.S. and M.S. degrees in aeronautical and astronautical engineering, and in 1970 a doctorate in biomedical engineering and neuroscience. In the 1970s he joined the Neurological Sciences Institute of what is today called Oregon Health and Sciences University, serving as its chair from 1980 to 1984. Nashner's continuing interest in astronautics is illustrated by the fact that in 1996, with Professor Laurence R. Young, he established the Apollo Program Prize at his alma mater, MIT, "to stimulate young people to follow in the path of the Apollo Program."[40]

An Oregon business magazine reported that "NeuroCom's initial support came largely from Good Samaritan Hospital, which incubated the company on-site and provided legal, accounting and payroll help."[41] In 1976, Nashner received a Research Career Development Award from the National Institutes of Health, followed by other NIH grants through 1985.[42] Nashner recently explained to an interviewer:

> I was an academic researcher funded by the National Institutes of Health and NASA in the area of human balance and movement. My own background has been multidisciplinary, including aeronautical engineering, biomedical engineering and neuroscience. The research of my colleagues and I provided a better understanding of the balance system, and through collaborative studies with healthcare professionals led to opportunities for proving better diagnostic and treatment tools to clinicians managing balance disorder patients. The additional startup funds for the company were provided primarily by three Phase I and two Phase II Small Business Innovation Research Grants (SBIR).[43]

Clearly, Nashner's inventions were partially supported by NASA, although also by NIH. It is clear that his work sprang from a long-established research

40. Marie Stuppard, "Aero and Astro Cites 19 Students and Faculty," *MIT Tech Talk*, 5 June 1996, *http://web.mit.edu/newsoffice/1996/aeroastro-awards-0605.html*.

41. Robin J. Moody, "Bioscience Company Finds the Perfect Balancing Point," *Portland Business Journal*, 16 May 2003, *http://www.bizjournals.com/portland/stories/2003/05/19/story3.html?page=2*.

42. Biosketch of Nashner for a speech given at the University of Pittsburgh, *http://www.shrs.pitt.edu/cms/School/News.asp?id=607*.

43. "What Is NeuroCom International?," interview with Lewis Nashner, *Twst.com*, 20 March 2006, *http://www.twst.com/notes/articles/tam212.html*.

program, although throughout his career some of his inspiration came from the space program. EquiTest and Balance Master were successful machines, although not very great contributions economically. In 2003, NeuroCom reported earnings "between $5 million and $10 million in annual revenue."[44] However, the machines have been used extensively as research instrumentation for studying human balance. For example, researchers at the University of Michigan and NASA's Johnson Space Center employed EquiTest to compare hypotheses about why astronauts sometimes continued to have balance problems long after landing.

A potentially important application area for the two instruments is research on populations whose systems of balance may be compromised by age or illness. For example, a study by Mark Redfern, Pamela Moore, and Christine Yarsky sought to understand what kinds of flooring surfaces helped elderly people keep their balance better.[45] The researchers tested elderly research subjects standing on seven different kinds of floors, using the EquiTest equipment. Findings were somewhat subtle, but there seemed to be a tendency for elderly people to keep their balance better on hard floors than soft ones. As many people understand, elderly people are especially susceptible to falling down, and the result can be a broken bone or even death. Of course, if they do fall, a hard floor may cause more damage than a soft one, so one must consider this trade-off when deciding whether to put carpeting in the home of an elderly person. These results may seem very simple yet, if substantiated by replication studies, could be very valuable precisely because they suggest a scientific basis for making decisions that significantly relate to human well-being. A literature research revealed more than two dozen articles referring to research with NeuroCom's instruments, in such journals as the *Journal of Vestibular Research* and *Aviation, Space, and Environmental Medicine*.

Case 3: Anti-Shock Garment

This spinoff concerns a method to counter hypotension, when the victim's blood pools in the legs and abdomen, by applying external pressure. Specifically, NASA's Ames Research Center had been aware for some years that methods developed to keep pilots from blacking out during extreme maneuvers might be applied to emergency response when ordinary citizens went into shock, whether from the trauma of an accident or a medical crisis,

---

44. Robin J. Moody, "Inovise Gets Funds Through Federal Science Grant," *Portland Business Journal*, 20 June 2003.
45. Mark S. Redfern, Pamela L. Moore, and Christine M. Yarsky, "The Influence of Flooring on Standing Balance Among Older Persons," *Human Factors* 39, no. 3 (1997): 445–455.

and needed to be transported quickly to a hospital. Each emergency-response ambulance could carry the anti-shock garment among its equipment, then apply it to the lower body of someone who had gone into shock, thereby increasing the supply of blood in the upper body. At the hospital the garment would be removed as soon as the emergency-room team was ready to address the fundamental problem in a more direct way.

A photograph in the *Spinoff* article shows a patient lying on a stretcher at the door of an ambulance as a team of two men works over her. One is giving her an IV as the other finishes applying the anti-shock garment. It appears to be in four connected sections—one for the lower part of the legs, one applied above the knees, one over the hips, and the final one on the stomach. The *Spinoff* story explains that the garment is not inflatable, but the picture shows it pressing tightly on the patient's lower body. The text says, "In addition to shunting blood from the patient's legs and abdomen to the heart, lungs and brain, the evenly and sequentially applied counter pressures help curb internal bleeding." The story says the garment was developed and distributed by two California companies, Zoex and Dyna Med, and that an Ames employee named Sheri Hillenga had joined Zoex to assist with marketing the item. The garment was called Dyna Med Anti-Shock Trousers (DMAST for short), and Zoex obtained patents for it in 1992.

The story actually begins much earlier than that. Ordway, Adams, and Sharpe report:

> In September 1969 Mrs. Mary Phillips, a housewife and mother of two children, was in the Stanford University Hospital bleeding uncontrollably after a minor operation. She received 46 pints of whole blood and 64 units of plasma in five weeks while doctors sought to control her hemorrhaging. One of the doctors recalled that a pressure suit had once been used to control bleeding during brain surgery at the Cleveland Clinic in Ohio. A call to NASA's nearby Ames Research Center produced a "g-suit," modified to fit the small patient. Placed on her and inflated for 10 hours, it brought the bleeding to a halt.[46]

The *Spinoff* article says Ames developed in 1971 a suit intended to control bleeding in the joints of hemophiliac children. A NASA Web site shows a picture from that year of the anti-hemophilia g-suit, apparently a full-body pressure suit.[47] The picture also shows a version of the "cool suit," a liquid-cooled garment that could help a person survive excessively hot conditions, or children suffering from hypohidrotic ectodermal dysplasia, an inherited disease

---

46. Ordway, Adams, and Sharpe, *Dividends from Space*, p. 47.
47. Ames image number AC71-8549, *http://ails.arc.nasa.gov/Images/Space/AC71-8549.html*.

among the consequences of which is a deficiency or lack of sweat glands. This is another spinoff from Ames, cited several times by NASA over the years.[48]

Sheri Hillenga had been involved with the pressure suit efforts at Ames as early as 1969 and continued to be interested in the potential of this idea until the opportunity came to promote it further. Zoex was a small company in Palo Alto, California, headed by Roxy Rapp. To assist in their marketing effort, Hillenga set up another small company named VMH Visual Communications. A Zoex corporation exists today in Houston, Texas, but it is a chemical company that does not appear to be related to the Palo Alto company. As of 2002, Hillenga and VMH Visual Communications had moved to Ashland, Oregon. Dyna Med was acquired by Galls, which is a subsidiary of Aramark Corporation, specializing in equipment for public safety and rescue apparel and equipment. It does not currently sell anti-shock trousers.[49]

Aside from the fact that DMAST is no longer on the market, there are two questions about this spinoff. First, how much credit can the space program claim for developing anti-shock garments? Second, how valuable are they? Within the field of medicine, the original credit for the invention is often assigned to George Washington Crile, who described his experiments with a rubber suit in the 1903 book *Blood-pressure in Surgery*.[50] Crile was a leading surgeon who won the Cartwright Prize of Columbia University for his research on shock, and his work was widely known. The reference made by Ordway, Adams, and Sharpe to the Cleveland Clinic speaks volumes about the source of the idea for Ames, because Crile was a founder of that clinic.[51]

When the Public Broadcasting System published Web pages for its 2002 program "Red Gold: The Epic Story of Blood," it included Crile among just 11 "innovators and pioneers" who deserved biographies. PBS recalled: "He devised and used clinically an ingenious 'pressure suit' which was capable of restoring blood to the circulation by the application of external pressure."[52] A 2000 article published in the student version of the *British Medical Journal* said, "Crile's design was later used in the second world war to prevent blackout

48. "Liquid Cooled Garments," *Spinoff 1979*, p. 100; "Cool Suit," *Spinoff 1987*, p. 104; "Spinoff from a Moonsuit (Cool Suit)," *Spinoff 1989*, p. 56.

49. Personal communication from Daniel P. Lockney, editor of NASA *Spinoff*, 16 January 2007.

50. George Washington Crile, *Blood-pressure in Surgery: An Experimental and Clinical Research* (Philadelphia: Lippincott, 1903).

51. See *http://www.case.edu/artsci/dittrick/cemetery/stop13.htm* and *http://www.clevelandclinic.org/aboutus/*.

52. "George Washington Crile," Public Broadcasting System, 2002, *http://www.pbs.org/wnet/redgold/innovators/bio_crile.html*.

of pilots subjected to high gravity forces. During the Vietnam war it was used to stabilize patients with haemorrhagic shock during transportation."[53]

The application for pilots may be a separate invention, usually attributed to the Canadian Dr. Wilbur Franks, who developed the first aviation g-suit (or anti-gravity suit) in 1941. Pilots fly in a seated position, and rapid turns of the aircraft can therefore force the blood away from the head, causing a pilot to lose consciousness. A cancer researcher at the Banting Institute, Franks has an honored place in the Banting Research Foundation Hall of Fame, which describes him as "inventor of the world's first anti-gravity suit for fighter pilots and the first high-speed human centrifuge, two innovations later used in the U.S. space program."[54] When the anti-shock trousers were inducted into the Space Foundation's Hall of Fame in 1996, the NASA press release said they "are an adaptation of the anti-gravity flight suits originally developed for pilots and astronauts."[55] However, by that time the medical use of such devices had a 93-year history of development.

Over the years many U.S. patents have been issued to inventors of anti-shock trousers. Many of these require inflation, including those described in Patent 1,608,239 granted in 1926; patent 3,933,150 in 1976; patent 4,039,039 in 1977; patent 4,270,527 in 1981; patent 4,355,632 in 1982; and patent 5,117,812 in 1992. One claimed advantage of the spinoff trousers is that they do not need to be inflated, but this was also true of the similar devices described in Patent 4,577,622, issued to Thomas Jennings in 1986; patent 4,848,324, issued in 1989 to Technion of Israel; and patents 5,146,932 and 5,259,397, issued to Francis McCabe in 1992 and 1993. Frustratingly, I have been unable to find the patent issued in 1992 mentioned in the *Spinoff* article, and I must assume one was applied for but not granted.

The second problem with this spinoff is that anti-shock trousers may not really be beneficial. In their 2003 book on spinoffs, Bijlefeld and Burke have noted that the effectiveness of anti-shock trousers has come under increasing criticism.[56] Writing in the U.S. Army's *Combat Casualty Care Guidelines: Operation Desert Storm* in 1991, Colonel Ronald Bellamy surveyed the recommendations of the Advanced Trauma Life Support (ATLS) course of the American College of Surgeons and noted: "The military anti-shock trouser

---

53. Karen Dickinson and Ian Roberts, "Medical Antishock Trousers (Pneumatic Antishock Garments) in Patients After Trauma: A Systematic Review of Randomised Controlled Trials," *Student British Medical Journal* 8 (February 2000): 25–27.

54. "Dr. Wilbur Franks," Banting Research Foundation Hall of Fame, *http://www.utoronto.ca/bantresf/HallofFame/Franks.html*.

55. "NASA Life-Saving Technologies Enter Space Hall of Fame," NASA press release 96-69, 10 April 1996, *http://www.nasa.gov/home/hqnews/1996/96-69.txt*.

56. Bijlefeld and Burke, *It Came from Outer Space*, p. 6.

(MAST) or PASG has been deleted from Army first- and second-echelon facilities. Therefore it is no longer a therapeutic option even though it is still a part of ATLS."[57]

The controversy over the value of anti-shock trousers has been very complicated, because there are many different conditions for which they might be used, each perhaps requiring a different judgment. For example, in a 1997 advisory the New York State Department of Health's Bureau of Emergency Medical Services downgraded its evaluation, listing eleven conditions for which the trousers may be harmful, four conditions for which they are probably at least not harmful, and only two where there is some evidence the trousers might help, although they remain of "uncertain efficacy"—namely "hypotension due to suspected pelvic fracture" and "severe traumatic hypotension (palpable pulse, blood pressure not obtainable)." The advisory pointedly notes that there are no conditions for which the trousers are definitely recommended, and reports that "the role of MAST (PAST) in the prehospital emergency medical care of adult and pediatric patients is extremely limited."[58]

In a text published in 2000, *Injury Prevention and Control*, Mathew Varghese reports: "The use of PASG or MAST is not recommended any more in the pre-hospital setting."[59] He especially stresses the danger that they can cause complications by restricting the patient's breathing if used over the abdomen; by cutting off the blood supply, thereby causing tissues to die; and by potentially damaging nerves and blood vessels. It is still possible to buy anti-shock trousers, and three different makes are advertised in supply catalogs for emergency response teams—inflatable models from David Clark, Armstrong Medical, and Life Support Products.[60] However, there is no evidence that the anti-shock trousers described in the 1996 spinoff are available; nor are any of the other noninflatable competitors that were patented. Inflated trousers may take longer to apply, but they adjust better to the contours of the body, take up less space in a crowded ambulance, and had achieved a solid position in the marketplace, from which it might have been difficult to dislodge them.

57. Ronald F. Bellamy, "The Nature of Combat Injuries and the Role of ATLS in Their Management," in *Combat Casualty Care Guidelines: Operation Desert Storm* (Washington, DC: Walter Reed Army Medical Center, 1991), available at *http://www.bordeninstitute. army.mil/cmbtcsltycare/default.htm*.

58. "Medical Anti-Shock Trousers," Advisory No. 97-04, 7 August 1997, New York State Department of Health Bureau of Emergency Medical Services.

59. Mathew Varghese, "Technologies, Therapies, Emotions, and Empiricism in Pre-hospital Care," in Dinesh Mohan and Geetam Tiwari, eds., *Injury Prevention and Control* (London: Taylor and Francis, 2000), 249–264, esp. p. 254.

60. David Clark, *http://www.davidclark.com/MAST/medical.shtml*; Armstrong Medical, *http:// www.armstrongmedical.com*; and Allied Healthcare Products, *http://www.alliedhpi.com*.

Case 4: Heart Pacemaker Technology

A rather more promising spinoff in *Spinoff 1996* is the "programmable pacemaker" for patients with slow or erratic heart rhythms. Over the years NASA has often publicized some of its connections to the development of heart pacemakers and the related technology of implantable defibrillators. The 1996 *Spinoff* article mentions several such connections but emphasizes recent developments by one company: "Introduced in 1995 by Pacesetter Systems, Inc., Sylmar, California, the Trilogy family of pacing systems represents a fourth generation advancement of the programmable unit first developed in the 1970s by NASA, Johns Hopkins Applied Physics Laboratory and Pacesetter Systems." Rather than focus narrowly on one example, here I consider the wider issue of a series of influences over the years. A range of people suffer from bradycardia, a heartbeat that is too slow, some of them in otherwise reasonably good health. Therefore, it makes sense to use artificial means to speed up their heart beat, and each year hundreds of thousands of Americans receive pacemakers to do this job. Thus the benefit to human survival and the quality of life is enormous.

A pacemaker works by stimulating the heart with a small electronic pulse, applied in just the right location. The original pacemakers emitted a constant, predetermined rate of pulses. However, this rate might be too fast or too slow for some patients, so the development of implantable pacemakers that could be programmed at a distance, without surgery, was a considerable step forward. Chemical batteries run down over time, so the ability to determine the condition of the battery, or to recharge it without removal, could reduce the need for further surgery. Ideally, a pacemaker should monitor the heart and adjust to the owner's changing needs. Research in this direction led to the possibility of an implantable defibrillator for patients whose hearts might unexpectedly beat too fast or in an uncoordinated manner that fails to pump the blood properly (called ventricular tachycardia or ventricular fibrillation). A brief historical survey is needed to identify the various points at which NASA may have had a positive input.

A recent history of pacemakers by Kirk Jeffrey traces the development of scientific knowledge about the heart, early experiments with pacemakers in the 1920s and 1930s, to Paul M. Zoll's demonstration of a medically successful pacemaker in 1952 at Beth Israel Hospital in Boston.[61] The device was hospital equipment for emergency use, external to the patient and using a substantial voltage because the electrodes were not placed inside the body,

---

61. Kirk Jeffrey, *Machines in Our Hearts: The Cardiac Pacemaker, the Implantable Defibrillator, and American Health Care* (Baltimore: Johns Hopkins University Press, 2001).

available commercially from 1954. A team at the University of Wisconsin led by C. Walton Lillehei, a pioneer in the field of open-heart surgery, first tried placing an electrode in the heart itself in 1957, initially using 110 volts. Seeking safer, low-voltage equipment, Lillehei commissioned Earl Bakken, an engineer and owner of a small company called Medtronic, who quickly created a battery-powered, transistorized pacemaker. Released commercially in 1958, Bakken's pacemaker was external to the human body but connected to internal electrodes and portable. By 1960 pacemakers were being widely used to sustain patients with chronic problems rather than only temporarily during surgery and the recovery period. Recognizing that long-term users could not easily tolerate wires through the skin to external equipment, a Swedish group created a fully implantable pacemaker in 1958; and by the end of 1960, eight different research groups had developed them.

The most important of the other groups was the partnership of surgeon William C. Chardack with electrical engineer Wilson Greatbatch. The two worked in simultaneous competition with Lillehei and Bakken and claimed the first "clinically successful" pacemaker implantation on 6 June 1960, considering the Swedish work to have been experimental and useless for patients.[62] On 22 July 1960, Greatbatch filed a patent application for a "medical cardiac pacemaker," acknowledging Lillehei's work, and he received Patent 3,057,356 on 9 October 1962. Beginning in 1960, Greatbatch and Chardack teamed up with Lillehei and Bakken to have Medtronic produce pacemakers based on Greatbatch-Chardack designs. By 1965 the technology was widely accepted, and Medtronic dominated the market for a decade.[63]

Greatbatch is often described as "the inventor of the implantable cardiac pacemaker," and he clearly played a key role—from his independent experiments with prototypes before collaborating with Chardack to his tireless work to improve pacemaker batteries after leaving Medtronic.[64] However, the idea that a complex device like the cardiac pacemaker had a single inventor promulgates false notions of how technological progress works. The National Academy of Engineering awarded its 2001 Russ Prize to "Earl E. Bakken and Wilson Greatbatch for their independent development of the implantable cardiac pacemaker."[65] A large number of other people also contributed,

---

62. Wilson Greatbatch, *The Making of the Pacemaker: Celebrating a Lifesaving Invention* (Amherst, NY: Prometheus, 2000), p. 16.

63. Alice M. Vollmar, "Medical Mechanic," *The World and I* 18, no. 12 (2003); and "Wilson Greatbatch, Electrical Engineer," an oral history conducted in 2000 by Frederik Nebeker, IEEE History Center, Rutgers University, New Brunswick, NJ, *http://www.ieee.org/portal/cms_docs_iportals/iportals/aboutus/history_center/oral_history/pdfs/Greatbatch396.pdf*.

64. John A. Adam, "Wilson Greatbatch," *IEEE Spectrum* 32, no. 3 (1995): 56–61.

65. See *http://www.nae.edu/NAE/awardscom.nsf/weblinks/DWHT-4T7KMA?OpenDocument*.

including Lillehei and Chardack, who were not eligible for the award because they were not engineers. Electrical engineers conceptualize a pacemaker as an electronic device that generates stimulation pulses, but the pulse generator is really only part of an entire system that includes the electrodes, the surgical techniques for implantation, and the medical knowledge of where to stimulate the heart in what way and when. Early electrodes tended to break from metal fatigue, and their contact with the heart tended to degrade. The first durable electrodes were developed by Chardack, not by Greatbatch, described in Chardack's Patents 3,198,195; 3,216,424; and 3,348,548. Much subsequent development work by many people was required to make pacemakers programmable, adaptive, and capable of handling problems other than bradycardia or heart block.

Granting Greatbatch's significance, he represents the first plausible influence of the space program on pacemaker development. While a student at Cornell, he had gotten a job building electronic equipment to monitor animals' heart rate and blood pressure, but what really got him interested in biomedical electronics was building amplifiers for monitoring animals' vital signs on suborbital test flights. He also was involved with building amplifiers for a predecessor of the Arecibo radio telescope.[66] Although these were conceptually related to space exploration, Greatbatch did not work for NASA. He refers to the Little Joe launches, but they did not come until after he had begun collaborating with Chardack. Also, he says the work he did supporting suborbital flights was for the Air Force, and Arecibo was a DARPA-NSF project. In early 1958, when Greatbatch's collaboration with Chardack began, Greatbatch used to call his experimental devices "Tikniks," in honor of the Russian Sputniks that had been launched over the previous several months.[67] Late in his life, Greatbatch became an advocate for economic development of the Moon, based on harvesting Helium-3 for use as a fuel in fusion reactors to solve the Earth's energy crisis.[68]

Greatbatch has clearly been inspired by space exploration but not through working with NASA, because NASA was only just being founded while he was working on his first pacemakers. Today, the head of Medtronic's pacemaker division is Stephen Mahle, who came to the company in 1972 from NASA's Manned Spacecraft Center in Houston and who negotiated agreements to use Greatbatch's lithium batteries when the latter left Medtronic.[69]

---

66. Greatbatch, *Making of the Pacemaker*, p. 233, 29.
67. Jeffrey, *Machines in Our Hearts*, p. 99.
68. Greatbatch, *Making of the Pacemaker*, pp. 223–228.
69. "Biography of Stephen Mahle," Medtronic.com, *http://wwwp.medtronic.com/Newsroom/ Biography.do?itemId=1108585002878&lang=en_US*.

The cases of Greatbatch and Mahle illustrate two of the mechanisms for technology transfer other than spinoffs. First, a receptive and talented person outside NASA can be inspired by its vision and accomplishments to undertake achievements in related areas. Second, an individual who works with NASA, or indirectly for a contractor or other NASA partner, and who moves on to a new job takes along experience and expertise of a general kind that may be quite valuable for technological innovation, quite apart from any specific invention NASA may have supported.

Having described the early history of pacemakers, we can now place the 1996 *Spinoff* story about programmable pacemakers in context. The story actually appears in three earlier issues as well—those for 1980, 1981, and 1990. Although Medtronic dominated the field in the early days, it did face some competition, including that from Pacesetter Systems. An obvious factor that limited the performance of implanted pacemakers was that precisely because they were enclosed within the human body, the physician could neither make adjustments nor check the battery. Often, patients suffered pacemaker replacement surgery simply because there was no good way to tell how much life the battery still had. Thus two-way communication with the pacemaker, including the ability to program its performance, would be an obvious advantage.

The first programmable heart pacemaker was manufactured by General Electric in 1961, and as is often the case for first-generation innovations, it was rather crude. The pacemaker included a magnetically operated switch that set the heart's beat rate at either 70 or 100 per minute, with the hope that the user himself could select the low rate for rest and the high rate for activity by moving a magnetic wand over his skin in a particular way. A more elaborate magnetically programmed model was marketed by Medtronic in 1972, but it also merely selected the speed.[70] Both involved using a magnet to physically move a component of the pacemaker, rather than being fully electronic, but Medtronic introduced a radio-control method in 1973.

The particular commercial product cited by the 1980 *Spinoff* story was called Programalith. Over the years new models were introduced—Synchrony and Trilogy—and Pacesetter Systems kept in contact with NASA, presumably for two reasons. First, it probably did find continuing inspiration in NASA work in telemetry and related technologies, even though the direct spinoff occurred in the late 1970s when pacemaker technology was still immature. Second, the company and the Agency could take pride in their informal relationship, generating publicity like the *Spinoff* stories that benefited NASA

---

70. Tarun Mittal, "Pacemakers: A Journey Through the Years," *Indian Journal of Thoracic and Cardiovascular Surgery* 19, no. 5 (2005): 236–249.

public relations and giving the company some added prestige with the medical profession.

Another potential benefit is more subtle and conjectural but interesting because it could apply to many other NASA spinoff technologies. In a competitive, high-technology market, there are pressures on each company to go in two opposite technical directions. As sociologists Paul DiMaggio and Walter Powell have influentially argued, there are strong pressures on competitors in a well-defined market to become more similar to each other than is technically necessary.[71] DiMaggio and Powell note that government or professional standards of quality could do this. In the case of pacemakers, there was an added factor in that patients, doctors, and health insurance plans constantly change—for example, patients moving to different cities or switching their health insurance plans. This means that as cardiologists and heart pacing specialists gain new patients these new patients will have different models of pacemakers from different manufacturers. As a profession, cardiologists may become resistant to a proliferation of models that they would understand less well. Although factors like this undoubtedly operate, Kirk Jeffrey has noted that pacemakers have not become uniform, commodity products for which low cost (given reliability) becomes the key market factor, but manufacturers have continued to innovate.[72] In its need to innovate, Pacesetter Systems required not only better products but different ones as well.

Depending on the industry, a significant fraction of earnings can come from license fees for patents. It can be difficult for a company to make a profit, if it must pay multiple heavy fees to other companies for the use of their patents. An innovative company may not be able to do business without paying some patent licensing fees, but if it has its own patents as well, it can trade licenses with the other companies. A current area of practical and scholarly debate concerns *patent thickets*. These are situations in which a given new technology must use several patents held by different firms, and economic or regulatory barriers to sharing patents prevent the technology from being developed. A traditional solution was *patent pools*—the major firms agreeing to cross-license all the relevant patents, perhaps charging fees only for companies that did not contribute significantly to the pool. However, in recent years

---

71. Paul J. DiMaggio and Walter W. Powell, "The Iron Cage Revisited: Institutional Isomorphism and Collective Rationality in Organizational Fields," *American Sociological Review* 48, no. 2 (1983): 147–160.

72. Kirk Jeffrey, "Cardiac Pacing and Electrophysiology at Millennium's End: Historical Notes and Observations," *PACE* 22, no. 12 (1999): 1713–1717.

the U.S. federal government has tended to treat patent pools as monopolistic practices, thus banning them.[73]

By innovating, an organization like Pacesetter not only could produce a better pacemaker but would also put itself in a better competitive position, for three reasons. First, as just noted, it would have patents to cross-license with other firms, so long as this did not lead to an illegal patent pool. Second, it would have sales advantages in competition with those of other companies that offered different innovations. In such cases, unfortunately, patients might need to wait until the patents expire to get pacemakers that combine all the beneficial innovations. Third, some of its patents would concern other ways of accomplishing the same benefits as patents held by the other companies. Often there is more than one route to a common goal. For example, there are at least two ways to avoid having to remove pacemakers often to change the batteries. One is simply longer-lasting batteries, perhaps coupled with telemetry to assess how much charge remains on it as it ages. The other is rechargeable batteries, which sound like a great idea but have problems of their own, including the frequent need for costly recharging; they also require replacement at some time.

What exactly was the programmable pacemaker a spinoff of? In NASA's *Spinoff* database the 1980 story says it "incorporated Apollo technology," whereas the 1981 story said it "originat[ed] from spacecraft electrical power systems technology." The 1990 story referred to "bidirectional telemetry developed for communication between earth stations and orbiting satellites." The 1996 story explained that Pacesetter drew on different aspects of space technology for three different aspects of pacemakers: rechargeable long-life batteries, perhaps inspired by spacecraft power systems; single-chip integrated circuits, supposedly developed for microminiaturization of spacecraft components; and programmability derived from NASA's two-way telemetry with satellites. Rechargeable batteries were not very much better than the best batteries that could not be recharged, at least in their early years. But they were valuable for Pacesetter because industry leader Medtronic had put so much emphasis on developing the very best nonrechargeable batteries, under Greatbatch's influence. Thus rechargeability was a strategy for competing with Medtronic, as much as it was an innovation that gives patients better service.

---

73. Gavin Clarkson and David DeKorte, "The Problem of Patent Thickets in Convergent Technologies," in William Sims Bainbridge and Mihail C. Roco, eds., *Progress in Convergence: Technologies for Human Wellbeing* (New York: New York Academy of Sciences, 2006), pp. 180–200.

The other technology—improvements of integrated circuit technology— is a vast field to which NASA undoubtedly contributed greatly and that deserves much more extensive treatment than I can provide in this chapter. However, as I address later in connection with the Optacon device, often several government agencies also made substantial contributions. So single-chip devices are probably not a narrowly defined spinoff but a broad area of progress where NASA's contribution, which is very significant, is difficult to separate out from those of other contributors, such as the Defense Department, computer manufacturers, and the chip makers themselves. Programmability is connected to rechargeability in at least two ways. First, both require some method for exchanging energy to and from the implanted pacemaker, thus potentially building on some of the same expertise. Second, programmability potentially places greater demands on the battery, especially if it involves extensive two-way communications. One could power programming and telemetry separately from the batteries that stimulate the heart—for example, through induction from the external programming device. But the fact that Pacesetter combined rechargeability with programmability emphasizes their technological connection.

In addition to citing Pacesetter Systems, the 1996 *Spinoff* story refers to work done in the 1970s at the Johns Hopkins Applied Physics Laboratory. A key person in that effort was Robert E. Fischell, who joined the laboratory in 1959.[74] His curriculum vitae lists: "Pacesetter Systems, Inc., Founder, Patent Licensor, Director from 1969 to 1986."[75] In the early 1960s he did work for the U.S. Navy's navigation satellite program that led to later NASA connections and to his work with heart pacemakers. In a 1998 autobiographical article, Fischell himself attributes his pacemaker work to his experience helping to develop the Navy's Transit satellite, not to his NASA work.[76] Dating from 1960, Transit was a system of satellites in low polar orbit, the first operational satellite navigation system and thus the precursor to the tremendously important Global Positioning System. Fischell's article explicitly underscores the importance of Transit as the source of the three advances he later claimed for his pacemakers: telemetry, power system, and microminiaturization. He specifically credits NASA with supporting later work that led to two distinct but technically related developments—a programmable implantable medication

74. Ron Perea, interview with Robert E. Fischell, 16 October 1993, *DukEngineer*, *http:// www.pratt.duke.edu/news/?id=354*.

75. See *http://www.psychoceramic.com/Fischell.html*.

76. Robert E. Fischell, "Applications of Transit Satellite Technology to Biomedical Devices," *Johns Hopkins APL Digest* 19, no. 1 (1998): 60–62.

system for insulin delivery and an implantable heart defibrillator, both of which NASA has counted as separate spinoffs.[77]

Fischell's early work on Transit exploited magnetism in a number of different ways to control and measure the rotation of an orbiting satellite. In 1963 and 1964, for example, he received Patents 3,104,080 ("energy absorption mechanism"); 3,114,518 ("magnetic despin mechanism"); and 3,118,637 ("magnetic attitude control"). Two other patents issued in 1969 and 1970 did not exploit magnetic phenomena: 3,424,907 ("satellite attitude detection system including cosine and spinstate detectors"), based on solar cells that located the sun; and 3,489,203 ("controlled heat pipe"), for managing heat produced by a satellite's electronics. Another set of patents issued from 1971 to 1974 again employed magnetism: 3,611,815 ("frictionless gyroscope"), based on magnetic bearings; 3,767,139 ("spacecraft spin stabilization system"), based on a flywheel that would need the magnetic bearings; and 3,785,595 ("system for sensing and compensating for the disturbance forces on a spacecraft"), inertial guidance using magnetism to levitate a mass. All these patents were assigned to the U.S. Navy. With Richard T. Ellis, Fischell also got Patent 3,489,372 ("satellite spin control system") in 1970, again assigned to the U.S. Navy and exploiting Earth's magnetic field. The application for his 1973 Patent 3,767,139 on the same topic says the earlier method was "utilized, for example, on the AE-B and DME-A satellites launched by NASA." Thus NASA was involved in these early efforts, but the Navy was the driver.

According to a very brief biography in MIT's "Inventor of the Week" archive and a University of Maryland alumni newsletter, Fischell first started thinking about pacemakers when he happened to see a notice in an academic journal about a pacemaker battery with a life of two years, presumably one of Greatbatch's creations.[78] Thus the first of his innovations in the field would be a system for recharging the batteries of implanted devices. In 1975 he got two patents based on charging by means of magnetic induction: 3,867,950 ("fixed rate rechargeable cardiac pacemaker") and 3,888,260 ("rechargeable demand inhibited cardiac pacer and tissue stimulator"). Both were assigned to Johns Hopkins University, and neither patent application referred to either the Navy or NASA.

---

77. "A Boon for the Diabetic (Insulin Pump)," *Spinoff 1987*, p. 76; and "Implantable Heart Aid," *Spinoff 1984*, p. 60.

78. "Inventor of the Week Archive," *http://web.mit.edu/invent/iow/fischell.html*; "On Alumnus Dr. Robert E. Fischell," *The Photon: Spotlight*, University of Maryland, physics research alumni newsletter, April 2000.

In 1979, Fischell received patent 4,275,739 ("charge control switch responsive to cell casing deflection") on behalf of the university, but the application included this notice: "The invention described herein was made in the performance of work under a NASA 5-23732 contract and is subject to the provisions of Section 305 of the National Aeronautics and Space Act of 1958, Public Law 85-568 (72 Stat. 435; 42 U.S.C. 2457)." This invention linked telemetry with battery charging, because it concerned a small sensor that would detect the pressure on a battery casing when it was completely charged. Such a system could be useful for satellites—for example, if they used solar power to charge storage batteries and needed to know when they were fully charged. When Fischell filed the patent application for his implanted drug-delivery system in 1979 (issued as Patent 4,373,527 in 1983), he made it programmable but did not cite NASA. In 1983, Arthur F. Hogrefe and Wade E. Radford filed for a patent—4,561,443 ("coherent inductive communications link for biomedical applications")—that involved telemetry and cited support from NASA under contract NDPR S-63983B. In 1985, Hogrefe and Radford obtained the patent for Johns Hopkins. Clearly, NASA was funding some of the work, even if it had not provided the original inspiration.

What appears to be a simple yet important spinoff, a "Programmable Pacemaker," turns out to be a complex story, with NASA providing some inspiration to at least a few leaders in the field and supporting research that contributed to improved pacemakers. However, other agencies contributed at least as much, and the medical community deserves a huge portion of the credit. Kirk Jeffrey's historical study cites Greatbatch frequently but argues that the electrical engineers have been given too much credit. To elaborate his argument slightly, we can consider the vagueness of the line between science and technology and the tendency of observers to hold a simplistic theory that progress consists of a series of patentable "inventions."

An electrical engineer would tend to conceptualize heart pacemaking in terms of an electronic device that needed to be invented by an electrical engineer. That is the concept behind the 2001 Russ Prize awarded to Bakken and Greatbatch. However, that device could never work until physicians and medical researchers had begun to specify when the heart should be stimulated, where, and how much. The heart surgeons and the electrical engineers collaborated, and then the annals of electrical engineering gave the credit to the electrical engineers. Pacemaking is really a system of many components, including the knowledge of the heart's complex response to electrical stimulation, the surgical procedures, proper designs for electrode leads that would neither break nor corrode, and finally the device that produced the electrical stimulation.

It is an odd fact of the history of science and technology that Western society has come to recognize the intellectual property of engineers but not that of scientists. An author or publisher can copyright a book, but doing so protects only its particular expression of ideas, not the ideas themselves. Patents protect ideas, if for a limited period of time. It might be said that scientists merely discover properties of nature, and nature cannot be patented. However, our society considers the ownership of land to be quite acceptable, and land is a part of nature. Economic or political historians may be able to tell us why this odd state of affairs came to be. If the U.S. patent system is merely a practical expedient that serves our economy by encouraging practical innovation, then one unhelpful consequence is that the system reinforces the partially false notion that inventions are isolated moments of innovation that exist apart from their human and technical environment.

In the case of programmable heart pacemakers, a vast army of individuals contributed a library full of ideas, some more important than others. NASA's direct contribution would seem to be relatively minor, providing some inspiration and support for one of the competitors in the market, but having only a small influence on the origins of pacemaker technology in general. The *Spinoff 1996* article correctly describes the spinoff as a set of improvements rather than the fundamental invention of pacemakers, but the public may not be attentive to such niceties. They may also not be clear that the case was one of diffusion of general technical ideas, and the involvement of one person (Fischell) who directly linked NASA with pacemakers, rather than an invention completed within NASA's research facilities and then transferred to a medical application. Thus one could debate whether the pacemaker is a spinoff at all.

Case 5: Cardiac Monitor

The accelerations of liftoff and reentry and the weightlessness experienced between these stressful events place the human heart under unusual conditions. Therefore, both for research purposes and to monitor the well-being of astronauts during their demanding missions, it is useful to be able to monitor the heart. *Spinoff 1996* reports the NASA-supported development of the *impedance cardiograph*. This is not the same thing as the familiar electrocardiograph, which measures the electrical activity of the heart itself as it beats. Rather, the aim is to estimate the volume of blood pumped through the heart, the so-called cardiac output. *Impedance* is the effective electrical resistance of an object in response to alternating current of a given frequency, and the pulsing flow of blood causes the impedance of the human thorax to vary. Impedance cardiography can be conceptualized as a special case of *plethysmography*, measuring the changing volume of a part of the body, usually as blood volume changes. *Spinoff 1996* explains that NASA's interest came quite early in the space program:

> In 1965, Johnson Space Center contracted with the University of Minnesota to explore the then-known but little-developed concept of impedance cardiography (ICG) as a means of astronaut monitoring. A five-year program led to the development of the Minnesota Impedance Cardiograph (MIC), an electronic system for measuring impedance changes across the thorax that would be reflective of cardiac function and blood flow from the heart's left ventricle into the aorta.[79]

The *Spinoff* database says that the work came out of Project Mercury, and indeed the hearts of Mercury astronauts were constantly monitored; but this was done by means of a conventional if miniaturized electrocardiograph, which could not measure the flow of blood directly, only the electrical signals generated by the heart as the muscle pumps. There are many ways to infer blood flow, but standard medical methods are extremely intrusive, involving catheterization and monitoring the dilution of a dye or actually heating the blood to use temperature differential to trace the flow in what is called *thermodilution*. Having experienced a similar dye procedure myself, I know it requires extensive equipment, trained staff, and can be done only for a short period of time. Thus it is totally unsuited for use on an astronaut during flight. Impedance cardiography is nonintrusive, involving a very slight electric current administered between electrodes on the surface of the skin. The challenge is that the data collected must be interpreted in terms of a complex mathematical model of the impedance of the human body, and thus development of the technology was far from simple.

As was the case for heart pacemakers, research in this area began long before a workable system could be built, at least as early as the 1940s. However, the point at which developments really took off does seem to have been 1965, when NASA began supporting the Minnesota work. The team was highly interdisciplinary. William G. Kubicek, often mentioned most prominently in scientific publications on the topic, was a professor of physical medicine and rehabilitation at the University of Minnesota, with a history of inventions employing pneumatic or hemostatic pressure. He was coinventor of a system to supply properly humidified air to patients who had received tracheotomies or comparable procedures, receiving patent 2,584,450 way back in 1952. Like the others mentioned here, the patent was assigned to the University of

---

79. *Spinoff 1996*, "Cardiac Monitor," p. 51.

Minnesota. In 1962 he received patent 3,050,050 for an "alternating pressure seat" for paraplegics that would avoid constantly pressing on the same spot, and patent 3,059,635 for a "respirator collar" for iron lungs.

In 1964, Kubicek and three colleagues filed a patent application for an "impedance plethysmograph," receiving patent 3,340,867 in 1967. This was clearly the predecessor of the impedance cardiograph. The patent application included a diagram of a transistorized electrical circuit, a diagram showing four beltlike electrodes applied to a patient (two around the neck and two around the abdomen), and two diagrams analyzing the structure and dynamics of the human torso. The very second sentence of the application states clearly: "The invention is particularly useful in determining cardiac output." The coinventors constituted all three other prominent members of the team that would develop the impedance cardiograph: Edwin Kinnen, Robert P. Patterson, and David A. Witsoe. Someone with a naive understanding of patents would conclude, from reading the application for patent 3,340,867, that the team had already invented the impedance cardiograph in 1964, a year before NASA provided support. However, a device for measuring the changing impedance of the human body is not yet a system for measuring blood flow. There must be a reasonably accurate mathematical model that translates the indicator (variations in impedance) into the real variable it indicates (blood flow), and the math must ultimately rest on empirical calibrations. Robert Patterson recalls that support from NASA was vital to his endeavors, stating that "starting in 1965, our laboratory performed an extensive series of animal experiments using dogs instrumented with electromagnetic flow meters on the pulmonary artery and aorta, and with catheters placed in the aorta and various heart chambers."[80] A key publication, still widely cited, is the team's extensive 1969 report from NASA Contract 101965.[81]

An indicator of continuing NASA interest is the fact that Patent 3,957,037 for a "readout electrode assembly for measuring biological impedance" was issued to NASA Administrator James Fletcher in 1976 on behalf of the Agency. *Spinoff 1996* says that the first operational use of impedance cardiography was during the STS-8 Space Shuttle flight in 1983. A 2004

---

80. R. P. Patterson, "Fundamentals of Impedance Cardiography," *IEEE Engineering in Medicine and Biology Magazine* 8, no. 1 (1989): 35–38, esp. p. 36.

81. W. G. Kubicek, D. A. Witsoe, R. P. Patterson, and A. H. L. From, eds., *Development and Evaluation of an Impedance Cardiographic System to Measure Cardiac Output and Other Cardiac Parameters*, NASA-CR-101965 (Houston, TX: NASA Manned Spaceflight Center, 1969).

report in *NASA Tech Briefs* about software to measure human physiological data included two channels of impedance cardiographic data as a matter of course.[82]

Around 1970, the results of this method were highly controversial, perhaps in part because exaggerated claims had been made for impedance measurements in earlier decades, but gradually a considerable scientific literature developed calibrating impedance methods against others and demonstrating its potential reliability.[83] In the cases of the bone density analyzer and balance evaluation system described earlier, the scientific literature supports their value, but neither has been the subject of very many studies. The situation with impedance cardiography is quite different. I found a huge literature, even overwhelming in its volume. This measurement approach is clearly very significant, both clinically and in terms of fundamental scientific research. One reason why there are so many publications is because the method is not straightforward but must infer blood flow from its electrical measurements, so there is huge scope for studies to evaluate and improve the performance of the technology. The subject is international in scope, with researchers active today not only in the United States but in nations as diverse as India, Israel, Italy, and the Netherlands.[84]

Impedance cardiographs based on the original Minnesota Impedance Cardiograph concept are being manufactured by a number of companies. Bio-Impedance Technology, a small company in Chapel Hill, North Carolina, makes them.[85] The BioZ line of impedance cardiographs is manufactured by

82. Bruce C. Taylor, Soumydipta Acharya, Patricia S. Cowings, and William B. Toscano, "User Interactive Software for Analysis of Human Physiological Data," *NASA Tech Briefs*, ARC-15287-1, Ames Research Center, May 2004, available at *http://www.techbriefs.com/ component/content/article/1265-arc-15287-1*.

83. Joseph M. Van De Water, Bruce E. Mount, James R. Barela, Roger Schuster, and Ferdinand S. Leacock, "Monitoring the Chest with Impedance," *Chest* 62 (1973): 597–603.

84. For example, see Paul A Nakonezny, Ray B. Kowalewski, John M Ernest, Louise C. Hawkley, David L. Lozano, Daniel A. Litvak, Gary G. Berntson, John J. Sollers III, Paul Kizakevich, John T. Cacioppo, and William R. Lovallo, "New Ambulatory Impedance Cardiograph Validated Against the Minnesota Impedance Cardiograph," *Psychophysiology* 38, no. 3 (2001): 465–473; Esther Raaijmakers, Th. J. C. Faes, Henk G. Goovaerts, Peter M. J. M. Vriew, and Robert M. Heethaar, "The Inaccuracy of Kubicek's One-Cylinder Model in Thoracic Impedance Cardiography," *IEEE Transactions on Biomedical Engineering* 44, no. 1 (1997): 70–76; Vinod K. Pandey and Prem C. Pandey, "Cancellation of Respiratory Artifact in Impedance Cardiography," *Proceedings of the 2005 IEEE Engineering in Medicine and Biology Annual Conference*, 1–4 September 2005, Shanghai, pp. 5503–5506; and G. Cotter, A. Schachner, L. Sasson, H. Dekel, and Y. Moshkovitz, "Impedance Cardiography Revisited," *Physiological Measurement* 27, no. 9 (2006): 817–827.

85. See *http://www.microtronics-nc.com/BIT/IFM.html*.

CardioDynamics of San Diego, California, which had net sales of $30.4 million in FY 2006.[86] Brochures for the Minnesota-based Väsamed company explicitly say their AcQtrac System is based on the work supported by NASA beginning in 1965.[87]

In 1995, Jaakko Malmivuo and Robert Plonsey summarized a considerable body of research on the accuracy of impedance cardiography, concluding that it is a harmless, easy way to measure the blood pumped by the hearts of normal individuals under normal conditions. They cautioned that the accuracy declined considerably when dealing with some heart diseases or when a normal person was under the influence of drugs, was breathing in an unusual manner, or suffering oxygen deficiency (hypoxia).[88] In 2004, CIGNA HealthCare reviewed impedance cardiography with the hope that its lower costs and noninvasive methods could allow it to replace conventional methods. However, the company concluded: "CIGNA HealthCare does not cover electrical bioimpedance for the measurement of cardiac output because such measurement is considered experimental, investigational or unproven."[89] Thus the success of this particular NASA spinoff is significant but incomplete.

Case 6: Telemedicine Program

The sixth medical example published in *Spinoff 1996* concerns a very specific example of telemedicine that represents much greater possibilities than the case itself. From the earliest days of the human space program, NASA needed to be able to monitor astronauts' conditions at a distance and prescribe remedial action when any problems arose. Already during the Mercury missions, telemetry had given physicians on the ground information about the astronaut's blood pressure, heart rate, and body temperature.[90]

Spinoff 1996 defines *telemedicine* as "the interactive transmission of medical images and data to provide better health care for people in remote or 'medically underserved' locations." This is actually a very narrow definition that describes NASA's own original needs and some of the earliest applications

---

86. See *http://www.cardiodynamics.com/index.html*; and "CardioDynamics Reports Fourth Quarter and Fiscal 2006 Results," press release, *http://markets.financialcontent.com/stocks/news/read?GUID=1107905*.

87. "AcQtrac System: Cardiovascular Hemodynamic Health" (Eden Prairie, MN: Väsamed), p. 4.

88. Jaakko Malmivuo and Robert Plonsey, *Bioelectromagnetism: Principles and Applications of Bioelectric and Biomagnetic Fields* (New York: Oxford University Press, 1995).

89. "Electrical Bioimpedance for the Measurement of Cardiac Output Coverage," position number 0200, effective date 15 October 2004, CIGNA HealthCare, p. 1.

90. Link, *Space Medicine in Project Mercury.*

outside the space program. But it falls far short of expressing the full potential of the technology. The particular example cited in *Spinoff 1996* concerned a cooperative effort begun the previous year to offer telemedicine in the underserved South Texas area, involving the University of Texas Health Sciences Center at San Antonio (UTHSCSA), South Texas Hospital in Harlingen, the Texas Department of Health, participating communities, VTEL Corporation in San Antonio, the telecommunications company Sprint, and Healthcare Open Systems and Trials (HOST) Consortium in Washington, DC. NASA provided expertise plus one-third of the first year's funding. Of great concern to health care researchers and practitioners was the lack of specialized care for children suffering from cancer, because their cases are sufficiently rare that low-population areas cannot afford local expertise to treat them. The program also addressed the increasing rates of tuberculosis in the area. The chief technology employed in this demonstration project was teleconferencing, including distance-learning classes and long-distance professional consultations.

This was far from the first NASA-supported demonstration of telemedicine in a real-world context. In fact, the first had taken place a quarter century earlier with the launch of the Applications Technology Satellite ATS-1 in 1971. Satellite communications equipment was set up in 26 villages in Alaska, and a pilot study was carried out using ATS-1 for radio communications. Paramedical personnel in the villages, who had been trained by the Public Health Service, received advice from doctors. The project supported not only advice-giving about the cases of particular patients but also education. The University of Washington provided a genetics course to medical students at the University of Alaska, and 22 nurses at remote locations received a class in coronary care.[91]

More extensive pilot efforts were carried out using the ATS-6 satellite, which was launched in 1974. In addition to carrying out a variety of scientific observations and televising curriculum to Rocky Mountain schoolchildren, ATS-6 expanded the Alaskan telemedicine experiments. Where ATS-1 had provided a single audio circuit, which did not even permit the two parties to speak at the same time, ATS-6 provided high-quality video with multiple audio channels that could also carry data. The telemedicine experiments were remarkably farsighted, addressing most of the major issues for an operational system in areas where research continues even today. In addition to talk and pictures, ATS-6 could simultaneously transmit a patient's pulse and electrocardiogram data. The data could be entered into the patient's data file in a

---

91. Albert Feiner, "Health Care and Education: On the Threshold of Space," *Science* 186, no. 4170 (1974): 1178–1186.

computer. To protect privacy, voice and audio could be scrambled.[92] ATS-6 was the last of the series, and Congress cancelled plans for further experiments in these directions.[93] This was not the end of telemedicine in Alaska, however. Efforts very much like the NASA-supported experiments of the early 1970s have continued until the present day, becoming progressively more operational and less experimental. In 1994 the University of Alaska launched a new Alaska Telemedicine Project, explicitly based on the ATS-1 and ATS-6 experiences. By the year 2000 the project was cooperating with the Arctic Council, working with representatives from Canada, Denmark (Greenland), Finland, Iceland, Norway, Russia, and Sweden. Satellite communications are an essential technology for telemedicine in the arctic, and the 2000 report of the project uses the word *satellite* 18 times.[94]

Similar telemedicine activities in Texas have continued over the decade following *Spinoff 1996*. The University of Texas Medical Branch in Galveston set up a Telehealth Center in 1998.[95] By the year 2000 the Center for Telemedicine at Texas Tech University had taken on a special responsibility for the state's widely dispersed prison population.[96] In 2005 the Texas legislature mandated a study to examine how the use of telemedicine for Medicaid patients could be facilitated.[97] The nature of telemedicine has evolved, and new technologies are constantly being introduced. For example, the University of Texas Health Sciences Center at San Antonio, which took the lead in the experiments described in *Spinoff 1996*, now offers to all the world a major Web-based digital library of streaming videos devoted to medical education called the South Texas Regional Family Medicine Grand Rounds Virtual Video Library.[98]

NASA's involvement has changed as well. In the year 2000, NASA signed a memorandum of understanding with the National Cancer Institute to share

---

92. F. W. Norwood, "The Satellite Technology Demonstration and Health-Education Telecommunications Experiments on ATS-6," *Proceedings of the Royal Society of London* A345, no. 1643 (1975): 541–556.

93. Daniel R. Glover, "ATS 6," *http://roland.lerc.nasa.gov/~dglover/sat/ats6.html*.

94. Carl M. Hild, "Arctic Telemedicine Project Final Report, Presented to the Sustainable Development Working Group of the Arctic Council," Institute for Circumpolar Health Studies, University of Alaska, Anchorage, 2000, p. 41.

95. See *http://www.utmb.edu/telehealth/about.asp*.

96. See *http://www.ttuhsc.edu/telemedicine/default.htm*; and Jennifer Proctor, "Medicine Behind Bars: Texas' Telemedicine Experiment," *Reporter of the Association of American Medical Colleges* 9, no. 13 (2000), *http://www.aamc.org/newsroom/reporter/oct2000/bars.htm*.

97. Texas Health and Human Services Commission, "Telemedicine in Texas Medicaid Pursuant to Senate Bill 1340, 79th Legislature, Regular Session, 2005," January 2006.

98. See *http://familymed.uthscsa.edu/grandrounds/virtual_lib/Virtuallib.htm*.

research on biomedical sensors. As NASA Administrator Daniel S. Goldin and his coauthors on a *Science* article explained:

> The NCI's need is for technologies that can couple minimally invasive sensing and signaling of early molecular signs of cancer in patients and that will have the capability for controlled and monitored intervention. NASA's requirements are for monitoring and maintaining the spacecraft environment, remote sensing of life on distant planets, and for diagnosis and treatment of injury and emerging disease in astronauts during long-duration space missions.[99]

Early detection is the key to curing cancer, and the hope is that it will be possible to detect cancers long before they would show up in x-rays or by touch during physical exams. Indeed, there is hope to identify precancerous cells that can be treated immediately. Thus the NASA-NCI collaboration fits in with the NCI's hope to prevent cancers; it also serves NASA's distinctive but overlapping purposes.[100] Ideally this would mean not waiting for the patient to visit the doctor's office but using some kind of simple equipment at home that could transmit data to the doctor over the Internet. Indeed, that is the new model of telemedicine we are gradually moving toward: patient-centered, prevention-oriented *telehealth*.[101]

Already today, citizens get much of their health and medical information over the Web. Increasingly, doctors and nurses carry tablets and smart phones. When *Spinoff 1996* was published, experiments were already in progress to give visiting nurses devices that were wirelessly connected to their organization's home office.[102] Especially in cases of chronic illness, the elderly, and postoperative care, constant electronic communication with the patient at home can be a great advantage.[103] In a book chapter describing the Texas

99. Daniel S. Goldin, Carol A. Dahl, Kathie L. Olsen, Louis H. Ostrach, and Richard D. Klausner, "The NASA-NCI Collaboration on Biomolecular Sensors," *Science* 292, no. 5516 (2001): 443–444.

100. Barbara K. Rimer, "Cancer Control Research 2001," *Cancer Causes and Control* 11, no. 3 (2000): 257–270.

101. Mary Chaffee, "A Telehealth Odyssey," *American Journal of Nursing* 99, no. 7 (1999): 26–33.

102. Rachel Wilson and Terry Fulmer, "Introduction of Wireless, Pen-Based Computing Among Visiting Nurses in the Inner City: A Qualitative Study," *Journal of Community Health Nursing* 14, no. 1 (1997): 23–37.

103. Mary Tellis-Nayak, "The Postacute Continuum of Care," *American Journal of Nursing* 98, no. 8 (1998): 44–49; and Ron Chepesiuk, "Making House Calls: Using Telecommunications to Bring Health Care into the Home," *Environmental Health Perspectives* 107, no. 11 (1999): A556–A560.

Telemedicine Project, Jane Preston notes that the problems of health-care delivery are not limited to rural populations:

> In the United States, increasing numbers of rural hospitals are closing.... The cost of medical care steadily rises. The gaps are spreading between the quality of care in urban state-of-the-art medical centers and the quality of and access to health care in the remainder of the country. Indeed, the picture is stark in rural communities, but it is equally bleak in urban areas where access is blocked by prisons walls, traffic, enclaves of poverty and ignorance. In short, our country does not provide quality health care equally to all citizens. Delivery of medical services is at crisis level.[104]

This paragraph is ambivalent. While asserting the value of telemedicine for patients who live in remote areas, it recognizes that the health and medical needs of many other people are not being met either. The trade-offs in providing benefits to some people at the cost of denying them to other people made the cost of widespread telemedicine prohibitive for many years. Pamela Whitten has argued that early telemedicine projects were all terminated after initial experiments and demonstrations because the computing and communications infrastructures were not really in place yet.[105] Except for some rural areas, the Internet is now ubiquitous in the United States, and increasing numbers of users have access to the wide-bandwidth connections needed for two-way video conferencing. Internet primarily relies on optical fiber cables and to a lesser extent on land-based microwave links and satellite communications. Thus NASA was a pioneer with ATS-1 and ATS-6 but may have a diminished role to play in the communications side of telemedicine today.

A major innovation is not merely technical but also social. For example, telemedicine spans governmental boundaries, raising serious questions about which jurisdictions will regulate medical practice. In the late 1990s the *American Journal of Nursing* carried several articles on the question. As Connie Helmlinger and Kathy Milholland asked in 1997: "In which state does a nurse need to be licensed when providing telehealth services to patients

---

104. Jane Preston, "Rural Health and the New Media," in Linda M. Harris, ed., *Health and the New Media: Technologies Transforming Personal and Public Health* (Mahwah, NJ: Lawrence Erlbaum Associates, 1995), pp. 65–86, esp. p. 65.

105. Pamela Whitten, "The State of Telecommunication Technologies to Enhance Older Adults' Access to Health Services," in Arthur D. Fisk and Wendy A. Rogers, eds., *Human Factors Interventions for the Health Care of Older Adults* (Mahwah, NJ: Lawrence Erlbaum Associates, 2001), pp. 121–146.

in another state?"[106] A widely discussed answer was that nursing and other medical professions would need to consider multistate licensure or some other form of regulation that was not tied to a particular geographic area.[107] Changing over to a new system that involved unprecedentedly large geographic areas would naturally require a good deal of negotiation. Regulation is not the only issue; and in the chaotic health-care system of the United States, the issue of who pays for what may be more important. Any government health-care program or geographically widespread health maintenance organization will undoubtedly employ telecommunications to an increasing extent, but it is not clear how different organizations will cooperate with each other, or how they will serve the uninsured who happen to live in rural areas.

One may question the historical significance of premature inventions and demonstration projects. Yet, to creative scientists, engineers, and entrepreneurs, they signal possible avenues for advance. For example, the first personal computer worthy of the name was probably the Altair, and the first personal computer with the modern mouse and Windows user interface was the Alto. Neither was commercially successful, and it was left to the Apple II and the MacIntosh to succeed where they had failed. Yet the Altair and the Alto directly inspired the later machines.[108] It is hard to say where telemedicine would be today without the ATS satellites or the mid-1990s Texas demonstration. Clearly, NASA has shown a direction toward improved health care of people in remote areas and, by stimulating development of telemedicine in general, has benefited all citizens. This is another example in which NASA's contribution to general progress was more important than any specific invention. A related example is the Digital Library Initiative (DLI), in which I was fortunate to participate across its entire history. The DLI began in 1994, through a partnership between the National Science Foundation (NSF), the Defense Advanced Research Projects Agency (DARPA), and NASA. DARPA had created the Internet for military researchers, and the NSF had civilized it. In 1994 the World Wide Web began its explosive growth on the basis of the Internet, and I can remember using a beta-test version that year of Mosaic, the first full-featured Web browser, which had been developed at the NSF-supported Illinois supercomputer center. NASA did not join the DLI with

---

106. Connie Helmlinger and Kathy Milholland, "Telehealth Discussions Focus on Licensure," *American Journal of Nursing* 97, no. 6 (1997): 61–62, esp. p. 61.

107. David Keepnews, "Emerging Issues in Licensure and Regulation," *American Journal of Nursing* 96, no. 10 (1996): 70–72; and Lucille A. Joel, "Multistate Licensure?," *American Journal of Nursing* 99, no. 1 (1999): p. 9.

108. William Sims Bainbridge, entries for "Altair" and "Alto," in William Sims Bainbridge, ed., *Encyclopedia of Human-Computer Interaction* (Great Barrington, MA: Berkshire, 2004).

telemedicine in mind, yet the results were valuable for telemedicine in often unexpected ways.

Of the first six major DLI projects, NASA was most interested in the Alexandria Digital Library, a consortium of universities headed by the University of California, Santa Barbara that specialized in "georeferenced materials," which means maps, geographic information systems, and the like. But all funding was pooled, so NASA really contributed to all the projects. The DLI project at Stanford, which was intended to develop new tools for collaborative use of the Internet, produced Google, among other things. The second phase of the DLI, which began in 1998, included the National Library of Medicine, and many of the projects over the years contributed technologies or ideas that can support or enhance telemedicine.[109]

3. Selected Other Spinoffs

To get a broader perspective on medical spinoffs and to stay within the historical focus of the work, I examined the five issues of *Spinoff* from 1976 through 1980 for examples that were in very different areas from the 1996 examples. In addition, I asked the editor of *Spinoff* to suggest an example that deserved to be included here on the basis of its importance.

Case 7: Tactile Reader for the Blind

The 1977 *Spinoff* described an electronic reading device for the blind as "new help for the sightless." The *Spinoff* database says: "Derived from NASA technology, the Optacon works by passing a mini-camera over a printed page with [the] right hand, the left hand senses a vibrating image of the letters the camera is viewing."[110] *Optacon* stood for Optical TActile CONverter. Electrical engineers associated with Stanford University developed this technology: John G. Linvill, James C. Bliss, James D. Meindl, and others. The Optacon was not merely an early step in the development of reading devices for the blind. It also contributed to the development of the field that today is called *haptics*—information technology that involves the human sense of touch.[111] Optacon was often used as a research instrument in haptics, and the field gained significantly from its heritage in a union of applications for the

---

109. Alexandria Digital Library, University of California, Santa Barbara, *http://alexandria.sdc. ucsb.edu/*; *http://www.dli2.nsf.gov/*; and "Stanford Digital Library Technologies," *http:// dbpubs.stanford.edu:8091/diglib/*.

110. *Spinoff 1977*, "New Help for the Sightless (Optacon)," p. 64.

111. Ralph L. Hollis, in Bainbridge, ed., *Encyclopedia of Human-Computer Interaction, s.v.* "Haptics."

blind and for pilots.[112] Tactile output is an important part of haptics, as currently defined, but equally important is technology that involves the human kinesthetic sense that responds to forces and movement. The historical origin of input-output devices involving the kinesthetic sense is probably the work in remote manipulation, notably the robot arms used in the nuclear industry to allow humans to handle radioactive materials without hazard.[113]

The idea of remote manipulators has a long history, and it is difficult to say where it originated. Decades ago, science fiction fans tended to attribute the invention to leading writer Robert A. Heinlein, whose story "Waldo" in the August 1942 issue of *Astounding Science-Fiction* concerned a disabled scientist who had invented a set of remote manipulators for himself, some tiny and precise, others huge and powerful. The fictional scientist's name was Waldo, and for a time fans and some engineers called manipulators *waldoes*. Heinlein himself disclaimed any credit for the idea, saying he got it from a 1918 issue of *Popular Mechanics* that described how a disabled engineer had made mechanical arms for himself.[114] Primary credit for developing the "master-slave manipulator" is usually given to Raymond C. Goertz, after whom the American Nuclear Society has named its robotics award.[115] Working at the Argonne National Laboratory around the 1950s, Goertz developed several manipulators for handling nuclear materials, and by 1955 the technology had been transferred to a corporation, Central Research Laboratories.[116] As the technology developed further, it became clear that remote manipulators would require a sense not only of pressure and movement but also of touch.

Haptics, broadly defined, is important for aviation and spaceflight. The old expression *flying by the seat of one's pants* recognizes that the physical sensation of acceleration or rotation is important for pilots. But it ignores what they feel through their hands. In the modern "fly-by-wire" days, when a computer stands between the control stick and the ailerons and elevators, force feedback can help the pilot feel the effect of his or her actions. Given all the complex instruments in the cockpit and the advantages of a heads-up display,

---

112. Z. Kuc, "A Bidirectional Vibrotactile Communication System: Tactile Display Design and Attainable Data Rates," *Proceedings of the Conference on VLSI and Computer Peripherals*," 8–12 May 1989, Hamburg, Germany, pp. 2/101–2/103.

113. Robert J. Stone, "Haptic Feedback: A Potted History, from Telepresence to Virtual Reality," *Proceedings of the First International Workshop on Haptic Human-Computer Interaction*, 31 August–1 September 2000, Glasgow, Scotland, pp. 1–7; available at *http://www.dcs.gla.ac.uk/~stephen/workshops/haptic/papers/stone.pdf*.

114. Robert A. Heinlein, "Science Fiction: Its Nature, Faults, and Virtues," in Basil Davenport, ed., *The Science Fiction Novel* (Chicago: Advent, 1969), pp. 14–48, esp. p. 26.

115. "American Nuclear Society," *http://www.ans.org/honors/va-goertz*.

116. "History of Telemanipulator Development," Central Research Laboratories, undated, *http://www.centres.com/nuclear/manip/maniphis.htm*.

it was reasonable for the Air Force and NASA to consider haptic information displays, including some that employ the tactile sense. Fundamental research in this area would also improve understanding of how the accuracy of the human tactile sense is affected by vibration.

The connection between aerospace interest in haptics and the possible applications for visually disabled people was made between the Ames Research Center and nearby Stanford University, where John Linvill and James Bliss formed a collaboration to attack the problem.[117] In a 1997 interview, Bliss explained that there were both accidental and personal dimensions to what happened:

> After finishing my master's degree in electrical engineering at Stanford, I received a fellowship from the National Science Foundation that would allow me to obtain a doctorate in circuit theory at MIT. When I arrived at MIT, however, I discovered that the professor who I had hoped would be my thesis advisor had switched his research interest to the application of electrical engineering to the problems of blind people. I also became interested in this subject after meeting a research associate at MIT who had been blinded in the Battle of the Bulge in World War II. After completing my doctoral thesis, "Communication via the Kinesthetic Sense," I returned to the San Francisco Bay Area, started a research group at SRI International to study the visual and tactile senses, and [became] an associate professor in electrical engineering at Stanford. Professor John G. Linvill, the chair of my department, had a daughter who was blind.[118]

In a 1996 issue of the newsletter of the Research Laboratory of Electronics at MIT, Linvill described his own path to the Optacon:

> In 1962, my family and I visited IBM's research lab near Stuttgart. We saw a high-speed printer driven against a fast-moving paper with carbon. It occurred to me that you could probably feel this. The fact was it could have drilled a hole in your hand. On the way back, I told my family, "I have a great idea to help Candy read." Our daughter Candy was blind and, ever since she was in kindergarten, my wife Marjorie had been her Braille teacher. She spent four hours a day preparing material for Candy, who went to regular school. I told Arnold Shostak at the Office of Naval Research about my idea. The Navy had

---

117. Elizabeth A. Muenger, *Searching the Horizon: A History of Ames Research Center, 1940–1976* (Washington, DC: NASA SP-4304, 1985), p. 94.

118. Paul J. Lewis, "An Interview with James C. Bliss," *RE:view* 29, no. 2 (1997): 62–63; the journal is named *RE:view* because the RE stands for the "rehabilitation and education" of those who have visual impairment.

people working underwater, and tactile communication seemed like a reasonable thing to work on. That was the beginning of the Optacon, and Candy became its principal guinea pig. Today, she's a clinical psychologist.[119]

Linvill joined the Stanford faculty in 1955, and Bliss joined SRI in 1956. SRI International was originally founded in 1946 by the university as the Stanford Research Institute and did not become a separate nonprofit organization until 1970. Bliss took the lead on the NASA contracts and on at least two from the Air Force.[120] The first projects for Ames involved air jet stimulators rather than vibrating reeds. Other support came from the Office of Naval Research, and once the possible applications for the blind became paramount, substantial grants came from the National Institutes of Health and the American Foundation for the Blind.

In 1964, Linvill filed a patent application, issued in 1966 as Patent 3,229,387, for a "reading aid for the blind." Applications often describe in detail one possible version of the invention, with statements to the effect that many other arrangements would also be covered by the patent. Linvill described a small, integrated device, hardly more than two inches across, with a lens at one end that could scan over the printed text, focusing on a matrix of tiny photoelectric light detectors. A circuit from each one would apply a signal to the corresponding output unit, when the part of the image it saw was dark, representing part of a printed letter. The patent application emphasized the tactile output device, which consisted of a matrix of piezoelectric reeds, each of which would vibrate when a small alternating current was applied to it. Thus, when the lens received the image of a capital letter *A*, output reeds arranged in that same shape would vibrate and be detected by one fingertip of the user.

Notice that this is not an example of modern optical character reading (OCR) because it does not interpret the letter. An OCR device for blind people could translate the images into Braille on a tactile display or a voice simulated by computer speech generation, but the many innovations required to make a device on these principles did not exist in the mid-1960s. In the same month that the patent was awarded, Linvill and Bliss published a detailed description of their research in the *Proceedings of the IEEE*. They stressed that it involved direct translation from visual to tactile, rather than interpretation.[121]

---

119. John G. Linvill, "Alumni Profile: John G. Linvill," *RLE Currents* 8, no. 1 (1996): 4–5.
120. NASA contracts included NAS 2-912, NAS 2-1679, NAS 2-5409, and NAS 2-2752, and NAS 2-4582; Air Force contracts included AF 33(657)-8824 and AF 33(615)-1099.
121. J. G. Linvill and J. C. Bliss, "A Direct Translation Reading Aid for the Blind," *Proceedings of the IEEE* 54, no. 1 (1966): 40–51.

Two features of this article deserve attention, because they reveal profound facts about the Optacon research program. First, the article is objectively of high quality, describing first-rate scientific research and fundamental engineering innovation. It includes mathematical formulas as well as circuit diagrams, and it reports the results of training research subjects to see how many words per minute they could learn to read. The Optacon is not the result of tinkering but of systematic research and development carried out by individuals of great talent and thorough training. The basic idea may sound simple, but success required fundamental research in electronics and in the psychology of human perception. For example, the contractor report Bliss filed with Ames for contract NAS 2-5409 is a serious study comparing perception and short-term memory between visual and tactile senses.[122]

Second, the Optacon harnesses science and technology for the benefit of human beings. Three human subjects volunteered for the pioneering research, but their identities are concealed in accordance with privacy standards in research. A photograph shows a girl reading from the tactile display, and the accompanying text explains: "Our initial subject was a 12-year-old girl who is in the seventh grade at a regular school. She is an avid Braille reader.... She has been partially blind since she was about 8 months old and totally blind since she was about two years old." The IEEE Web site offers a photograph of the same girl demonstrating the Optacon at the 1969 International Solid-State Circuits Conference, and she is identified there as Candace Linvill.[123]

In 1967, Linvill was chair of the Electrical Engineering Department at Stanford, and he recruited James D. Meindl from the U.S. Army Signal Research and Development Laboratories, in Fort Monmouth, New Jersey, to come to Stanford and help develop the integrated circuit chips that would mediate between the light sensors and the tactile display.[124] Up to this point, the Optacon had relied on a computer-controlled instrumentation system and was not yet a workable, self-contained, hand-held device. Integrated circuits were still new at this point in time. Credit for inventing the first integrated circuits is split between Jack Kilby of Texas Instruments and Robert Noyce of Fairchild Semiconductor. Kilby filed a patent application on 6 February 1959 and received patent 3,138,743 on 23 June 1964. Noyce filed 30 July 1959 and received patent 2,981,877 on 25 April 1961. Several

122. J. C. Bliss, J. W. Hill, and B. M. Wilber, "Tactile Perception Studies Related to Teleoperator Systems," NASA CR-1775 (Washington, DC: NASA, 1971).

123. See *http://www.spectrum.ieee.org/print/3646*.

124. Tekla S. Perry, "Wizard of Watts," biography of James D. Meindel on the *IEEE Spectrum* Web site, *http://www.spectrum.ieee.org/print/3646*.

patents were required to fully develop the idea, and the technology was still in its infancy when the Optacon was developed.

Bliss had received patent 3,353,027 in 1967 for a photocell aiming device and patent 3,385,159 in 1968 for a ranging instrument. Both of these devices use physical vibration to scan the environment, locating an object within the field of view. It is worth noting that, like the Optacon, these patents combine a lens, photocells, and vibration to achieve the desired result. It is also worth noting some similarities between these two patents and the dominant technical approach to television in the 1920s, before Philo Farnsworth—whom one biographer has called the "last lone inventor"—developed the fully electronic system that became the basis of the television industry.[125] The semimechanical systems of the 1920s employed a rotating disk through which a spiral of tiny holes had been punched to scan the image with a photocell in the camera and to produce the picture in the display. By the late 1960s all-electronic television had a four-decade history and was commonplace, bulky, and expensive. Integrated circuits introduced the possibility of creating an entire array of photoreceptors at low cost and small size, thus rendering mechanical scanning systems obsolete and expensive television cameras unnecessary in the Optacon.[126]

In 1970, with Stanford University as one of their investors, the team created a corporation named Telesensory Systems to manufacture Optacon. Linvill and Meindl stayed at Stanford, and Bliss went to the new company. Around 1980 the Optacon was one of three competing reading machines available on the market. The Stereotoner converted printed letters not to tactile images but to sound patterns. Both the Optacon and Stereotoner were difficult to learn, and reading speed remained far lower than practiced Braille readers could achieve. Only a very small fraction of blind people ever used either device. The Kurzweil Reading Machine used a computer to turn printed words into recognizable speech. Although underpowered and expensive, it was a harbinger of the future.[127]

In 1970, Bliss had coauthored a comprehensive survey of sensory aids for the blind, with physicist Patrick W. Nye, who was then at the California

---

125. Evan I. Schwartz, *The Last Lone Inventor: A Tale of Genius, Deceit, and the Birth of Television* (New York: HarperCollins, 2002).

126. Roger D. Melen and James D. Meindl, "A Transparent Electrode CCD Image Sensor for a Reading Aid for the Blind," *IEEE Journal of Solid-State Circuits* SC-9, no. 2 (1974): 41–49.

127. Harvey Lauer, "The Reading Machine That Hasn't Been Built Yet," *AFB Access World* 4, no. 2 (2003), *http://www.afb.org/afbpress/pub.asp?DocID=aw040204.*

Institute of Technology.[128] The next year, Nye moved to the Haskins Institute in New Haven, Connecticut, which was loosely connected to Yale University and already had a long tradition of research on human speech and reading. There his team was able to develop the first functional reading machine that combined optical character reading with computerized speech synthesis. But this was not yet a commercial product. Samuel J. Mason, the MIT professor who had interested Bliss in work for the blind, continued to do research in this area and produced a reading machine about the same time.[129] Innovator, entrepreneur, and MIT graduate Ray Kurzweil founded Kurzweil Computer Products in 1974 to develop optical character reading, and in 1976 he demonstrated his text-to-speech reading machine.[130]

Over the years, Telesensory gradually gave greater emphasis to technologies that assisted people with partial vision; in 1996 it finally ceased making Optacon, and the company went out of business in 2005.[131] Thousands of blind people benefited from the Optacon, and a search of the World Wide Web in January 2007 revealed a number of testimonials to how it had improved individual lives along with many laments that it was no longer available. Canon had introduced the technology to Japan, and arguably it was better adapted for the many characters of Japanese and Chinese than for English, just as text-to-speech technology is more difficult for Chinese than for English.[132] Quite apart from the direct benefit for blind users, Optacon played an important role in stimulating other innovations across the field of assistive technologies, and the individuals who developed it continued to make contributions to technical progress.

In 1994, Bliss received Grant 9362053 from the NSF to develop an image-processing system based on a personal computer for visually disabled people and founded his own small company, JBliss Imaging Systems. In 2000 and 2002 he received a two-stage Small Business Initiation Research grant from the NSF (0132058, 0060386) to develop a complete information-handling system for such users. The description on the NSF online database describes

---

128. Patrick W. Nye and James C. Bliss, "Sensory Aids for the Blind: A Challenging Problem with Lessons for the Future," *Proceedings of the IEEE* 58, no. 12 (1970): 1878–1898.

129. Robert W. Mann, letter to the editor, *Journal of Rehabilitation Research and Development* 38, no. 1 (2001): xvii.

130. "Raymond Kurzweil," official biography, *http://www.kurzweilai.net/bios/bio0005.html?printable=1.*

131. Deborah Kendrick, "From Optacon to Oblivion: The Telesensory Story," *AFB Access World* 6, no. 4 (2005): *http://www.afb.org/afbpress/pub.asp?DocID=aw060403.*

132. See "1974 A business group is established to deal with communication support products," in the official company history on the Canon Web site, *http://www.canon.com/about/history/episode06b.html.*

the goal: "The research objective is to combine optical character recognition (OCR), speech synthesis and recognition technologies, together with displays based on the latest vision research to provide an integrated system with a consistent, easy to learn, command structure."[133] In 2005 he retired and licensed his technology to a new company that still carries his name, JBliss Low Vision Systems—many of whose products use electronics to magnify printed text so that people with poor but intact vision can read it.[134]

In a 2005 survey of research and development of tactile user interfaces, Lilly Spirkovska of the Ames Research Center cited the Optacon research extensively and reported that the general area remains one of interest for NASA.[135] Research on vibrating, tactile input for the blind continues. For example, Francis Quek at Virginia Tech recently led an NSF-funded group, examining the potential for a haptic system to help blind children learn mathematics.[136] It is based on the premise that gesture is one of the ways that humans shape and communicate concepts, and observations of mathematics classes reveal that gesture is often used between teachers and students. Of course, blind students cannot see these gestures and thus miss one of the dimensions of the mathematical discourse. The hope is that future technology could combine sophisticated computer vision with a haptic device such as a glove with a matrix of vibrators to communicate the movement of the gestures to the blind students. Thus, while the Optacon may have passed from the scene, it helped many blind people until better technology was available and contributed to the development of a tradition of research and invention that endures today.

The final question, however, concerns how crucial NASA's contribution to the Optacon may have been. The *Spinoff* database overstates when it says that the Optacon was "derived from NASA technology." NASA contributed some funding in the early days, especially for the haptic side of the work, but funding also came from several other sources. A historical account of the development of the Optacon research indicates that the first funding came from the Office of Naval Research (ONR); then, between 1966 and 1971, the Office of Education in the Department of Health, Education, and Welfare

---

133. Abstract of NSF grant 0132058, *http://www.nsf.gov/awardsearch/*.

134. See *http://www.jbliss.com/overview.htm*.

135. Lilly Spirkovska, "Summary of Tactile User Interfaces Techniques and Systems," NASA/TM-2005-213451, Ames Research Center, Moffett Field, CA, 2005.

136. Francis Quek and Madiv McNeill, "Embodiment Awareness, Mathematics Discourse, and the Blind," in William Sims Bainbridge and Mihail C. Roco, eds., *Managing Nano-Bio-Info-Cogno Innovations: Converging Technologies in Society* (Berlin: Springer, 2006), pp. 266–279.

invested $1.8 million because of its interest in technologies for the disabled.[137] Although Linvill's original patent application does not mention NASA, applications before the late 1970s tended to focus narrowly on the technical claims for the invention; so that omission is not solid evidence. However, in their reminiscences Linvill and Bliss never mention NASA. Neither got the inspiration for a machine to help the blind from NASA, and the Optacon was not itself designed to serve an aerospace purpose. Thus NASA played a supporting role, which may have been significant when Linvill and Bliss first needed support for their research, but it was not the primary source for the technological innovation.

Case 8: Dental Use of Nitinol Wire

A 1979 spinoff was the use of Nitinol wire in dentistry.[138] The popular Web site *www.space.com* summarizes the story thus:

> A nickel and titanium alloy known as Nitinol, originally developed by NASA for aerospace application, is used in a type of dental arch wire, which has orthodontic application. In contrast to the traditional steel arch wire, the Nitinol arch wire reduces the number of times braces require adjustment, since it returns to its original shape as teeth are pulled (the alloy is a type of 'memory metal' which does not kink when bent.).[139]

An online registry of orthodontists says, "Nitinol (thanks NASA!) is a new metal that retains its shape and is strong enough to withstand the force exerted by orthodontic appliances."[140] Comments such as these convey the impression that NASA invented Nitinol, something that NASA itself does not claim. We shall see that NASA did indeed contribute to progress in understanding and using this remarkable material, but it was discovered elsewhere.

The name Nitinol explains what this metal is made of and where it originated. The first two syllables are the symbols for nickel (Ni) and titanium (Ti). The last syllable is the acronym for the Naval Ordnance Laboratory, the place where Nitinol's qualities and first applications were discovered. The Naval Ordnance Laboratory was established in White Oak, Maryland, in 1944.[141] As its name implies, its chief purpose was research on naval gun-

---

137. Christophe Lécuyer, What Do Universities Really Owe Industry? The Case of Solid State Electronics at Stanford," *Minerva*, 43 (2005): 51–71.

138. "Dental Arch Wire," *Spinoff 1979*, p. 74.

139. See *http://www.space.com/adastra/adastra_spinoffs_050127.html.*

140. See *http://www.1stbraces.com/orthodontic-braces-article8.shtml.*

141. "White Oak, Maryland," *http://www.globalsecurity.org/military/facility/white-oak.htm.*

nery technologies, which required much expertise in metallurgy. In 1974 it merged with the Naval Weapons Laboratory, and the site was relinquished by the Navy in 1997. Its current successor is the Carderock Division of the Naval Surface Warfare Center in West Bethesda, Maryland.[142]

In 1959 experienced metallurgist William J. Buehler was carrying out research on a number of substances when he observed that alloys of nickel and titanium were more ductile—flexible but resistant to impact—than other alloys. He then focused his research on nickel-titanium alloys that contained approximately equal numbers of atoms of the two elements. Because nickel has a greater atomic weight than titanium, about 55 percent of the mass of an equal mixture of atoms will consist of nickel. Therefore the alloy he was investigating was called 55 Nitinol. The first big clue that it was something very special came when Buehler and his assistant had just cast six bars of 55 Nitinol. To get a quick sense of the properties of the material, Buehler intentionally dropped one bar on the concrete floor, and it made what he later called a "very dull thud," similar to the sound of dropping a lead bar. Not expecting this sound, he dropped another, and it rang like a bell. After repeating this simple experiment several times, he discovered that hotter bars made a bell-like sound, while cooler bells made a thud. Something about the internal structure of the alloy apparently depended sensitively on temperature. A series of experiments, both systematic and accidental, led to the full realization in 1961 that 55 Nitinol is a "memory metal." It can be formed into a shape at one temperature, formed into a second shape at another temperature, then returned to the first temperature, where it spontaneously resumes the original shape.

The next year, Buehler gained expertise on the physics of crystals when Frederick E. Wang joined the lab and developed the theory of how Nitinol works. In 1961, Buehler and the Navy filed an application for a patent explaining methods for making alloys like Nitinol and describing their properties, and patent 3,174,851 was granted in 1965. In 1966, again on behalf of the Navy, Buehler filed a patent application for methods to convert heat energy to mechanical energy, describing the principles that could power an engine based on Nitinol's ability to change shape in response to changing temperatures, receiving patent 3,403,238 two years later. Other patents issued to the Navy based on the research included patent 3,351,463 ("high strength nickel-based alloys") in 1967 and patent 3,753,700 ("heat recoverable alloy") in 1973.

---

142. Bob Hardy, "Wacker Reflects on Structures; Materials," Heritage Lecture Series, September 1998, *http://www.dt.navy.mil/pao/excerpts%20pages/1998/heri9.html*.

According to two brief histories of Nitinol's discovery by George Kauffman and Isaac Mayo, who interviewed both Buehler and Wang, the first successful product was a coupling to join together the hydraulic-fluid lines on the Navy's F-14 jet fighter.[143] Nitinol's ability to return to a former shape was used to make it clamp over the ends of two sections of the pipe, holding them together. In 1969, John D. Harrison of the Raychem Corporation began collaborating with Buehler and Wang, and the resulting idea was called "Cryofit" because the low temperature of liquid nitrogen was used to reform the couplings. Raychem does not appear to carry the product today, but the Aerofit company does, and Cryofit couplings have been used on a variety of both military and commercial aircraft.[144]

It is worth noting two facts about Raychem. First, the company has been involved in many patents, and a search for the word *Raychem* in the U.S. Patent and Trademark Office's patent database turns up fully 2,758 that were either issued to Raychem or in some other way cite the company explicitly. Coincidentally, the most recent one I found was related to both medicine and Nitinol: patent 7,160,322 issued on 9 January 2007 to an individual for an "implantable cardiac prosthesis for mitigating prolapse of a heart valve." The base of the device is a mesh, and "the mesh may be formed of a shape memory alloy material, such as a nitinol (nickel-titanium alloy) wire" to facilitate installation and adjustment. The only publication mentioned in the patent application is Raychem's 1999 product brochure, "Nitinol Solutions."

Second, an important part of Raychem's business has long been couplings, and a number of patents unconnected to Nitinol describe couplings that change shape to seal the connection tightly. Patent 3,574,313 issued in 1971 concerns a "wraparound closure sleeve" that is "heat recoverable." This patent concerns the physical shape of the sleeve rather than its material but cites earlier patents about plastics with shape-memory qualities, not Nitinol. Patent 3,379,218 ("closure sleeve for pipes or the like") issued in 1968 similarly assumes the material would be some kind of polymer with a temperature-related memory property. In 1971, Buehler and Wang got patent 3,558,369 on behalf of the Navy for a "method of treating variable transition temperature alloys," describing alloys in which some or all of the nickel in Nitinol is replaced by iron or cobalt. In 1973 patent 3,753,700 was issued to Raychem for an alloy in which about 7 percent of Nitinol's nickel atoms are replaced

---

143. George B. Kauffman and Isaac Mayo, "The Story of Nitinol: The Serendipitous Discovery of the Memory Metal and Its Applications," *The Chemical Educator* 2, no. 2 (1996): 1–21; and George B. Kauffman and Isaac Mayo, "The Metal with a Memory," *Invention and Technology Magazine* 9, no. 2 (1993): 18–23.

144. Raychem, *http://www.raychem.com/*; and Aerofit, *http://www.aerofit.com/sma/partnos.htm*.

by iron, claiming superior qualities and reporting: "For example, a hydraulic coupling made of the alloy was provided with a heat unstable diameter of 8% greater than the heat stable diameter." The 1973 patent 3,759,552 ("hydraulic coupling with metallic sealing member") requires a metal having Nitinol's properties and even discusses a nickel-titanium alloy but does not explicitly refer to the Navy, Buehler, or Nitinol. Raychem's 1975 patent 3,872,573 specifically centers on using the Nitinol family of alloys for couplings.

The work that led to dental applications began in 1968, when George B. Andreasen in the Department of Orthodontics of the School of Dentistry at the University of Iowa read about Nitinol. Andreasen contacted Buehler, but at first the Navy was reluctant to release information about this strategically valuable innovation. In 1969, however, Buehler was allowed to send Andreasen a three-foot section of Nitinol wire for evaluation. It proved to be greatly superior to stainless steel, and Andreasen quickly published articles in the leading dentistry journals publicizing this fact. In 1972 he made an initial attempt to file for a patent, making a successful application in 1973 and receiving patent 4,037,324 in 1977. This patent describes exactly the innovation reported in the NASA *Spinoff* publication. Soon, after a little more development, the Unitek division of 3M was supplying Nitinol wire to orthodontists, and a 1997 company brochure explains that it is easier to use than stainless steel, requires less adjustment, and often straightens teeth more quickly. The product's slogan makes this point cutely: "because treatment efficiency always comes down to the wire."[145]

To this point in the saga, I have not even mentioned NASA. Is the spinoff claim entirely spurious? No. Andreasen's patent application cites a very important contribution NASA had made in 1969: "A description of the materials and certain of their properties also may be found in the brochure entitled 'Nitinol Characterization Study' dated September, 1969. This document [is] identified as N-69-36367 or NASA CR-1433." NASA's Langley Research Center had become interested in the potential of Nitinol, and in 1967 it gave a contract to Goodyear Aerospace Corporation to study the material systematically. Buehler provided samples of Nitinol rods and foil, and Battelle Institute, a subcontractor in Columbus, Ohio, processed some of the rod into wires of varying diameters: 100, 20, 15, and 10 mils (a mil is $\frac{1}{1,000}$ of an inch). Each kind of sample was studied to determine such things as the force involved when it changed shape, its responses to tension and compression at various temperatures, how many times it could change shape before degrading, and its variable electrical resistance. One series of

---

145. "Nitinol: Heat Activated Wire," product brochure (Monrovia, CA: 3M Unitek, 1997).

experiments determined how much Nitinol wire would spring back after being bent around mandrels of differing diameters.

After 19 months of research, Goodyear provided Langley with the data that became the report. It contains many graphs showing the performance of Nitinol under various conditions, which would be useful to engineers in determining how to design reliably functional components from the material. Earlier tests in Buehler's lab had given inconsistent results, in part because the samples may not have been consistently made; so it was important to begin to understand the exact sources of unreliability. The report explained that the behavior of a Nitinol object depended very much on the detailed history that formed the microscopic grains of which any large piece was composed. The report's final sentence states: "From these results it is recognized that 55-Nitinol has many potential applications in advanced space structures, especially where requirements for expandable and erectable structures or self-actuating devices in space are needed."[146]

Given the difficulty of manufacturing Nitinol reliably, especially in its early days, it was an expensive material. Thus, it could find its first cost-effective applications only in fields where there was a premium on high performance. When the first major report on the societal implications of nanotechnology was published in 2001, one chapter gathered together these demanding application areas: space exploration, national security, and medicine.[147] As we have just seen, these were the first areas of application for Nitinol, including dental wires and the crucial couplings for the F-14's hydraulic system, plus the space applications foreseen by the NASA report. This observation highlights one reason why spinoffs from NASA to medical applications may be especially common. In both spaceflight and medical care, high performance has a much higher priority than in most other sectors of the economy.

The history of Nitinol in dentistry illustrates the important point that major inventions in modern technology often require a vast amount of development research that would be far beyond the capability of an individual inventor. The search engine of the U.S. Patent and Trademark Office finds fully 6,154 patents whose applications contain the word *Nitinol*, up through 9 January 2007. Of these, 114 mention NASA.

Twelve Nitinol-related patents were the direct result of work done for NASA. Three early Nitinol patents belonged to the U.S. government: 4,553,393 for "memory metal actuator"; 4,665,334 for "rotary stepping

146. William B. Cross, Anthony H. Kariotis, and Frederick J. Stimler, "Nitinol Characterization Study," NASA CR-1433 (Washington, DC: NASA, 1969), p. 58.

147. Mihail C. Roco and William Sims Bainbridge, eds., *Societal Implications of Nanoscience and Nanotechnology* (Dordrecht, Netherlands: Kluwer, 2001), pp. 203–237.

device with memory metal actuator"; and 4,765,139 for "thermocouple for heating and cooling of memory metal actuators." In 1991 the Tini Alloy Company, whose name evokes titanium and nickel, earned patent 5,061,914 for "shape-memory alloy micro-actuator," and in 1999 it got patent 5,903,099 for "fabrication system, method and apparatus for micro-electromechanical devices"—both based on work done for NASA. Five patents belonging to NASA itself describe using microwaves to treat prostate enlargement (6,289,249; 6,512,956; 6,592,579; 6,675,050; and 6,944,504), with some assistance from Nitinol. SRI International was issued two patents based on applications that mentioned Nitinol in describing a range of possible developments: 6,617,963 for "event-recording devices with identification codes" and 6,806,808 for a "wireless event-recording device with identification codes." The remaining 102 patents referred to work at NASA, usually by citing a publication. Eight referred to the Goodyear Nitinol characterization study, and 66 referred to a 1972 NASA publication based on it that had wider distribution: "55-Nitinol—The Alloy with a Memory."[148]

Case 9: Springback Foam

Perhaps the best-known NASA spinoff, temper foam, is featured in the FAQ on the *Spinoff* Web site and in fully six editions of *Spinoff*, from 1976 through 2005.[149] In the 1979 edition it is only three pages away from that other materials development, Nitinol wire.[150] Both can be described as memory materials that return to their original shape after being deformed, but the two were results of totally separate development efforts. The medical applications of temper foam are chiefly cushioning for patients, such as hospital pads, wheelchair seats, and the like. Because the material forms gently to the shape of the weight it is supporting, it is less likely to put undue pressure on any given spot. This may seem like a very humble application, unrelated to cure of disease, but treating hospital patients gently is an essential part of cure, reducing stress and injury.

---

148. C. M. Jackson, H. J. Wagner, and R. J. Wasilewski, "55-Nitinol—The Alloy with a Memory: Its Physical Metallurgy, Properties, and Applications" (Washington, DC: NASA SP-5110, 1972).

149. "Spinoff Frequently Asked Questions," *http://www.sti.nasa.gov/tto/spinfaq.htm*; "Versatile Padding (Temper Foam)," *Spinoff 1976*, p. 84; "Shock Absorbing Helmets," *Spinoff 1977*, p. 100; "Temper Foam," *Spinoff 1981*, p. 76; "Foam Cushioning," *Spinoff 1988*, p. 68; "Foot Comfort for the Fashionable," *Spinoff 2002*, p. 88; and "Forty-Year-Old Foam Springs Back with New Benefits," *Spinoff 2005*, pp. 46–49.

150. "Springback Foam," *Spinoff 1979*, p. 77.

It is generally agreed that temper foam was invented by Charles A. Yost and Charles Kubokawa, although they did not apparently receive a patent for the idea. Yost worked for various NASA contractors before starting his own company, Dynamic Systems, in 1969. Kubokawa worked at NASA Ames from 1963 to 1989.[151] *Spinoff 2005* says, "As an aeronautical engineer with the Systems Dynamics Group at North American Aviation, Inc., Charles Yost helped to build a recovery system for the Apollo command module in 1962." While working for Stencel Aero Engineering in Asheville, North Carolina, Yost carried out research for NASA on how to improve the survivability of airplane accidents.

The 1968 report Yost coauthored with Ronald W. Oates covered a very wide range of topics, such as the rate of survivability when an aircraft hits the ground at varying angles and speeds, the main injuries caused when the upper bodies of unrestrained passengers flail around during the impact, and the difficulty of designing for the protection of passengers of different sizes. One section concerned aluminum honeycomb materials for seats to absorb the shock of impact, including this criticism of other substances: "Materials such as sponge, solid rubber, cork, and paper wadding generally exhibit spring characteristics with an attendant rebound problem."[152] The rebound would exacerbate the flailing of the upper body, causing injuries when it strikes other objects. Aluminum honeycomb crushes somewhat gracefully on impact, absorbing much of the shock and not rebounding. The report does not mention temper foam but does identify two characteristics that an ideal cushioning material should have: 1) adapting to passengers of different shapes and sizes, and 2) absorbing shock without immediately rebounding. It also announced plans for Yost to work with Ames in a broadly based effort to develop a more protective seat system for aircraft.

The company Yost founded says, "Dynamic Systems began research in 1969 to perfect foam cushion materials having both high energy absorption and soft pressure properties. In cooperation with NASA, the materials were applied to seating systems, such as wheelchair cushions, ejection seats, and crash safety seats for aircraft."[153] The 2005 edition of *Spinoff* further reports: "The Leicester, North Carolina–based company sold the rights to the technology in 1974, but later returned to market second- and third-generation derivatives that were less temperature-sensitive and more environmentally

---

151. "Living on the Ocean Floor," National Japanese American Historical Society, 1999, *http://www.nikkeiheritage.org/nh/fvxin4.html*.

152. Charles A. Yost and Ronald W. Oates, "Human Survival in Aircraft Emergencies," NASA CR-1262 (Washington, DC: NASA, 1968), p. 29.

153. See *http://www.sunmatecushions.com/about_dynamic_systems_inc.php*.

friendly than the original version."[154] It currently markets a variety of products, including Sunmate and Pudgee, which are varieties of temper foam, and Liquid Sunmate, which allows medical personnel to create form-fitting cushions for particular patients. The applications include the health-care field, sports equipment, and a sculpture foam for art projects. In addition to the original application for aircraft seating, the company also sells the products for cars used in mines, racecars, and horse saddle pads.

The great fame that temper foam has achieved did not come through the Dynamic Systems company, despite the good work it has done for society since Yost founded it in 1969. Rather, its fame has come through the "Swedish sleep system" marketed by the Tempur-Pedic company. The investors' FAQ linked from the Tempur-Pedic corporation Web site says:

> Tempur-Pedic was originally founded in 1992 after nearly a decade of research and development of a product formulation originally designed for use by NASA…. In the early 1970's, NASA engineers developed a viscoelastic memory foam to relieve astronauts of the incredible G-forces experienced during lift-off. Tempur-Pedic's Swedish partners began experiments to perfect the NASA formula for consumer use and after nearly a decade and millions of dollars of research; the company introduced our improved version—the proprietary TEMPUR® pressure-relieving material…. In 1998 Tempur-Pedic was awarded the "Certified Technology Seal" from NASA. This seal verifies that the underlying product technology was derived from the United States' efforts and experiments in space.[155]

A bloglike Web site devoted to information about foam mattresses, *MyFoamMattress.net*, disputes this spinoff and argues that bedding companies use the supposed connection to NASA to promote their products, with the implication that NASA goes along with this alleged charade in order to boost its own reputation in the spinoff area. The core of the argument is this:

> The original NASA foam was never suitable for sleeping because it broke down in time and lacked the comfort needed to make a good mattress. But the Program never intended to use space foam for bedding. Remember, it was developed for astronauts' seats, not for sleeping. Once released to the industrial world, progress was imminent. Years of research and development by the Swedish company Tempurpedic did turn this early recipe into an unique material now commonly known as memory foam. Some people still like to

---

154. "Forty-Year-Old Foam Springs Back with New Benefits," *Spinoff 2005*, pp. 46.
155. See *http://phx.corporate-ir.net/phoenix.zhtml?c=176437&p=irol-faq*.

call it NASA foam. This earned them the Certified Space Technology status usually given to a "product that is the direct result of technology developed for space."[156]

In a sense these antagonistic information sources agree that the foam originally made by Yost's company was not suitable for ordinary bedding, and of course the original spinoff claims concerned specialized medical applications, not mass consumer products like bed mattresses. Millions of dollars of research, invested over a decade by the Swedish company, Fagerdala World Foams, was required to develop the mattress now distributed in the United States under the name Tempur-Pedic. Arguably, Yost's work for NASA Ames was an essential step in the development but only one among many. There is nothing wrong with that. Fundamental research and prototyping is often much less costly and time-consuming than the later stages of development of a commercial product. However, the publicity of the mattress company appears to exaggerate NASA's contribution and to identify it with the space program rather than recognizing NASA's broader mandate.

The entries in the Spinoff database from 1976 and 1979 say that temper foam's origin was "improvement of aircraft seats" and "protective covering for aircraft seats." The 1988 entry says the origin was the "space shuttle." The 2005 entry says the origin of the spinoff was "improved airline seating for crash and vibration protection." Except for the 1988 entry, which may simply have recognized that the material was being applied to the astronaut seats in the Space Shuttle, NASA has consistently acknowledged that the origin was in its aviation technology work, not specifically the space program. The public tends to think of NASA as "the space program," however, and the Tempur-Pedic company apparently found the futuristic aura of space exploration provided a useful connotation for its advertising campaign.

Temper foam is primarily a spinoff of NASA's work to develop better technologies for commercial aviation. As a Tempur-Pedic brochure explains: "In 1998, at a press conference held at its headquarters in Washington, D.C., NASA saluted Tempur-Pedic for 'significant contributions to transferring aeronautical and space research technology into the private sector to save lives, promote economic opportunity and help improve the quality of life for humankind.' In turn, Tempur-Pedic presented Daniel S. Goldin, NASA administrator at the time, with the one-millionth Swedish Neck Pillow

---

156. Daniel Burrows and Liz Hoffman, "The Secret of NASA Mattress Foam," 2006, *http://www.myfoammattress.net/memory/nasa.html.*

produced by the company."[157] The phrase "aeronautical and space research technology" allows historians to say it was a spinoff from aeronautical research, and the company is free to claim it was a spinoff from space research.

Temper foam can claim some inspiration from the space program. During his many years at Ames, Kubokawa was involved with space-related as well as aviation-related research. Yost had been involved in some work for Apollo, but his interest in spaceflight dates from about 1950, when he decided that actual flying saucers could be propelled by electrostatic fields rather than rockets. At age 17 he built a Tesla coil seven feet high that literally electrified the environment. Late in his life, Yost wrote: "Through my R&D work on the Apollo project (1962–1966), I became totally disgusted by the limitations of rocket technology. I became totally focused on discovering a means of electric propulsion, or flying saucer technology."[158] When Yost died in 2005, he was editor of a visionary periodical, *Electric Spacecraft*, that explored the possibility of propelling space vehicles by means of magnetic or electric fields.[159]

Despite its fame as a material for commercial bed mattresses, and its application in various sports-related products, temper foam continues to have respectable medical applications. To its credit, and also its profit, Tempur-Pedic has campaigned to reduce the serious problem of bedsores in American nursing homes through the use of its materials.[160] Thus temper foam is a legitimate NASA spinoff, although practical applications have required considerable development work, and the credit is due to NASA's aeronautics rather than space research.

Case 10: Cochlear Implants

This case was suggested to me by Daniel P. Lockney, former editor of NASA's *Spinoff*. Sometimes called a bionic ear, a cochlear implant substitutes for portions of the human ear that are not functioning properly in deaf people. As the National Institute on Deafness and Other Communication Disorders explains, a cochlear implant physically consists of two main parts—one positioned outside the head, usually perched at the back of the ear, and the implant itself, which is placed deep inside the structures of the inner ear. Functionally, the system consists of four parts. A microphone acts just like the one in a hearing aid, picking up the sounds. A speech processor translates

---

157. "Tempur-Pedic Pressure Relieving Swedish Mattresses and Pillows," Tempur-Pedic International, 2004, p. 3.

158. Charles Yost, "A Biographical Snapshot," *Electric Spacecraft* 39 (5 August 2005): 4.

159. See *http://www.electricspacecraft.com/journal.htm*.

160. See the history of the Tempur-Pedic company at FundingUniverse.com, *http://www.fundinguniverse.com/company-histories/TempurPedic-Inc-Company-History.html*.

the sounds into a form that can be meaningfully handled by the human nervous system. A transmitter and receiver pass the signals to the implanted part of the system. Deep inside the ear, the electrode array stimulates different regions of the auditory nerve. Modern cochlear implants allow many users to interpret speech and other sounds usefully, but the effect is not a clear duplicate of what people with normal hearing perceive.[161]

Spinoff 2003 attributes the invention of the cochlear implant to NASA employee Adam Kissiah and offers an unusually clear explanation of the circumstances under which he developed it:

> Driven by his own hearing problem and three failed corrective surgeries, Kissiah started working in the mid-1970s on what would become known as the cochlear implant, a surgically implantable device that provides hearing sensation to persons with severe-to-profound hearing loss who receive little or no benefit from hearing aids. Uniquely, the cochlear implant concept was not based on theories of medicine, as Kissiah had no medical background whatsoever. Instead, he utilized the technical expertise he learned while working as an electronics instrumentation engineer at NASA's Kennedy Space Center for the basis of his invention. This took place over 3 years, when Kissiah would spend his lunch breaks and evenings in Kennedy's technical library, studying the impact of engineering principles on the inner ear.[162]

This paragraph is packed with interesting information. The fact that Kissiah lacked a medical background—and the story mentions no medical collaborator—is striking. But it is also interesting to see that the story does not seem to describe the stereotype of a spinoff. This is not a case in which an invention was made with NASA support to accomplish something for the progress of aviation or space exploration, then transferred to other applications outside NASA. Rather, a motivated individual draws on expertise acquired through his NASA work to develop something outside his NASA responsibilities. Thus, like several other cases considered in this chapter, this one calls into question the very concept of spinoff.

In 1977, Kissiah obtained patent 4,063,048 for an "implantable electronic hearing aid," revised in 1982 as RE31,031. The patent application does not cite NASA support but acknowledges: "The invention described herein was made by an employee of the United States Government and may be manufactured and used by or for the Government for governmental purposes without the payment of any royalties thereon or therefor." The patent

---

161. See *http://www.nidcd.nih.gov/health/hearing/coch.asp.*
162. *Spinoff 2003*, p. 14.

is somewhat unusual in that only a very short period of time passed between the application and the awarding of the patent, from 16 March 1977 to 13 December 1977, rather than the usual two or three years. It does not cite any scientific or technical literature, and the only publication mentioned is a popular magazine: "A surgical process for implanting such electrodes is discussed more fully in an article entitled 'The Electric Ear' appearing in the April 1974 issue of *Newsweek* magazine." This sole citation suggests that other people were already attempting to develop cochlear implants, and this was in fact the case.

Kissiah's application cites one previous patent, 3,751,605 "method for inducing hearing," issued in 1973 to Robin P. Michelson on behalf of Beckman Instruments, Inc. Michelson's patent also described an implantable device to stimulate the auditory nerve. The difference is that Michelson's device would stimulate the auditory nerve with a single signal, whereas Kissiah's would stimulate different areas with different signals, which is much closer to the way the human ear usually works. So Kissiah's patent is really for a significantly improved cochlear implant, rather than for the very first such device.

Michelson's patent application cited three earlier patents. In 1969, James H. Doyle received patent 3,449,768 for an "artificial sense organ," on the basis of a series of applications filed over the previous eight years that would stimulate the auditory nerve with a series of pulses the inventor believed were suitable for simulating hearing. In 1965, Behrman A. Docotte and Louis E. Adin received patent 3,209,081 for a "subcutaneously implanted electronic device" that was like a conventional hearing aid but had an implanted sound generator; it did not stimulate the auditory nerve electronically. Way back in 1939, Héctor Pescador had received patent 2,164,121 for a "hearing apparatus for the deaf" that entered the ear canal rather more aggressively than conventional hearing aids but, again, did not stimulate the nerve directly.

A Web site called *HearAgain.org* "was established to create greater awareness about Adam Kissiah and his work on the implantable hearing device and the cochlear implant."[163] This site makes a very clear claim for the importance of his invention: "The patent is considered the first patentable design for digital electronics stimulation of the acoustic nerve in humans."[164] Indeed, the cochlear implants invented by Michelson and Doyle could be described as analog devices, rather than digital.

The Web site's brief biography of Kissiah notes that he received a B.S. degree in physics from the University of North Carolina, Chapel Hill, and worked for RCA and Pan American World Airways at Cape Canaveral in support of

163. See *http://www.hearagain.org/about_us.htm.*
164. See *http://www.hearagain.org/biography.htm.*

the testing of Redstone, Jupiter, Mercury, Pershing, and Minuteman rockets. In 1963, he joined NASA and worked on all the piloted rocket programs at Kennedy Space Center until the end of 1989. The site also lists honors that Kissiah received, notably "the prestigious NASA Space Act Award which included a signed certificate from NASA Administrator Sean O'Keefe and $21,000, the largest monetary award ever given to a single inventor in Kennedy's history."[165]

As both *Spinoff 2003* and *HearAgain.org* report, an unsuccessful effort was made to develop cochlear implants through a company called Biostim, based on Kissiah's patent. After the company's dissolution, the implication is that the technology was widely taken up by other companies. However, one gets a very different story if one starts searching for information on the history of cochlear implants, rather than working from Kissiah's patent. Invention of a well-functioning, multiple electrode cochlear implant is widely attributed to an Australian, Graeme Clark, who successfully implanted his first device in 1978. This was the year after Kissiah's patent, but a functioning device is very different from an idea described on paper. One way to assess the relative influence of Clark versus Kissiah is to see what the four major manufacturers of cochlear implants say today. The manufacturer called Cochlear offers a Flash-enabled timeline on its Web site explaining that Clark began work in 1967, inspired by his own father's deafness; became chair of the Department of Otolaryngology of the University of Melbourne in 1970; and received a research grant in 1977. His first 1978 patient received a 10-channel cochlear implant that allowed him to recognize the tune of the beloved Australian song "Waltzing Matilda." The Web site does not mention Kissiah.[166] The Web sites of the Austrian MED-EL and American Advanced Bionics companies do not have history sections and mention neither Clark nor Kissiah.[167] The French manufacturer Neurelec stakes its own claim to invention of the cochlear implant: "In the early 1970's French scientists and clinicians had been the first to design and develop an implantable hearing aid and in 1986, MXM launched the Digisonic® programme. This was to lead to the provision of the first multi-channel cochlear implant which was entirely digital, re-programmable and which transmitted the whole sound spectrum."[168]

In 1995, the Institute of Medicine of the National Academy of Sciences published a collection of historical essays titled *Sources of Medical Technology*. Stuart S. Blume's chapter on the early years of cochlear implantation, up

---

165. See *http://www.hearagain.org/biography.htm*.
166. See *http://www.cochlearamericas.com/index.asp*.
167. See *http://www.medel.com/* and *http://www.bionicear.com*.
168. See *http://www.neurelec.com/en/neurelec_company.html*.

through 1982, does not mention Kissiah. It properly places great emphasis on the fundamental research in biology and medicine that provided the knowledge necessary for any invention and cites early clinical experiments carried out in France in 1957. It described the resistance that pioneers like William House and Blair Simmons faced when they began research in the 1960s, then gives Michelson much credit. An extended section describes the important French work of the 1970s. Blume does mention Clark but only as a somewhat late participant in the development of the cochlear implant who played a role in finally bringing the technology to the point at which it could be exploited commercially.[169]

One is reminded of the public relations competition between the United States and the U.S.S.R. over the rights to claim the inventor of liquid-fuel rockets for their nations. Konstantin Eduardovitch Tsiolkovsky was the Soviet candidate, and Robert H. Goddard was the American one. Tsiolkovsky began earlier but never got beyond the stage of theoretical writing, whereas Goddard was later but built working rockets. Clark is Australia's "inventor," while Kissiah can play this role for the United States and for NASA. Kissiah's patent cannot be ignored. A total of 45 later U.S. patents cite it, most recently patent 7,010,354 ("sound processor for cochlear implants"), filed in 2001 by Graeme Clark himself and issued to Clark's Bionic Ear Institute in 2006. The fact that Kissiah worked completely outside the conventional medical science community may have been a disadvantage for him, as scientific marginality was a disadvantage for Tsiolkovsky. Kissiah deserves honor for his contribution, but cochlear implants were developed by dozens of researchers in several nations, making them a collective rather than individual invention.

Case 11: Fast Neutron Cancer Tumor Treatment

The 10 previous cases were originally framed as classical spinoffs, but the 11th is different. *Spinoff 1979* placed it in the category of *community service*, although it deals with an experimental medical technology. Fast neutrons are a variety of radiation treatment used to destroy cancer cells that are different from x-rays or radioisotopes in that fast neutrons must be produced by a particle accelerator such as a cyclotron. NASA's Lewis Research Center (renamed the Glenn Research Center in 1999) in Cleveland, Ohio, had one. The Cleveland Clinic teamed up with Lewis to provide fast neutron to some of its cancer patients.

---

169. Stuart S. Blume, "Cochlear Implantation: Establishing Clinical Feasibility, 1957–1982," in Nathan Rosenberg, Annetine C. Gelijns, and Holly Dawkins, eds., *Sources of Medical Technology* (Washington, DC: National Academies Press, 1995), pp. 97–124.

The cyclotron was invented by Ernest O. Lawrence in 1929, who received patent 1,948,384 for it in 1934 and won a Nobel Prize for it in 1939. The cyclotron was the first "atom-smasher" research tool that accelerated ions to high speed by spinning them in a circle with a fluctuating electric field. Neutrons cannot be accelerated directly in a cyclotron because unlike ions they are not electrically charged. To make a neutron beam, Lawrence accelerated ions of deuterium, the isotope of hydrogen that contains one proton and one neutron, then smashed them into a beryllium target. In 1938, *Time* magazine publicized the idea of fighting cancer using a cyclotron-produced neutron beam, which Lawrence had developed in collaboration with his physician brother.[170]

A cyclotron with the power to treat tumors is a large device, expensive both to build and to operate. Thus it is not surprising that this method of treatment never became commonplace. From 1972 until 1977, about 700 patients received fast-neuron treatment at three facilities that already had cyclotrons for other purposes. Two were educational institutions—the University of Washington and Texas A&M University—and the third was the Naval Research Laboratory in Washington, DC. The Cleveland Clinic began offering this kind of treatment only after the three original facilities had already gained five years of experience.[171] *Spinoff 1979* describes the treatment as experimental, intended to evaluate the benefit for patients as much as to provide those benefits. The costs of converting one of the cyclotron's target areas into a treatment room, and of the necessary ancillary equipment, came from a grant from the National Cancer Institute.

The use of cyclotron-generated neutrons to fight cancer continues to be a valuable treatment option, but facilities offering it are rare. An information sheet dated 2003 from the Wayne State University Physicians Group in Michigan claims: "The neutron cyclotron is now the standard of care for institute patients, offering them a treatment success rate that is 10 percent higher than that of patients treated with standard, conformal external beam radiation therapy."[172] A Web search found other advanced centers offering comparable treatment, including the National Superconducting Cyclotron Laboratory at Michigan State University, the Clatterbridge Neutron Cancer Therapy Cyclotron at Merseyside in the United Kingdom, and the University

---

170. "Cyclotrons for Cancer," *Time*, 28 November 1938.

171. Robert G. Parker, Herbert C. Berry, Jess B. Caderao, Arthur J. Gerdes, David H. Hussey, Robert Ornitz, and Charles C. Rogers, "Preliminary Clinical Results from U.S. Fast Neutron Teletherapy Studies," *Cancer* 40 (1977): 1434–1438.

172. Wayne State University Physicians Group in Michigan, "Radiation/Oncology Services Available," 2003, *http://www.med.wayne.edu/wsupg/specialty/radiation_oncol/services.htm*.

of Washington Medical Center, which was one of the original three sites. As of May 2006, the related Seattle Cancer Care Society said that only three such facilities exist in the United States.[173] The University of Washington cyclotron treats only about 100 patients a year, most of whom suffer from salivary gland tumors.

Why would NASA's Lewis Research Center have needed a cyclotron in the 1970s? Lewis was the laboratory where much of the research on advanced propulsion systems took place, and the cyclotron was part of the program to develop nuclear reactor power for spacecraft.[174] Specifically, it was operated by the Radiation Physics Branch directed by James Blue. In the 1960s, considerable progress was made developing the technology to build nuclear rocket engines that were twice as efficient as chemical engines and rivaled them in thrust. Science fiction writers like Robert A. Heinlein had long assumed that only nuclear rockets would be capable of supporting a high level of activity in space.[175] When the nuclear propulsion efforts were cancelled in 1972, it is hard to know how important widespread public opposition to nuclear energy was versus the Nixon administration's desire to reduce near-term costs at the expense of humanity's future in space. In any case, the cyclotron lost its original purpose and was available for public service.

Fortuitously, a NASA oral history project interviewed June C. Bahan-Szucs about her time at Lewis, and she mentioned the original medical use of the cyclotron. In the interview, she expressed great distress at the "RIF" (reduction in force) firings that occurred at that time and a sense of futility that their efforts were not appreciated. About the nuclear energy program, she reported:

> As a result of that, something beautiful happened, because Dr. James Blue worked there, and he was extremely upset to think that the cyclotron that we had was no longer in use, and eventually he got Cleveland Clinic to come out, and we gave cancer treatments at NASA, NASA Lewis Research Center, at that time. [The patients] would be brought out in [the Cleveland Clinic] ambulance, and they would get the horizontal beam of a cancer treatment that could [reach cancer that nothing else could reach]. There were only three places in the whole world that could give them at that time. You could never say that [anyone was] cured from cancer, but after x number of years, seven

---

173. See *http://www.seattlecca.org/patientsandfamilies/adultCare/treatmentOptions/radiation/NeutronTherapy/*.

174. Mark D. Bowles and Robert S. Arrighi, *NASA's Nuclear Frontier: The Plum Brook Reactor Facility* (Washington, DC: NASA SP-2004-4533, 2004).

175. Robert A. Heinlein, *Rocket Ship Galileo* (New York: Scribner, 1947).

or whatever the [number] was at that time, they had some wonderful success stories, and all because of one man.[176]

Although in principle nuclear reactor rockets could achieve orbit on their own power, among the very most promising propulsion systems for space vehicles that have already achieved orbit is ion drive, largely developed over the same period at Lewis. This kind of engine uses a high-voltage electrical field to accelerate ions to a high velocity, much faster than the jet from a chemical rocket engine. The source of the energy is usually not chemical but can be electricity from solar cells—giving essentially free power so long as the spacecraft has not left the inner solar system—or from a nuclear generator. Lewis supported a considerable amount of high-quality ion drive research in the 1970s, much of it guided by Harold R. Kaufman of Colorado State University.[177] A cyclotron demonstrated methods that can be used in an ion drive. The chief drawback of ion drives is that their thrust is low, so some other kind of propulsion is needed to achieve orbit. Kaufman built his first working ion drive at Lewis in 1960, but the first spacecraft to use ion propulsion was the highly successful Deep Space 1 probe, which was not launched until 1998.[178]

Perhaps it is appropriate to conclude this consideration of spinoffs with this poignant example. Our civilization was apparently not ready to plunge forward into outer space, whether with nuclear rockets or by a heavy investment in chemical rockets and ion drive. Thus, one piece of equipment that was intended to help humanity reach the stars was diverted to humane medical uses.

4. Conclusion

Historical scholarship can contribute to the storehouse of knowledge needed to make decisions about current issues and planning for the future. This relatively modest study of spinoff from NASA to biomedical technology has presented some analytical ideas that may be useful in that context, but it also raises questions for future research. Importantly, what is true for spinoffs to the medical area may not be true for other categories of science and technology,

---

176. Interview with June C. Bahan-Szucs by Sandra Johnson, 29 September 2005, *http://www.jsc.nasa.gov/history/oral_histories/NACA/Bahan-SzucsJC_9-29-05.pdf*.

177. Harold R. Kaufman, "Charge-Exchange Plasma Generated by an Ion Thruster," annual report for NASA CR-135318, Colorado State University, Fort Collins, December 1977.

178. "Ion Propulsion: Over 50 Years in the Making," NASA press release, 6 April 1999, *http://science.nasa.gov/newhome/headlines/prop06apr99_2.htm*.

where the number of reported NASA-related patents is much greater than in the medical area.[179]

At the present time, debates are raging about the plans announced by NASA in 2004 for new directions in the space program.[180] Critics charge that both science and new technology are being downgraded in favor of achieving goals that may be interesting to groups in the general public but have only limited potential to contribute new technical ideas of broad application.[181] A group claiming to speak for the international space exploration community has specifically argued that much more attention must be given to spinoffs, improving both the publicity about them and the means for transferring technology from space to applications that have terrestrial benefits.[182] This will be harder to do, the fewer innovations the world's space programs actually develop. However, this spinoff-centric view of the value of space neglects to take account of the very problematic nature of spinoffs.

The clearest conclusion of this chapter is that the historians and social scientists who advised NASA around 1970 were right. Only rarely does a distinct innovation arise completely within the space program and then find applications outside. Rather, the development of space technology is intertwined with many other fields of science and engineering advancement. Again and again, we have seen that NASA helped improve a technology that already existed or supported development of one among many competing versions of an innovation. NASA's most important contribution to technical progress, outside space technology itself, has been as an active, general partner with other high-tech institutions of society, pushing our scientific knowledge and technical abilities forward along a broad front. *Spinoff* may properly be what grammarians call a *mass noun* or *uncountable noun*—like water, sand, and space itself. Outside of metaphorical uses (e.g., the waters of Mesopotamia or the sands of time), without the addition of a measuring unit (e.g., cubic light years of space) water, sand, and space cannot be counted. Similarly, *spinoff* is a general flow of technical progress, and to be grammatically correct one might want to avoid speaking of "a spinoff" or "spinoffs."

179. Adam B. Jaffe, Michael S. Fogarty, and Bruce A. Banks, "Evidence from Patents and Patent Citations on the Impact of NASA and Other Federal Labs on Commercial Innovation," *Journal of Industrial Economics* 46, no. 2 (1998): 183–205.

180. *The Vision for Space Exploration* (Washington, DC: NASA, 2004).

181. National Research Council, *An Assessment of Balance in NASA's Science Programs* (Washington, DC: National Academies Press, 2006); and Joseph N. Pelton, "Revitalizing NASA? A Five-Point Plan," *Space Policy* 22 (2006): 221–225.

182. Robert A. Goehlich, Chris Blanksby, Gérardine M. Goh, Yuko Hatano, Bojan Pečnik, and Julielynn Wong, "Space Spin-offs: Making Them Known, Improving Their Use," *Space Policy* 21 (2005): 307–312.

This is really an epistemological, ontological, or social-scientific problem more than a grammatical one, but it is aggravated by journalistic goals. In 1981 and again in 1986, I carried out brief pilot research at the Jet Propulsion Laboratory looking at how the scientists and the scientifically sophisticated journalists there attempted to make sense for the general public of the discoveries made moment by moment as Voyager 2 encountered the planets Saturn and Uranus. Every morning, in the Von Kármán Auditorium, a panel of experts presented the latest findings to a large audience of journalists, some of whom were famous people but understood little about science. Every afternoon, there was an informal discussion between perhaps a dozen very sophisticated journalists and a few scientists, in which the two professional groups struggled to find the right metaphors to communicate ideas broadly but accurately. By the time the news was widely disseminated, it often resembled what the public likes to think more than the technical reality.

News is about *stories*. A news story has protagonists who are specific individuals with names and faces. This is the way ordinary people think about innovation. An individual artist or inventor innovates by creating a specific thing, easy to see and describe. This prejudice reinforces the false stereotype that the space program benefits society through a series of distinct inventions. Unfortunately, the system of government patents and the goals of corporate public relations also reinforce this dubious model of innovation. How important really was Kissiah's cochlear implant patent, or Yost's work on temper foam? It is hard to say, but it was interesting to see Australian and French cochlear implant inventors promoted by competing companies, and a mattress maker claiming its product came from the glamorous space program rather than from a Swedish company's research or efforts to improve the crash-worthiness of airliner seats.

A corollary of spinoffs' problematic nature is that the extent of NASA's contribution in each case is open to debate. If inventions came nicely packaged in boxes, it would be easier to say who owned them. To NASA's credit, the annual *Spinoff* reports usually make it clear that a story is about a particular kind of heart pacemaker, associated with a particular company, rather than claiming that NASA somehow invented the pacemaker itself. But it is also true that some of the innovations described throughout this chapter have not been very influential. The anti-shock trousers were a valiant attempt to help people, and the people who worked on them deserve credit for their efforts. But the particular innovation was not commercially successful, and the value of applying general pressure to a shock victim's lower body has come into serious question in medical circles. Three of the examples involve measurement techniques: the bone density analyzer, the balance evaluation system, and the impedance cardiograph. All three must be described as successes, but only the third appears to be very important. Even it has not gained full acceptance in the medical community.

The Optacon reader for the blind helped thousands of people but, like so many other technologies, was eventually superceded by better technologies. This observation places the social science issues in an historical context again. In many cases, the propitious time to develop an innovation and to use it may come and go. Often, there are alternate ways of solving a problem, and one or two gain temporary market dominance while others are kept on the shelf. Awareness that there is one way of solving a problem may encourage engineers to see another way. Thus a space-derived innovation that was technically sound but could not compete with better-established methods can still be judged a success by contributing to the general store of technical knowledge.

This raises the much more general question of the timeliness of medical innovations coming from the space program. Here we may apply the astronautical metaphor of a *launch window*. To reach certain goals in space, one must launch neither too early nor too late but only during a limited span of time. One historical example is the development of large liquid-fuel launch vehicles in the 1950s mentioned earlier. Although best suited for spaceflight, these engines were primarily developed for intercontinental ballistic missiles. At the time, nuclear warheads were heavy, and the social movement promoting spaceflight convinced government leaders that liquid fuels were the right technology for missiles. So, large liquid-fuel vehicles were built that could be adapted for spaceflight. But the early 1960s saw a shift back to solid fuel rockets for military purposes as the warheads became lighter, the electronics and other technologies became more efficient, and the lesser maintenance issues with solid fuels was given higher priority for military uses. Thus there was a launch window roughly spanning the 1950s during which the military was motivated to develop large liquid-fuel rocket technology that then could transfer to space applications.[183]

Many of the examples discussed here date from the early years of the space program. During that launch window, NASA was developing space technology almost from scratch, so it needed to innovate in many different areas. That means that the opportunities for widely applicable innovations were probably at their historical maximum. This chapter has been historical in nature, so it has naturally emphasized earlier examples. But recent issues of *Spinoff* cite some very old cases, so this observation is not entirely an artifact of my research approach. Clearly, it would be interesting to see a quantitative

---

183. William Sims Bainbridge, *The Spaceflight Revolution* (New York: Wiley Interscience, 1976); and Bainbridge, "Beyond Bureaucratic Policy: The Spaceflight Movement," in James Everett Katz, ed., *People in Space* (New Brunswick, NJ: Transaction, 1985).

study of the dates—for example, of patents or publications—of all spinoffs in the NASA database. Logically, as NASA's research emphasis changes over time, the fields benefiting from technology transfer should shift as well.

One other question that arose in my research and probably deserves future study concerns the role of fundamental biomedical research. The world's space programs have conducted a considerable amount of biological or medical research in orbit, but I did not happen to encounter an example of a spinoff that came from the results of these studies. A systematic study of the consequences of orbital biomedical research would be interesting in its own right, but one might need to test the hypothesis that such research has drawn on a wide range of fields of biology but really has not contributed to them in return. The medical spinoffs listed in the spinoff database, and the subset of them described here, are only indirectly related to biology. The chief categories they belong to are electrical engineering, mechanical engineering, and materials science.

As part of my work at the National Science Foundation from 1999 through the present, I have had the opportunity to take a leading role in the examination of the societal implications of nanotechnology and the remarkable convergence of many fields that is uniting nanotechnology with biotechnology, information technology, and new technologies based on cognitive science. In partnership with Mihail C. Roco and others, I have organized conferences and edited books summarizing the insights of about 200 leading scientists and engineers in these fields.[184] Their primary observation—that most branches of technology are converging—applies well to several of the innovations described in this chapter. They concern a biological organism, the human body, but they involve not only knowledge of anatomy, physiology, and neuroscience but also electronics, computing, and the proper selection of materials for electrodes, power supplies, and structural components. This awareness reverses the problematic evaluation of the spinoff concept, rendering it a virtue.

---

184. Roco and Bainbridge, *Societal Implications of Nanoscience and Nanotechnology*; *Converging Technologies for Improving Human Performance* (Dordrecht, Netherlands: Kluwer, 2003); *Nanotechnology: Societal Implications—Maximizing Benefit for Humanity* (Berlin: Springer, 2006); *Nanotechnology: Societal Implications—Individual Perspectives* (Berlin: Springer, 2006); Mihail C. Roco and Carlo D. Montemagno, eds., *The Coevolution of Human Potential and Converging Technologies* (New York: New York Academy of Sciences, 2004); William Sims Bainbridge and Mihail C. Roco, eds., *Managing Nano-Bio-Info-Cogno Innovations: Converging Technologies in Society* (Berlin: Springer, 2006); *Progress in Convergence: Technologies for Human Wellbeing* (New York: New York Academy of Sciences, 2006); William Sims Bainbridge, *Nanoconvergence* (Upper Saddle River, NJ: Prentice Hall, 2007).

In the biomedical area, NASA's chief contribution cannot be packaged neatly in distinct spinoff inventions. Rather, NASA has been an early and active promoter of convergence itself. Bringing the separate branches of science and engineering together strengthens all of them, at the small cost of making it harder to tell a coherent story to the general public. If spinoff stories are one of the best journalistic ways to communicate the benefits of the space program, then it will need to be done responsibly. In my personal judgment, NASA has been responsible over the years. Either in the *Spinoff* stories themselves, in technical NASA publications on the same topics, or in the help offered me by the current editor of *Spinoff*, I always found correct information to place the innovation in a broader context, including the clues that allowed me to find the other information reported here.

The general public has several misconceptions about NASA spinoffs, including having the impression that NASA developed innovations that it either was not involved with or where it developed customized version of things that already existed. On the Frequently Asked Questions page of its "Scientific and Technical Information" Web site, NASA has attempted to correct some of these misperceptions.[185] It denies any credit for Tang, Teflon, and Velcro. For its own needs, NASA developed improved barcodes, quartz clocks, and smoke detectors, which may have had some wider application, but it claims no credit for their invention.

In 1980, a Harris poll asked about 640 adults living in Kentucky to judge whether several things were results of "the space program." Fully 48.1 percent believed that Teflon was, compared with 37.4 percent who felt it was not; the remaining 14.5 could not make up their minds. Overwhelmingly, Kentuckians believed two direct applications were results of the space program: weather prediction methods (86.3 percent) and satellite communications (88.1 percent). Seven other possible spinoffs got the following levels of belief in ascending order: latex paint (27.0 percent), synthetic rubber (33.4), microwave ovens (53.1 percent), push-button telephones (53.7), hand-held calculators (57.6 percent), and rechargeable pacemakers (61.9). Clearly, the public connects the space program with advances in electronics, and the rechargeable pacemaker scores higher than any of the seven other spinoffs.[186]

In conclusion, historians should appreciate spinoff stories because they are one of the best ways to communicate the history of the space program to members of the public who otherwise may be unaware of it. Quite apart

---

185. See *http://www.sti.nasa.gov/tto/spinfaq.htm*.
186. Analysis of original data from Harris study no. 20038, a 26 March–17 April 1980 poll of 671 Kentucky residents, data provided by the Odom Institute, University of North Carolina, Chapel Hill.

from what they say about the spinoffs themselves, these stories inform people about the early days of space exploration, when heavily instrumented animals were sent into space before the Mercury astronauts, when Gemini prototyped methodologies for the Apollo program, and when the vast complexity of the Space Shuttle was being developed. If high-school students study American history as a sequence of presidents, wars, and political reforms, in college they can be introduced to scholarly history of the United States in all its breadth and depth. So, too, spinoff stories are a good introduction to the history of space exploration, especially if many students and members of the public can be enticed to study the subject more deeply later on.

Chapter 3

NASA's Role in the Manufacture of Integrated Circuits

Andrew J. Butrica

Introduction: The Legend

This chapter addresses a specific question: what was the role of NASA in improving the manufacture of integrated circuits (ICs) during the Apollo era? The answer to that question already is a well-known and accepted fact. Indeed, not only has the Agency's role in advancing integrated-circuit production become recognized as fact, it has achieved the status of urban legend through the attention showered on the Apollo program and, more particularly, on a single project—the Apollo guidance and navigation computers—and through the retelling of the story, most recently on the Internet,[1] where the Apollo guidance computer has its own history Web site.[2]

The key to the legendary impact of NASA electronics is the large number of integrated circuits that the Agency bought for the Apollo program, or at the very least for the Apollo guidance and navigation computers. Because of the extraordinary number of integrated circuits that NASA bought for that project, the Agency stimulated the commercialization of the integrated

---

1. A search of the Internet produced the following selection of results: Steven J. Dick, "Why We Explore," *http://www.nasa.gov/mission_pages/exploration/whyweexplore/Why_We_04.html* (accessed 17 November 2005); IEEE Virtual Museum, "Missiles, Rockets, and the Integrated Circuit," *http://www.ieee-virtual-museum.org/collection/event.php?id=3457010&lid=1* (accessed 17 November 2005); IEEE Virtual Museum, "Let's Get Small: The Shrinking World of Microelectronics," *http://www.ieee-virtual-museum.org/exhibit/exhibit.php?id=159270&lid=1&seq=3* (accessed 17 November 2005); John Roach, "Apollo Anniversary: Moon Landing 'Inspired World,'" 16 July 2004, *http://news.nationalgeographic.com/news/2004/07/0714_040714_moonlanding_2.html* (accessed 17 November 2005); Jim Grichar, "Wielding the Budget Axe: It's Time To Abolish NASA," *http://www.lewrockwell.com/grichar/grichar33.html* (accessed 17 November 2005); Hum Mandell, "On a Mission," *http://utopia.utexas.edu/articles/alcalde/mandell.html?sec=science&sub=astronomy* (accessed 17 November 2005); and Newsgroups: sci.space.history, *http://yarchive.net/space/politics/nasa_and_ICs.html* (accessed 17 November 2005).
2. "The Apollo Guidance Computer," *http://hrst.mit.edu/hrs/apollo/public/index.htm* (accessed 12 April 2006).

circuit by bringing down their unit cost. According to Eldon C. Hall—who helped to create the Apollo guidance computer—NASA bought more than one million integrated circuits for the Apollo program between 1962 and 1967. As a result, between 1961 and 1965, the Apollo program became the largest single consumer of integrated circuits. At least one Internet site echoes Hall's sanguine declaration by claiming that the first few prototypes of the Apollo computers contained about two-thirds of all the integrated circuits in the world.[3]

Other sources closer to the time period, however, assert that the procurement numbers were smaller. For example, a November 1964 article in *Aviation Week* stated that the number of integrated circuits that NASA had purchased up to that date for the entire Apollo program (including more than the guidance computers) was 200,000.[4] Furthermore, a 1965 internal NASA Headquarters report stated that the Apollo navigation and guidance computers used 150,000 "microcircuits."[5] Although these numbers from 1964 and 1965 do not represent the total purchased by 1967, when more Apollo spacecraft and their computers went into production, one wonders nonetheless how the total number of integrated circuits purchased for the project reached a million. In the end, the number of integrated circuits that NASA bought for the Apollo guidance computers—or for the entire Apollo program—is not known with any certainty. The enormous procurement of integrated circuits for the Apollo computers, Hall states, "provided the semiconductor industry with an incentive to develop the technologies that gave birth to the integrated circuits common in modern electronics. Today, all electronic equipment depends on the descendants of these semiconductor chips. They are so common and plentiful that their origins are lost in a forgotten history."[6] The large procurement spurred industry to increase production, which brought down the unit price paid by commercial consumers. NASA's role was limited to that of a buyer of a large number of integrated circuits for the Apollo computers.

---

3. Eldon C. Hall, *Journey to the Moon: The History of the Apollo Guidance Computer* (Reston, VA: American Institute of Aeronautics and Astronautics, 1996), p. 19; and Newsgroups: sci.space.history, *http://yarchive.net/space/politics/nasa_and_ICs.html* (accessed 17 November 2005).

4. Barry Miller, "Microcircuitry Production Growth Outpaces Applications," *Aviation Week*, 16 November 1964, 79. Hall, *Journey to the Moon*, p. 141, states that procurement quantities of integrated circuits for the Block I computers approached 200,000 by the summer of 1964 but does not provide a source for the statement.

5. NASA Headquarters, *Electronic Systems Program Review, April 27, 1965* (Washington, DC: NASA Office of Program and Special Reports, 1965), p. 5. The term "microcircuit" usually referred to an integrated circuit.

6. Hall, *Journey to the Moon*, p. 1.

Long before the publication of Hall's rendition of the Apollo guidance computer story, Herbert S. Kleiman laid out the same argument but in more detail in his 1966 dissertation. Kleiman focused on the entire Apollo program rather than just the guidance computers. While other parts of the Apollo program used integrated circuits, he reflected, the guidance computer "has, for several reasons, carried the greatest impact." Citing a published 1964 statement that NASA so far had purchased 200,000 integrated circuits for the Apollo program, and making a conservative price estimate of $20 each, he concluded, "It is easy to realize the salutary effects which this purchase conveyed for the other IC [integrated circuit] products being offered by the firms involved."[7]

In the spring of 1964, Kleiman points out, Fairchild Semiconductor—the supplier of the "major share" of the Apollo program integrated circuits— announced "the first off-the-shelf integrated-circuit product line directly aimed at stimulating the non-military, non-space market." He argues that it was highly unlikely that the firm could have made this move if it did not have NASA support for its higher-priced integrated circuits. At the least, the firm's ability to offer the commercial products "was facilitated by the significant NASA support it had."[8] In short, NASA's large procurements of integrated circuits for the Apollo program accelerated the acceptance of integrated circuits as a commercial product. Kleiman's argument for NASA's role goes beyond just the influence of falling unit prices into the area of psychology. The Agency's decision to use integrated circuits "must have been a powerful stimulus on systems designers who were still 'on the fence' whether to include the IC devices in their own designs," especially because NASA was using them "for the most important mission of the whole space program" and "in a critical area where electronic failure was probably equivalent to mission failure or at least a diminution of mission effectiveness." But, Kleiman admits, the influence of this decision is "impossible to measure." The psychological impact of NASA's decision was to allay the fears of those in industry who were hesitant about the future of the technology.[9]

---

7. Herbert S. Kleiman, "The Integrated Circuit: A Case Study of Product Innovation," D.B.A. (business administration doctorate) thesis, George Washington University, June 1966, pp. 210–211. The article was Barry Miller, "Microcircuitry Production Growth Outpaces Applications," *Aviation Week*, 16 November 1964, p. 79.

8. Kleiman, "Integrated Circuit," p. 212. The Fairchild company history Web site bears out his claim: "1964: Fairchild introduces the industry's first linear integrated circuit. It is the first IC whose operation is dependent upon matched active and passive components." Fairchild Semiconductor, "A History of Innovation—1964," *http://www.fairchildsemi. com/company/history_1964b.html* (accessed 10 April 2006).

9. Kleiman, "Integrated Circuit," pp. 210–211.

More recently, historian Leslie Berlin has looked again at the initial resistance to the use of integrated circuits. Many engineers, designers, and purchasing agents feared that integrated circuits would put them out of work because they perceived their jobs as depending on circuits created from off-the-shelf transistors, resistors, and capacitors. Also, if the manufacturers of integrated circuits designed and built the circuits themselves, engineers at the customer companies feared that they would have nothing to do. If the customer for the integrated circuit was the design engineer, why would a design engineer with 25 years of experience want a circuit designed by a 30-year-old employee of an integrated-circuit manufacturing firm? Finally, she notes, some detractors were concerned that even though silicon was ideal for making transistors, better materials than silicon were available for making resistors and capacitors. Making them out of silicon, they worried, might degrade a circuit's overall performance.[10]

Along with Apollo, an equally pioneering application of integrated circuits was the Air Force's Advanced Minuteman Missile program, which was the earliest military use of integrated circuits. The decision to build the Apollo guidance computers with integrated circuits preceded the announcement in December 1962 that the Minuteman II missile would incorporate integrated circuits. Its guidance and control system used about 3,000 integrated circuits, with almost 2,000 of them going into the computer.[11] Jerome Kraus, in his 1973 study of the semiconductor industry, characterized the Minuteman decision to use integrated circuits as being crucial for promoting their commercial use because it meant paying Texas Instruments $1.2 million, Westinghouse $300,000, and RCA $300,000 for integrated circuits. By 1965, shortly after the first flight of the Minuteman II in September 1964, the missile production rate was up to six or seven per week—a schedule that called for Texas Instruments, Westinghouse, and RCA to supply more than 4,000 integrated circuits every week.[12] The total number of integrated circuits purchased for the Minuteman II is not known, yet one wonders whether the Minuteman or Apollo used more of them.

---

10. Leslie Berlin, *The Man Behind the Microchip: Robert Noyce and the Invention of Silicon Valley* (New York: Oxford University Press, 2005), p. 136.

11. Kleiman, "Integrated Circuit," p. 210; Donald MacKenzie, *Inventing Accuracy: A Historical Sociology of Nuclear Missile Guidance* (Cambridge: MIT, 1990), pp. 206–207; and Roy Neal, *Ace in the Hole: The Story of the Minuteman Missile* (Garden City, NY: Doubleday, 1962).

12. Jerome Kraus, "An Economic Study of the U.S. Semiconductor Industry," Ph.D. thesis, New School for Social Research, 1973, p. 45; and Paul E. Ceruzzi, *A History of Modern Computing*, 2nd ed. (Cambridge: MIT Press, 2003), p. 187.

More to the point, was the military or NASA a larger consumer of integrated circuits during this crucial period in the history of the integrated circuit? Their relative market shares—and by extension their relative influence on the development of the integrated circuit—is not knowable because of the organization of the available data. When the Department of Commerce's Business and Defense Services Administration tabulated the quantity and dollar value of semiconductors consumed in the United States, it used only two categories: "defense" and "non-defense." Starting in 1959, the "defense" category included the Defense Department, the Atomic Energy Commission, the Central Intelligence Agency, the Federal Aviation Administration, and NASA. Furthermore, errors inevitably crept into the numbers. The administration's quarterly survey asked manufacturers to indicate whether shipments were for defense or nondefense customers, but producers did not know the ultimate consumer of their products.[13] Despite the difficulties in providing a quantitative measure of the relative roles of the military (mostly through the Minuteman II) and the civilian space Agency (mainly the Apollo program), the historian Paul E. Ceruzzi has provided a qualitative assessment. He wrote, "The current 'revolution' in microelectronics thus owes a lot to both the Minuteman and the Apollo programs. The Minuteman was first; it used integrated circuits in a critical application only a few years after they were invented. Apollo took the next and equally critical step; it was designed from the start to exploit the advantages of integrated logic."[14]

In order to understand better the argument that NASA (or the military for that matter) primed the production pump that would cause integrated circuits to begin flowing into commercial applications, we first must consider the history of the price paid for microcircuits. Manufacturers, of course, had begun producing integrated circuits in 1961, before either the Apollo or Minuteman decisions to use them. In 1961, only two companies made integrated circuits: Fairchild Semiconductor and Texas Instruments. At first, both companies offered integrated circuits to only their military customers, but in March 1961, Fairchild introduced a series of six compatible Micrologic Elements and began selling them to NASA and commercial clients for $120 each. By summer, the company was manufacturing hundreds of units per week as their unit price dropped below $100 for lots of more than a thousand. In October 1961, Texas Instruments brought out a comprehensive array of its Series 51 Solid Circuits and sold them at even lower prices. By the end of 1961, Fairchild had sold fewer than $500,000 of its Micrologic devices at about $100 apiece. Texas Instruments was having such problems selling

---

13. Kraus, "Economic Study of the U.S. Semiconductor Industry," pp. 89–91.
14. Ceruzzi, *History of Modern Computing*, p. 188.

integrated circuits that it cut prices from $435 to $76 in 90 days, but with little effect. Integrated circuits bought for the Apollo guidance computer in 1962 in lots of a thousand or more from Fairchild fell from $31.10 to $20.00 to $11.25 per unit between May and October, but a purchase of 3,000 in April 1963 cost $15.00 each.[15]

Despite these dramatic price reductions from the 1961 highs, the widespread commercial application of integrated circuits did not take place. As Berlin has explained, customers did not object to these prices per se. The real issue was their cost compared to the cost of buying the individual components that each integrated circuit virtually contained. That cost often was as much as 50 times more than just buying the parts. The size, weight, and volume advantages of integrated circuits did not outweigh this cost factor.[16] The breakthrough, according to Berlin and historian Christophe Lécuyer, took place in the spring of 1964, when Fairchild began selling Micrologic flip-flop integrated circuits for less than the cost of the discrete components needed to build an equivalent circuit and for less than the manufacturing cost of the integrated circuit. The goal of the price cut was to stimulate demand for integrated circuits and, consequently, to create a commercial market for them by making them cheaper than equivalent circuits made of individual diodes, transistors, capacitors, and resistors. The price cuts would lure businesses that were more aware of parts prices than military contractors. Lécuyer points out additionally that Fairchild already had used this price-cutting approach rather effectively in the transistor business just a few years before. By assembling components in Hong Kong, Fairchild Semiconductor could lower its transistor prices, which in turn enabled the company to sell them for computer and consumer-electronics applications.[17]

As a result of Fairchild's bold move, in less than a year the demand for microcircuits expanded dramatically, and Fairchild received a single order for 500,000 integrated circuits—the equivalent, according to Berlin, of 20 percent of the entire industry's output for the previous year. One year later, in 1966, computer maker Burroughs placed an order for 20 million integrated

---

15. Ernest Braun and Stuart MacDonald, *Revolution in Miniature: The History and Impact of Semiconductor Electronics* (New York: Cambridge University Press, 1978), p. 114; Michael Riordan and Lillian Hoddeson, *Crystal Fire: The Birth of the Information Age* (New York: W. W. Norton & Company, 1997), pp. 271–272; Berlin, *Man Behind the Microchip*, p. 136; and Hall, *Journey to the Moon*, 17 and 80.

16. Berlin, *Man Behind the Microchip*, p. 136.

17. Berlin, *Man Behind the Microchip*, p. 137; and Christophe Lécuyer, *Making Silicon Valley: Innovation and the Growth of High Tech, 1930–1970* (Cambridge: MIT Press, 2006), p. 241. He traces the idea back to Robert Freund, "Competition and Innovation in the Transistor Industry," Ph.D. dissertation, Duke University, 1971.

circuits with Fairchild.[18] In comparison to these quantities, the procurements of integrated circuits for the Apollo and Minuteman programs seem diminutive. Instead of counting acquisition numbers, Lécuyer and Berlin define the takeoff point for the commercialization of integrated circuits as a business decision. The Apollo and Minuteman decisions are clearly not part of their argument.

In all of these accounts (and others not mentioned here), NASA's role was limited to that of a consumer of industrial products. Is this an accurate portrayal of the Agency's role? Did NASA, for example, contribute to the improvement of manufacturing processes through research and development? Kleiman, for one, concludes that as a sponsor of research and development, NASA had "the least impact upon the advancement of the IC technology" for two reasons: the level of the funding and the nature of the programs being sponsored. Looking at funding levels, NASA's role as a sponsor of integrated-circuit research and development was "insignificant" and "slight" compared with that of the Air Force. Holding back the Agency's ability to have more than just a minor impact were its "special and peculiar needs." NASA put a high value on reliability as well as reduced size, low weight, and light power consumption. The Armed Forces desired these characteristics to a lesser degree, and they were of only slight concern in industrial or consumer applications.[19]

This chapter hopes to transcend the legend and the literature by taking a fresh look at NASA's Apollo-era electronics research. It begins with background sections on integrated circuits and their societal impact, on the history of the transistor and the integrated circuit, and finally on the rise of the integrated-circuit industry.

Background

What Is an Integrated Circuit?

Integrated circuits are tiny electronic devices about one centimeter square that contain at least two electronic components (such as transistors, diodes, resistors, and capacitors) and the connections required to form a circuit. A typical integrated circuit today might contain millions of interconnected components. Integrated circuits are manufactured in large batches on a wafer traditionally consisting of a slice of pure crystalline silicon. The best-known example of an integrated circuit, also known as a chip, is the microprocessor that is at the heart of every modern computer. They serve in a

---

18. Berlin, *Man Behind the Microchip*, p. 139.
19. Kleiman, "Integrated Circuit," 209.

spectrum of electronic hardware from computers and cellular phones to digital microwave ovens.

The integrated circuit challenged the rising supremacy of the transistor, which, following its invention, began to vie with existing electronic technology—namely the vacuum tube. Because the integrated circuit contained multiple components (each the equivalent of a vacuum tube)—plus their associated capacitors, resistors, and interconnecting conductors—in a single miniature device, the new technology quickly made vacuum tubes obsolescent. Integrated circuits offered small size, reliability, fast-switching speeds, low power consumption, mass production capability, and ease of adding complexity that exceeded the features and advantages of vacuum tubes and eventually transistors as well.

The Societal Importance of Integrated Circuits

The advent and subsequent ubiquity of the integrated circuit have become an integral part of what has been called the third industrial revolution.[20] Because they are found almost everywhere in electronic devices as well as in home appliances (dishwashers, refrigerators), automobiles (airbag sensors, engine management, and controls for doors, lighting, seats, heating, air-conditioning, and emissions), cellular telephones, and numerous other applications, the integrated circuit has achieved enormous social, economic, and technological importance. Their usage in computers as tiny microprocessors made possible tremendous reductions in computer size and cost, which consequently made computers far more available for previously unimagined uses. The gradual shrinking of computers until they fit in a pocket-size cellular telephone has been just one of the enormous societal impacts of the integrated circuit.

Integrated circuits, however, have done more than just solve technological problems; they actually have changed the way that engineers designed electrical circuits. As historian Berlin has argued, many engineers, designers, and purchasing agents feared that integrated circuits would put them out of work, and rightly so. Eventually, the integrated circuit and computer-aided design would revolutionize both how electronics engineers performed their

---

20. One of the oldest works on the so-called third industrial revolution is G. Harry Stine, *The Third Industrial Revolution* (New York: Putnam, 1975). See also Joseph Finkelstein, ed., *Windows on a New World: The Third Industrial Revolution* (New York: Greenwood Press, 1989); Joseph Finkelstein, *The American Economy from the Great Crash to the Third Industrial Revolution* (Arlington Heights, IL: Harlan Davidson, 1992); and Jeremy Greenwood, *The Third Industrial Revolution: Technology, Productivity, and Income Inequality* (Washington, DC: AEI Press, 1997).

jobs and the very nature of electrical engineering education.[21] Engineers had grown accustomed to creating circuits with a minimum of transistors and diodes because they were relatively more expensive than resistors and capacitors. But, as it turned out, transistors and diodes were both smaller and easier to put on an integrated circuit than resistors or diodes. The result was that the circuits most adaptable to integration were digital circuits, with many transistors performing "yes-no" or "on-off" logic functions. Because these logic circuits were fundamental to building computers, the integrated circuit not only made truly small computers possible, it actually encouraged engineers to look for digital solutions to design problems.[22]

History

Invention of the Transistor

To understand the history of the integrated circuit, one first must consider the invention of the transistor because so many of the techniques developed to manufacture transistors were the same techniques used later to fabricate integrated circuits. Before the integrated circuit came along, the chief semiconductor products were transistors and diodes. Although solid-state diodes were manufactured at least as early as World War II, the semiconductor industry did not begin its rapid development until the invention of the transistor. The invention of the transistor, once believed to have had a single point of origin, is no longer a straightforward story. Indeed, the pioneer of the transistor appears to have been a professor at the University of Leipzig working well before World War II, while investigators in France following the war also have an apparently valid claim to the title of inventor of the transistor.[23]

During the early 1920s, Julius E. Lilienfeld, a Polish-born professor of physics at the University of Leipzig, conducted experiments with roentgen radiation that contributed to the development of the x-ray tube and collaborated with Count Ferdinand von Zeppelin on the design of hydrogen-filled dirigibles. As early as 1926, Lilienfeld had applied for several patents for a

---

21. On this point, see John G. Linvill, James B. Angell, and Robert L. Pritchard, "Integrated Electronics vs. Electrical Engineering Education," *Proceedings of the IEEE* 52, no. 12 (December 1964): 1425–1429.

22. "The Micro World" in *A Century of Electricals: An Exhibit by the IEEE History Center (1984),* http://www.ieee.org/organizations/history_center/general_info/century_menu.html (accessed 13 April 2006).

23. Kraus, "Economic Study of the U.S. Semiconductor Industry," p. 2; and Riordan, "The Lost History of the Transistor," *IEEE Spectrum* 41, no. 5 (May 2004): 44–49.

complex device remarkably akin to a transistor,[24] a multilayer structure consisting of metallic and semiconductor layers. Whether the device could have worked is an interesting question that has cropped up from time to time. Lilienfeld, however, appears to have lacked the resources to develop his ideas. He became a U.S. citizen after leaving Germany for the Virgin Islands in 1935, at which time he retired from active research. Still, his patents had an impact on future developments, for they hindered the patenting of transistor technology by researchers at Bell Telephone Laboratories many years later.[25]

The invention of the point-contact transistor in 1947 by John Bardeen and Walter H. Brattain with William Shockley is a story that has been told many times and in many ways. Beginning in the 1930s, researchers at the Bell Telephone Laboratories in Murray Hill, New Jersey, started looking for a solid-state amplifier to replace the multitude of electromechanical relays that formed the backbone of the telephone company's nationwide network of telephone lines. In the course of that research, Bardeen, Brattain, and Shockley jointly discovered the so-called transistor effect while studying the properties of the semiconductor germanium. As a result, in 1947, Bardeen and Brattain constructed a crude transistor apparatus (a so-called point contact transistor) that demonstrated the transistor effect and earned them the Nobel Prize in 1956.[26]

Bell Telephone Laboratories sent samples of this so-called Type A transistor to military, government, corporate, and university laboratories, while Western Electric offered to license all comers for its transistor patents on payment of a $25,000 advance royalty, an offer made as government antitrust lawyers filed a suit in 1949 seeking to separate American Telephone

---

24. Lilienfeld, "Rectifying Apparatus for Alternating Current," U.S. Patent #1,611,653, filed 27 March 1926, issued 21 December 1926; Lilienfeld, "Method and Apparatus for Controlling Electric Currents," U.S. Patent #1,745,175, filed 8 October 1926, issued 28 January 1930. Virgil E. Bottom, "Invention of the Solid State Amplifier," *Physics Today* 17, no. 2 (February 1964): 24–26, discusses these patents in some detail.

25. "Obituaries: Julius E. Lillienfeld [*sic*]," *Physics Today* 16, no. 11 (November 1963): 104; Bottom, "Invention of the Solid State Amplifier," p. 24; Braun and MacDonald, *Revolution in Miniature*, pp. 29–30 and 48; G. W. A. Dummer, *Electronic Inventions, 1745–1976*, 1st ed. (New York: Pergamon Press, 1977), p. 82; David Morton and Joseph Gabriel, *Electronics: The Life Story of a Technology* (Westport, CN: Greenwood Press, 2004), p. 26; J. B. Johnson, "More on the Solid State Amplifier and Dr. Lilienfeld," *Physics Today* 17, no. 5 (May 1964): 60–62; and "Solid State Devices," *Electronic Design* 24 (23 November 1972): 72.

26. Riordan and Hoddeson, *Crystal Fire*, pp. 70 and 115–141; Michael Eckert and Helmut Schubert, *Crystals, Electrons, Transistors: From Scholar's Study to Industrial Research*, translated by Thomas Hughes (New York: American Institute of Physics, 1990), pp. 157–166; Morton and Gabriel, *Electronics*, pp. 27–34; and Braun and MacDonald, *Revolution in Miniature*, pp. 41–42 and 46–48.

& Telegraph from its manufacturing arm—Western Electric—while also dealing with the Bell telephone patents.[27] The transistor's technological problem was how to manufacture a more rugged version in large numbers. The solution came with the invention of the junction transistor, a more practical form of the transistor announced by Bell Telephone Laboratories on 4 July 1951. It featured many advantages over the point-contact transistor: lower power consumption, more efficient signal amplification, and less waste heat. Meanwhile, the point-contact transistor went into production at Western Electric, and in 1952 it entered service in telephone switching equipment.[28]

Throughout the 1950s, the company manufactured both point and junction transistors, but junction transistors showed more potential. Bell transistor patents at the same time came under legal and technological fire. The Radio Corporation of America (RCA) had a cross-licensing deal in electronics with Western Electric. Like Bell, RCA was under a Justice Department investigation that culminated in the filing of an antitrust suit in 1954. RCA—thanks to its cross-licensing arrangement with Western Electric—offered the Bell patents without requiring deposits as advance royalty payments. The Bell transistor also faced a technological challenge from John Saby's fabrication of an alloy-junction transistor at General Electric's Schenectady, New York, laboratory in 1951, which RCA quickly adapted for mass production.[29]

The transistors in question were made out of germanium, not silicon. Transistors, diodes, and integrated circuits eventually would be created out of silicon, not germanium. Silicon was much harder to work with than germanium because of its higher melting point, higher chemical reactivity, and other challenging characteristics. In February 1951, Gordon Teal of Texas Instruments achieved a breakthrough when he managed to grow individual silicon crystals and form p-n junctions, but he did not succeed in creating

---

27. The suit was *U.S. v. Western Electric Co. and American Telephone & Telegraph Co.*, Civil Action 17–49, U.S. District Court of New Jersey, 14 January 1949. "Semiconductors," *Business Week* no. 1595 (26 March 1960): 93; and Kraus, "Economic Study of the U.S. Semiconductor Industry," p. 26.

28. Riordan and Hoddeson, *Crystal Fire*, pp. 168, 169, 170–186 and 192–194; Braun and MacDonald, *Revolution in Miniature*, p. 54; and Eckert and Schubert, *Crystals, Electrons, Transistors*, p. 166.

29. Braun and MacDonald, *Revolution in Miniature*, p. 55; "Semiconductors," 94; Kraus, "Economic Study of the U.S. Semiconductor Industry," p. 27; and Riordan and Hoddeson, *Crystal Fire*, pp. 199–200. Subsequently, in a consent decree of 24 January 1956, AT&T signed a decree leaving Western Electric in the Bell family but freeing the Bell patents issued up to that time to one and all. In 1958, RCA signed a consent decree ending the Government suit and releasing nearly all of its existing patents royalty-free. "Semiconductors," 94–95; and Kraus, "Economic Study of the U.S. Semiconductor Industry," pp. 27–28.

silicon transistors until 1954. These became the first commercially available grown-junction silicon transistors.[30]

A key new transistor manufacturing process developed around this time by both General Electric and Bell Telephone Laboratories simultaneously was the diffusion technique. Later it would be vital to the invention and manufacture of integrated circuits. The diffusion process allowed an impurity to diffuse into a semiconductor when they were in a vapor state, and the degree of diffusion into the semiconductor could be controlled by regulating the furnace time and temperature of the process. The addition of complex photolithographic procedures allowed manufacturers to imprint intricate mask patterns on the semiconductor so that diffusion took place only in the selected areas. The resulting transistor performed better at higher frequencies and was more reliable.[31]

Fairchild Semiconductor, Motorola, and Texas Instruments also produced transistors using the so-called "mesa" technique developed at Bell Telephone Laboratories. The technique was so named because the transistors looked like miniature mesas of the American Southwest. It involved etching a tiny plateau (the "mesa") on the surface of a germanium or silicon wafer. After diffusing a layer or two of dopants just beneath this surface, technicians applied a patch of inert material (such as wax) on it and treated the surface with a strong acid. The acid dissolved the semiconductor everywhere except under the patch. Assemblers attached two fine, closely spaced wires to the top of the resulting flat-topped protrusion, and a third lead was fastened to the bottom layer.[32] The mesa transistor had many advantages in addition to the ease with which operators could regulate the thickness of its base region. It worked in the high-frequency range. It was rugged. It dissipated heat readily. Because one could manufacture them in batches to a degree, they were inexpensive to produce. Some serious disadvantages common to transistors remained,

---

30. Riordan and Hoddeson, *Crystal Fire*, pp. 207 206–207, 208; David C. Brock, "Useless No More: Gordon K. Teal, Germanium, and Single-Crystal Transistors," *Chemical Heritage* 24, no. 1 (Spring 2006): 33–35; Riordan, "Lost History of the Transistor," pp. 44–49; Dummer, *Electronic Inventions*, p. 111; Eckert and Schubert, *Crystals, Electrons, Transistors*, pp. 173–174; and Braun and MacDonald, *Revolution in Miniature*, pp. 62–63.

31. Eckert and Schubert, *Crystals, Electrons, Transistors*, pp. 178–179; Braun and MacDonald, *Revolution in Miniature*, pp. 63; Ross Knox Bassett, *To the Digital Age: Research Labs, Start-up Companies, and the Rise of MOS Technology* (Baltimore, MD: Johns Hopkins University Press, 2002), pp. 146–147; and "Semiconductors," 96 and 101.

32. Braun and MacDonald, *Revolution in Miniature*, pp. 83–84; "Semiconductors," 101; and Riordan and Hoddeson, *Crystal Fire*, p. 262.

however. For example, electrical connections still had to be made by hand—a procedure that was both slow and expensive.[33]

To resolve the shortcomings of the mesa technique, Jean Hoerni, one of the founders of Fairchild Semiconductor, invented the planar process. The word "planar" suggested flatness, in contrast to the mesa's elevated profile. Hoerni suggested creating a protective icing layer of silicon dioxide around the transistor. This layer allowed manufacturers to control the dopants that reached below the silicon dioxide icing. For instance, it would allow gallium to diffuse through the layer while stopping phosphorus and other select dopants. The planar process entailed repeating three basis steps. First, oxidation created a mask over the wafer. Next, a photolithographic procedure opened up "windows" in selected parts of the oxide layer. Finally, impurities were diffused into the exposed silicon. Operators repeated the sequence so that diffusion layers of impurities could be inserted exactly where needed.[34]

The planar process worked with silicon but not germanium, which was incapable of maintaining a silicon oxide layer. When combined with photolithography—which provided a means for creating extremely fine, delicate patterns with tiny features smaller than one-thousandth of an inch across—the planar process offered a wealth of new manufacturing possibilities. As a result, by 1963, 30 firms were using the planar process to make silicon transistors or integrated circuits.[35]

Another key transistor fabrication process was epitaxial deposition. Developed in 1960 by Bell Telephone Laboratories, it involved depositing a single crystal layer using silicon vapor on a crystalline substrate. Components could be formed in the deposited silicon by planar diffusion without interfering with the substrate. As a result, the substrate contributed mechanical strength to the device without undesirable electrical characteristics. Before the advent of epitaxial deposition, the semiconductor industry started with a crystal as pure as needed in the initial stage, then added impurities at each step in a controlled manner. The new Bell method of manufacturing transistors used single crystals grown from the gas phase with controlled impurity levels—that is, the desired impurities were introduced as the silicon crystals

---

33. Braun and MacDonald, *Revolution in Miniature*, p. 84.

34. Riordan and Hoddeson, *Crystal Fire*, p. 262; Braun and MacDonald, *Revolution in Miniature*, p. 85. Bell Telephone Laboratories had used a silicon layer as a mask, but that practice did not prevent diffusion of impurities, such as phosphorus, through the mask to the areas presumed to be protected beneath the silicon dioxide. Fairchild discovered that when the oxide mask itself contained phosphorus, passivation of lower layers was complete. Braun and MacDonald, *Revolution in Miniature*, p. 85.

35. Riordan and Hoddeson, *Crystal Fire*, p. 263; Morton and Gabriel, *Electronics*, p. 77; and Bassett, *To the Digital Age*, p. 147.

were being formed. It also offered the unique advantage of the ability to grow very thin regions of controlled purity.[36]

One of the most important and pervasive transistor—and, later, integrated-circuit—fabrication processes was photolithography. This multiple-stage process is repeated over and over as many times as necessary to make a given transistor or integrated circuit. Essentially, it involves transferring a pattern—an intricate design of minute geometric shapes representing electronic components and circuit connections on a template (called a "mask")—to the surface of a silicon wafer. The term "photolithography" combines the concepts and processes of both photography and lithography. Lithography (literally "writing on stone"), a printing process invented by the Bavarian actor and playwright Alois Senefelder in 1798, is based on the fundamental fact that oil and water repel each other. A lithographer draws or paints designs on specially prepared limestone with greasy ink or crayons. When the stone is moistened with water, the areas of the stone not covered by ink or crayon absorb moisture. Then an oily ink is applied with a roller, and this ink adheres only to the drawing because the damp parts of the stone repel the ink. Finally, a print is made by pressing paper against the inked drawing.[37]

Photolithography, as the name implies, combines lithography with the technology of photography. This multiple-stage process is repeated as many times as necessary to make a given chip. As in lithography, the technique transfers a pattern—in this case, an intricate design of minute geometric shapes on a template (called a mask)—to the surface of the silicon wafer. A different mask might be used each time that the process repeats. The first step in photolithography is deposition, in which an insulating layer (usually of silicon dioxide) is grown or deposited on the slice of silicon. Its purpose is to create a barrier layer on the wafer's silicon substrate that can be patterned to form circuit elements using photolithography. After the formation of this silicon-dioxide layer, a chemical called a photoresist is applied to the wafer's surface. The photoresist acts much like the oily crayons and ink (or the water) in lithography, depending on whether they are positive or negative.

A positive photoresist is exposed to ultraviolet (UV) light wherever the underlying material is to be removed. The ultraviolet light changes the chemical structure of the photoresist so that it becomes more soluble in the developer. The action of sunlight on photographic emulsion is similar. The developer solution then washes away the exposed photoresist, leaving

---

36. Braun and MacDonald, *Revolution in Miniature*, p. 86; Dummer, *Electronic Inventions*, p. 127; and Richard L. Petritz, "Contributions of Material Technology to Semiconductor Devices," *Proceedings of the IRE* 50, no. 5 (May 1962): 1030–1031.

37. See Wilhelm Weber, *A History of Lithography* (New York: McGraw-Hill, 1966).

windows of the bare underlying material. With the use of a positive photoresist, the mask contains an exact copy of the pattern that is to remain on the wafer. Negative photoresists operate in just the opposite way. Exposure to ultraviolet light causes the negative photoresist to become polymerized and consequently more difficult to dissolve. The negative photoresist remains on the surface wherever it is exposed, and the developer solution removes only the unexposed portions. Negative photoresist masks, as a result, contain the inverse (or photographic negative) of the pattern to be transferred. Negative photoresists were popular in the early history of integrated circuit processing, but positive photoresists gradually became more widely used because they offered better control of the process for small geometrical features.

Following the application of the photoresist, the integrated circuit is heated in a process called soft-baking. This process is critical because it causes the photoresist layer to become photosensitive. Soft-baking for too long or for not long enough renders the wafer incapable of being made into a batch of integrated circuits. The next step—called mask alignment—is equally crucial. A mask (also called a photomask) is a square glass plate with a patterned emulsion of metal film on one side. The mask is aligned with the wafer so that the pattern can be transferred onto the wafer surface. Once the mask is aligned accurately with the pattern on the wafer's surface, the photoresist is exposed through the pattern on the mask with a high-intensity ultraviolet light.

There are three primary exposure methods: contact, proximity, and projection. In contact printing, the photoresist-coated silicon wafer comes into physical contact with the glass mask and is exposed to ultraviolet light. This physical contact permits very high resolutions—finely detailed circuit designs—but debris trapped between the photoresist and the mask can damage the mask and cause defects in the pattern. Proximity printing attempts to minimize those defects. It is similar to contact printing except that a small gap, 10 to 25 microns wide, is maintained between the wafer and the mask during exposure. The gap minimizes (but does not necessarily eliminate) mask damage. However, it yields lower resolutions, with the result that fewer components can be placed on a given chip. Projection printing avoids mask damage entirely. An image of the pattern on the mask is projected onto the resist-coated wafer, which is many centimeters away. To achieve high resolution, only a small portion of the mask is imaged at a time. This small image field is scanned over the surface of the wafer. Resolutions can be competitive (but not equal to) contact printing.

The final stages of photolithography are development (much like photographic development) followed by hard-baking. This last step is necessary to harden the photoresist and to improve adhesion of the photoresist to the wafer surface. The various stages of photolithography are repeated numerous times, building up the wafer microscopic layer by microscopic layer, until the final integrated circuit design is achieved.

Transistor Industry

Thanks to continual development of fresh manufacturing methods, the price of transistors fell throughout the 1950s and 1960s. The transistor that Fairchild Semiconductor sold for $150 in February 1958 (the 2N697), for example, sold for $28.50 in September 1959 and sold for less than 10 cents in the mid-1960s. Simultaneously, the number of transistors sold ballooned. Between 1954 and 1956, for instance, 17 million germanium and 11 million silicon transistors were sold in the United States, altogether worth about $55 million. In 1957, the 10th anniversary of the invention of the transistor, U.S. production of the device was at 30 million per year, with 5 million being made by Western Electric alone. The average cost had fallen to a dollar or two each, and annual sales topped $100 million.[38] Data gathered by the Electronic Industries Association bears out the dramatic expansion of the transistor industry. From 1954 to 1960, the compounded yearly growth rate of the transistor business was nearly 100 percent—that is, each year, on average, the total revenue from transistor sales nearly doubled. For the electronics industry in toto, the rate was only about half that at nearly 50 percent. In 1954, transistors brought in one-eighth the amount derived from the sales of diodes and rectifiers, but in 1960 that ratio changed in favor of the transistor to about 1.3:1. The change was apparent already in 1959, when transistors alone accounted for slightly more than half of the industry's $400 million sales. The electronics industry now was advancing on the back of the transistor.[39]

The number of companies making transistors also increased during the 1950s, and many formed solely to manufacture transistors. For example, William Shockley and Arnold Beckman, the latter already wealthy from the manufacture of a pH meter of his invention, formed the Shockley Semiconductor Laboratory in February 1956. The next year, eight Shockley employees left to set up their own company—Fairchild Semiconductor— with financing from Fairchild Camera and Instruments of New York City, which was winning a growing number of satellite and missile contracts. The

---

38. Kraus, "Economic Study of the U.S. Semiconductor Industry," pp. 153–154; Kleiman, "Integrated Circuit," p. 81; Bassett, *To the Digital Age*, p. 147; Braun and MacDonald, *Revolution in Miniature*, p. 69; and Riordan and Hoddeson, *Crystal Fire*, p. 254.

39. Kleiman, "Integrated Circuit," p. 78; and "Semiconductors," 78.

pattern of employees leaving one company to set up their own repeated throughout the semiconductor industry (but especially at Fairchild, giving birth to the term "Fairchildren"), multiplying the number of manufacturers. In 1951, four companies in the United States made transistors for the commercial market. By 1952, there were 8; by 1953, 15; and by 1956, no fewer than 26. In 1953, eight of the companies making transistors had been major manufacturers of vacuum tubes: General Electric, RCA, Sylvania, Raytheon, Philco, Columbia Broadcasting System (CBS), Tung-Sol, and Westinghouse. The remaining seven were Western Electric, Motorola, Texas Instruments, and four other new companies founded specifically to make transistors, such as Germanium Products.[40]

The largest consumer of transistors was the Defense Department. When Western Electric began manufacturing transistors for Bell Telephone Laboratories, the company had four wealthy customers: the three military services and AT&T. In 1952, for example, the military purchased almost the entire output of 90,000 transistors from Western Electric. These early transistors were very expensive and did not contribute to any kind of electronic miracle. The history of the integrated circuit would mirror this early growth phase. Even though transistor unit prices were at their highest, manufacturers could count on the military as a reliable customer at prices that would not find takers in the civilian sector. The military was an ideal customer for transistors under these market conditions because they were concerned more with availability, reliability, and technical performance than with price. In addition, the armed forces were voracious consumers of transistors. A single missile might carry as many as 6,000 of them, and its control equipment might contain tens of thousands more.[41]

Between 1952 and 1964, the Defense Department injected some $50 million into the U.S. semiconductor industry, most of it applied to improve production conditions. As early as 1951, to improve the manufacture of transistors, the three services assigned responsibility for overseeing the improvement of military transistor production to the Army's Signal Corps. The goal was to increase the availability of the transistor, reduce its cost, and improve its performance and reliability. Thus, most of the defense electronics money distributed to industry was in the form of production improvement contracts

---

40. Riordan and Hoddeson, *Crystal Fire*, pp. 233–237 and 251–252; Braun and MacDonald, *Revolution in Miniature*, pp. 66–67, 124, and 125; "Transistors: Growing Up Fast," *Business Week*, 5 February 1955, p. 86; Kleiman, "Integrated Circuit," p. 78; and Bassett, *To the Digital Age*, p. 45.

41. Riordan and Hoddeson, *Crystal Fire*, pp. 233–234; Eckert and Schubert, *Crystals, Electrons, Transistors*, pp. 168–169; Braun and MacDonald, *Revolution in Miniature*, pp. 57 and 79; and "Semiconductors," 78.

rather than research and development funds. The first of these contracts were signed in 1952 with Western Electric, General Electric, Raytheon, Sylvania, and RCA. By the early 1960s, when overproduction was eroding manufacturing prices and profits, the only protection that many firms had was to use the extra capacity imposed by the military to manufacture yet more devices, compensating for low unit prices by raising supplies.[42]

With the creation of NASA in 1958, the civilian space Agency soon became a significant buyer of transistors. The start of the lunar landing effort provided the semiconductor industry with a user whose demands for quality as well as quantity resulted in further growth. Still, in 1963 the armed services continued to dominate the transistor marketplace. The greatest demand came from the Air Force, followed by NASA; both were concerned particularly with reliability and reducing size and weight. The combined value of transistors purchased for aircraft ($22.8 million) and missiles ($20.3 million) exceeded the value of transistors bought for space applications ($33.0 million). The military market peaked around 1960, when the armed forces consumed nearly half the value of all semiconductor shipments made. But although military purchases outdistanced other purchases in dollar amounts, they did not represent the largest quantities bought. This discrepancy arose from the military's demand for the newest components and for those that would meet rigorous specifications, which would have been the most expensive items available.[43]

Despite the military's preponderant position in the transistor and solid-state electronics marketplace, commercial demand for transistors outside of AT&T began to build during the middle of the 1950s. The first commercial use of transistors, of course, was in such company equipment as rural telephone carrier amplifiers and headset amplifiers for operators. Another early application was the transistorized hearing aid, which first appeared in 1952. In honor of their namesake, Alexander Graham Bell, a lifelong advocate for the hearing impaired, Bell Telephone Laboratories waived the patent royalties for these hearing aids. Selling transistors for hearing aids was the application that launched Texas Instruments—future manufacturer and inventor of integrated circuits—into the commercial market.[44] In 1954, however, that firm's military business was beginning to falter, and Texas Instruments sought

42. Eckert and Schubert, *Crystals, Electrons, Transistors*, pp. 168; and Braun and MacDonald, *Revolution in Miniature*, pp. 80, 81, and 94.

43. Kraus, "Economic Study of the U.S. Semiconductor Industry," p. 6; and Braun and MacDonald, *Revolution in Miniature*, p. 91.

44. Riordan and Hoddeson, *Crystal Fire*, pp. 205 and 210–211; and Braun and MacDonald, *Revolution in Miniature*, pp. 54 and 55–56.

a new commercial market for its transistors. Through a joint venture with the Regency Division of Industrial Development Engineering Associates (IDEA), the Texas-based company entered production of so-called transistor radios using germanium components. As a result, the first commercial transistor radio, the Regency TR1, hit the market in October 1954, and Texas Instruments subsequently became the main supplier of transistors to the major radio manufacturers: Admiral, Motorola, RCA, and Zenith.[45]

Shortly after the appearance of the transistor radio, new commercial applications for the transistor emerged. In 1954, IBM announced that it would no longer use vacuum tubes in its computers. Nonetheless, the first commercial use of semiconductors in computers was probably the Remington Rand Univac in 1953. The first completely transistorized general purpose digital computer—IBM's 608—used thousands of transistors, which reduced its volume by 50 percent and decreased its power consumption by an impressive 90 percent. Eventually even small computers would have a thousand or more transistors. Phonographs, dictating machines, pocket pagers, automobile radios and fuel-injection systems, clocks, watches, toys, and transistorized TV sets soon followed, as did such NASA applications as the Explorer and Vanguard satellites.[46]

The Integrated Circuit

Despite the proliferation of transistor applications and sales, these tiny devices still had serious flaws that held back their adoption. Early on, a critical problem was the large number of useless transistors on each wafer—commonly about 20 percent of a batch. With time, however, the transistor's Achilles' heel became apparent. Each transistor had to be connected manually—soldered in place—with two or three tiny wires. Other components—diodes, resistors, and capacitors—required manual soldering as well. Mistakes, imperfections, and electrical shorts inevitably resulted. As long as each part had to be made, tested, packed, shipped, unpacked, retested, and connected to other parts,

---

45. Riordan and Hoddeson, *Crystal Fire*, pp. 211–213; and Dummer, *Electronic Inventions*, p. 116. The perception of the transistor radio as a Japanese product has persisted nonetheless. Japanese transistor radios began with a company known familiarly as Totsuko (Tokyo Tsushin Kogyo), founded in 1946, which manufactured a range of electrical and electronic equipment from heating pads to voltmeters for the domestic market. After developing their own germanium transistor, the company began selling its own radio model under the name Sony. Riordan and Hoddeson, *Crystal Fire*, pp. 214 and 217.

46. Braun and MacDonald, *Revolution in Miniature*, p. 78; Kraus, "Economic Study of the U.S. Semiconductor Industry," pp. 34–35; Kleiman, "Integrated Circuit," p. 79; "Semiconductors," 78; Riordan and Hoddeson, *Crystal Fire*, p. 254; and Braun and MacDonald, *Revolution in Miniature*, p. 78.

it would be the sheer individuality of components rather than technical or production limitations that would constrain semiconductor improvement. In the end, no matter how reliable the individual components were, ultimately they were only as reliable as the joints connecting them and the manual methods used for wiring circuits. Of course, the need for interconnections grew as circuitry complexity increased, with the result that the chance of failures caused by soldering also escalated.[47]

The solution was the integrated circuit. Integrated circuits perform the functions of several discrete components (transistors, diodes, capacitors, resistors) and incorporate all necessary interconnections into a single device. The term "monolithic" integrated circuit—common in the 1960s—referred to the fabrication of the entire device from a single ("mono") crystal ("lith"). With the invention of the integrated circuit, manufacturers could go from batch production of individual components to batch fabrication of entire circuits on a single wafer. Even more importantly, microcircuits required no postproduction soldering of wires between components.

The invention of the integrated circuit as both a concept and a device can be understood only within the context of a larger research framework— namely, the search for miniaturized electronics that began after World War II with the invention of the transistor. Miniaturization subsumed a range of technological solutions that sought to make everything smaller, lighter, and less hungry for power. Miniaturization meant diminutive light bulbs and switches; tiny probes to record a person's pulse, blood pressure, respiration, and temperature; and even little vacuum tubes.[48] Throughout the 1950s, researchers searched for a variety of solid-state solutions to electronics miniaturization. One of the major contenders was thin-film technology, in which deposited films of controlled properties and thickness were used to form multilayer electronic circuit elements such as capacitors and resistors. A limiting factor to the technology was the inability to form transistors. The focus of the armed services was on manufacturing methods (packaging, automation) that yielded prefabricated (prefab) modules. Strictly speaking, they were not forerunners to the integrated circuit, whose invention addressed the central challenge of the interconnection of solid-state components, but scholars nevertheless include them in their histories of that device. Their description, therefore, is included in the following discussion of the origins of the integrated circuit.

---

47.   Braun and MacDonald, *Revolution in Miniature*, pp. 76 and 113; Riordan and Hoddeson, *Crystal Fire*, p. 255.

48.   Horace D. Gilbert, "Introduction: Miniaturization as a Concept," in Horace Gilbert, ed., *Miniaturization* (New York: Reinhold Publishing Corporation, 1961), pp. 1–12.

Military Concepts

One of the oldest of these military programs was the Navy Bureau of Aeronautics' Project Tinkertoy, started in 1950 at the National Bureau of Standards. The goal of Project Tinkertoy was the automatic assembly and inspection of circuit components in a modular package. Silk-screen printing techniques formed resistors and capacitors (but not diodes or transistors) over printed wiring on steatite[49] ceramic wafers 22 millimeters (⅞ inch) square and 1.5 millimeters (¹⁄₁₆ inch) thick. Four to six of these wafers were automatically selected, stacked, and joined mechanically and electrically in a stack. Next, machine-soldered riser wires were attached at notches along the sides of each wafer. The resulting module generally had a tube socket on the top wafer. The program was directed toward high production of electronic equipment, and a disproportionate amount of the available funds went toward developing mass-production machinery, leaving a number of technical and reliability problems unsolved.[50]

The Navy had spent almost $5 million on Tinkertoy when the project came to a halt upon the disclosure in September 1953 that the entire scheme was founded on vacuum tube technology, not transistors. At one point, ACF Industries attempted to develop a commercial version of Tinkertoy in a program called Compac (for component package) and invested more than $1 million of its own money. Although the Tinkertoy modular approach to electronics packaging found its way into production items, it faded in the late 1950s as the transistor began to replace the vacuum tube. Transistors and printed wiring came into widespread use toward the end of the program and offered far greater potential volume reduction.[51]

---

49. Steatite is a type of soapstone consisting almost entirely of talc as well as a type of ceramic material made from soapstone with small amounts of additives and heated into a ceramic material. Steatite often has been used as an insulator or encasing for electrical components, but its use to form beads and seals dates back thousands of years.

50. Kleiman, "Integrated Circuit," p. 47; Riordan and Hoddeson, *Crystal Fire*, p. 255; Jack S. Kilby, "Invention of the Integrated Circuit," *IEEE Transactions on Electron Devices* ED-23, no. 7 (July 1976): 648; Michael F. Wolff, "The Genesis of the Integrated Circuit," *IEEE Spectrum* 13, no. 8 (August 1976): 45–53, esp. 45; Robert L. Henry and C. C. Rayburn, "Mechanized Production of Electronic Equipment," *Electronics* 26 (December 1953): 160–165; Eckert and Schubert, *Crystals, Electrons, Transistors*, p. 179; and Jack J. Staller and Arthur H. Wolfson, "Miniaturization in Computers," in Gilbert, *Miniaturization*, pp. 157–158.

51. Braun and MacDonald, *Revolution in Miniature*, p. 107; Kleiman, "Integrated Circuit," p. 48; "Microelectronics Today," *Electronic Industries* 21, no. 12 (December 1962): 92; Kleiman, "Integrated Circuit," pp. 49–50; Edmund L. Van Deusen, "Electronics Goes Modern," *Fortune* 51, no. 5 (June 1955): 146; Dummer, *Electronic Inventions*, pp. 108–109; R. L. Goldberg, "Solid State Devices: Packaging and Materials," *Electronic Design* 24 (23 November 1972): 126; and Staller and Wolfson, "Miniaturization in Computers," pp. 157.

Attempting to retain the good features of the Tinkertoy approach and to create a transistorized version of Tinkertoy, the Army's Signal Corps initiated its Micromodule[52] program in partnership with RCA. The Micromodule concept was an approach to miniaturization that included transistors along with printed circuits, dip soldering, and modular construction. The Micromodule program attempted to make connections in a uniform, reliable, mass-producible fashion akin to the printed-circuit boards that became commonplace following World War II. Utilizing multiple component logic wafers, groups of 12 to 18 wafers were mounted vertically with a thin insulating material between them. The Micromodule process combined high-density packaging, machine assembly, and modular design. Even more money went into the new program—$26 million between 1958 and 1963—with the bulk of funding going to the prime contractor, RCA.[53]

The Micromodule program shifted the emphasis of Tinkertoy from automating to miniaturizing electronics packaging. RCA suggested an approach that was similar to Tinkertoy but used smaller wafers that were 310 millimeters square and spaced 10 millimeters apart. RCA encapsulated the assembled module with an epoxy resin to increase mechanical strength and to provide environmental protection. The company's idea was based on the use of micro-elements—extremely small components of uniform size and shape that one could combine into tiny modules, each capable of performing a complete circuit function (for example, an oscillator, amplifier, or gate) according to it circuit design. In turn, one could interconnect the modules in various ways to constitute a wide variety of electronic assemblies.[54]

The Signal Corps promoted its Micromodule process as a standard package. Indeed, it was the first attempt at functional modular replacement—that is, the treatment of the entire module as a single component. The Signal Corps anticipated that by 1964, a million units per year would be rolling

---

52. Alternatively spelled "Micro-Module."
53. Kleiman, "Integrated Circuit," pp. 51–52; Staller and Wolfson, "Miniaturization in Computers," pp. 158–159; Braun and MacDonald, *Revolution in Miniature*, p. 107; "Microelectronics Today," p. 93; Kilby, "Invention of the Integrated Circuit," p. 648; Stanley F. Danko, Willie L. Doxey, and J. P. McNaul, "The Micro-Module—A Logical Approach to Miniaturization," *Proceedings of the IRE* 47 (May 1959): 894–904; George F. Senn and Rudolph C. Riehs, "Miniaturization in Communications Equipment," in Gilbert, *Miniaturization*, pp. 105–106; Riordan and Hoddeson, *Crystal Fire*, p. 255; Wolff, "Genesis of the Integrated Circuit," p. 45; Dummer, *Electronic Inventions*, p. 123; and Goldberg, "Solid State Devices," pp. 126–127.
54. Braun and MacDonald, *Revolution in Miniature*, p. 107; "Microelectronics Today," 93; Dummer, *Electronic Inventions*, p. 123; Kleiman, "Integrated Circuit," p. 52; and Senn and Riehs, "Miniaturization in Communications Equipment," p. 106.

off production lines. Those hopes did not materialize, however. The project ended as ignominiously as its predecessor. Just as they were starting to gain popularity in the early 1960s, Micromodules became dinosaurs, overtaken by more successful miniaturization techniques. The integrated circuit in particular punctured its chances of achieving sufficient production volume to support a competitive price.[55]

The Air Force was not without its electronics miniaturization program, but it started much later. The basic goal of its Molecular Electronics (or Molectronics) Program was to rearrange the basic molecular structure of materials in a controlled manner to cause the material to perform circuit functions. In more practical terms, it initially attempted to create circuits from single crystals of solid-state materials—the essential concept behind the integrated circuit. But by then the invention of the integrated circuit was history. Nevertheless, the Molecular Electronics program got under way through an Air Force contract with Westinghouse Electric, which engaged in molecular electronic research and to that end even formed a Molecular Electronics Division located in Elkridge, Maryland. The program succeeded in creating an amplifier no larger than a dime by 1960.[56]

The Air Force was especially interested in reducing the size, weight, and power consumption of its electronics, probably more so than the Navy or Army. The service also hoped that miniaturization would increase component reliability. Still, the greater reliability of small semiconductor components remained unproven even by the end of the 1950s. The obvious advantages of smaller size—reduced size, weight, and power consumption—were sufficient incentives to drive miniaturization, but the possibility of greater reliability at no greater cost made doubly sure that the miniaturization movement was self-sustaining.[57] Of all the military miniaturization efforts, that of the Air Force most closely resembled (and actually morphed into) integrated circuit

55. Dummer, *Electronic Inventions*, p. 123; Goldberg, "Solid State Devices," pp. 126–127; Braun and MacDonald, *Revolution in Miniature*, p. 107; and Kleiman, "Integrated Circuit," pp. 51–55.

56. Staller and Wolfson, "Miniaturization in Computers," pp. 116 and 119; Riordan and Hoddeson, *Crystal Fire*, p. 255; Kilby, "Invention of the Integrated Circuit," pp. 649–650; and Smithsonian Science Service, "Molecular Electronic Amplifier Smaller than a Dime," 22 January 1960, *http://americanhistory.si.edu/collections/scienceservice/052092.htm* (accessed 13 April 2006).

57. Braun and MacDonald, *Revolution in Miniature*, pp. 103–104; Kleiman, "Integrated Circuit," pp. 115–119; Dummer, "Miniaturization and Micro-Miniaturization," *Wireless World* 65 (1959): 545–549; and L. J. Ward, "Microminiaturization," *Journal of the Institute of Electronic Engineers* 8 (1962): 200.

technology. The Air Force's Molecular Electronics work lived on,[58] and today it is a nanotechnology program funded largely through the Defense Sciences Office of the Defense Advanced Research Projects Agency (DARPA).[59]

Geoffrey W. A. Dummer

The earliest formulation of the integrated-circuit idea, however, came from not the Pentagon but Geoffrey W. A. Dummer, who worked for the United Kingdom's Royal Radar Establishment at Malvern, England, founded in 1940 as the Telecommunications Research Establishment to develop radar applications for the Royal Air Force.[60] On 5 May 1952, in a talk he gave at a symposium sponsored by the Institute of Radio Engineers (IRE) in Washington, DC, Dummer described the concept of the integrated circuit. He declared, "With the advent of the transistor and the work in semiconductors generally, it seems now possible to envisage electronic equipment in a solid block with no connecting wires. The block may consist of layers of insulating, conducting, rectifying and amplifying materials, the electrical functions being connected directly by cutting out areas of the various layers."[61]

---

58. One of the most recent and useful works on the subject is Gianaurelio Cuniberti, Giorgas Fagas, and Klaus Richter, eds., *Introducing Molecular Electronics* (New York: Springer, 2005). Many of the earliest books on molecular electronics appeared during the 1980s, as the field evolved toward nanotechnology. See, for instance, Forrest L. Carter, ed., *Molecular Electronic Devices* (New York: M. Dekker, 1982); Forrest L. Carter, ed., *Molecular Electronic Devices II* (New York: M. Dekker, 1987); Forrest L. Carter, ed., *Molecular Electronics: Beyond the Silicon Chip*, 1st ed. (Fort Lee, NJ: Technical Insights, 1983); Forrest L. Carter, ed., *Molecular Electronics: Beyond the Silicon Chip*, 2nd ed. (Fort Lee, NJ: Technical Insights, 1985); and Felix T. Hong, *Molecular Electronics: Biosensors and Biocomputers* (New York: Plenum Press, 1989). The pace of publication (and research) picked up during the 1990s. Some examples are Pavel I. Lazarev, ed., *Molecular Electronics: Materials and Methods* (Boston: Kluwer Academic Publishers, 1991); Geoffrey J. Ashwell, ed., *Molecular Electronics* (New York: Wiley, 1992); Kristof Sienicki, ed., *Molecular Electronics and Molecular Electronic Devices* (Boca Raton, FL: CRC Press, 1993); Michael C. Petty, Martin R. Bryce, and David Bloor, *An Introduction to Molecular Electronics* (New York: Oxford University Press, 1995); J. Jortner and M. Ratner, eds., *Molecular Electronics* (Malden, MA: Blackwell Science, 1997); and Ari Aviram and Mark Ratner, eds., *Molecular Electronics: Science and Technology* (New York: New York Academy of Sciences, 1998).

59. "Molectronics: Overview," *http://www.darpa.mil/mto/mole/* (accessed 23 May 2006).

60. D. H. Tomlin, "The RSRE (Royal Signals & Radar Establishment): A Brief History from Earliest Times to Present Day," *IEE Review* 34 (1988): 403–407.

61. Riordan and Hoddeson, *Crystal Fire*, pp. 255–256; Braun and MacDonald, *Revolution in Miniature*, pp. 108; Morton and Gabriel, *Electronics*, p. 76; Dummer, *Electronic Inventions*, p. 109; Dummer, "Solid Circuits: Glimpses into the Future at Malvern Components Symposium," *Wireless World* 63, no. 11 (November 1957): 516–517; Dummer, "A History of Microelectronics Development at the Royal Radar Establishment," *Microelectronics and Reliability* 4, no. 2 (1965): 194–195; Dummer, "Integrated Electronics Development in

Five years later, Dummer convinced his superiors at the Royal Radar Establishment to award a contract to the Plessey Company to pursue this concept. The result was a metal model that demonstrated how a transistorized switching circuit (known as a "flip-flop") used in computers and other applications might be fashioned from silicon crystals. The device was placed on display during the September 1957 International Components Symposium held at the Royal Radar Establishment. The flip-flop circuit, contained within a tiny piece of silicon approximately ¼ inch (about 6.35 millimeters) square by ⅛ inch (3.2 millimeters) thick, consisted of four transistors, seven resistors, and three capacitors. The silicon had various sections removed to leave thin bridges of material (about 2 centimeters long by ½ millimeter square) with relatively high resistances (around 1,000 ohms [Ω]). Other resistors were created by depositing films of resistive material on the surface of the silicon, while capacitors were made in a similar manner from thin layers of evaporated or plated gold with insulators between them.[62]

In addition, Dummer managed to convince his employer to place a contract with the Plessey Company in April 1957 for the development of a model demonstrating the technique of shaping silicon crystals to control their resistance. This device, too, was on display at the symposium as an illustration of the possibilities of solid-circuit techniques. Despite this success in demonstrating the concept, the British government was not forthcoming with additional developmental funding. Consequently, Dummer's work on conceiving and developing the integrated circuit was "quietly shelved."[63]

Jack Kilby and Robert Noyce

Jack S. Kilby of Texas Instruments is a long-acknowledged inventor of the integrated circuit, but the idea that sparked his work hardly predated that of Dummer. On 24 July 1958, Kilby, a recent hire at Texas Instruments, while

---

the United Kingdom and Western Europe," *Proceedings of the IEEE* 52, no. 12 (December 1964): 1412–1425.

62. Dummer, "A Review of British Work on Microminiaturization Techniques," in *Electronics Reliability and Microminiaturization*, vol. 1 (New York: Pergamon Press, 1962), pp. 39–41; Braun and MacDonald, *Revolution in Miniature*, p. 108; Riordan and Hoddeson, *Crystal Fire*, p. 256; Wolff, "Genesis of the Integrated Circuit," p. 45; Dummer, "A History of Microelectronics Development," p. 195; Dummer, "Solid Circuits: Glimpses into the Future at Malvern Components Symposium," *Wireless World* 63, no. 11 (November 1957): 516; and Dummer, "Integrated Electronics Development," p. 1415.

63. Dummer, "A History of Microelectronics Development," p. 195; Dummer, "Solid Circuits: Glimpses into the Future at Malvern Components Symposium," p. 516; Dummer, "Integrated Electronics Development," p. 1415; and Braun and MacDonald, *Revolution in Miniature*, p. 108.

working in the miniaturization laboratory, wrote in his laboratory notebook that if circuit elements such as transistors, resistors, and capacitors could be made of the same material, they could be included in a single integrated circuit. By 12 September 1958, he had built a simple oscillator integrated circuit with five components that were not connected by wires. On 6 February 1959, Texas Instruments, on behalf of Kilby, sought a patent for the invention, and on 23 June 1964, it received U.S. patent # 3,138,743 for "miniaturized electronic circuits." The patent showed a circuit with only two transistors, but also with resistors and two capacitors. After building a prototype device, Kilby began work on fabricating more of them using photolithography.[64] His invention subsequently was considered to have been of such value and importance that in 2000, he shared the Nobel Prize in physics for "his part in the invention of the integrated circuit."[65]

All the same, though, Kilby's initial integrated circuit was a kludge. Its most serious drawback was the need for individual gold wires to connect the components, thereby making the concept difficult to scale up to any useful complexity. Meanwhile, Robert N. Noyce, a founder of Fairchild Semiconductor, was thinking about new uses for the planar process. Where Kilby and Dummer had focused on making different components (diodes, transistors, resistors) from the same material, Noyce focused on the electrical connections. Instead of wires manually soldered in place, the company would use photolithography to deposit fine lines of metal, such as aluminum. The narrow metal lines running atop the protective glass layer would be insulated completely from the electrical activity taking place just beneath it. Finally, in a separate fabrication step, one could insert external contact wires through tiny holes in the silicon-dioxide layer. The next step would be to create multiple devices inside the silicon slice and link them together in a single miniature circuit.[66] With this new process, one could manufacture

---

64. Riordan and Hoddeson, *Crystal Fire*, pp. 256–261; Braun and MacDonald, *Revolution in Miniature*, pp. 102–103; Morton and Gabriel, *Electronics*, p. 77; Kilby, "Invention of the Integrated Circuit," pp. 650–651; Eckert and Schubert, *Crystals, Electrons, Transistors*, pp. 180–181.

65. The prize acknowledged those whose work was fundamental to modern information technology, especially high-speed transistors, laser diodes, and integrated circuits. Kilby received one half of the prize money, while the other half was shared by Zhores I. Alferov and Herbert Kroemer "for developing semiconductor heterostructures used in high-speed- and opto-electronics." "Jack St. Clair Kilby (1923–2005): Engineering Monolith," *IEEE Spectrum* 42, no. 8 (August 2005): 10.

66. Riordan and Hoddeson, *Crystal Fire*, pp. 263–264; Morton and Gabriel, *Electronics*, p. 77; Bassett, *To the Digital Age*, p. 46; and Eckert and Schubert, *Crystals, Electrons, Transistors*, pp. 181–183.

hundreds of self-contained circuits—indeed, entire electronic devices—on just one silicon wafer.

Knowing that Texas Instruments already had filed for a patent, but not knowing its contents, Fairchild Semiconductor began drawing up its own patent papers. Their strategy was to focus on a detailed technical description that concentrated on the use of the company's planar techniques to make monolithic integrated circuits.[67] On July 30, 1959, Fairchild's attorneys filed for their "Semiconductor device-and-lead structure," and the company received U.S. Patent # 2,981,877 on 25 April 1961. That patent showed a circuit consisting of only one transistor plus two diodes, two capacitors, and four resistors.

Development of the IC Industry

The integrated circuit, like the transistor before it, eventually drove the electronics industry to new heights and changed the very geography of the semiconductor industry. In the early 1950s, the East Coast was the heart of the semiconductor business because that was the location of the existing large electronics corporations. Later that decade, as many small firms burst into the market, the industry remained centered there but coalesced in those areas that best suited the new industry's demands. Long Island and the Boston region, especially around the Route 128 ring road, proved most attractive to new companies, and one still can find many semiconductor businesses there. A second major electronics hub grew up in the Santa Clara Valley south of San Francisco, the location of both Shockley Semiconductor and Fairchild Semiconductor. The manufacture of integrated circuits above all else was the driving force behind the creation of this new West Coast electronics center. By 1969, no fewer than 25 semiconductor firms were located there within a few miles of each other.[68] The rise of the nation's second semiconductor hub was under way.[69]

The semiconductor industry proved to be no place for large, established firms, especially those that had been engaged in making vacuum tubes. During the early and mid-1950s, a handful of large, established electronics firms dominated the semiconductor industry. Their position eroded rapidly as new companies entered the business. By the end of the 1950s,

---

67. Riordan and Hoddeson, *Crystal Fire*, p. 264.
68. Braun and MacDonald, *Revolution in Miniature*, pp. 123–124.
69. The reader will understand that the positing of these two semiconductor manufacturing regions does not ignore the presence of isolated but essential manufacturers in outlying areas, such as Texas Instruments in Dallas and Motorola's semiconductor division in Phoenix. These exceptions, in fact, prove the rule.

when dozens of firms were in the semiconductor business, the market leaders were no longer the old established companies but rather the new ones. In 1960, in fact, the nation's two largest transistor manufacturers were Texas Instruments and Transitron Electronic Corporation, neither of which had ever manufactured a tube.[70] Established electronics companies, such as Motorola and Westinghouse, desperately tried to catch up with Fairchild Semiconductor and Texas Instruments, the pioneers of integrated-circuit technology. One company, Teledyne, in its attempt to compete in the new market, lured key employees from Fairchild in 1961 to start Amelco, a subsidiary devoted to making integrated circuits. Still, in 1971, the top five companies were IBM, Texas Instruments, Motorola, Western Electric, and Fairchild Semiconductor—not one of which had been a leader in the commercial semiconductor market in the mid-1950s.[71]

The first integrated circuits were expensive. As a result, the sales pattern of the integrated circuit replicated that of the transistor. The first customers again were the armed forces, followed this time by the civilian space Agency. One of the first applications was in a small working computer that Texas Instruments delivered to the Air Force in October 1960. It had a few hundred bits of solid-state memory. NASA, for its part, purportedly supported a portion of the Texas Instruments effort to develop the so-called Series 51 computer. In 1962, the company received a large contract to design and build a family of 22 special circuits for the Minuteman II missile. Meanwhile, Fairchild Semiconductor entered into mass production (a relative term) of its chips in 1961 and 1962, and it won substantial contracts from NASA and a number of commercial equipment makers.[72]

To understand the rising importance of the integrated circuit, one can compare shipments of integrated circuits and shipments of transistors from 1963 to 1971. In 1963, 302.9 million transistors were shipped versus only 4.5 million integrated circuits. The number of transistors shipped rose over this period to a high of 1,249.1 million in 1969 then began to decline, while the number of integrated circuits shipped increased far more steadily, reaching 635.2 million in 1971 compared with 880.7 million transistors in the

70. Braun and MacDonald, *Revolution in Miniature*, p. 122; "Semiconductors," 83. Leo and David Bakalar were the founders and primary (87 percent) owners of Transitron, located in Wakefield, Mass., and founded in 1952. William B. Harris, "The Company That Started with a Gold Whisker," *Fortune* 60 (August 1959): 98–101; Braun and MacDonald, *Revolution in Miniature*, pp. 110 and 122; Eckert and Schubert, *Crystals, Electrons, Transistors*, pp. 174.

71. Riordan and Hoddeson, *Crystal Fire*, p. 273; "Semiconductors," 74; and Braun and MacDonald, *Revolution in Miniature*, p. 123.

72. Kilby, "Invention of the Integrated Circuit," p. 653; and Hall, *Journey to the Moon*, 141.

same year. These numbers are more impressive when one considers the transistors integrated into microcircuits compared to the total number of individual transistors shipped. The percentage of transistors in integrated circuits, only 10.4 percent in 1963, skyrocketed to 94.5 percent in 1971.[73] At the same time, both the average price of integrated circuits and the percentage of integrated circuits purchased by the military fell. Advances in fabrication techniques brought down integrated-circuit prices throughout the 1960s. Units sold in 1962 cost $100 for small lots and $50 for larger quantities, but integrated circuits in 1975 went for a mere 80 cents each. When unit prices were the highest, the armed services bought the largest quantities of integrated circuits. According to one source, the Defense Department accounted for 100 percent of all integrated circuits purchased in 1962 and 94 percent in 1963, but that portion slipped to 72 percent in 1965 and only 37 percent in 1968 as commercial orders rose.[74]

The first commercial application of the integrated circuit (as had been the case for the transistor) was in hearing aids beginning in December 1963. Other new nonmilitary uses followed, particularly in computers and space applications. Shortly after Fairchild introduced a series of six compatible Micrologic Elements and began selling them to NASA and others in March 1961, a new market for integrated circuits opened up when, only two months later, President John Kennedy announced on 25 May 1961 that the United States intended to put an astronaut on the Moon by the end of the decade. NASA engineers, already concerned about every gram of weight on their spacecraft, welcomed the integrated circuit for use in the computer, communication, and other electronic systems required for human spaceflight.[75]

If the integrated circuit replicated the history of the transistor in many ways, it was unquestionably unique in one way. Unlike the transistor, the integrated circuit consisted of not one component but many. And the number of components that one could fit on a single integrated circuited kept growing. The devices made for computers in 1962 had two to four logic "gates" per unit, but those produced in 1975 featured more than 2,000 gates (the equivalent of 4,000 bits of memory) each. In addition, the number of integrated circuits that one could manufacture in a single batch grew as the size of silicon wafers expanded to 1 inch in 1964 and to 1.5 inches in 1966.[76] The

---

73. Kraus, "Economic Study of the U.S. Semiconductor Industry," pp. 80 and 209; and Braun and MacDonald, *Revolution in Miniature*, p. 112.

74. Kilby, "Invention of the Integrated Circuit," p. 653; and Braun and MacDonald, *Revolution in Miniature*, p. 113.

75. Braun and MacDonald, *Revolution in Miniature*, p. 105; and Riordan and Hoddeson, *Crystal Fire*, pp. 271–272.

76. Kilby, "Invention of the Integrated Circuit," p. 653.

giddy excitement generated by these and other key advances in integrated circuit fabrication undoubtedly induced Gordon Moore, director of research and development at Fairchild Semiconductor, to write his famous 1965 paper titled "Cramming More Components onto Integrated Circuits." In the paper, Moore observed, "The complexity for minimum component cost has increased at a rate of roughly a factor of two per year." The observation, now known as Moore's law, subsequently evolved into the rule that the number of components contained on an integrated circuit doubles every year.[77]

Electronics and NASA

The Challenges

Clearly, NASA had a more than passing interest in integrated circuits. The Agency was a major consumer of electronics of all kinds; they were as central to the Agency's mission as launchers. As NASA became involved in supersonic transport; satellites for meteorology, navigation, and communication; and human spaceflight, the rockets to launch those payloads—and the Mercury, Gemini, and Apollo spacecraft themselves—depended massively on electronics. Indeed, electronics were indispensable to all of the Agency's programs. As NASA Administrator James E. Webb explained to an audience in 1964: "Our accomplishments in space have stemmed from two principal sources: the first, of course, is the modern rocket ...; the second—whose importance must not be forgotten—is modern electronics. This combination has given us a tool whose boundaries are essentially unlimited, except for the finite imaginations of the users."[78] Electronics costs represented a substantial portion of NASA's outlays. "Electronics components," Webb told Congress, "account for over 40 per cent [*sic*] of the cost of our boosters, over 70 per cent of the cost of our spacecraft, and over 90 per cent of the cost of the resources going to tracking and data acquisition."[79] These same electronics also played a crucial part in enabling the United States to compete with its Cold War rival the Soviet Union in the arenas of both military struggle and international affairs and prestige.

---

77. Gordon E. Moore, "Cramming More Components onto Integrated Circuits," *Electronics* 38, no. 8 (April 1965): 114–117.

78. James E. Webb, "Electronics in the Space Age," address to the National Electronics Conference, Chicago, IL, 19 October 1964, pp. 6–7, Folder 3937, NASA Historical Reference Collection, NASA History Division, Washington, DC (hereafter, NHRC).

79. James E. Webb to George P. Miller, Chairman, Committee on Science and Astronautics, House of Representatives, 21 March 1963, Folder 4884, NHRC.

It was common knowledge that the United States did not possess the large boosters used by the Soviet Union to launch their bigger, heavier satellites. Thanks to its superior launchers, between 1957 and 1965 the Soviet Union piled up one accomplishment after another in space. Miniaturization made it possible for the United States to deliver satellites into space that were lighter in weight than Soviet satellites. Reduced size, weight, and power requirements also were becoming increasingly vital in developing new weapon systems. Every pound of missile weight shaved off through miniaturization meant greater range, more economical use of fuel, and, in certain cases, heavier pay-loads. By one estimation, for every pound of weight eliminated, one could reduce a missile's fuel load by 100 pounds. Moreover, the extremely high ratio of total vehicle weight to useful payload weight for U.S. launchers in 1960 was greater than 1,000 to 1, making weight reduction a top priority. Electronics also made missiles more accurate. For example, the United States needed fewer Minuteman I missiles because its microelectronics significantly increased their accuracy.[80]

The public exhibition of a Sputnik spacecraft at the 1958 Brussels World's Fair (also known as Expo '58) revealed the backward state of Soviet space-craft electronics by U.S. standards—or at least provided an opportunity for U.S. engineers and managers to disparage Soviet accomplishments. The satel-lite might have contained some semiconductors, U.S. observers conjectured; vacuum tubes were more common in Soviet spacecraft. Many of the Sputnik components displayed appeared large and roughly comparable to commercial components available in the United States a decade earlier. Moreover, the same observers remarked on the extensive use of hand-wiring and the "awk-ward" character of Sputnik assembly techniques compared with those found on U.S. satellites.[81]

Whether expressed as a "missile gap" or a "space race," electronics estab-lished itself as the sine qua non for defense and space. Electronics pro-vided the weight that counterbalanced the Soviet Union's lead in heavy-lift launchers. But electronics also was the deadweight that held back launch

---

80. Olin B. King, "Miniaturization in Missiles and Satellites," in Gilbert, *Miniaturization*, pp. 58–59; J. D. Hunley, ed., *The Birth of NASA: The Diary of T. Keith Glennan* (Washington, DC: NASA SP-4105, 1993), pp. 23, 31–32 and 40; Roger D. Launius, "Introduction," in Roger Launius and Dennis R. Jenkins, eds., *To Reach the High Frontier: A History of U.S. Launch Vehicles* (Lexington: University Press of Kentucky, 2002), p. 8; Gilbert, "Introduction," pp. 1–3; and Jeffrey Zygmont, *Microchip: An Idea, Its Genesis, and the Revolution It Created* (Cambridge, MA: Perseus, 2003), p. 72.

81. King, "Miniaturization in Missiles and Satellites," p. 59. In his discussion of the design of Sputnik I, Asif A. Siddiqi, *Challenge to Apollo: The Soviet Union and the Space Race, 1945–1974* (Washington, DC: NASA SP-4408, 2000), pp. 161–164, does not mention anything about the electronics hardware.

successes. "A vast majority of our flight failures, not to mention flight delays," Administrator Webb apologetically told a reporter, "arise from electronic failures." Moreover, that the highest percentage of flight failures occurred because of electronic components was a "well advertised" fact, according to Albert J. Kelley, who was head of the Electronics and Control Directorate at NASA Headquarters.[82] Kelley knew firsthand how a diode—a small, inexpensive solid-state part—could ruin a mission because he was head of the 1962 Ranger Board of Inquiry that investigated the failure of Ranger 5. The spacecraft, which was to photograph the lunar surface, instead entered an orbit around the Sun. The Kelley board found over a half-dozen problems with the Ranger program, and a major program shakeup as well as a postponement of the launches of Rangers 6 through 9 resulted. Later, in September 1963, as Ranger 6 began its final round of qualifying tests at the Jet Propulsion Laboratory, a new problem—discovered by the Lewis Research Center— derailed all launch plans. Short circuits had led to two acute test failures in the General Electric guidance components. Loose gold flakes in certain diodes were at fault, and hundreds of the same diodes had already been installed in Rangers 6 and 7.[83]

A new investigation determined that the incidence of gold-flake contamination was so high that most equipment containing it was unsuitable for flight. The flaking originated from poorly bonded excess gold cement at the attachment of a silicon wafer that supported the post inside the diode. The only fix was to replace all the suspect diodes (purchased from Continental Devices) and to postpone the flights of Ranger 6 and Ranger 7 until all the diodes had been replaced. The "famous and troublesome diodes" escaped detection because system and environmental testing could not measure the reverse-current resistance of every diode continuously. If a temporary short happened while forward voltage was applied, no measurable effect resulted. Of course, no tests could simulate the zero gravity of outer space, where the gold flakes would float inside the diodes.[84]

82. C. Wendel, "Blame Electronics for Space Failures," *Electronic News* (25 May 1964): 1; NASA Headquarters, *Space Vehicles, Electronics, and Control Research Program Review, March 9, 1963* (Washington, DC: NASA Office of Programs, 1963), p. 132. Hereafter, *1963 Program Review*.

83. All previous Ranger mishaps were the result of electronics failures: the Agena second stage shut down prematurely on Ranger 1 and Ranger 2; on Ranger 3, the radio command system of the Atlas rocket malfunctioned and the Ranger's digital sequencer failed, as it did on Ranger 4 as well. R. Cargill Hall, *Lunar Impact: A History of Project Ranger* (Washington, DC: NASA SP-4210, 1977), pp. 173–176, 185–186 and 197.

84. Hall, *Lunar Impact*, p. 197; NASA Headquarters, *Electronics & Control Program Review, February 29, 1964* (Washington, DC: NASA Office of Programming, 1964), p. 31. Hereafter, *1964 Program Review*.

NASA Headquarters, and Kelley's Electronics and Control Directorate in particular, found the diode setback of interest for two reasons: it highlighted the need for both future research and better component standards. The Agency required "common electronics performance standards which can be applied by all." By measuring newly developed components against adequate standards, NASA could create a list of qualified parts that the entire Agency would use, instead of qualifying parts for each and every project. Headquarters viewed standardization as an important factor in achieving component reliability, too.[85] Electronics breakdowns were serious setbacks in the country's efforts to compete against the Soviet Union in both the Cold War and the Space Race, just as electronics (especially miniaturization) was looked to as the technological means for bridging the launch-weight capability disparity between the two powers. Coincidentally, a new technology emerged—at the same time as the civilian space Agency—that revolutionized electronics by furnishing a new and more efficacious method for achieving miniaturization. That technology was the integrated circuit. Sharing equally in the microelectronics spotlight were thin films, an older technology that involved depositing material on a ceramic or glass substrate to form resistors and capacitors, and hybrids that combined thin-film and integrated-circuit technologies.[86]

NASA planned to make extensive use of microelectronics technologies, which, Al Kelley noted, were having "a very, very significant impact on all the component technology."[87] A survey of NASA centers conducted during the summer of 1965 indicated that by 1970, an estimated 70 percent of all NASA spacecraft electronics hardware "would be buil[t] in microelectronic form."[88] Already the Apollo Navigation and Guidance computer was using 150,000 microcircuits, and the Agency foresaw its more complex spacecraft eventually using microcircuits "by the hundreds of thousands."[89] Integrated circuits were attractive to NASA for a number of reasons. For one, they incorporated transistors and diodes, unlike thin-film applications. They also were smaller, cheaper, more reliable, and less power-hungry than vacuum tubes. One of the biggest drawbacks of integrated circuits, however, was their basic

85. *1964 Program Review*, p. 31; *1963 Program Review*, pp. 132 and 134.

86. ARINC Research Corporation, *Microelectronic Device Data Handbook*, vol. 1 (Washington, DC: NASA, July 1968), pp. 2-1, 2-2, 2-16, 2-17, 2-55, 2-56, 2-57.

87. *1963 Program Review*, p. 134.

88. NASA Headquarters, *Electronic Systems Program Review, March 22, 1966* (Washington, DC: NASA Office of Program and Special Reports, 1966), p. 49. Hereafter, *1966 Program Review*.

89. NASA Headquarters, *Electronic Systems Program Review, April 27, 1965* (Washington, DC: NASA Office of Program and Special Reports, 1965), p. 5. Hereafter, *1965 Program Review*.

incompatibility with conditions encountered in space, such as the long times that electronic systems had to operate unattended or such physical stresses as radiation (which distorted silicon's crystal structure) and temperature extremes.[90] Silicon integrated circuits, for example, were not qualified to operate at temperatures higher than 150 degrees centigrade. No suitable substitute material for silicon was available, but alternate technologies were promising, including thin-film technology (with transistors added in somehow), integrated circuits made from new materials, and even vacuum "microtubes."[91]

Integrated circuitry was reducing packaged electronics to the size of a "speck" and changing the way NASA was thinking about how it designed things. "We can no longer get off the shelf and put together any components we want," Al Kelley explained. "We have to plan in advance what the integrated circuit will be. In one of the simplest circuits you can think of, with, say, three elements, there are ten billion combinations that you can put together."[92] The solution, again, was standardization. By creating standardized integrated circuits, one could minimize the number of microcircuits that NASA would have to build "so we can get some kind of uses out of them at reasonable costs."[93]

A critical barrier to using integrated circuits was the manufacturing process. Only a small portion of the integrated circuits fabricated on a wafer were suitable for NASA's demanding requirements. To begin with, 25 percent of the devices on a completed wafer were unusable for any purpose. Further handling of the wafer damaged another 25 percent. In fact, handling and assembling electronic devices were major sources of electronics failures in general. Of the remaining integrated circuits, another 25 percent were suitable for low-quality commercial applications, 15 percent were good enough for high-quality industrial or military uses, and only 10 percent were usable in aerospace applications.[94] These percentages indicated that if NASA were to increase its use of integrated circuits dramatically, the Agency would have to learn how to improve integrated-circuit fabrication processes, just as the diode debacle had demonstrated the need for standardization and research

---

90. NASA Headquarters, *Space Vehicles, Electronics, and Control Research Program Review, March 9, 1963* (Washington, DC: NASA Office of Programs, 1963), p. 132. Hereafter, *1963 Program Review.*

91. NASA Headquarters, *Electronic Systems Program Review, February 14, 1967* (Washington, DC: NASA Office of Program and Special Reports, 1967), p. 24. Hereafter, *1967 Program Review.*

92. NASA Headquarters, *Space Vehicles, Electronics, and Control Research Program Review, March 9, 1963* (Washington, DC: NASA Office of Programs, 1963), p. 134.

93. *1963 Program Review*, pp. 134–135.

94. *1963 Program Review*, 31; and *1967 Program Review*, 20.

(especially in new testing methods). Above all else, the Agency needed reliable solid-state electronics.

NASA Electronics Research

NASA did not lack for electronics research personnel or facilities or programs. At Headquarters, electronics research was the concern of the Office of Advanced Research and Technology (OART). That office, however, did not have a separate electronics division until the Agency's 1 November 1961 reorganization. The formation of the office was less a response to President John Kennedy's 25 May 1961 mandate to land an astronaut on the Moon by the end of the decade than a consequence of the arrival of the new NASA Administrator, James Webb.[95] The charter of the new Electronics and Control Directorate, according to its director, Albert Kelley, was, "Get NASA into its proper role in space electronics research." The directorate itself was "a one-man operation to start," according to Kelley, "with essentially the instructions to get NASA into the electronics business." After a year, the staff size grew to 35 members.[96]

Funding for electronics research expanded, too. Between 1963 and 1967, the electronics research and development portion of the OART budget increased from 6.3 percent to 12.5 percent, while the office's overall budget remained at about the same level. The Agency's spending on just microcircuit research in 1963 was at an annual rate of about $1 million,[97] and in 1964, for the first time, NASA's funding of integrated-circuit research surpassed that of the Army and Navy.[98]

These numbers tell only part of the story. Additional electronics funding came out of the budgets of other Headquarters organizations, such as the Office of Space Science and Applications, as well as the various centers. Each center had its own electronics research and development program, in fact. The Marshall Space Flight Center, the Jet Propulsion Laboratory, and the Goddard Space Flight Center (all classified as flight centers) tended to contract out the work, while the research centers, as they were known (the

---

95. Arnold S. Levine, *Managing NASA in the Apollo Era* (Washington, DC: NASA SP-4102, 1982), pp. 34–43; *1963 Program Review*, p. 85; and *1964 Program Review*, p. 1.

96. *1964 Program Review*, p. 41; and Albert J. Kelley, interview by Neil Furst, transcript, 20 May 1968, Boston, 2 and 3, Folder 1168, NHRC. According to Kelley (*1964 Program Review*, p. 1), the Electronics and Control Division formed on 1 November 1961 with three people.

97. *1967 Program Review*, p. 1; and Philip Trupp, "NASA Stress on Semicon Microcircuits," *Electronic News*, 21 January 1963, p. 1.

98. But not that of the Air Force. Kleiman, "Integrated Circuit," p. 201.

Ames Research Center, the Langley Research Center, and the Lewis Research Center), tended to conduct far more internal research. As a result, in 1963, for example, Langley had both the largest professional staff engaged in electronics research and development and the largest research and development budget. Marshall, however, spent as much as Langley on electronics research ($5 million) but had a significantly smaller professional research staff.[99]

These numbers still do not paint the whole picture of NASA's integrated-circuit research, especially at the centers. In 1962, Al Kelley's Electronics and Control Directorate surveyed the strengths and weaknesses of the various centers' electronics research and technology programs. His task group evaluated the programs in a variety of specific research areas, such as astrophysics, biophysics, engineering instrumentation, communications and tracking, advanced computing devices, information theory, advanced electronic devices, piloted and automatic flight control, advanced control theory, advanced component technology, optical and infrared techniques, inertial and electromagnetic guidance, display technology, flight mechanics, and trajectory analysis. For the most part, the survey found that center research projects tended to be "specific rather than basic, more technological than fundamental in nature, and of more immediate application than those that are supported by headquarters."[100] This research orientation reflected the project-oriented interests of the centers' researchers. "A substantial fraction of the capable electronics personnel at the centers are concerned primarily with project management or space flight project engineering" rather than with long-range research. Instead of carrying out long-range internal studies, NASA routinely conducted long-range research by awarding grants and contracts to universities, nonprofit institutions, and industry. Not surprisingly, then, the survey concluded that "a substantial fraction of the capable electronics personnel at the centers are concerned primarily with project management or space flight project engineering."[101]

This propensity to orient management and engineering toward projects and flight missions rather than basic long-range problems was more pronounced at the flight centers than at the research centers. Goddard, for example, was involved largely in advancing spacecraft technology associated with its primary mission: scientific exploration of the area between Earth and

99. *1964 Program Review*, p. 23.

100. NASA, *Electronics Research Center: Need for Space Electronics Research, Organization, and Implementation Plans* (Washington, DC: NASA, January 1964), p. 25; and NASA, *Electronics Research Center* (Washington, DC: NASA, January 1964), p. 25, Folder 4881, NHRC.

101. NASA, *Electronics Research Center*, p. 25.

the Moon. Its electronic research focuses on areas such as spin-stabilization techniques, flight sensors, and antenna control. Similarly, the Jet Propulsion Laboratory worked to develop better spacecraft to explore the Moon, the planets, and interplanetary space, and it was involved heavily in such applied research fields as antennas, video systems, photo-scan techniques, and deep-space communications. The Marshall Space Flight Center, concerned with developing large chemical-propulsion rocket launchers such as the Saturn booster, specialized in guidance, telemetry techniques, lasers, and antennas.[102]

As a result of the importance of spaceflight projects at these centers, the "greatest proportion of NASA's in-house technical capability in electronics and control is vitally enmeshed in flight development and operations," the Kelley study concluded. "Electronics research professionals at these centers, reacting naturally to problems that arise on flight projects, rarely have the time or the opportunity to take the long-range view, identify problems that will hamper future missions, and undertake research designed to lay the foundation for the eventual solution of these problems. In view of the primary missions of the centers, this is appropriate. But it yields a climate different from that required for a research organization."[103] Electronics work at the NASA research centers, according to the survey, was somewhat more basic than that carried on at the flight centers but still related to the center's main mission. Ames worked on control and information systems, gravity-oriented satellites, and visual displays, while the (Dryden) Flight Research Center—because of its heavy commitment to aircraft—supported research projects in optical devices and airborne infrared temperature sensors for piloted flight. The Lewis electronics research program centered on the hardware needs for that center's historical role in propulsion, while Langley devoted much of its research effort to such electronics areas as pilot control of spacecraft, rendezvous, docking, radar transponders as tracking aids, and navigation and control for glider vehicles.[104]

Research into the fundamentals of integrated circuits—or electronic components in general—was simply not part of program-oriented work, but the centers *did* have facilities, personnel, and funding for applied work on integrated circuits. In 1965, for example, Langley, Marshall, and Goddard had silicon integrated-circuit laboratory facilities, but they were engaged primarily in questions of component testing, reliability, and quality control, including studies of thin-film and thick-film technologies. They frequently contracted

---

102. NASA, *Electronics Research Center*, p. 33.
103. NASA, *Electronics Research Center*, pp. 25–26.
104. NASA, *Electronics Research Center*, p. 33.

with industry to perform needed research.[105] Goddard, for example, began with a thin-film laboratory, then added another for making integrated circuits not available commercially. The Center soon developed three different microcircuits that ended up bound for use in spacecraft. Goddard staff, however, left the development of microelectronic processes to contractors, such as the Philco Corporation.[106] Langley, in its film facility,[107] developed a new process for depositing films of transistors one-thousandth of an inch thick on ceramic wafers. Such so-called thick-film transistors previously had to be inserted by hand. Center personnel also developed specialized integrated circuits and other electronics through contracts with Texas Instruments and other firms.[108]

Marshall had one of the more ambitious Apollo-era electronics programs, at least until the Electronics Research Center became operational. One of its two chief goals was the conversion of the Saturn launch vehicle electronics to integrated circuits to reduce component size and weight and to increase their life and reliability. The work took place largely through contracts with industry and academia, although some internal research and development was undertaken. The center's Astrionics Laboratory, Marshall's "center of gravity" for microelectronics research, focused on both thin-film and microcircuit fabrication processes, using internal as well as contractor studies.[109]

---

105. Larry Ramsdell, "NASA Quest: Ultimate in IC Reliability," *Electronic News*, 8 January 1968, p. 5; Research Triangle Institute, *Microelectronics in Space Research* (Washington, DC: NASA SP-5031, August 1965), p. 43. The Research Triangle Institute did not include the Electronics Research Center.

106. John C. Lyons and David R. Dargo, "Goddard Space Flight Center Microelectronics Program," in Electronics Research Center, *Proceedings of Second NASA Microelectronics Symposium*, 1 June 1967 (Washington, DC: NASA-TM-X-55834, 1967), pp. 1–5. Hereafter, ERC, *Proceedings*.

107. The terms "thick film" and "thin film" referred to microelectronic technologies other than integrated circuits. Thick film structures generally contained only conductors, resistors, and capacitors deposited on a ceramic substrate; other components, such as transistors, had to be added separately.

108. Charles Husson, "Langley Research Center Microelectronics Program," in ERC, *Proceedings*, pp. 7–14; NASA, *Fifteenth Semiannual Report to Congress, January 1–June 30, 1966* (Washington, DC: NASA, 1966), p. 88; and Ramsdell, "NASA Quest," p. 5.

109. James C. Taylor, "Microelectronics Program at Marshall Space Flight Center," in Marshall Space Flight Center, Research and Development Operations, *Research Achievements Review Series No. 5, Electronics Research at MSFC* (Huntsville: Marshall Space Flight Center, 1965), pp. 21 and 23–24; *1967 Program Review*, p. 17; Taylor, "Microelectronics Program at Marshall Space Flight Center"; George D. Adams, Salvadore V. Caruso, L. L. Folsom, Robert F. DeHaye, and George L. Filip, "Thin-Film Microcircuit Development at Marshall Space Flight Center," in ERC, *Proceedings*, pp. 41–58. For its part, the Jet Propulsion Laboratory studied advanced techniques for interconnecting electronic circuits. *1967 Program Review*, p. 19.

Microelectronics Reliability

Given the centers' predilection for mission-oriented research, establishing the reliability of integrated circuits and other electronics component would seem to be a logical task for them. In fact, each center had its own reliability office and programs and—independently of Headquarters and other centers—managed the qualification of flight and other components within the organization established for each individual program. Each center and its contractors created their own specifications, their own vendor surveys, and their own circuit qualifications without any coordination across the Agency. As a result, for instance, 750 variations in specification applied to a single type of transistor. Of those, 58 variations applied to use in NASA high-reliability projects. In addition, a vendor might be surveyed multiple times by different centers for the same integrated circuits.[110]

Headquarters, of course, had a reliability office, too. An organization chart dated 17 January 1961 shows the NASA Office of Reliability and Systems Analysis, headed by Landis S. Gephart and reporting to the Office of the Associate Administrator, who was then Robert C. Seamans, Jr. General Management Instruction 4-2-1, "Reliability Policy as Applied to NASA Programs," dated 1 February 1961, instituted policies and procedures for achieving reliable systems and defined reliability as "the probability that a system, subsystem, component or part will perform its required functions under defined conditions at a designated time and for a specified operating period."[111] With the arrival of James Webb, the reliability office became linked administratively to quality assurance on 13 October 1961 as the Office of Reliability and Quality Assurance within the Office of Programs. A new policy statement numerated the responsibilities of the office and charged center directors with establishing "a single organizational point for quality assurance responsibility and authority" at each field installation.[112] The

---

110. *1966 Program Review*, p. 50.

111. Levine, *Managing NASA in the Apollo Era*, p. 32; NASA Management Manual, General Management Instructions, "Reliability Policy as Applied to NASA Programs," 1 February 1961, Folder "NASA SR & QA (Gen.) through 1966," Box 18,153, NHRC; and NASA Management Manual, General Management Instructions, "Functions and Authority—Office of Reliability and Systems Analysis," 16 January 1961, Folder "NASA SR & QA (Gen.) through 1966," Box 18,153, NHRC.

112. Howard M. Weiss, "NASA's Quality Program: Achievements and Forecast," 12 September 1973, 2, Folder "NASA SR & QA (Gen.) through 1966," Box 18,153, NHRC; and NASA Management Manual, General Management Instructions, "Quality Assurance Policy as Applied to NASA Programs," 13 October 1961, Folder "NASA SR & QA (Gen.) through 1966," Box 18,153, NHRC.

Agency also cooperated with the Defense Department in matters of parts and system reliability.[113]

The Columbia Accident Investigation Board has claimed that the reliability office vanished around this time: "Although a NASA Office of Reliability and Quality Assurance existed for a short time during the early 1960s, it was funded by the human space flight program. By 1963, the office disappeared from the Agency's organizational charts. For the next few years, the only type of safety program that existed at NASA was a decentralized 'loose federation' of risk assessment oversight run by each program's contractors and the project offices at each of the three Human Space Flight Centers."[114] A cursory look at agency organizational charts dated 1 November 1963, 2 January 1966, 15 March 1967, and 1 May 1968 appears to support this assertion. Nonetheless, NASA Management Instruction 1136.5, dated 17 August 1965, indicates that the Agency still had a Reliability and Quality Assurance Office, but its director reported to the Deputy Associate Administrator for Industry Affairs.[115] Contrary to the Columbia Accident Investigation Board, the office continued to exist but at a level too low for detection in these upper-level Agency organizational charts.

The Headquarters reliability office played a central role in the Agency's quest for reliable electronics components through an organization called the NASA Parts Steering Committee. Its membership included representatives from each center and the Headquarters offices of space science, human spaceflight, advanced research and technology, tracking and data acquisition, and reliability and quality assurance. The director of the Office of Reliability and Quality Assurance served as chairman, with the head of the Electronics and Control Directorate assisting as vice chairman. The committee's major functions included 1) providing advice and assistance on planning and policy regarding parts program, especially regarding technical and administrative matters; 2) recommending policies and procedures for centralizing parts management; 3) determining the extent of NASA participation in parts activities outside of the Agency; 4) recommending research and development on issues relating to parts and material; 5) advocating ways of exchanging parts information and data; 6) proposing standard terminology

---

113. On this point, see, among others, W. Fred Boone, *NASA Office of Defense Affairs: The First Five Years* (Washington, DC: NASA HHR-32, 1970), p. 265.

114. Columbia Accident Investigation Board, *Report*, vol. 1 (Washington, DC: NASA, August 2003), p. 178.

115. Levine, *Managing NASA in the Apollo Era*, pp. 44, 51, 56 and 58; and NASA Management Instruction (NMI) 1136.5, "Functions and Authority—Reliability and Quality Assurance Office," 17 August 1965, Folder 15,903, NHRC.

and criteria for parts specifications and standards; and 7) reviewing and suggesting changes in proposed or existing NASA-wide documents pertaining to parts.[116]

Virtually little else is known about the committee and its activities; however, we are lucky to have some sketchy documentation regarding an ad hoc subgroup of the committee that took on the formal name of the Microelectronics Subcommittee. The subgroup included representatives from all NASA centers and was chaired by the heads of the Office of Reliability and Quality Assurance and the Electronics and Control Directorate. Its formation reflected the recognition by NASA managers and engineers that microelectronics technology was changing rapidly and that the Agency needed to take steps to accommodate the new technology through an Agency-wide approach.[117]

The Microelectronics Reliability Program aimed to create just such an Agency-wide approach to reliability. It resulted specifically from the efforts of the Electronics Research Center's Qualifications and Standards Laboratory to "define its role within the NASA complex in the area of Q&S [Qualifications and Standards] efforts." The laboratory "deliberately aimed" their proposal at "the microelectronic component field" for several reasons. Among them were the fact that the Agency lacked a systematic approach to parts and components, the anticipated rising use of microelectronic parts over the next five years to between 50 and 75 percent of the dollar amount spent on components (including integrated circuits), and the belief that "a major proportion of the component research and development dollar over the next 10 years will be devoted necessarily to the microelectronic component field."[118]

The Microelectronics Reliability Program had its roots in the first NASA-wide Reliability and Quality Assurance Meeting held at NASA Headquarters on 27 October 1965 at the instigation of James O. Spriggs and Robert F. Garbarini, who were in the Headquarters Office of Space Science and Applications. Participants included representatives from the offices of industrial affairs, space science and applications, human spaceflight, and advanced research and technology, as well as the Electronics Research Center. The

---

116. "Office of Reliability and Quality Assurance Functions for Potential Transfer to ERC," n.d., Folder "Qualifications & Standards Lab," Box 1, Record Group (hereafter RG) 255, Accession Number (hereafter AN) 71A3002, Washington National Record Center, Suitland, MD (hereafter WNRC).

117. *1966 Program Review*, p. 50.

118. Appendix B, "Microelectronics Reliability Program," Memorandum, Robert L. Trent to W. Crawford Dunlap, "Microelectronics Reliability Program," 1 November 1965, Folder "Qualifications & Standards Lab," Box 1, RG 255, AN 71A3002, WNRC.

meeting became a forum for discussing preliminary plans for what would come to be called the NASA Microelectronics Reliability Program.[119]

The essence of the program was to realize reliability by working toward three broad interrelated goals: the establishment of NASA-wide standards and general specifications for microelectronic parts, the coordination of vendor surveys, and the qualification of vendor production lines for general classes of microelectronic circuits. NASA could not attain any of these goals—or make the NASA Microelectronics Reliability Program work—unless it obtained the cooperation of industry. In recognition of that need, the space Agency asked the Council of Defense and Space Industry Associations (CODSIA)[120] to study the Microelectronics Reliability Program and to make comments and suggestions. The CODSIA critique also would provide insight into how best to present the program to both industry and the Defense Department in order "to give it the best chance of being understood and accepted."[121]

The NASA Microelectronics Reliability Program endorsed utilizing three modalities to achieve its goals: symposia, a data bank, and research. Three such symposia took place during the Apollo era in 1964,[122] 1967,[123] and 1968.[124]

119. Memorandum, Robert L. Trent to W. Crawford Dunlap, "Microelectronics Reliability Program," 1 November 1965, Folder "Qualifications & Standards Lab," Box 1, RG 255, AN 71A3002, WNRC; *Reliability and Quality Assurance* (Washington, DC: NASA CR-156261, 31 October 1965); and John E. Condon, "Introductory Remarks," in *Second NASA-Wide Reliability and Quality Assurance Meeting* (Washington, DC: NASA-TM-102990, 1 January 1966), pp. 5–6.

120. CODSIA was formed in 1964 by industry associations having common interests in the defense and space fields. The Department of Defense encouraged the establishment of this organization as a vehicle for obtaining broad industry reactions to new or revised procurement regulations, policies, and procedures. CODSIA serves as a focal point for the Federal Government to submit questions that address policies, regulations, directives, and procedures relating to the supplier-purchaser relationship between government and industry. These areas have primarily included government acquisition, contracting, and management system requirements. Among its member associations was the American Electronics Association and the Electronics Industry Alliance. "CODSIA," *http://www.codsia.org/* (accessed 7 April 2006).

121. *1966 Program Review*, pp. 50–51; Memorandum, Robert L. Trent to NASA Headquarters, "Industrial Comment on Microelectronics Reliability Program," 11 March 1966, Folder "TECH: Electronic Components Research Office," Box 6, RG 255, AN 71A2309, WNRC.

122. No published proceedings appear to have issued from the first symposium. Coincidentally, the Office of Naval Research organized a somewhat similar meeting in Washington, DC, at the Department of the Interior auditorium. Samuel J. Mathis, Jr., Richard E. Wiley, and Lester M. Spandorfer, eds., *Symposium on Microelectronics and Large Systems* (Washington, DC: Spartan Books, 1965).

123. ERC, *Proceedings*.

124. "Microelectronics Meeting Breaks Attendance Record," *ERC News*, 1 March 1968, p. 4, Folder "ERC History House Organ 1966–1969," Box 8, RG 255, AN 71A-2309, WNRC. No proceedings of this third symposium appear to have been printed.

Their stated purpose was the interchange of ideas among NASA, the military, industry, and academia. The organizers hoped that better component specifications and better end products would result from coupling user experience with NASA research results. Papers read at the 1967 Microelectronics Symposium covered a diversity of topics, such as thin-film transistors, metal-oxide semiconductor field-effect transistors (MOSFETs), power integrated circuits in the Saturn, microcircuits in the Apollo TV camera, and assuring integrated-circuit reliability through the application of scanning electron and fast-scan infrared microscopy. The third symposium featured panel discussions on the "Goals of Microelectronics at NASA," "Methods for Obtaining High Reliability Microelectronics," integrated-circuits applications, analyzing integrated circuits, component reliability, film formation and devices, and computer-aided circuit design.[125]

The second group of measures that made up the Microelectronics Reliability Program involved organizing and disseminating data on electronics parts through a series of handbooks and a technical data bank.[126] Over a period of 10 years, the handbook effort yielded more than 60 publications. Typical of these was the two-volume *Microelectronic Device Data Handbook* published in July 1968.[127] Compiled by the ARINC Research Corporation of Annapolis, Maryland, for the Electronics Research Center, and reviewed by the Army Electronics Command, Ft. Monmouth, New Jersey, it provided engineers and circuit designers with guidance for selecting microelectronic devices for space systems, with an emphasis on their reliability. Volume I contained five sections of text on such topics as system-design considerations, testing, the physics of failure, and procurement specifications. Volume II gave the characteristics of about 2,000 devices manufactured by 32 companies from the Alpha Microelectronics Corporation in Beltsville, Maryland, to the Westinghouse Electric Corporation's Molecular Electronics Division in Elkridge, Maryland. The microelectronics covered included a wide variety of

125. *1965 Program Review*, p. 10; ERC, *Proceedings*; "Research Highlights," *ERC News*, 1 March 1968, n.p., Folder "ERC History House Organ 1966–1969," Box 8, RG 255, AN 71A2309, WNRC.

126. *1966 Program Review*, p. 51; *1965 Program Review*, p. 10.

127. Even earlier was the *Microelectronics Device Data Handbook* completed in May 1964 under contract NASw-831 and developed for use by NASA project engineers, design engineers, technicians, and parts specialists as well as by NASA contractors as a ready reference for device information. "Office of Reliability and Quality Assurance Functions for Potential Transfer to ERC," n.d., Folder "Qualifications & Standards Lab," Box 1, RG 255, AN 71A3002, WNRC.

integrated-circuit types, thin-film devices, and even metal-oxide silicon (now called metal-oxide semiconductor) transistors.[128]

Subsequently, in April 1969, NASA issued a four-volume handbook on the reliability of silicon integrated circuits prepared by Texas Instruments for the Marshall Space Flight Center—namely the Quality and Reliability Assurance Laboratory of its Parts and Microelectronics Technology Branch. The key difference between this *Reliability Handbook for Silicon Monolithic Microcircuits* and the 1968 *Microelectronic Device Data Handbook* was that the previous guide had dealt with microelectronics in general, while the Marshall-sponsored effort focused solely on integrated circuits. The first volume,[129] an introduction to integrated-circuit technology, discussed typical problems experienced with certain applications; the most reliable and trouble-free methods of using each type of circuit; how to obtain maximum information from manufacturers' data sheets; the characteristics, advantages, and disadvantages of the various package types available; and methods for interconnecting packages and assembling circuits into components and systems. The second volume[130] focused on failure mechanisms, including defects introduced during the manufacturing process, their causes, and screening procedures. Volume three[131] considered failure analysis exclusively and ranged from methods for evaluating integrated circuits prior to opening their packaging to techniques for opening packaging to evaluating integrated circuits after opening their packaging.[132] The fourth volume[133] was the most theoretical. It discussed various methods for assessing the reliability of integrated circuits and included lengthy sections on statistical and other definitions of reliability.

---

128. ARINC Research Corporation, *Microelectronic Device Data Handbook, Vol. I: Test*, CR-1110 (Washington, DC: NASA, July 1968), p. iii; ARINC Research Corporation, *Microelectronic Device Data Handbook, Vol. II: Manufacturer and Specific Device Information*, CR-1111 (Washington, DC: NASA, July 1968), p. 1-1.

129. W. C. Weger et al., *Reliability Handbook for Silicon Monolithic Microcircuits, Vol. 1: Application of Monolithic Microcircuits*, CR-1346 (Washington, DC: NASA, April 1969).

130. Wilburn O. Shurtleff, *Reliability Handbook for Silicon Monolithic Microcircuits, Vol. 2: Failure Mechanisms of Monolithic Microcircuits*, CR-1347 (Washington, DC: NASA, April 1969).

131. Wilton L. Workman, *Reliability Handbook for Silicon Monolithic Microcircuits, Vol. 3: Failure Analysis of Monolithic Microcircuits*, CR-1348 (Washington, DC: NASA, April 1969).

132. For a discussion of the range of packaging then in use, see J. L. Easterday, D. A. Kaiser, and C. H. Burley, *Industrial Survey of Electronic Packaging* (Redstone Arsenal, AL: Redstone Research and Development Directorate, Army Missile Command, November 1966), p. ii.

133. Jim D. Adams et al., *Reliability Handbook for Silicon Monolithic Microcircuits, Vol. 4: Reliability Assessment of Monolithic Microcircuits*, CR-1349 (Washington, DC: NASA, April 1969).

While these handbooks appeared to duplicate some of the information available through manufacturers' guides to their products, they suffered the temporal limitations of all such publications. Microelectronic components came and went rapidly, sometimes becoming outmoded in two or three years. Therefore, a key element of the NASA Microelectronics Reliability Program was the creation of the Technical Data Bank. The data bank was a pilot program managed by Chauncey W. Watt, Chief of the Components Standards Branch at the Electronics Research Center. Access was granted to NASA and contractor circuit designers to this "comprehensive, up-to-date, and easily accessible file of microelectronic technical and qualification data" that contained a full spectrum of technical specifications necessary for buying and using qualified parts. It also identified those manufacturers who had been qualified to produce specified blocks of microcircuits.[134]

The reliability and other technical information collected on microelectronics components by all NASA centers and their contractors went into the database. When a center or contractor needed a new microcircuit, they could consult the Technical Data Bank to see if an approved circuit already existed. If not, the center would proceed—using qualified vendors and the NASA-wide specification format—to procure the required circuits, then would feed the results into the data bank. The availability of a wide range of microelectronics data, NASA hoped, would avoid duplication in the search for qualified parts and vendors, and the standardization of integrated circuits would save the Agency money.[135]

The data bank, developed by the Information Dynamics Corporation of Reading, Massachusetts, under contract to the Electronics Research Center, contained NASA and manufacturer information on 1,000 specific circuits of key interest to NASA. In addition to the Electronics Research Center, those centers participating in the program included the Goddard Space Flight Center, the Marshall Space Flight Center, the Jet Propulsion Laboratory, and the Ames Research Center. On the Defense Department side, the Air Force Rome Air Development Command also participated. The data bank offered

---

134. *1966 Program Review*, pp. 50–51; "Research Highlights," *ERC News*, 1 September 1967, n.p., Folder "ERC History House Organ 1966–1969," Box 8, RG 255, AN 71A2309, WNRC; Memorandum, Franklyn W. Phillips, Acting Director, ERC, to NASA Headquarters, Office of Advanced Research and Technology, "Information Required Concerning Data Centers," 25 April 1966, Attachment, C. W. Watt, "The NASA Microelectronics Reliability Program," 10 March 1966, Folder "Procedures—Controlled Mail ERC #500," Box 9, RG 255, AN 71A2309, WNRC.

135. *1966 Program Review*, pp. 50–51; "Research Highlights," *ERC News*, 1 September 1967, n.p., Folder "ERC History House Organ 1966–1969," Box 8, RG 255, AN 71A2309, WNRC.

users three levels of information. One level was the published manual that listed the various circuits available and the names of their manufacturers. More detailed information was available on microfilm. The third level was a computer system that helped engineers find information on peculiar characteristics of a given part.[136]

The Electronics Research Center—through its Qualifications and Standards Laboratory—played a critical role in defining and implementing the NASA Microelectronics Reliability Program. The center hoped to participate in the program by acting as the lead center for establishing NASA-wide microelectronics standards and specifications, for centralizing data storage and retrieval programs for electronic parts, and for qualifying programs for flight equipment. In the end, the Qualifications and Standards Laboratory did take on the role of Agency-wide program coordinator.[137]

NASA Electronics Research Center

Electronics Research Task Group

The third prong of the Microelectronics Reliability Program was a striking program of basic and applied research undertaken in the Agency's laboratories, especially those located at the Electronics Research Center (ERC). The center was at the heart of the Agency's efforts to obtain reliable integrated circuits and other solid-state components. Located in Cambridge, Massachusetts, the ERC was within walking distance of MIT. The greater metropolitan area abounded in electronics resources and talent: Harvard, the industries along Route 128, the Air Force's Cambridge Research Laboratory and Electronics Systems Division at Hanscom Field, MIT's Lincoln Laboratory and Instrumentation Laboratory, and the Mitre Corporation. The Electronics Research Center formally opened in September 1964, taking over the administration of contracts, grants, and other Agency business in New England from the antecedent North Eastern Operations Office, which had been created in July 1962.[138] The center—and the research it conducted—began

---

136. "Research Highlights," *ERC News*, 1 September 1967, n.p., Folder "ERC History House Organ 1966–1969," Box 8, RG 255, AN 71A2309, WNRC.

137. Memorandum, Robert L. Trent to W. Crawford Dunlap, "Microelectronics Reliability Program," 1 November 1965, Folder "Qualifications & Standards Lab," Box 1, RG 255, AN 71A3002, WNRC; Memorandum, Robert L. Trent to Distribution, "Presentation of Proposed Microelectronics Reliability Program," 11 January 1966, Folder "Qualifications & Standards Lab," Box 1, RG 255, AN 71A3002, WNRC.

138. "Need for Establishment of an Eastern Operations Office," 31 January 1962, Folder 4884, NHRC; and NASA News Release, "NASA to Establish Northeastern Operations Office," Release No. 62-155, 3 July 1962, Folder 4884, NHRC.

before its formal opening, operating in rented quarters in Cambridge as well as out of Headquarters as the Electronics Research Task Group (ERTG).

The idea of creating the group began with a memorandum of 2 March 1962, in which Dr. Robert C. Seamans Jr., then NASA's Associate Administrator, instructed the Office of Advanced Research and Technology to present "a plan to strengthen NASA's capability in the electronics and guidance and control field to support current and long range programs" for inclusion in the fiscal 1964 budget. In response, Al Kelley's newly formed Electronics and Control Directorate undertook a detailed study of the Agency's electronics research resources and capabilities, studied the Agency's long-range electronics research needs, and recommended a plan to meet those needs. By June 1962, Kelley had drafted a plan. He found no lack of electronics expertise within NASA in the areas of guidance and navigation, control and stabilization, communications and tracking, and instrumentation and data processing. That expertise, however, was "widely diffused throughout the Centers with spotty emphasis, in short a heterogeneous group of bits and pieces."[139] The plan that Kelley developed and refined proposed creating an entirely new NASA center, the Electronics Research Center.

The activity that would evolve into the new center started shortly thereafter when the Electronics Research Task Group came into being within the Office of Advanced Research and Technology on 6 February 1963. The purpose and intent in establishing the group was to conduct the necessary technical facilities and administrative planning in readiness for such time as Congress endorsed the proposed Electronics Research Center.[140] In the words of Al Kelley, "We wanted to have a technical cadre who could then move up to Boston and merge with the Northeast Office and have the nucleus of a center. So it was really to get us off and running, and get some people who could translate the plans into action."[141] The electronics task group, attached to Kelley's directorate and under his leadership, began organizing and instituting electronics research in advance of the center's creation. The group's initial efforts reflected its separation into specialized areas denoted organizationally as the Electromagnetic Division, the Instrumentation and Data Processing

---

139. [Electronics and Control Directorate], "Plan to Implement NASA Electronic Systems Technology Capability," [June 1962], n.p., Folder 4886, NHRC; Albert J. Kelley, interview by Cargill Hall and Richard Dowling, transcript, 28 August 1968, Washington, DC, pp. 3–9 and 15–17, Folder 4877, NHRC.

140. *1964 Program Review*, pp. 2 and 35.

141. Kelley, interview, by Hall and Dowling, p. 36.

Division, and the Components and Technology Division. Within a year of its founding, the group had 20 employees located in Washington, DC.[142]

Until the task group could move into rental quarters in Cambridge and build up the level of its internal research staff, the electronics protocenter would have to rely on contracting for research. "While it is necessary to strengthen considerably the in-house capability for electronics systems R&D," a 1962 internal report read, "it is unrealistic to assume that the lag in in-house personnel staffing in this technical area will be corrected before approximately 1966. Therefore, it is planned that a substantial portion of the electronics and control R&D load will be assumed by industry and university contracts during this buildup period."[143]

One source of work for the electronics task group was the Office of Reliability and Quality Assurance, which considered transferring a number of studies to the protocenter. One such study (contract NASw-919) was a survey of packaging and interconnection problems ($49,000), while another was a follow-on review of a microelectronics standardization program that included specifications for selection, procurement, and qualification criteria and test methods standards. The office also considered transferring oversight of the joint program carried out with the Air Force Rome Air Development Center that conducted long-term life and stress testing of various types of high-reliability, high-usage electronics parts and analyzed failed components. A contract with General Electric's Valley Forge Space Technology Center in King of Prussia, Pennsylvania, implemented the program. Another candidate was the Electronics Components Reliability Center (ECRC), a program conducted by the Battelle Memorial Institute that included participation by NASA, the Air Force Rome Air Development Center, and 12 aerospace contractors. The program had a twofold objective: 1) the collection and dissemination of parts and component reliability and test data among the members and 2) research in such areas as parts reliability methodology and the physics of aging. NASA funded its participation through the Office of Reliability and Quality Assurance (contract NASr-9) at $19,500 annually.[144]

---

142. NASA Headquarters, *Electronics & Control Program Review, February 29, 1964* (Washington, DC: NASA Office of Programming, 1964), pp. 35–36. The ERTG was structured functionally under the Office of a Manager with two basic research groups, Electronics and the other being guidance and control. Also included were the Technical Services group for the facilities planning, and a group for administrative functions. Ibid., 36.

143. [Electronics and Control Directorate], "Plan to Implement NASA Electronic Systems Technology Capability," [ca. June 1962], n.p., Folder 4886, NHRC.

144. "Office of Reliability and Quality Assurance Functions for Potential Transfer to ERC," n.d., Folder "Qualifications & Standards Lab," Box 1, RG 255, AN 71A3002, WNRC.

By August 1964, immediately prior to the official start of Electronics Research Center operations in Cambridge, the electronics task group was overseeing contracts for 1) an investigation of radiation-resistant device phenomena, 2) the physics of failure and reliability for microelectronic and thin-film devices, and 3) research and development on ferrite memory technology. Other research that transferred to the rented Cambridge facilities included 1) a study of the feasibility of high-temperature thin films, 2) a study of glass passivation of integrated circuits, and 3) the development of a thin-film-space-charge-limited triode (a type of transistor more tolerant of space radiation). In addition, the protocenter had 16 pending patent applications submitted by researchers working on various grants and contracts. Most (all but four) resulted from research carried out by either Pratt & Whitney or the Geophysics Corporation of America (GCA).[145]

The ERC Laboratories

Effective 1 September 1964, the administrative functions and staff of NASA's North Eastern Operations Office merged with the technical research personnel of the Electronics Research Task Group to form the Electronics Research Center. Assuring the reliability of NASA's microelectronics was at the heart of the center's mission. Center research took place in 10 different laboratories: space guidance, systems, computers, instrumentation research, space optics, power conditioning and distribution, microwave radiation, electronics components, qualifications and standards, and control and information systems. Researchers in these laboratories worked in such areas as laser and microwave communications, the miniaturization and radiation resistance of electronic components, guidance and control systems, photovoltaic energy conversion, information display devices, instrumentation, and computers and data processing. The computer-related work encompassed the spectrum of software and hardware needs.[146]

---

145. NASA, *Fourteenth Semiannual Report to Congress, July 1–December 31, 1965* (Washington, DC: NASA, 1965), pp. 99–100; Manager, Electronics Research Task Group to Office of Legislative Affairs, "Request for Information on Acceleration of Research Contracts," 24 August 1964, "Proposals/Contracts," Box 11, RG 255, AN 71A3002, WNRC; "Preliminary History of the National Aeronautics and Space Administration during the Administration of President Lyndon B. Johnson: Final Edition," Manuscript, Vol. 1 (Washington, DC: NASA, 1969), 1:V-11, 1:V-34, and 1:V-35, NHRC; "National Aeronautics and Space Administration Electronics Research Center Patent Docket," 25 July 1966, Folder "Reading File—July thru Dec. 1966," Box 11, RG 255, AN 71A2309, WNRC.

146. Associate Administrator for Advanced Research and Technology, "Administrator's Briefing Memorandum," 23 August 1965, Folder 4884, NHRC; Albert J. Kelley, "Staff Report on the Electronics Research Center," draft, no date, Folder 4883, NHRC.

Much of the effort relating to integrated circuits took place in either the Electronics Components or the Qualifications and Standards Laboratories, which together comprised the Electronic Components Research Division (renamed the Components Division after the 1968 restructuring). In general, the Qualifications and Standards Laboratory addressed (among others) problems associated with the processes for fabricating integrated circuits to make them more reliable, while the Electronics Components Laboratory focused on developing and improving electronic components, including those found in integrated circuits. The Component Technology Laboratory had branches devoted to advanced research, materials, devices, physical electronics, and microelectronics. The Qualifications and Standards Laboratory had two major functions. It was the coordinating center ("lead center") for research on qualifications and standards—that is, reliability work—for all of NASA on behalf of the Microelectronics Subcommittee. Its second function was to perform the basic research needed for reliability work. The Qualifications and Standards Laboratory had branches involved in design criteria and component standards, but the laboratory's largest effort was that of the Failure Mechanisms Branch.

An example of the high caliber of the ERC laboratory staff was the head of the Electronic Components Research Division, W. Crawford Dunlap. Along with Robert N. Hall (of the Hall Effect fame), Dunlap invented General Electric's germanium diode and transistor. He spent 11 years with the General Electric Company, first as a member of their Schenectady research laboratories, then as a consultant on semiconductors at their Syracuse laboratory. He also had been director of solid-state electronics research at the Raytheon Corporation and supervisor of solid-state research at the Bendix Research Laboratories. He held 20 patents, including a German patent that covered all semiconductor p-n junction devices produced by the alloying or diffusion technique and that resulted in a multimillion-dollar licensing payment to the General Electric Company.[147]

Douglas M. Warschauer directed the Component Technology Laboratory (later called the Electronic Materials Laboratory). Before joining the ERC, Warschauer was Manager of the Physics Laboratory at the Itek Corporation[148] in Lexington, Massachusetts. Previously, he had been a research scientist at the Aeronautical Research Laboratory, Wright-Patterson AFB, and MIT's

---

147. NASA News Release, "Van Atta, Dunlap Head Divisions at Electronics Laboratory," Release No. 64-317, 21 December 1964, Folder 4884, NHRC.

148. The Itek Corporation is best known as the manufacturer of lenses and cameras for the CORONA and other satellite spying systems. Jonathan E. Lewis, *Spy Capitalism: ITEK and the CIA* (New Haven, CT: Yale University Press, 2002).

Lincoln Laboratory before becoming Principal Research Scientist at the Raytheon Corporation. The author of two books, one of which dealt with semiconductors and transistors, his research interests included high-pressure physics, crystal growth, and the optical and electrical properties of semiconductors and lasers.[149]

Robert L. Trent directed the Qualifications and Standards Laboratory and its three branches concerned with the physics of failure mechanisms, design criteria, and component standards. Before joining NASA, Trent was Vice President and Resident Manager of CTS Microelectronics, Inc., in Ridgefield, Connecticut; Vice President of Research and Development at the National Semiconductor Corporation in Danbury, Connecticut; and Technical Director of the Sperry Semiconductor Division in Norwalk, Connecticut. Earlier, from 1941 to 1957, he worked at the Bell Telephone Laboratories supervising systems and circuit development as well as advanced development in semiconductor devices and circuits.[150]

The PREDICT Facility

The heart of the Qualifications and Standards Laboratory was the PREDICT facility. PREDICT was an acronym for Process Reliability Evaluation and Determination of Integrated Circuit Techniques. More professional and technical staff worked in that facility than in any other part of the Qualifications and Standards Laboratory.[151] The facility was a pilot plant—outfitted with the same production equipment found in industry—that could make complete integrated circuits in any quantity or configuration desired from beginning to end, except for making the silicon crystals themselves. The intention was not to compete with industry, which had neither the time nor the incentive to conduct such research because of a lack of market demand, according to industry representatives who examined and commented on the facility's plans.[152]

The mission of the PREDICT facility was to study failure mechanisms and advanced integrated-circuit fabrication processes in depth as well as the

---

149. NASA ERC, "Biographical Information on Dr. Douglas Warschauer," no date, Box 8, RG 255, AN 71A2309, WNRC; and Douglas Warschauer, *Semiconductors and Transistors* (New York: McGraw-Hill, 1959).

150. NASA ERC, "Biographical Information on Robert L. Trent," no date, Box 8, RG 255, AN 71A2309, WNRC.

151. Based on a count of ERC employees listed in the 1968 and 1969 telephone directories. NASA Electronics Research Center, *Telephone Directory* (Cambridge, MA: NASA Electronics Research Center, January 1968); and NASA Electronics Research Center, *Telephone Directory* (Cambridge, MA: NASA Electronics Research Center, July 1969).

152. *1967 Program Review*, pp. 27–30.

processes for making thin-film and hybrid[153] circuits in order to improve their reliability. Research also would investigate manufacturing methods for achieving high reliability for long life, sterilization capability, radiation resistance, and low-power operation. The basic PREDICT fabrication technology was the planar diffusion process for creating silicon integrated circuits, although the facility also had a complete line of equipment for making thin-film devices. The fabrication processes studied included diffusion, photolithography, etching, oxidation, masking, isolation, vacuum deposition, passivation, scribing, resistor adjustment, interconnection, and packaging. Photolithography took place in the Photo-Resist laboratory, while the Diffusion Room had two furnaces for performing diffusion operations. The function of the Interconnection laboratory was self-evident, while studies of various packaging techniques occurred in the assembly area.[154]

PREDICT personnel also applied the knowledge gained in the pilot plant to the problems encountered by manufacturers. In one case, a small, unnamed NASA subcontractor was having a few problems with process controls. A PREDICT specialist checked out the company's visual inspection techniques. The technician not only improved those inspection techniques but also caught a problem in the scribing operation that prepared wafers for separation into individual integrated circuits. In another instance, a small, unnamed firm was experiencing problems with a complicated processing technique that the company had developed to produce a highly stable device. After a detailed study, PREDICT workers were able to tell the manufacturer the specific processing areas to emphasize in order to obtain the specific electrical characteristics needed. With the aid of an electron microscope, the PREDICT technicians detected cracks in an aluminum lead that had been bonded ultrasonically to an integrated circuit. The minute cracks were significant in causing the device to fail in the long term.[155]

---

153. Hybrid microcircuits involved marrying (or cross-breeding, to continue the analogy) thin film and silicon-based integrated-circuit technologies. One could deposit thin films on a silicon substrate to improve the capabilities of a silicon integrated-circuit device but at the cost of additional fabrication processing steps, increased cost, and (potentially) decreased reliability. Another hybrid technology involved depositing a thin film on a substrate to create passive components (capacitors, resistors) and interconnections, then inserting silicon integrated circuits. This approach appeared to combine the best attributes of both silicon and thin-film technologies. Some of the television cameras developed for the Apollo program utilized Motorola hybrid devices. Research Triangle Institute, *Microelectronics in Space Research* (Washington, DC: NASA SP-5031, August 1965), pp. 13 and 124.

154. *1967 Program Review*, pp. 27–30; Albert J. Kelley to Mac C. Adams, Associate Administrator, OART, 4 November 1965, Folder "DD/Reading File Sept–Dec 1965," Box 4, RG 255, AN 71A2309, WNRC.

155. Ramsdell, "NASA Quest," p. 5.

Research Case Studies

A complete examination of the integrated-circuit research carried out at the Qualifications and Standards Laboratory, the Electronics Components Laboratory, and the PREDICT facility of the NASA Electronics Research Center is beyond the scope of this work, as is an assessment of the parallel efforts carried out at other NASA centers. Instead, this section uses a series of case studies to consider the efforts made to improve processes for fabricating integrated circuits, thin-film devices, and hybrid circuits. The addition of thin-film devices stems from their inclusion along with integrated circuits and so-called hybrid circuits as microelectronic devices. Furthermore, the case studies include some research conducted by other centers, especially the Marshall Space Flight Center, the Goddard Space Flight Center, and the Langley Research Center, as well as research performed by industry and academia through grants and contracts administered by the Electronics Research Center and other centers.

The presentation of the case studies is in four parts. The first deals with understanding the causes of component failure and reliability. The second concerns the creation of components impervious to the harsh conditions of space, especially radiation and temperature extremes. The third part discusses advances in specific fabrication processes, such as ion implantation and separating integrated circuits from the wafer. The final section tackles improvements in testing methods, such as the use of scanning electron microscopes.

Failure and Reliability

NASA's need for small quantities of high-quality electronics subsystems made the statistical approach to testing large numbers of parts uneconomical. As Yasushi Sato has shown in his study of Apollo program reliability, NASA officials in charge of reliability at the highest levels, such as Nicholas E. Golovin and Landis S. Gephart, were engineers with backgrounds in statistics, and they argued for the indirect use of statistical techniques.[156] Lower failure rates, with the knowledge that an inherently failure-free design procedure had not yet been discovered, only compounded the problem by requiring longer testing times. NASA's approach was to gain more knowledge on new components through both research and experience, to achieve high confidence and reliability. Better knowledge of materials, processes and their controls, and the physics of both success and failure was obtainable only through research.[157]

---

156. Yasushi Sato, "Reliability in the Apollo Program: A Balanced Approach Behind the Scenes," *Quest* 13, no. 1 (2006): 22–29.

157. *1965 Program Review*, p. 7.

Rensselaer Polytechnic Institute

Typical of the general investigations of failure and reliability carried out by NASA was that of Professor Sorab K. Ghandi[158] of Rensselaer Polytechnic Institute in Troy, New York, initiated in June 1965. His school received a noncompetitive NASA contract (NAS 12-34) for research on a "Study of Reliability in Microcircuits," sponsored by the Qualifications and Standards Laboratory of the Electronics Research Laboratory.[159] One of the three lines of investigation that he pursued was the replacement of gold in the fabrication of silicon devices. Jean Hoerni at Fairchild Semiconductor had developed the new technique of gold doping to improve the switching speeds of silicon transistors. Going against the conventional wisdom that saw gold as a contaminant reducing transistor gain, Hoerni diffused gold on the back of the silicon wafer. His gold-doped transistors turned off faster, and David Allison and other Fairchild engineers soon applied the technique to new switching transistor products used in computing.[160] By 1965 gold had found extensive use in silicon microelectronics. Usually gold diffusion (Hoerni's doping technique) was the final process in fabricating the devices prior to metallization and packaging. The introduction of gold resulted in a number of problems, however. Gold has an extremely high diffusion constant, about five to six orders of magnitude greater than that of boron and phosphorous, the usual semiconductor dopants. Consequently, gold atoms cannot be considered immobile at temperatures 300°C or higher. Also, because gold is a noble metal, one could not diffuse elemental gold.[161]

The diffusion process proceeded from a gold-silicon alloy, which resulted in damage to the surface of the wafer to a depth of many microns. This was of no consequence when the entire integrated circuit was to be doped with gold. In that case, gold was applied to the side of the wafer that constituted the substrate, and one removed the damaged layer mechanically prior to packaging. In other cases, however, it was necessary to use the actual face of the wafer on which the microcircuit was fabricated. Because the fabrication of integrated circuits used the first few microns of wafer material, the usage of gold was not feasible in these instances. Another problem was the metallurgical

158. Sorab Khushro Ghandi was the author of *The Theory and Practice of Microelectronics* (New York: Wiley, 1968).

159. Electronics Research Center, Monthly Report, Month of June 1965, Folder "Chron File 1965," RG 255, AN 71A-2309, WNRC.

160. Lécuyer, *Making Silicon Valley*, p. 154.

161. S. K. Ghandi, D. N. Arden, F. L. Thiel, E. Henry, and R. Wooley, *Final Report on Study in Optimization of Microcircuit Design* (Troy, NY: Electrical Engineering Department, Rensselaer Polytechnic Institute, 1 July 1966), pp. 2–3.

incompatibility of gold and aluminum, the material most commonly used for integrated-circuit connections (metallization). At elevated temperatures, gold, in the presence of aluminum and silicon, gave rise to the common complaint called "purple plague," a serious mechanism for device failure.[162]

Ghandi and his fellow Rensselaer Polytechnic Institute researchers considered four candidate metals to replace gold in integrated circuits: iron, nickel, cobalt, and copper. They focused on nickel because of the existence of a large body of knowledge concerning some of its properties in silicon as well as Rensselaer's past experience using nickel in microwave devices. Following a literature search on nickel in silicon, the metal seemed to be promising. Their subsequent research included diffusing nickel at different temperatures and attempted to achieve uniform doping. They concluded that nickel, like gold, could be used in high-speed switching devices, but that the behavior of nickel in silicon was considerably more complex than that of gold because of effects at both extremes of the diffusion temperature range.[163]

Librascope

Professor R. E. Back of Northeastern University in Boston carried out a similar reliability study titled "Study of Reliable Solid-State Circuits" through NASA grant (NGR-22-011-007).[164] In addition, the Electronics Research Center awarded a research contract to the Librascope division of General Precision, Inc., for a "Study of Failure and Reliability in Microelectronic Devices" in 1965. The goal of the Librascope study was to acquire a better understanding of basic failure mechanisms and to identify methods for detecting them through testing. Ultimately, NASA hoped, it would lead to methods for

---

162. Ghandi et al., *Final Report on Study in Optimization of Microcircuit Design*, p. 3. A common failure in lead wire connections was called "black plague." This defect was the result of a chemical reaction between gold, aluminum, and silicon. NASA Headquarters, *Electronic Systems Program Review, April 27, 1965* (Washington, DC: NASA Office of Program and Special Reports, 1965), p. 7. The aluminum used for metal connections between components also suffered from "spike-over," whose cause often was excessive voltage or a continuity interruption resulting from a cracked wafer. *1965 Program Review*, p. 8.

163. Ghandi et al., *Final Report on Study in Optimization of Microcircuit Design*, pp. 3, 4, 6–8, and 16; and S. K. Ghandi, K. E. Mortenson, J. N. Park, "Impact Ionization Devices," *IEEE Transactions on Electron Devices* ED-13, no. 6 (1966): 515–519.

164. NASA, *Twelfth Semiannual Report to Congress, July 1–December 31, 1964* (Washington, DC: NASA, 1965), p. 228; memorandum, Director, Electronics Research Center, to NASA Headquarters, Director, Office of Grants and Research Contracts, "Extension of Grant NGR-22-011-007," 27 October 1965, Folder "Qualifications & Standards Lab," Box 1, RG 255, AN 71A3002, WNRC.

predicting failure rates through an understanding of failure mechanisms and to tests for screening out potentially weak devices.[165]

The effort focused on aluminum, including the bonding of gold and aluminum. Librascope performed lifetime and heat stress testing of a specific type of commercial integrated circuit (known as a NAND/NOR gate) made by Fairchild, Motorola, Signetics, and ITT that included both diode transistor logic (DTL) and complementary transistor logic (CTL) circuits. The test consisted of temperature cycling from –55°C to +125°C, temperature-step stress testing from +200°C nonoperating until failure of units, and operating life tests at +125°C.[166]

After thousands of hours of tests on hundreds of integrated circuits from the five manufacturers, Librascope technicians concluded that "the bulk of the failures in microelectronic devices at the present time are process failures and not random failures." Process failures, in fact, had beset the research effort. During the first year of the study, for example, Librascope had to eliminate Motorola from further consideration because of early incidences of purple plague. They rejected ITT completely during the study's second year because of a fundamental fabrication problem, while a different Motorola microcircuit exhibited a high incidence of failures in a given region of the device studied.[167]

Librascope found three fabrication processes to be under inadequate control. These were the minimization of the rate of growth of purple plague, the minimization of the rate and extent of interaction between thin-film aluminum and the underlying dielectric, and possibly incomplete removal of photoresist in wafer regions that had "windows" in the dielectric for thin-film electrode connections. The researchers proposed several possible screening tests to isolate select problems, including visual inspection prior to sealing and vibration and x-ray testing.[168]

---

165. Electronics Research Center, Monthly Report, Month of June 1965, Folder "Chron File 1965," RG 255, AN 71A-2309, WNRC; and Librascope Group, *Study of Failure and Reliability in Microelectronic Devices: First Quarterly Report* (Glendale, CA: Librascope, 30 November 1965), 3.

166. Librascope Group, *Study of Failure and Reliability*, 1965, pp. 4, 24, 25, 43, and "Appendix: Test Procedure," p. 1; Librascope Group, *Study of Failure and Reliability in Microelectronic Devices: Third Quarterly Report* (Glendale, CA: Librascope, May 1966), pp. 1–2.

167. Librascope Group, *Study of Failure and Reliability*, May 1966, p. 23.

168. Librascope Group, *Study of Failure and Reliability*, May 1966, pp. 23–24; Librascope Group, *Study of Failure and Reliability in Microelectronic Devices: Fifth Quarterly Report* (Glendale, CA: Librascope, November 1966).

Tritium Tracer

One of the causes of failures in integrated circuits was believed to have been hydrogen entrapped between layers of silicon dioxide, but no experimental evidence existed to support that belief. In order to investigate the presence of hydrogen as a cause of failure, the Electronics Research Center turned to the Autonetics Division of North American Aviation, with which it already had a contract to study improved process techniques for making silicon integrated circuits. The center had the company develop a tritium[169] tracer and obtain quantitative information on the entrapment of hydrogen in grown silicon dioxide layers.[170]

The Autonetics team, under J. E. Meinhard, began by studying 30 transistors of the type used on the Minuteman II missile, half of which were processed with deuterium, a stable isotope of hydrogen. Prior to bringing the tritium tracer to bear on their work, the team performed gas analyses of the transistors that indicated the presence of mainly nitrogen, not hydrogen, but they admitted that hydrogen may have leaked or diffused into the sealed packages during baking. The presence of nitrogen was normal.[171] Upon investigation with the tritium tracer, Meinhard's group discovered that significant amounts of hydrogen had been retained in the silicon oxide matrix when steam had been present in the gas used to oxidize silicon. They carried out a subsequent examination of 75 silicon wafers at the company's General Atomic facilities. In addition, they selected 30 general-purpose amplifier microcircuits (10 each made by Texas Instruments, Westinghouse, and Norden) to investigate the effects of ambient hydrogen on transistor gain. The Norden circuits showed nitrogen as the major constituent, and the same gas was the major constituent in the Texas Instruments devices. Nonetheless, five contained small amounts

---

169. Tritium is a radioactive isotope of hydrogen.

170. Contract NAS 12-4 was with the Autonetics Division of North American Aviation, Inc., for $69,000 to study "process techniques to study integrated circuits." "ERC R&D Accomplishments," no date, "Headquarters—OART," Box 6, RG 255, AN 71A3002, WNRC; "Research Mission Prompts New Thinking on Procurement Methods," *Missiles and Rockets* 16, no. 22 (31 May 1965): 30; J. E. Meinhard, *Process Techniques Study of Integrated Circuits, Quarterly Report No. 7* (Anaheim, CA: Autonetics, February 1967); and "Component Technology Has 'Core' Role," *Missiles and Rockets* 16, no. 22 (31 May 1965): 37.

171. Meinhard, *Process Techniques Study of Integrated Circuits, Quarterly Report No. 3* (Anaheim, CA: Autonetics, 15 February 1966), pp. 7–8; and Meinhard, *Process Techniques Study of Integrated Circuits, Quarterly Report No. 5* (Anaheim, CA: Autonetics, 15 May 1966), pp. 8–9.

of carbon dioxide, and four contained helium, which probably stemmed from their being pressurized in helium prior to leak testing.[172]

In contrast, the results obtained from testing the Westinghouse devices were "erratic." Three showed hydrogen contents above 85 percent with a smaller percentage of nitrogen, while two devices contained no detectable hydrogen but noteworthy amounts of nitrogen and oxygen. The explanation appeared to lie in the package-sealing process. The sealing of the integrated circuits took place in a continuous-belt furnace that had a nitrogen blanket on both ends and a small region of hydrogen at the center of the furnace. As a result, the gas composition of the packaged device depended on where in the furnace the package became sealed. This study—and the work conducted in parallel internally at the Electronics Research Center—provided the first direct evidence of hydrogen contamination, a potential failure mechanism in microelectronic devices.[173]

Components

Research carried out at the ERC and other centers on certain components was relevant to the advancement of integrated-circuit reliability insofar as the improved components would be integrated into those microcircuits and would be more resistant to such adverse space conditions as radiation and high temperatures. After development of a transistor or diode that met these rigorous conditions, the task of manufacturing them as part of an integrated circuit would follow. One example of this class of research was the development of a radiation-resistant transistor using thin-film techniques.

Silicon solid-state devices were less tolerant of radiation than conventional, nonsilicon resistors and capacitors, and unquestionably less tolerant than thin-film devices, because they contained no silicon. Silicon semiconductors were susceptible to radiation damage because that material consists of a single crystalline structure, and dislocations or disturbances in its crystal lattice would cause it to deteriorate. Proton radiation and other energetic particles such as that found in space could cause distortions in the silicon crystal lattice, primarily through the displacement of atoms from lattice points into interstitial regions, and cause a degradation in performance because of changes induced in the material's physical properties.[174]

---

172. Meinhard, *Quarterly Report No. 5*, p. 13; Meinhard, *Quarterly Report No. 7*, pp. 3, 7 and 10.
173. Meinhard, *Quarterly Report No. 7*, p. 12; and "ERC R&D Accomplishments."
174. *1964 Program Review*, pp. 15–16; and Battelle Memorial Institute, Columbus Laboratories, *A Study of the Effect of Space Radiation on Silicon Integrated Circuits: Final Report*, vol. 1 (Columbus, OH: Battelle Memorial Institute, 1965), p. 39.

Battelle Memorial Institute

In order to understand the impact of radiation on specific commercially available solid-state devices, NASA's Goddard Space Flight Center paid the Columbus, Ohio, laboratories of the Battelle Memorial Institute to investigate the effect of space radiation on silicon integrated circuits in 1964 and 1965. The study looked at 16 types of integrated circuits representing various prevalent logic configurations,[175] functions, and fabrication techniques (such as planar epitaxial or triple-diffusion planar processes) made by five different companies. Battelle exposed them to a 3-Mev electron environment under different electrical conditions until they failed. The researchers came to several conclusions. Among those was that no one class of microcircuits was inherently superior to another, but those using faster transistors (usually epitaxial transistors) were more resistant to failure.[176] In a subsequent leg of the study conducted in 1965 and 1966, the Battelle technicians examined four other types of integrated circuits, including digital MOS (metal-oxide semiconductor) circuits, manufactured by nine companies,[177] by exposing them to electron radiation until failure. Again, they exposed all devices to 3 Mev of electron energy, except the MOS circuitry, which they exposed to 1.5 Mev electrons. Again, they reached several conclusions about circuit types, transistor gain, and the failure mechanisms. They also concluded that currently available MOS microcircuits resisted lower radiation exposures better than the other three genres of circuits studied.[178]

The Mead Triode

Another critical component research area was in the development of thin-film devices capable of resisting space radiation. The earliest of these was a so-called Mead triode, named after the microelectronics pioneer Professor Carver A. Mead of the California Institute of Technology (later Gordon and Betty Moore professor emeritus), who first proposed making an all-evaporated thin-film triode in a particular manner in 1961. Mead's triode was a new

---

175. The logic configurations were Resistor-Capacitor-Transistor Logic (RCTL), Resistor-Transistor Logic (RTL), Diode-Transistor Logic (DTL), Emitter-Coupled Logic (ECL), and Transistor-Transistor Logic (T2L). Battelle Memorial Institute, *Study of the Effect of Space Radiation*, pp. iv, 5, and 7.

176. Battelle Memorial Institute, *Study of the Effect of Space Radiation*, pp. ii, 1, 2, and 162.

177. The manufacturers were Amelco, Fairchild, General Instrument, Motorola, Philco, Radiation Incorporated, Signetics, and Westinghouse. The MOS devices were the General Instrument 7531 and 7532 and the Fairchild µM400. Battelle Memorial Institute, *Study of the Effect of Space Radiation*, p. 6.

178. Battelle Memorial Institute, *Study of the Effect of Space Radiation*, pp. ii and iii.

type of device that used the principle of tunnel emission, a concept from quantum mechanics that was just then being explored in solid-state electronics following the discovery of the tunneling effect in 1958 by Leo Esaki, a Japanese physicist then working for the Sony Corporation. For his discovery Esaki received the Nobel Prize in physics in 1973.[179] Tunneling phenomena provided a path for electrons, and the voltage applied to the Mead triode determined the number of electrons in motion and that, in turn, allowed one to turn the triode into an amplifying device.[180]

Mead showed that one could create a controlled electron source by using a metal-insulator-metal diode structure in which the second metal layer was very thin. By adding an additional insulator and a metal collector layer, one also could devise a triode. Mead built and studied both diode and triode devices made from several kinds of materials but found that he obtained the best triodes when making them out of aluminum oxide (Al_2O_3) insulating films as well as tantalum pentoxide (Ta_2O_5).[181] In 1962, with funding from the Air Force and NASA, Gulu T. Advani and two other researchers at MIT's Electronic Systems Laboratory attempted to construct a Mead triode out of aluminum and aluminum oxide, but utilizing thin-film fabrication techniques, and succeeded.[182]

University of Virginia

Beginning in 1962, Professor Robert L. Ramey and a group of graduate students at the Research Laboratories for the Engineering Sciences at the University of Virginia, in Charlottesville, performed a more intensive study of these thin-film devices under a grant from the Langley Research Center. The initial idea was to develop thin-film devices for use in various instruments, but the state of the art required substantial theoretical investigation to develop the basic knowledge of the physics of thin films to create practical devices. During the first year of the study, from December 1962 to December 1963, Ramey focused on developing the accuracy of the equipment used to deposit the films and evaluating and calibrating the equipment.[183]

---

179. Berlin, *Man Behind the Microchip*, p. 66.

180. *1964 Program Review*, p. 16.

181. Carver A. Mead, "Operation of Tunnel Emission Devices," *Journal of Applied Physics* 32 (1961): 646–652.

182. Gulu T. Advani, James G. Gottling, and Martin S. Osman, *Thin Film Triode Research* (Cambridge: MIT Electronic Systems Laboratory, March 1962).

183. Robert L. Ramey, *A Study of Thin Film Vacuum Deposited Junctions, Annual Status Report* (Charlottesville: University of Virginia, January 1964), pp. 1, 2, and 27 (hereafter Ramey, January 1964); Ramey, *A Study of Thin Film Vacuum Deposited Junctions, Semiannual Status Report* (Charlottesville: University of Virginia, June 1967), p. 2 (hereafter Ramey, June 1967).

The next phase called for research into three basic areas of thin-film phenomena—conductivity, Hall effect, and photoelectric effects—utilizing both single films and rectifying junctions between dissimilar films. In 1965, during the study's third year, the laboratory entered various theoretical and analytical areas of research on thin films as well as some developmental work that eventually led to the issuance of two U.S. patents.[184] Having achieved control over the physical and electronic properties of the deposited films, Ramey devised a method for depositing germanium films with controlled-hole mobility that led to the first NASA patent to emerge from the research.[185] Their research into the design and fabrication of thin-film electronic devices had placed them in a position to capitalize on the abilities they developed through the Langley grant. The laboratory, moreover, had developed a "boat" for the evaporation of insulating films (such as those made from silicon monoxide, SiO).[186]

Hughes Aircraft Research

While the University of Virginia study contributed to knowledge of the physics of thin-film devices, including space charge effects,[187] NASA needed practical applications of that knowledge to the solution of space electronics problems. The Electronics Research Center therefore decided to develop a thin-film space-charge-limited triode, specifically for NASA spacecraft with orbits in the Van Allen radiation belt, based on the theoretical concepts of G. T. Wright, published in 1963.[188] The development effort took place at both

---

184. Robert Ramey, Hugh S. Landes, and Eugene A. Manus, "Active Microwave Irises and Windows," U.S. Patent #3,649,935, filed 18 August 1970, issued 14 March 1972, assigned to NASA; and Ramey, Landes, and Manus, "Thin Film Microwave Iris," U.S. Patent #3,676,809, filed 28 May 1970, issued 11 July 1972, assigned to NASA.

185. Robert Ramey and William D. McLennan, "Depositing Semiconductor Films Utilizing a Thermal Gradient," U.S. Patent #3,420,704, filed 19 August 1966, issued 7 January 1969, assigned to NASA.

186. Ramey, January 1964, p. 27; and Robert Ramey, *A Study of Thin Film Vacuum Deposited Junctions, Semiannual Status Report* (Charlottesville: University of Virginia, July 1965), pp. 2–3 (hereafter Ramey, July 1965).

187. Ramey, July 1965, p. 18; and Robert Ramey, *A Study of Thin Film Vacuum Deposited Junctions, Annual Status Report* (Charlottesville: University of Virginia, December 1965), p. 2.

188. G. T. Wright, "Space-Charge-Limited Solid-State Devices," *Proceedings of the IEEE* 30 (November 1963): 1642–1652; Hughes Aircraft Company, Solid State Research Center, *Development of a Thin-Film Space-Charge-Limited Triode, Final Report* (Newport Beach, CA: Hughes Aircraft Company, June 1966), p. 7. On the development of the device, see also Rainer Zuleeg and Peter Knoll, "A Thin-Film Space-Charge-Limited Triode," *Proceedings of the IEEE* 54, no. 9 (September 1966): 1197–1198; and Knoll and Zuleeg, "A Thin-Film Space-Charge-Limited Triode," *Proceedings of the IEEE* 55, no. 2 (February 1967): 249.

the ERC and the Microelectronics Division of the Hughes Aircraft Company in Newport Beach, California. Hughes started work in March 1965 and produced results by June 1966 under the leadership of Rainer Zuleeg.[189]

Space charge is the electrical current that results when a metal object is heated to incandescence in a vacuum. When a metal object is placed in a vacuum and is heated to incandescence, the energy is sufficient to cause electrons to "boil" away from the surface atoms and surround the metal object in a cloud of free electrons. Because the resulting electron cloud has a negative charge, any nearby positively charged object will attract it, thus producing an electrical current that passes through the vacuum. Space charges also can occur within a solid, liquid, or gas.

The Hughes study considered both experimental and theoretical aspects of thin-film space-charge-limited triodes, including their development, their limitations, and the feasibility of manufacturing them via photolithography and related processes. Zuleeg's team created several versions of the thin-film space-charge-limited triode. In fact, prior to winning the NASA contract, the team already had fabricated one consisting of silicon layered on sapphire. The production of the device for NASA, however, utilized more sundry materials and fabrication processes. The original device had a mesa structure, not unlike early transistors, but in the course of development Hughes replaced it with a planar structure.[190]

Learning how to make batches of this special device was a rocky road. The batches of devices processed under the NASA contract initially yielded very few operable devices as a result of manufacturing problems in the etching and mask design steps. Subsequent lots underwent a number of fabrication variations in order to determine approaches to solving these problems. As a result, the last five batches yielded higher portions of usable devices. The Hughes researchers concluded that the availability of better silicon films in the near future would result in better fabrication results.[191]

The results of a range of radiation tests were promising. In addition, it appeared that the space-charge-limited triode had considerable potential as a microwave amplifier. Zuleeg's team believed that, among other advantages,

---

189. "Preliminary History of the National Aeronautics and Space Administration during the Administration of President Lyndon B. Johnson: Final Edition," Manuscript, Vol. 1 (Washington, DC: NASA, 1969), V-34; *1967 Program Review*, p. 18; and NASA, *Fourteenth Semiannual Report to Congress*, p. 99.

190. Hughes Aircraft Company, *Development of a Thin-Film Space-Charge-Limited Triode*, pp. 4 and 7.

191. Kenneth G. Aubuchon, Peter Knoll, and Rainer Zuleeg, *Research for Development of Thin-Film Space-Charge-Limited Triode Devices* (Newport Beach, CA: Hughes Aircraft Company, May 1967), pp. 27–29, 31, and 65.

the devices appeared to be capable of straightforward integration into high-speed and high-frequency (3 to 4 GHz) circuits. The thin-film space-charge-limited triode offered a specific advantage over regular integrated circuits built on silicon wafers. The disadvantage of the wafer arose from the fact that all the electronic parts in the integrated circuit were coupled electrically via the conductive substrate, and this coupling limited the operation of the device at high frequencies. From NASA's point of view, the thin-film triode had several advantages. Because it was a thin-film device, it was compatible with other thin-film technologies, thereby simplifying their usage over silicon wafer microcircuits. Additionally, the space-charge-limited triode offered high-speed, radiation-resistance, and high-temperature capabilities.[192]

Texas Instruments

That the space-charge-limited triode could operate at high frequencies (3 to 4 GHz) was encouraging, because NASA wanted to take advantage of the many benefits gained by operating at higher frequencies, especially in the microwave range. It was in this vein that the Electronics Research Center's Microwave Radiation Laboratory commissioned Texas Instruments to develop a microwave integrated circuit as part of its millimeter and submillimeter circuits and component program. The first step (R&D 65-45, "Solid State Integrated Microwave Circuits") was a study to define the specific problem areas associated with integrated circuits at microwave frequencies. The goal was to help to solve the disadvantages of integrated circuits at microwave frequencies, such as circuitry restrictions, tight tolerances, element isolation, and low reactance Q's.[193]

Throughout the first phase of the study program, Texas Instruments made extensive use of the results of company-sponsored research and, especially, the work done by the firm on the Molecular Electronics for Radar Applications (MERA) program under contract with the Air Force Systems Command

192. Aubuchon, Knoll, and Zuleeg, *Research for Development of Thin-Film Space-Charge-Limited Triode Devices*, pp. 1, 57, 62, and 65; Hughes Aircraft Company, *Development of a Thin-Film Space-Charge-Limited Triode*, p. 47; and *1967 Program Review*, p. 25.

193. NASA Research and Technology Coordinated Documentation Form, 26 October 1964, Folder "FY1965 Program Actions," Box 9, RG 255, AN 71A3002, WNRC; G. M. Trafford to W. E. Kock, "Procurement Request for a Research Study on Solid State Integrated Microwave Circuits," 19 March 1965, Folder "Official File (R&D) Calendar Year 1964," Box 9, RG 255, AN 71A3002, WNRC; ERC, Task Approval Document, 19 February 1965, Folder "Official File (R&D) Calendar Year 1964," Box 9, RG 255, AN 71A3002, WNRC.

at Wright-Patterson Air Force Base.[194] Subsequently, under Contract NAS 12-75, Texas Instruments conducted analytical studies and completed preliminary designs of a solid-state silicon, integrated-circuit telemetry transmitter capable of operating in the S band (2 to 4 GHz)—which was used for air-traffic control and long-range weather applications—as well as in the meteorological telemetry band of 1700 to 1710 MHz (L band). The design goal was to create an FM transmitter with an output of 2 watts at 15 percent efficiency. The estimated size of the completed device was 2 inches × 2 inches × 3 inches and its weight about 2 pounds. Hardware was available as well as a report. Laboratory devices currently under evaluation were capable of 1 watt at 2 GHz.[195] The Texas Instruments team—actually Albert E. Mason, Jr., project engineer; and Louis I. Farber, engineer—initially studied solid-state microwave devices, techniques, and components capable of functioning in the frequency range of 1 to 6 GHz. They looked at a number of devices, including transistors, thin-film devices, and so-called Schottky barrier diodes. They eventually would use this information to design a simple hypothetical microwave FM telemetry transmitter.[196]

The leap to frequencies above 1 gigahertz was considerable, as integrated circuits in 1964 and 1965 could achieve a bandwidth of 100 MHz, a considerably small fraction of 1 GHz. Laboratory transistors had been built capable of continuous-wave power outputs of about 1 watt at frequencies above 2 GHz. Part of the problem was that the small size of integrated circuits gave rise to inductance and parasitic capacitance. For several years prior to the NASA study, Texas Instruments had been working toward extending the frequency response and power-handling capability of transistors and had succeeded in developing a new generation of UHF silicon transistors—namely, the TI3016A silicon planar transistor. Under laboratory conditions Texas Instruments was able to create devices that developed an output of 30 mW at 4.5 GHz with a pulsed input of 10 mW at 2.25 GHz. They also experimented with power amplifiers under the military's MERA program, and they built an entire single-stage 500 mW preamplifier as a single integrated circuit through selective epitaxial deposition.[197]

---

194. Texas Instruments Incorporated, *Study of Solid-State Integrated Microwave Circuits* (Washington, Government Printing Office, 31 December 1965), p. iv.

195. The contract actually called for a transmitter with wider specifications: 1 to 5 watts of output power at 1 to 2 GHz. "ERC R&D Accomplishments"; and Texas Instruments Incorporated, *Study of Solid-State Integrated Microwave Circuits*, p. 3.

196. Texas Instruments Incorporated, *Study of Solid-State Integrated Microwave Circuits*, pp. iii, iv, and 145.

197. Texas Instruments Incorporated, *Study of Solid-State Integrated Microwave Circuits*, pp. 1, 21, 24, 27, and 32.

The Texas Instruments engineers produced a microwave integrated circuit, but their conclusion at the end of the NASA study was that the knowledge gained about the devices, techniques, and components studied was fleeting because of the rapid progress being made in microwave integrated circuits. Before finalizing the design of the transmitter, they argued, it would be necessary to update the study by examining the latest work in microwave integrated circuits.[198] Texas Instruments was so certain that the NASA work would lead to useful new microwave integrated circuits that it asked NASA upfront for a waiver of the Agency's patent rights in 1965 in advance of signing a contract. The company claimed to have made a substantial investment of its own funds as well as those of the MERA program. However, NASA pointed out that the internal funds had underwritten work on devices that did not operate at microwave frequencies, and the MERA dollars only confirmed NASA's position that the work in microwave integrated circuits had been done with government money, which precluded the issuance of a waiver. In fact, the Agency concluded that communications in the microwave range (1 GHz to 6 GHz) had no "general commercial application" and that the government had been the principal developer of these devices.[199]

Improving the IC Fabrication Process

The ERC's Electronics Component and Qualifications and Standards laboratories conducted substantial research internally as well as through grants and contacts with industry and academia aimed at improving the processes for fabricating integrated circuits. The centers equally were involved in the effort, but the Electronics Research Center distinguished itself as the leader in basic research intended to increase fundamental understanding of the physics and chemistry of those processes. The goal, as always, was to fabricate better, more reliable integrated circuits.

Ion Implantation

One of the most promising techniques for improving the fabrication of integrated circuits was ion implantation.[200] Pioneered at Bell Telephone

---

198. Ibid., p. 147.
199. Memorandum, R. E. Walsh, Contract Negotiator, to J. T. Dennison, Director, Technology Utilization, "Request by Texas Instruments, Incorporated, for Waiver of Title under New Technology Clause—Proposed Contract ERC-R&D 65-45," 11 August 1965, "(T) Chronological File—Mr. Dennison (1965)," Box 15, RG 255, AN 71A-2309, WNRC.
200. For a short history of ion implantation, see Richard Fair, "History of Some Early Developments in Ion-Implanted Technology Leading to Silicon Transistor Manufacturing," *Proceedings of the IEEE* 86 (1998), pp. 111–137.

Laboratories in the late 1940s and early 1950s for the manufacture of transistors, ion implantation was one of the oldest methods for introducing the impurities—called dopants—that created the so-called n and p areas in silicon or germanium to form transistors or diodes.[201] As a result of this pioneering work with transistors and diodes, ion implantation in silicon and germanium was a given. But in light of the vulnerability of silicon to space radiation, NASA hoped to utilize other materials in integrated circuits. Its research therefore focused on the possibility of using ion implantation with alternative materials.

At the same time, industry—with financial assistance from the armed forces (the Naval Avionics Facility at Indianapolis, for instance)—was investigating silicon-based ion implantation techniques. In 1964, CBS Laboratories was looking at the feasibility of creating active semiconductor p-n junction and tunnel devices via electron-beam technology. They already had built backward and tunnel diodes. The CBS researchers successfully doped silicon substrates with indium using an ion gun controlled by electronic raster circuits similar to those that manipulate the electron beam in television picture tubes. Because the researchers could control the ion gun quite finely, industry hoped that the technique could lead to the eventual elimination of the banks of costly diffusion furnaces that were part and parcel of the integrated-circuit manufacturing process and might solve some of the problems encountered in forming transistors and diodes on thin-film substrates. Indeed, the ultimate goal of the CBS research was a thin-film technique that would have had all the advantages of planar silicon technology.[202]

The ion implantation research overseen by the Electronics Research Center appears to have resulted from an unsolicited proposal submitted in 1965 by the Sprague Electric Company. That company, through the National Research Corporation acting as its subcontractor, conducted a 12-month study of techniques that used a focused ion beam, instead of the conventional masking and diffusion techniques, to create reliable, radiation-resistant integrated circuits. Initially they produced a wafer of n-p-n transistors doped via ion implantation. Heat treatment for radiation damage followed in the same vacuum machinery as the ion implantation. Eventually, they extended

201. J. Stephen, "Ion Implantation in Semiconductor Device Technology," *Radio and Electronic Engineer* 42, no. 6 (June 1972): 265–283; Dummer, *Electronic Inventions*, p. 106; and William Shockley, "Forming Semiconductive Devices by Ionic Bombardment," U.S. Patent #2,787,564, filed 28 October 1954, issued 7 April 1957.

202. Frank Leary, "Microelectronics," *Space/Aeronautics* (February 1964): 71. The Westinghouse Central Research Laboratory also was using a rastered electron beam, but to polymerize photoresist material preparatory to diffusing p-n junctions in silicon, in order to ease mask requirements and reduce costs. Ibid.

the wafer pattern to add resistors and capacitors to complete the integrated circuit design.[203]

A far more ambitious study soon ensued. Between 1967 and 1970 the Hughes Research Laboratories in Malibu, California, examined ion implantation on behalf of the Electronics Research Center. During the first phase of the study, completed in October 1967, the laboratory demonstrated that one could form p-n junctions in gallium arsenide (GaAs) and silicon carbide (SiC) using ion implantation. The next step was an evaluation of the devices resulting from implanted zinc, cadmium, tin, and sulfur in gallium arsenide and antimony and bismuth in silicon carbide.[204] The Hughes researchers presented their results at two professional meetings held in 1968.[205]

As the research of the Hughes team progressed into 1969, they achieved a first: doping phosphorus and antimony in silicon carbide. The potential advantage of ion implantation over other fabrication processes, the Hughes team reported, was demonstrated best in the formation of silicon carbide p-n junctions. Industry normally grew the hexagonal form of silicon carbide at temperatures of 2,500°C or above and normally carried out the diffusions for forming junctions at 2,000°C to 2,500°C. Typically one had to maintain these high temperatures for several hours. In contrast, the Hughes investigators succeeded in producing p-n junctions in silicon carbide by ion-implanting

---

203. Electronics Research Center, Task Approval Document, Research and Development, 17 June 1965, "FY1965 Program Actions," Box 9, RG 255, AN 71A3002, WNRC; Memorandum, Director, Electronics Research Center, to Associate Administrator for Advanced Research and Technology, "Request for Approval of FY 1965 R&D Task," and attachment, "Research and Technology resume," "Official File (R&D) Calendar Year 1965," Box 9, RG 255, AN 71A3002, WNRC. Contract R&D 66-60, "Ion Implantation Technique Study," $55,000, "Procurement Request Status Report," attached to Memorandum, Chief of Procurement to Director, "Monthly Report—November 1965," 3 December 1965, "Code AA Activity reports—1965," Box 15, RG 255, AN 71A-2309, WNRC.

204. R. G. Hunsperger, H. L. Dunlap, and O. J. Marsh, *Development of Ion Implantation Techniques for Microelectronics* (Washington, DC: NASA, October 1968), pp. 1–2.

205. The meetings were those of the Electrochemical Society Meeting, held in Boston, 7 May 1968, and the IEEE Solid State Device Research Conference, Boulder, 17 June 1968. Hunsperger, Dunlap, and Marsh, *Development of Ion Implantation Techniques for Microelectronics*, p. 3. An abstract of the Boulder paper appeared as R. G Hunsperger, Marsh, and Carter A. Mead, "The Presence of Deep Levels in Ion Implanted p-n Junctions in GaAs and Their Effect on the Electrical Characteristics," *IEEE Transactions on Electron Devices* 15, no. 9 (September 1968): 687–687. A more complete version appeared as Hunsperger, Marsh, and Mead, "The Presence of Deep Levels in Ion Implanted Junctions," *Applied Physics Letters* 13, no. 9 (1968): 295–297.

nitrogen, phosphorus, or antimony, then annealing them at temperatures as low as 1,000°C for only 1 to 2 minutes.[206]

Further successes came out of the Hughes research. They continued to improve their processing and fabricating techniques, and by the end of the study in August 1970 they demonstrated the ability to form good quality p-type layers by implanting cadmium or zinc at room temperatures, while producing n-type layers with sulfur ions. Ultimately the Hughes effort contributed to the development of new techniques for doping semiconductors through ion implantation instead of diffusion and masking processes. They showed, too, that one could dope materials through ion implantation that one could not dope using standard fabrication processes. Another key advantage to implantation was the ability to use low temperatures in place of the high temperatures required for the diffusion process. Implantation also permitted one a far better degree of control of the formation of patterns as integrated circuits became smaller and smaller still.[207]

Separation Anxiety

One of the most critical stages in the fabrication of integrated circuits was their separation from the wafer. A significant portion (upwards of 75 percent in some cases) of them were damaged or destroyed routinely during their separation from the wafer mainly by scratching or breaking them.[208] Any solution to this problem would have obvious benefits for industry, because it would mean reaping more useful integrated circuits per batch. Two researchers in the Failure Mechanisms Branch of the Electronics Research Center—Irving Litant and Anthony J. Scapicchio—came up with a solution, which subsequently received a U.S. patent and earned the inventors employee recognition awards.[209]

---

206. Dunlap, Hunsperger, and Marsh, *Development of Ion Implantation Techniques for Microelectronics* (Washington, DC: NASA, October 1969), pp. 1 and 4.

207. Dunlap, Hunsperger, and Marsh, October 1969, pp. 5–6; Marsh, Dunlap, R. Hart, and Hunsperger, *Development of Ion Implantation Techniques for Microelectronics* (Washington, DC: NASA, 1970), pp. xi and 1; and *1967 Program Review*, p. 26.

208. Irving Litant and Anthony J. Scapicchio, "Apparatus and Method for Separating a Semiconductor Wafer," U.S. Patent #3,493,155, filed May 1969, issued 3 February 1970, p. 4.

209. "Research Highlights," *ERC News*, 1 June 1969, n.p., Folder "ERC History House Organ 1966–1969," Box 8, RG 255, AN 71A2309, WNRC; NASA, *Twenty-Second Semiannual Report to Congress, July 1–December 31, 1969* (Washington, DC: NASA, 1970). Upon the closing of the Electronics Research Center in 1970, both Litant and Scapicchio transferred to the Department of Transportation's Transportation Systems Center. Boyd C. Myers II, *A Report on the Closing of the NASA Electronics Research Center* (Washington, DC: NASA, 1 October 1970), pp. 201–202.

Typically one scribed silicon wafers, and the lines so formed determined the boundaries of the integrated circuits. Then one flexed the wafer in such a manner as to separate the integrated circuits from the wafer. Operators used saws or ultrasonic cutters to separate the microcircuits. In some cases they might employ a curved cylindrical anvil or a roller. Technicians had to flex the wafer in one direction then in another before cutting, and they had to be careful to minimize damaging and contaminating the circuits. Several other techniques were in use, but they all suffered from the same difficulties.

The Litant-Scapicchio technique entailed using an open-ended cylindrical chamber that contained a convex hemisphere on which one placed the scribed wafer. An operator placed a flexible diaphragm over the wafer and convex hemisphere and positioned a second chamber over the convex hemisphere. Fluid pressure introduced into the upper chamber forced the flexible diaphragm downward and onto the wafer. As the pressure increased, the flexible diaphragm "walked" across the surface of the wafer. The pressure continued to increase until the entire wafer was broken. After the turning off of the pressure and the removal of the upper chamber and flexible diaphragm, the wafer was in the form of numerous chips. During the entire operation the integrated circuits did not come in contact with each other, thereby avoiding damage to the microcircuits. The separation of the integrated circuits was clean, clear-cut, and orderly along the scribed lines, and little if any flakes or dust formed as a result. In actual practice the inventors achieved very high yields.[210]

Irving Litant also invented a method for locating leaks in hermetically sealed containers intended for use in testing packaged electronics. Sensitive electronic devices, such as transistors and integrated circuits, frequently were packaged in metal containers that had been sealed hermetically in small protective metallic or ceramic enclosures to protect them from a variety of environmental contaminants. The problem was that once these packages were sealed, one had to test to ensure that the seals were tight.[211] Of course, various testing methods already existed. One method—prescribed by the Defense Department (MIL-STD-202C)[212]—consisted of immersing the device in a bath of ethylene glycol or glycerol heated to 150°C and watching for bubbles of gas escaping. This method was highly unreliable, however, and several other methods were available. A method in use at the Electronics Research Center

---

210. Litant and Scapicchio, "Apparatus and Method," p. 4.
211. Litant, "Method for Detecting Leaks in Hermetically Sealed Containers," U.S. Patent #3,548,636, filed 30 September 1968, issued 22 December 1970, assigned to NASA; and Litant, *Leak Detection in Hermetically Sealed Devices* (Washington, DC: NASA, October 1969), p. 1.
212. L. D. Hanley, J. Partridge, and E. C. Hall, *The Application of Failure Analysis in Procuring and Screening of Integrated Circuits* (Cambridge: MIT, October 1965), pp. 48–49.

employed the standard helium leak detector but modified significantly by the addition of a sample container having a controlled orifice. The particular advantage of this method was that leaks intermediate between gross and fine could be detected reliably. Litant's improved, patented method basically involved introducing a low-boiling fluorocarbon liquid (Freon 11, for example) into the sealed container through any leak, and the detection of the resultant vapor escaping from the leak by means of a halogen leak detector. The technique allowed one to pinpoint leaks and to determine the leak rate.[213]

Researchers at the Electronics Research Center developed several additional methods for improving the integrated-circuit fabrication process, such as the use of lasers to scribe wafers. Laser-scribing eliminated the tiny cracks that often formed at the intersection of scribe lines, minimized or eliminated the preparation of the scribe channels, and increased yield during the breaking process. NASA touted the laser technique as being able to compete successfully with existing methods as a high-speed production tool for scribing wafers.[214]

The Electronics Research Center also was in the middle of the push toward computer-aided design of circuits, including those that would become integrated circuits.[215] Universities and the armed services were the key players in the development of the new techniques. The Joint Services Electronics Program,[216] for example, invested heavily in the computer design of integrated circuits at the University of California, Berkeley. The most widely used design software developed was a circuit simulation program known as the Simulator Program with Integrated Circuits Emphasis (SPICE).[217]

The Electronics Research Center had an "extremely active" program in computer-aided circuit design. Its focus was on circuit analysis techniques to the design of microelectronic circuits. The center contracted out much of the research (as had the military) through various grants and contracts to a number of universities,[218] such as Villanova University in Pennsylvania[219]

213. Litant, *Leak Detection*, pp. 1, 2, 3, and 5–6.

214. The work was carried out by C. Calihan of Spacerays, Inc., Burlington, MA, under contract to the Electronics Research Center. "Laser Scribing of Silicon Wafers," NASA Tech Brief, Brief 70-10437, October 1970.

215. *1967 Program Review*, p. 30.

216. On the JSEP, see A. L. Gilbert and B. D. McCombe, "The Joint Services Electronics Program: An Historical Perspective," in Arnold Shostak, ed., *Fortieth Anniversary of the Joint Services Electronics Program* (Arlington, VA: ANSER, 1986), pp. 2–5.

217. Gary Akin, "Consider the Epoch of Electronics," *Leading Edge* (August 1990): 9; and W. Oldham and D. Angelakos, "University of California, Berkeley," Shostak, *Fortieth Anniversary of the Joint Services Electronics Program*, pp. 9–11.

218. *1967 Program Review*, p. 30.

219. Tsute Yang and Henry T. Koonce, *Semi-Annual Report on Computer-Aided Circuit Analysis* (Villanova: Villanova University, December 1965); Yang and Koonce, *Semi-Annual*

and MIT in Massachusetts.[220] The Electronics Research Center took center stage on 11–12 April 1967 when it hosted its computer-aided circuit design seminar at MIT's Kresge Auditorium. Participants came from IBM and its Watson Research Center, Bell Telephone Laboratories, Los Alamos Scientific Laboratory, the Sandia Corporation, Rice University, the Boeing Corporation, the Air Force Weapons Laboratory (Kirtland, New Mexico), the Bendix Radio Division, and of course the Electronics Research Center—the only NASA center participating in the seminar. The event concluded with a panel discussion on time-sharing versus batch processing.[221]

The total NASA computer-aided design program comprised more than the design and analysis of integrated circuits. The computer-aided "efforts include techniques for diagramming circuits and modeling the components, the automated production of the masks and layouts from which the actual circuits are made, and the reduction and analysis of test data. To date [March 1969], these efforts have produced a series of computer programs suitable for analyzing digital circuit designs and laying out the precision masks used in device fabrication. In Fiscal Year 1970, these efforts will continue with emphasis on a broader range of circuit types, the development of automated test instrumentation, and the elimination of hand-drawn layouts in the device fabrication process. Our aim in this area is to capitalize on the facilities and competence of industrial and university research and direct selected investigations of particular interest to meeting future aerospace operational needs."[222]

Testing Methods

An essential and critical stage in the manufacture of reliable integrated circuits was their testing. Microscopes had long been a standard tool for the nondestructive testing of transistors and later microcircuits. The screening

---

Report on Computer-Aided Circuit Analysis (Villanova: Villanova University, June 1966); Yang and Koonce, *Semi-Annual Report on Computer-Aided Circuit Analysis* (Villanova: Villanova University, December 1966).

220. MIT Electronic Systems Laboratory, *Computer-Aided Electronic Circuit Design* (Cambridge: MIT Electronic Systems Laboratory, June 1965); MIT Electronic Systems Laboratory, *Computer-Aided Electronic Circuit Design* (Cambridge: MIT Electronic Systems Laboratory, June 1966); and MIT Electronic Systems Laboratory, *Computer-Aided Electronic Circuit Design* (Cambridge: MIT Electronic Systems Laboratory, January 1967).

221. NASA Electronics Research Center, *Computer-Aided Circuit Design Seminar* (Cambridge, MA: Electronics Research Center, 1967), passim.

222. "Oral Statement of Francis J. Sullivan, Director of Electronics and Control, National Aeronautics and Space Administration, Before the Subcommittee on Advanced Research and Technology, Committee on Science and Astronautics, House of Representatives," 11 March 1969, Folder 2243, NHRC.

procedures for the Apollo program, carried out by the Raytheon Corporation's Space and Information Systems Division, called for the use of microscopes in testing integrated circuits for the guidance and navigation computers. In general, technicians performed inspections at determined minimum magnifications—for example, at a minimum magnification of 150 times to discover scratches on the chip's surface or corrosion of the metallic connections. Other checks required minimal magnifications of only 80 times and some as low as 20 times.[223]

Microscopes also served as a critical tool in the three tests prescribed for detecting packaging leaks. These were known as the helium or "radiflo" leak, the nitrogen bomb, and the hot glycerol bubble tests, the latter being performed in accordance with specifications laid out by the Defense Department (MIL-STD-202C). The tritium tracer was not available yet. For these tests the Apollo program specified the use of binocular microscopes capable of magnification of only 7 to 10 times in order to observe the stream of bubbles that would indicate a package leak.[224]

Such visual observations were standard practice for both NASA and the military, but were inadequate for testing for a range of failure mechanisms. The Apollo guidance and navigational computer integrated circuits underwent tests to determine whether their leads were too long or came in contact with each other. Microscopes were incapable of penetrating their packaging, so x-raying proved an excellent screening procedure. However, the technique was useless for devices that employed aluminum leads (the material of choice for integrated-circuit connections) because that metal is transparent to x rays. A major improvement came out of the ERC's Qualifications and Standards Laboratory, when the staff succeeded in using an x-ray spectrograph to measure the thickness of aluminum deposited on silicon to within ± 10Å. The spectrograph operated in conjunction with a vacuum evaporator for depositing aluminum film and a diffusion furnace to study aluminum interfaces.[225]

Scanning Electron Microscope

One of the routine instruments utilized by industry today for testing integrated circuits on the production line is the electron microscope. Electron microscopes already had been around for decades when the integrated circuit

---

223. L. D. Hanley, J. Partridge, and E. C. Hall, *The Application of Failure Analysis in Procuring and Screening of Integrated Circuits* (Cambridge: MIT, October 1965), pp. 2 and 36–39.

224. Hanley, Partridge, and Hall, *Application of Failure Analysis*, pp. 48–49.

225. Ibid., p. 15; and "ERC R&D Accomplishments."

came along, as they date back to the 1930s and 1940s.[226] Adapting them to integrated-circuit testing on a production line was not a straightforward proposition for a number of reasons. To begin with, early in the development of the electron microscope, a question arose concerning two different types of lens systems. It was not at all obvious whether the electromagnetic lens or the electrostatic design would prove to be superior. Their application to the manufacture of integrated chips would appear to have had to wait until 1965, when Cambridge Instruments introduced the first commercial scanning electron microscope (the Stereoscan).[227]

The electron microscope held out promise because of the extremely small sizes involved with integrated circuits. They had an area of about 1 square millimeter, and some contained dozens of transistors, diodes, and resistors. The electrical interconnections between them often were but a few microns wide, which made physical contact with them for testing not only difficult but also dangerous to their mechanical and electrical integrity.[228]

Already during the 1950s, some experimental work had taken place in the scientific study of semiconductors with electron microscopes.[229] Still, given the variety of electron microscopes available (at least for scientific research), which one would be the most appropriate instrument for integrated-circuit testing was not clear. Therefore, to determine the electron microscope type

---

226. For the history of the electron microscope, see Ladislaus Marton, *Early History of the Electron Microscope* (San Francisco: San Francisco Press 1968); Peter W. Hawkes, ed., *The Beginnings of Electron Microscopy* (Orlando, FL: Academic Press, 1985); John Reisner, "An Early History of the Electron Microscope," *Advances in Electronic and Electron Physics* 73 (1989): 134–230; Ernst Ruska, "The Development of the Electron Microscope and of Electron Microscopy," *Reviews of Modern Physics* 59 (July 1987): 627–638; and O. H. Griffith and E. Engel, "Historical Perspective and Current Trends in Emission Microscopy, Mirror Electron Microscopy and Low-Energy Electron Microscopy," *Ultramicroscopy* 36 (May 1991): 1–28.

227. Gregory C. Kunkle, "Technology in the Seamless Web: 'Success' and 'Failure' in the History of the Electron Microscope," *Technology and Culture* 36, no. 1 (January 1995): 81; "Major Events in the Development of the Electron Microscope and Its Application to Cell Biology," *http://www.ncbi.nlm.nih.gov/books/bv.fcgi?rid=mboc4.table.1751* (accessed 11 May 2006). This article is from an online version of Bruce Alberts, Alexander Johnson, Julian Lewis, Martin Raff, Keith Roberts, and Peter Walter, *Molecular Biology of the Cell*, 4th ed. (New York: Garland Science, 2002).

228. Leon C. Hamiter, Jr., "Fast Scan Infrared Microscope for Improving Microelectronic Device Reliability," 1 in Research and Development Operations, Marshall Space Flight Center, *Research Achievements Review*, Vol. 2, Report No. 5 (Huntsville, AL: Marshall Space Flight Center, 1966).

229. See, for example, J. W. Allen and K. C. A. Smith, "Electron Microscopy of Etched Germanium Surfaces," *Journal of Electronics* 1 (1956): 439–443; and C. W. Oatley and T. E. Everhart, "The Examination of p-n Junctions in the Scanning Electron Microscope," *Journal of Electronics* 2 (1957): 568–570.

best suited to the inspection of integrated circuits, Dr. James E. Cline and colleagues in the ERC's Failure Mechanisms Branch undertook a study of two candidate types.[230] Although historians have focused on the development of scanning electron microscopes by General Electric and RCA,[231] the application of electron microscopy to the production of integrated circuits appears to have begun at the Westinghouse Research Laboratories in Pittsburgh well before the availability of the Cambridge Instruments microscope. In 1960, Oliver C. Wells, Thomas E. Everhart, and R. K. Matta built an advanced scanning electron microscope (called the Micro-Scan) specifically to study solid-state devices and to improve the fabrication of integrated circuits. Researchers at Marshall's Astrionics Laboratory purchased a Micro-Scan and used it starting in 1966 as a research tool to investigate failure mechanisms in integrated circuits.[232]

About the same time, a different research team at the Goddard Space Flight Center began examining how to use scanning electron microscopy in the manufacture of integrated circuits. Robert J. Anstead, John W. Adolphsen, and Samuel R. Floyd in the Quality Assurance Division, who published their initial results in 1968, attempted to analyze failure modes in semiconductors and wire plating. As the Goddard team continued its investigation of solid-state reliability physics, they focused on specific problems whose solution was not

---

230. James E. Cline, James M. Morris, and Seymour Schwartz, "Scanning Electron Mirror Microscopy and Scanning Electron Microscopy of Integrated Circuits," *IEEE Transactions on Electron Devices* 16, no. 4 (April 1969): 371–375.

231. See, especially, Gregory C. Kunkle, "Technology in the Seamless Web: 'Success' and 'Failure' in the History of the Electron Microscope," *Technology and Culture* 36, no. 1 (January 1995): 80–103. Cyrus C. M. Mody, "Corporation, Universities, and Instrumental Communities: Commercializing Probe Microscopy, 1981–1996," *Technology and Culture* 47, no. 1 (January 2006): 56–80, discusses the scanning tunneling microscope, a later development. Mody dismisses the importance of the scanning electron microscope in the history of nondestructive testing because its use required "breaking and discarding expensive silicon wafers." Ibid., 64. Both articles suffer or benefit—depending on one's perspective—from their overwhelming historiographical bent. A sorely needed history of the application of scanning electron microscopy to the production of solid-state devices remains to be written.

232. G. Berryman and T. R. Edwards, Marshall Space Flight Center, "Evaluation of Semiconductor Devices Using the Scanning Electron Microscope," in ERC, *Proceedings*, p. 424; Thomas E. Everhart, Oliver C. Wells, and R. K. Matta, "A Novel Method of Nondestructive Semiconductor Device Measurements," *IEEE International Electron Devices Meeting* 9 (1963): 72–72G; Thomas E. Everhart, Oliver C. Wells, and R. K. Matta, "A Novel Method of Nondestructive Semiconductor Device Measurements," *Proceedings of the IEEE* 52, no. 12 (December 1964): 1642–1647. See also Oliver C. Wells, Thomas E. Everhart, and R. K. Matta, "Automatic Positioning of Device Electrodes Using the Scanning Electron Microscope," *IEEE Transactions on Electron Devices* 12, no. 10 (October 1965): 556–563; and Oliver C. Wells, *Scanning Electron Microscopy* (New York: McGraw-Hill, 1974).

readily at hand through optical microscopy. The electron microscope allowed them to see conditions previously unseen. Among these problems were the discontinuities in integrated-circuit aluminum connections. By 1970 they had completed developing the application of scanning electron microscopy to integrated-circuit production lines for the detection of metallization faults.[233]

Fast-Scan Infrared Microscope

A different line of attack was taken up by researchers at the Marshall Space Flight Center under Leon C. Hamiter Jr. through a contract with the Raytheon Corporation. Their microscope operated in the infrared spectrum. Infrared illumination of integrated circuits would furnish inspectors with several advantages over optical microscopy. For one, silicon is essentially transparent at infrared wavelengths. Of course, one of the most obvious advantages is that infrared inspection avoided contact with the microcircuit, thereby reducing the likelihood of damage during examination. NASA considered this ability important to Agency research into what made electronic parts fail and hoped that the microscope one day might ease the inspection load in the assembly of spacecraft systems.[234]

The fast-scan infrared microscope that Marshall and Raytheon developed measured small temperature differences in the integrated circuit that pointed to possible trouble areas. Wherever electric current flows, a portion of it turns into heat (called power dissipation), so that wherever current flows the temperature of the part through which it flows rises. Experience showed that electronic parts that operated significantly hotter or colder than average tended to fail first.[235]

---

233. Robert J. Anstead, "Failure Analysis Using a Scanning Electron Microscope," [IEEE] *Sixth Annual Reliability Physics Symposium, Los Angeles, CA, November 6–8, 1967* (New York: Electron Devices and Reliability Societies of the Institute of Electrical and Electronics Engineers, 1968); Robert J. Arnstead and Samuel R. Floyd, "Thermal Effects on the Integrity of Aluminum to Silicon Contacts in Silicon Integrated Circuits," *IEEE Seventh Annual Reliability Physics Symposium, Washington, DC, December 2–4* (New York: Electron Devices and Reliability Societies of the Institute of Electrical and Electronics Engineers, 1969); John W. Adolphsen and Robert J. Anstead, "The Use of the Scanning Electron Microscope as a Semiconductor Production Line Quality Control Tool," in Charles P. Marsden, ed., *Silicon Device Processing*, NBS SP-337 (Washington, DC: National Bureau of Standards, 1970), pp. 384–397.

234. Contract NAS 8-11604 for the "design, development, and evaluation of a fast-scan infrared detection and measuring instrument to assess the true electrical performance and thermal characteristics of microelectronics devices." "Office of Reliability and Quality Assurance Functions for Potential Transfer to ERC," n.d., "Qualifications & Standards Lab," Box 1, RG 255, AN 71A3002, WNRC; and *1965 Program Review*, p. 9.

235. *1965 Program Review*, p. 8.

The fast-scan infrared microscope consisted of a pedestal located inside on the left side of the infrared microscope unit system. The operator mounted the integrated-circuit device to be examined on this pedestal. A binocular optical microscope aided the operator in positioning and inspecting the device. The heat from the microcircuit undergoing testing was scanned by the motion of a mirror, and various mirrors reflected and focused the heat on the cryogenically cooled infrared detector. The output of the detector was recorded as a function of the scanning mirror's position and by inference the location of the heat source.[236]

The optics of the infrared microscope provided magnification, while the scanning-mirror system provided a fast rate of scanning not unlike that used in television. Each infrared "snapshot" depicted an area 10 microns across. Scanning speed was 10 frames per second (at 100 lines per frame, 1,000 lines per second). The instrument's design simplified signal processing as much as possible, but still providing a single frame image of the scanned target as well as a read-out of infrared radiation amplitude over time on a continuous basis. The microscope furnished output in two forms. One was an analog signal with a maximum frequency of 100 Hz, and another was an indexed video signal, which made it ideal for recording output on a conventional videotape recorder. One then could reproduce the data from the recorder as video images on an oscilloscope, and the line scans could be recorded directly on a strip chart recorder. Furthermore, one could use the recorder data as input to an analog-to-digital converter whose output was fed into a buffer unit for storage and future computer processing.[237]

The Marshall researchers foresaw uses for the fast-scan infrared microscope beyond the inspection of production-line solid-state devices. For example, one could place prototype integrated circuits in the unit, producing a "thermal map" that would enable design engineers to verify its operation, in particular to eliminate excessive power dissipation and unwanted heating of sensitive components. In addition, one could compare the thermal maps of production circuits with standards established for each basic device. Any significant variations during the production process would become apparent. From these thermal maps, then, both circuit designers and production personnel could institute changes in order to correct defects or other anomalous

---

236. Ibid.

237. Leon C. Hamiter, Jr., "Fast Scan Infrared Microscope for Improving Microelectronic Device Reliability," in ERC, *Proceedings*, pp. 436–439; and Hamiter, "Fast Scan Infrared Microscope for Improving Microelectronic Device Reliability," in Research and Development Operations, Marshall Space Flight Center, *Research Achievements Review*, vol. 2, Report No. 5 (Huntsville: Marshall Space Flight Center, 1966), pp. 2–3.

qualities. Hamiter and his Marshall colleagues looked forward to placing the instrument in a microcircuit manufacturer's plant for about three months of testing to establish the relationship between the effectiveness and efficiency of infrared testing of integrated circuits.[238]

Impacts and Conclusions

The purpose of this chapter was to determine what, if any, role NASA had in improving the manufacture of integrated circuits during the Apollo era. It tackled that question on two fronts. The first was a reconsideration of the role of the decision to utilize integrated circuits for the Apollo guidance and navigation computers on the growth of the integrated-circuit industry, especially the stimulus of commercial—as opposed to governmental—applications. The second front was a broader examination of NASA's integrated-circuit research and development efforts during that same time period.

The Legend Revisited

The long-standing argument is that the integrated-circuit industry took off because NASA purchased a large number of them for the Apollo guidance and navigation computers (or for the entire Apollo program in alternate versions of the story). Those hefty purchases induced industry to introduce changes that led to commercial applications of the integrated circuit.[239] Some scholars also have pointed to the role of military buys of integrated circuits for the Minuteman II; however, those followed the initial NASA procurements for the Apollo program.[240] Paul Cerruzi perhaps has best summed up the relative contributions of these two significant procurements of the first integrated circuits, when he wrote: "The current 'revolution' in microelectronics thus owes a lot to both the Minuteman and the Apollo programs. The Minuteman was first; it used integrated circuits in a critical application only a few years after they were invented. Apollo took the next and equally critical step; it was designed from the start to exploit the advantages of integrated logic."[241]

---

238. Hamiter, "Fast Scan Infrared Microscope," in *Research Achievements Review*, pp. 4–5; and Hamiter, "Fast Scan Infrared Microscope," in ERC, *Proceedings*, pp. 439–440 and 443.

239. Hall, *Journey to the Moon*; Kleiman, "Integrated Circuit"; Eli Ginzbert, James W. Kuhn, Jerome Schnee, and Boris Yavitz, *Economic Impact of Large Public Programs: The NASA Experience* (Salt Lake City, UT: Olympus Publishing Company, 1976), pp. 57–59, just to name three sources.

240. Ceruzzi, *History of Modern Computing*, p. 188; Neal, *Ace in the Hole*; MacKenzie, *Inventing Accuracy*; Kleiman, "Integrated Circuit," p. 210; and Kraus, "Economic Study of the U.S. Semiconductor Industry," p. 45.

241. Ceruzzi, *History of Modern Computing*, p. 188.

Leslie Berlin and Christophe Lécuyer have provided an alternative account that shifts the focus from the number of integrated circuits bought by the government to the price that customers paid for them. As long as the price of integrated circuits exceeded the price of the individual components represented in the integrated circuit, Berlin has argued, commercial consumers stayed away from them. But in 1964, when Bob Noyce of Fairchild Semiconductor decided to lower the unit price of certain integrated circuits below that of the aggregated components, businesses began to buy them. Although the price of integrated circuits had been declining since their introduction, commercial sales did not take off until Noyce lowered prices below the price paid for the equivalent components. Furthermore, the lowered prices were *below* what it cost Fairchild Semiconductor to make them. The large government procurements were not a motivating factor, and in fact they are not part of the narrative that Berlin and Lécuyer weave.[242]

Because the legend relies on the large size of government procurements prior to 1964, one must wonder the extent to which those purchases actually were large, especially compared with the commercial buys that began following the 1964 price drop. Eldon Hall generally cites 200,000 as the number of integrated circuits that NASA bought for the Apollo guidance and navigation computers. The total number of integrated circuits that the Air Force bought for the Minuteman II is unknown. We *do* know, however, that each missile's guidance and control system used about 3,000 integrated circuits, and that by 1965, shortly after the first flight of the Minuteman II in September 1964, missile production rate was up to six or seven per week—a schedule that called for Texas Instruments, Westinghouse, and RCA to supply over 4,000 integrated circuits every week.[243]

Even though the size of these procurements seems big, they pale in comparison to those made by industry once Noyce lowered the price of certain integrated circuits. Within a year of the price cuts, Fairchild Semiconductor received a single order for 500,000 microcircuits, a number equal to a fifth of the entire industry's production output for the preceding year, according to Berlin, and the computer manufacturer Burroughs put in an order for several millions in 1966. Fairchild Semiconductor now was selling integrated circuits in lots of 500,000,[244] more than double the number of integrated

---

242. Lécuyer, *Making Silicon Valley*, p. 241; Berlin, *Man Behind the Microchip*, pp. 136–139. The small exception is that Fairchild worked hard to overcome customers' objections to the IC. One thing they did was to participate "in several widely publicized reliability tests sponsored by the federal government, including experiments for the Apollo project." Berlin, *Man Behind the Microchip*, p. 137.

243. Ceruzzi, *History of Modern Computing*, p. 187.

244. Berlin, *Man Behind the Microchip*, p. 139.

circuits in all of the Apollo guidance and navigation computers. The commercial demand for standardized integrated circuits for computers outpaced government demand immediately and established a new meaning of "big" for integrated-circuit purchases. One wonders, therefore, the extent to which the significantly smaller NASA and Air Force procurements actually had an impact on the commercialization of integrated circuits. The question demands new research.

Both the Apollo legend and the Berlin-Lécuyer thesis focus on large-scale purchases of standardized integrated circuits that Fairchild Semiconductor and Texas Instruments churned out in large numbers. Mass production is a relative term in this context. But the production and consumption of standardized integrated circuits on a large scale did not constitute the entire market for integrated circuits, but only one part of it. Alongside the large-scale standardized market was another that comprised smaller lots of specialized integrated circuits with electrical characteristics designed to meet a consumer's particular needs. As one NASA publication characterized the market: "In space research, requirements exist for small quantities of special microelectronic devices and for large quantities of standardized devices."[245]

NASA had a voracious appetite for these specialized microelectronic products. Scholars, moreover, have provided a number of examples in which the production of small lots of specialty integrated circuits played a crucial role in the history of electronics firms, but without overtly making this point. For example, Berlin tells us that when Fairchild Semiconductor was a newly created company hungry for its first order, it obtained one from IBM for 100 transistors with special characteristics (capable of switching 150 milliamps while operating at 60 volts and 50 MGz) for which IBM agreed to pay $150 each. In Berlin's words, "The IBM order made Fairchild Semiconductor."[246]

The example of Signetics shows a conscious effort to make lots of specialized integrated circuits instead of large numbers of standardized devices. With little capital, the company reasoned, they could not compete with Fairchild in the low-cost, high-volume transistor market, so they decided to concentrate their resources on microcircuits. As a result, Signetics became the first corporation to specialize in integrated circuits in the United States. When the founders discovered that microcircuits were in little demand, they had their engineers design logic circuits of the type known as diode transistor logic (DTL) and familiar to most system firms. After six months the firm realized that the demand for custom integrated circuits was more limited

245. Research Triangle Institute, *Microelectronics in Space Research* (Washington, DC: NASA SP-5031, August 1965), p. 127.
246. Berlin, *Man Behind the Microchip*, pp. 92–93.

than they originally had thought, so they switched to a family of standard integrated circuits based on a DTL configuration in the belief that these had certain inherent advantages over the other types of logic configurations. Still, they were getting nowhere with sales, so the company hired a military sales specialist whose only job was to garner military research and development contracts. Success arrived in 1963, when Signetics won a $159,000 contract to develop a high-speed DTL family for the Signal Corps. Signetics' fortunes began to change. Signetics' DTL rapidly became the standard in the market. That same year, the Pentagon began to require that defense contractors utilize microcircuits to the maximum extent possible, thereby creating a large market for integrated circuits "that went well beyond the Apollo and Minuteman II programs."[247]

NASA, then, was a consumer of both standardized integrated circuits produced en masse and small-lot specialty integrated circuits. The common portrayal of the space agency as just a buyer of integrated circuits is far from the actuality, however. Both consumption patterns drove the space agency to undertake its own research and development programs—not to mention the construction and operation of its own fabrication facilities—not just to be a better customer of industry but to guide the direction of scientific research and the development of technology and manufacturing processes. The case studies provided above bear witness to NASA's internal and external research and development efforts related to the manufacture of better integrated circuits, whether to achieve greater reliability or devices capable of withstanding the rigorous conditions of space.

The case studies also portray a government research effort that was on the cutting edge along with industry, academia, and other government laboratories. NASA combined internal research with contracted work to some degree as a response to the 1962 *Report to the President on Government Contracting for Research and Development* by David Bell, Director of the Bureau of the Budget, and others.[248] More importantly, under Administrator James Webb, NASA research was woven into a web of institutional relationships, a so-called university-industry-government complex, for waging "war" on the technological frontier. For Webb, this complex was not a necessary evil but a positive boon.[249] To understand NASA research during the Apollo era, we therefore must understand it as taking place within this institutional web.

---

247. Lécuyer, *Making Silicon Valley*, pp. 140, 212, 229, 230, 231, 233, 236.

248. Levine, *Managing NASA in the Apollo Era*, pp. 75–84.

249. Walter A. McDougall, … *The Heavens and the Earth: A Political History of the Space Age* (Baltimore, MD: Johns Hopkins University Press, 1985), p. 381.

Technology Utilization

NASA was not only committed to accomplishing improvements in the manufacture of integrated circuits but also made a concerted effort to gather information internally on those accomplishments. The Electronics Research Center, for example, regularly generated documents that fed the Agency's hunger for news of its own success. The monthly reports compiled in 1965, for example, included information on expenses for research and development programs, the construction and rental of facilities, personnel growth, visitors to the center, and research news.[250] With the start of weekly reports, such as those for 1967, the focus shifted to appointments of new senior-level personnel, meetings and symposia, and research and development results (not to mention the ongoing news of facility construction).[251]

The activities and successes documented in the weekly and monthly reports flowed upward through the NASA bureaucracy to Headquarters, where the information supported a number of efforts, not the least of which was technology utilization (or "spinoff"). The Headquarters reorganization of 1963 included the creation of the Office of Technology Utilization in order to work out means by which the technical byproducts of space research and development could be disseminated most effectively to private industry and other users. This effort was a major undertaking, according to Levine, because it involved identifying useful technology, evaluating its potential, supporting research on technology transfer (often through grants to universities and research institutes), and matching data collected with potential users.[252] The office[253] promoted the use and diffusion of the technological innovations that developed from NASA research activities. These innovations stemmed from work conducted internally within the Agency's laboratories or in academic or industrial facilities under contract with NASA. One of the chief, but certainly not the sole, outlet for the dissemination of NASA technical successes was the Tech Brief. Each Tech Brief discussed one innovation in detail with appropriate diagrams and technical drawings and included information on how businesses could license the technology. A list of selected Tech Briefs that

---

250. The 1965 monthly reports are in "Chron File 1966," Box 3, RG 255, AN 71A-2309, WNRC.

251. Weekly reports for 1967 are in "Chron File-September 1967," "Chron File-October 1967," "Chron File-November 1967," and "Chron File-December 1967," Box 11, RG 255, AN 71A-2309, WNRC.

252. Edward E. Furash, "The Problem of Technology Transfer," in Raymond A. Bauer and Kenneth J. Gergen, eds., *The Study of Policy Formation* (New York: Free Press, 1968), pp. 281–328; and Levine, *Managing NASA in the Apollo Era*, pp. 42–43.

253. Also known as the Office of Technology Utilization and Policy Planning.

issued from the integrated-circuit work of the Electronics Research Center is offered in Appendix A.

As a means of improving industrial practices, the diffusion of technical knowledge through publications with technical illustrations harkens back to the 18th century and the *Encyclopédie ou Dictionnaire raisonné des sciences, des arts et des métiers par une société de gens de lettres.*[254] Published under the direction of Denis Diderot and Jean le Rond d'Alembert, the work consisted of 21 volumes of text, 11 volumes of plates, and some 72,000 articles written by over 140 contributors. Among articles on philosophy, religion, history, and other topics, the compilers included a substantial number of scientific and technical entries relating to what today would be called physics, chemistry, astronomy, and medicine as well as automatons, glassblowing, metallurgy, and papermaking, not to mention articles devoted to discovery (*"découverte"*) and invention (*"invention"*) itself.

The *Encyclopédie*, like the NASA Tech Briefs, spread technical and scientific knowledge in a passive manner. That is, the publication made the information available, but it did not engage an end user of the knowledge in any active manner. Consequently, gauging the effectiveness of the *Encyclopédie* approach is difficult, if not impossible. The case of the NASA Tech Briefs, however, is not so daunting because of the evolution of the legal context of innovation. Unlike the technological practices spread by the *encyclopédistes* of the 18th century, NASA operates in a society that legalizes and regulates technological change. Thus, if a business chose to use one of the innovations described in a Tech Brief, it would ask NASA for a license to use it. The records of the licensing of Tech Brief innovations therefore provide a valuable index of the transfer of knowledge from NASA research programs to the commercial sector. Unfortunately, such records are unavailable because they are either off-limits to researchers or have been destroyed as part of the Agency's records management policy.

Patent waivers provide a second legalistic trace of technology transfer, but only for those innovations created by industry under contract to NASA. When work under a NASA contract or grant led to a new discovery or invention, the rights to it belonged to NASA under the National Aeronautics and Space Act of 1958. The argument was that the invention had been funded by

---

254. Denis Diderot and Jean le Rond d'Alembert, *Encyclopédie ou Dictionnaire raisonné des sciences, des arts et des métiers par une société de gens de lettres* (Paris: Briasson [etc.], 1751–1765), republished as Charles Coulston Gillispie, ed., intro. and notes, *A Diderot Pictorial Encyclopedia of Trades and Industry; Manufacturing and the Technical Arts in Plates,* Selected from *L'Encyclopedie; ou, Dictionnaire raisonné des sciences, des arts et des métiers,* 2 vols. (New York: Dover Publications 1959). Selections from the work have been translated into English several times.

"the people" and should belong to them through the government. Granting a patent to the contractor would create a monopoly (the intent of patent law) that could hinder technological development within the aerospace industry. Contractors, of course, opposed this policy and claimed that it would discourage industry from bidding on NASA contracts, but clearly the number and size of the companies placing bids appears to belie that point.

As a way of allowing industry to enjoy the rights to such inventions, Section 305, "Patent Rights in Inventions," of the NASA Act of 1958 gave the NASA Administrator the right to waive "all or any part" of those patent rights under conditions outlined in the Act. However, the waiver did not give away the Agency's rights to use the invention at no charge. These waiver provisions applied to inventions resulting from work carried out by industry and academia as well as by NASA employees. From time to time, therefore, a firm or university would request a patent waiver, so that they could exploit the invention for their own advantage. Whether NASA granted the waiver or not, behind the *intent* to request a waiver was the belief that the invention had commercial utility as well as potential economic impact. One cannot calculate what that economic impact was or might have been, but one can document the *perception* that an invention made with NASA funding had commercial potential through these requests for patent waivers. A list of the patent waiver requests and their disposition is included in Appendix B.

Technology Utilization and Congress

While technology utilization (and the gathering of internal information for Tech Briefs) functioned to diffuse technical knowledge gained through NASA-sponsored research programs, it also served an auxiliary political purpose. Technology utilization had become a political issue by August 1964, when Congress authorized the establishment of a National Commission on Technology, Automation, and Economic Progress. As spelled out in Public Law 88-444, the commission's four primary functions were to identify and assess the past effects and the current and prospective role and pace of technological change; to identify and describe the impact of technological and economic change on production and employment and the social and economic effects on the nation's economy, workforce, communities, families, social structure, and human values; to define those areas of unmet community and human needs toward which application of the new technologies might most effectively be directed; and to assess the most effective means for channeling new technologies into promising directions, including civilian industries where accelerated technological advancements will yield general benefits, and assess the proper relationship between governmental and private investment in the application of new technologies to large-scale human and community needs.[255]

Politicians, scholars, and others were trying to come to grips with society's relationship with technology.[256] What followed from this intellectual trend and the congressional commission in particular was the need of NASA to not just conduct successful research programs, but also to demonstrate that the research results had some positive benefit to society. From that moment forward, technology utilization was no longer just about technical knowledge, but also had socioeconomic and political implications ("societal impacts").

The evaluation of how particular technologies affected society came into vogue as "technology assessments," particularly following congressional debate on the question.[257] NASA programs did not escape the critical eye of these technology assessments. They became the subject of a myriad of studies (still ongoing) into the "impact" of NASA research on society. They had a variety of foci. Some looked at NASA's impact on biomedical research[258] that often responded to political pressures being brought to bear on NASA. For example, one major study looked at the use of NASA technology by industry in the Midwest,[259] where the Agency had no centers, but where congressional members, such as Rep. Donald Rumsfeld (R-IL), pressured NASA to spend

---

255. Richard Lesher and George J. Howick, *Assessing Technology Transfer* (Washington, DC: NASA SP-5067, 1966), preface.

256. Not coincidentally, this was also the period that saw the emergence of science technology and society studies and the first U.S. edition of the writings of Jacques Ellul on technology and society (*La technique; ou, L'enjeu du siecle*): Ellul, *The Technological Society*, translated by John Wilkinson, introduction by Robert K. Merton, 1st ed. (New York: Knopf, 1964). It appeared more recently in English as Ellul, *The Technological Bluff*, translated by Geoffrey W. Bromiley (Grand Rapids, MI: W. B. Eerdmans, 1990). For Ellul's ideas on technology and society, see also the translation of his *Entretiens avec Jacques Ellul: Jacques Ellul on Religion, Technology, and Politics*, translated by Joan Mendès France (Atlanta, GA: Scholars Press, 1998).

257. An excellent source on the early technology assessment movement is the collection of essays in George Washington University, Program of Policy Studies in Science and Technology, *Readings in Technology Assessment: Selections from the Publications of the Program of Policy Studies in Science and Technology* (Washington, DC: George Washington University, 1975).

258. See, for instance, Southwest Research Institute, *Southwest Research Institute Assistance to NASA in Biomedical Areas of the Technology Utilization Program, Quarterly Progress Report No. 4* (San Antonio, TX: Southwest Research Institute, 19 September 1967); and United Aircraft Corporation, Biosciences and Technology, Space and Life Sciences Department, Hamilton Standard Division, *Medical and Biological Applications of Space Telemetry* (Washington, DC: NASA SP-5023, July 1965).

259. Howard M. Gadberry, *Utilization of NASA-Generated Space Technology by Midwestern Industry, Quarterly Progress Report No. 1* (Kansas City, MO: Midwest Research Institute, 1962), and subsequent reports.

more money in the region.[260] Also along these same lines of regional politics, NASA always listed the distribution of its grants and contracts by state.

These impact studies are of particular interest to an investigation of NASA's role in the manufacture of integrated circuits because several of them that date from the Apollo era and immediately thereafter attempted to address that very question. Nonetheless, for several reasons they left much to be desired in terms of an answer to the question of NASA's impact. For example, a 1976 study purporting to deal with "the NASA experience" repeatedly gave information on not just NASA, but on NASA and the Defense Department combined, referring in several places to "space-defense."[261] The crux of the problem in separating NASA from military economic impacts lies in the categories devised by the Department of Commerce's Business and Defense Services Administration to tabulate data. The only categories were "defense" and "non-defense." Starting in 1959, the "defense" category included NASA, along with the Atomic Energy Commission, the Central Intelligence Agency, and the Federal Aviation Administration (not to mention, of course, the several components of the Defense Department).[262]

Attempts to assess retrospectively the impact of NASA research programs on society have suffered additional problems. One of the major suppliers of these studies has been the Denver Research Institute of the University of Denver, which has conducted impact studies for NASA since almost the founding of the space agency.[263] During the 1970s the institute prepared and updated from time to time a *Benefits Briefing Notebook* for NASA's Office of Technology Utilization. Although the overt purpose of the *Notebook* was to "provide the Agency with accurate, convenient, and integrated resource information on the transfer of aerospace technology to other sectors of the U.S. economy" for use in "speeches, articles, or other purposes,"[264] at least one

---

260. Ken Hechler, *Toward the Endless Frontier: History of the Committee on Science and Technology, 1959–1979* (Washington, DC: U.S. House of Representatives, 1980), pp. 220–229.

261. Ginzbert et al., *Economic Impact of Large Public Programs*, pp. 57, 60, 68.

262. Kraus, "Economic Study of the U.S. Semiconductor Industry," pp. 89–91.

263. According to John G. Welles, Lloyd G. Marts, Robert H. Waterman Jr., John S. Gilmore, and Robert Venuti, *The Commercial Application of Missile/Space Technology, Part 1* (Denver, CO: Denver Research Institute, September 1963), p. iii, the institute began working on a NASA impact-study grant in November 1961, just two years after the Agency's creation.

264. Denver Research Institute, *NASA Benefits Briefing Notebook* (Denver, CO: Denver Research Institute, 1976), preface. See also Martin D. Robbins, John A. Kelley, and Linda Elliott, *Mission-Oriented R & D and the Advancement of Technology: The Impact of NASA Contributions*, 2 vols. (Washington, DC: NASA, 1972).

unmentioned purpose likely was to provide NASA with fodder for congressional hearings.

Even though electronics was not one of the "benefit cases" investigated by the Denver Research Institute, they did not overlook success stories related to the manufacture of integrated circuits. The one that stands out was titled "microelectronics production quality assurance" and classified under the rubric "manufacturing capital goods." The *Notebook* states that supporting documentation for the technology transfers was available from NASA's own Office of Technology Utilization. The basis for the claims made in the "microelectronics production quality assurance" entry, however, provided no NASA source but rather an earlier study conducted by the Denver Research Institute's own 1973 study, *Industrial Products and Practices*.[265] One wonders about the ultimate source of information because of the nature of the claims made in the *Notebook*.

The "microelectronics production quality assurance" entry discusses two supposed impacts of NASA relevant to the manufacture of integrated circuits: certified production lines and the use of scanning electron microscopes. It begins with this statement: "NASA established reliability program and procurement standards for microelectronic products in 1964 to assure best manufacturing practices would be used by suppliers."[266] Several sources contradict this statement. Although NASA Headquarters, at least, had a reliability office at least as early as January 1961, and a General Management Instruction relating to reliability policy appeared at least as early as February 1961,[267] these did not deal specifically with microelectronics. In fact, the Microelectronics Reliability Program did not begin until October 1965 at the earliest.[268]

The "microelectronics production quality assurance" next states that NASA "introduced [the] concept of [a] Certified Production Line (CPL) so that entire production lines, rather than products themselves, were certified for NASA procurement after [the] vendor had established quality control practices and in-house evaluation methods to satisfy NASA inspection teams."

---

265. A search for this study has turned up nothing.

266. Denver Research Institute, *NASA Benefits Briefing Notebook*, entry B-1.

267. Levine, *Managing NASA in the Apollo Era*, pp. 32; NASA Management Manual, General Management Instructions, "Reliability Policy as Applied to NASA Programs," 1 February 1961, Folder "NASA SR & QA (Gen.) through 1966," Box 18,153, NHRC; NASA Management Manual, General Management Instructions, "Functions and Authority—Office of Reliability and Systems Analysis," 16 January 1961, Folder "NASA SR & QA (Gen.) through 1966," Box 18,153, NHRC.

268. Memorandum, Robert L. Trent to W. Crawford Dunlap, "Microelectronics Reliability Program," 1 November 1965, Folder "Qualifications & Standards Lab," Box 1, RG 255, AN 71A3002, WNRC; *Reliability and Quality Assurance* (Washington, DC: NASA CR-156261, 31 October 1965); and Condon, "Introductory Remarks," pp. 5–6.

The Defense Department, according to the Denver Research Institute, later adopted the Certified Production Line approach, and industry, as a result of adopting it, increased productivity significantly, with one vendor boosting yields by 20 percent.[269] What is most ambiguous about the wording of the text is that NASA appears to have introduced the Certified Production Line in 1964.

The Certified Production Line statements do not jibe entirely with an article on the history of spare parts at NASA written by Leon Hamiter, who led the Marshall group that developed the original line certification standards, which became the basis for the military's production line certification standards and which applied to the manufacture of integrated circuits. The individual who introduced these changes at NASA actually came to the Agency from the military. Hamiter states that in 1964, J. L. "Larry" Murphy came to NASA from the Navy, and over the following years, he planned and implemented the NASA Standard Parts Program as well as NASA's use of the military's parts specification system, line certification program, and other reliability and quality assurance measures.[270]

As a result, the NASA Microcircuit Line Certification Program began in 1968, and NASA certified the first microcircuit lines in 1969. In August 1968, Marshall Space Flight Center was conducting a pilot effort of its specification 85-M-03877 for line certification, mainly for the Skylab Program (but eventually for the Shuttle and its associated space station), to prove the value of the concept.[271] However, as late as 1971 a NASA-wide line certification requirements document was awaiting publication. In the military's procurement specifications system, manufacturers must follow the practices outlined in MIL-M-38510 as well as the certification requirements for microcircuits laid out in MIL-STD-976, which originated as the NASA Microcircuit Line Certification Program.[272] The origins of NASA's certified production line, then, do not date back to 1964, but the close connection between NASA and

269. Denver Research Institute, *NASA Benefits Briefing Notebook*, entry B-1.

270. Leon Hamiter, "The History of Space Quality EEE Parts in the United States," paper presented at the ESA Electronic Components Conference, ESTEC, Noordwijk, The Netherlands, 12–16 November 1990, ESA SP-313 (March 1991), *http://www.cti-us.com/HistoryEEESpacePartsinUSA.pdf* (accessed 24 May 2006).

271. M. F. Nowakowski and F. Villella, "Microelectronics Research for Shuttle and Space Station," in Marshall Space Flight Center, Research and Development Operations, *Research Achievements Review* 3, Report 11 (Huntsville, AL: Marshall Space Flight Center, 1970), pp. 59–63, describes the steps to be controlled in the manufacturing process.

272. Leon Hamiter and Howard Weiss, "NASA's Quality Program—Achievements and Forecast," 21 May 1971, pp. 9–10, Folder "NASA SR & QA (Gen.) through 1966," Box 18,153, NHRC. According to Weiss, p. 9, who was Deputy Director of Reliability and Quality Assurance at NASA Headquarters, NASA and the Defense Department

the armed forces in the development of these manufacturing specifications is rather certain.

The Denver Research Institute entry also refers to the application of scanning electron microscopes to the production of integrated circuits. Their statements are close to being on firm ground, but not entirely grounded in the facts. Indeed, NASA played a far more significant role in applying scanning electron microscopy to the manufacture of integrated circuits than their entry acknowledges. The entry begins by stating that "Marshall [Space Flight Center] obtained one of [the] first scanning electron microscopes ever built."[273] The statement is not quite accurate. Scanning electron microscopes had been around for a few decades by the 1960s. Moreover, they probably did not buy one of the earliest ones sold by Westinghouse under the name Micro-Scan and first developed in 1960, as detailed earlier in the section "Scanning Electron Microscope." By their own admission, the researchers in Marshall's Astrionics Laboratory did not start to experiment with the Micro-Scan as a tool for studying integrated-circuit failure mechanisms until 1966.[274] The span of time between 1960 and 1966 suggests that this was not the first Micro-Scan that Westinghouse sold.

The Denver Research Institute entry further states that Marshall "developed SEM [scanning electron microscopy] inspection techniques to analyze failure modes for microelectronics [and] provided failure analysis reports to microelectronic vendors and encouraged manufacturers to use SEM for [the] same purpose."[275] As discussed in the section "Scanning Electron Microscope," Marshall researchers did study device failure mechanisms with their Micro-Scan, but I also noted the work of the team at the Goddard Space Flight Center, which certainly outpublished the Marshall researchers on the subject of applying scanning electron microscopy to the manufacture of microcircuits. These two groups were not the full extent of the effort underwritten by NASA dollars, however.

Three investigators at the University College of North Wales in Bangor—with underwriting from both the Navy Office of the Ministry of Defence

---

collaborated on creating "General Specification For Microcircuits," MIL-M-38510, which called for line certification of items of the highest reliability level.

273. Denver Research Institute, *NASA Benefits Briefing Notebook*, entry B-1.

274. G. Berryman and T. R. Edwards, Marshall Space Flight Center, "Evaluation of Semiconductor Devices Using the Scanning Electron Microscope," in ERC, *Proceedings*, p. 424; Everhart, Wells, and Matta, "A Novel Method of Nondestructive Semiconductor Device Measurements," *IEEE International Electron Devices Meeting* 9 (1963): 72–72G; and Everhart, Wells, and Matta, "A Novel Method of Nondestructive Semiconductor Device Measurements," *Proceedings of the IEEE* 52, no. 12 (December 1964): 1642–1647.

275. Denver Research Institute, *NASA Benefits Briefing Notebook*, entry B-1.

and NASA—also looked into the use of the scanning electron microscope to study solid-state failure mechanisms. Despite some positive results, they concluded, however, that the instruments currently available were not versatile enough to exploit completely the utility of the scanning electron microscope and that the irradiation from the microscope actually altered the properties of the device being examined.[276]

Yet another group working on the use of scanning electron microscopy in the production of solid-state devices with sponsorship from NASA's Electronics Research Center was at Texas Instruments. Building at least partially on the work of Westinghouse researchers Oliver C. Wells, Thomas E. Everhart, and R. K. Matta (who designed the Micro-Scan), they found that the instrument furnished useful information quickly and, in some cases of corrosion failure, provided answers that other methods were unable to deliver.[277] The Denver Research Institute entry elaborates on the Texas Instruments study of the scanning electron microscope. "[The] largest producer of microelectronics," they declare, "Texas Instruments (Texas), with annual sales over $1 billion." The company was "certified by NASA and used 2 [SEMs] for quality assurance and failure analysis."[278] The company did conduct carry out work for NASA under the rubric of the "Development of quality standards inspection criteria and reliability screening techniques for large-scale integrated circuits" because they requested a patent waiver for the techniques developed under the NASA contract in 1968.[279]

The Interplanetary Monitoring Platform

Returning to the earliest technology utilization studies, one finds that NASA was using integrated circuits in a substantial number of applications that required small lots of specialized devices. A survey of electronics in NASA space programs carried out by the Research Triangle Institute, in Durham,

---

276. P. R. Thornton, D. V. Sulway, and D. A. Shaw, "Scanning Electron Microscopy in Device Diagnostics and Reliability Physics," *IEEE Transactions on Electron Devices* ED-16, no. 4 (April 1969): 360–370. See also D. A. Shaw, D. V. Sulway, and R. C. Wayte, *The Further Development of Scanning Electron Microscopy in the Microelectronics Field, Semiannual report, 1 January–30 June 1968* (Washington, DC: NASA, 30 June 1968); and I. G. Davies, D. A. Shaw, D. V. Sulway, P. R. Thornton, and R. C. Wayte, *Device Failure Analysis by Scanning Electron Microscopy* (Washington, DC: NASA, 1 January 1968).

277. Ronald H. Cox, Delbert L. Crosthwait, Jr., and Robert D. Dobrott, "The Application of the Scanning Electron Microscope to the Development of High-Reliability Semiconductor Products," *IEEE Transactions on Electron Devices* ED-16, no. 4 (April 1969): 376–380.

278. Denver Research Institute, *NASA Benefits Briefing Notebook*, entry B-1.

279. NASA, *Nineteenth Semiannual Report to Congress, January 1–June 30, 1968* (Washington, DC: NASA, 1968), p. 176.

North Carolina, in 1965 found integrated circuits in a range of electronic equipment aboard spacecraft and on the ground, from the amplifier circuits in a digital tape recorder to the data processor on the Orbiting Geophysical Observatory, from analog-to-digital converters to the ground support equipment for the Apollo guidance computer.[280]

To these quotidian applications of the integrated circuit one must add a rather impressive, if little known, one. Launched on 26 November 1963, the Interplanetary Monitoring Platform (IMP)—not the Apollo guidance and navigation computers and not the Minuteman II—carried the first integrated circuits into space.[281] Part of the Explorer series[282] of NASA probes, the Interplanetary Monitoring Platform was Goddard's proposal to establish a network of 11 satellites that would collect data on space radiation in support of the Apollo program. Its instruments investigated plasma (ionized gas), cosmic rays, and magnetic fields in interplanetary and cislunar space from a variety of solar and terrestrial orbits.

The first Interplanetary Monitoring Platform (called variously IMP 1, IMP A, and Explorer 18, among other names) functioned normally until 30 May 1964, then intermittently until 10 May 1965, when it was abandoned. The integrated circuits had not contributed to any of the spacecraft's anomalous performance. The mission was a scientific success. The data transmitted to Earth provided the first direct evidence of a collisionless magnetohydrodynamic shockwave surrounding Earth and its magnetosphere. The spacecraft also provided information on the nature of the transition region between the magnetopause and shock front; the magnitude, direction, and variations of the interplanetary magnetic field; and the energy and fluxes of the solar wind and solar and cosmic rays.[283]

The existence of the Van Allen Radiation Belts (evidenced by earlier Explorer probes) forced scientists to revise their theories of how the Sun's particle radiations affected Earth's atmosphere. In the process they made

---

280. Research Triangle Institute, *Microelectronics in Space Research* (Washington, DC: NASA SP-5031, August 1965), pp. 101–114 and 118–121.

281. Specifically, they were SN510 and SN514 integrated circuits manufactured by Texas Instruments. Research Triangle Institute, *Microelectronics in Space Research*, p. 95; Edgar G. Bush, *The Use of Solid Circuits in Satellite Instrumentation* (Greenbelt, MD: Goddard Space Flight Center, July 1964), esp. I and 1–6; and John Rhea, "ERC Is Focal Point of Future Efforts," *Aerospace Technology* 21, no. 11 (20 November 1967): 54.

282. The IMP series consisted of Explorers 18, 21, 28, 33, 34, 35, 41, 43, 47, and 50.

283. Goddard Space Flight Center, *Interplanetary Monitoring Platform, IMP I—Explorer XVIII, Second Interim Status Report* (Greenbelt, MD: Goddard Space Flight Center, July 1965), pp. 1–4 (hereafter, Goddard, *IMP I*); and Goddard Space Flight Center, *Interplanetary Monitoring Platform, IMP II—Explorer XXI, Flight Report* (Greenbelt, MD: Goddard Space Flight Center, August 1965), p. 1 (hereafter, Goddard, *IMP II*).

another fundamental discovery that occupied much of space physicists' satellite research in the 1960s: the existence of a previously unpredicted region above the ionosphere called the magnetosphere. The discovery of the Van Allen Belts also focused attention on the interaction between the Sun and Earth's magnetic field. Gradually a new picture began to emerge. The particles in the so-called solar wind were deflected around Earth by the planet's magnetic field in a manner similar to the way the bow of a boat turns aside water. The first Interplanetary Monitoring Platform measured a distinct "bow shock" area where the solar wind encountered Earth's magnetic field. In fact, IMP A crossed the magnetopause and the bow shock many times.[284]

The second platform, launched 3 October 1964, continued the radiation research of its predecessor on the radiation environment of cislunar space and the quiescent properties of the interplanetary magnetic field but also initiated monitoring that would enable NASA to predict solar flares for the Apollo program. One of the Agency's concerns was the possible radiation hazard that solar cosmic ray events might pose for astronauts. NASA was concerned primarily with solar cosmic rays. When events such as solar flares occur, a greater number of solar cosmic rays are released into space. One of the missions of the Interplanetary Monitoring Platform network was to determine how great a hazard these events might create for humans traveling in space.[285]

The Interplanetary Monitoring Platform program did not just demonstrate how integrated circuits were advancing scientific knowledge. Later platforms also tested cutting-edge integrated circuit technology. One of the newest form of solid-state devices was the metal-oxide semiconductor field-effect transistor (MOSFET), a form of field-effect transistor composed of a channel of n-type or p-type semiconductor material (usually silicon, hence many writers once used the term "metal-oxide silicon"). The "metal" in the name refers to the metallic-oxide gate element used to control the transistor's operation.

The MOS decision is astoundingly daring in light of the novelty of the technology. The MOS transistor had been invented only in 1960, by M. M. "John" Atalla and Dawon Kahng, two engineers at Bell Telephone Laboratories, and the formal announcement of the MOS transistor's existence as a potentially useful technology occurred only in 1963. In 1963 the MOS transistor still was not ready to be sold as a product, but IBM and Fairchild Semiconductor had research programs, even though they would not

284. Homer E. Newell, *Beyond the Atmosphere: Early Years of Space Science* (Washington, DC: NASA, 1980), pp. 179–180; and Lane E. Wallace, *Dreams, Hopes, Realities: NASA's Goddard Space Flight Center: The First Forty Years* (Washington, DC: NASA, 1999), pp. 84–85.

285. Goddard, *IMP II*, pp. 1–2; and Wallace, *Dreams, Hopes, Realities*, p. 86.

be the firms to bring the MOS transistor to market. MOS technology was much slower than the standard transistors found in integrated circuits, but its simplicity of fabrication allowed manufacturers to incorporate more transistors into integrated circuits.[286]

The new MOS technology promised to solve one of spacecraft designers' growing predicaments, the need for greater electronic capability on board for communications and other functions. One can measure this capability in terms of the number of communication channels and the number of transistors used. For example, Explorer 12 (launched 16 August 1961) utilized 20 channels and 200 transistors, while Ariel 1 (an ionospheric probe launched 26 April 1962) made use of 90 channels and 600 transistors. The first three Interplanetary Monitoring Platforms, in contrast, employed had 175 channels and 1,200 transistors, surpassed quickly by the next two platforms, which would have 256 channels and over 2,000 transistors.[287]

The central core of electronics for the Interplanetary Monitoring Platforms was supposed to be a standardized design, but in reality the electronic systems underwent redesign for each successive flight, while still keeping down weight and size.[288] To address the growing need for more channels and electronic circuitry that demanded an increasing number of transistors, the designers of the Interplanetary Monitoring Platforms electronics turned to MOS technology for the IMP D (Explorer 33), IMP E (Explorer 34), and IMP F (Explorer 35). IMP D launched on 1 July 1966, while IMP E and IMP F went up on 24 May 1967 and 19 July 1967. The electronics on the IMP D and IMP E were far complex than those on the first platforms. One difference was the greater functional complexity of its encoding system. IMP A had about 5,000 electrical parts, of which 3,000 were not resistors (that is, capacitors, diodes, transistors, and MOSFETs). If the Goddard Space Flight Center engineers had used the same circuit design and fabrication technique as on the IMP A, they estimated that the encoding system would have had about 9,400 electrical parts, of which 5,600 would not be resistors. Ironically, the IMP A encoding system had pushed the state of the art using conventional components, but now the IMP D would have to go another route.[289]

---

286. Dawon Kahng, "A Historical Perspective on the Development of MOS Transistors and Related Devices," *IEEE Transactions on Electron Devices* ED-23, no. 7 (July 1976): 655–657; Bassett, *To the Digital Age*, pp. 3, 12–13; and Lécuyer, *Making Silicon Valley*, p. 256.

287. Arthur D. Little, *An Examination of the Applicability of Microelectronic Circuits to the Telemetry and Command Subsystems of Several Applications Spacecraft* (Washington, DC: NASA, May 1965), II.2 and III.1.

288. Ibid., III.1.

289. Donald C. Lokerson, *IMP D&E (AIMP) PFM Encoding System Interface Document* (Greenbelt, MD: Goddard Space Flight Center, August 1966), p. 25.

The designers took several steps to reduce the number of parts going into IMP D. One was to eliminate thousands of parts by redesigning several circuits. The engineers rejected using regular integrated circuits as building block circuits, because the integrated circuits were still under development and would not be ready for the IMP D and E flights. The other approach was to use MOS field effect transistors in a basic building block circuit. The Goddard designers saw two promising things happening as a result. One was that about 93 percent of the system could be built as MOSFET blocks or resistors, and the other was that the total number of parts used would dip to less than 4,000, with less than 1,000 of them not being resistors. This approach resulted in the IMP D employing about the same number of parts as the IMP A encoding system, but with fewer nonresistor parts, even though the IMP D had about twice the functional complexity as the IMP A.[290]

Goddard decided to go with the "new animal" for a number of reasons, including its electrical properties and the reduction of parts, not to mention that using the MOS technology would keep the spacecraft on schedule. "Its disadvantage," Goddard engineers declared, "is that 'SUPPOSEDLY' less is known about its long term reliability than is known about the conventional approach." The engineer in charge had been assured "by competent personnel" that the manufacturing processes required to make an all MOSFET monolithic chip were considerably fewer than those for conventional circuits.[291]

The two basic MOSFET blocks for the IMP D were manufactured by the General Micro Electronics Company in California. Surprisingly, they did not buy them from semiconductor leader Fairchild, which had announced in October 1964 its MOS transistor called the FI-100. The announcement was timed to coincide with the release of reliability tests on the devices to show a lack of reliability and stability problems with MOS transistors. Already, however, Frank Wanlass had left Fairchild Semiconductor in December 1963 to join the start-up General Micro Electronics, formed in the summer of 1963 by former Fairchild technical and marketing people. The firm hired him specifically to get into the MOS business. General Micro Electronics, not Fairchild, was the first to introduce an MOS transistor in May 1964.[292]

Wanlass provided the driving force behind the first major push to sell MOS large-scale integrated circuits. Wanlass held many meetings with prospective

---

290. Ibid., pp. 25–26.
291. Ibid., pp. 26 28.
292. Ibid., p. 28; Bassett, *To the Digital Age*, pp. 110, 116, 117, 150–151; Ginzbert et al., *Economic Impact of Large Public Programs*, p. 64; and Lécuyer, *Making Silicon Valley*, p. 256.

customers to promote MOS technology; however, the government was General Micro Electronics' most important early customer. Key to garnering these contracts was the firm's personnel. Art Lowell, the company's founder, was a retired Marine colonel who previously headed the avionics branch of the Navy's Bureau of Weapons and while there had championed the use of integrated circuits in naval systems in the early 1960s. The company's "first MOS contract was for a chip with six to seven MOS transistors for NASA's Interplanetary Monitoring Platform satellite."[293] No wonder, then, that the Goddard engineers, regarding their experience working with General Micro Electronics, reported: "The cooperation received from that plant has been outstanding in that 'prototype' blocks of both types were delivered within a month after they received our drawings." The MOS technology fit the bill, and, in the words of the engineers: "The blocks worked very well."[294]

Final Thoughts

This chapter has reexamined the role of NASA's Apollo program—in particular the Apollo guidance and navigation computer project—as a catalyst in the commercial success of the integrated circuit. While the procurements of integrated circuits may have seemed big, subsequent commercial purchases dwarfed them. Yet the relatively smaller NASA buys would have had a larger impact in the initial integrated-circuit market whose size was considerably small. Furthermore, any discussion of the impact of NASA's procurements must also take into account the near simultaneous purchases of integrated circuits for the Minuteman II missile. One must question whether the injection of government dollars into the two major manufacturers of integrated circuits really had an impact on their overall business viability. Both Fairchild Semiconductor and Texas Instruments were engaged profitably in the production and sale of transistors, diodes, and other electronic devices. The problem was not one of stimulating the producers but of stimulating the consumers.

Scholars Lécuyer and Berlin have provided a different picture of how the integrated circuit achieved commercial success, one in which governmental procurements are absent. The key was Robert Noyce's decision to lower the price of certain integrated circuits below the price of the individual components represented by the microcircuit and below the cost of manufacturing the integrated circuit. The same strategy had worked before but with transistors.

---

293. Bassett, *To the Digital Age*, pp. 110 and 150–151; and Lécuyer, *Making Silicon Valley*, p. 240. Philco-Ford acquired General Micro Electronics in 1966. Ginzbert et al., *Economic Impact of Large Public Programs*, p. 66.

294. Lokerson, *IMP D&E (AIMP) PFM Encoding System Interface Document*, p. 28.

Their interpretation therefore reshapes the problem in terms of stimulating demand rather than production. Furthermore, it precludes NASA's purchases of integrated circuits from having had an impact on the manufacture of integrated circuits for a commercial market.

This chapter also looked at NASA research programs that dealt with the manufacture of integrated circuits and found a substantial amount of activity. The driving force behind the activity was the Agency's struggle to achieve component reliability. Much of the effort took place at the Electronics Research Center, the Agency's lead center for electronics research during the Apollo era, as well as at other Centers such as the Goddard Space Flight Center and the Marshall Space Flight Center. What distinguished the Electronics Research Center from their other centers in this field of endeavor was its possession of a pilot plant for studying the fabrication of integrated circuits called the PREDICT facility.

The research programs discussed in the case studies presented here covered a range of aspects of the manufacture of integrated circuits. Some were investigations of the underlying chemical and physical mechanisms that caused integrated circuits and other solid-state devices to fail, such as gold and hydrogen contamination. Other research involved testing commercial integrated circuits under various physical conditions, such as heat and radiation. In some cases, such as the thin-film space-charge-limited triode and microwave integrated circuits, NASA efforts focused on creating new components based on recent scientific discoveries or on perfecting thin-film technologies that promised to complement or perhaps even replace integrated circuits in hostile space environments.

The space agency's efforts to improve the integrated-circuit fabrication process seem more impressive, but only because the processes and testing methods developed have become familiar aspects of semiconductor manufacturing. Researchers studied the use of ion implantation and invented new patented techniques for separating wafers into chips and for discovering leaks in semiconductor packaging, while others collaborated with universities and the armed forces in the development of computer-aided circuit design. X-ray and infrared radiation became the basis for new testing procedures, and the scanning electron microscope became a quality control tool on the integrated-circuit production line. Finally, one needs to recall that the first integrated circuit to fly in space was not in a Minuteman II missile or an Apollo capsule, but on Goddard's Interplanetary Monitoring Platform.

Still, these successes beg the question of whether or not NASA's research programs had any impact on integrated circuits outside the Agency. Mainly because of the lack of documentation, demonstrating the transfer of a discovery or innovation from a laboratory conducting work on behalf of NASA to its application in industry, commerce, or agriculture is like trying to nail Jell-O to the wall. However, uncommon circumstances allow one to perform a sort of "thought experiment" regarding NASA's impact on the manufacture of integrated circuits. A thought experiment (*Gedankenexperiment*) is a well-structured hypothetical question that uses "What if?" reasoning—an imagined scenario, if you will—to understand the way things actually are. One can perform such a thought experiment in the case of integrated circuits because NASA's Electronics Research Center, which performed the lion's share of the Agency's electronics research during the Apollo era, shut its doors in 1970. Among many other activities, the center's Qualifications and Standards Laboratory performed important analyses of components for a number of NASA space programs. One such investigation carried out by the laboratory for the Jet Propulsion Laboratory involved integrated-circuit failure modes and mechanisms for the Mariner '69 program.[295]

Qualifications and Standards also conducted failure analysis on a group of defective plastic-encapsulated transistors furnished by a system contractor working for the Manned Spacecraft Center (now the Johnson Space Center). The transistors were a part of the Apollo lunar scientific equipment package. The failure modes were determined and a report was sent to the contractor involved together with a compilation of guidelines established by users of high-reliability devices on the suitable application of plastic encapsulated microelectronics. In another instance the laboratory evaluated a specification (S45OP3) for a semiconductor screening technique that Goddard had developed for the Nimbus weather satellite because the Center wanted to apply the specification across all of Goddard's programs.[296]

With the closing of the Electronics Research Center in 1970, did NASA suddenly experience incidents involving defective integrated circuits? In May 1971, Centaur-Mariner H was lost because of an integrated circuit that functioned as an operational amplifier (part RN709) in the pitch channel rate gyro in the flight control system. The associated zener diode also was suspect, but in the end the cause of the mission failure was a $5 integrated circuit. The cost of the lost spacecraft was about $60 million. In addition, the discovery of

---

295. The use of integrated circuits on Mariner '69 is discussed in Edward Clinton Ezell and Linda Neuman Ezell, *On Mars: Exploration of the Red Planet, 1958–1978* (Washington, DC: NASA, 1984), p. 243; and "Research Highlights," *ERC News*, 1 August 1968, n.p., Folder "ERC History House Organ 1966–1969," Box 8, RG 255, AN 71A2309, WNRC. Integrated circuitry and packaging techniques were directly borrowed from Mariner Venus 67 (aka Mariner 5) and the 1969 Mars craft. Ezell and Ezell, *On Mars*, p. 167.

296. "Research Highlights," *ERC News*, 1 January 1969, n.p., Folder "ERC History House Organ 1966–1969," Box 8, RG 255, AN 71A2309, WNRC.

cracks in the silicon integrated circuits in the new Centaur computer resulted in a cost of $500,000 for replacement. Next, four Skylab system test failures occurred because of shorts in microcircuits caused by "gold-ball" contamination. As a result, NASA had to retrofit some 3,000 integrated circuits at a cost of about $4 million. Finally, a metallization problem in integrated circuits caused a complete shutdown of the manufacturer's production line, thereby holding up delivery of some 40,000 integrated circuits for NASA projects at Goddard, JPL, and Marshall.[297]

Despite the efforts to establish the Microelectronics Reliability Program during the 1960s, a draft NASA Management Instruction instituting it was not drawn up until late in 1970, and the Microelectronics Reliability Program was established by NMI 5300.4 on 4 June 1972. Its purpose was to ensure necessary reliability and quality of procured microcircuits in a cost-effective manner. The Management Instruction established the microcircuit reliability program, which culminated the efforts of the Microelectronics Subcommittee. The program provided a uniform approach for the specification, testing, and procurement of microelectronics. Where NASA utilized military specifications, NASA requirements were merged with those of the Defense Department. As for fulfilling the testing and analysis role played by the Qualifications and Standards Laboratory, NASA now utilized the Quality Engineering and Evaluation Laboratory of the Naval Ammunition Depot (later the Naval Weapons Support Center) in Crane, Indiana.[298]

297. J. L. Murphy to Members of the Microelectronics Subcommittee, "Minutes of the Sixteenth Meeting of the Microelectronics Subcommittee," 22 July 1971, Folder "Chron R&QA 1970–1973," Box 18,153, NHRC; Virginia P. Dawson and Mark D. Bowles, *Taming Liquid Hydrogen: The Centaur Upper Stage Rocket, 1958–2002* (Washington, DC: NASA, 2004), p. 120; Memorandum, Director, Reliability & Quality Assurance to Assistant Administrator for Industry Affairs & Technology Utilization, "Resource Requirements for Implementation of the NASA Microelectronics Reliability Program," 24 August 1972, Folder "Chron R&QA 1970–1973," Box 18,153, NHRC.

298. Memorandum, Director, Reliability & Quality Assurance, to Assistant Administrator for Industry Affairs & Technology Utilization, "Resource Requirements for Implementation of the NASA Microelectronics Reliability Program," 24 August 1972, Folder "Chron R&QA 1970–1973," Box 18,153, NHRC; Memorandum, Director, Reliability & Quality Assurance to Distribution List, "Request for Issue Clearance on Draft NASA Management Instruction NMI 5320.(X) 'NASA Microelectronics Reliability Program,'" 11 December 1970, Folder "Chron R&QA 1970–1973," Box 18,153, NHRC; Associate Administrator for Organization and Management to NASA Installations, "18th Meeting of the NASA Parts Steering Committee," 19 March 1973, Folder "Chron R&QA 1970–1973," Box 18,153, NHRC.

Appendix A

NASA Tech Briefs

The following are Tech Briefs issued by the Electronics Research Center that deal with integrated-circuit technology. The list in table 3.1 is not comprehensive but is limited by the availability of documentation.

TABLE 3.1. NASA Tech Briefs for Integrated Circuits Issued by the Electronics Research Center

	Title of Tech Brief	Brief Number	Date Issued
1	Thermal and bias cycling stabilizes planar silicon devices	67-10176	June 1967
2	Failure rates for accelerated acceptance testing of silicon transistors	68-10541	December 1968
3	Reliable method for testing gross leaks in semiconductor component packages	68-10562	December 1968
4	Improved method of dicing integrated circuit wafers into chips	69-10441	September 1969
5	Microelectronic device data handbook	69-10687	December 1969
6	Reducing contact resistance at semiconductor to metal or aluminum to metal interfaces	69-10689	December 1969
7	Controlled substrate cooling improves reproducibility of vapor deposited semiconductor composites	69-10732	December 1969
8	Technique for depositing silicon dioxide on indium arsenide improves adhesion	70-10475	September 1970
9	New method for photoresist stripping	70-10497	September 1970
10	Laser scribing of silicon wafers	70-10437	October 1970
11	Growth of single-crystal gallium nitride	70-10473	October 1970
12	Glass-to-metal bonding process improves stability and performance of semiconductor devices	70-10477	October 1970
13	Copper-titanium eutectic alloy improves electrical and mechanical contact to silicon carbide	70-10444	November 1970
14	Aluminum-silicon eutectic alloy improves electrical and mechanical contact to silicon carbide	70-10445	November 1970
15	p-n junctions formed in gallium antimonide	70-10500	November 1970

Appendix B

Patent Waiver Requests

This appendix provides information on the patent waivers requested by industry and academia for work relating to integrated circuits (and thin-film technology in some instances) conducted with financial support from NASA. The discussion is limited because the source of data, NASA's semiannual reports to Congress, lists patent waiver information only from 1963 (the 9th semiannual report) to 1969 (the 22nd semiannual report). A total of 27 such waiver requests appear in these reports. With three exceptions, the requests came from companies. Some were major aerospace firms, such as General Dynamics, North American Rockwell, and the Lockheed Missiles and Space Company, while others were key electronics businesses that supplied the aerospace industry, such as IBM and TRW. Those leading the pack in requesting waivers were Texas Instruments, Westinghouse, and CBS Laboratories. Outside of industry, only the Massachusetts Institute of Technology and the California Institute of Technology requested waivers. Table 3.2 indicates the number of patent waiver requests reported by company and institution.

TABLE 3.2. Patent Waiver Requests by Company/Institution

Company/Institution	Waiver Request Number
Texas Instruments	5
Westinghouse	4
CBS Laboratories	3
Hughes Aircraft Company	2
Philco Corporation	2
Electro-Optical Systems, Inc.	1
General Dynamics, Electronics Division	1
International Business Machines (IBM)	1
General Precision, Librascope Group	1
Lockheed Missiles and Space Company	1
North American Rockwell, Autonetics Division	1
Thompson Ramo Wooldridge (TRW)	1
Tyco Laboratories	1
California Institute of Technology (CalTech)	1
Massachusetts Institute of Technology (MIT)	2

Table 3.3 comprises the list of waiver requests found through a search of the semiannual reports. It indicates the year of the disposition of the waiver request, the description of the "invention," and the name of the "petitioner" as listed in the semiannual reports, and whether the waiver was denied. All but a few were granted. The citation information denotes in condensed form the number of the semiannual report and the page number. For example, a reference in NASA, *Ninth Semiannual Report to Congress, January 1–June 30, 1963* (Washington, DC: NASA, 1963), page 190, would read "9:190."

TABLE 3.3. Patent Waiver Requests by Invention and Petitioner Name

Year	Invention	Petitioner	Granted/ Denied	Reference
1963	Modular-circuit package	IBM	Granted	9:190
1966	Integrated circuit	Electro-Optical Systems	Granted	15:177
1966	Delivery, development, fabrication, and delivery of improved MOS transistors	Philco	Denied	16:235
1966	R&D of a thin-film space charge limited triode	Hughes Aircraft	Granted	16:236
1967	Contacts for semiconductor devices, particularly integrated circuits, and methods for making the same	Westinghouse	Granted	17:191
1967	Microelectronic S-band receiver	General Dynamics	Granted	17:193
1967	R&D thin-film space-charge limited-triode device	Hughes Aircraft	Granted	17:193
1967	Silicon carbide semiconductor junction device	Tyco Laboratories	Granted	18:198
1967	Semiconductor fabrication technique permitting examination of epitaxially grown layers	Westinghouse	Granted	18:199
1967	Study of cold substrate deposition of thin-film passive elements	Librascope Group	Granted	18:201
1967	Study of failure modes in silicon solid-state devices	Philco	Granted	18:201
1968	Solid state triode	CalTech	Granted	19:174
1968	Integrated circuit thin-film magnetometer	Lockheed	Granted	19:175
1968	Investigation of single-crystal ferrite thin films	North American Autonetics	Denied	19:176
1968	Development of quality standards inspection criteria and reliability screening techniques for large-scale integrated circuits	Texas Instruments	Granted	19:176

(continued)

Year	Invention	Petitioner	Granted/ Denied	Reference
1968	(Complementary Micropower Transistor) Complementary J-K Flip-Flop using transistor logic	Texas Instruments	Granted	20:210
1968	(Complementary multiple emitter micropower) high-speed, low-power magic gate	Texas Instruments	Granted	20:210
1968	Process for fabricating integrated circuits having matched complementary transistors and product	Texas Instruments	Granted	20:210
1968	Monolithic circuit and high Q capacitor	Texas Instruments	Granted	20:210
1968	Shielding method and device for polycrystalline and epitaxy growths	TRW	Granted	20:210
1968	Method for producing dimensionally stable photosensitive resist patterns	CBS Laboratories	Granted	20:211
1968	Method for producing reliable contacts b/n resistors & low sensitivity materials	CBS Laboratories	Granted	20:211
1968	Method for restoring the electrical properties of ion bombarded semiconductor devices	CBS Laboratories	Granted	20:211
1969	Solid state device	MIT	Granted	21:208
1969	Fabrication of compound semiconductor films	MIT	Granted	21:209
1969	Integrated circuit with multiple collection current	Westinghouse	Denied	21:209
1969	Semiconductor integrated circuit having complementary MIS and Darlington bipolar transistor elements	Westinghouse	Denied	21:209

Chapter 4

NASA's Role in the Development of MEMS (Microelectromechanical Systems)

Andrew J. Butrica

Introduction

In various venues, NASA has claimed to have been instrumental in the development of microelectromechanical systems (MEMS). For example, a recent Agency publication stated: "MEMS applications are directly traceable to the miniature accelerometers [devices that detect changes in acceleration] NASA developed in the 1970s to measure changes in speed of small objects or activity levels of people or animals during human space flight. MEMS technology is used now in consumer products to trigger automobile airbags, regulate pacemakers and even keep washers and dryers balanced. MEMS-based products have grown into a $3 billion per year industry. The original NASA-sponsored work on an MEMS accelerometer is referenced in 83 patents; the earliest reference was made in 1975 and the latest in 2003."[1]

This particular passage suggests a rather limited role for NASA in promoting MEMS research—namely in the specific area of "miniature accelerometers"—but the question really is rather broader. This chapter looks at the particular case of NASA's subsidy of accelerometer development during the 1970s, including the claim of references to the "original NASA-sponsored work on an MEMS accelerometer" in 83 patents dating from 1975 to 2003. However, this investigation necessarily must cast a broader net into the field of biomedical instrumentation to capture more faithfully the context within which NASA subsidized the creation of the MEMS accelerometer.

This exploration of NASA's early contributions to MEMS research will be somewhat untidy out of necessity. To begin with, no scholarly history

---

1. NASA Public Affairs Office, *NASA Hits: Rewards from Space*, NP-2004-04-349-HQ, p. 5, *http://www1.nasa.gov/pdf/54862main_hits.pdf* (accessed 9 August 2005); NASA, *Rewards from Space: How NASA Improves Our Quality of Life*, p. 4, *http://www.nasa. gov/pdf/54862main_benefits.pdf* (accessed 9 August 2005); and Newsletter, *Metropolitan Detroit Science Teachers Association,* no. 9 (Winter 2004): 13, *http://www.mdsta.org/ MDSTA_News.1203.pdf* (accessed 20 March 2006).

of MEMS exists to date. Very little MEMS research took place prior to the 1980s, the period of this study, and the term "MEMS" itself did not exist until 1987,[2] as MEMS began to take off both commercially and as an organized discipline. Determining where within NASA this inquiry ought to focus is similarly fraught with ambiguity. Integrated-circuit technologies and manufacturing processes are the sine qua non of MEMS research, and several NASA Centers not only performed research in integrated circuits but also had their own integrated-circuit fabrication facilities. Without ignoring this wide-ranging activity within NASA, this chapter focuses on the early MEMS work carried out at Stanford University's Integrated Circuits Laboratory, because the development of the "miniature accelerometer" in question took place there. Likewise, NASA's Ames Research Center takes center stage because of its role as underwriter of the accelerometer project and other MEMS research. Appendix A provides an overview of the MEMS research that NASA's Ames Research Center underwrote at Stanford University.

Finally, in addition to looking into NASA's claim, this inquiry also asks whether the NASA-funded MEMS work had any social, technological, or economic impact. By focusing on NASA's relationship with Stanford University, this chapter points out the need for further research on the space agency's role in the development of the electronics industry in Silicon Valley. Historians have examined the development of Silicon Valley from the perspective of industry's influence on Stanford[3] as well as relations between Stanford and the federal government.[4] Their silence on the role of NASA suggests that the

---

2.   The first use of the term MEMS was in conjunction with a series of workshops sponsored by the National Science Foundation (NSF) that took place in 1987 and 1988. See the Report of the NSF Workshop on Microelectromechanical Systems Research: Kaigham J. Gabriel, J. F. Jarvis, and William S. Trimmer, "Small Machines, Large Opportunities: A Report on the Emerging Field of Microdynamics," in William S. Trimmer, ed., *Micromechanics and MEMS: Classic and Seminal Papers to 1990* (New York: IEEE Press, 1997), pp. 117–144.

3.   Christophe Lécuyer, *Making Silicon Valley: Innovation and the Growth of High Tech, 1930–1970* (Cambridge: MIT Press, 2006); Leslie Berlin, *The Man Behind the Microchip: Robert Noyce and the Invention of Silicon Valley* (New York: Oxford University Press, 2005).

4.   Stuart W. Leslie, *The Cold War and American Science: The Military-Industrial-Academic Complex at MIT and Stanford* (New York: Columbia University Press, 1993); Stuart W. Leslie, "Profit and Loss: The Military and MIT in the Postwar Era," *Historical Studies in the Physical and Biological Sciences* 21 (1990): 59–85; Stuart W. Leslie, "Playing the Education Game to Win: The Military and Interdisciplinary Research at Stanford," *Historical Studies in the Physical and Biological Sciences* 18 (1987): 55–88; Rebecca S. Lowen, "'Exploiting a Wonderful Opportunity': Stanford University, Industry, and the Federal Government, 1937–1965," Ph.D. dissertation, Stanford University, 1990; and Rebecca S. Lowen, *Creating the Cold War University: The Transformation of Stanford* (Berkeley: University of California Press, 1997).

Agency was a missing actor in the Silicon Valley story, despite the location of the Ames Research Center in the middle of that region. In the end, however, one question remains: Was NASA's role so critical that the Agency should get the credit it claims for MEMS development? Before directly investigating that claim, this chapter first reviews what MEMS are and sketches an outline of their early history.

MEMS History

What Are MEMS?

The term "microelectromechanical system" in the United States (other terms are used elsewhere in the world) denotes a specialized class of integrated circuits. The manufacture of MEMS devices entails the same techniques utilized in the fabrication of integrated circuits, such as photolithography, etching, and doping. However, the idea behind MEMS is not to make tiny electronic circuits but to create miniature devices capable of performing mechanical or other functions. The term "micromachining" broadly refers to the use of integrated-circuit and other precision fabrication techniques to make MEMS, and it connotes the distinct nature of the devices thus formed. As with integrated circuits, MEMS are made not individually but in large batches on wafers, and this batch production lowers unit cost. Today, a MEMS chip might contain both the MEMS device—such as a sensor or accelerometer—and the electronic circuitry associated with it.[5] MEMS are extremely small, but they are not as tiny as nanotechnologies. MEMS range in size from below 1 μm (1 micrometer or micron,[6] equal to 1×10^{-6} meters) to above 1 mm (1 millimeter, equal to 1×10^{-3} meters). In other words, they are smaller than a drop of water or the thickness of a human hair. Even smaller are nanotechnologies. They are measured in nanometers, which are increments equal to one billionth of a meter (1×10^{-9} meters).

Although MEMS devices did not achieve wide commercial success until the 1980s, their origins extend back into the 1950s and 1960s, when researchers began to appreciate the fact that one could make discrete electronic components—transistors and diodes as well as entire integrated electronic circuits—from silicon, germanium, and other semiconductors. At the same time, some researchers realized that one might create miniature sensors and

---

5. Gregory T. A. Kovacs, *Micromachined Transducers* (Boston: WCB McGraw-Hill, 1998), p. 1; and David Morton and Joseph Gabriel, *Electronics: The Life Story of a Technology* (Westport, CT: Greenwood Press, 2004), p. 146.

6. "Micron" is the term used by manufacturers of integrated circuits in place of micrometer, which is one millionth of a meter. A human hair is about 50 micrometers wide.

other devices out of these same materials. By the 1980s the devices fabricated on wafers included gears, shafts, belts, pulleys, valves, springs, mirrors, nozzles, and connectors as well as a wide spectrum of sensors.[7] They quickly found their way into ink-jet printers, automobiles, mobile telephones, and a multitude of medical and scientific applications, such as the instrumentation used in a recent study of glacier activity. Indeed, today MEMS devices are nearly ubiquitous.[8]

Feynman's Challenge

For many, the moment that marked the beginning of MEMS was a 1959 talk given by the influential physicist Richard P. Feynman (1918–1988). For others, it is the defining moment for the birth of nanotechnology.[9] His scientific and societal contributions spanned the theory of quantum electrodynamics, the development of the atomic bomb, and the investigation of the Space Shuttle Challenger disaster. In 1965, for his work on quantum electrodynamics, Feynman received the Nobel Prize along with Julian Schwinger and Sin-Itiro Tomonaga.[10]

Feynman spent much of his career teaching at the California Institute of Technology (Caltech). On 29 December 1959 at the annual meeting of the

---

7. Kovacs, *Micromachined Transducers*, p. 1; Kurt Petersen, "Prologue," in Kovacs, *Micromachined Transducers*, p. iii; Edward Regis, *Nano: The Emerging Science of Nanotechnology: Remaking the World Molecule by Molecule* (Boston: Little, Brown, 1995), p. 209; and Sharon Begley, "Welcome to Lilliput: Devices Smaller Than a Pinhead Explore the Newest Scientific Frontier," *Newsweek* 117, no. 15 (15 April 1991): 61.

8. General Motors Corporation's Cadillac Division used chips with tiny embedded accelerometers to help dissipate bumps in the 1993 model of its Seville Touring Sedan and other automobiles. Peter Coy, "Mighty Mites Hit It Big," *Business Week*, April 26, 1993, p. 93; Erico Guizzo, "Into Deep Ice," *IEEE Spectrum* 42, no. 12 (December 2005): 28–35, especially 32–33; James D. Meindl, interview by Jennifer Ross-Nazzal, Houston, TX, and Atlanta, GA, 11 January 2005, transcript, p. 19; Stephen Terry, interview by Jennifer Ross-Nazzal, Moffett Field, CA, 4 October 2004, transcript, NASA History Office, NASA Headquarters, Washington, DC (hereafter NHO), pp. 26–27; and Kurt W. Petersen, interview by Jennifer Ross-Nazzal, Moffett Field, CA, 5 October 2004, transcript, NHO, pp. 11–12.

9. For those hailing Feynman's challenge as the start of nanotechnology, see, for instance, John Gribbin and Mary Gribbin, *Richard Feynman: A Life in Science* (New York: Dutton, 1997), 170: "hailed today as the first clear statement of the possibilities of nanotechnology"; Regis, *Nano*, pp. 72 and 77; and Jagdish Mehra, *The Beat of a Different Drum: The Life and Science of Richard P. Feynman* (New York: Oxford University Press, 1994), 446.

10. For Feynman's biography, see James Gleick, *Genius: The Life and Science of Richard Feynman* (New York: Pantheon Books, 1992); Gribbin and Gribbin, *Richard Feynman*; Mehra, *Beat of a Different Drum*; and Richard P. Feynman with Christopher Sykes, *No Ordinary Genius* (New York: W.W. Norton, 1994).

American Physical Society—held that year at Caltech—Feynman gave a presentation titled "There's Plenty of Room at the Bottom." In it, he considered the ability to manipulate matter on an atomic scale and some interesting ramifications, particularly the possibility of denser computer circuitry and microscopes capable of seeing things much smaller than were visible through scanning electron microscopes. Feynman proposed that it might be possible to manipulate material at the atomic level using a top-down approach. One would use ordinary machine shop tools to develop and operate a set of ¼-scale machine shop tools, then employ those to develop ¹⁄₁₆-scale machine tools, including miniaturized hands to operate them. This reduction in scale would continue until the tools were able to manipulate atoms directly.

Feynman recognized that this process periodically would require redesigning the tools as different physical forces and effects came into play. For example, gravity would diminish, but the effects of surface tension and van der Waals attraction[11] would increase. Feynman concluded his talk with a pair of challenges that turned out to be remembered a lot longer than the details of his presentation. The physicist offered prizes of $1,000 each to the first person to create an electrical motor smaller than .04 cm (¹⁄₆₄ of an inch) and to the first person to write the information from a book page on a surface ¹⁄₂₅,₀₀₀ smaller in linear scale.[12]

The prize offers spurred innovation but failed to generate any fundamentally new fabrication techniques. Feynman's motor challenge was met by a meticulous craftsman—William H. McLellan, an engineer at Electro-Optical Systems in Pasadena, California—using conventional tools. In 1985, Tom Newman, a Stanford University graduate student, collected the second Feynman prize by succeeding in reducing the first paragraph of Charles Dickens's *A Tale of Two Cities* by ¹⁄₂₅,₀₀₀.[13] These responses met the conditions

11. The phrase "van der Waals attraction" refers to the forces of attraction or repulsion between molecules or between parts of the same molecule other than those caused by covalent bonds or the electrostatic interaction of ions with one another or with neutral molecules.

12. Richard Feynman, "There's Plenty of Room at the Bottom," in Horace D. Gilbert, ed., *Miniaturization* (New York: Reinhold Publishing Corporation, 1961), pp. 282–296; and Feynman, "There's Plenty of Room at the Bottom," pp. 3–9, in Trimmer, *Micromechanics and MEMS*. The talk also is available at *http://www.zyvex.com/nanotech/feynman.html* (accessed 22 February 2006). For an account of the talk and how people reacted to it, see Regis, *Nano*, pp. 72–77; and Mehra, *Beat of a Different Drum*, pp. 441–446.

13. Newman used an electron beam to write the opening page in an area of 5.9 × 5.9 µm. T. H. Newman, K. E. Williams, and R. F. W. Pease, "High Resolution Patterning System with a Single Bore Objective Lens," *Journal of Vacuum Science Technology* B5, no. 1 (January–February 1987): 88–91. William McLellan built an electric motor ¹⁄₆₄ of an inch on a side in 1960. Regis, *Nano*, pp. 72–77; Mehra, *Beat of a Different Drum*, p. 445. The first micromotor (.0001 of an inch across) was made at the University of California at Berkeley in 1988. The creation of a tiny, nanotechnology-scale motor was achieved

that the physicist had laid out, but they did not stimulate the rise of either MEMS or nanotechnology.

First MEMS Devices

Despite the failure of Feynman's challenges to stimulate technological innovation, the development of MEMS devices already was under way when he gave his 1959 American Physical Society presentation. The pressure sensor (also called a strain gauge) was the first MEMS invention and success story. They became a MEMS cash cow for decades because of the wide commercial demand for them, the demand to make them smaller and more reliable, and their relative simplicity as a MEMS technology.[14] Pressure sensors measure strain or pressure because of the piezoresistive property of the material (traditionally, metallic foil or a very fine wire arranged in a grid pattern) used in their construction. The piezoresistive effect, first reported in 1856 by the British scientist Lord Kelvin, is the phenomenon in which the resistance of a metallic wire increases with increasing strain and decreases with decreasing strain.[15] Thus, as the strain on a pressure sensor increases, the electrical resistance of its metallic element increases in direct proportion to the amount of strain, and the change can be detected and measured as a change in voltage.

These sensors remained essentially wire gauges until the 1950s, when a key discovery of one of silicon's many physical properties occurred. In 1953, Charles S. Smith, a physics professor at Case Institute of Technology,[16] took a sabbatical to work at Bell Telephone Laboratories, where he studied the piezoresistivity of the semiconductors germanium and silicon. Smith published

---

later by Mehran Mehregany at Case Western Reserve University. Begley, "Welcome to Lilliput," p. 61; William F. Allman, "Shrinking the Future," *U.S. News & World Report* no. 112 (9 March 1992): 53; and Gary Stix, "Frothing a Raindrop," *Scientific American* 266, no. 5 (May 1992): 128.

14. Terry, interview, pp. 26 and 27; Kurt E. Petersen, "Silicon as a Mechanical Material," *Proceedings of the IEEE* 70, no. 5 (May 1982): 439; Obert N. Tufte, Paul W. Chapman, and Donald Long [all three of the Honeywell Research Center, Hopkins, MN], "Silicon Diffused-Element Piezoresistive Diaphragm," *Journal of Applied Physics* 33, no. 11 (1962): 3322–3327; Wolf D. Frobenius, A. C. Sanderson, and Harvey C. Nathanson [of Westinghouse], "A Microminiature Solid-State Capacitive Blood Pressure Transducer with Improved Sensitivity," *IEEE Transactions on Biomedical Engineering* 20, no. 4 (July 1973): 312–314.

15. B. E. Noltingk, "History of Electrical Devices for Measuring Strains and Small Movements," *Journal of Scientific Instruments* 32 (1955): 157–158; and Kovacs, *Micromachined Transducers*, p. 211.

16. Case Institute of Technology became Case Western Reserve University after its 1967 merger with Western Reserve University.

the results of his research in the prominent *Physical Review* in 1954.[17] The measurements that he obtained indicated that one could use germanium or silicon, instead of metal, to construct strain gauges. In fact, Smith discovered that sensors using germanium and silicon potentially could be far more sensitive than traditional metal gauges.

Not long after Smith's discovery, commercial production of silicon-based pressure sensors began. One of the pioneers in this field was the New Jersey–based firm Kulite Semiconductor Products, founded in 1960 by Anthony D. Kurtz specifically to manufacture silicon-based strain gauges. Kurtz, who had undergraduate and doctoral degrees in physics from the Massachusetts Institute of Technology, worked in laboratories at MIT and Honeywell in Boston before starting his own company.[18] As the company's patent stated, silicon-based gauges had "recently … come into widespread use." The silicon strain gauges were integrated on a thin silicon substrate and employed resistors created by a diffusion process. The thin silicon substrate was mounted on a base to act as a diaphragm for sensing pressure changes.[19]

Kurtz's former employer, Honeywell, also fabricated silicon sensors as early as 1962.[20] Their proprietary method for making thin diaphragms involved mechanically milling a cavity into a silicon substrate. The firm did not begin to apply so-called micromachining techniques until 1970, when Honeywell employed isotropic etching to micromachine silicon diaphragms, and in 1976 started making use of anisotropic etching. Etching is isotropic or anisotropic depending on whether or not the etching takes place along the crystallographic lines of the silicon wafer.[21] In 1971 researchers at Case Western Reserve University developed a pressure sensor as a part of their biomedical applications program. The development of miniature sensors and probes for a gamut of biomedical uses was an important branch of MEMS sensor evolution. When National Semiconductor began selling the first high-volume

17. Charles S. Smith, "Piezoresistance Effect in Germanium and Silicon," *Physical Review* 94, no. 1 (April 1954): 42–49.

18. "Biography: Dr. Anthony D. Kurtz, Chief Scientist and CEO," *http://www.kulite.com/biography.asp* (accessed 28 August 2005).

19. David J. Frist, Anthony D. Kurtz, and Jean-Pierre Pugnaire, "Temperature Compensated Semiconductor Strain Gage Unit," U.S. patent 3,245,252, filed 18 November 1961, issued 12 April 1966, assigned to the Kulite-Bytrex Corporation, Newton, MA.

20. Tufte, Chapman, and Long, "Silicon Diffused-Element Piezoresistive Diaphragm," pp. 3322–3327.

21. For a discussion of the history of anisotropic etching, see Kenneth E. Bean, "Anisotropic Etching of Silicon," *IEEE Transactions on Electron Devices* 25, no. 10 (October 1978): 1185–1192.

pressure sensor in 1974, piezoresistive pressure sensor technology became a low-cost, batch-fabricated manufacturing technology.[22]

Among the many new types of pressure sensors was the MEMS capacitive pressure sensor, developed and demonstrated first at Stanford University in 1977 by Craig S. Sander, James W. Knutti, and James D. Meindl in the Integrated Circuits Laboratory. Instead of measuring resistance to detect pressure changes, capacitive sensors use electrical capacitance—that is, the storage of electrical charge. Capacitive pressure sensors are inherently more sensitive to pressure changes than piezoresistive sensors, but the signal from a capacitive sensor easily can be lost in electronic noise. The diaphragm of the Stanford sensor, which was recessed about 5 micrometers into the silicon wafer, served as one plate of a capacitor. The other capacitor plate was a metal film on a sheet of glass anodically bonded to the wafer.[23]

Major Early MEMS Advances

One of the most important early MEMS achievements was the invention of surface machining in 1965 by researchers at the Westinghouse Research Laboratories in Pittsburgh, Pennsylvania. More precisely, Harvey C. Nathanson and Robert A. Wickstrom created a device called a resonant gate field effect transistor, or resonant gate transistor for short. It was no ordinary transistor but a type of metal-oxide semiconductor transistor. A metal-oxide semiconductor field-effect transistor (MOSFET) consists of a channel of N-type or P-type semiconductor material (usually silicon). The "metal" in the name referred to the metallic-oxide gate element used to control the transistor's operation.[24]

---

22. Eugene M. Blaser and D. A. Conrad, "Active Integrated Pressure Transducer," *Biomedical Sciences Instrumentation* 8 (1971): 83–85; and "Historical Background," *http://mems. cwru.edu/shortcourse/part1_1.html* (accessed 2 March 2006).

23. James B. Angell, Stephen C. Terry, and Phillip W. Barth, "Silicon Micromechanical Devices," *Scientific American* 248, no. 4 (April 1983): 55; and Cheng-Hsien Liu, PowerPoint presentation for course titled "Micro-Electro-Mechanical Transducers," Fall 2004, "MEMS Technology: The Past, Present, and Future," p. 13, *http://memsliu. pme.nthu.edu.tw/~liuch/transducer/Lecture2_History_ho.pdf* (accessed 9 January 2006). Although more than one source—including one from Stanford—gives priority to Stanford for the first capacitive sensor built using integrated-circuit processes, Anthony D. Kurtz, "Pressure Sensitive Transducers Employing Capacitive and Resistive Variations," U.S. patent 3,748,571, filed 7 September 1972, issued 24 July 1973, assigned to Kulite Semiconductors Products, preceded the Stanford work.

24. Harvey C. Nathanson and Robert A. Wickstrom, "A Resonant-Gate Silicon Surface Transistor with High-Q Band-pass Properties," *Applied Physics Letters* 7 (1965): 84–86; Harvey C. Nathanson, William E. Newell, Robert A. Wickstrom, and John Ransford Davis, Jr., "The Resonant Gate Transistor," *IEEE Transactions Electron Devices* ED 14, no. 3 (March 1967): 117–133; Ross Knox Bassett, *To the Digital Age: Research Labs, Start-up Companies, and the Rise of MOS Technology* (Baltimore, MD: Johns Hopkins

The Westinghouse MOSFET consisted of a plated-metal cantilever beam suspended over the channel region of the transistor. The cantilever beam served as the gate electrode. The device probably was the earliest micromechanical cantilever beam experiment. When the device operated as a high-Q ("high quality")[25] electromechanical filter, a DC voltage applied to the beam biased the transistor at a convenient operating point while the input signal electrostatically attracted the beam through the input force plate, thereby effectively increasing the capacitance between the beam and the channel region of the transistor. When the device operated as an analog filter, the input signal caused the beam to vibrate. Only when the signal contained a frequency component corresponding to half the beam's mechanical resonant frequency were the beam motions large enough to induce an output from the underlying transistor structure.[26]

Practical, commercial use of the Westinghouse resonant gate transistors never materialized for a number of reasons. Some reasons arose from technical problems, while others had to do with overall trends in electronics. The most serious technological difficulties Nathanson and Wickstrom encountered were 1) reproducibility and predictability of resonant frequencies, 2) temperature stability, and 3) device failure from fatigue. The main trend in electronics was toward digital circuits, higher frequencies of operation, higher accuracies, and lower voltages—all of which militated against the resonant gate transistor and similar devices.[27]

The importance of this electromechanical curiosity in the history of MEMS was the process devised for creating it—namely, the so-called sacrificial layer technique. As a result, the concept of "surface" micromachining—an essential MEMS fabrication technique—was born. Surface micromachining indicated that the silicon substrate served primarily as a mechanical support on which the micromechanical elements were fabricated. The beam's fabrication began with deposition of a spacer layer, followed by application of a photoresist and removal of those areas where the beam was to be plated. After application of the plating, one stripped the photoresist and etched away the spacer layer,

University Press, 2002), pp. 147 and 148. The work was done under contract with the Air Force. Harvey C. Nathanson, John Ransford Davis, and Terence R. Kiggins, "Resonant Gate Transistor with Fixed Position Electrically Floating Gate Electrode in Addition to Resonant Member," U.S. patent 3,590,343, filed 31 January 1969, issued 29 June 1971, assigned to Westinghouse Electric Corporation, p. 4.

25. "High-Q" or "high-quality" factor in physics and electrical engineering rates the decay (energy dissipation) of anything oscillating. A "high-Q" means a lower rate of energy dissipation and that the oscillations die off at a slower rate.

26. Petersen, "Silicon as a Mechanical Material," pp. 443–444.

27. Ibid., p. 444.

leaving the plated beam suspended above the surface by a distance corresponding to the thickness of the spacer film.[28]

On the other hand, a commercially successful MEMS application initiated during the mid-1970s was the micromechanical ink-jet nozzle array, the foundational technology for today's ink-jet printers. Ernest Bassous, Larry Kuhn, and their colleagues at the IBM Thomas J. Watson Research Center did the pioneering work on these MEMS devices, developing many different nozzle systems.[29] Although ink-jet nozzles per se dated back to the 1960s, their creation did not entail the use of integrated-circuit fabrication techniques. Subsequently in 1979 Hewlett-Packard developed its own MEMS ink-jet nozzles.[30]

The 1970s also saw the rise of MEMS applications—as well as an array of other electronic devices—in automobiles. During those years of awareness of oil consumption imposed by international events—not to mention the first federal regulations on automotive fuel economy and pollution—the pressures on automotive engineers for engine economy and emissions were at their highest. At the same time, however, tooling costs and customer openness to the required changes curbed the rate at which automotive engineers and managers could meet the new economic and pollution requirements by downsizing. A major part of the solution unavoidably involved embracing more electronics, and sensors came to play a central role in the new electronic automobiles. As Lee Iacocca, the chairman of the Chrysler Corporation, declared in 1976: "The world will beat a path to the door of a man with a better MAP [manifold absolute pressure] sensor."[31]

---

28. R. T. Howe, "Surface Micromachining for Microsensors and Microactuators," *Journal of Vacuum Science and Technology* 16 (1988): 1809–1813; and Petersen, "Silicon as a Mechanical Material," p. 443.

29. Ernest Bassous, Howard H. Taub, and Lawrence Kuhn, "Ink Jet Printing Nozzle Arrays Etched in Silicon," *Applied Physics Letters* 31 (1977): 135–137; Ernest Bassous and E. F. Baran, "The Fabrication of High Precision Nozzles by the Anisotropic Etching of (100) Silicon," *Journal of the Electrochemical Society* 125 (1978): 1321; Petersen, "Silicon as a Mechanical Material," pp. 431–432; Angell, Terry, and Barth, "Silicon Micromechanical Devices," pp. 50–51; Ernest Bassous, "Nozzles Formed in Mono-Crystalline Silicon," U.S. patent 3,921,916, filed 31 December 1974, issued 25 November 1975, assigned to IBM; and Bassous, Kuhn, Arnold Reisman, and Taub, "Ink Jet Nozzle," U.S. patent 4,007,464, filed 23 January 1975, issued 8 February 1977, assigned to IBM.

30. See, for example, Richard G. Sweet, "High Frequency Recording with Electrostatically Deflected Ink Jets," *Review of Scientific Instruments* 36 (January 1965): 131; and Morton and Gabriel, *Electronics*, p. 146. Sweet was affiliated with the Systems Techniques Laboratory, Stanford Electronics Laboratories, Stanford University.

31. William G. Wolber and Kensall D. Wise, "Sensor Development in the Microcomputer Age," *IEEE Transactions on Electron Devices* ED-26, no. 12 (December 1979): 1864–1865.

Automotive engine control sensor art became the pacing factor in the introduction of microcomputer-based engine controls. By 1979 about 15 percent of the automobiles produced in the United States had sensor-controlled electronic engine systems, and about 25 percent of the automobiles built around the world had at least one electronic engine control loop. During 1982 and 1983, among the more successful mass-produced MEMS products produced for the automobile industry were manifold absolute pressure sensors. The next MEMS device to make an appearance in the automobile was the airbag system, which used MEMS accelerometers to detect a crash by sensing the change in acceleration. The MEMS devices, the size of a dime and costing but a few dollars each, replaced an electromechanical accelerometer that was roughly the size of a soda can, weighed several pounds, and cost about $15 each. The airbag accelerometer has been one of the real MEMS success stories over time.[32]

The Growth of the MEMS Field

MEMS eventually came to have a number of new automobile applications, including their use to check tire inflation levels and to control skidding. The takeoff of the MEMS field during the 1980s and 1990s was remarkable. The first start-up companies committed to commercializing applications other than pressure sensors—Microsensor Technology, Inc., and Transensory Devices, Inc.—started up in 1981 and 1982. A reflection of the growth of MEMS was the coining of the term around 1987, when a series of three workshops on Microdynamics and MEMS took place in July 1987 in Salt Lake City, Utah; in November 1987 in Hyannis, Massachusetts; and in January 1988 in Princeton, New Jersey. By the mid-1990s MEMS device revenue already was measured in billions and stood at about $10 billion in 2000—more than the $8 billion predicted in 1993. In 2001 the market-research firm Frost & Sullivan estimated that the total MEMS market, then at $1.4 billion, would increase at a compound annual growth rate of 17 percent through 2004, when they expected the market to exceed $3 billion. Automotive applications such as airbag sensors composed one-third of the total market, while

---

32. William G. Wolber, "Automotive Engine Control Sensors '79—An Overview Update," in *Second International Conference on Automotive Electronics, 29 October–2 November 1979* (London: Institution of Electrical Engineers, 1979), p. 125; Wolber and Wise, "Sensor Development in the Microcomputer Age," p. 1865; Allman, "Shrinking the Future," p. 52; Coy, "Mighty Mites Hit It Big," pp. 92–94; Charles P. Poole, Jr., and Frank J. Owens, *Introduction to Nanotechnology* (Hoboken, NJ: J. Wiley, 2003), 333; Petersen, "Prologue," p. iii; Terry, interview, p. 30; James B. Angell, interview by Jennifer Ross-Nazzal, San Francisco, CA, 5 October 2004, transcript, NHO, p. 19; and Morton and Gabriel, *Electronics*, p. 146.

the medical market was the second largest industry using MEMS in products such as disposable blood pressure sensors.[33]

Supporting this surprising growth in the United States was fundamental research undertaken largely at such research universities as the University of California at Berkeley, MIT, the University of Wisconsin, and Georgia Institute of Technology. By about 1983 every leading university in the United States had established major research programs. MEMS industrial research also took off during the 1980s and 1990s. In the early 1980s only about five Fortune 500 companies had research programs in MEMS, while by the end of the 1990s, more than 25 such companies were contributing to the technology. Government funding of research also played a role. In 1991 the National Science Foundation (NSF) and the Defense Advanced Research Projects Agency (DARPA) together spent $2.5 million per year on micromachines. At the same time, private funding of research was between $1.5 million and $2.5 million a year.[34]

One measure of the growth of MEMS was its development as a discipline reflected in the emergence of specialized journals and conferences. A pioneering work was Kurt Petersen's "Silicon as a Structural Material," a review article of the state of the field at that time, which included micromachining and device applications.[35] Before 1980 articles on solid-state sensors were scattered in the journals of various fields, such as electronic devices, automobiles, instrumentation, materials, physics, and analytical chemistry. The journal *Sensors and Actuators*, first published in 1980, aimed to provide a forum for papers in the field. The American Institute of Physics began publishing its quarterly *Journal of Micromechanics and Microengineering* in 1991, while the *Journal of Microelectromechanical Systems*, a quarterly joint publication of the Institute of Electrical and Electronics Engineers (IEEE) and the American Society of Mechanical Engineers (ASME), started in 1992.[36]

MEMS conferences began much earlier, and again sensor research led the way. The International Conference on Solid State Sensors and Actuators, established in 1981, is a biannual (odd years) meeting that rotates between

33. Coy, "Mighty Mites Hit It Big," p. 92; Petersen, interview, pp. 11–12; Petersen, "Prologue," p. iii; "Historical Background"; Technical Insights, *MEMS: Powerhouse for Growth in Sensors, Actuators, and Control Systems*, 2nd edition (Englewood, NJ: John Wiley & Sons, 2001); and "MEMS: Smaller Is the Next Big Thing," NASA Tech Briefs, August 2001, *http://www.findarticles.com/p/articles/mi_qa3957/is_200108/ai_n9002755* (accessed 27 March 2006).

34. Coy, "Mighty Mites Hit It Big," p. 93; Petersen, "Prologue," p. iii; and Begley, "Welcome to Lilliput," p. 61.

35. Petersen, "Silicon as a Mechanical Material," pp. 420–457.

36. Petersen, "Prologue," p. iii.

the United States, Japan, and Europe. The conference publishes a technical digest, and some of the papers appear in special issues of *Sensors and Actuators*. The International Meeting on Chemical Sensors is a biannual conference (held in even years) that also rotates between the United States, Japan, and Europe. It publishes a technical digest as well, along with selected papers in *Sensors and Actuators B*. In addition, in the United States a Workshop on Solid-State Sensors and Actuators has taken place at Hilton Head, South Carolina (in even years), and publishes a technical digest of the papers presented. Furthermore, the international IEEE Workshop on Micro Electro Mechanical Systems met six times beginning in 1987 and ending in 1993, and each workshop published a proceedings digest. Of course, because of the interdisciplinary interest in MEMS, a number of conferences in other fields also offered sessions on MEMS.[37]

Biomedical Instrumentation

To understand why NASA might be interested in funding MEMS research, one first must consider the Agency's interest in biomedical instrumentation that followed logically from its missions to send animals and humans into space. The focus of this effort at NASA Headquarters was the Office of Life Science Programs. The creation of tiny biomedical instruments for taking cardiovascular, neurological, and other measurements and transmitting the data wirelessly to recording stations was a vital piece of this endeavor, at a time when continuous monitoring of this physiological information was a rather novel idea.[38]

The sensors and other biomedical instruments that NASA undertook to build and improve assisted in conducting tests on animals and plants in a space environment—that is, to determine the effects of weightlessness on them. For that purpose, for example, NASA launched three so-called biosatellites between 1966 and 1969 that transported fauna and flora into space and performed various experiments on them.[39] Additional animal trials took place in Earth-bound laboratories as well. Yet another application for these

---

37. Petersen, "Prologue," p. iii.
38. John A. Pitts, *The Human Factor: Biomedicine in the Manned Space Program to 1980* (Washington, DC: NASA Special Publication-1985-4213, 1985), pp. 40 and 73–74; and Stanley M. Luczkowski, "Bioinstrumentation," in Richard S. Johnston, Lawrence F. Dietlein, and Charles A. Berry, eds., *Biomedical Results of Apollo* (Washington, DC: NASA SP-1975-368, 1975), p. 485.
39. On this program, see Thomas P. Dallow, Dennis W. Jenkins, and Joseph F. Saunders, "The NASA Biosatellite Program," *Astronautics and Aeronautics* 4 (January 1966): 48–52; or the more elusive *Biosatellite Project Historical Summary Report*, TM-X-72394

tiny sensors was to look for life on other planets. For that purpose the Jet Propulsion Laboratory created the Voyager Biological Laboratory for a future, but later canceled, undertaking to seek out signs of life on Mars.[40] The twin Viking spacecraft, however, took up that mission in 1975.

Astronauts, too, benefited from the sensor and biotelemetry research. Research on human biomedical instrumentation took place at both the Electronics Research Center (ERC) in Cambridge, Massachusetts, and the Ames Research Center. Researchers in Cambridge cooperated with Dr. Harold Klein, Assistant Director for Life Sciences at Ames,[41] on projects of mutual interest, and starting in 1964, Gene A. Vacca, in the NASA Headquarters Office of Advanced Research and Technology, coordinated bioinstrumentation research at both centers.[42] Research in life sciences started at Ames in 1961 with the arrival of scientists Richard S. Young and Vance Oyama, who built a small "penthouse" laboratory atop the instrument research building, and in November 1961 the Center hired Webb E. Haymaker, a world-renowned neuropathologist, to direct new activities in that field.[43]

The Ames-Stanford Biomed Link

The size and scale of the life sciences effort at Ames echoed in the number and scope of organizational departments that Klein's Life Sciences Office oversaw. Each of the three divisions—Exobiology, Environmental Biology, and Biotechnology—in turn consisted of several branches. For example, the Environmental Biology Division included branches dealing with

---

(Washington, DC: NASA, 1969), which focuses more on mission planning, launchers, and spacecraft design.

40. NASA, *Sixteenth Semiannual Report to Congress, July 1–December 31, 1966* (Washington, DC: NASA, 1967), pp. 83, 86, and 88–90. On the Mars Voyager mission, see Curtis Peebles, "The Original Voyager: A Mission Not Flown," *Journal of the British Interplanetary Society* 35 (1982): 9–15.

41. Ames Research Center (ARC), *Telephone Directory* (Moffett Field, CA: ARC, November 1964), p. 56.

42. Chief, Program & Resources Office, Electronics Research Center, to Chief, Instrumentation Research Laboratory, Electronics Research Center, "Details on IRL Contact with NASA Centers," 30 December 1965, Folder "December [1965]," Box 16, Record Group (hereafter RG) 255, Accession Number (henceforth Acc. No.) 71A2309, Washington National Records Center, Suitland, MD (hereafter WNRC); William Rambo, memo for the record, "NASA Study Program File," regarding "Mr. Gene Vacca, NASA Project Officer," 30 October 1964, Folder 5, "Chron File 1964 Oct.-Dec.," Box 2, ACCN 97-093, Papers of William R. Rambo, SC 132, Department of Special Collections, Stanford University Libraries, Stanford, CA (hereafter Rambo Papers).

43. Glenn E. Bugos, *Atmosphere of Freedom: Sixty Years at the NASA Ames Research Center*, SP-4314 (Washington, DC: NASA SP-2000-4314, 2000), p. 57.

Neurobiology, Experimental Pathology, Physiology, and Biochemical Endocrinology.[44] Splicing these new research disciplines into a laboratory whose historical foundation was aerodynamics and wind tunnels was not a trouble-free operation, as Glenn Bugos, the official NASA Ames historian, has pointed out. Researchers in the life sciences, he wrote, "seemed grafted onto the Center. They used different disciplines, procedures and language. Many of the leading biologists were women, at a time when women were still sparse in the physical sciences. The biologists looked for success from different audiences."[45]

The idea at both Ames and the ERC was to combine sensors and transmitters, so that NASA could implant them in the body of an astronaut permanently for continuous monitoring, while the astronaut moved about freely without having to be attached by cables. The emphasis, a NASA Headquarters report stated, was on "small, light-weight instruments that do not require an umbilical cord." The sensor readings would travel to an onboard telemetry transmitter via "a small, low-powered radio link." The wireless sensor system allowed the astronaut "freedom of motion and small weight so that the instruments will not impede his motion." The effort also involved replacing current sensors and data links—all of which were analog—with digital equivalents.[46]

NASA's thinking was that because "extended manned space missions require the astronaut to perform demanding functions, complete freedom of motion is essential. Hence, attached leads, wires and cables are considered an impediment. In addition, it is desirable to eliminate skin chafing and irritation that would result from long-term application of sensors applied directly to the skin." As Albert J. Kelley, at the time Deputy Director of the ERC, explained: "We visualize teams of scientist-astronauts spending months in outer space conducting exacting scientific investigations including extravehicular research…. It will be important to medically monitor them with devices which are totally remote rather than attached to their skin. Unattached and unimplanted sensors … will allow complete freedom of movement and eliminate discomfort."[47]

Equally integral to NASA's biomedical instrumentation programs was a major attempt to transfer the technology to society ("spinoff" or, in the language of the period, "technology utilization"). Since 1963, NASA Headquarters had

---

44. ARC, *Telephone Directory* (1964), pp. 56–57.

45. Bugos, Atmosphere of Freedom, p. 58.

46. NASA Headquarters, *Space Vehicles, Electronics, and Control Research Program Review* (Washington, DC: Office of Programs, Management Reports, 9 March 1963), p. 113.

47. NASA, Electronics Research Center, news release No. 65-12, 7 April 1965, Folder "Procurement Branch," Box 9, RG 255, Acc. No. 71A3002, WNRC.

the Office of Technology Utilization[48] to formulate and implement the means by which NASA could disseminate the technical byproducts of space research and development most effectively to private industry and other users.[49] The office also involved itself in evaluating how successfully the space agency had passed on its research results to society, a process that came into general vogue as "technology assessments," particularly following congressional debate on the question.[50] NASA's impact on biomedical research was the subject of at least three such assessments.[51]

The School of Medicine at Stanford University was an important player in spinning off NASA biomedical technology with the support of the space agency. Explicitly for that purpose, the School's Cardiology Division set up the Biomedical Technology Transfer team with NASA funding and cooperation in July 1970. Under the direction of Donald C. Harrison, the team consisted of members drawn from the Medical School faculty and NASA engineers, including James A. White, whose expertise was instrumentation and who was head of the Ames Research Center's Instrumentation Division. In addition, individuals from the School of Medicine and Stanford University at large participated as consultants. The institutions participating in the Stanford program included medical centers, hospitals, and private physicians located throughout the West.[52]

---

48. Also known as the Office of Technology Utilization and Policy Planning.

49. Edward E. Furash, "The Problem of Technology Transfer," in Raymond A. Bauer and Kenneth J. Gergen, eds., *The Study of Policy Formation* (New York: Free Press, 1968), pp. 281–328; and Arnold S. Levine, *Managing NASA in the Apollo Era* (Washington, DC: NASA SP-1982-4102, 1982), pp. 42–43.

50. An excellent source on the early technology assessment movement is the collection of essays in George Washington University, Program of Policy Studies in Science and Technology, *Readings in Technology Assessment: Selections from the Publications of the Program of Policy Studies in Science and Technology* (Washington, DC: George Washington University, 1975).

51. See, for instance, Bio-Dynamics, Inc., *Study of the Transferral of Space Technology to Biomedicine: Final Report* (Cambridge, MA: Bio-Dynamics, 21 February 1964); United Aircraft Corporation, Biosciences and Technology, Space and Life Sciences Department, Hamilton Standard Division, *Medical and Biological Applications of Space Telemetry* (Washington, DC: NASA SP-1965-5023, July 1965); and Southwest Research Institute, *Southwest Research Institute Assistance to NASA in Biomedical Areas of the Technology Utilization Program, Quarterly Progress Report No. 4* (San Antonio, TX: Southwest Research Institute, 19 September 1967).

52. NASA, "NASA-Stanford Research Team," news release, 19 August 1971, box 18,422, NASA Historical Reference Collection, NASA History Office, NASA Headquarters, Washington, DC (hereafter NHRC); ARC, *Telephone Directory* (1964), p. 52; and Stanford University School of Medicine, Cardiology Division, *Biomedical Technology Transfer: Applications of NASA Science and Technology: Final Report, July 1, 1971–September 30, 1972* (Palo Alto, CA: Stanford University, 1972), pp. 2–3, 5, and 8–12.

Cardiology led the program because heart illnesses were "by far the leading single cause of death in the United States" and the Cardiology Division was "a nationally recognized authority" in the field. The Cardiology Division reported to NASA. "One major technological need" in cardiology as well as in other areas of medicine was "the development of improved electrodes. Over the past 15 years, there has been little improvement in generally available electrodes." According to one report, "the Stanford team has considered electrode technology a problem of major medical importance." These electrodes became the initial focus of MEMS research at Stanford, but in the Department of Electrical engineering, not the School of Medicine. In addition to electrodes, the program tackled the creation of an apnea monitor, a miniature electrocardiogram (ECG) telemetry unit for ambulatory patients, and a temperature telemeter for gastrointestinal tract diagnoses as well as the digital transmission of medical data.[53]

Stanford was not the only school to participate in such NASA biomedical technology transfer efforts. As early as March 1966, the Solid State Laboratory of the Research Triangle Institute in Research Triangle Park, North Carolina, near Durham, Raleigh, and Chapel Hill, undertook a major effort to apply "NASA science and technology" to biomedical problems and enlisted the support of the Duke University Medical Center, the Bowman-Gray School of Medicine Department of Biomedical Engineering at Wake Forest College, the University of North Carolina Medical School as well as Rockefeller University, Monte Fiore Hospital, and the Albert Einstein Institute in New York.[54]

Stanford University did not act just as a middleman by fostering the adoption of NASA biomedical technologies but also performed research to originate such technologies. As early as the fall of 1961, NASA had research contracts with a number of universities, businesses, and military laboratories for research in the life sciences. Many of these dealt with methods for studying blood flow, and a contract with Corbin-Farnsworth, Inc., called for the company to develop a remote blood pressure monitoring transducer.[55]

Stanford University received a substantial sum of funding from NASA for biomedical research thanks to the efforts of Joshua Lederberg, who in 1958 won the Nobel Prize in Physiology or Medicine with Edward L. Tatum

---

53. *Biomedical Technology Transfer*, pp. 4 and 21–22; and NASA, "NASA-Stanford Research Team."

54. J. N. Brown, Jr., "Biomedical Applications of NASA Science and Technology," *Quarterly Progress Report* 3, 15 December 1966–14 March 1967, Research Triangle Institute, Research Triangle Park, NC, 1967, pp. ii and 1–3.

55. NASA, *Research Grants and Contracts: Quarterly Program Report* (Washington, DC: NASA, 1 October 1961), pp. 19–20.

and George Beadle.[56] In the spring of 1962, Lederberg submitted a proposal to NASA Headquarters for the construction of a $600,000 Biomedical Instrumentation Laboratory at Stanford. The Agency provided $535,000 for its construction between September 1962 and March 1965. Even earlier, beginning 1 April 1960, NASA Headquarters underwrote a long-term research project undertaken by Lederberg on microorganisms. By 1980, NASA had furnished $4,673,087 for the effort.[57] This was the largest life sciences research effort that NASA funded at Stanford, judging from the totality of NASA life sciences grants and research contracts extended to Stanford researchers between 1960 and 1980.[58]

The Ames Research Center began funding biomedical research at Stanford University on 1 October 1961, when the facility commissioned Frank Morrell, M.D., and Lenore K. Morrell, Ph.D., to undertake a study of the physiological mechanisms involved when humans adapted to monotonous environments. Ames continued to underwrite the Morrells through September 1967, providing a total of $316,614, making this one of the most high-priced efforts paid for by Ames.[59] Dozens of grants and research contracts followed, too numerous for repetition here. However, a description of the largest contracts, those totaling over $200,000, gives an idea of what Ames was funding at Stanford.

The largest contract, totaling $1,229,032 between 1968 and 1980, went to Donald C. Harrison, head of the Stanford School of Medicine Cardiology

---

56. Nobelprize.org, "Joshua Lederberg: Biography," *http://nobelprize.org/nobel_prizes/medicine/laureates/1958/lederberg-bio.html* (accessed 17 August 2006).

57. NASA, *Research Grants and Contracts: Supplementary Program Report* (Washington, DC: NASA, 1 March 1962), p. 13; NASA, *NASA's University Program* (Washington, DC: NASA, June 1969), p. 26; and NASA, *NASA's University Program* (Washington, DC: NASA, 1979), p. 27.

58. This statement is based on an examination of life sciences grants and contracts in the space agency's reports. These reports include 1) the quarterly reports issued during the 1960s alternatively as NASA, *Research Grants and Contracts: Quarterly Program Report*, and NASA, *NASA University Research Grants and Contracts: Quarterly Status Report*, and located in boxes 18,414–18,416 at the NHRC; and 2) the annual reports published as NASA, *NASA's University Program*, available at the NHRC and in library government document collections. Information on the life sciences grants and research contracts was extracted from these sources and constituted as a set of spreadsheets that included separate listings by Stanford Principle Investigator, Ames contract overseer, and total dollar amounts. These spreadsheets collectively are referred to in this chapter as the Ames Life Sciences Research at Stanford Database.

59. NASA, Grants and Research Contracts (1 February 1962), p. 4; NASA, Research Grants and Contracts: Quarterly Program Report (Washington, DC: NASA, 1 July 1964), p. 84; NASA, Research Grants and Contracts: Quarterly Program Report (Washington, DC: NASA, 1 January 1966), p. 146; and NASA, Research Grants and Contracts: Quarterly Program Report (Washington, DC: NASA, 1 April 1967), p. 79.

Division, for "evaluation of the cardiovascular system during various circulatory stresses."[60] The second largest research contract went to James D. Meindl, a professor of electrical engineering who was a key player in Stanford's early MEMS research. The funding for the research, for "ultrasonic Doppler measurement of renal artery blood flow," amounted to $695,154 from 1972 to 1979.[61] Max Anliker, a professor of aeronautics and astronautics in the School of Engineering from 1958 to 1971, received $482,486 from Ames for a study of "biomechanics within the field of cardiovascular physiology" between September 1967 and June 1973.[62]

Ames paid for a study of the "effects of altered gravitational stress on fluid balance and the circulation of the blood," by John A. Luetscher, Jr., who led the transformation of the Stanford University Medical Center into an academic medical school from a hospital-based institution located in San Francisco, disbursing a total of $392,090 between 1970 and 1980.[63] In order to design and develop special experimental hardware, the Ames Biosatellite program paid out $372,123 to Colin S. Pittendrigh, known as the "father of the biological clock," for his pioneering work between 1969 and 1974.[64] Lastly, Thomas R. Kane, a professor of mechanical engineering, conducted research on the "dynamics of the human body in free fall" from 1966 to 1971 for Ames, which paid him $245,396.[65]

60. *NASA's University Program* (1969), p. 24; NASA, *NASA's University Program* (Washington, DC: NASA, May 1972), p. 28; NASA, *NASA's University Program* (Washington, DC: NASA, 1973), p. 24; NASA, *NASA's University Program* (Washington, DC: NASA, 1974), p. 20; and *NASA's University Program* (1979), p. 27.

61. *NASA's University Program* (1979), p. 27.

62. *NASA's University Program* (1973), p. 23. For biographical information on Anliker, see *Stanford Magazine*, November–December 2002, "Obituaries: Faculty/Staff," *http://www.stanfordalumni.org/news/magazine/2002/novdec/classnotes/obituaries.html* (accessed 18 August 2008).

63. *NASA's University Program* (1979), p. 27. For biographical information on Luetscher, see Myron H. Weinberger, "John Arthur Luetscher, Jr. (1913–2005)," *Hypertension* 47 (2006): 627–628, *http://hyper.ahajournals.org/cgi/content/full/47/4/627* (accessed 14 August 2008).

64. *NASA's University Program* (1974), p. 18. On Pittendrigh, see Stanford University, "Colin Pittendrigh, 'Father of Biological Clock,' Dies at 77," news release, *http://news-service.stanford.edu/pr/96/960325pittendrig.html* (accessed 15 August 2008).

65. *NASA's University Program* (1972), p. 29. On Kane, see his Web site, Stanford Engineering Faculty, "Thomas R. Kane," *http://me.stanford.edu/faculty/facultydir/kane.html* (accessed 22 August 2008).

Instrument Design

This varied research represented an extension of the work being conducted at the Ames Research Center by its own staff. Within the Ames organization, researchers in the life sciences divisions worked in tandem with technicians from James White's Instrumentation Division. In contrast, an examination of the grants and research contracts in the life sciences that Ames awarded to Stanford faculty indicates that the overseers of the funding (the NASA contract representatives) were all in the life sciences divisions.[66] The collaboration of life scientists and instrument technicians echoes the relationship that Thomas Edison had with instrumentmakers such as John Kreuzi in his laboratory.[67] It also raises questions about the friction between the scientists of the life sciences divisions and the Instrumentation Division that Bugos describes. Indeed, some of the earliest biomedical work at Ames grew out of the laboratory's aerodynamic culture and its need for testing instruments.

Such was the case of the miniature capacitive accelerometer that Grant W. Coon, a member of the Measurements Research Branch of the Ames Instrumentation Division,[68] developed and patented in 1959, originally for testing model aircraft and missile parts in wind tunnels. Patent records show that Coon's work soon entered the domain of biomedical instrumentation.[69] In order to measure blood pressure, one mounted his tiny ("ultraminiature") diaphragm-type capacitance transducer on the end of a cardiac catheter. The device allegedly was small enough to be inserted into a baby.[70] By 1967, Ames staff members were testing the probes on anesthetized dogs. The probes were less than 0.13 cm (.05 inches) in diameter, and after they entered an artery, technicians maneuvered them into the left ventricle to measure the blood pressure inside the dog's artery.[71]

---

66. Ames Life Sciences Research at Stanford Database.

67. William Kennedy-Laurie Dickson and Antonia Dickson, *The Life and Inventions of Thomas Alva Edison* (New York: T.Y. Crowell, 1894), p. 107.

68. ARC, *Telephone Directory* (15 October 1963), pp. 7 and 44.

69. Grant W. Coon, "Diaphragm Type Capacitance Transducer," U.S. patent 3,027,769, filed 3 March 1959, issued 3 April 1962, assigned to NASA; Grant W. Coon and Donald R. Harrison, *Miniature Capacitive Accelerometer Especially Applicable to Telemetry*, ARC 72 (Moffett Field, CA: Ames Research Center, 1 November 1966); Grant W. Coon, *A Capacitive Accelerometer Suitable for Telemetry*, NASA TM-X-2644 (Moffett Field, CA: Ames Research Center, 1 September 1972). See as well John Dimeff, James W. Lane, and Grant W. Coon, "New Wide-Range Pressure Transducer," *Review of Scientific Instruments* 33 (1962): 804–807.

70. Grant W. Coon, *Ultraminiature Manometer-Tipped Cardiac Catheter*, ARC 10054 (Moffett Field, CA: Ames Research Center, 1 December 1967).

71. NASA Tech Brief, "Ultraminiature Manometer-Tipped Cardiac Catheter," December 1967, Folder 6911, NHRC. Coon later developed a capacitive pressure transducer

Coon was not the only member of the Ames Instrumentation Division investigating biomedical devices for the space program. Indeed, as Edwin P. Hartman pointed out in his history of Ames: "Although the life-science and the physical-science groups at Ames had little in common, mutual benefits occasionally arose from their being together…. One area in which the physical scientists were able to help the life scientists was in instrument design." Thomas Benton Fryer and Gordon J. Deboo, he added, "were particularly active in this field."[72] Fryer, born 17 July 1925, spent the first 13 years of his life in China, where he was born. When his family returned to the United States in the late 1930s, Fryer attended high school, and during World War II he joined the Marine Corps, which sent him to the universities of Colorado and Michigan. Upon his graduation as an electrical engineer from the University of Michigan at the end of the war, Fryer worked for the Naval Research Laboratory for two years and, in 1949, began work at Moffett Field, one of several laboratories operated by the National Advisory Committee for Aeronautics (NACA) and which became NASA's Ames Research Center in 1958.[73] Fryer's career at Ames thus spanned from the NACA wind tunnels years into the era of biomedical research.

Fryer, then assistant chief of the Electronics Research Branch, and Gordon Deboo worked with Charles M. Winget from the Physiology Branch (Environmental Biology Division).[74] Before becoming a research scientist at Ames, Winget, who received a Ph.D. in comparative physiology from the University of California at Davis in 1957, had a postdoctoral fellowship at the National Institutes of Health and a teaching position at the University of Guelph in Canada. Starting in 1977, Winget taught a graduate course in biorhythms at UC Davis.[75] Avian biorhythms were the subject of his research with Fryer and Deboo.

---

specifically for high-temperature environments, which built on his earlier work. Grant W. Coon, "Trielectrode Capacitive Pressure Transducer," U.S. patent 3,948,102, filed 17 July 1975, issued 6 April 1976, assigned to NASA; and Grant W. Coon, *Trielectrode Capacitive Pressure Transducer*, ARC-10711 (Moffett Field, CA: Ames Research Center, 1 February 1975).

72. Edwin Hartman, *Adventures in Research: A History of Ames Research Center, 1940–1965* (Washington, DC: NASA SP-1970-4302, 1970), p. 483.

73. Michael Taylor, "Thomas Benton Fryer—NASA Engineer," *San Francisco Chronicle*, 9 July 2005, p. B-7; electronic edition, SFGate.com, *http://www.sfgate.com/cgi-bin/article.cgi?file=/c/a/2005/07/08/BAGHJDKORC1.DTL&type=printable* (accessed 30 September 2008).

74. ARC, *Telephone Directory* (December 1967), pp. 9, 13, 40, 46, and 50. As late as 1963, the Electronics Research Branch was known as the Measurements Research Branch. ARC, *Telephone Directory* (1963), p. 44.

75. A. H. Smith, E. M. Bernauer, A. L. Black, R. E. Burger, J. H. Crowe, J. M. Horowitz, G. P. Moberg, and E. M. Renkin, "History of Physiology at University of California, Davis,"

Winget, Fryer, and Deboo developed an instrument that made measurements inside living subjects and transmitted the information to an external source that recorded the data. Hartman described it as "particularly useful."[76] Applying the device to research on various birds, they obtained quantitative measurements of fowl ovulation and oviposition, determined their heart rates and deep body temperatures, and analyzed the circadian cycles of those temperatures.[77] Winget also oversaw a research contract on circadian rhythms undertaken by Stanford University professor Colin Pittendrigh, whose specialty was biological clocks.[78]

The system that Fryer and Deboo developed with Winget—battery included—was no larger than a penny, but it used individual electronic components rather than an integrated circuit.[79] In 1965, Fryer and Deboo actually considered using integrated circuits because of their small size and potentially higher reliability. They rejected the idea, however, because of what they perceived as the many disadvantages of integrated-circuit technology: high cost, relatively high power requirements, poor component tolerances, low yields, limited component value ranges, the fact that some components—such as tantalum capacitors—could not be integrated, and the experimental nature at that time of radio-frequency integrated circuitry.[80]

Cardiovascular studies at Ames also drew on the collaboration of instrument and life sciences researchers. For example, in 1967, Deboo and Jack M. Pope in the Electronics Research Branch developed a so-called cardiotachometer with David B. D. Smith in the Human Performance Branch (Biotechnology

---

The Physiologist 29, no. 5 supplement (1986): 52, *http://www.the-aps.org/publications/tphys/legacy/1986/issue5/46.pdf* (accessed 22 August 2008).

76. Hartman, Adventures in Research, p. 483.

77. Charles M. Winget, E. G. Averkin, and Thomas B. Fryer, "Quantitative Measurement by Telemetry of Ovulation and Oviposition in the Fowl," *American Journal of Physiology* 209, no. 4 (October 1965): 853–858; and Charles M. Winget and Thomas B. Fryer, "Telemetry System for the Acquisition of Circadian Rhythm Data," *Aerospace Medicine* 37, no. 8 (August 1966): 800–803.

78. NASA's University Program (1972), p. 30.

79. Thomas B. Fryer, Gordon J. Deboo, and Charles M. Winget, *Miniature Long Life Temperature Telemetry System*, TM-X-56104 (Washington, DC: NASA, 1964); Gordon J. Deboo and Thomas B. Fryer, "Miniature Biopotential Telemetry System," *American Journal of Medical Electronics* 13 (July–September 1965): 138–142; Thomas B. Fryer, Gordon J. Deboo, and Charles M. Winget, "Miniature Long-Life Temperature Telemetry System," *Journal of Applied Physiology* 21, no. 1 (January 1966): 295–298; and Thomas B. Fryer, "Survey of Biomedical Instrumentation Developments at Ames Research Center," *National Biomedical Sciences Instrumentation Symposium, Biomedical Sciences Instrumentation; Proceedings*, vol. 3 (1967): 103–112.

80. Deboo and Fryer, "Miniature Biopotential Telemetry System," p. 140.

Division).[81] The instrument detected and displayed the human heart rate during physiological studies.[82] Later, in 1972, Fryer and Pope devised and patented a digestible temperature transmitter. This endoradiosonde measured human deep body temperature to indicate general health, but it also was useful for investigating animal circadian rhythms.[83] Fryer, Pope, and Winget also developed a multiple-channel transmitter capable of measuring several body temperatures simultaneously for a more complete study of the circadian rhythm of deep body temperatures.[84] In addition, Pope worked with John Dimeff, who had become the head of the Instrument Division, in designing an apparatus that automatically detected if hospital patients with surgically implanted tracheotomy tubes failed to breathe. Like many of these biomedical devices, it featured a miniature radio transmitter that allowed technicians to monitor patients from a remote location and freed the patient of extraneous wires.[85]

By 1970, Fryer could point to a substantial array of tiny biomedical instruments that he and his fellow members of the Instrument Division had devised for wirelessly transmitting data on live subjects. Some devices telemetered electroencephalogram (EEG) and electrocardiogram (ECG) signals (known also as biopotential data) to recording equipment, while others measured pressure using commercially available solid-state strain gauges planted in animal subjects. Yet other instruments recorded internal body temperatures in humans and animals. However, if one looks at the overview of Ames progress in this field that Fryer provided in 1970, one does not encounter any reference to integrated circuits or the processes for making them.[86]

81. ARC, *Telephone Directory* (1967), pp. 30, 34, and 50.

82. Gordon J. Deboo, Jack M. Pope, and David B. D. Smith, "Cardiotachometer with Linear Beat-to-Beat Frequency Response," *Psychophysiology* 4 (1967): 486–492.

83. Jack M. Pope, Thomas B. Fryer, and Harold Sandler, *An Ingestible Temperature-Transmitter*, ARC-10583 (Moffett Field, CA: Ames Research Center, 1 July 1972); Jack M. Pope and Thomas B. Fryer, "Miniature Ingestible Telemeter Devices to Measure Deep-Body Temperature," U.S. patent 3,971,362, filed 27 October 1972, issued 27 July 1976, assigned to NASA.

84. Thomas B. Fryer, Jack M. Pope, and Charles M. Winget, *An Implantable Multi-Channel Temperature Transmitter*, TM-X-61199 (Washington, DC: NASA, 1967).

85. John Dimeff and Jack M. Pope, *Automatic Patient Respiration Failure Detection System with Wireless Transmission*, ARC-10174 (Moffett Field, CA: Ames Research Center, 1 October 1968); and Jack M. Pope, John Dimeff, and S. Abraham, "A Wireless Respiration Failure Detection System," *Medical and Biological Engineering* 12, no. 3 (May 1974): 348–354, which states: "Research supported by the Anita Oliver Lunn Foundation."

86. A good review of these instruments and techniques is provided in Tom Fryer, *Implantable Biotelemetry Systems* (Washington, DC: NASA SP-1970-5094, 1970).

The one exception is a telemeter that Fryer himself devised in 1966 and patented for implanting in animals.[87] He offered that one could build it either from conventional discrete parts or as an integrated circuit.[88] Fryer, as an electrical engineer and assistant chief of the Electronics Research Branch (which became the Electronic Instrument Development Branch in 1971),[89] was knowledgeable about integrated-circuit technology, but he rejected it in 1970 because of its excessive cost and the state of the art. He explained that "the making of the necessary masks and the processing of monolithic integrated circuits" was "very expensive"; as a result, "custom-made circuits for specific requirements of biotelemetry are impractical. The standard off-the-shelf devices are inexpensive because of the economics of mass production, but to date they have not met the power and voltage requirements of this application.... The present state of the art makes it impractical to construct a circuit, such as the temperature transmitter, on one monolithic integrated circuit; the circuit is too complex for one chip without very poor production yields and resultant high cost."[90]

Fryer's thinking may have represented the "state of the art" in thinking at the Ames Research Center regarding the application of integrated-circuit technologies to biomedical instrumentation in 1970. That thinking, however, did not stop him from being the first Ames employee to oversee a MEMS project at Stanford University starting in November 1969.[91] By then, Stanford already had produced its first MEMS device. The exact details of how Ames came to initiate MEMS research at Stanford are not clear; however, a probable narrative emerges from the recollections undertaken for this chapter in conjunction with extant information on grants and research contracts let to Stanford by the Ames Research Center.

Ames and MEMS

According to Kensall D. "Ken" Wise, one of the leading figures of Stanford's pioneering MEMS research, Nigel C. Tombs, a NASA Ames employee, "may have provided the spark" that instigated that MEMS project.[92] There is reason

---

87. Tom Fryer, "Telemeter Adaptable for Implanting in an Animal," U.S. patent 3,453,546, filed 5 November 1966, issued 1 July 1969, assigned to NASA.

88. Tom Fryer, *Miniature Bioelectric Device Accurately Measures and Telemeters Temperature*, ARC-52 (Moffett Field, CA: Ames Research Center, 1 February 1966).

89. ARC, *Telephone Directory* (October 1971), p. 48.

90. Fryer, Implantable Biotelemetry Systems, p. 83.

91. NASA, NASA's University Program (1972), p. 28.

92. Kensall D. Wise, telephone interview by Jennifer Ross-Nazzal, Houston, TX, and Ann Arbor, MI, 6 December 2004, transcript, NHO, p. 16.

to believe that Tombs may have ignited the interest of the Ames Research Center in the Stanford MEMS work, for he was intimately involved in the research that Ames funded at Stanford and he had specialized knowledge of integrated-circuit fabrication techniques. Tombs was a recent arrival at Ames, having come from the ERC as part of NASA's closing of that facility. The Center's termination entailed merging the ERC's bioinstrumentation projects, including studies in bionics, bioinstrumentation, human spacecraft monitoring, the bioinstrumentation of flight experiments, and advanced biosensors with those at Ames. Transferred to Ames effective on 27 June 1970, Tombs worked for the Measurement Sciences Branch under Boris Ragent.[93] Although new to Ames and NASA as a whole, Tombs brought with him a long career of laboratory work and innovation that stretched back to his native country of England.

Born in Swindon, England, a small town about 80 miles west of London, on 27 November 1925, Tombs earned a B.S. from the Royal College of Science in 1945 and a Ph.D. from Imperial College in 1949 before working for the General Electric Company, Ltd., at their Hirst Research Laboratories. Established in 1919 in Wembley, the Hirst was the first industrial research laboratory in England.[94] During his 10 years at General Electric, Tombs was the group leader of their magnetic material research laboratories[95] and received six British patents relating to the manufacture of permanent magnets and assigned to General Electric.[96] He also developed a new technique for making

---

93. Boyd C. Myers II, *A Report on the Closing of the NASA Electronics Research Center, Cambridge, Massachusetts* (Washington, DC: NASA, 1 October 1970), pp. 232 and 234; and Glenn Bugos, e-mail message to author, 1 September 2005.

94. The General Electric Company, Ltd., was not at all associated with the U.S. company of the same name but was founded by two German immigrants, Hugo Hirst and Gustav Binswanger (later Byng). On the GE laboratories, see "Physics in Industry at the Wembley Laboratories," *Nature* 111 (10 March 1923): 344–345; and Sir Robert Clayton and Joan Algar, "GEC Hirst Research Centre," *Physics in Technology* 16 (1985): 76–84; as well as Robert Clayton and John Algar, *The GEC Research Laboratories 1919–1984* (London: P. Peregrinus in association with the Science Museum, 1989).

95. *Who's Who in Government*, 1st edition (Chicago: Marquis Who's Who, 1972), p. 511; and *Who's Who in Government*, 2nd edition Chicago: Marquis Who's Who, 1975), p. 624.

96. The patents were found through a search of Nigel C. Tombs as inventor in the database maintained by the European Patent Office at *http://ep.espacenet.com/* (accessed 17 January 2006). In chronological order by filing date, the following are the patents in question: Tombs, "Improvements in or relating to the manufacture of permanent magnets," GB patent 761,459, filed 8 September 1953, issued 14 November 1956, assigned to General Electric; Tombs, "Improvements in or relating to the production of gamma ferric oxide," GB patent 755,852, filed 8 September 1953, issued 29 August 1956, assigned to General Electric; Tombs, "Improvements in or relating to apparatus for the manufacture of powdered materials," GB patent 749,265, filed 10 September 1953, issued 23 May 1956, assigned to General Electric; Tombs, "Improvements in or relating to the manufacture of

the ceramic dielectric material used to make capacitors for radios and other electronic equipment and assigned the patent to General Electric in Britain as well as to Hazeltine Research, Inc., of Chicago in the United States.[97]

Upon leaving General Electric in 1959, Tombs and his wife moved to the United States, where he worked on semiconductor devices in RCA's Semiconductor and Materials Division in Somerville, New Jersey, which at the time was engaged in fabricating a novel type of device known as a metal-oxide semiconductor (MOS).[98] Next, briefly in 1961, Tombs was head of General Precision's solid-state laboratory, and carried on solid-state research at the Sperry Rand Research Center in Sudbury, Massachusetts, before working for NASA.[99] At Sperry Rand, Tombs's research turned toward the improvement of integrated-circuit fabrication processes, as witnessed by his patent applications. His new technique involved depositing an insulating film of silicon nitride (Si_3N_4) on a silicon chip in an atmosphere of silane (SiH_4) and ammonia at temperatures ranging from 600°C to 1,000°C. The silicon nitride replaced the silicon oxide commonly used to form p-n junctions (the building blocks of solid-state circuitry) in integrated circuits and as the insulating dielectric in MOS transistors and diodes. By baking at temperatures below the norm (1,000°C), the process was far less likely to cause faults in the resulting

permanent magnets," GB patent 723,496, filed 5 October 1953, issued 9 February 1955, assigned to General Electric; Tombs, "Improvements in or relating to the manufacture of permanent magnets," GB patent 723,497, filed 5 October 1953, issued 9 February 1955, assigned to General Electric; Tombs, "Improvements in or relating to the manufacture of magnetizable powder cores," GB patent 811,935, filed 1 June 1955, issued 15 April 1959, assigned to General Electric. The search revealed several additional patents relating to magnets issued in West Germany and the United States but assigned to the Société d'Electro-Chimie d'Electro-Métallurgie et des Aciéries Electriques d'Ugine of Paris. That company specialized in, among other electrochemical products, steel alloys. See Charles Le Menestrel, *Ugine: Histoire des aciéries électriques* (Lyons: Editions Lyonnaises d'Art et d'Histoire, 1994).

97. The original patent was Tombs, "Improvements in or relating to ceramic dielectric composition," GB patent 698,946, filed 26 May 1950, issued 28 October 1953, assigned to General Electric. The same patent then was assigned to Hazeltine in the United States: Tombs, "Ceramic dielectric materials and methods of producing the same," U.S. patent 2,768,901, filed 22 May 1951, issued 30 October 1956, assigned to Hazeltine. Tombs, "Improvements in or relating to the manufacture of ceramic bodies composed of one or more oxides," GB patent 828,066, filed 21 November 1955, issued 17 February 1960, assigned to General Electric, represents an improvement on the manufacturing process.

98. Kenyon Kilbon, "Pioneering in Electronics: A Short History of the Origins and Growth of RCA Laboratories, Radio Corporation of America, 1919 to 1964," unpublished manuscript, revised August 1964, pp. 317–318, available at the David Sarnoff Library, "Pioneering in Electronics," *http://www.davidsarnoff.org/kil.html* (accessed 2 August 008); and Bassett, *To the Digital Age*, pp. 42–43.

99. *Who's Who in Government*, 1st edition (Chicago: Marquis Who's Who, 1972), p. 511; and *Who's Who in Government*, 2nd edition Chicago: Marquis Who's Who, 1975), p. 624.

solid-state products. Sperry Rand considered the technique so important that it filed for patents in Canada, Europe, and the United States.[100]

Thus, before working at NASA, Tombs possessed intimate knowledge of integrated-circuit fabrication techniques and had contributed to the advancement of that knowledge. Furthermore, published sources indicate that while at Ames, Tombs personally assisted in the MEMS research undertaken at Stanford. In two journal articles published in IEEE journals, Stanford graduate student Samaun Samadikun acknowledged the "cooperation and guidance" of "N. Tombs."[101] He also noted that Tombs had assisted him in testing the sensors on animals.[102] Given this social propinquity and his familiarity with integrated-circuit fabrication techniques, he very well "may have provided the spark" that Ken Wise attributed to him.[103]

In acknowledging the "cooperation and guidance" of Ames personnel, Samadikun also thanked "J. [John] Dimeff, and B. [Benjamin] Beam of the NASA Ames Research Center."[104] John Dimeff put forth his own claims as the initiator of the Ames MEMS research at Stanford in a letter dated 29 August 2005 and faxed to Ames historian Glenn Bugos.[105] Dimeff's claims are dubious at best; they appear to present a cloudy and at times counterfactual recollection of events, suggesting a tangential role at best. Dimeff asserted that the MEMS contracts let to Stanford "were both the result of concepts I

---

100. Tombs, "Silane Method for Making Silicon Nitride," U.S. patent 3,573,096, filed 23 June 1965, issued 30 March 1971, assigned to Sperry Rand Corporation. A search of the European Patent Office database reveals that Tombs also received a patent for the invention in Britain, Canada, France, West Germany, the Netherlands, and Sweden. In all instances Sperry Rand was the assignee.

101. Samaun Samadikun, Kensall D. Wise, E. Nielsen, and James B. Angell, "An IC Piezoresistive Pressure Sensor for Biomedical Instrumentation," *1971 IEEE International Solid-State Circuits Conference Digest of Technical Papers* 14 (February 1971): 104; and Samaun Samadikun, Kensall D. Wise, and James B. Angell, "An IC Piezoresistive Pressure Sensor for Biomedical Instrumentation," *IEEE Transactions on Biomedical Engineering* BME-20, no. 2 (March 1973): 109.

102. Samadikun, Wise, and Angell, "An IC Piezoresistive Pressure Sensor for Biomedical Instrumentation," pp. 108 and 109; and Samaun Samadikun, "An Integrated Circuit Piezoresistive Pressure Sensor for Biomedical Instrumentation," Ph.D. dissertation, Stanford University, August 1971, pp. 39–42.

103. The author was unable to reach Tombs for comment. Tombs's telephone number is unlisted, and he did not respond to two letters sent to him by the author.

104. Samadikun, Wise, Nielsen, and Angell, "An IC Piezoresistive Pressure Sensor for Biomedical Instrumentation," p. 104; and Samadikun, Wise, and Angell, "An IC Piezoresistive Pressure Sensor for Biomedical Instrumentation," p. 109.

105. Letter, John Dimeff to Glenn Bugos, 29 August 2005, attachment to e-mail, from Glenn Bugos, 7 September 2005, furnished to author by Bugos, historian to NASA's Ames Research Center.

suggested."[106] He lacked Tombs's background in the fabrication of integrated circuits; rather, he was steeped in the laboratory's tradition of aerodynamics and wind-tunnel testing as one of its major instrument designers. According to Ames historian Edwin Hartman: "The development of wind-tunnel instrumentation was an activity which grew in complexity and volume, and before the end of 1958 represented a rather large part of the total instrument development effort." With the formation of the new Instrument Research Division, Dimeff became its chief. Hartman also noted Dimeff's contribution to the improvement of the magnetometer, a device that measures the direction and intensity of a magnetic field, and Dimeff patented a few magnetometer and electrometer improvements during the 1960s.[107] His patents from the 1950s included a wire strain gauge, a device for pairing several underground circuits, and a type of potentiometer—an apparatus for varying voltage in a circuit.[108] Dimeff clearly understood electrical instruments, but that knowledge did not imply a grasp of or familiarity with integrated-circuit fabrication techniques.

Dimeff also claimed that the contract for "an improved pressure transducer" "was placed with Professor [James B.] Angell at Stanford and monitored by my Assistant Division Chief, [Benjamin H.] Ben Beam."[109] However, NASA records state that the monitor on the pressure sensor contract was Tom Fryer, not Ben Beam, who was not listed as a supervisor of any MEMS contracts.[110] In 1969, Beam was the Assistant Chief of the Instrumentation Division

---

106. Ibid.

107. Hartman, *Adventures in Research*, pp. 246, 279, and 478. Dimeff obtained several patents for improvements to magnetometers as well as electrometers: Dimeff and Grant W. Coon, "Vibrating element electrometer with output signal magnified over input signal by a function of the mechanical Q of the vibrating element," U.S. patent 3,384,820, filed 17 May 1965, issued May 1968, assigned to NASA; Dimeff, "Electrostatic charged particles analyzer having deflection members shaped according to the periodic voltages applied thereto," U.S. patent 3,532,880, filed 30 January 1968, issued 6 October 1970, assigned to NASA; Dimeff, "Cryogenic apparatus for measuring the intensity of magnetic fields [a cryogenic magnetometer]," U.S. patent 3,437,919, filed 1 July 1965, issued April 1969, assigned to NASA.

108. Dimeff, "Directly strained, capacitance strain gauge," U.S. patent 2,933,665, filed 5 April 1956, issued 19 April 1960, assigned to NASA; Dimeff, "Apparatus for coupling a plurality of ungrounded circuits to a grounded circuit," U.S. patent 3,059,220, filed 2 July 1959, issued 16 October 1962, assigned to NASA; Dimeff, "Servomotor capacitance-coupled potentiometer wiper circuit," U.S. patent 2,844,776, filed 15 November 1955, issued 22 July 1958, assigned to NASA.

109. Letter, John Dimeff to Glenn Bugos, 29 August 2005.

110. The names of the six contract monitors for the four MEMS contracts can be found in *NASA's University Program* (1972), pp. 25 and 28; *NASA's University Program* (1973), p. 26; *NASA's University Program* (1974), p. 23; NASA, *NASA's University Program* (1977), p. 25; and *NASA's University Program* (1979), p. 28.

directly under Dimeff,[111] but during essentially all of the pressure sensor contract period—that is, from November 1969 to September 1972, he was Assistant Chief of the Research Facilities and Instrumentation Division but reporting to Angelo Giovannetti, not Dimeff, who had moved up the organizational ladder to be the Assistant Director for Advanced Instrumentation in the Office of the Director of Research Support.[112] Still, it is probable that Beam played some part in the Stanford work given Samadikun's acknowledgment of Beam's assistance, as cited earlier, but the nature of that work he left unspecified.

Dimeff's letter also offered a muddied recollection of a second MEMS contract, again based on concepts that he suggested. The contract was with "someone (?) at Stanford" and called for that person to "machine" a groove into a block of silicon to be "covered to form a column for a mass chromatograph."[113] The contract in question did not relate to a "mass chromatograph" but rather to a *gas* chromatograph—an entirely different genre of scientific instrument. One would expect the author of the concept, especially a person whose entire career was spent on instrumentation, to recall such a fundamental difference. The chromatograph was perhaps the most important of the MEMS devices created at Stanford with NASA funding and certainly received the most funding.

Dimeff correctly identified the (initial) Ames contract monitor as "Ralph Donaldson of my Division."[114] Ralph W. Donaldson, Jr., was in the Measurement Sciences Branch of the Instrumentation Division in 1969 and therefore in Dimeff's division at that time. But during the period of the chromatograph contracts—that is, from 1971 to 1980—Donaldson did not report to Dimeff. Rather, he was in the Electronic Instrument Development Branch and reported to Boris Ragent and Tom Fryer (Deboo, starting in 1977).[115] Dimeff's understanding of the purpose of the gas chromatograph contract seems muddled, to say the least. He stated that the intent was to "support [the] biological research of Dr. Joan Dannellis [*sic*] of the Life

---

111. ARC, *Telephone Directory* (August 1969), p. 46.

112. ARC, *Telephone Directory* (November 1970), p. 46; ARC, *Telephone Directory* (1971), pp. 47 and 48; ARC, *Telephone Directory* (November 1972), pp. 46 and 47.

113. Letter, John Dimeff to Glenn Bugos, 29 August 2005.

114. Ibid.

115. Information on their positions is from ARC, *Telephone Directory* (1969), pp. 9 and 46; ARC, *Telephone Directory* (1971), pp. 9 and 48; ARC, *Telephone Directory* (1972), pp. 9 and 47; ARC, *Telephone Directory* (May 1974), pp. 9 and 50; ARC, *Telephone Directory* (July 1975), pp. 9 and 50; ARC, *Telephone Directory* (January 1977), pp. 10 and 52; ARC, *Telephone Directory* (1977–1978), pp. 9 and 51; ARC, *Telephone Directory* (April 1979), pp. 9 and 50; and ARC, *Telephone Directory* (October 1980), pp. 10 and 54.

Sciences Directorate."[116] Joan Vernikos more recently, from 1993 to 2000, was Director of Life Sciences at NASA Headquarters. Before joining Ames in 1966, she held several academic posts at, among other places, the Ohio State University Medical School, the Stanford University Medical Center's Department of Psychiatry, and the University of London. Her research focused on understanding the hormonal and behavioral mechanisms that trigger responses to stress and applying that knowledge to aviation and space applications as well as developing a framework for discerning how spaceflight and terrestrial gravitation affect the human body.[117]

Joan Vernikos Danellis began her career at Ames in the Physiology Branch (Environmental Biology Division) under Jiro Oyama, and from 1972 to 1975 she was the Chief of the Human Studies Branch within the Biomedical Research Division, whose head was Harold "Hal" Sandler. Subsequently, until at least 1980, she had an unspecified position in Sandler's Biomedical Research Division.[118] During her tenure at Ames, Vernikos Danellis oversaw a number of contracts at Stanford. She did not monitor the first two of those contracts by herself, but with Charles Winget, the more senior Ames employee, and they reflected his interest in circadian rhythms and her concern with hormone levels. The subsequent contracts that she supervised covered research carried out from 1969 to 1977 and investigated such issues as adrenaline activity during bed rest, the effects of changing gravitational stress on body fluids and blood circulation, and the influence of chronic and repeated stress on the pituitary adrenal system. The gas chromatograph, Vernikos has affirmed, was not relevant to her research.[119]

A preponderance of evidence, on the other hand, indicates that the objective of the gas chromatograph research—at least from NASA's perspective—was for the Mars Viking mission. Such was the recollection of Ken Wise, who was the principle investigator on the first gas chromatograph contract.

---

116. Letter, John Dimeff to Glenn Bugos, 29 August 2005.

117. The Victory Speakers' Bureau, "Dr. Joan Vernikos," *http://artofvictory.com/Vernikos.htm* (accessed 4 September 2008); and "Dr. Joan Vernikos," *http://www.joanvernikos.com/about/bio.php* (accessed 4 September 2008).

118. ARC, *Telephone Directory* (1964), pp. 8 and 50; ARC, *Telephone Directory* (1969), pp. 8 and 50–51; ARC, *Telephone Directory* (1971), pp. 8 and 45; ARC, *Telephone Directory* (1972), pp. 7 and 44; ARC, *Telephone Directory* (1974), pp. 8 and 47; ARC, *Telephone Directory* (1975), pp. 8 and 51; ARC, *Telephone Directory* (1977), pp. 8 and 51; ARC, *Telephone Directory* (1977–78), pp. 8 and 50; ARC, *Telephone Directory* (1979), pp. 9 and 55; and ARC, *Telephone Directory* (1980), pp. 9 and 53.

119. *NASA's University Program* (1972), pp. 28 and 30; *NASA's University Program* (1973), pp. 20 and 21; *NASA's University Program* (1975), p. 27; *NASA's University Program* (1976), p. 30; and *NASA's University Program* (1977), pp. 22 and 24; and notes, telephone conversation with Joan Vernikos, 24 November 2008.

In addition, one of the graduate students who benefited from the chromato-graph funding asserted in his dissertation that "the idea of a miniature GC [gas chromatograph] column etched in silicon was conceived by NASA for a small, lightweight, and rugged chromatograph to be landed on Mars in the Viking 75 probe." Elsewhere, the same student declared that the chromato-graph potentially could be used on "planetary probes."[120]

The person who emerges as a key individual from a look at the extant grant and research contract records[121] as well as publications by Stanford graduate students—not to mention his own account of events—is Harold "Hal" Sandler. Born in 1929 and raised in Cincinnati, Ohio, Sandler majored in chemistry and mathematics at the University of Cincinnati and graduated from its medi-cal school with a degree in internal medicine in 1955. After being drafted and assigned to the Navy in 1961, he set off on a career in aerospace medical research at the Navy's medical aviation laboratory in Johnsville, Pennsylvania, where Neil Armstrong and other astronauts trained.[122] He next took a position at Ames in 1965 and participated in various studies related to weightlessness; however, his main interest was in biomedical instrumentation, a sine qua non for understanding bodies in a state of real or artificial weightlessness.[123]

From the start Hal Sandler was in the Biomedical Research Division, which was in John Billingham's Biotechnology Division under Harold Klein, and by 1971 he was branch chief. Shortly after arriving at Ames, Sandler told a NASA interviewer that he contacted John Dimeff, who led the Instrumentation Division and was, in Sandler's words, "the godfather, or the enabler, of the bioinstrumentation project." He also struck up a "lifelong friendship" with Tom Fryer.[124]

120. *NASA's University Program* (1972), p. 25; Wise, interview, p. 16; Stephen Clark Terry, "A Gas Chromatography System Fabricated on a Silicon Wafer Using Integrated Circuit Technology," Ph.D. dissertation, Stanford University, 1975, p. 1; Stephen Clark Terry, John Hallock Jerman, and James Angell, "A Gas Chromatographic Air Analyzer Fabricated on a Silicon Wafer," *IEEE Transactions on Electron Devices*, ED-26, no. 12 (December 1979): 1886.

121. Ames Life Sciences Research at Stanford Database.

122. James R. Hansen, *First Man: The Life of Neil A. Armstrong* (New York: Simon & Schuster, 2005), pp. 54–55.

123. Finding Aid, Harold Sandler Papers, MS-352, Special Collections and Archives, Wright State University Libraries, Dayton, OH, pp. 3–4; Harold Sandler, telephone interview by Jennifer Ross-Nazzal and Rebecca Wright, Houston, TX, and Bainbridge Island, WA, 15 October 2004, transcript, NHO, pp. 3–4.

124. ARC, *Telephone Directory* (1967), pp. 33 and 50; ARC, *Telephone Directory* (December 1968), pp. 32 and 50; ARC, *Telephone Directory* (1969), pp. 31 and 50; ARC, *Telephone Directory* (1971), pp. 29 and 44; and Sandler, interview, 2004, p. 5.

Sandler states that he fostered and encouraged the Ames effort to develop miniature sensors that one could implant in the bodies of subjects, be they humans or animals, as well as telemetry systems for broadcasting those sensor readings for use by researchers. As a result, his name appeared regularly in Stanford dissertations and as a MEMS contract supervisor. For example, Samaun Samadikun acknowledged his help in testing his device on animals. Stephen C. Terry, whose 1975 dissertation dealt with the development of a miniature gas chromatograph, also recognized Sandler's help. In addition, Sandler was the sole monitor during the first year of the gas chromatograph contract, beginning on 1 September 1973.[125]

At the same time, Sandler became aware that implantable biomedical sensors had become commercially available with the founding in 1967 of Konigsberg Instruments, Inc., in Pasadena, California, by former Ames employee Eph Konigsberg.[126] Sandler and others at Ames realized that they needed to make their sensors, such as those designed by Fryer and Deboo, much smaller. According to Sandler, the shift from dogs to monkeys as research subjects demanded smaller instruments. "No one wanted to use the dog, which was my favorite subject for study for long-term space effects. An animal I had little experience with, the monkey, was being touted as the only thing that they would consider, and they weren't even very eager about that. Well, if I was going to start working in monkeys, everything had to be reduced by a factor of ten."[127]

Ames, however, lacked the capability to achieve significant reductions in scale through the application of integrated-circuit technology, the need for which Sandler characterized as "critical." Dimeff, Sandler recalled, attempted to start an integrated-circuit laboratory at Ames, but "it sputtered and it failed. Costs were way too high, and talent wasn't available," because of the growing demands of Silicon Valley businesses. Fryer, as noted earlier, also lamented the high cost of integrated-circuit technology. Ames lacked an integrated fabrication facility, such as the PREDICT laboratory (Process Reliability Evaluation and Determination of Integrated Circuit Techniques)

---

125. Samadikun, "An Integrated Circuit Piezoresistive Pressure Sensor," p. iii; Terry, "Gas Chromatography System," p. v; *NASA's University Program* (1974), p. 23; and *NASA's University Program* (1980), p. 34.

126. Sandler, interview, 2004, p. 6; Konigsberg Instruments, "Company Overview," *http://www.konigsberginc.com/* (accessed 24 November 2008); and notes, telephone conversation with Harold Sandler, 24 November 2008.

127. Sandler, interview, 2004, p. 13.

built at the ERC, which allowed that center's staff to create experimental integrated circuits.[128]

Ames, lacking its own integrated-circuit fabrication capability, needed to find a job shop where they could have experimental devices manufactured to order. According to Sandler, Dimeff suggested that he visit "several individuals starting an IC [integrated circuit] lab over at Stanford and see what could be worked out." At that point he met James A. Meindl, and the MEMS collaboration between Ames and Stanford began.[129] Subsequently, however, Sandler recalled that Fryer, with whom he worked closely on implantable sensors and telemetry, probably had the idea to exploit the pioneering MEMS research that Stanford was carrying out.[130] Shading his recollections is Sandler's self-admitted desire to ensure that Dimeff received credit for the biomedical accomplishments of Ames. "If any good has come from all these efforts," he told NASA interviewers, "I would really like to share them with him."[131]

Could Tom Fryer have played a role in instigating the Ames MEMS projects at Stanford?[132] In addition to Hal Sandler's recollection, one can point to the fact that Fryer was heading Ames's biomedical instrumentation effort prior to Sandler's arrival at Ames in 1965, as witnessed by his writings in this area[133] and his position as Assistant Chief in the Measurements Research

128. Ibid. The PREDICT laboratory is discussed in chapter 3, "NASA's Role in the Manufacture of Integrated Circuits," in this volume.

129. Sandler, interview, 2004, p. 13.

130. Notes, telephone conversation with Harold Sandler, 24 November 2008.

131. Sandler, interview, 2004, p. 5. He shared this sentiment. Notes, telephone conversation with Harold Sandler, 24 November 2008.

132. Fryer could not be contacted, having died in 2005.

133. The list of Tom Fryer's writings prior to 1965, usually coauthored with Deboo, is rather long. The following can be found by searching the NASA Technical Reports Server, *http://ntrs.nasa.gov/search* (accessed 14 November 2005): Gordon J. Deboo and Tom Fryer, "An Improved, Miniaturized Bio-Medical Amplifier," TM-X-51506, 1963, 13-page manuscript submitted for publication in *IEEE Transactions on Biomedical Electronics*; Gordon J. Deboo and Tom Fryer, *A Miniature Bio-Potential Telemetry System*, TM-X-54931 (Moffett Field, CA: Ames, March 1964); Gordon J. Deboo and Tom Fryer, "A Solid State Preamplifier for Micro-Electrode Studies," TM-X-51731, May 1964, 4-page manuscript submitted for publication in the *Journal of Physiology*; Tom Fryer and Gordon J. Deboo, "A High-Performance Miniature Biopotential Telemetry System," in Robert Plonsey, ed., *Engineering in Medicine and Biology: Proceedings of the 17th Annual Conference on Engineering in Medicine and Biology*, vol. 6 (Cleveland, OH: Conference Committee, 1964), p. 39; Tom Fryer and Charles W. Beck, "A Low-Power Analog-to-Digital Converter," TM-X-51840, 21 July 1964, 12-page manuscript submitted for publication in *Electronics Magazine*; Winget, Averkin, and Fryer, *Quantitative Measurement by Telemetry of Ovulation and Oviposition in the Fowl*. In addition, two manuscripts prepared prior to 1965 appeared as Tom Fryer and Gordon J. Deboo, "A High-Performance Miniaturized Preamplifier for Biological Applications," *Medical*

Branch and Electronics Research Branch.[134] Also in support of Fryer as the instigator of the Stanford collaboration is the fact that when he retired in 1979 and formed Biomedical Monitoring Systems, Inc.,[135] the collaboration ended.[136] Dimeff left in 1975, when he formed Dimeff Associates of San Jose, California,[137] but Sandler remained at Ames until January 1987,[138] several years after the Ames MEMS projects were over.

The Rashomon effect variations in these narratives likely reflects the different hierarchical levels at which these individuals participated. Within the Stanford integrated-circuit laboratory, Ken Wise and Samaun Samadikun interacted with Nigel Tombs, who helped with the animal testing. It is probable that this social propinquity—combined with Tombs's knowledge of integrated-circuit fabrication methods and his role as monitor on a different MEMS contract[139]—led to the perception that Tombs was the principal behind Ames's interest in MEMS research. Dimeff, Sandler, and Fryer were at the opposite end of the hierarchy, being part of upper management with Dimeff above Fryer as Chief of the Instrumentation Division. Dimeff undoubtedly played a role in this story, but his dubious claims and counterfactual recollection of events suggests a tangential role at best. His rise within the Ames bureaucracy at this time probably took him farther and farther away from the details, but he assuredly was in a position to approve

---

Electronics and Biologic Engineering 56 (April 1965): 203–204; and Gordon J. Deboo and Tom Fryer, "Miniature Biopotential Telemetry System," *American Journal of Medical Electronics* 4 (July–September 1965): 138–142.

134. For example, ARC, *Telephone Directory* (1963), p. 44; and ARC, *Telephone Directory* (1964), p. 53.

135. Taylor, "Thomas Benton Fryer—NASA Engineer."

136. *NASA's University Program* (1980), p. 34, lists the contract, although it is not to be found among the Stanford contracts in NASA, *NASA'S University Program Active Projects Fiscal Year 1982* (Washington, DC: NASA, 1983), pp. 29–41.

137. Manta, "Dimeff Assoc," *http://www.manta.com/coms2/dnbcompany_9018r* (accessed 17 October 2008). Further corroboration of his departure is found in the Ames telephone directories, which listed him in May 1974 but not in July 1975. See ARC, *Telephone Directory* (1974), p. 49; and ARC, *Telephone Directory* (1975). On the other hand, a search of U.S. patents on both Google and the U.S. Patent Office database, *http://patft.uspto.gov/* (accessed 23 October 2005) for "John Dimeff" reveals that the last two patents that he received and assigned to NASA had filing dates in December 1975. Dimeff, "Optically selective, acoustically resonant gas detecting transducer," U.S. patent 4,055,764, filed 22 December 1975, issued 25 October 1977, assigned to NASA; and Dimeff, "Self-calibrating radiometer," U.S. patent 4,030,362, filed 22 December 1975, issued 21 June 1977, assigned to NASA. The filing dates, however, reflected when the NASA patent attorney had completed his paperwork, irrespective of Dimeff's participation in the patent application process.

138. Sandler, interview, 2004, p. 30.

139. *NASA's University Program* (1972), p. 25.

and encourage Sandler's ventures. As James Meindl of Stanford recalled, Dimeff was "more at an administrative or management level in those days and, as far as I was aware, not doing research the way Hal Sandler and some of his friends were."[140] Nonetheless, it is not conceivable that Dimeff may have advised Sandler to look into the possibility of using Stanford facilities. Finally, Tombs, not Dimeff, was a more believable candidate as the inventor of the MEMS projects.

Once Sandler arrived at Ames, he unquestionably was the prime mover in expanding the Center's biomedical instrumentation effort. That effort, however, already was in progress, and Fryer was leading the way. The published grant and research contract records are more puzzling than clarifying of Sandler's role. He told NASA interviewers that the gas chromatograph was not his project but belonged rather "to help people who were on the third floor, above us," meaning those searching for extraterrestrial life.[141] Ironically, as stated earlier, he was the first monitor on the gas chromatograph contract.[142]

Fryer, who supervised the first MEMS contract, was an electrical engineer and an instrumentmaker in search of solutions to the specific problems related to the development of small-scale sensors and allied telemetry systems. The Stanford integrated-circuit laboratory, given the lack of a comparable capability at Ames, could serve him capably as a job shop for experimenting with what he characterized as the costly advanced technology of integrated-circuit fabrication techniques. Dimeff may have told him about the Stanford facility, or he may have learned on his own.

The sensor work carried out under the first MEMS contract, monitored by Fryer, fit perfectly with his research agenda, but it readily could have dovetailed with Sandler's instrument needs. Indeed, Samadikun acknowledged his help in animal testing his devices.[143] Therefore, Tom Fryer may have played an important role in Ames's entry into MEMS research. This suggested role for Fryer in no way is meant to disparage Sandler's vital role in spearheading the development of state-of-the-art biomedical instruments nor his probable principal role (within the Ames Life Sciences Divisions) in leading the Center into carrying out MEMS research at Stanford University. I now turn to look at a better-documented account, the emergence of the Stanford Integrated Circuits Laboratory and its research agenda.

---

140. Meindl, interview, p. 4.

141. Sandler, interview, 2004, p. 28.

142. Ames Life Sciences Research at Stanford Database.

143. Samadikun, "An Integrated Circuit Piezoresistive Pressure Sensor for Biomedical Instrumentation," p. iii.

Stanford MEMS Research

Stanford Integrated Circuits Laboratory

The Stanford University Electrical Engineering Department had a long-standing interest in the possibilities that integrated circuits opened up and authorized the construction of their own integrated-circuits laboratory. Having a laboratory where one could fabricate chips was a prerequisite for exploring MEMS because the creation of MEMS devices resulted directly from the application of integrated-circuit fabrication techniques. Initially, integrated-circuit technology research took place in the Solid-State Electronics Laboratory, a part of the Stanford Electronics Laboratories (SEL) managed by the university's Electrical Engineering Department. The main architect of Stanford's solid-state electronics program was Professor John G. Linvill, who had been hired from Bell Telephone Laboratories in 1955. During the mid-1960s he incorporated integrated circuits into Stanford's electrical engineering courses and research.[144]

One of Linvill's key hires in that effort—and for the eventual entry into MEMS research—was James B. Angell. Angell received his S.B. and S.M. degrees in 1946 and his Sc.D. degree in 1952, all in electrical engineering, from MIT. From 1951 to 1960 he had been with the Research Division of the Philco Corporation in Philadelphia, Pennsylvania, where he managed solid-state circuit research and led a group studying the principles of evaluating and applying transistors and other solid-state devices, especially to computing and high-frequency applications. Previously, from 1946 to 1951, Angell had been a research assistant in the MIT Research Laboratory of Electronics, where he studied noise in tracking radars. In 1960 he joined the Stanford faculty, becoming a full professor in 1962 and then director of the Solid-State Electronics Laboratory in 1964, replacing Linvill, who now was executive head of the Electrical Engineering Department. Angell also directed the Solid-State Industrial Affiliates program from 1964 to 1970.[145]

In 1965, as the Integrated Circuits Laboratory was becoming involved in MEMS work, the size of its faculty, staff, and facilities were rather modest. The laboratory initially consisted of about four rooms in the basement of the McCullough Building, built in 1964 by H. J. Brunnier Associates but

---

144. Christophe Lécuyer, "What Do Universities Really Owe Industry? The Case of Solid State Electronics at Stanford," *Minerva* 43, no. 1 (Spring 2005): 52 and 53.

145. William P. Rambo to "SEL Staff," 21 September 1964, Folder 4, "Chron File 1964 Aug.-Sept.," Box 2, ACCN 97-093, Rambo Papers; Lécuyer, "What Do Universities Really Owe Industry?," p. 56; Dawn Levy, "James Angell, Electrical Engineer and Former Carillonneur, Dies," *Stanford Report* 38, no. 19 (8 March 2006): 2; and Angell, interview, p. 9.

dedicated in October 1965.[146] The firm also had built the Varian Laboratory for physics research and teaching in 1961.[147] The National Science Foundation underwrote the building's construction costs in part. The principal occupants of the McCullough Building included the Center for Materials Research, which involved faculty from engineering, physics, chemistry, and other departments as well as groups from the Stanford Electronics Laboratories and the Department of Applied Physics. The idea was to promote interdisciplinary research by fostering "communication" among researchers in several fields. Associated with the Integrated Circuits Laboratory was the Integrated Circuit Fabrication Laboratory, where faculty and students could make experimental integrated circuits.[148] The collocation of the two laboratories in the basement of the McCullough Building created a fruitful synergy between integrated-circuit fabrication, research, and teaching.

During its earliest years the Integrated Circuits Laboratory received outside funding from a variety of sources. The National Institutes of Health (NIH) provided its main support, with the National Institute of General Medical Sciences providing more than a million dollars a year in funding. Later, the National Institute for Occupational Safety and Health (NIOSH) began underwriting the work, providing at least a half-million dollars. In contrast, some of those associated with the Integrated Circuits Laboratory recalled that NASA funding at Stanford University (not just the Integrated Circuits Laboratory) was usually about $100,000 per year. A small amount of additional funding came from the National Science Foundation.[149] This picture of NASA funding, as discussed later in this chapter, does not portray the Agency's underwriting of Stanford research accurately.

An initial grant that was crucial in the history of the Integrated Circuits Laboratory came from the U.S. Office of Education. The $1,800,000 grant funded the miniaturization of the Optacon, a device recently developed at

146. Angell, Terry, and Barth, "Silicon Micromechanical Devices," p. 49; Lécuyer, "What Do Universities Really Owe Industry?," p. 58; Terry, interview, pp. 4 and 13; Meindl, interview, p. 10; and Memo, Rambo to Angell et al., 4 October 1965, Folder 11, "Chron File 1965 Sept.-Oct.," Box 2, ACCN 97-093, Rambo Papers.

147. "H. J. Brunnier Associates," *http://www.hjbrunnier.com/projects.html?Catalog=Universities* (accessed 7 October 2005).

148. Rambo to Angell and Prof. Melvin Chodorow, "Research Equipment Funds," 23 February 1966, Folder 13, "Chron File 1966 Jan.-Feb.," Box 2, ACCN 97-093, Rambo Papers; [Rambo], "The Jack A. McCullough Building," no date, but location in folder dates it to 4 October 1965, Folder 11, "Chron File 1965 Sept.-Oct.," Box 2, ACCN 97-093, Rambo Papers; Meindl, interview, p. 15; Terry, interview, p. 4; and Angell, interview, p. 9.

149. Meindl, interview, pp. 9 and 17; Terry, interview, pp. 4 and 24; and Petersen, interview, pp. 14 and 15.

Stanford University that scanned printed pages and communicated a vibrating facsimile of each letter to the reader's finger. Linvill was interested in building the device because he hoped that it would help his blind daughter, Candace, to read printed materials. Thanks to the grant, according to historian Christophe Lécuyer, Linvill had the financial resources to transform the Integrated Circuits Laboratory from a teaching facility into a research operation.[150]

This characterization of the laboratory's funding, though, ignores the critical importance of military underwriting that historian Stuart W. Leslie has emphasized.[151] In 1947, Stanford began receiving contract funding from the Joint Services Electronics Program for work carried out in the Stanford Electronics Research Laboratories (known as the Stanford Electronics Laboratories after 1955). The Joint Services Electronics Program came into being in March 1946 initially to take over financial support of the MIT Research Laboratory of Electronics[152] and similar laboratories at Harvard and Columbia Universities. In 1947 the Joint Services Electronics Program consolidated four separate contracts that Stanford had with the Office of Naval Research into a single contract and added supplementary financial support from the Army and the Air Force.[153]

The solid-state program of the Stanford Electronics Laboratories reflected changing military priorities for more compact, reliable, and durable electronics for guided missiles, communications, and so-called smart weapons. The school's contract with the military electronics program increasingly emphasized solid-state research, as did its contracts with the individual services. A $250,000 Air Force contract for studies of adaptive systems (for improving the reliability of military electronics systems) supported about half of the

---

150. Lécuyer, "What Do Universities Really Owe Industry?," pp. 59–60. Linvill set up a company, Telesensory, to manufacture the Optacon. See Deborah Kendrick, "From Optacon to Oblivion: The Telesensory Story," *AccessWorld* 6, no. 4 (July 2005), *http://www.afb.org/afbpress/pub.asp?DocID=aw060403* (accessed 27 March 2006); and Harvey Lauer, "The Reading Machine That Hasn't Been Built Yet," *AccessWorld* 4, no. 2 (March 2003), *http://www.afb.org/afbpress/pub.asp?DocID=aw040204* (accessed 27 March 2006).

151. Leslie, *Cold War and American Science*; Leslie, "Profit and Loss," pp. 59–85; and Leslie, "Playing the Education Game to Win," pp. 55–88.

152. On the MIT Research Laboratory of Electronics, see Andrew Butrica, *To See the Unseen: A History of Planetary Radar Astronomy* (Washington, DC: NASA, 1996), p. 28.

153. A. L. Gilbert and B. D. McCombe, "The Joint Services Electronics Program: An Historical Perspective," in Arnold Shostak, ed., *Fortieth Anniversary of the Joint Services Electronics Program* (Arlington, VA: ANSER, 1986), pp. 2–5; Leslie, *Cold War and American Science*, pp. 25 and 58; and S. E. Harris and J. S. Harris, "Stanford University Electronics Laboratory and Microwave-Ginzton Laboratory," in Shostak, *Fortieth Anniversary of the Joint Services Electronics Program*, p. 105.

Stanford Electronics Laboratories' total solid-state electronics effort. Between 1948 and 1960 the program represented a sizable fraction of the Stanford Electronics Laboratories support, including some of the early seed money to establish the first integrated-circuits laboratory at Stanford and the first work on processing and fabrication. In 1960 military dollars continued to dominate the Stanford Electronics Laboratories' external funding sources. In that year it received $330,000 from the joint services contract plus another $2.3 million from individual contracts with the three armed services. Meanwhile, the laboratories received only $200,000 from all other sources, including just $32,000 from the National Science Foundation. NASA's contribution, according to Leslie, increased dramatically from virtually nothing to roughly a third of the laboratories' budget over the following decade (the 1970s), but Defense Department money and interests continued to dominate.[154] However, as for the Integrated Circuits Laboratory, the 1970s witnessed a dramatic and pivotal shift in support away from military dollars, as discussed below.

For several years what would become the Integrated Circuits Laboratory operated informally and without a name within the Solid State Laboratory. According to Lécuyer, the Integrated Circuits Laboratory originated in 1964, when Professor Robert L. Pritchard developed a laboratory course on microcircuits and established the Integrated Circuits Laboratory as a component of the department's Solid State Laboratory. Pritchard's microcircuits course familiarized students with the complex processes—such as photolithography and thin-film deposition techniques—employed in the making of integrated circuits.[155] Nonetheless, Stanford Electronics Laboratories documents reveal a different and later origin for the Integrated Circuits Laboratory.

The key event was the recruitment of James D. Meindl by Linvill in 1967 to direct the laboratory as its founding director and to undertake research in medical electronics. Meindl, who received a Ph.D. in electrical engineering in 1958 from Carnegie Mellon University in Pittsburgh, first learned about integrated circuits in 1959, while working at the Army's Signal Corps Research and Development Laboratories in Fort Monmouth, New Jersey, where he became the founding director of the Integrated Electronics Division in 1965. Upon arriving at Stanford, Meindl turned to nearby Shockley Semiconductor for the expertise to process advanced integrated circuits. He hired Jacques Beaudoin, an experienced Shockley research engineer, and made him the

---

154. Leslie, *Cold War and American Science*, p. 73; and Harris and Harris, "Stanford University Electronics Laboratory and Microwave-Ginzton Laboratory," pp. 106 and 107.
155. Lécuyer, "What Do Universities Really Owe Industry?," p. 58.

laboratory's chief engineer.[156] In this way the Integrated Circuits Laboratory continued to build its strengths in integrated-circuit technology and processing by borrowing industry talent.

In 1967, when Meindl joined the electrical engineering faculty, the Stanford campus was in the throes of antiwar protests. Meindl intended to submit two proposals. One was a joint proposal with the medical school to the NIH; the other was a submission to the Central Intelligence Agency (CIA). When Meindl witnessed campus reaction to the CIA recruiting effort, he wondered about the propriety of the CIA proposal and possible repercussions with the Research Policy Committee, even though the proposed research would be unclassified and carried out in the open—in short, within the conditions approved by the Academic Council.[157]

The decision to steer in the direction of biomedical research was momentous, as the field of biomedical electronics was growing. The professional movement had started already during the 1950s, with the formation of the Professional Group on Medical Electronics within the Institute of Radio Engineers (IRE) and the annual conferences on electronic instrumentation run by the American Institute for Electrical Engineering (AIEE). Following the 1963 merger of the AIEE and the IRE to form the Institute of Electrical and Electronics Engineering (IEEE), the IRE Professional Group became the IEEE Professional Group on Bio-Medical Engineering and later the IEEE Engineering in Medicine and Biology Society.[158] Meanwhile, also during the early 1960s, the NIH took significant steps to support biomedical engineering. First, it created a committee under the National Institute of General Medical Sciences to evaluate program and project applications. Then, it set up a biomedical engineering training study section to evaluate training-grant applications, many of which served biophysics and biomedical engineering.

---

156. Meindl, interview, pp. 2–4; "Microelectronics Pioneer Awarded IEEE Medal of Honor," *The Institute* (March 2006), p. 4; and Lécuyer, "What Do Universities Really Owe Industry?," p. 59. Meindl was John M. Fluke Professor of Electrical Engineering and Associate Dean for Research, School of Engineering. In 2006 the IEEE recognized Meindl's contributions to microelectronics—including the sensors that he developed for a portable electronic reading aid for the blind—by awarding him its Medal of Honor, the organization's highest award. "Microelectronics Pioneer Awarded IEEE Medal of Honor," p. 4.

157. Rambo to Prof. W. F. Baxter, "Research Proposal Prepared by Professor Meindl," 8 November 1967, Folder 8, "Chron File 1967 Oct.-Dec.," Box 3, ACCN 97-093, Rambo Papers.

158. "The Whitaker Foundation: A History of Biomedical Engineering," *http://www.whitaker. org/glance/history.html* (accessed 27 March 2006).

A special "floating" study section processed applications in bioacoustics and biomedical engineering.[159]

Meindl went ahead with the NIH proposal in a very big way. His preparations led to a major funding proposal to the National Institute of General Medical Sciences (application GM 17940-01) for its Integrated Electronics for Medical Applications Program. The research proposal, titled "Integrated Electronics for Medical Applications," asked for $1,396,400 for the five-year period from 1 September 1970 to 31 August 1975. The proposal represented a milestone in the evolution of the Integrated Circuits Laboratory, not to mention the Stanford Electronics Laboratories and Stanford University.

In its "broad statement of research objectives," the proposal laid out the rationale for why the NIH should invest in integrated-circuit technology. It asserted that the "promising potential of this technology in medical research and practice is largely unfulfilled." The objective of the proposed program was "to bring to bear the combined talents of integrated electronics engineers and medical personnel in an intensive effort to advance medical research and practice through the application of integrated electronics." The program consisted of four projects, each of which aspired to apply integrated-circuit technology "primarily toward innovative solutions to generic problems in medical research and practice, and secondarily toward immediate solutions of particular problems."[160]

The four projects were labeled "large-scale integration," "transducers," "telemetry systems," and "advanced technology." "Large-scale integration" was the term in current use for the creation of "hundreds, thousands, and tens-of-thousands of transistors and associated circuit elements in a single silicon substrate" and was, in the language of the proposal, "the most important force in modern electronics." Among the actual devices mentioned were ones to allow the blind to read texts, "unique multi-element biopotential sensors and stimulators based on integrated circuit technology" for neurological applications, and "totally implantable telemetry systems...for accurate measurement of instantaneous blood flows and related variables" to be used in cardiovascular physiology and urology.[161]

The range of faculty committed to the program of research reflected the interdisciplinary nature of the proposed work as well as the efforts of Meindl to marry the Electrical Engineering Department and the School of Medicine. The electrical engineering faculty involved in the program comprised Meindl,

---

159. Ibid.
160. P. I. Meindl, "Integrated Electronics for Medical Applications," application GM 17940-01, March 1970, Folder 5, "N.I.G.M.S.," Box 6, ACCN 97-093, Rambo Papers.
161. Ibid.

then associate professor of electrical engineering; John G. Linvill, chairman of the department; James B. Angell, associate chairman of the department; and James C. Bliss, associate professor of electrical engineering. The School of Medicine faculty consisted of Arnold Starr, assistant professor of medicine (neurology); William W. Angell, assistant professor of cardiovascular surgery; Eugene Dong, Jr., assistant professor of cardiovascular surgery; and Duncan E. Govan, assistant professor of surgery (urology).[162]

To sell the program to the National Institute of General Medical Sciences, Meindl organized a one-day site visit for the institute's personnel on 16 March 1970. Meindl gave an overview of the program; talks by William Rambo, associate dean for research at the School of Engineering, and Dr. Robert J. Glaser, vice president for medical affairs and dean of the School of Medicine, followed. The medical school and electrical engineering faculty then gave presentations on the proposed research program, much of it based on research recently carried out by doctoral students. Several, of course, dealt with sensory aids for the blind, while others involved brain probes and telemetry systems linked to implanted blood flowmeters.[163] In addition, a brochure for graduate studies at Stanford had been prepared for a new program on "bioengineering."[164] All of the work soon paid off; the school won the grant.[165]

The grant fostered a funding shift of the Integrated Circuits Laboratory's budget from military to health research. This shift is borne out by funding data contained in Meindl's grant application. It indicates that laboratory faculty already had military dollars through the Joint Services Electronics Program for research on "testing of LSI [Large Scale Integration] circuits," "micropower integrated circuits," "micropower circuits for medical electronics," and "models in integrated circuits." Those funds supported just 5 percent of each professor's time. NASA, too, was underwriting "research on pressure sensors for biomedical instrumentation" (since October 1969), which accounted for 10 percent of one professor's time. It was the first MEMS project underwritten by NASA. In addition, the laboratory already received Office of Education funds, which supported 10 percent of one faculty position,[166] and it planned

162. Ibid.

163. "National Institute of General Medical Sciences Site Visit," 16 March 1970, Folder 5, "N.I.G.M.S.," Box 6, ACCN 97-093, Rambo Papers.

164. Stanford University School of Engineering, "Engineering in Biology and Medicine: Preliminary Brochure on Graduate Programs for Stanford Engineering Students," March 1970, Folder 5, "N.I.G.M.S.," Box 6, ACCN 97-093, Rambo Papers.

165. Meindl to Rambo, "NIH Grant," 1 October 1970, Folder 31, "Solid State Correspondence Miscellaneous," Box 6, ACCN 97-093, Rambo Papers.

166. An August 1969 listing of Solid-State Electronics Laboratory research indicates that the Office of Education money (OEG 0-8-071112-2995) was supporting even more

to submit grant proposals to both the Public Health Service and the National Institute of Neurological Disorders and Stroke. The goal, the proposal stated, was for the NIH to become the primary supporter of Stanford research in biomedical integrated-circuit applications: "It is intended that this NIH program will increase substantially the research on integrated electronics for medical applications at Stanford. It is anticipated that the proposed program will be the major source of support for this research at Stanford."[167]

The shift to medical applications also became visible in the work carried out for the Joint Services Electronics Program. An obvious example is Meindl's work on "micropower integrated circuits for medical electronics," which started in October 1968. The program's broad objective was to investigate the application of integrated-circuit technology to medical electronics by looking at such issues as low-power drain uses and small size. Specifically, Meindl investigated two problems. One was the design of a battery-operated optical-tactile reading aid for the blind that translated the optical image of a printed character to a tactile facsimile that a reader sensed with his or her finger. The second problem was the design of a totally implantable telemetry system for monitoring instantaneous blood flow within an animal or a human being. The device used a Doppler flow probe to obtain an FM signal whose frequency was proportional to the blood's velocity.[168]

Contrary to Lécuyer's assertion, as stated earlier, it was only *after* Meindl successfully won this key NIH grant for biomedical research that the appellation Integrated Circuits Laboratory came into use, not earlier in 1964. For some time SEL Director William Rambo had been planning "to recognize the momentum in our Integrated Circuits research separate from the Solid-State Electronics activities, through the formal identification of an Integrated Circuits Laboratory, with Professor James Meindl as Director."[169] In 1970, Rambo informally organized the Integrated Circuits Laboratory with Meindl as its head, but no formal announcement came forth until March 1971. The initial integrated-circuits faculty consisted of Meindl, Linvill, Angell, and Malcolm McWhorter, all of whom had been associated with the Solid-State Electronics Laboratory. With the official announcement of the formation of

---

faculty time. See Attachment, Stanford Electronics Laboratory, "Program and Abstracts of Papers for Stanford University Electronics Research Review," August 1969, Folder 32, "Electronics Research at Stanford-1970," Box 6, ACCN 97-093, Rambo Papers.

167. Meindl, "Integrated Electronics for Medical Applications."

168. Angell to Rambo and Dr. D. C. Bacon, "Recommendation Regarding ONR–JSEP Funding for 1970–1971," 21 April 1970, Folder 31, "Solid State Correspondence Miscellaneous," Box 6, ACCN 97-093, Rambo Papers.

169. Rambo to SEL Personnel, "Organizational Changes," 26 March 1971, Folder 29, "SEL Laboratory Reorganization," Box 5, ACCN 97-093, Rambo Papers.

the Integrated Circuits Laboratory in 1971, Angell relinquished his responsibilities in the Solid-State Electronics Laboratory because of his "substantial time commitment" in the Electrical Engineering Department, where he had been chair of graduate admissions since 1969, handling some 1,200 applications per year.[170]

By the early 1980s the Integrated Circuits Laboratory was outgrowing the McCullough Building basement. The school undertook an important fundraising campaign to construct and equip a new building that would enable the expansion of MEMS and integrated-circuit research. Stanford succeeded in convincing some 20 key Silicon Valley–based electronics companies to contribute to the new facility, including the Hewlett-Packard Development Company, Xerox's Palo Alto Research Center, and the Intel Corporation. As a result, the new Integrated Circuits Laboratory found a new home inside the Center for Integrated Systems, which Linvill established in 1980. The Center brought together the activities of four laboratories (Solid-State, Integrated Circuits, Computer Systems, and Information Systems) and included a new fabrication facility. Today, the McCullough Building houses the Laboratory for Advanced Materials Research.[171] The Integrated Circuits Laboratory became much larger and grander. It is currently the home to 14 faculty members, 9 research associates, 100 Ph.D. students, and 10 full-time staff members. Specific areas of research include gate dielectrics, diffusion, deposition, etching, ion implantation, and associated thermal processing. One of the laboratory's areas of expertise is the simulation of complex fabrication sequences and the ability to predict accurately the resulting device structures.[172]

The Brain Probe

The rise of MEMS work at Stanford University took place within this larger context of a research and funding shift from the military to medicine. The shift also entailed a cross-pollination between the Electrical Engineering Department and the medical school. The first MEMS device to issue from the integrated-circuit laboratory reflected this mix of military and medical

---

170. Rambo to Dr. D. C. Bacon, Linvill, Dr. Ralph Smith, "SEL Laboratory Organization," 16 December 1970, Folder 29, "SEL Laboratory Reorganization," Box 5, ACCN 97-093, Rambo Papers; Rambo to Prof. Gerald Pearson, "Organization of the Solid-State Electronics Laboratory," 15 March 1971, Folder 29, "SEL Laboratory Reorganization," Box 5, ACCN 97-093, Rambo Papers; and Levy, "James Angell, Electrical Engineer and Former Carillonneur, Dies," 2.

171. Meindl, interview, pp. 10–11; Angell, interview, p. 10; Lécuyer, "What Do Universities Really Owe Industry?," p. 65; and "Stanford Materials Science and Engineering," *http://mse.stanford.edu/* (accessed 27 March 2006).

172. "Integrated Circuits Laboratory," *http://cis.stanford.edu/icl/* (accessed 27 March 2006).

elements. Specifically, the project received funding from the Joint Services Electronics Program and involved creating a biomedical instrument whose conception came from the medical school faculty.

The device was the project of a graduate student, Kensall D. Wise, who arrived at the Stanford Electrical Engineering Department in 1965. Subsequently, Wise became a key figure in the development of MEMS at Stanford. From 1964 to 1965 he worked at Bell Telephone Laboratories in the Digital Device Integration Department, where he was concerned with the design, integration, and testing of high-speed circuitry. Wise became a graduate research assistant for Angell, whose research at the time centered on the reliability of integrated circuits, so Wise spent his first year at Stanford dealing with reliability and redundancy in integrated circuits.[173]

The details of how the biomedical project came about are not clear. The two accounts given by Angell and Wise are not mutually exclusive and involve a similar timeline. The dissimilarities between the two versions may reflect the differing perspectives of a professor and a graduate student. According to Angell, in the spring of 1966 representatives from the Departments of Neurology and Electrical Engineering met to discuss projects on which they could cooperate, particularly ones appropriate for the newly furnished integrated-circuits laboratory. In view of the difficulties encountered in manufacturing conventional microelectrodes, Angell recalled, the group discussed the application of integrated-circuit technology to the manufacture of microelectrodes. As a result, in September 1966 a project started to examine the question in detail, and it also became Ken Wise's doctoral dissertation. Wise carried out the research under Angell in the Electrical Engineering Department and Starr in the Medical School's Department of Neurology. The main research interests of Starr, who had been a research associate in neurobiology at the National Institutes of Health from 1960 to 1962, were the physiological processes related to hearing.[174]

Wise recalled a somewhat different series of events. The starting points of the project from his perspective were a seminar given in May of 1966 by the Stanford University head of neurology before the Electrical Engineering

---

173. Angell, Terry, and Barth, "Silicon Micromechanical Devices," pp. 49–50; Kensall Wise, James Angell, and Arnold Starr, "An Integrated-Circuit Approach to Extracellular Microelectrodes," *IEEE Transactions on Bio-Medical Engineering* BME-17, no. 3 (July 1970): 247; and Wise, interview, pp. 3 and 4.

174. James Angell, "Study of Hearing with a New Multielectrode Microprobe," grant application NS 08834-01A1, Public Health Service, for the period 1 July 1970 to 30 June 1973, for a total of $169,103, in Folder 31, "Solid State Correspondence Miscellaneous," Box 6, ACCN 97-093, Rambo Papers; and Wise, Angell, and Starr, "An Integrated-Circuit Approach to Extracellular Microelectrodes," p. 247.

Department and a proposal written by Professor John L. Moll in the Electrical Engineering Department. In his dissertation, Wise acknowledged Moll "for his original ideas in the conception of this project" but also Starr for his "suggestions and encouragement."[175]

Moll played a seminal role in semiconductor development over his long career. While working as a recently minted Ph.D. (in electrical engineering from Ohio State University) at Bell Telephone Laboratories from 1952 to 1958, Moll argued for using silicon in the manufacture of semiconductor switches instead of germanium. Daunting technical problems had to be overcome, but Moll and his group succeeded with the help of Carl Frosch in the laboratory's chemistry department. Moll left Bell Telephone Laboratories in 1958 to teach and research the physics of silicon devices at Stanford. Moll also was the recipient of at least one NASA grant. In 1964, for example, he investigated the metallurgical, electrical, and optical properties of gallium phosphide—a compound utilized in various semiconductor devices, including light-emitting diodes (LEDs)—with funding from NASA's Lewis Research Center in Cleveland, Ohio.[176]

In response to the neurology seminar, Wise recalled, Moll suggested using photolithographic techniques similar to those he had seen under development at Bell Telephone Laboratories to create a probe capable of recording electrical impulses from multiple sites in the central nervous system. The idea was radically different from Linvill's Optacon, which did not plug directly into the brain; however, Linvill's concerns may have spurred interest in Moll's project. A couple of proposals submitted to the NIH failed to find support; the brain probe instead received underwriting from the Joint Services Electronics Program.[177]

---

175. Wise, interview, p. 5; and Kensall Wise, "A Multielectrode Microprobe for Biopotential Recording," Ph.D. dissertation, Stanford University, Department of Electrical Engineering, May 1969, p. v.

176. Michael Riordan and Lillian Hoddeson, *Crystal Fire: The Birth of the Information Age* (New York: W.W. Norton & Company, 1997), pp. 221 and 222; Solid-State Electronics Laboratory, *Fundamental Studies of the Metallurgical, Electrical, and Optical Properties of Gallium Phosphide: Quarterly Progress Report for Period April 1–June 30, 1964*, NASA Lewis Research Center Grant NsG-555 (Stanford, CA: Stanford Solid-State Electronics Laboratory, 1964); Solid-State Electronics Laboratory, *Fundamental Studies of the Metallurgical, Electrical, and Optical Properties of Gallium Phosphide: Quarterly Progress Report for Period July 1–September 30, 1964*, NASA Lewis Research Center Grant NsG-555 (Stanford, CA: Stanford Solid-State Electronics Laboratory, 1964). The research began in February 1964 and continued until at least 1972. NASA, Office of University Affairs, *NASA's University Program: Interim Report of Active Grants & Research Contracts* (Washington, DC: NASA, May 1972), p. 27.

177. Wise, interview, pp. 4, 5, and 21; Wise, "Multielectrode Microprobe for Biopotential Recording," p. v; Wise, Angell, and Starr, "An Integrated-Circuit Approach to Extracellular

Moll's project—a device to record electrical pulses from multiple sites in the brain—became not just the topic of Wise's doctoral thesis but his introduction to MEMS as well. Furthermore, it was the first Stanford project that had to do with the fabrication of so-called micromachined devices (the term "micromachined" did not come into currency until later). Although Stanford researchers were carrying out various sensor projects, this was the first that involved precision silicon etching. The goal was to etch individual silicon islands containing transistors precisely and to connect them together with tiny gold beams. The faculty and staff of the integrated-circuits laboratory did not possess such capabilities, so Wise spent the summer of 1966 back at the Bell Telephone Laboratories in Murray Hill, New Jersey, talking to the people who were doing precision etching work, and then he brought that technical knowledge back to Stanford.[178]

The probe that Wise designed was intended to measure electrical potentials (voltages) in animal brains. Its electrodes were small enough to make contact with individual neurons. Wise chose gold for the electrodes because of its resistance to protein poisoning and its proven suitability for high-resolution electroforming. He formed each gold electrode on a silicon structure that provided the electrode needed material strength, because gold lacked the requisite rigidity for the probe electrodes. A thin coating of silicon oxide (glass) over the silicon structure insulated it from surrounding tissue. Removing the coating just at the electrode tip formed the recording area, and the insulated electrode projected beyond the silicon structure a short distance (10 to 50 microns) to minimize tissue damage at the recording site.[179]

Wise created the tiny silicon device using photolithography, a fundamental technique for manufacturing integrated circuits. He employed an acid-resistant, light-sensitive lacquer as a photoresist. The probe manufacturing process took place in four steps: 1) substrate preparation, 2) metallization, 3) insulation, and 4) finishing. Starting with a thermally oxidized silicon wafer only 50 microns thick, he defined an oxide pattern on the wafer. Wise used the pattern as an etching mask as he formed mesas by removing the silicon to a depth of 25 to 35 microns. He then oxidized the wafer again, defining

---

Microelectrodes"; Contract Nonr-225(83) in Attachment, Stanford Electronics Laboratory, "Program and Abstracts of Papers for Stanford University Electronics Research Review," August 1969, Folder 32, "Electronics Research at Stanford-1970," Box 6, ACCN 97-093, Rambo Papers, pp. 20–21.

178. Wise, interview, pp. 4 and 5; and Angell, Terry, and Barth, "Silicon Micromechanical Devices," pp. 49–50.

179. Wise, "Multielectrode Microprobe for Biopotential Recording," pp. 25–26; and Wise, Angell, and Starr, "An Integrated-Circuit Approach to Extracellular Microelectrodes," p. 238.

new oxide patterns on both sides of the wafer. The next step, metallization, created the gold electrodes. He then deposited a thin layer of silicon-dioxide insulation over the wafer using a radio-frequency glow-discharge system.[180]

This technique allowed thin, uniform films of high-quality glass to deposit at temperatures below 300°C. Another photoresist operation selectively removed the deposited silicon dioxide to form the recording site at the electrode tip. The probe was now ready for the finishing operation, in which Wise separated the individual probes from the wafer, attached the output wires, and mounted each probe in a suitable handle. The complete manufacturing sequence represented between one and two man-weeks of work. Typically, Wise realized from 20 to 30 probes from a single silicon wafer, and two to three wafers constituted a typical processing run. He did not consider the fabrication time to be excessive.[181]

The brain probe underwent evaluation by testing it in five anesthetized living cats. The probes went into an opening of about 1 cm in their skull. In all, 40 probes went through this testing procedure. Of them, 18 recorded the activity of several cells, and 7 successfully recorded from individual cells. Only five electrode tips broke in over 60 passes through the brain with the probes. The silicon structure was strong enough to penetrate the dura, a tissue layer surrounding the brain, but the projecting electrodes lacked that strength. Despite their advantages over conventional probes, these tiny probes had their disadvantages and required further development. Eventually, however, Wise's work led to the development of a variety of brain probes.[182]

After completing his dissertation in 1969, Wise remained at Stanford as a research associate in the Electrical Engineering Department, continuing to apply integrated-circuit technology to biomedical sensors. In 1974, however, he began teaching at the University of Michigan, where he resumed his brain probe efforts, this time with funding from NIH, in particular from the Neural Prosthesis Program of the National Institute of Neurological Disorders and Stroke (NINDS). Today, the electrodes are used all over the world for research in neuroscience in such areas as deafness, epilepsy, and Parkinson's disease.[183]

180. Prof. Robert H. Weissman provided the glow-charge system. Wise, "Multielectrode Microprobe for Biopotential Recording," p. v.

181. Wise, Angell, and Starr, "An Integrated-Circuit Approach to Extracellular Microelectrodes," pp. 239–241; and Wise, "Multielectrode Microprobe for Biopotential Recording," pp. 26–35.

182. Wise, Angell, and Starr, "An Integrated-Circuit Approach to Extracellular Microelectrodes," pp. 244–246; Wise, "Multielectrode Microprobe for Biopotential Recording," pp. 36–37; and Angell, Terry, and Barth, "Silicon Micromechanical Devices," p. 50.

183. Wise, interview, pp. 8–9; Kensall Wise, NIH grant 3N01NS012384-00181, "Intracortical Recording Electrode Arrays," NINDS, from September 1981 to September 1984; James

The Biomedical Pressure Sensor

The first MEMS project that the Stanford Electrical Engineering Department and its associated Integrated Circuits Laboratory carried out with funding from the Ames Research Center was a biomedical pressure sensor. The sensor was the creation of Samaun Samadikun in collaboration with Ken Wise and James Angell, and it was the subject of Samadikun's doctoral dissertation undertaken in the Department of Electrical Engineering and completed in August 1971.[184]

Wise's experience with the brain probe and his knowledge of MEMS techniques were integral to the success of Samadikun's doctoral work. Indeed, Wise was the one who guided Samadikun into that particular area of research. His dissertation project was to integrate both the pressure sensor and its associated circuitry on a single silicon chip. The student had to solve three problems. One was how to fashion the diaphragms out of silicon. A second was how to integrate the pressure-sensitive sensors on the diaphragms, and the last problem was how to separate the finished chips from the wafer.[185]

How the Sensor Worked

The heart of the device was a thin-diaphragm piezoresistive pressure sensor. The piezoresistive effect provided an observable resistance change that changed in direct proportion to the amount of pressure applied. The diaphragm magnified the stress placed on it (even small amounts of stress), and the magnification was proportional to the square of the ratio of the diaphragm diameter to its thickness. Piezoresistors interconnected to form a so-called bridge sensed the stresses in the diaphragm caused by pressure. The term "bridge" refers to an electrical arrangement invented by the English scientist Charles Wheatstone to measure resistance. It traditionally consists of four resistors arranged in such a way that one can determine the value of an unknown resistance when the resistances of the others are known. In the case

Angell, "Integrated Circuit Transducers," NIH grant 5P01GM017940-030002, National Institute of General Medical Sciences, fiscal 1972, 1973, and 1974, which appears to have involved electrodes in cats. On the Neural Prosthesis Program, see "Neural Prosthesis Program," *http://www.ninds.nih.gov/funding/research/npp/index.htm* (accessed 27 March 2006).

184. Samadikun, "An Integrated Circuit Piezoresistive Pressure Sensor for Biomedical Instrumentation," Ph.D. dissertation, Stanford University, August 1971.

185. Samadikun, e-mail message to author, 1 February 2006; and Samadikun, "An Integrated Circuit Piezoresistive Pressure Sensor for Biomedical Instrumentation," p. iii.

of the sensor, the unknown value was the piezoresistance of the sensor, which changed in direct proportion to the amount of pressure exerted on it.[186]

For the sensor to work with small pressures, a pressure magnification scheme was necessary. Nearly all sensors based on the piezojunction effect utilized some sort of stylus to magnify the effect of the pressure. By concentrating the force on a small area at the point of a needle, large stresses affected a small area. Another approach used a cantilever-beam structure in which one formed the stress-sensitive devices on one side of the beam, one end of which was clamped to a rigid supporting structure. One applied force at the free end of the beam. A third method involved a diaphragm structure consisting of a circular diaphragm mounted on the tip of a catheter. One could achieve large pressure magnification, if one ensured a large ratio between the diaphragm diameter and its thickness. This is the route that Samadikun chose.[187]

Samadikun evaluated the effectiveness of his sensors by measuring the blood pressure of laboratory dogs with them. He and Nigel Tombs first tested them in a pressurized water bath before trying them on animals. The trials on live dogs went well. "Everything went smooth without a hitch during that first experiment," Samadikun recalled. After sterilizing the sensors in a chemical solution, Samadikun and Tombs inserted them into the left descending aorta of an anesthetized canine, then they guided the probe to the desired upstream location with the help of a fluoroscope. The tests appeared to demonstrate that the device could withstand rough treatment during insertion into a dog, and no noticeable reaction occurred either with the silicon or the epoxy sealant during measurements that lasted 90 minutes in the animal's cardiovascular system.[188]

The Etching Technique

Silicon pressure sensors, of course, were not novel. Rather, the novelty of the Stanford work lay in the etching process used to create the diaphragm. Rather than electrochemical etching or another known method, Samadikun used anisotropic etching. During discussions with Wise, who had experience

---

186. James Angell, "Research on Pressure Sensors for Biomedical Instruments," Final Report NASA-Ames Grant #05-020-401, NASA CR-142351 (Washington, DC: NASA Center for AeroSpace Information, 19 February 1975), p. 1.

187. Samadikun, Wise, and Angell, "An IC Piezoresistive Pressure Sensor for Biomedical Instrumentation," p. 102; and Samadikun, "An Integrated Circuit Piezoresistive Pressure Sensor for Biomedical Instrumentation," pp. 17–25.

188. Samadikun, e-mail message to author, 1 February 2006; Samadikun, Wise, and Angell, "An IC Piezoresistive Pressure Sensor for Biomedical Instrumentation," pp. 108 and 109; and Samadikun, "An Integrated Circuit Piezoresistive Pressure Sensor for Biomedical Instrumentation," pp. 39–42.

with both isotropic and anisotropic etching, Samadikun and Wise decided to use a wet etching technique to make the diaphragms and to explore different methods of stopping the etching process to produce diaphragms with thicknesses that were independent of the etch rate. Their discussions also gave rise to the idea of using an anisotropic etching of a V-groove in (100) silicon for two distinct purposes: to control the thickness of the diaphragm and to separate the sensor chips from the finished wafer.[189]

In isotropic etching, material is removed from the wafer substrate in a nondirectional manner (in relationship to the crystal planes) via chemical etching. Anisotropic (or nonisotropic) etching is rather different and an important commercial process in manufacturing integrated circuits. When photolithography is used to print resist lines on silicon wafers, to adequately reproduce very tiny lines (below 0.1 micrometers) into underlying silicon and metal layers on a wafer positioned horizontally, the direction of the etching must be vertical only, so that the etching compound does not spread in the horizontal plane. Anisotropic etching requires a substrate with a well-defined crystalline structure, which silicon typically exhibits. The anisotropic etch is directional along the exposed plane in the crystal lattice. As atoms are removed from the crystal lattice, different planes are exposed to the etching substance. Since the density of atoms on the planes varies, the etching rate varies significantly. Potassium hydroxide (KOH) dissolves the silicon that has been left exposed by the photolithography masking step. Such alkali solvents dissolve the silicon in a highly anisotropic way, with some crystallographic directions dissolving up to a thousand times faster than others. This is a common method for creating V-shaped grooves in silicon wafers.[190]

Several etching chemicals were available to Samadikun, including hydrazine, pyracatechnol, and potassium hydroxide. He chose potassium hydroxide, which was relatively inexpensive and easy to handle. The way in which he used this anisotropic etching technique made possible a novel thickness-monitoring scheme that also acted as a chip separation etch. Samadikun produced the first devices himself as a "proof of concept" of the procedure, then fabricated

---

189. Samadikun, e-mail message to author, 1 February 2006.

190. The invention of anisotropic etching at Bell Telephone Laboratories in 1967 was a key milestone in the history of MEMS. See Herbert A. Waggener, Roger C. Kragness, and A. L. Tyler, "Anisotropic Etching for Forming Isolation Slots in Silicon Beam Leaded Integrated Circuits," *International Electron Devices Meeting Abstracts* 11, no. 1 (October 1967): 68; Herbert A. Waggener, Roger C. Kragness, and A. L. Tyler, *Electronics* 40 (1967): 274; and Roger C. Kragness and Herbert A. Waggener, "Precision Etching of Semiconductors," U.S. patent 3,765,969, filed 13 July 1970, issued 16 October 1973, assigned to Bell Telephone Laboratories.

batches of sensors with diaphragm diameters of 0.5 mm and thicknesses of only 5 microns, surrounded by a 0.15-mm-wide ring of thick silicon.[191]

Several years after Samadikun left Stanford for his home in Indonesia, Wise applied for a patent to cover the diaphragm-forming technique and gave credit to Samadikun as his fellow inventor. Issued in 1975, the patent provides a rather detailed discussion of a generalized technique for forming thin regions of predetermined thickness in a silicon wafer, rather than its specific application to form diaphragms. The process exploited the anisotropic properties of silicon and the fact that certain anisotropic etching compounds had etch rates several hundred times larger in the (100) crystallographic direction than along other directions.[192] For a discussion of relevant silicon crystallography, see the technical essay (Appendix B) at the end of this chapter.

The Source of Funding

Samadikun's work on the sensor—including the development of the specific anisotropic etching process—was underwritten unequivocally by NASA, although not his initial studies at Stanford. Support for his first five quarters of study came from the United Nations International Atomic Energy Agency, with a NASA grant taking care of the remainder of his schooling. Samadikun specifically cited the NASA grant (NGL-05-020-401) that supported his sensor research in his dissertation[193] as well as in subsequent publications—namely, a paper he read at the 1971 IEEE International Solid-State Circuits Conference[194] and an article published in the *IEEE Transactions on Biomedical Engineering*,[195] which Stanford simultaneously duplicated as a NASA Contractor Report.[196] After Samadikun left Stanford in 1971, NASA continued to underwrite the sensor research until September 1973, when

191. Samadikun, Wise, and Angell, "An IC Piezoresistive Pressure Sensor for Biomedical Instrumentation," p. 105; Samadikun, e-mail message to author, 1 February 2006; and Angell, "Research on Pressure Sensors for Biomedical Instruments," p. 1.

192. The following discussion is from Wise and Samadikun, "Method for Forming Regions of Predetermined Thickness in Silicon," U.S. patent 3,888,708, filed 10 April 1974, issued 10 June 1975.

193. Samadikun, "An Integrated Circuit Piezoresistive Pressure Sensor for Biomedical Instrumentation," p. iii.

194. Samadikun, Wise, Nielsen, and Angell, "An IC Piezoresistive Pressure Sensor for Biomedical Instrumentation," p. 104.

195. Samadikun, Wise, and Angell, "An IC Piezoresistive Pressure Sensor for Biomedical Instrumentation," p. 101.

196. Samadikun, Wise, and Angell, *An IC Piezoresistive Pressure Sensor for Biomedical Instrumentation*, NASA-CR-142604 (Washington, DC: NASA Center for AeroSpace Information [CASI], 1 March 1973).

NIH picked up its funding with a contract for a study of the "Physiology of the Oviduct" and a grant from the Integrated Electronics for Medical Applications Program.[197] Thus, looking at the full range of funding for Samadikun and the sensor, although NASA funded a portion of the research, one also must recognize the assistance from the United Nations International Atomic Energy Agency and the National Institutes of Health.

The Gas Chromatograph

Another early MEMS device developed in the Stanford Integrated Circuits Laboratory was a gas chromatograph on a silicon wafer. A gas chromatograph is an instrument that identifies and measures gases found in an unknown mixture by separating, identifying, and quantifying each gas in the mixture. First, a sample of the mixture is injected by a valve into a long capillary column, through which it is flushed by an inert carrier gas (usually helium). The walls of the capillary column are lined with a thin layer of a material, such as silicone oil or a polymer, so that different gases have different degrees of solubility with the lining material. As the gases passed through the column, they were adsorbed and desorbed repeatedly in the lining. Because the time a component gas remained adsorbed depended on its solubility, each gas traveled through the column at a different rate. As a result, the gases emerged at different but specific and predictable times. The output stream passed over a detector that measured a given property of the gas (its thermal conductivity, for example). The detector's output signals, then, corresponded to the individual gases.[198]

The creation of the miniature gas chromatograph was the subject of dissertation research carried out by Stephen Clark Terry, who completed his degree in 1975.[199] Ultimately, his work relied on two other Stanford dissertations on thermal-conductivity detectors for gas chromatographs. The first, completed in 1973, was that of Frederick Andrew Perner, who compared thermistor and pyroelectric types of detectors. Perner acknowledged receiving financial help from IBM and Dr. Glenn C. Carle of NASA's Ames Research Center in the form of a grant (NGL 05-020-543) specifically for the miniature gas chromatograph project.[200] Carle was in the Life Detection

197. Angell, "Research on Pressure Sensors for Biomedical Instruments," pp. 1, 3, and 6.

198. Angell, Terry, and Barth, "Silicon Micromechanical Devices," p. 51.

199. Terry, "Gas Chromatography System."

200. Frederick Andrew Perner, "A Pyroelectric Thermal Conductivity Detector for a Miniature Gas Chromatography System," Ph.D. dissertation, Department of Electrical Engineering, Stanford University, October 1973, p. iv. The published abstracts of NASA grants,

Systems Branch, whose chief was Vance Oyama, who is perhaps best known for the Viking mission search for life on Mars.[201]

Based on Perner's results, Terry took a different approach. He later used the thermal-conductivity detector developed by John Hallock "Hal" Jerman, whose dissertation was underwritten not by NASA but NIOSH.[202] Jerman's contribution was to design a detector that could be fabricated using the planar process. Its basic configuration emerged during a major redesign of the entire integrated-circuit gas chromatograph.[203] The chromatograph also was representative of the interdisciplinary effort that the McCullough Building was supposed to foster, having benefited from the support of Assistant Professor L. M. "Bill" Stephenson of the Chemistry Department. More than that, the device grew out of the growing expertise and pioneering MEMS endeavors of the Integrated Circuits Laboratory. For example, it relied on the laboratory's experience with photolithography and the special anisotropic etching techniques had been developed for the fabrication of tiny brain probes on Wise's dissertation and Samadikun's sensors.[204]

Description of the Chromatograph

Terry's gas chromatograph was extremely small. Indeed, the silicon wafer on which it was built was only 5 cm (2 inches) in diameter. The device consisted of a capillary column, a sample-injection valve, and a thermal-conductivity detector—all of which fit on a wafer. One of the technical challenges was the volume of the capillary column, which was much smaller than that of a conventional column. Chromatographs operate properly only if the volume of the injected sample gas is much smaller than the volume of the column. Therefore, it was necessary to design a miniature sample-injection valve to accompany the miniature capillary column. Because of the necessity of minimizing the internal volume between the valve and the column as well as between the column and the detector, the valve and column were fabricated on the same wafer, and the detector was a silicon chip mounted on the wafer.[205]

---

however, state that Ralph W. Donaldson Jr. was the contract monitor. *NASA's University Program* (1973), p. 26.

201. ARC, *Telephone Directory* (1974), pp. 5 and 48.

202. John Hallock Jerman, "A Miniature, Thin-Film Thermal Conductivity Detector for an Integrated Gas Chromatograph," Ph.D. dissertation, Stanford University, May 1981, p. vii.

203. Ibid., p. 33.

204. Terry, "Gas Chromatography System," pp. v and 1.

205. Angell, Terry, and Barth, "Silicon Micromechanical Devices," p. 51; and Kovacs, *Micromachined Transducers*, p. 869.

The capillary column was a groove 1.5 meters long, wound into a spiral so that it fit on the wafer. A glass plate bonded to the wafer formed the top surface of the column. At the input end of the column, a hole led to the bottom surface of the wafer. The helium carrier gas entered the column through this hole. A short distance away, another hole in the column channel led to a valve on the back surface of the wafer. The sample gas entered a separate channel through yet another hole and flowed through the channel to the valve, where the valve injected it into the capillary column.[206]

The valve seat on the back surface of the wafer consisted of a silicon sealing ring, which surrounded both the input and the output openings, and a silicon seating ring, which surrounded only the output opening that led to the capillary column. A diaphragm of nickel and Teflon was clamped against the sealing ring. Normally, a spring held the plunger of a solenoid against the diaphragm, pushing it against the seating ring. When the solenoid was actuated, the plunger withdrew and the diaphragm relaxed, allowing gas to flow from the input opening over the seating ring and into the output opening. The effective dead volume of the valve was the volume of the capillary-column orifice, which was only 4 nanoliters.[207]

At the output end of the capillary column, another etched hole led to a gas channel etched into the bottom surface of the wafer. The chip on which the thermal-conductivity detector was built was inverted over this channel and clamped to the wafer. The detector was a thin-film metal resistor on a thermally isolating membrane of Pyrex glass in the middle of the chip. A constant electric current passed through the resistor, and changes in its resistance were monitored. Sample gases had thermal conductivities lower than that of the carrier gas. Therefore, they removed less heat from the resistor and so increased its resistance and created voltage peaks. The amplitude of each voltage peak was proportional to the quantity of a given gas in the mixture.[208]

Micromachining the Gas Chromatograph

The micromachining of the chromatograph wafer began with etching the valve seat. The valve well, the sealing ring, and the seating ring were defined by isotropic etches through concentric circular openings in the oxide layer on the back of the wafer. The holes in the valve and at the ends of the capillary columns and the sample-gas channels resulted from an anisotropic etch through square openings defined in the silicon oxide on the front surface of

---

206. Angell, Terry, and Barth, "Silicon Micromechanical Devices," pp. 51–52.
207. Ibid., p. 52.
208. Ibid., p. 52.

the wafer. The capillary column and the carrier-gas channels were delineated by an isotropic etch through a silicon oxide pattern on the front wafer surface.[209] At this point, the capillary column was a shallow spiral groove about 200 micrometers across at the surface of the wafer and 40 micrometers deep. The open spiral was made into an enclosed channel by stripping the silicon oxide off the surface and bonding the wafer to a plate of Pyrex glass using a technique called anodic bonding (sometimes called electrostatic bonding or the Mallory process as well). The possibility that extruded glue or solder might block the shallow capillary column precluded both as bonding agents.[210]

Next, one oxidized the wafer and covered it with a layer of Pyrex glass deposited by a technique called sputtering. After the depositing of thin-film metal resistors on top of the glass, an anisotropic etch through square openings in the silicon oxide on the back of the wafer removed the entire thickness of silicon, leaving a membrane of thermally isolating glass under each set of resistors. One then sawed the wafer into chips, attached wires to each device, and the front surface of each chip was clamped over the gas channel at the end of the capillary column.[211] Because the capillary column had such a minute volume, the miniature gas chromatograph required much smaller amounts of carrier gas than a conventional chromatograph. This economy of carrier gas and the device's small size made it possible to build lightweight portable chromatographs. In fact, in 1983 a portable instrument consisting of five miniature chromatographs and a microcomputer was under development by Microsensor Technology, a company founded by Jerman to commercialize his MEMS-based integrated gas chromatograph.[212]

The Source of Funding

The Ames Research Center unquestionably was behind the development of the gas chromatograph on a silicon wafer. In his dissertation, Stephen Terry

---

209. Terry, "Gas Chromatography System," pp. 41–51; Angell, Terry, and Barth, "Silicon Micromechanical Devices," p. 52; and Petersen, "Silicon as a Mechanical Material," pp. 434–435.

210. Angell, Terry, and Barth, "Silicon Micromechanical Devices," pp. 52–53. For a discussion of anodic bonding, see G. Wallis and D. I. Pomerantz, "Field Assisted Glass-Metal Sealing," *Journal of Applied Physics* 40, no. 10 (October 1969): 3946; Timothy Alan Nunn, "A Silicon Absolute Pressure Transducer for Biomedical Applications," Ph.D. dissertation, Stanford University, Department of Electrical Engineering, October 1977, pp. 91–92.

211. Angell, Terry, and Barth, "Silicon Micromechanical Devices," p. 53.

212. "About Iolon," *http://www.iolon.com/mteam.html* (accessed 28 March 2006). Microsensor Technology still exists. See "Microsensor Technology," *http://www.microsensortech.com* (accessed 27 March 2006).

acknowledged NASA funding and specifically thanked Ralph Donaldson, Hal Sandler, and Glenn Carle of Ames. NASA also funded the earlier work on miniature thermal-conductivity detectors built on a chip by Perner, whose studies received additional support from IBM.[213] On the other hand, Jerman's dissertation was not underwritten by NASA but rather by NIOSH.[214] Joint publications based on their gas chromatograph theses did not report a single or consistent funding source. For instance, a 1977 technical report authored by Terry and Jerman stated that their funding source was NIOSH under contract NIOSH-210-76-0140,[215] but their 1979 article written with Jim Angell and published in the peer-reviewed *IEEE Transactions on Electron Devices* reported financial support from both NIOSH *and* NASA.[216] The dual funding of the research showed up in the projected uses of the gas chromatograph. The apparatus was "expected to find application in the areas of portable ambient air quality monitors, implanted biological experiments, and planetary probes."[217]

Clearly, the NASA grant preceded the NIOSH contract. Aside from the earlier underwriting of Perner's dissertation, a NASA contractor report submitted by James Angell in 1974—prior to the completion of Terry's dissertation in 1975—stated specifically that the space agency had funded the technology for fabricating the very small valves whose function was to introduce a small sample of the gas to be analyzed into the main carrier gas stream flowing through the chromatograph's column.[218] NASA appeared to

---

213. Terry, "Gas Chromatography System," p. v; Perner, "Pyroelectric Thermal Conductivity Detector for a Miniature Gas Chromatography System," p. iv. Perner specified NASA grant NGL 05-020-543, while Terry's dissertation acknowledged NASA grant NGR 05-020-690.

214. Jerman, "Miniature, Thin-Film Thermal Conductivity Detector," p. vii.

215. Stephen Clark Terry and John Hallock Jerman, *Feasibility Study of a Pocket-Sized Gas Chromatographic Air Analyzer*, report SEL-77-027 (Stanford, CA: Stanford University, Stanford Electronics Laboratories, July 1977), cover and p. 1. The final report on the project, filed with the National Institute for Occupational Safety and Health in 1981, indicates that the work was supported by the agency's Division of Physical Sciences and Engineering in Cincinnati, Ohio. James B. Angell, John Hallock Jerman, Stephen Clark Terry, and S. Saadat, *A Prototype Gas Analysis System Using a Miniature Gas Chromatograph*, NIOSHTIC #00113750 and #PB83-105122 (Cincinnati: NIOSH, April 1981).

216. Terry, Jerman, and Angell, "Gas Chromatographic Air Analyzer Fabricated on a Silicon Wafer," p. 1880. This source also provided a different NIOSH contract number: 210-77-0159. The possibility that the two contracts were for different fiscal years might explain the discrepancy in NIOSH contract numbers.

217. Terry, Jerman, and Angell, "Gas Chromatographic Air Analyzer Fabricated on a Silicon Wafer," p. 1886.

218. James Angell, "Interim Status Report on Research on Miniature Gas Analysis Systems," NASA CR-138138, 2 January 1974, p. 1.

be the sole underwriter of the research at that point, with the NIOSH money coming later as Jerman joined the effort.

The NIOSH money was available because the Occupational Safety and Health Act of 1970 had authorized the Secretary of Health, Education, and Welfare to prescribe regulations requiring employers to measure, record, and make reports on the exposure of employees to gases and substances that might endanger their safety or health. Chronic exposure to a large number of the gases found in industrial environments is potentially dangerous to workers. To record worker exposure and to enforce the law, the federal government needed to be able to provide or to suggest to employers monitoring methods of suitable accuracy for making measurements of employee exposure. The miniature gas chromatograph fit the bill for this regulatory need.[219]

Silicon Accelerometer

A third pioneering MEMS device developed at the Stanford Integrated Circuits Laboratory was an accelerometer, first described in the 1977 Stanford dissertation of electrical engineering graduate student Lynn Michelle Roylance.[220] Accelerometers measure, display, and analyze acceleration (a change in the amount or direction of velocity) but also can measure vibrations. In 1977 accelerometers were not entirely new. Several MEMS lectures posted on the Internet claim that Kulite Semiconductor Products demonstrated a silicon accelerometer in 1970, but they do not present any evidence or cite any sources for their assertion.[221] Amid a number of patents issued to Kulite at the time for transducers,[222] devices that convert energy from one form into

---

219. Terry and Jerman, *Feasibility Study of a Pocket-Sized Gas Chromatographic Air Analyzer*, p. 1.

220. Lynn Michelle Roylance, "Miniature Integrated Circuit Accelerometer for Biomedical Applications," Ph.D. dissertation, Stanford University, November 1977.

221. See, for example, Cheng-Hsien Liu, PowerPoint presentation, "Micro-Electro-Mechanical Transducers," p. 13; Kristofer S. J. Pister, "Introduction to MEMS Design and Fabrication," slide 2, *http://www-bsac.eecs.berkeley.edu/~pister/245/2005S/Lectures/Intro.ppt* (accessed 9 January 2006); and Haixia Zhang, "Lecture 1: Introduction to MEMS," for course MEMS Devices and Design, Spring 2004, p. 14, *http://ime.pku.edu.cn/mems/courses/device&design/Lecture_01_Introduction_to_MEMS_ZHX.pdf* (accessed 9 January 2006).

222. See, for example, Anthony D. Kurtz and Joseph R. Mallon, Jr., "High Temperature Transducers and Housing Including Fabrication Methods," U.S. patent 3,800,264, filed 14 March 1972, issued 26 March 1974, assigned to Kulite Semiconductors Products; Anthony D. Kurtz and Charles L. Gravel, "Semiconductor Transducers Having H-Shaped Cross-Sectional Configurations," U.S. patent 3,739,315, filed 18 May 1972, issued 12 June 1973, assigned to Kulite Semiconductor Products; Anthony D. Kurtz, "Pressure Sensitive Transducers Employing Capacitive and Resistive Variations," U.S. patent 3,748,571, filed 7 September 1972, issued 24 July 1973, assigned to Kulite Semiconductors Products.

another, a search of European and United States patents reveals two, filed in 1970 and 1971, for an accelerometer that one could insert into a subject's blood stream.[223] Surprisingly, in light of the implied priority for the Kulite accelerometer given by Internet sources, the 1970 patent presents a rather different picture of the current state of the art. "Presently," the patent reads, "a great number of such devices are manufactured by using monolithic integrated circuit techniques."[224] This language infers that Kulite was not first and calls into question our knowledge of MEMS history.

The heart of Roylance's accelerometer was a flexible beam micromachined out of silicon. In 1972, Wolf Dietrich Frobenius and other Westinghouse researchers announced the creation of an accelerometer that used a flexible *metal* beam. It was compatible with integrated-circuit technology and could be fabricated in batches.[225] The originality of the work carried out at Stanford is attested to by a recent work, one whose author provided footnotes and sources. Gregory T. A. Kovacs wrote: "One of the earliest examples of a micromachined piezoresistive strain-gauge accelerometer (or, in fact, *any* micromachined accelerometer) is the device made by Roylance and Angell (1979) for use in biomedical implants to measure heart wall accelerations."[226]

Description of the Accelerometer

The first descriptions of the miniature accelerometer appeared in Roylance's dissertation and an associated technical report, both dated November 1977.[227] Later, she and her professor, James Angell, discussed the accelerometer in a presentation given at the 1978 IEEE Solid-State Circuits Conference,[228] and

---

223. Anthony D. Kurtz, Joseph R. Mallon, and Charles L. Gravel, "Electromechanical Transducers and Housings," U.S. patent 3,654,579, filed 11 May 1970, issued 4 April 1972, assigned to Kulite Semiconductor Products; and Anthony D. Kurtz, Charles L. Gravel, and Joseph R. Mallon, "Transducers Employing Integral Protective Coatings and Supports," U.S. patent 3,753,196, filed 5 October 1971, issued 14 August 1973, assigned to Kulite Semiconductors.

224. Kurtz, Mallon, and Gravel, "Electromechanical Transducers and Housings," col. 1.

225. Wolf Dietrich Frobenius, S. A. Zeitman, M. H. White, D. D. O'Sullivan, and R. G. Hamel, "Microminiature Ganged Threshold Accelerometers Compatible with Integrated Circuit Technology," *IEEE Transactions on Electron Devices* 19, no. 1 (January 1972): 37–40.

226. Kovacs, *Micromachined Transducers*, p. 229.

227. Roylance, "Miniature Integrated Circuit Accelerometer"; and Lynn Michelle Roylance, *Miniature Integrated Circuit Accelerometer for Biomedical Applications*, SEL-77-044 (Stanford, CA: Stanford Electronics Laboratories, Integrated Circuits Laboratory, November 1977).

228. Lynn Michelle Roylance and James B. Angell, "A Miniature Integrated Circuit Accelerometer," *1978 IEEE International Solid-State Circuits Conference Digest of Technical Papers* 21 (February 1978): 220–221.

described it in some detail in an article published in the *IEEE Transactions on Electron Devices* in 1979.[229] The accelerometer built on previous work carried out in the Stanford Integrated Circuits Laboratory on silicon sensors for biomedical applications (the Wise and Samadikun dissertations) and the gas chromatograph. Its fabrication combined standard integrated-circuit photolithography techniques with controlled etching processes, the same methods that had been used to fabricate the sensors and gas chromatograph. The impetus for its development arose from discussions with biomedical researchers, which suggested two specific applications. The first addressed problems encountered by obstetricians and gynecologists during late pregnancy and labor. It assumed that motion within the uterus correlated with the movements of the fetal heart, because the force of blood being pumped to the extremities tended to move the fetus within its buoyant environment. If one could measure motion in the uterus, theoretically it might signal to the doctor that the fetus was in difficulty early enough to prevent serious injury or death. The other more feasible application was to measure the motion of the heart wall over the course of the cardiac cycle. A matrix of miniature sensors sutured to the heart muscle would provide detailed information about the force of contraction of various sections of the heart wall and, eventually, information about velocity and displacement as well. A better understanding of the mechanics of the heart also might enable such a matrix of sensors to signal the early phases of coronary occlusion, the prelude to a heart attack.[230]

The accelerometer was a major advance in the size and weight of such devices. Roylance described it as "more than an order of magnitude reduction in volume and mass compared to commercially available accelerometers with equivalent sensitivity." It was light enough (less than 0.02 grams) to allow highly accurate, highly sensitive measurements of changes in heart muscle motion. It also was small enough (2 mm × 3 mm × 0.06 mm) for several of them to fit inside a pill that, when swallowed, could monitor the magnitude and direction of the pill's movement through the intestinal tract, while telemetry circuitry inside the pill transmitted the signals to an external receiver.[231]

Essentially, the device was a glass-silicon-glass sandwich. The center silicon layer—the heart of the device—was a very thin (15 micron) cantilevered beam of silicon surrounded by a 200-micron-thick silicon supporting rim. The rim provided a rigid support for one end of the beam, an area for contact

229. Lynn Michelle Roylance and James B. Angell, "Batch-Fabricated Silicon Accelerometer," *IEEE Transactions on Electron Devices* 26, no. 12 (1979): 1911–1917.

230. Roylance, "Miniature Integrated Circuit Accelerometer," pp. iv, 1, and 3.

231. Ibid., p. 6; Angell, Terry, and Barth, "Silicon Micromechanical Devices," p. 45; and Petersen, "Silicon as a Mechanical Material," p. 441.

pads, and mounting surfaces parallel to the plane of the beam. The beam widened at its free end into a rectangular paddle that supported a mass composed of a dense substance such as gold or silicon.[232]

The resistor diffused into the top surface of the beam changed its value in proportion to the amount of acceleration because of the stress produced in the beam by changes in velocity. A second resistor, placed in an area free of stress, compensated for temperature changes that the device otherwise might sense incorrectly as acceleration. Wells etched into the glass caps, which were sealed to the supporting rim using anodic bonding, formed a cavity completely enclosing and protecting the beam, yet allowed it to deflect freely up to a given distance determined by the depth of the wells.[233]

Fabrication of the Silicon Accelerometer

In general, Roylance fabricated the accelerometer in a batch process that used standard integrated-circuit techniques (photolithography and diffusion) as well as special anisotropic etching techniques to shape the silicon and glass. The starting material was n-type (100) silicon, chosen because the preferred (110) direction for p-type piezoresistors coincided with the pattern orientation of anisotropic etchants (such as potassium hydroxide) in silicon. Roylance could achieve precise dimensional control even with a large etch depth, because etching the (111) planes took place two orders of magnitude slower than (100) and (110) surfaces.[234] The first step in creating the accelerometers was to etch half a dozen widely spaced alignment holes completely through the wafer. Next, Roylance grew a 1.5-micron layer of thermal silicon oxide over the silicon wafer, then performed two photolithographic and diffusion steps in order to form the 10 Ω/square p+ contacts and the 100 Ω/square P piezoresistors. The photolithography stripped away the front silicon oxide but left intact the back oxide layer as a mask for the final etching. She did not apply a metal coating to the silicon because of the lack of electrical connections with those components.[235]

---

232. Roylance and Angell, "Miniature Integrated Circuit Accelerometer," p. 220; and Roylance and Angell, "Batch-Fabricated Silicon Accelerometer," p. 1911.

233. Roylance and Angell, "Miniature Integrated Circuit Accelerometer," p. 220; and Roylance and Angell, "Batch-Fabricated Silicon Accelerometer," pp. 1911–1912.

234. Roylance, "Miniature Integrated Circuit Accelerometer," p. 19–46; Roylance and Angell, "Miniature Integrated Circuit Accelerometer," p. 220; Roylance and Angell, "Batch-Fabricated Silicon Accelerometer," p. 1911; and Kovacs, *Micromachined Transducers*, p. 229.

235. Roylance, "Miniature Integrated Circuit Accelerometer," pp. 19–46; and Kovacs, *Micromachined Transducers*, p. 229.

The remaining processing steps shaped the cantilevered beam, the silicon mass (if gold was not used), and the window where the glass overhung the silicon. After opening windows in the backside oxide layer, Roylance etched away the silicon around the beam and mass as well as in the area where the beam was to be thinned. She stopped the etching when the beam area was twice the desired final thickness. A photolithography on the top surface of the partially etched wafer defined the air gap around the beam and the window opening. The final etch engraved the beam from the bottom plus the air-gap regions from both top and bottom. The moment the silicon disappeared from the large window openings signaled the end of the etching. The final step was to strip away the remaining silicon oxide.

The materials from which she prepared the glass cover plates were 200-micron-thick pieces of #7740 Pyrex glass polished optically flat on one side. The silicon-glass bonding process dictated the type of glass. The glass had to be slightly conductive at the bonding temperature and its thermal expansion rate had to match that of silicon. Unfortunately, the Pyrex glass was not nearly as easy to etch as silicon. Roylance etched wells in the top and bottom glass covers using a chrome-gold etch mask and a mixture of 30 percent nitric acid (HNO_3) and 70 percent hydrogen fluoride (HF) at 48°C. With the masking layer stripped away, she deposited aluminum on the top glass to form the metal bonding pads.[236]

The next steps were to sandwich the silicon wafer between the two glass wafers, bond them, then separate the individual devices from the wafer. Assembling the accelerometer sandwich involved attaching gold masses (if used) and aligning and bonding the glass covers to the silicon. Only after completion of the bonding process did Roylance break apart the individual accelerometers and begin packaging them. The anodic bonding process produced a hermetic and irreversible seal between silicon and glass. It consisted of aligning the glass and silicon, raising the temperature to about 400°C, then applying 600 V between the silicon and the glass. The advantages of anodic bonding included its simplicity, the lack of glue or solder, and visual inspection of the bonding results. Bonded areas appeared dark gray, while other areas were much lighter in color and showed interference fringing. Finally, with the top and bottom covers bonded, Roylance separated the individual devices using a dicing saw and attached the leads.[237]

---

236. Roylance, "Miniature Integrated Circuit Accelerometer," pp. 19–46; and Kovacs, *Micromachined Transducers*, p. 230.

237. Roylance, "Miniature Integrated Circuit Accelerometer," pp. 19–46; and Kovacs, *Micromachined Transducers*, p. 230.

The Source of Funding

Determining the source of funding for Roylance's research on the silicon accelerometer is not a straightforward proposition. An article about the device published in a 1979 issue of *IEEE Transactions on Electron Devices*, for example, provided no acknowledgment of the research's funding source.[238] Nonetheless, an unspecified amount of initial support evidently came from the Joint Services Electronics Program under grant N000-14-67-A-0112-0004.[239] The semiannual report submitted to the Joint Services Electronics Program for the first half of 1973 promised that the Integrated Circuits Laboratory would undertake "a new project dealing with charge-coupled devices for analog delay lines" later in the year.[240] The following report mentioned no work on charge-coupled devices, but it *did* include a new undertaking called Project 4606, "A Solid-State Accelerometer." The principal investigator was James Angell; Lynn Roylance was listed as "staff." The project's objective was "to fabricate a miniature accelerometer, based on acceleration-induced strain in a piezoresistor, using integrated-circuit technology." This was the very device that Roylance created for her dissertation. The report contained a photograph of the first such device, about 2.1 mm long and shaped like a bone.[241]

The subsequent semiannual report—for the first half of 1974—included a resume of progress made on the accelerometer,[242] but the next report failed to mention the project. Instead, Project 4606 had become "Precision Chemical Machining of Single-Crystal Silicon," which Angell was carrying out as principal investigator with Stephen Terry and Phillip Barth. It involved perfecting techniques for fabricating very thin layers of silicon on insulated substrates, such as silicon dioxide or on a glass matrix.[243] One can conclude therefore that underwriting for Roylance's dissertation research during the

---

238. Roylance and Angell, "Batch-Fabricated Silicon Accelerometer," pp. 1911–1917.

239. Angell, *Interim Report*, p. A-1.

240. Stanford Electronics Laboratories, *Semi-Annual Status Report No. 124, 1 January through 30 June 1973*, Report SU-SEL-73-028 (Stanford, CA: Stanford Electronics Laboratories, June 1973), p. 18.

241. Stanford Electronics Laboratories, *Semi-Annual Status Report No. 125, 1 July through 31 December 1973*, Report SU-SEL-73-051 (Stanford, CA: Stanford Electronics Laboratories, December 1973), pp. 20–22.

242. Stanford Electronics Laboratories, *Semi-Annual Status Report No. 126, 1 January through 30 June 1974*, Report SU-SEL-74-043 (Stanford, CA: Stanford Electronics Laboratories, June 1974), pp. 27–29.

243. Stanford Electronics Laboratories, *Semi-Annual Status Report No. 127, 1 July through 31 December 1974*, Report SU-SEL-74-056 (Stanford, CA: Stanford Electronics Laboratories, December 1974), pp. 33–35.

second half of 1973 and the first half of 1974 came from the Joint Services Electronics Program.

The process by which NASA began funding the research is uncertain. In January 1974, when James Angell reported to NASA regarding the gas chromatograph effort under grant NGR 05-02-690, he included a separate appendix, titled "Solid State Accelerometer," that furnished the space agency with developmental information on Roylance's silicon accelerometer. The work, he explained, was underwritten by the Joint Services Electronics Program and was being "reported here for informational purposes only."[244] It is possible that Angell's description of the accelerometer interested Ames researchers, who agreed to have Roylance continue her research under the gas chromatograph grant. Acknowledgment of the NASA funding, namely grant NGR 05-020-690 (which funded the gas chromatograph), appeared in the first two published descriptions of the silicon accelerometer: Roylance's dissertation and its associated technical report, both of which appeared in November 1977.[245] The same NASA grant also received recognition in the published précis of a paper on the accelerometer that Roylance and Angell delivered at the 1978 IEEE Solid-State Circuits conference.[246] Angell continued as the grant's principal investigator until 1980, three years after Roylance completed her dissertation, and it still was for "miniature gas analysis systems."[247]

Impact Evaluations

The Biomedical Pressure Sensor

NASA and NIH

NASA funding paid for development of this MEMS device and its testing on dogs, and the sensor successfully tested their blood pressure.[248] The Stanford sensor effort continued after the NASA underwriting ended in September 1973, with money from the National Institutes of Health. The NIH support was indicative of the laboratory's funding shift away from its traditional sponsor, the Joint Services Electronics Program, to its new backers within the health research establishment. From 1972 to 1974, while NASA was funding the pressure sensor, James Angell was principal investigator on a

---

244. Angell, *Interim Report*, p. A-1.
245. Roylance, "Miniature Integrated Circuit Accelerometer," p. iii; and Roylance, *Miniature Integrated Circuit Accelerometer*, cover and p. iii.
246. Roylance and Angell, "Miniature Integrated Circuit Accelerometer," p. 220.
247. NASA's University Program (1980), p. 28.
248. Samadikun, Wise, and Angell, "An IC Piezoresistive Pressure Sensor for Biomedical Instrumentation," p. 109.

National Institute of General Medical Sciences grant to develop integrated-circuit sensors that Stanford tested on cats. Later, he was principal investigator on a major grant between 1977 and 1986 from the National Center for Research Resources to develop a "resource" for creating biomedical sensors and transducers using integrated-circuit fabrication techniques.[249] The successful development of the pressure sensor with NASA dollars thus helped to propel Stanford University and its Integrated Circuits Laboratory in their quest for patronage from the NIH. Any statements that NASA makes about its contribution to the development of the sensor must concede the role of the NIH money as well.

MEMS Techniques

In addition, NASA supported the development of a new integrated-circuit fabrication technique—a variety of anisotropic etching—that made possible the measured production of the diaphragm. An additional application of this technique took place at Stanford even before approval of the Wise-Samadikun patent for the process in the dissertation research of Timothy A. Nunn. Finished in October 1977, Nunn's dissertation involved creating a miniature pressure transducer on a silicon wafer. The effort was underwritten by both the NIH and NASA—through the gas chromatograph grant.[250] In many ways, Nunn's work continued that of Wise and Samadikun, and it represented NASA's continuing support of biomedical sensor development at the Integrated Circuits Laboratory. Nunn's device measured absolute pressure using a thin silicon diaphragm and a piezoresistive bridge circuit, much like that of Wise and Samadikun. Nunn later applied for a patent, and in it he specifically proposed using the Wise-Samadikun anisotropic etching process to control the thickness of the diaphragm.[251] The diffusion of the technique was a technological success and laid the foundation for future technological growth, but no indication of its commercial application has appeared yet.

---

249. These were grants no. 5P01GM017940-020002 (for 1972), 5P01GM017940-030002 (for 1973), and 5P01GM017940-040002 (for 1974). These grants were found via a search on NIH Computer Retrieval of Information on Scientific Projects (CRISP), *http://crisp.cit.nih.gov/* (accessed 11 March 2006), for "Angell, James" as PI.

250. Nunn, "Silicon Absolute Pressure Transducer for Biomedical Applications," pp. iv–vi.

251. Timothy Alan Nunn, "Miniature absolute pressure transducer assembly and method," U.S. patent 4,079,508, filed 13 May 1976, issued 21 March 1978, assigned to Stanford University, especially p. 7.

Bandung High Tech Valley

Perhaps one of the longest lasting and most direct contributions of this MEMS work was not in the United States at all but overseas, and not technological but social. After receiving his doctoral degree from Stanford, Samaun Samadikun joined the Department of Electrical Engineering of the Bandung Institute of Technology in Bandung, Indonesia. There, he became professor of electronics in 1974 and subsequently the first director of the Inter-University Center for Microelectronics (1984–1989), where he also was a senior researcher. Samadikun later served his government as director general of Energy, Ministry of Mining and Energy (1978–1983), and as chairman of the Indonesian Institute of Sciences (LIPI) (1989–1995). He has received several awards and honors, including the 1998 Award of the Association of South Eastern Asian Nations in recognition of his meritorious service to science and technology, and he was a Founding Fellow of the Islamic Academy of Sciences and a Founding Fellow of the Indonesian Academy of Sciences.[252]

Samadikun attempted to raise interest in creating something like Silicon Valley in Indonesia in an area now dubbed the Bandung High Tech Valley. It benefits from the same ingredients as Silicon Valley: a confluence of businesses, universities, and government agencies engaged in the production and exploitation of science and technology. Upon his return from Stanford, where he acquired firsthand knowledge of how Silicon Valley operated, Samadikun brought that knowledge back to Indonesia, where he pushed the idea of creating a similar high-tech area based on semiconductors and telecommunications. Although that dream still remains unrealized, Samaun Samadikun is considered one of its key proponents.[253]

The Gas Chromatograph

The miniature gas chromatograph efforts of Terry, Perner, and Jerman, although funded initially by NASA, received important additional support from the National Institute for Occupational Safety and Health. This additional financing allowed the gas chromatograph on a chip to evolve into a prototype device and eventually a commercial product. Terry and Jerman received two patents for the improved gas chromatograph in 1984—one of which was assigned to Stanford University and the other to Microsensor

252. "Prof. Samaun Samadikun," *http://www.ias-worldwide.org/profiles/prof63.htm* (accessed 6 January 2006).

253. Budi Rahardjo, "A Story of Bandung High-Technology Valley," 2002, *http://64.233.179. 104/search?q=cache:p3vqaj9OOgcJ:budi.insan.co.id/articles/a-portrait-of-BHTV2. doc+%22Samaun+Samadikun%22&hl=en&start=6* (accessed 6 January 2006).

Technology, Inc., in Fremont, California,[254] a manufacturer of gas analysis equipment. The firm became MTI Analytical Instruments in 1981 and in 1998 was purchased by the Hewlett-Packard Company, which incorporated it into its Chemical Analysis Group.[255] NASA's funding of the Stanford gas chromatograph therefore served as an early stimulus to the development of a product taken up by the gas-analysis industry.

In his discussion of the gas chromatograph, Kurt Petersen, in his ground-breaking 1982 article on MEMS technologies and techniques, declared it to be "one of the more ambitious, practical, and far-reaching applications of silicon micromechanical techniques."[256] More recently, Gregory Kovacs described the contribution of the gas chromatograph to the development of MEMS techniques in the following terms: "Perhaps the first active micromachined valve was that demonstrated by Terry (1975) and Terry et al. (1979) as a component of an integrated gas chromatography system. While external actuation was used, *the basic concepts of the micromachined silicon valve seats spelled out in this work continue to be used in subsequent micromachined valves.* Two designs were tested, one using a silicon diaphragm/silicon valve seat arrangement, and the other using a Teflon-coated polyimide (Kapton) membrane/silicon valve seat design. Both designs required hand assembly (as is presently still the case, in general) and used external actuation, but *demonstrated principles that would later be applied to truly batch-fabricated micromachined valves*"[257] (emphasis added).

The original intent of NASA funding for the chromatograph was for the Viking mission to Mars, which launched twin probes to the planet in search of evidence of life in August and September 1975.[258] A good number of records support this notion. For example, according to Ken Wise, the Integrated Circuits Laboratory worked on a silicon-chip gas chromatograph

254. Jerman and Terry, "Gas Chromatography System and Detector and Method," U.S. patent 4,471,647, applied 22 March 1982, issued 18 September 1984, assigned to Stanford University; Jerman and Terry, "Miniature Gas Chromatograph Apparatus," U.S. patent 4,474,889, applied 26 April 1982, issued 2 October 1984, assigned to Microsensor Technology. Patent 4,471,647 states that the U.S. government had rights to the invention because the NIOSH (not NASA) had funded its development.

255. "HP to Acquire MTI Analytical Instruments; Purchase Targeted to Expand HP's GC Analysis Business," *Business Wire*, 17 February 1998, *http://www.findarticles.com/p/articles/ mi_m0EIN/is_1998_Feb_17/ai_20296835* (accessed 9 January 2006).

256. Petersen, "Silicon as a Mechanical Material," p. 434.

257. Kovacs, *Micromachined Transducers*, p. 827.

258. For a history of the Mars Viking mission, see Edward Clinton Ezell and Linda Neuman Ezell, *On Mars: Exploration of the Red Planet, 1958–1978* (Washington, DC: NASA SP-1984-4212, 1984).

for the Mars Viking project.[259] In addition, in his dissertation Terry wrote that "the idea of a miniature GC [gas chromatograph] column etched in silicon was conceived by NASA for a small, lightweight, and rugged chromatograph to be landed on Mars in the Viking 75 probe."[260]

The development of the miniature gas chromatograph for the Viking mission, however, did not guarantee that it flew to Mars as part of the Viking mission. The timing of the initial development of the gas chromatograph by Terry (who completed his dissertation in 1975) and the earlier work by Perner (completed in October 1973) preceded the launch of the two Viking probes in August 1975. While this timeline appears to support the possibility that Viking used the Stanford chromatograph, textual descriptions of the chromatographs that went to Mars point in a different direction. These devices were part of the Viking biology instrument that between 1970 and the spring of 1975 the TRW Systems Group—under contract to Martin Marietta Corporation—designed, built, and tested to search for signs of life on Mars.[261]

The Viking biology instrument was actually three systems in a single integrated package, each of which performed a distinct experiment intended to reveal the presence of life on Mars. The complexity of the instrument arose from the large number of different functions that had to be performed in order to conduct biological analyses of the Martian surface material. Only one of the scientific instruments included in the biology package was a miniaturized gas chromatograph capable of measuring parts per million concentrations of such metabolic gases as methane.[262]

The gas chromatograph was part of the gas exchange experiment for the detection of biological activity in the Martian soil. Its function was to measure changes in the concentration of gases caused by the metabolism and growth of microorganisms. The experiment—devised by Vance I. Oyama of NASA's Ames Research Center along with Bonnie J. Berdahl and Glenn C. Carle, also at Ames—was based on a common characteristic of life on Earth—namely that all organisms produce and/or consume various gases, such as hydrogen, nitrogen, methane, and carbon dioxide. In the gas exchange experiment, the

259. Wise interview, p. 16.

260. Terry, "Gas Chromatography System," p. 1. Also, Terry, Jerman, and Angell, "Gas Chromatographic Air Analyzer," p. 1886, states that the gas chromatograph potentially could be used on "planetary probes."

261. F. S. Brown, H. E. Adelson, M. C. Chapman, O. W. Clausen, A. J. Cole, J. T. Cragin, R. J. Day, C. H. Debenham, R. E. Fortney, R. I. Gilje, D. W. Harvey, J. L. Kropp, S. J. Loer, J. L. Logan, Jr., W. D. Potter, and G. T. Rosiak, "The Biology Instrument for the Viking Mars Mission," *Review of Scientific Instruments* 49, no. 2 (February 1978): 140.

262. Ibid., p. 140.

metabolism and growth of the microorganisms were stimulated either by humidifying the soil or contacting the soil with a watery solution containing a variety of nutrients and growth factors. The gas chromatograph periodically analyzed the resulting (if any) changes in gas composition.[263]

Thus, the gas chromatograph column was but a small part of the complex mechanism designed to conduct the gas exchange experiment. The lower portion of the gas exchange experiment module contained what was called the gas chromatographic analysis assembly, a specialized gas chromatograph for the analysis of gases of biological interest. The assembly basically was a conversion to flight hardware of the chromatographic system used in the laboratory.[264]

The TRW general description of the biology instrument stated that "gases and liquids move through tiny plumbing network controlled by 39 miniature solenoid valves."[265] These valves were complex mechanisms that controlled the flow of gases, liquids, and vapors within the chromatograph assembly and the two other experimental apparatus. The valves could not have been those developed at Stanford, because TRW characterized them as solenoids—devices that consist of wire-wound coils—not the silicon-based valves created by Terry. And diagrams of the valves plainly establish that they were not constructed from silicon on a chip.[266]

Glenn Carle described the laboratory hardware as early as 1970,[267] probably as Perner was starting his comparison of thermistor and pyroelectric types of thermal-conductivity detectors with a grant from Carle. The description of the flight hardware also strongly suggests that the chromatograph column was not built on a silicon wafer. It states that the entire gas chromatographic analysis assembly, which was 9 cm square by 6 cm high (larger than the 5 cm of the silicon wafer), consisted of "a gas sampling system, a pair of matched porous polymer bead chromatographic columns, and a thermistor thermal-conductivity detector."[268] Although Perner's research considered thermistor detectors, he demonstrated that a detector with a pyroelectric crystal as a

263. Ibid., p. 158.

264. Ibid., p. 161.

265. TRW Systems Group, "Biology Instrument," no date, Folder 5496, NHRC.

266. Brown et al., "Biology Instrument for the Viking Mars Mission," p. 175; H. E. Adelson, F. S. Brown, and R. I. Gilje, "Looking for Life on Mars," *Quest* 1, no. 2 (Fall 1977): 41–42.

267. Glenn Carle, *Journal of Chromatographic Science* 8 (1970): 550–556. For a description of the process, see Vance Oyama and Glenn Carle, "Pyrolysis Gas Chromatography Application to Life Detection and Chemotaxonomy," *Journal of Gas Chromatography* 5 (March 1967): 151–154.

268. Brown et al., "Biology Instrument for the Viking Mars Mission," p. 161.

sensing element was 500 times more responsive than a thermistor.[269] This evidence does not indicate that Perner's research impacted the design of the Viking gas chromatographs.

The apparatus that went to Mars incorporated "a pair of matched porous polymer bead chromatographic columns." This description precludes the construction of the columns as grooves in a silicon wafer like the Stanford device. The size of the gas chromatograph and its column were too large, moreover. TRW described the gas chromatograph as "only three and a half inches in diameter" and containing "coiled tubes" with an inside diameter "about the size of a pencil lead" and "filled with porous polymer beads."[270] In addition, a published photograph of the gas chromatographic analysis assembly makes the fact unequivocally clear that the chromatograph was not constructed on a silicon wafer.[271] The Stanford miniature gas chromatograph therefore did not fly on the Viking Mars mission. Rather, its destiny appears to have lain in the field of medicine.

The Accelerometer

Although the Joint Services Electronics Program initially funded the accelerometer work, NASA subsequently picked up the tab—and for a longer period of time. Thus, the statement made by NASA that the Agency developed tiny accelerometers during the 1970s "to measure changes in speed of small objects or activity levels of people or animals during human space flight" is essentially true for the most part, even though the purported intention of the grant under which the work was performed was to develop a miniature gas chromatograph. Other NASA claims regarding the MEMS accelerometer are harder to corroborate—namely that "the original NASA-sponsored work on an MEMS accelerometer is referenced in 83 patents; the earliest reference was made in 1975 and the latest in 2003."[272]

This nature of this claim gives rise to a number of difficulties. A search of the online U.S. patent database[273] for "Roylance, Lynn" in all fields for

---

269. Terry, "Gas Chromatography System," p. 102.

270. Adelson, Brown, and Gilje, "Looking for Life on Mars," p. 45.

271. Brown et al., "Biology Instrument for the Viking Mars Mission," p. 161.

272. NASA, "Big Functions in a Small Package," p. 5; NASA Public Affairs Office, *NASA Hits: Rewards from Space*; NASA, "Big Functions in a Small Package," p. 4; and NASA Public Affairs Office, *NASA Hits: Rewards from Space*.

273. "United States Patent and Trademark Office," *http://www.uspto.gov/patft/index.html* (accessed 13 March 2006). A search of the European patent database, which includes U.S. patents, produced a single result; however, this database does not permit users to search all patent fields, such as the reference sections. European Patent Office, home page *http://ep.espacenet.com/* (accessed 13 March 2006).

the period 1976 to the present (August 2008) revealed only 31 patents that referred to the work of Lynn Roylance, not 83.[274] These references, moreover, appeared no earlier than 1984. Without exception, they all cited her 1979 IEEE article,[275] while four cited her dissertation as well. The first three patents that referenced Roylance's work were inventions by researchers at Honeywell. They also cited published works by Stanford researchers Petersen and Terry.[276]

An examination of these patents showed that Roylance's article and dissertation were cited in those sections of the patent text called "references cited" and "other references" as being part of the relevant technical literature consisting of all publications plus issued patents. Cited patents fall in a separate category of "prior art." In patent parlance, her publications were provided in the body of the patent to indicate "the background of the invention"—that is, the known "art," in order to differentiate the patented invention (a new thing) from the art (the established knowledge).

References to Roylance's accelerometer necessarily had to fall in this category of prior art, because the patent office never issued a patent for the device. With the disclosure of the invention in a technical publication, obtaining a patent was out of the question. The device, according to Roylance, had its own technological problems, some of which were indigenous to all strain-gauge accelerometers.[277] In addition, because of the beam thickness minimum imposed by the micromachining technology as well as the small volume available for the mass, the lowest accelerations it could detect (from ±10 to ±15 g) were above the lowest range detectable by commercial units (±5 g).[278] The device's biggest failing, Roylance opined, was that it sensed movement along only one axis, not three.[279] On the other hand, the good news was that one could fabricate the accelerometer inexpensively in batches. A 5 cm

---

274. The original search, conducted in March 2006, produced only 14 hits. The oldest patent was Robert G. Johnson and Robert E. Higashi, "Method of Making Semiconductor Device," U.S. patent 4,472,239, filed 8 July 1983, issued 18 September 1984. The most recent reference to Roylance's work is Jian Liu, Carl L. Hansen, and Stephen R. Quake, "Microfluidic Rotary Flow Reactor Matrix," U.S. patent 7,413,712, filed 30 April 2004, issued 19 August 2008.

275. Roylance and Angell, "Batch-Fabricated Silicon Accelerometer," p. 1911–1917.

276. Johnson and Higashi, "Method of Making Semiconductor Device"; Philip J. Bohrer, "Flow Sensor," U.S. patent 4,478,076, filed 30 September 1982, issued 23 October 1984; and Bohrer and Johnson, "Flow Sensor," U.S. patent, 4,478,077, filed 30 September 1982, issued 23 October 1984.

277. Roylance, "Miniature Integrated Circuit Accelerometer," pp. 153, 154, and 158.

278. Ibid., p. 158.

279. Notes, telephone conversation with Lynn Roylance, 27 September 2008.

(2 inch) wafer cost about $40 but yielded approximately 200 accelerometers (about 20 cents each).[280]

Did the miniature accelerometer have any societal impacts? The patent literature search furnishes only a fragile link between invention and the open literature. The results of that search should not lead inexorably to the assumption that the accelerometer (and its associated fabrication techniques) did not have an impact on later work. Once a discovery enters the arena of public knowledge, whatever applications follow potentially may never materialize in any patent or technical literature search, but they still may be part and parcel of a manufacturing trade secret or other form of "art."

Nonetheless, the search for societal impacts is disappointing. For example, an article published in the *Scientific American* in 1983 by associates of the Stanford Integrated Circuits Laboratory mentioned the accelerometer in passing but did not state that it was under commercial development, unlike other MEMS devices described.[281] Subsequently, a wide variety of more complex piezoresistive accelerometers were fabricated, but not until the 1990s.[282] Perhaps the lack of issue from the accelerometer invention was the immature nature of the technology itself. In 1982 chip-sized accelerometers still had no commercial importance. Technological barriers seem to have remained. Such cantilever-beam accelerometers made by etching clear through the wafer had to address serious packaging problems. Special top and bottom motion-limiting plates, for example, had to be included in the assembly to prevent beam damage during possible acceleration overshoots.[283]

In the end, the sum of Roylance's achievement was to demonstrate the feasibility of the concept of creating a miniature accelerometer using integrated-circuit fabrication techniques as well as the viability of the techniques used in its construction.[284] However, if one looks for a linear development of the technology—that is, one in which each advance depends directly on the progress realized by the preceding discovery—the type of technological progression implicit in NASA's claims regarding its accelerometer investment such a linear sequence does not exist. The first three patents that reference Roylance's publications, for improvements in sensors and their manufacture by researchers at Honeywell, do not link their work to that of Roylance or anyone else

---

280. Roylance, "Miniature Integrated Circuit Accelerometer," pp. 158–159.

281. Angell, Terry, and Barth, "Silicon Micromechanical Devices," p. 45.

282. Kovacs, Micromachined Transducers, p. 231.

283. Petersen, "Silicon as a Mechanical Material," p. 449.

284. Notes, telephone conversation with Lynn Roylance, 27 September 2008.

at Stanford in this fashion.[285] The claim of a direct technological evolutionary line from the Ames-Stanford device to modern MEMS accelerometers is not valid.

Conclusions

NASA indeed got something for its money from the NASA Ames investment in MEMS research at Stanford's Integrated Circuits Laboratory. And so did society. The sensor work—continued under NIH funding—produced a patented (that is, technologically new) method for anisotropic etching in a silicon wafer, and an improved version of the sensor contributed to research carried out under the auspices of the NIH. The Wise-Samadikun patented technique for anisotropic etching had an immediate use in the Nunn patent and, as part of the patent literature, it achieved a modicum of technological success but no apparent commercialization. However, subsequent patents incorporating the Wise-Samadikun technique may have seen commercial use. NASA's investment in the gas chromatograph also paid off as a commercialized product.

The paucity of MEMS research carried out at the Stanford Integrated Circuits Laboratory with NASA funding compared to that carried out under the sponsorship of the armed forces or the health establishment appears to reinforce the impression that NASA was not a major player in the development of Silicon Valley, despite the presence of the Ames Research Center in its midst. Nonetheless, historian Stuart Leslie has stated that NASA's financial contribution to the Stanford Electronics Laboratories rose from nothing to about a third of its budget during the 1960s, although Defense Department money continued to dominate.[286] Where did that NASA money go?

Perhaps the recipient of the largest amount of NASA funding within the Stanford Electronics Laboratories was the Stanford Center for Radar Astronomy. The center was a joint venture of Stanford University and the Stanford Research Institute created in 1962 to foster scientific and engineering efforts and to provide graduate student training in radar astronomy and space science. A NASA grant underwrote the center itself, while additional military and civilian awards supported a range of theoretical and

---

285. These are the three patents cited above—namely, Robert G. Johnson and Robert E. Higashi, "Method of Making Semiconductor Device," U.S. patent 4,472,239, filed 8 July 1983, issued 18 September 1984; Philip J. Bohrer, "Flow Sensor," U.S. patent 4,478,076, filed 30 September 1982, issued 23 October 1984; and Bohrer and Johnson, "Flow Sensor," U.S. patent, 4,478,077, filed 30 September 1982, issued 23 October 1984—all assigned to Honeywell.

286. Leslie, *Cold War and American Science*, p. 73.

experimental radio and radar research on space, ionospheric, and communication theory topics.[287] It was the umbrella organization for Von Eshleman, Robert A. Helliwell, and Oswald G. Villard, Jr., who had NASA grants totaling $1,285,751 in 1969. In addition, NASA paid for the construction of research laboratory facilities as part of the Space Engineering Building to the tune of over $2 million.[288]

In contrast, integrated circuits and MEMS failed to receive a substantial amount of NASA sponsorship. The time was ripe for the Stanford Electronics Laboratories and its Integrated Circuits Laboratory to shift the brunt of its financial support away from the armed forces. As William Rambo, associate dean for research at the School of Engineering, wrote to the electrical engineering faculty, students, and staff in 1970, in a rather understated manner: "Important questions are underway concerning our research, particularly those projects supported by the Department of Defense."[289] NASA dollars certainly could have provided a civilian haven for research, but they did not flow into integrated-circuit research with the same volume as they did elsewhere within the Electrical Engineering Department. But did NASA want to spend such sizable sums on MEMS? Exobiology at Stanford, on the other hand, received ample funding. In 1969, for example, as Ames began funding the first MEMS project, NASA Headquarters awarded Joshua Lederberg in the Stanford School of Medicine a $535,000 grant for the construction of biomedical instrumentation facilities as well as a $410,000 grant for "cytochemical studies of planetary microorganisms."[290]

By this measure at least, NASA's funding level of MEMS research was low alongside the sums the Agency was investing in the Stanford Medical School or the Stanford Center for Radar Astronomy. Still, for this relatively small investment, NASA and society reaped a not insignificant return. NASA's investment was still insufficient to realize the full potential of the research; money from other sources aided in developing these MEMS projects. In short, NASA was not the sole underwriter of these MEMS advances. Should NASA, in making pronouncements about its contribution, acknowledge the role of other agencies in encouraging the Stanford effort? Indeed, NASA does not deserve sole credit for fostering the origins of MEMS at Stanford, and its financial contribution was far from representing the lion's share of support.

---

287. Butrica, *To See the Unseen*, p. 155.

288. *NASA's University Program* (1969), pp. 22–26.

289. Memo, Rambo to EE Students, Staff, and Faculty, "Electronics Research at Stanford," 11 May 1970, Folder 32, "Electronics Research at Stanford-1970," Box 6 [no accession number], Rambo Papers.

290. *NASA's University Program* (1969), pp. 22 and 26.

The Agency also should acknowledge that the Stanford MEMS effort already was under way, when NASA's Ames Research Center began to work with Stanford on MEMS projects. If one shifts one's perspective from that of NASA to that of the Stanford MEMS program, the space agency definitely did not play a central or defining role.

Appendix A

NASA Grants and Research Contracts to Stanford

Ames MEMS Research Only

Ames contract monitor	Ames branch	Subject of research	Dates	Total dollar amount	Stanford principal investigator	Contract or grant number
Thomas B. Fryer	Electronics Research Branch	Research on pressure sensors for biomedical instrumentation	11/1969– 9/1972	$80,000	James B. Angell	NGR 05-020-401
Ralph W. Donaldson, Jr., and Nigel C. Tombs	Measurement Sciences Branch	Research on integrated circuits for gas chromatography	6/1971– 10/1971	$3,725	Kensall D. Wise	NAS 2-6491
Ralph W. Donaldson, Jr.	Electronic Instrument Development	Research miniature gas analysis systems	5/1971– 2/1973	$35,000	James B. Angell	NGR 05-020-543
Harold Sandler, Ralph W. Donaldson, Jr., and Glenn C. Carle	Biomedical Research Division	Miniature gas analysis systems	9/1973– 11/1980	$282,632	James B. Angell	NGR 05-02-690

Appendix B

Technical Essay[291]

Silicon has a distinctive cubic crystal structure. At each corner of the cube as well as in the center of each of the cube's faces sits an atom of silicon. The cubes interlock in such a way that several atoms from neighboring cubes are located in each individual cube. The axes of the individual cubes form an orthogonal coordinate system (x, y, z) that allows one to specify directions (vectors) and planes within the crystal. A crystalline direction is labeled by three coordinates called the Miller indexes, which are whole number (integer) multiples of the length of one edge of a single cube. The same set of indexes designates the planes perpendicular to that direction.

For instance, a [110] direction or vector points diagonally across the face of a given cube. The coordinates of the vector also designate the complete set of atomic planes perpendicular to that direction. The notation (100) describes all the planes perpendicular to the x axis. Because the crystal structure is symmetrical, the x, y, and z directions are interchangeable. As a result, one can describe all equivalent directions and sets of planes. The notation <110> designates the diagonals across any face of a unit cube, and the notation [110] designates the sets of planes perpendicular to all <110> vectors. The different bracket styles thus distinguish planes from directions as well as generalized planes and directions from specific planes and directions.

The orientation of the crystalline structure is a fundamental property that the fabrication of MEMS devices exploits, because some etching chemicals attack different crystal directions at diverse rates. Most anisotropic etching chemicals move quickly in the crystalline direction perpendicular to the (110) plane but less speedily in the direction perpendicular to the (100) plane. The direction perpendicular to the (111) plane etches rather slowly, if at all. Anisotropic etching chemicals create holes composed of flat walls, while isotropic etching consistently produces a rounded hole.

What determines the specific shape of an anisotropically etched hole is the crystalline orientation of the wafer surface along with the shape and orientation of the openings in the photolithographic mask on that surface and the dependence of the etching chemical itself on the orientation of the silicon. A square opening leaning along the <110> directions of a <100> wafer yields a pyramid-shaped pit with [111] side walls. If the opening is larger, the point of intersection of the [111] planes is deeper, and one can create a flat-bottomed pit by stopping the etching before it reaches that depth. On the other hand,

---

291. Abstracted from Angell, Terry, and Barth, "Silicon Micromechanical Devices," pp. 46–48.

a rectangular opening made on the same wafer yields a V-shaped pit. One also can create holes with parallel side walls by etching a wafer with a dissimilar surface orientation. A <110> wafer has two sets of [111] planes that are perpendicular to the surface but not to each other. If one orients an oxide opening on a <110> wafer in a certain fashion, the etching process creates a hole with vertical side walls. The side walls that intersect at angles less than 90 degrees (acute angles) are linked to still other planes. These are just some of the various shapes that one can form using this method.

Part III

THE WORLD AT LARGE

Chapter 5

Powering Space Exploration: U.S. Space Nuclear Power, Public Perceptions, and Outer Planetary Probes

Roger D. Launius

1. Introduction

Since the dawn of the Space Age more than 50 years ago, the United States has pursued a variety of methods for delivering electrical power to spacecraft in flight. Nuclear power systems are the only ones that have been found acceptable for deep space missions. While these technological systems have made possible a myriad of accomplishments in space, especially the successful flights to the outer planets, the details of space nuclear power generation are virtually unknown to even the most knowledgeable observers. What *is* known, furthermore, is frequently limited to the often-incomplete reporting of controversies over the propriety of using nuclear systems for space power. This chapter traces the development of this technology from its origins in the 1960s to the present. It describes the evolution of the systems involved and the decision-making process whereby NASA chose to adopt one approach over another. Finally, it analyzes the public debate over the employment of these technologies for spaceflight.

2. Satellite Power Systems in Summary

Flying in space requires reliable, uninterrupted, stable electrical power, not only for engines to maneuver and navigate but also for onboard systems that perform a range of functions.[1] One of the critical components of any

---

1. There are several general works on space power systems. These include Mukund R. Patel, *Spacecraft Power Systems* (New York: CRC, 2004); A. K. Hyder, R. L. Wiley, G. Halpert, D. J. Flood, and S. Sabripour, *Spacecraft Power Technologies* (London: Imperial College Press, 2000); Martin J. L. Turner, *Rocket and Spacecraft Propulsion: Principles, Practice, and New Developments* (Chichester, UK: Springer Praxis Books, 2004); Donald B. Mackay, *Design of Space Powerplants* (Englewood Cliffs, NJ: Prentice-Hall, 1963); Peter Frankel, "Space Electric Power System," Lockheed Technical Report LR 17558, February 1964; and S. Lieblein and H. O. Slone, "Electric Power Generation Systems for Use

satellite, either in Earth's orbit or dispatched elsewhere, is the power system that allows the operation of its many systems. There are only four methods of providing the electrical power needed for spacecraft, all of them with positives and negatives. The first method, and the one used on the first spacecraft launched into orbit, was batteries. Their wattage was limited, but even more limited was their longevity. Within a few weeks they always ran down and the spacecraft's systems no longer operated. For example, about three weeks after the launch of Sputnik 1 on 4 October 1957, its batteries ran down and it ceased to broadcast telemetry, although it remained in orbit for about 90 days after launch.[2]

Second, to help resolve that problem, NASA pioneered in the 1960s fuel cell technology, which generated more electricity for the size of the cell and had a longer effective life. Even so, fuel cells have an effective life of less than two months.[3] Of course, this may change in the future as NASA pursues more efficient fuel cells for its human exploration program that could have remarkably long lives.[4] Third, photovoltaic solar cells emerged in the 1960s as a useful alternative to batteries and fuel cells. They have a long life measured in years rather than weeks or months, and with additional refinement they have become the critical power generation technology for most spacecraft.[5] They have one important drawback, however: they require the Sun's powerful light source to be effective. For spacecraft traveling into deep space beyond Mars, where the Sun becomes much less intense, photovoltaic systems up to this point have proven insufficient. This may change in the future

in Space," in *Advances in Aeronautical Sciences*, Vol. 4, *Second International Congress in the Aeronautical Sciences, Proceedings, Zurich, Switzerland, Sept. 12–16, 1960* (New York: Pergamon Press, 1962), pp. 1131–1152.

2.   Hyder et al., *Spacecraft Power Technologies*, p. 1.

3.   A. John Appleby, *Fuel Cell Handbook* (New York: Van Reinhold Co., 1989); Leo Blomen and Michael Mugerwa, *Fuel Cell Systems* (New York: Plenum Press, 1993); Barton C. Hacker and James M. Grimwood, *On the Shoulders of Titans: A History of Project Gemini* (Washington, DC: NASA Special Publication-1977-4203, 1977), pp. 148–152, 208–215; Karl Kordesch and Günter Simader, *Fuel Cells and Their Applications* (New York: VCH, 1996); David L. Douglas and Herman A. Liebhafsky, "Fuel Cells: History, Operation, and Applications," *Physics Today* 13, no. 6 (1960): 26–30; Karl V. Kordesch, "25 Years of Fuel Cell Development (1951–1976)," *Journal of the Electrochemical Society* 125, no. 3 (1978): 77C–91C;Marvin Warshay and Paul R. Prokopius, *The Fuel Cell in Space: Yesterday, Today, and Tomorrow* (Cleveland, OH: Lewis Research Center, NASA-TM-102366, 1989); and Hyder et al., *Spacecraft Power Systems*, pp. 223–234.

4.   John H. Scott, "Fuel Cell Development for NASA's Human Exploration Program: Benchmarking with 'The Hydrogen Economy,'" *Transactions of the ASME* (2007): 8 pp., *http://ntrs.nasa.gov/archive/nasa/casi.ntrs.nasa.gov/20070023719_2007023572.pdf* (accessed 24 October 2007).

5.   Hyder et al., *Spacecraft Power Systems*, pp. 71–149.

as new technologies increase the efficiency of energy collection and power management, but past and present capabilities have not allowed their use.[6] Accordingly, when requirements are for short mission times or systems do not require high power, chemical and/or solar energy may be used effectively to make electricity. But for the generation of high power levels over longer periods of time, especially farther away from the Sun, nuclear energy has thus far been the only way to satisfy mission requirements.

For this reason, as well as others of a more sublime nature, many spacecraft designers have adopted nuclear power technology as a means of powering spacecraft on long deep space missions. As the chief of NASA's nuclear electric power program remarked in 1962:

> Basically, radioisotopes are of interest because they represent a compact source of power. The energy available in radioisotopes is many orders of magnitude larger than that available in batteries, and thus they constitute a unique, concentrated energy source that may be used for space purposes if design requirements are met. Radioisotope power is inherently reliable. It cannot be turned on or off. There are no moving parts of oriented arrays. It will provide heat energy in accordance with the fixed laws of radioactive decay. This heat is absorbed in a device that converts the heat directly into electricity.[7]

There are several types of nuclear power that could be employed, everything from small reactors to nuclear heaters to the dominant technology of radioisotope thermoelectric generators (RTGs). In those small space nuclear reactors, energy could be generated through controlled fission of uranium. The heat created through this process would then be used to power either a thermoelectric or a dynamic turbine or alternator conversion system. While excess heat would be dissipated through a radiator, electricity generated through this process would serve to power the spacecraft. These reactors had the capability to generate more than 100 kilowatts of electricity, making them much more powerful than other forms of energy generation in space, including RTGs. The simpler process of allowing the natural decay of an isotope and harnessing its heat to generate electricity with an RTG, however, has become the preferred method for supplying the power needs of American deep space probes; it has also been used on some Earth orbital and lunar spacecraft (see figures 5.1 and 5.2). It operates by releasing heat during the decay process

6.   "Why the Cassini Mission Cannot Use Solar Arrays," JPL Fact Sheet, November 1996.

7.   Fred Shulman, "The NASA Nuclear Electric Power Program," in Morris A. Zipkin and Russell N. Edwards, eds., *Power Systems for Space Flight* (New York: Academic Press, 1963), pp. 15–27, quotation from pp. 17–18.

of a suitable radioactive material which is then converted into electricity by means of an array of thermocouples, with the outer end of each thermocouple connected to a heat sink. Radioactive decay of the fuel produces heat that flows through the thermocouples to the heat sink, generating electricity in the process. The thermocouples are then connected through a closed loop that feeds an electrical current to the power management system of the spacecraft. Indeed, all U.S.-launched systems have used plutonium-238 for this purpose.[8]

In addition to its longevity, space nuclear power offers a significant saving in terms of mass associated with an individual mission when compared to the other possibilities. As policy analyst Steven Aftergood reports: "For all

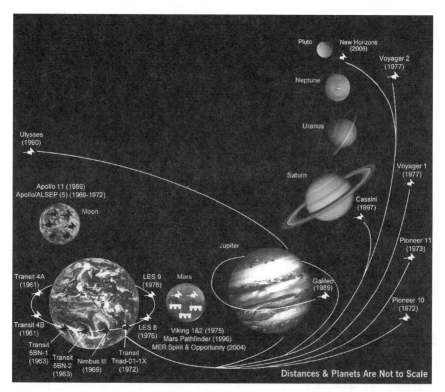

Credit: NASA

FIGURE 5.1. Forty-one RTGs have been used successfully on 23 spacecraft since 1961. This graphic shows the range of missions that have been powered by nuclear power sources.

---

8. Steven Aftergood, "Background on Space Nuclear Power," *Science & Global Security* 1, nos. 1–2 (1989–1990): 93–107.

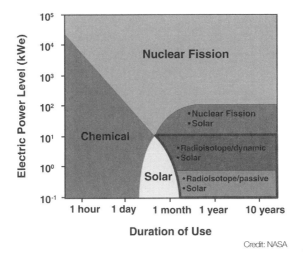

Credit: NASA

FIGURE 5.2. This figure compares the various types of power available for space-craft. The longer the duration of the mission, the more necessary space nuclear power becomes.

practical purposes, nuclear reactors are required when moderate to high levels of continuous power are required for an extended period."[9] Another observer, admittedly one committed to exploiting space nuclear power much more aggressively than has been done thus far, writes:

> Nuclear power has been used for deep space vehicles for over 40 years. RTGs have been used for spacecraft electrical power since 1961. All RTGs have operated as designed, both in normal operations and accident conditions. RTGs were designed carefully with consideration for the accident environments that might be experienced during every phase of the launch. The design requirement is to protect public and worker health and safety during all phases of operations during launch and accident conditions.[10]

These systems have provided power ranging from 2.7 watts on the very early systems to 500 watts on more recent flights.[11] Even so, while RTGs have been a proven, reliable technology, they have fostered only relatively low power

9. Ibid., p. 94.

10. Beverly A. Cook, "Making Space Nuclear Power a Reality," 2005, p. 3, AIAA-2005-0101.

11. James H. Lee, "Aerospace Nuclear Safety: An Introduction and Historical Overview," paper presented at the International Topical Meeting: Advanced Reactor Safety, Pittsburgh, PA, 17–21 April 1994.

efficiency—only about 7 percent at the beginning of the mission—and researchers have consistently sought to improve on that fact.[12]

Of course, the conversion of heat into electricity as done by an RTG is not a new concept. It was discovered more than 150 years ago by the German physicist Thomas Johann Seebeck, who first observed that electric voltage might be produced when two dissimilar but conductive materials were joined in a closed circuit and the two junctions were kept at different temperatures. These junctions gained the name thermocouples, and they generated electricity through the movement and interactions of the electrons the two materials. The thermocouples in RTGs therefore use heat from the natural decay of radioactive plutonium-238 to heat one material of the thermocouple, while the other remains cold from the temperatures in space, and electricity results from the interactions.[13]

Beginning in the late 1940s, several threads converged to make it possible to develop and use radioisotope thermoelectric generators. First, the Atomic Energy Commission (AEC) began to develop radioisotopes for atomic weapons. This prompted scientific research to understand the nature of the half-lives of various isotopes, decay processes, and charge separation. Second, scientists began to experiment with the development of small nuclear power generators for a variety of uses on Earth, especially at the poles and under the seas where scientific instruments could be placed and left alone for months at a time. The first bench-test RTGs emerged from the Mound Laboratory (operated for the AEC by the Monsanto Research Corporation) in 1953 and quickly found application in Antarctica to power scientific research stations.[14] Indeed, Mound scientists Kenneth Jordan and John Birden had hit upon the RTG as a possibility almost by accident. They had been frustrated in their efforts to use decaying radioactive materials to boil water to drive a steam turbine. They then decided to apply the thermocouple principle to harness heat

---

12. David J. Anderson, Wayne A. Wong, and Karen L. Tuttle, "An Overview and Status of NASA's Radioisotope Power Conversion Technology NRA," p. 1, 2007, AIAA-2007-0022221.

13. DOE Fact Sheet, "Space Radioisotope Power Systems: Multi-Mission Radioisotope Thermoelectric Generator," April 2002.

14. B. C. Blanke and J. H. Birden, "Nuclear Battery—Thermocouple Type, Final Report," Mound Laboratory, MLM-1106, AD 251119, 1960; M. Benedict and T. H. Pigford, *Nuclear Chemical Engineering* (New York: McGraw-Hill, 1957); B. C. Blanke, J. H. Birden, and K. C. Jordan, "Nuclear-Battery-Thermocouple Type Summary Report," Monsanto Research Corp., MLM-1127, 1962; B. C. Blanke, "Nuclear-Thermocouple Conversion," *Proceedings of the 12th Annual Battery R&D Conference*, 1958; K. C. Jordan and J. H. Birden, "Thermal Batteries Using Po-210," Mound Laboratory, MLM-984, 1954; and J. G. Morse, "Energy for Remote Areas," *Science* 139, no. 3560 (1963): 1175–1180.

from decaying isotopes and, after working out the calculations quickly, built a successful model of an RTG. The Jordon/Birden principle soon became the basis for all radioisotope thermoelectric generators.[15] Third, advances in thermoelectricity and semiconductors for the first time made the type of power source offered by RTGs feasible. As William R. Corliss and Douglas G. Harvey commented about this heady time in a 1964 textbook on the subject: "The right ingredients were present and the dough began to rise."[16]

3. Origins of Nuclear Power Systems for Spaceflight

In the latter part of the 1940s, several engineers began to consider the possibility of using nuclear power sources for space exploration. The seminal document in this consideration appeared in 1946 from the newly established RAND Corporation on a "preliminary design of an experimental world-circling spaceship," exploring the viability of orbital satellites and outlining the technologies necessary for their success.[17] It did not take long for scientists and engineers to graft nuclear power sources onto their considerations, and 1947 brought the first publications concerning the subject.[18] By 1949, a full-scale analysis by RAND had sketched out the large-scale use of nuclear power sources for satellites in Earth's orbit.[19] Beginning in 1951, at the request of the Department of Defense (DOD), the AEC sponsored research into nuclear power for spacecraft to support the Project Feedback study of the United States Air Force (USAF), leading to the development of a reconnaissance satellite. By June 1952, as reported in an early classified study of the effort, "preliminary results of the reactor analyses were available; all were favorable to the feasibility of the proposal." This extensive and positive discussion of radioisotopic power for space application led to an exponential growth in interest in isotopic power for space satellites. A year later, in May 1953, USAF Headquarters took the next step by authorizing development

---

15. Carol Craig, *RTG a Source of Power: A History of the Radioisotopic Thermoelectric Generators Fueled at Mound* (Miamisburg, OH: Mound Public Relations, 1983).

16. William R. Corliss and Douglas G. Harvey, *Radioisotope Power Generation* (Englewood Cliffs, NJ: Prentice-Hall, 1964), p. 9.

17. Project RAND, Douglas Aircraft Company's Engineering Division, "Preliminary Design of an Experimental World-Circling Spaceship," SM-11827, 2 May 1946, RAND Corporation.

18. M. A. Greenfield and C. Starr, "Studies on Nuclear Reactors: Temperature Stability of an Epithermal Reactor," NAA-SR-6, November 1947, North American Aviation, Inc.

19. S. L. Gender and H. A. Kock, "Auxiliary Power Plant for the Satellite Rocket: A Radioactive Cell-Mercury Vapor System to Supply 500 Watts for Durations of Up to One Year," RAND Report, February 1949.

work on a nuclear power source for satellites. This research effort led directly to the nuclear power systems used on spacecraft in the early 1960s.[20] The AEC oversaw this effort, pursuing two related avenues. The first led to a small nuclear reactor and the second to the RTG. Codenamed SNAP (for Systems for Nuclear Auxiliary Power), these systems were numbered with the odd numbers designating RTGs and even numbers for the reactors. For the RTGs, SNAP-1 was built at the Mound Laboratory under AEC supervision in 1954. It used a thermocouple heated by polonium (Po)-210 for fuel. Exceeding all expectations, SNAP-3 used advanced thermoelectric conversion devices with the first Po-210 fuel capsules; the capsules would soon become a standard in future RTGs. In the reactor arena, the SNAP-2 system used a 50-kilowatt reactor system weighing about 600 pounds and employing liquid NaK—a sodium (Na) and potassium (K) alloy—as a coolant to transfer heat through a mercury loop. This reaction—basic chemistry, really—produced 3 kilowatts of electricity. This led to the research on two additional space power units, SNAP-8 and SNAP-10, which emphasized a metal hydride reactor technology first used in SNAP-2.[21]

These efforts led to a long-standing record of success in meeting the electrical needs of deep space vehicles while offering both reliable and safe operations. As Richard Engler commented:

> The history of the radioisotope power program is basically a success sto[r]y, although it is certainly not one of linear success. The program was initiated by the AEC under impetus from the Department of Defense but first went public late in that decade as part of the "Atoms for Peace" movement, with President Eisenhower showing an atomic battery to the world and extolling its peaceful potential uses. Subsequently, while the Defense Department supported mostly test applications of the radioisotopic power devices in space,

---

20. Robert L. Perry, "Origins of the USAF Space Program, 1945–1956" (Space Systems Division Supplement), in *History of Deputy Commander (AFSC) for Aerospace Systems, 1961*, Vol. 5, ch. 3, Andrews Air Force Base, MD; available at *http://www.fas.org/spp/eprint/origins/part08.htm* (accessed 26 October 2007). See also and Corliss and Harvey, *Radioisotope Power Generation*, pp. 10–11.

21. G. M. Anderson and F. H. Featherstone, "The SNAP Programme: U.S. AEC's Space-Electric Power Programme," *Nuclear Engineering* 5 (October 1960): 460–463; William R. Corliss, "Nuclear Power in Outer Space," *Nucleonics* 18 (August 1960): 58–60; William R. Corliss, "Parameters for Radioisotope Generator Design," American Institute of Electrical Engineers, preprint CP 62-1239, 1962; R. C. Hamilton, "Auxiliary Power for Space Probes," *Astronautics* 4 (August 1959): 30–34; and K. P. Johnson, "Power from Radioisotopes," Atomic Energy Commission, AECU-4373, 1958; *SNAP Nuclear Generator Press Kit* (Washington, DC: Atomic Energy Commission, 26 June 1961), "Attachment 2: SNAP Fact Sheet," note, p. 2.

the program reached its pinnacle of success through uses by the civilian space agency, NASA.

This technology proved exceptionally quiet for most of its history, until the latter 1970s when concerns about all things nuclear erupted in the public consciousness. This was in part because it involved neither explosive power nor a human-built reactor to operate.[22] Even so, it has been discussed at the highest levels of national discourse. President John F. Kennedy in 1961 believed that nuclear power would be used to send an American into space, while "Nuclear Power will sustain him there."[23]

The possibilities of space nuclear power first entered the public sphere in January 1959 when President Dwight D. Eisenhower posed for a photo op with an RTG in the Oval Office of the White House (see figure 5.3). It was SNAP-3, the AEC-developed power source that so many involved in the

Credit: Department of Energy

FIGURE 5.3. On 16 January 1959 this photograph appeared in Washington, DC, newspapers to show the president's interest in radioisotope thermoelectric generators. President Dwight D. Eisenhower with officials from the Atomic Energy Commission examines in the Oval Office an RTG developed specifically to provide spacecraft with the power necessary to operate.

---

22. Richard E. Engler, "Atomic Power in Space: A History," DOE/NE/32117-H1, Department of Energy, March 1987, p. 2.
23. Kennedy's quotation appears in "Special Report on Nuclear Energy in Space," *Nucleonics* 19 (April 1961): cover.

space program pinned their hopes on for exploration of the solar system. AEC officials hailed this RTG as a "significant breakthrough"—one that was reliable, simple, flexible, and safe. Just as importantly, they said, "We can tailor the product to fit the customer."[24] In the context of the post-Sputnik high-technology competition with the Soviet Union in the late 1950s, Eisenhower undoubtedly viewed this showing of the first RTG as a useful propaganda device, graphically demonstrating American technological verisimilitude. He emphasized that this nuclear device was not destructive; rather, it was a means of supporting peaceful scientific expeditions for ramifications for the positive development of humanity. Accordingly, the SNAP-3 served as a proof-of-concept for Eisenhower's Atoms for Peace initiative, a positive use of nuclear technology around the globe. Its small size, inconspicuousness, and nonthreatening nature served Eisenhower well in helping to defuse the caustic international confrontations between the United States and the Soviet Union (see figure 5.4).[25]

RTGs have evolved over the past fifty years from the early SNAP systems to the current General Purpose Heat Source (GPHS) system that has flown on a wide range of NASA deep space missions. For instance, the Galileo mission to Jupiter contained two RTGs, while Ulysses had one RTG to power its systems. The GPHS had a thermal power of 4.4 kilowatts and contained a total plutonium mass of 9.4 kilograms.[26] Because of the low wattage of these systems, all space probes flown by NASA have been power constrained. For example, the Cassini spacecraft launched to Saturn in 1997, with power supplied by three GPHS RTGs, has the most electrical power on board any deep space vehicle. But they produced only 900 watts of on-board electrical power at the time of launch. When compared to the number of 60-watt light bulbs in a normal home, the power for Cassini paled in comparison. To help resolve this constraint, a two-pronged effort has been pursued to enhance the

---

24. "President Shows Atom Generator," *Evening Star* (Washington, DC), 16 January 1959, p. 1.
25. "First Atom Battery Developed by U.S.," *Washington Post*, 17 January 1959, p. 1; "5-Lb. Device Hailed as Big Breakthrough," *Evening Star*, 16 January 1959, p. 1; and "Hearings Before Subcommittee of the Joint Committee on Atomic Energy Congress of the United States, Eighty-Fifth Congress Second Session on Outer Space Propulsion by Nuclear Energy," 22 and 23 January and 6 February 1958.
26. Gary L. Bennett et al., "The General-Purpose Heat Source Radioisotope Thermoelectric Generator: Power for the Galileo and Ulysses Missions," *Proceedings of the 21st Intersociety Energy Conversion Engineering Conference*, August 1986, Vol. 3, pp. 1999–2011.

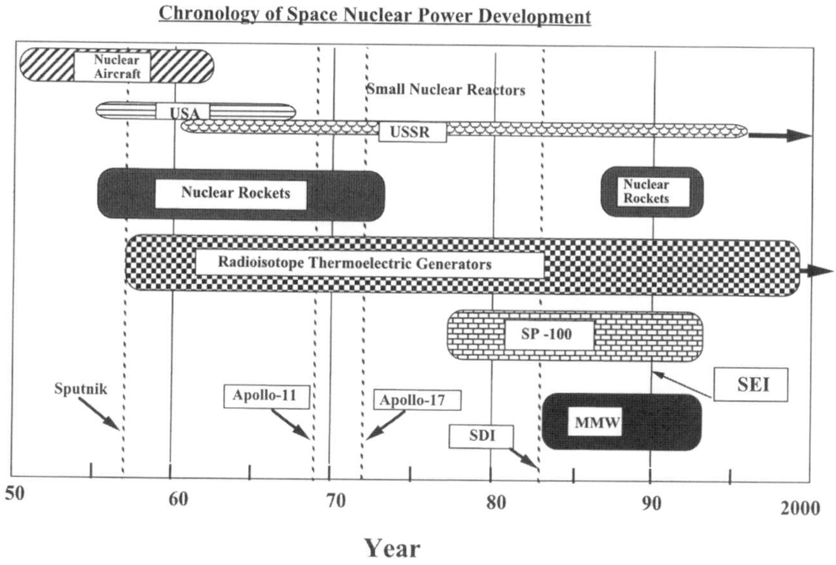

Chronology of Space Nuclear Power Development

Credit: NASA

FIGURE 5.4. The figure provides a chronology of the major elements of space nuclear power since the 1950s. Radioisotope thermoelectric generators have satisfied the power generation need of deep space probes to explore the outer planets since the 1960s.

wattage of RTGs and to economize on what may be accomplished with a limited amount of electricity.[27]

This research has led to a coordinated R&D effort, under the direction of the AEC (renamed the Department of Energy in 1977) and NASA, over many years to advance technology along a broad range of space power areas.

Basic research:
- Photovoltaic energy conversion
- Chemical energy conversion
- Thermal (nuclear) energy conversion
- Power management
- Thermal management

Focused research:
- Space nuclear power
- Surface power and thermal management

27. Cook, "Making Space Nuclear Power a Reality," p. 1; and Robert E. Gold, Ralph L. McNutt Jr., Paul H. Ostdiek, and Louise M. Prockter, "Radioisotope Electric Propulsion As an Enabler of Comprehensive Reconnaissance of Small Bodies," *Spacecraft Reconnaissance of Asteroid and Comet Interiors* (2006).

- Earth-orbiting platform power and thermal management
- Deep space probe power and thermal management
- Laser power beaming
- Mobile surface power systems

As Gary L. Bennett, the dean of space nuclear power at NASA, and electrical engineer Ronald C. Cull remarked in the context of NASA's Space Exploration Initiative (SEI) in 1991:

> The ongoing NASA research and technology program in space energy conversion provides a foundation from which to build the focused technology programs to meet the SEI power requirements. An augmented program focusing on space nuclear power, high capacity power, surface power and thermal management, Earth- orbiting platform power and thermal management, spacecraft power and thermal management for deep-space vehicles, laser power beaming, and mobile surface systems power has been defined to develop the specific focused technologies for SEI applications.[28]

4. Space Nuclear Power and the Early Satellite Efforts

The application of nuclear power to spaceflight really began in the 1950s, when the Navy, through its contractor, the Applied Physics Laboratory (APL) of Johns Hopkins University, developed the first RTGs for space applications. Specifically, Transit, the first navigation satellite, flew an RTG in 1965. Intended as a method of ensuring the capability of the inertial navigation systems of the U.S. Navy's Polaris ballistic missile submarines, the Transit system promised 80–100 meter accuracy. Accordingly, it supported one-third of the nation's strategic triad in enabling targeting and ensuring that the deterrent threat posed to the Soviet Union was real.[29] It originated on 18 March 1958, when the APL's Frank T. McClure wrote two memoranda to APL Director Ralph E. Gibson: "Yesterday I spent an hour with Dr. [William H.] Guier and Dr. [George C.] Weiffenbach discussing the work they and their colleagues have been doing on Doppler tracking of satellites. The principal problem

---

28. Gary L. Bennett and Ronald C. Cull, "Enabling the Space Exploration Initiative: NASA's Exploration Technology Program in Space Power." (Washington, DC: NASA-TM-4325, 1991), p. 11.

29. William H. Guier and George C. Weiffenbach, "Genesis of Satellite Navigation," *Johns Hopkins APL Technical Digest* 1, no. 2 (1997): 178–181; and H. D. Black, R. E. Jenkins, and L. L. Pryor, "The TRANSIT System, 1975," JHU/APL TG 1306 (Laurel, MD: Johns Hopkins University Applied Physics Laboratory, December 1976).

facing them was the determination of the direction in which this work should take in the future. During this discussion it occurred to me that their work provided a basis for a relatively simple and perhaps quite accurate navigation system." Most important, McClure noted, it offered the solution to a vexing problem of genuine military significance during the Cold War.[30]

The first Transit satellite, Transit 1A, took off from the Space Operations Center at Cape Canaveral, Florida, on 17 September 1959 but failed during launch. A second satellite, Transit 1B, was launched on 13 April 1960 and operated for 89 days. There followed a succession of Transit satellites, with a general development of greater capability and longevity interspersed with failures of missions. An important and troubling issue was how to maximize the spacecraft's useful service life on orbit—the best that the Navy could achieve seemed to be about a year with batteries and solar arrays.[31] RTGs offered a ready alternative. As John Dassoulas of APL recalled: "I had been looking into the possibilities of isotopic power since we first began the Transit program. We had a five-year goal for the life of the operational Transit, and we weren't confident that the hermetic seals on batteries would hold up for five years."[32]

Dassoulas attended a space technology symposium in 1959 that prompted his conversion to the belief that nuclear space power had real potential for Transit. By happenstance, he sat on the airplane back to Washington, DC, near Col. G. M. Anderson of the Atomic Energy Commission (AEC). Their conversation led to a visit to the Martin Nuclear Division in Baltimore to learn more about the RTG program then under way. While the bench-test RTGs at Martin used polonium (Po-210) as a fuel source, with its relatively short half-life of 138 days, it led to longer-lived systems using plutonium (Pu-238) as the isotope of choice for the heat source. As two veterans of this project recalled:

> As word spread about a possible flight opportunity, many proposals, including some not so credible ones, were put forward. It was clear to those working the

---

30. Frank T. McClure, quoted in Vincent L. Pisacane, "The Legacy of Transit: Guest Editor's Introduction," *Johns Hopkins APL Technical Digest* 19, no. 1 (1998): 5–10, quotation from p. 7.

31. Robert J. Danchik, "An Overview of Transit Development," *Johns Hopkins APL Technical Digest* 19, no. 1 (1998): 18–26; Patrick W. Binning and Jay W. Middour, "A Brief History of NRL's Early Firsts in Spaceflight," AAS 07-327, 10 August 2007; and Bradford W. Parkinson, Thomas Stansell, Ronald Beard, and Konstantin Gromov, "A History of Satellite Navigation," *Navigation: Journal of the Institute of Navigation* 42, no. 1 (1996): 109–164.

32. John Dassoulas as quoted in Engler, "Atomic Power in Space," p. 22.

spacecraft design that most of these proposals had not been developed with the entire system in mind. One alternative scheme proposed Strontium-90 (90Sr) as the isotope of choice; however, it involved implementing a shield of mercury around the SNAP device (to protect the workers) that could presumably be drained off prior to launch. This would have imposed significant design constraints in safety, reliability, and weight that were clearly unacceptable. It should be noted that none of these suggestions came from the SNAP office of the AEC.[33]

In the end the Transit 4A and 4B satellites were provided with SNAP-3B power sources by the AEC. Both satellites also used solar cells supplying 35 watts at the start of the mission in addition to the RTG.

An early and persistent issue in the use of nuclear power sources for spacecraft was that of safety. It took time and energy to acquire approval to launch these nuclear systems, however. The first tests to assure the safety of RTGs for Transit spacecraft were conducted in the fall of 1960, and the DOD formally requested that the AEC initiate a program in February 1961 "to provide two plutonium-238 isotope-fueled generators for TRANSIT satellites to be launched in June and July."[34] A detailed safety analysis conducted under AEC auspices in March 1961 focused on potential hazards that might result from launch or reentry failures. It concluded that, because of the shielding developed for the RTG and the nature of the system itself, "if the radioisotope generator considered is launched in the trajectory proposed for Transit vehicles, it will not produce a significant radiation hazard."[35]

The AEC's Glenn Seaborg proved a persistent advocate for this mission. He officially asked the president on 6 May 1961 to approve the launch, citing the findings of a hazards study that "any danger to the public is extremely unlikely." He added, "I call this to your attention since this first application of a nuclear auxiliary power source in space is likely to have a wide public impact."[36] The Department of State resisted this launch, in no small part because of its international implications, but the DOD and the AEC persisted and eventually succeeded in obtaining approval. Before the Transit

33. John Dassoulas and Ralph L. McNutt Jr., "RTGs on Transit," paper presented at Space Technology and Applications International Forum, February 2007, Albuquerque, NM.

34. John Graham, Acting Chairman of AEC, to Vice President Lyndon B. Johnson, 10 May 1961, DOE archives, Washington, DC.

35. D. G. Harvey and T. J. Dobry, *Safety Analysis of the Transit Generator* (Baltimore: The Martin Company, MND-P-2479, March 1961), p. vii; and AEC General Manager memorandum for Glenn Seaborg and the Commissioners, "Impact Test Results for the Transit Generator," 22 April 1961, DOE archives.

36. Glenn Seaborg to Chet Holifield, JCAE Chair, date uncertain, DOE Archives.

launch there had been no AEC protocol for delivery of Pu-238 for the integration tests; AEC officials hand-carried the RTG to the Applied Physics Laboratory in the trunk of a private automobile. Security stood guard over the system, and engineers completed their integration tests as quickly as possible. Thereafter, they delivered the RTG to Martin Nuclear in Baltimore, where it underwent shipment to Cape Canaveral.[37]

As this took place, the public learned of the impending launch of a nuclear power plant and organized a protest. Picking up on the high-level discussions inside the Kennedy administration, on 16 May 1961, the *New York Times* broke the story, suggesting that the "problem confronting the Administration … is not so much a technical decision as one of diplomatic, political, and psychological considerations."[38] Three days later, the *New York Times* pressed the issue, highlighting concerns from State Department officials "that in the event of an unsuccessful launching, the satellite, with its radioactive parcel, could fall on Cuba or some other Latin-American country." They feared, in the politically charged involvement over the failed Bay of Pigs invasion of Cuba, that this would add fuel to any international incident that might result. Some even expressed concern that other nations might "take offense about having radioactive materials flown over their territory."[39]

Accordingly, the DOD reconfigured Transit 4A to fly without the RTG, reluctantly accepting a lesser capability on orbit. The story differs as to how the approval finally came down to fly the RTG on Transit 4A. Some believed that it was the culmination of a month-long set of internal negotiations between the DOD and the State Department to proceed with the June 1961 launch of Transit 4A, with final approval clearing the spacecraft for launch on 23 May 1961.[40] Others claimed that it contained the RTG only because of the intervention of President Kennedy, who personally gave an approval to proceed during a small dinner party at which Seaborg pled the case for the mission. Regardless, about two days before the scheduled liftoff, a military team flew the RTG from Baltimore to Patrick Air Force Base in Florida, where the launch team destacked the payload and inserted the SNAP-3 system. The vehicle then launched from Launch Complex 17 on 29 June 1961, and operated for 15 years before the satellite was finally shut down.

---

37. Dassoulas and McNutt, "RTGs on Transit."

38. "Nuclear Power Is a Space Issue," *New York Times*, 16 May 1961.

39. "U.S. Hesitates to Use Atom Device in Satellite Flight Across Cuba," *New York Times*, 19 May 1961, p. 2.

40. Glenn T. Seaborg to DOE General Manager, 8 June 1961, Washington, DC; and Howard C. Brown Jr. to Glenn T. Seaborg, 23 June 1961, as cited in Engler, "Atomic Power in Space," pp. 25, 125.

Transit 4B followed on 15 November 1961 and operated until June 1962, when a thermoelectric converter in the power unit failed. The satellite ceased communications on 2 August 1962, but there were some reports of picking up telemetry from it as late as 1971.[41]

The launch of Transit 4A made headlines. The *New York Journal-American* offered a positive story. It reported: "The successful orbiting of the nuclear device … gives American scientists a significant lead over Russia in the race to harness atomic power for space exploration."[42] Previous concerns voiced by officials from the State Department withered with the success of this flight, and serious intergovernmental opposition never found traction thereafter. By October, Seaborg was promoting the use of atomic power as the logical technology to power spacecraft. He asserted:

> The presence of the "atomic battery" in the satellite is a symbol of a "marriage" that was bound to occur—between Space and the Atom. We have known for some time that the two were made for each other. No one would be tempted, at the present time, to abandon other sources of energy for space. However, the atom has made greater strides toward coming of age for space application in the past few years than many of us could have hoped. The day is not far off when atomic energy will be available in many different packages for practical use in space vehicles.[43]

At the same time, he lobbied with Vice President Lyndon B. Johnson, the chair of the Space Council, for greater use of space nuclear power. He argued that the success of the first mission could be replicated over and over, providing efficient power systems for spacecraft.[44]

The initial successes prompted the development of the Transit 5B series of satellites containing nuclear power sources. Launched atop Thor Able-Star

41. S. J. De Amicis, *Artificial Earth Satellites Designed and Fabricated by the Johns Hopkins University Applied Physics Laboratory*, JHU/APL SDO 1600 (rev.) (Laurel, MD: Johns Hopkins University Applied Physics Laboratory, 1987); National Research Council (NRC), "Past U.S. Nuclear Power and Propulsion Programs," *Priorities in Space Science Enabled by Nuclear Power and Propulsion*, Committee on Priorities for Space Science Enabled by Nuclear Power and Propulsion, Space Studies Board, National Research Council (Washington, DC: National Academies Press, 2006), Appendix A; and Dassoulas and McNutt, "RTGs on Transit."

42. "3-in-1 Satellite Is World's First," *New York Journal-American*, 29 June 1961, p. 1.

43. Glenn T. Seaborg, "Nuclear Power and Space," presentation at International Symposium on Aerospace Nuclear Propulsion, Hotel Riviera, Las Vegas, NV, AEC press release, 24 October 1961, DOE archives.

44. Glenn T. Seaborg to Vice President Lyndon B. Johnson, 4 November 1961; and Lyndon B. Johnson to Glenn T. Seaborg, 6 November 1961, both in DOE archives.

rockets, Transit 5BN-1 reached orbit on 28 September 1963; but it achieved gravity-gradient stabilization upside down, which limited its signal output to the ground. Transit 5BN-2 was launched with an RTG power source on 5 December 1963 and operated for approximately one year. The last RTG-powered navigation satellite, Transit 5BN-3, was launched on 12 April 1964 but failed to achieve orbit, and its failure prompted widespread concern. As a U.S. GAO report noted in the latter 1990s: "In 1964, a TRANSIT 5BN-3 navigational satellite malfunctioned. Its single RTG, which contained 2.2 pounds of plutonium fuel, burned up during reentry into Earth's atmosphere. This RTG was intended to burn up in the atmosphere in the event of a reentry."[45]

It did, and this sent shock waves through the world community. The Atomic Energy Commission tried to assuage the public's fears, reporting that "from previous safety analysis and tests it had been concluded the reentry will cause the plutonium-238 fuel to burn up into particles of about one-millionth of an inch in diameter. These particles will be widely dispersed... and would not constitute a health hazard."[46] This proved too optimistic. One study concluded that "a worldwide soil sampling program carried out in 1970 showed SNAP-9A debris present at all continents and at all latitudes."[47] As reported in *New Scientist*, within a decade after its reentry, atmospheric measurements "showed that about 5 percent of its plutonium-238 remained in the atmosphere. The activity of the release is about 10 percent of that of plutonium-239 released in all tests of nuclear weapons in the atmosphere up to now. It is the main source of plutonium-238 in the environment."[48]

NASA's economic impact statement, conducted in advance of the Cassini space launch in 1996, added:

> Since 1964, essentially all of the SNAP-9A release has been deposited on the Earth's surface. About 25 percent... of that release was deposited in the northern latitudes, with the remaining 75 percent settling in the southern hemisphere.... The release into the atmosphere was consistent with the RTG design philosophy of the time. (Subsequent RTGs, including the RTGs on

45. U.S. GAO Report, "Space Exploration: Power Sources for Deep Space Probes," U.S. GAO/NSIAD-98-102, May 1998, p. 18.

46. AEG press release, 22 April 1964, attached to a letter from Glenn Seaborg to the Executive Secretary, National Aeronautics and Space Council, 23 April 1964, DOE archives.

47. "Emergency Preparedness for Nuclear-Powered Satellites" (Paris: Organization for Economic Cooperation and Development and Swedish National Institute for Radiation Protection, 1990), p. 17.

48. "Ariadne," *New Scientist*, 2 March 1991, p. 88. See also William J. Broad, "Fallout from Nuclear Power in Space," *Science* 219, no. 4580 (1983): 38–39.

the Cassini spacecraft, have been designed to contain the Pu-238 fuel to the maximum extent possible, recognizing that there are mass and configuration requirements relative to the spacecraft and its mission that must be considered with the design and configuration of the power source and its related safety requirements.)[49]

Such reports, and the concerns that they engendered, led to the development of a very rigorous testing and safety program and to the restriction of space nuclear power to only those missions for which it was absolutely critical.

Its immediate result was to prompt the Navy to rely thereafter on solar-powered satellites because of the many high-level approvals necessary to launch a nuclear power system and the safety hazard inherent in failure. Of the six objectives for this series of satellites listed here, only three were fully met (3, 4, and 5), while the remainder were at best partially resolved:

1. Provide a means by which U.S. Navy ships may navigate anywhere in the world.
2. Demonstrate satisfactory operation of all satellite subsystems.
3. Demonstrate satisfactory operation and potential long life capability of the SNAP 9-A power supply.
4. Improve our understanding of the effects of ionospheric refraction on radio waves.
5. Demonstrate satisfactory operation of the satellite-borne data injection memory system.
6. Increase knowledge of the Earth's shape and gravitational field.

Each of these satellites contained a SNAP-9A power source: a cylinder 30.48 centimeters in diameter by 20.32 centimeters high with four radiating fins, weighing 12.3 kilograms. They provided, when working correctly, 25 watts at 6 volts for a projected satellite lifetime of five years in space.[50]

At the same time, the U.S. military flew one nuclear reactor in space, solely as a test program, in the mid-1960s. Designated SNAPSHOT, this mission was launched from Vandenberg Air Force Base, in California, on 3 April 1965, with the SNAP-10A reactor. A heritage project based on earlier SNAP reactors, the 435-kilogram system produced 500 watts of energy for

---

49. NASA Office of Space Science, "Cassini Final Environmental Impact Statement," June 1995, pp. 3–44, *http://saturn.jpl.nasa.gov/spacecraft/safety/chap3.pdf* (accessed 1 July 2008).
50. Gary L. Bennett, "Space Nuclear Power: Opening the Final Frontier," AIAA 2006-4191, presentation at Fourth International Energy Conversion Engineering Conference and Exhibit (IECEC), San Diego, CA, 26–29 June 2006, pp. 3–4.

one year. Precautions abounded for this test: for example, the reactor was not started until the spacecraft reached orbit. The test was successful until 43 days into the mission, when a voltage regulator on the carrier vehicle, an Agena upper stage, failed, and the test had to be terminated.[51] Thereafter, as Canadian nuclear policy analyst Michael Bein commented:

> The only U.S. satellite thus far to carry a nuclear fission reactor failed in 1965 after 43 days aloft and was subsequently boosted into a 4,000-year orbit in order that its radioactivity might have time to decay to safer levels before it descends to Earth. Injection into higher orbit is the method of reactor "disposal" preferred by both the American and Soviet programs.[52]

5. Space Nuclear Power at High Tide

The period between the flights of the Transit navigational satellites and the flights of NASA's outer planetary probes, Voyagers 1 and 2, in the late 1970s may best be characterized as the high tide of space nuclear power. During that time NASA flew no fewer than 14 RTGs, and the DOD operated another 11, while the Soviet Union launched 20 on various spacecraft. These included RTGs on the Apollo lunar missions, the flights of Pioneers 10 and 11, the Viking missions to Mars, and the so-called "Grand Tour" of the solar system made by Voyagers 1 and 2. Throughout this period, furthermore, the technology evolved and became increasingly capable. Table 5.1 depicts the total number of RTGs launched to date by the United States on space missions.

Viewing the efforts of the DOD, NASA officials determined that, although the possible use of reactors for space power was rejected, RTGs would be helpful in its planetary exploration program. For example, as reported in the Atomic Energy Commission's SNAP Fact Sheet: "NASA's inquiries about using RTGs for Project Surveyor—the unmanned soft lunar exploration program—had led to work at the AEC on SNAP-11. This device, to be filled with curium-242, would weigh 30 pounds, and would provide a minimum of 18.6 watts of power continuously for 90-day lunar missions."[53] While

---

51. Ibid., p. 5; D. W. Staub, "SNAP 10A Summary Report," Atomics International Report NAA-SR-12073, 25 March 1967; S. S. Voss, "SNAP (Space Nuclear Auxiliary Power) Reactor Overview," June 1982–December 1983, Air Force Weapons Lab, Kirtland AFB, NM, Final Report, copy in DOE History Office; and Steven Aftergood, "Nuclear Space Mishaps and Star Wars," *Bulletin of Atomic Scientists* 42, no. 8 (1986): 40–43, esp. p. 40.

52. Michael Bein, "Star Wars and Reactors in Space: A Canadian View," 1986, *http://www.animatedsoftware.com/spacedeb/canadapl.htm#ref28* (accessed 1 July 2008).

53. "AEC SNAP Fact Sheet," 1 September 1963, quoted in Engler, "Atomic Power in Space," p. 34.

TABLE 5.1. U.S. Spacecraft Using Radioisotope Systems

Spacecraft	Power Source	No. RTGs*	Mission Type	Launch Date	Status
Transit 4A	SNAP-3	1	Navigational	29 Jun 1961	Currently in orbit
Transit 4B	SNAP-3	1	Navigational	15 Nov 1961	Currently in orbit
Transit 5BN-1	SNAP-9A	1	Navigational	28 Sep 1963	Currently in orbit
Transit 5BN-2	SNAP-9A	1	Navigational	5 Dec 1963	Currently in orbit
Transit 5BN-3	SNAP-9A	1	Navigational	12 Apr 1964	Aborted; burned up
Nimbus B-1	SNAP-19	2	Meteorological	18 May 1968	Aborted; retrieved
Nimbus III	SNAP-19	2	Meteorological	14 Apr 1969	Currently in orbit
Apollo 11	ALRHU	Heater	Lunar	16 Jul 1969	On lunar surface
Apollo 12	SNAP-27	1	Lunar/ALSEP	14 Nov 1969	On lunar surface
Apollo 13	SNAP-27	1	Lunar/ALSEP	11 Apr 1970	Aborted in Pacific
Apollo 14	SNAP-27	1	Lunar/ALSEP	31 Jan 1971	On lunar surface
Apollo 15	SNAP-27	1	Lunar/ALSEP	26 Jul 1971	On lunar surface
Pioneer 10	SNAP-19	4	Planetary	2 Mar 1972	Heliopause
Apollo 16	SNAP-27	1	Lunar/ALSEP	16 Apr 1972	On lunar surface
Triad-01-1X	Transit-RTG	1	Navigational	2 Sep 1972	Currently in orbit
Apollo 17	SNAP-27	1	Lunar/ALSEP	7 Dec 1972	On lunar surface
Pioneer 11	SNAP-19	4	Planetary	5 Apr 1973	Heliopause
Viking 1	SNAP-19	2	Mars Lander	20 Aug 1975	On Martian surface
Viking 2	SNAP-19	2	Mars Lander	9 Sep 1975	On Martian surface
LES 8, LES 9	MHW-RTG	2, 2	Communication	14 Mar 1976	Currently in orbit
Voyager 2	MHW-RTG	3	Planetary	20 Aug 1977	Heliopause
Voyager 1	MHW-RTG	3	Planetary	5 Sep 1977	Heliopause
Galileo	GPHS-RTG	2	Planetary	18 Oct 1989	Intentionally deorbited into Jupiter
Ulysses	GPHS-RTG	1	Planetary	6 Oct 1990	Sun's polar regions
Mars Pathfinder	LWRHU	Heater	Mars Lander	4 Dec 1996	Operated on Mars
Cassini	GPHS-RTG	3	Planetary	15 Oct 1997	Operating at Saturn
New Horizons	GPHS-RTG	1	Planetary	19 Jan 2006	En route to Pluto

*All U.S. RTGs are fueled by plutonium-238; the SNAPSHOT reactor was fueled by uranium-235.

Source: Gary L. Bennett, James J. Lombardo, and Bernard J. Rock, "Development and Use of Nuclear Power Sources for Space Applications," *Journal of the Astronautical Sciences* 29, no. 4 (1981): 321–342; Nicholas L. Johnson, "Nuclear Power Supplies in Orbit," *Space Policy* 2, no. 3 (1986): 223–233; and Gary L. Bennett, "Space Nuclear Power: Opening the Final Frontier," AIAA 2006-4191, presentation at Fourth International Energy Conversion Engineering Conference and Exhibit (IECEC), San Diego, CA, 26–29 June 2006, p. 2.

NASA chose to forego RTG usage for Surveyor, the Agency adopted it for the Apollo lunar landing program. It had a willing partner in Glenn Seaborg and the AEG. A report advocating the use of RTGs emerged from the AEC in February 1964, emphasizing the appropriateness of space nuclear power for extended and deep space missions because the "performance of ambitious space missions will require amounts of reliable power so large that they can be achieved only from nuclear systems."[54] A similar report from NASA's Jet Propulsion Laboratory in 1964 advocated the employment of RTGs to power deep space probes in cases where solar power would be insufficient to meet the needs of the spacecraft.[55] In June 1965, NASA and the AEC reached agreement on the establishment of a joint Space Nuclear Systems Division. Harry Finger, the senior official working on space nuclear issues, emphasized the need "to develop systems that bracket as broad a range of potential mission uses as possible, and parallel with this, continue to push the technology into more advanced areas in order to try to improve the performance and life capability of these systems."[56]

Even with this impetus, it took five years after the loss of Transit 5BN-3 for another RTG to reach orbit, and the effort to achieve it was slow and prickly. As never before, NASA weighed in to ensure the safety of the RTGs from any conceivable accident. The AEC was a willing accomplice, of course, and took this charge seriously. The management structure evolved to carry out this mission. First, the two organizations used the joint office to coordinate all efforts, giving it both authority and responsibility to conduct the program effectively and safely. Like the larger Apollo program, the joint office pursued RTG efforts with the same top-down leadership style that was so successful elsewhere, emphasizing configuration control and project management as the only true means of achieving acceptable results. This centralization of design, engineering, procurement, testing, construction, manufacturing, spare parts, logistics, training, and operations worked well. The approach was lauded in the November 1968 issue of *Science* magazine, the publication of the American Association for the Advancement of Science:

54. U.S. Atomic Energy Commission, "Systems for Nuclear Auxiliary Power: A Report by the Commission," TID-20103, February 1964.

55. Eric S. Pedersen, "Heat-Sterilizable Power Source Study for Advanced Mariner Missions," JPL Technical Memorandum No. 33–780, 1 July 1964, NASA Center for Aerospace Information, Linthicum, MD.

56. "Agreement Between the Atomic Energy Commission and the National Aeronautics and Space Administration on Space Nuclear Systems," draft prepared by AEC General Manager, 22 January 1965; and "AEC Creates Division of Space Nuclear Systems: Harold B. Finger Named Director," AEC press release, 17 June 1965, both in NASA Historical Reference Collection.

In terms of numbers of dollars or of men, NASA has not been our largest national undertaking, but in terms of complexity, rate of growth, and technological sophistication it has been unique.... It may turn out that [the space program's] most valuable spin-off of all will be human rather than technological: better knowledge of how to plan, coordinate, and monitor the multitudinous and varied activities of the organizations required to accomplish great social undertakings.[57]

Finger employed the same approach in building and flying the RTGs used in the Apollo program and other missions of NASA. The AEC's Bernard Rock reflected on this approach to overseeing the RTG program and its influence on other activities of his organization: "My background was technical, but I soon saw how important management was in the NASA scheme of things and I sensed that this concern with management was correct. I went out and enrolled in some courses in engineering administration.... Apollo was many orders of magnitude greater in size and complexity than" other AEC programs and it was successful largely because of its rigorous management.[58]

The SNAP-27 RTG became the power supply for the Apollo Lunar Surface Experiments Package (ALSEP) that was left on the Moon by all Apollo missions but the first one (see figure 5.5). This was largely because of the scientific objectives of the Apollo program. Of course, the reasons for undertaking Apollo had little to do with furthering scientific understanding. Its impetus rested almost solely on Cold War rivalries and the desire to demonstrate technological verisimilitude to the peoples of all the nations of the world. Even so, a great deal of good scientific knowledge emerged from the exercise as scientists gained entrée to the program and maximized the scientific return on this investment. They succeeded in having established at each of the landing sites a self-contained experiments package that would measure, record and send data back to Earth on a variety of factors, such as seismic occurrences, surface vibrations, responses of the Moon to fluctuations in solar and terrestrial magnetic fields, and changes in the low concentrations of gas in the virtually nonexistent lunar atmosphere.[59]

Ongoing debates about the size and mass of experiments, as well as their power requirements, roiled the mission planning efforts throughout the mid-1960s. The scientists agreed that the first investigations should relate to geology (especially sample collection), geochemistry, and geophysics. They also

---

57. Dael Wolfe, "The Administration of NASA," *Science* 163 (15 November 1968): 753.
58. Bernard Rock, as quoted in Engler, "Atomic Power in Space," p. 59.
59. Edgar M. Cortright, ed., *Apollo Expeditions to the Moon* (Washington, DC: NASA SP-1975-350, 1975), pp. 240–241.

Credit: NASA

FIGURE 5.5. On the Moon in 1969, Apollo 12 astronaut Alan Bean prepares to load the plutonium-238 heat source in the SNAP-2 thermoelectric generator. This generator produced 73 watts of power for the Apollo Lunar Surface Experiments Package for nearly eight years.

agreed that the early landings should focus on the return of as many diverse lunar rock and soil samples as feasible, the deployment of long-lasting surface instruments, and the geological exploration of the immediate landing areas by each crew. These could be expanded later to include surveys of the whole Moon and detailed studies of specific sites in the equatorial belt.[60]

The scientific "geeks" exploited this opportunity to place more than 50 experiments on the various Apollo missions and, in the case of the last landing mission, to have one of their own, Harrison Schmitt, undertake fieldwork on the Moon. The science packages deployed on the Moon included the following types of experiments:

---

60. *NASA 1965 Summer Conference on Lunar Exploration and Science, Falmouth, Massachusetts, July 19–31, 1965* (Washington, DC: NASA SP-88, 1965), pp. 7–12, 16–19.

- *Soil Mechanics Investigation* studied the properties of the lunar soil (Apollo 11, 12, 14, 15, 16, and 17).
- *Solar Wind Composition Experiment* collected samples of the solar wind for analysis on Earth (Apollo 11, 12, 14, 15, and 16).
- *Passive Seismic Experiment* detected lunar "moonquakes" and provided information about the internal structure of the Moon (Apollo 11, 12, 14, 15, and 16).
- *Laser Ranging Retroreflector* measured very precisely the distance between the Earth and the Moon (Apollo 11, 14, and 15).
- *Lunar Dust Detector* studied the effects of lunar dust on the operation of the experiment package (Apollo 11, 12, 14, and 15).
- *Lunar Surface Magnetometer* measured the strength of the Moon's magnetic field (Apollo 11, 12, 15, and 16).
- *Lunar Portable Magnetometer* measured the strength of the Moon's magnetic field (Apollo 14 and 16).
- *Cold Cathode Gauge* measured the abundance of gases in the lunar atmosphere (Apollo 12, 14, and 15).
- *Suprathermal Ion Detector Experiment* studied the lunar ionosphere (Apollo 12, 14, and 15).
- *Solar Wind Spectrometer* measured the composition of the solar wind (Apollo 12 and 15).
- *Active Seismic Experiment* provided information about the structure of the upper 100 meters of the lunar regolith (Apollo 14 and 16).
- *Charged Particle Lunar Environment Experiment* measured plasmas around the Moon (Apollo 14).
- *S-Band Transponder Experiment* measured regional variations in the Moon's gravitational acceleration (Apollo 14, 15, 16, and 17).
- *Bistatic Radar Experiment* measured scattering of radar waves from the lunar surface (Apollo 14).
- *Heat Flow Experiment* measured the amount of heat coming off of the Moon (Apollo 15, 16, and 17).
- *Metric and Panoramic Cameras* provided systematic photography of the lunar surface (Apollo 15, 16, and 17).
- *Laser Altimeter* measured the heights of lunar surface features (Apollo 15, 16, and 17).
- *X-ray Fluorescence Spectrometer Experiment* measured the composition of the lunar surface (Apollo 15, 16, and 17).
- *Gamma-ray Spectrometer Experiment* measured the composition of the lunar surface (Apollo 15, 16, and 17).
- *Alpha Particle Spectrometer Experiment* measured radon emission from the lunar surface (Apollo 15, 16, and 17).

- *Orbital Mass Spectrometer Experiment* measured the composition of the lunar atmosphere (Apollo 15, 16, and 17).
- *Bistatic Radar Experiment* measured the scattering of radar waves from the lunar surface (Apollo 15, 16, and 17).
- *Subsatellite* measured regional variations in the Moon's gravitational acceleration and magnetic field and the distribution of charged particles around the Moon (Apollo 15, 16, and 17).
- *Far Ultraviolet Camera/Spectrograph* took pictures and spectra of astronomical objects in ultraviolet light (Apollo 16).
- *Cosmic Ray Detector* measured very high-energy cosmic rays from the Sun and other parts of our galaxy (Apollo 16 and 17).
- *Active Seismic Experiment* provided information about the structure of the upper 100 meters of the lunar regolith (Apollo 16 and 17).
- *Lunar Surface Magnetometer* measured how the strength of the Moon's magnetic field varied over time (Apollo 16).
- *Traverse Gravimeter Experiment* measured how the Moon's gravitational acceleration varied at different locations near the landing site, which helped to measure the thickness of the basalt layer in this region (Apollo 17).
- *Lunar Neutron Probe* measured the penetration of neutrons into the lunar regolith, which helped to measure the overturn rate of the regolith (Apollo 17).
- *Surface Electrical Properties* measured the propagation of electrical waves through the lunar crust (Apollo 17).
- *Lunar Seismic Profiling Experiment* provided information about the structure of the upper kilometer of the lunar crust (Apollo 17).
- *Lunar Atmospheric Composition Experiment* measured the composition of the Moon's tenuous atmosphere (Apollo 17).
- *Lunar Ejecta and Meteorites Experiment* measured the impact of small meteorites on the Moon (Apollo 17).
- *Lunar Surface Gravimeter* attempted to detect gravity waves (Apollo 17).
- *Apollo Lunar Sounder Experiment* used radar to study the structure of the upper kilometer of the lunar crust (Apollo 17).
- *Ultraviolet Spectrometer Experiment* studied the composition of the lunar atmosphere (Apollo 17).
- *Infrared Radiometer* measured the cooling of the Moon's surface at night as a way to determine the physical properties of the lunar soil (Apollo 17).

Collectively, these experiments yielded more than 10,000 scientific papers and a major reinterpretation of the origins and evolution of the Moon.[61]

The Bendix Aerospace Systems Division developed the Apollo Lunar Surface Experiments Package (ALSEP) to support these activities, to be powered by a Pu-238-fueled, 75-watt isotopic power unit built by General Electric. This unit later had to be downsized to 50 watts, and ultimately the SNAP-27 was useful only when armed by an astronaut during extravehicular activity (EVA). It would therefore provide power on the Moon throughout the long (14-Earth-day) lunar night for the ALSEP but could not be used on robotic missions.[62] All but Apollo 11 used the SNAP-27, and that first mission used a smaller, nuclear heating unit. George E. Mueller, NASA's Associate Administrator for Manned Space flight, explained why:

> We have sharpened the focus on some of the problems involved. The first landing mission represents a large step from orbital operations.... The 1/6 g lunar surface environment will be a new experience. We cannot simulate it completely on Earth. We find that we simply do not have as much metabolic data as we would like in order to predict with high confidence, rates in a 1/6 g environment. Only educated guesses are possible on the difficulties the astronaut will have in maneuvering on the surface or the time it will take him to accomplish assigned tasks.... The decision not to carry ALSEP on the first mission is due to the time necessary for deployment and not to any concern of operating with the RTG. You have the strongest advance assurance I can give

61. Donald A. Beattie, *Taking Science to the Moon: Lunar Experiments and the Apollo Program* (Baltimore, MD: Johns Hopkins University Press, 2001), pp. 252–253; Bevan M. French, "The New Moon: A Window on the Universe," n.d., unpublished paper, Bevan M. French Biographical Files, NASA Historical Reference Collection; Sen. Frank Moss, "The Value of Lunar Science," *Congressional Record*, 29 January 1976, p. S797; William K. Stevens, "Lunar Science Thrives in Wake of Apollo," *New York Times*, 17 July 1979, pp. C-1, C-3; Bevan M. French, *What's New on the Moon* (Washington, DC: NASA EP-131, 1977); Carlton C. Allen, Richard V. Morris, and David S. McKay, "Oxygen Extraction from Lunar Soils and Pyroclastic Glass," *Journal of Geophysical Research* 101, no. E11 (1996): 26,085–26,095; William H. Gregory, "Data Show Moon As Ever More Complex," *Aviation Week & Space Technology*, 16 April 1973, pp. 38–43; Billy Goodman, "Apollo's Geology Lesson," *Air & Space* (June–July 1994): pp. 42–51; and Louis Varricchio, "Inconstant Moon—A Brief History of U.S. Lunar Science from 1840 to 1972," unpublished paper in NASA Historical Reference Collection.

62. J. R. Bates, W. W. Lauderdale, and H. Kernaghan, "*ALSEP Termination Report*," NASA Reference Publication 1036 (Washington, DC: NASA, 1979); A. A. Pitrolo, B. J. Rock, W. C. Remini, and J. A. Leonard, "SNAP-27 Program Review," in *Proceedings of the Fourth Intersociety Energy Conversion Engineering Conference*, 22–26 September 1969, Washington, DC.

that ALSEP will be carried on the second mission. I also foresee significant RTG use in the future as lunar exploration progresses.[63]

The first use of the RTG on Apollo 12 proved a moment of truth for its proponents. Not that they were unconcerned for the safety of the astronauts—all precautions were taken, but the crew had trouble deploying the system. Astronaut Alan Bean easily deployed the ALSEP as intended, but he could not activate the RTG. As Mission Commander Pete Conrad relayed to Mission Control: "It really gets you mad, Houston.... Al put the tool on, screwed it all the way down and the fuel element would not come out of the kit. He's taking the tool off and working it again." Bean added: "I tell you what worries me, Pete. If I pull on it too hard, it's a very delicate lock mechanism ... just get the feeling that it's hot and swelled in there or something. It doesn't want to come out.... Come out of there, rascal." Bean used a common technique when frustrated by a mechanical device—he got a hammer. That sent the RTG staff into a spin, but his light taps were sufficient to dislodge the fuel capsule and activate the RTG. The SNAP-27 then began to produce power for the ALSEP as intended and operated thereafter without any problem. The first use of an RTG on a human mission was successful.[64]

In no small measure because of the ALSEP's capacity for sustained operation, the Apollo program proved one of the most significant scientific expeditions ever undertaken. Lunar geologist Don Wilhelms commented on the state of knowledge about the Moon resulting from Apollo in his outstanding 1993 recollection, *To a Rocky Moon*:

> I think that to a first approximation we can summarize the geologic style of the Moon very simply. Primary and secondary impacts, helped by a little lava and minor faulting, have created almost the entire range of lunar landforms. The cosmic impact catastrophes have alternated with gentle volcanic extrusions and an occasional fire fountain originating deep in the Moon's interior. Horizontal plate motions like those of Earth are unknown on the Moon. Vertical motions are more important, but only in the settling of the mare mascons and in the rise of crater floors that are not loaded with mare basalt. The Moon's face has been molded by the rise of basaltic magmas into receptacles

---

63. George E. Mueller to Glenn T. Seaborg, 13 November 1968, NASA Historical Reference Collection.

64. Cortright, *Apollo Expeditions to the Moon*, p. 225; "Exuberance Sets Tone of First EVA," *Aviation Week & Space Technology*, 24 November 1969, pp. 19–21; and "Scientists Concede Value of Man in Lunar Experiment Deployment," *Aviation Week & Space Technology*, 1 December 1969, pp. 20–21.

dug in plagioclase-rich terra material by impacts. The Moon is neither cosmic exotica nor a little Earth.[65]

As reported in *Science* in 1973: "Man's knowledge of the Moon has been dramatically transformed during the brief 3½ years between the first and last Apollo landing[s]."[66]

The only other difficulty with the RTGs used in the Apollo program came as a result of the failure of Apollo 13 and the near-loss of the crew. Like its sister missions, Apollo 13 had an ALSEP on the lunar module powered by an RTG that would have been left on the Moon had the mission not been aborted. As it was, the lunar module returned to Earth with the capsule and crew and was jettisoned over the Pacific Ocean prior to the crew's reentry into the atmosphere. It was targeted for the Tonga Trench, one of the deepest points in the Pacific, and all evidence suggests that the RTG impacted the ocean as intact. Crews trolled the area in search of debris, measuring radio-activity in the area. They found none. Everyone involved in the investiga-tion agreed that the lunar module had broken up on reentry, as anticipated, but that the graphite-encased plutonium-238 fuel cask survived the breakup and went down intact to more than 20,000 feet in the depths of the Tonga Trench. Some RTG insiders went on television to reassure the public that no one was in danger from the RTG. Even so, there was not much public outcry.

One AEC engineer close to the program recalled that he received only two questions about this potential safety issue; he assured them that there was no reason for concern. The AEC issued a statement about two weeks after the Apollo 13 mission, indicating that "air sampling over the predicted impact area of the SNAP-27 fuel cask freed from the Apollo 13 lunar module showed no traces of radiation above that already present in the atmosphere. The absence of additional radiation indicates that the cask containing the plu-tonium fuel survived as designed the heat of reentry, impacted in the South Pacific intact and sank to the ocean bottom."[67] Some antinuclear power activ-ists never accepted this conclusion, but no compelling evidence to the con-trary has ever been brought forward.

A major shift in the use of space nuclear power came with NASA's decision to pursue outer-planetary exploration. In the early 1960s several scientists

---

65. Don E. Wilhelms, *To a Rocky Moon: A Geologist's History of Lunar Exploration* (Tucson: University of Arizona Press, 1993), p. 344.

66. Allen L. Hammond, "Lunar Science: Analyzing the Apollo Legacy," *Science* 179, no. 4080 (1973): 1313–1315.

67. Director, AEC Public Information, "Response to Queries On SNAP-27 Reentry," 28 April 1970, DOE History Office; and Beattie, *Taking Science to the Moon*, pp. 130–131.

realized that once every 176 years, both the Earth and all the giant planets of the solar system gather on one side of the Sun, making possible close-up observation of all the planets in the outer solar system in a single flight. This geometric line-up made up Voyager's so-called "Grand Tour." Moreover, the flyby of each planet would bend the spacecraft's flight path and increase its velocity enough to deliver it to the next destination. This would occur through a complicated process known as "gravity assist," something like a slingshot effect, whereby the flight time to Neptune could be reduced from 30 to 12 years. Such a configuration was due to occur in the late 1970s, and it led to one of the most significant space-probe efforts undertaken by the United States.[68]

For such a lengthy mission, NASA would need a long-lasting power source. Solar power would not work because of the distance from the Sun; the logical, perhaps the only realistic, answer was to use RTGs to generate power on the spacecraft. To prepare the way for a more extensive Grand Tour, NASA conceived Pioneer 10 and Pioneer 11 as outer solar system probes stripped bare through successive budgetary constraints that forced a somewhat less ambitious effort than originally intended (see figure 5.6). Both were small nuclear-powered, spin-stabilized spacecraft that Atlas–Centaur launched. Pioneer 10 was launched on 3 March 1972. It arrived at Jupiter on the night of 3 December 1973, and although many were concerned that the spacecraft might be damaged by intense radiation discovered in Jupiter's orbital plane, the spacecraft survived, transmitted data about the planet, and continued on its way out of the solar system, away from the center of the Milky Way galaxy.

In 1973, NASA launched Pioneer 11, providing scientists with their first close-up view of Jupiter. The close approach and the spacecraft's speed of 107,373 miles per hour, by far the fastest ever reached by an object launched from Earth, hurled Pioneer 11 across the solar system some 1.5 billion miles toward Saturn, where it encountered the planet's south pole within 26,600 miles of its cloud tops in December 1974. In 1979, Pioneer 11 again encountered Saturn, this time passing within 13,000 miles of the planet, where it discovered two new moonlets and a new ring, and charted Saturn's magnetosphere, its magnetic field, its climate and temperatures, and the general

---

68. Henry C. Dethloff and Ronald A. Schorn, *Voyager's Grand Tour: To the Outer Planets and Beyond* (Washington, DC: Smithsonian Institution Press, 2003); Richard L. Dowling, William J. Kasmann, Michael A. Minovitch, and Rex W. Ridenoure, "The Origin of Gravity-Propelled Interplanetary Travel," in J. D. Hunley, ed., *History of Rocketry and Astronautics: Proceedings of the Twenty-fourth History Symposia of the International Academy of Astronautics*, Vol. 19, AAS History Series (San Diego: Univelt, Inc., 1997), pp. 63–102; Gary A. Flandro, "Discovery of the Grand Tour Voyager Mission Profile," in Mark Littmann, ed., *Planets Beyond* (New York: John Wiley and Sons, 1988), pp. 95–98.

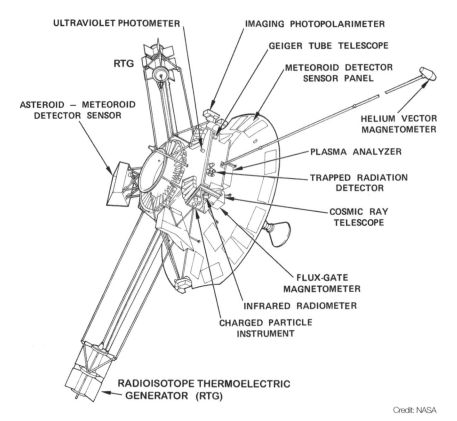

Credit: NASA

FIGURE 5.6. The Pioneer 10 and 11 spacecraft, launched in 1972 and 1973, respectively, used RTGs to power them on their voyages to Jupiter and Saturn. Four RTGs are mounted on two booms extending from the spacecraft, shown here in tandem.

structure of its interior. In 1990, Pioneer 11 officially departed the solar system by passing beyond Pluto and heading toward interstellar space at the center of the Milky Way galaxy. Pioneer 11 ended its mission 30 September 1995, when the last transmission from the spacecraft was received.[69]

---

69. Mark Wolverton, *The Depths of Space: The Pioneers Planetary Probes* (Washington, DC: Joseph Henry Press, 2004); Robert Hotz, "The Last Liftoff," *Aviation Week & Space Technology*, 11 December 1972, p. 17; "Pioneer 10 Blazes Path to Jupiter, Beyond," *Aviation Week & Space Technology*, 10 December 1973, p. 21; "Pioneer 11 Retargeted for Saturn Encounter," *Aviation Week & Space Technology*, 25 March 1974, p. 18; "Use of Isotopic Nuclear Systems on the NASA PIONEER Spacecraft" (staff paper), 30 December 1971, DOE Archives; "Pioneer's Success Buoys Saturn Hopes," *Aviation Week & Space Technology*, 9 December 1974, p. 16; Palmer Dyal, "Pioneers 10 and 11 Deep Space Missions," NASA Technical Memorandum 102269, February 1990, NASA Center for Aerospace Information, NASA HQ, Washington, DC; Charles F. Hall, "Pioneer 10 and Pioneer 11," *Science* 188, no. 4187 (1975): 445–446; and J. W. Dyer, "Pioneer Saturn," *Science* 207, no. 4429 (1980): 400–401.

Earth received Pioneer 10's last, very weak signal on 22 January 2003. The last time a Pioneer 10 contact actually returned telemetry data was 27 April 2002. At last contact, Pioneer 10 was 7.6 billion miles from Earth, or 82 times the nominal distance between the Sun and Earth. At that distance it takes more than 11 hours and 20 minutes for the radio signal, traveling at the speed of light, to reach Earth. The spacecraft will continue to coast silently as a ghost ship into interstellar space, heading generally toward the red star Aldebaran, which forms the eye of the constellation Taurus (The Bull). Aldebaran is about 68 light years away. It will take Pioneer 10 more than two million years to reach it. "From Ames Research Center and the Pioneer Project, we send our thanks to the many people at the Deep Space Network (DSN) and the Jet Propulsion Laboratory (JPL) who made it possible to hear the spacecraft signal for this long," said Pioneer 10 Flight Director David Lozier at the time of the last contact.[70]

Both Pioneer 10 and 11 were remarkable space probes, stretching from a 30-month design life cycle into a mission of more than 20 years and returning useful data not just about the Jovian planets of the solar system but also about some of the mysteries of the interstellar universe.[71] The program—perhaps this is an understatement—was a huge success. Such success would not have resulted without the four RTGs on each spacecraft providing power. Each Pioneer spacecraft employed four SNAP-19 generators as the sole power source mounted in tandem pairs on extendable booms 120 degrees apart. As stated in the SNAP-19 final report:

> For this first all-nuclear power application in outer space, each RTG is required to produce 30 watts at high probability (0.995) at Jovian encounter, which is specified to occur up to 36 months after delivery to NASA. This performance is to be achieved in accord with the constraints of 38 to 42.5 watts at delivery and a maximum weight of 30.5 pounds. The flight time through the asteroid belt and up to encounter with Jupiter is between 20 and 30 months. Thus, the 36-month specification includes six months operation (most with RTG output shorted) prior to launch and the maximum travel time.

---

70. "Pioneer 11 to End Operations after Epic Career," NASA press release, 29 September 1995, NASA Historical Reference Collection.

71. Richard O. Fimmel, James A. Van Allen, and Eric Burgess, *Pioneer: First to Jupiter, Saturn, and Beyond* (Washington, DC: NASA SP-1980-446, 1980); Richard O. Fimmel, William Swindell, and Eric Burgess, *Pioneer Odyssey* (Washington, DC: NASA SP-1977-396, 1977); and "Farewell Pioneer 10: Spacecraft Sends Its Last Signal," *Space Times: Magazine of the American Astronautical Society* 42 (May–June 2003): 19.

This report added: "The fuel is in the form of pucks, about 2 inches in diameter and 0.2 inch thick, of plutonium moly cermet (PMC). Eighteen pucks comprise a complete fuel stack for the capsule."[72]

The spacecraft also had a dozen radioisotope heater units (RHUs), each generating 1 watt, to heat components in space. They were strategically placed in the Thruster Cluster Assembly, near the Sun sensor, and at the magnetometer. There was no problem with the long-term power capabilities of these RTGs. As one account of the mission noted: "The spacecraft continued to make valuable scientific investigations in the outer regions of the solar system until routine tracking of the probe was stopped on 31 March 1997, for budgetary reasons, and NASA formally decommissioned it."[73]

In the meantime, NASA technicians prepared to launch what became known as the Voyager probes. Even though the four-planet mission was known to be possible, it soon became too expensive to build a spacecraft that could go the distance, carry the instruments needed, and last long enough to accomplish such an extended mission. Thus, the two Voyager spacecraft were funded to conduct intensive flyby studies of Jupiter and Saturn only— in effect repeating on a more elaborate scale the flights of the two Pioneers. Nonetheless, the engineers designed as much longevity into the two Voyagers as the $865-million budget would allow. NASA launched them from the Kennedy Space Center, Florida: Voyager 2 lifted off on 20 August 1977, and Voyager 1 entered space on a faster, shorter trajectory on 5 September 1977. The three RTGs on the two Voyagers each weighed 56 kilograms, had a diameter of 42.2 centimeters, and a length of 114 centimeters. Like the SNAP-27 that served as a power source on the Moon during the Apollo mission, this RTG consisted of a cylindrical fuel supply surrounded by rings of thermocouples. Again, there were cooling fins attached to the cold shoes of the thermocouples. Using plutonium-238 as the fuel source, as in previous missions, these elements were shaped so that each pellet produced approximately 250 watts of thermal power.

The fuel modules were encased in a heat- and impact-resistant shell designed to prevent a vehicle accident from releasing plutonium. The testing on this power source showed that the RTG containers would remain intact even in a launch-vehicle explosion or a reentry accident.[74] As the mission

72. Teledyne Isotopes, "SNAP 19 Pioneer F & G Final Report," IESD 2873-172, DOE Report, June 1973, pp. I-1-I-5.

73. "Pioneer 10: Last Signal Sent from RTG-powered Spacecraft," *Nuclear News* (April 2003): 65–67.

74. Francis de Winter, Gerhard Stapfer, and Enrique Medina, "The Design of a Nuclear Power Supply with a 50 Year Life Expectancy: The JPL Voyager's SiGe MIIW RTG," paper presented at 1999 IECEC, Vancouver, British Columbia, Canada; Martin Marietta

progressed, having successfully accomplished all its objectives at Jupiter and Saturn by December 1980, additional flybys of the two outermost giant planets, Uranus and Neptune, proved possible—and irresistible—to mission scientists. Accordingly, as the two spacecraft flew across the solar system, remote-control reprogramming was used to redirect the Voyagers for the greater mission. Eventually Voyager 1 and Voyager 2 explored all the giant outer planets, 48 of their moons, and the unique systems of rings and magnetic fields those planets possess.[75]

The two spacecraft returned information to Earth that has revolutionized solar-system science, helping resolve some key questions while raising intriguing new ones about the origin and evolution of the planets. The two Voyagers took well over 100,000 images of the outer planets, rings, and satellites as well as millions of magnetic, chemical spectra, and radiation measurements. They discovered rings around Jupiter, volcanoes on Io, shepherding satellites in Saturn's rings, new moons around Uranus and Neptune, and geysers on Triton. The last imaging sequence was Voyager 1's portrait of most of the solar system, showing Earth and six other planets as sparks in a dark sky lit by a single bright star, the Sun. Now traveling out of the solar system in the early 21st century, Voyager 2 has reached the heliopause and sent back the first information ever received from the outer boundary of our solar neighborhood. It revealed that at a distance of 83.7 astronomical units the spacecraft had five encounters with the termination shock, something unexpected as it passed into interstellar space.[76]

As of 2015, Voyager has continued to return scientific data. Communications remain possible until its nuclear power sources can no longer supply enough electricity to power critical subsystems. Originally built to explore Jupiter and Saturn, Voyager 1 is today farther from Earth than any other human-made object and speeding outward at more than 38,000 miles per hour. Both spacecraft carry phonograph records (primitive DVDs) that contain sounds and images portraying the diversity of life and culture on Earth. Perhaps 40,000 years from now, when the Voyager spacecraft come within the vicinity of nearby stars, these records will be discovered and played by an intelligent alien being, if such exists. On the 22 April 1978 broadcast

---

Corporation, Denver Division, "Viking 75 Project Mission Planning Handbook," Contract No. NAS1-9000, June 1974, HB-3720438, JPL 49, Box 9, folder 126, JPL Archives, Pasadena, CA; and Langley Research Center, Viking Project Office, "Viking 75 Project Plan," n.d. [1975], M75-131-1, JPL 49, Box 3, folder 36, JPL Archives, Pasadena, CA.

75. Dethloff and Schorn, *Voyager's Grand Tour.*
76. "Scientific Exploration: What a Long, Strange Trip It's Been," *Nature* 454 (3 July 2008): 24–25; and Ron Cowen, "Postcards from the Edge," *Science News,* 5 July 2008, pp. 8–10.

of the television program *Saturday Night Live*, comedian Steve Martin breathlessly announced that extraterrestrials had found the record and sent back the message "Send more Chuck Berry."[77] Again, such success would not have resulted without the RTGs on each spacecraft providing electrical power.

One observer has called the 1970s the "golden age" of planetary science, perhaps a bit of an overstatement, but appropriate in certain ways in part because of the power capabilities of the RTGs placed on planetary probes.[78] Virtually every year in that decade brought the launch of at least one major planetary probe and the start of several others that were not launched until the late 1980s.[79] Indeed, 12 planetary probes launched during the 1970s visited all of the planets of the solar system, some landing on such bodies as Mars. The solar system exploration program of the 1970s was the stuff of legend and myth, in some measure because of its success. Yet it was also much more. It represented a rich harvest of knowledge about Earth's neighboring planets, a transformation of our understanding of the solar system's origin and evolution, and a demonstration of what might be accomplished using limited resources when focusing on scientific goals rather than large human spaceflight programs aimed at buttressing American prestige.[80]

6. Reconsideration and Retrenchment

From the very first conceptualization of space, nuclear power engineers worried about its safety. Even more than nuclear reactor power plants and submarines powered by nuclear reactors, the challenges of ensuring the safety of individuals in the event of catastrophe consumed the designers and builders of RTGs. The AEC used plutonium-238 as the fuel of choice for RTGs primarily because it emits "alpha" particles, known to be the least-penetrating type of radiation, incapable of supporting a chain reaction, and sustaining a

---

77. David Samuels, "Alien Tongues," in Debbora Battalia, ed., *E.T. Culture: Anthropology in Outerspaces* (Durham, NC: Duke University Press, 2005), pp. 112–113.

78. Robert S. Kraemer, *Beyond the Moon: A Golden Age of Planetary Exploration, 1971–1978* (Washington, DC: Smithsonian Institution Press, 2000).

79. There were launches in 1971, 1972, 1973, 1975, 1977, and 1978.

80. As Representative George E. Brown Jr. (D-CA) remarked in a speech to the National Academy of Sciences in 1992, "it is also important to recall that some of our proudest achievements in the space program have been accomplished within a stagnant, no-growth budget. The development of ... the Viking lander, Voyagers I and II, Pioneer Venus ... were all carried out during the 1970s, when the NASA budget was flat. It would be wise to review how we set priorities and managed programs during this productive time." (George E. Brown Jr., Remarks, 9 February 1992, copy in NASA Historical Reference Collection.)

long half-life.[81] This meant that the danger to living organisms rested with ingesting radioactive particles dispersed in the atmosphere should the capsule containing the fuel be breached in an explosion on launch or during a reentry. The key to safety therefore relied on redoubling efforts to ensure successful launches and hardening the containers in the event of catastrophe.

Extensive and ongoing tests by the AEC/DOE on a successive generation of plutonium-238 fuel capsules served to lessen this danger to the extent that nuclear space power's advocates have argued it had little risk. As one statement from an engineering firm working on this technology stated:

> The potential hazard is essentially zero. The fuel modules are unlikely to be bre[a]ched in any accident, but even if all of the coatings and containers were to fail, there is little chance that any person would consume enough material to cause any health consequences. Plutonium oxide is a dense and relatively nonreactive material; it is most likely that it would rapidly fall out of the air and sink to the bottom of the ocean.[82]

The Atomic Energy Commission, later the DOE, also enforced a rigorous process of reviews and approvals to obtain permission to launch an RTG on a mission. Its regulators forever questioned every aspect of the construction of the hardware, the safety of the transporting and handling, the placement of the RTGs on the spacecraft, the reliability of the launch vehicles, and the conduct of the mission as a whole.[83]

For the first decade and a half of space nuclear power, the public—even though it had an interest in the risk that RTGs and space nuclear reactors portended—did not register serious misgivings about the use of this technology in space. This changed rather dramatically in the later 1970s in response to two incidents, one directly bearing on space operations and the other a dramatic ground accident. On 24 January 1978, the Soviet Cosmos 954 reentered the atmosphere, spreading thousands of pieces of radioactive debris over more than 100,000 square kilometers of northwest Canada. A few of the recovered fragments showed a high degree of radioactivity. "The Cosmos 954 reactor included 110 pounds of highly enriched uranium 235," according to the *Time* story reporting on the incident. "This is a long-lived fuel whose 'half-life'—the time it takes for half the material to lose its radioactivity—is

---

81. "What Is Plutonium-238?" *APP RPS Pu-238 FS 1210-12.pdf.*
82. Adams Atomic Engines, Inc., "Cassini," *Atomic Insights* 2, no. 6 (1996), *http://www.atomicinsights.com/sep96/Cassini.html* (accessed 4 July 2008).
83. Engler, "Atomic Power in Space," pp. 55–62; James H. Lee and Dave Buden, "Aerospace Nuclear Safety: An Introduction and Historical Overview," paper presented at Advanced Reactor Safety Conference, Pittsburgh, PA, 17–21 April 1994.

an astonishing 713 million years."[84] These reports prompted U.S. President Jimmy Carter to propose a moratorium on the use of nuclear power for space-flight. "If we cannot evolve those fail-safe methods, then I think there ought to be a total prohibition against [nuclear-powered] Earth-orbiting satellites," he said. A permanent ban, of course, did not take place, but what did result was a more strict control regime that emerged in the aftermath of the accident, recompense for the government of Canada and its citizens, and a delay of more than a decade in the launch of RTGs on U.S. space probes, and then exclusively for outer planetary missions.[85]

The Cosmos 954 incident raised the consciousness of the public about the potential hazards of nuclear power in space. Coupled with the public's intense reaction to the serious accident at Unit 2 of the Three Mile Island nuclear power plant in Pennsylvania in March 1979, the Cosmos 954 event quickly eroded support for the use nuclear power in any setting. By October 1981, according to one study, a majority of Americans opposed nuclear power for the first time since the advent of the atomic age: "In fact, over the last four surveys spanning 7 years [through 1988], opposition has exceeded support by a margin of about 2:1, a complete flip-flop from the earliest Harris survey."[86]

The significance of the Three Mile Island accident to public perceptions of risk tied to the technology cannot be overestimated. Although most ana-lysts had believed prior to that accident that public perceptions of risk were related to serious loss of life and destruction of property, this accident defied the model. "Despite the fact that not a single person died, and few if any latent cancer fatalities are expected," wrote Paul Slovic in *Science* magazine, "no other accident in our history has produced such costly societal impacts." It stampeded the public toward more expensive and arguably more envi-ronmentally destructive power sources. It made virtually impossible the

---

84. "Cosmos 954: An Ugly Death," *Time*, 6 February 1978, *http://www.time.com/time/magazine/article/0,9171,945940-1,00.html* (accessed 4 July 2008).
85. Glenn H. Reynolds and Robert P. Merges, *Outer Space: Problems of Law and Policy* (Boulder, CO: Westview Press, 1998), pp. 179–186; W. K. Gummer, F. R. Campbell, G. B. Knight, and J. L. Richard, *Cosmos 954: The Occurrence and Nature of Recovered Debris* (Hull, Quebec: Canadian Government Publishing Center, INFO-0006, 1980), p. 11; Philip J. Klass, "Technical, Political Concerns Impede Space Nuclear Power," *Aviation Week & Space Technology*, 1 February 1988, p. 59; and Carl Q. Christol, "International Liability for Damage Caused by Space Objects," *American Journal of International Law* 74 (April 1980): 346–371.
86. Eugene A. Rosa and Riley E. Dunlap, "Poll Trends: Nuclear Power, Three Decades of Public Opinion," *Public Opinion Quarterly* 58, no. 2 (1994): 295–324, quotation from p. 296. See also Connie de Boer and Ineke Catsburg, "A Report: The Impact of Nuclear Accidents on Attitudes Toward Nuclear Energy," *Public Opinion Quarterly* 52, no. 2 (1988): 254–261.

continuation of the nuclear power capability of the nation and the advancement of that technology. "It may even have led to a more hostile view of other complex technologies, such as chemical manufacturing and genetic engineering," Slovic added. This increasing public concern was not mirrored in the scientific and technical communities, which contended that the risk was manageable.[87] For the next two decades this opposition to nuclear power would be manifested in direct confrontations with antinuclear activists on all launches of spacecraft using RTGs for on-board power.

7. Direct Resistance to the Use of RTGs: The Galileo and Cassini Missions

After the Three Mile Island accident the use of RTGs in space missions met direct opposition from the antinuclear community. Five missions employing nuclear material have been flown since that accident—Galileo (1989), Ulysses (1990), Mars Pathfinder (1996), Cassini (1997), and New Horizons (2006)—and all of them received some form of public opposition. Also, with the loss of Challenger in a fiery explosion on 28 January 1986, any probe with nuclear material to be deployed from the Space Shuttle received serious scrutiny from the public. The Galileo and Cassini missions were what most concerned antinuclear activists, and efforts to prevent both launches took extravagant turns. On 18 October 1989, NASA's Galileo spacecraft, again with RTGs to provide on-board power, began a gravity-assisted journey to Jupiter, where it sent a probe into the atmosphere and observed the planet and its satellites for several years beginning in December 1995.

Jupiter was of great interest to scientists because it appeared to contain material in an original state left over from the formation of the solar system, and the mission was designed to investigate the chemical composition and physical state of Jupiter's atmosphere and satellites. A mission in the planning since the late 1970s to be deployed from the Shuttle, after the loss of the Challenger, Galileo brought together government officials and the public to force a review of what was proposed. Representative Edward Markey (D-MA) persuaded the Department of Energy to release its risk analysis of the Galileo and Ulysses mission launches, which contained the disturbing conclusion that launch failures on those flights could result in between 202 and 386 cancer deaths, more than quintuple the national average.

The proposed use of the Centaur liquid hydrogen–liquid oxygen upper stage attached to the space probe as the vehicle that would propel Galileo

---

87. Paul Slovic, "Perception of Risk," *Science* 236, no. 4799 (1987): 280–285.

and Ulysses from the Shuttle's cargo bay on their journeys ensured that an explosion on launch would be more destructive than any ever experienced before. The post-Challenger accident probability estimates and the high volatility of the Centaur upper stages persuaded NASA Administrator James C. Fletcher to scrap plans to use Centaurs from the payload bay. This decision led to a reconsideration of the manner in which NASA would send Galileo on its way beyond Earth's orbit. An inertial upper stage (IUS), though much less powerful than the Centaur, was called upon to do the job of sending the spacecraft to Jupiter; but it would require the use of complex orbital mechanics, including flyby gravity assists of Venus and Earth, to reach Jupiter. The same process took place with Ulysses, a mission dedicated to solar astronomy, which deployed in 1990.[88]

NASA also considered replacing the two RTGs on the spacecraft with solar arrays because of the political issues associated with a launched nuclear-powered satellite in an environment of considerable public opposition. The project team eventually rejected this replacement because of several technical factors. As the study team for the alternate Galileo power system noted: "In view of the insurmountable mass and schedule difficulties associated with a solar retrofit of Galileo, the study team concluded that the only alternative to an RTG-powered Galileo mission would be to cancel the Galileo mission and design a completely new, solar-powered spacecraft for the late 1990s." Based on this conclusion, NASA pursued and eventually received permission to deploy the RTG-powered Galileo spacecraft from the payload bay of the Space Shuttle in 1989.[89]

The space agency's environmental impact statement analyzed the physical hazards of the mission and concluded that, as in all such space missions, the launch sequence held the most potential hazard to living things on the Earth's surface:

---

88. H. Josef Hebert, "Questions Raised About Launching Spaceships with Nuclear Payloads," Associated Press, 19 February 1986; Eliot Brenner, "Study Estimates Plutonium Risks from Shuttle Blast," United Press International, 18 March 1986; "Space Agency Readies Environmental Statement," United Press International, 23 November 1987, all in NASA Historical Reference Collection. See also Philip M. Boffey, "Nuclear Reactor Hazard Is New Concern in Shuttle Flights," *New York Times*, 18 March 1986, p. C1; Kathy Sawyer, "NASA Junks Rocket Plan Because of Safety Concerns," *Washington Post*, 20 June 1986, p. A1; and Virginia P. Dawson and Mark D. Bowles, *Taming Liquid Hydrogen: The Centaur Upper Stage Rocket, 1958–2002* (Washington, DC: NASA SP-2004-4230, 2004), pp. 207–212.

89. Jet Propulsion Laboratory, "Jupiter Solar Mission Alternative Feasibility Study," JPL D-6679, 15 August 1989, p. iii, NASA Center for Aerospace Information.

An intensive analysis of the proposed action indicates that the possible health and environmental consequences of launch or mission anomalies pose small risks…. The accident estimated to be most probable would pose very small health risks and very small probability of detectable environmental contamination. The maximum credible accident (having a probability of one in 10 million) would be an accidental reentry into the Earth's atmosphere during a planned VEEGA [Venus-Earth-Earth-Gravity Assist] flyby, releasing Pu238 upon impact with the ground. The very low probability "maximum case" would lead to an increase of an estimated 9.8 cancer fatalities over a 70-year period among a population of 83,000 persons, which normally would have an estimated 16,000 cancer fatalities over the same period.[90]

Antinuclear groups filed lawsuits to prevent the launch, alleging that the spacecraft's two RTGs posed an unacceptable risk to the residents of Florida and making a connection to the Challenger Space Shuttle accident in 1986 as an unacceptable worst-case scenario with nuclear material aboard. Such a potent carcinogen as plutonium-238, they argued, would render large areas of Florida uninhabitable and infect the bones and lungs of millions of people along the coast. Because the plutonium-238 would be encased in hardened graphite containers, NASA's engineers insisted that the risk was minimal, even in a Challenger-like explosion. Tests on those containers, they argued, had absorbed shocks and concussions more than 10 times as severe as a rocket explosion. The antinuclear activists refused to accept these arguments, noting that 3 of 22 U.S. missions with nuclear material aboard had failed, and asserting that the nation should not take that chance again. There the matter rested until adjudicated.

A centerpiece of antinuclear concern—and this may have been one of the driving forces in catalyzing opposition to the launch—was the unique mission profile of the Galileo probe. Because of Galileo's deployment from the Space Shuttle, it would be able to reach Jupiter only by using a gravity-assist trajectory that required it to pass close to Venus and have two swings past Earth before slingshotting it to Jupiter. This Venus–Earth–Earth Gravity Assist (VEEGA) mission profile was ingenious in many ways, and it allowed Galileo to encounter many interesting objects in space, including Venus, asteroid 243 Ida, and asteroid 951 Gaspra.[91] Concerned not only about the

---

90. NASA Office of Space Science, "Draft Environmental Impact Statement for the Galileo Mission (Tier 2)," December 1988, NASA TM-102926, NASA Center for Aerospace Information, pp. ii–iii.

91. Jet Propulsion Laboratory, "The Galileo Tour Guide," p. 3, *http://www2.jpl.nasa.gov/galileo/tour/2TOUR.pdf* (accessed 5 July 2008).

explosion of the Space Shuttle Challenger in 1986 and the possibility of such an accident with nuclear material aboard but also with this VEEGA trajectory, antinuclear activists redoubled efforts to prohibit the launch. What if the trajectory calculations were slightly off? The possibility for Galileo's uncontrolled reentry into the Earth's atmosphere on one of its flybys only added to larger concerns about the use of nuclear power in space. Protesters had a point, Carl Sagan allowed. Although a strong supporter of the Galileo mission, he weighed in with this opinion before the launch in 1989; he also allowed that "there is nothing absurd about either side of this argument."[92]

The lawsuit that went before Judge Oliver Gasch in the U.S. District Court in the District of Columbia arguing that NASA had violated the National Environmental Protection Act (NEPA) by failing to fully document the launch dangers in its environmental impact statement. Just two days before the scheduled launch on 12 October 1998, Judge Gasch rendered his decision on this case, ruling that NASA had fulfilled the letter of the law concerning NEPA and that the launch could move forward. "The court will not substitute its own judgment regarding the merits of the proposed action for that of the government agencies," he wrote. "NEPA itself does not mandate particular results, but simply prescribes the necessary process." Having followed the NEPA process, Gasch noted, NASA had appropriately discharged its responsibilities under the law, and he rejected the plaintiff's request for a restraining order, directing that NASA continue the Galileo launch on the Space Shuttle.

NASA had finally received permission to proceed; but just as this took place, the launch had to be delayed because of a technical malfunction on the Shuttle.[93] Delaying the launch allowed the antinuclear opposition time to file an appeal, but the U.S. Circuit Court of Appeals rejected this appeal on 16 October 1989. The launch was rescheduled for 18 October, and despite several protests at Kennedy Space Center in the days leading up to the launch,

---

92. Carl Sagan, "Benefit Outweighs Risk: Launch Galileo Craft," *USA Today*, 10 October 1989, inquiry page.

93. *Florida Coalition for Peace & Justice v. Bush*, Civil Action No. 89-2682-OG, United States District Court for the District of Columbia, 1989 U.S. Dist. Lexis 12003, filed and decided on 10 October 1989; W. Mitchell Waldrop, "Rifkin Tried To Stop Galileo Launch," *Science* 246 (6 October 1989): 30; W. Mitchell Waldrop, "Stop the Plutonium Shuttle!" *Science* 246 (22 September 1989): 1328; "The Lethal Shuttle," *Nation*, 22 February 1986, p. 16; "Launch Hazards of Atomic Satellites Studied," *Washington Post*, 19 May 1986, p. A11; "NASA Says a Nuclear Generator on Satellite Will Pose Little Risk," *New York Times*, 7 January 1989, p. 9; Jay Malin, "Galileo Survives Suit, but Mishap Delays Liftoff," *Washington Times*, 11 October 1989, p. A3; and Kathy Sawyer, "Faulty Engine Computer Delays Launch of Shuttle," *Washington Post*, 11 October 1989, p. A3.

the Shuttle finally launched. It deployed without incident once in Earth's orbit and began its lengthy journey to Jupiter.[94]

Because of a unique orbital inclination, Galileo passed both Venus and Earth and made the first close flyby of asteroid Gaspra in 1991, providing scientific data on all before reaching Jupiter in 1995. Until 2003, Galileo continued to transmit scientific measurements back to Earth for analysis. Galileo's mission has led to a reinterpretation of understanding about Jupiter, its moons, and the outer solar system. A short list of Galileo's most important discoveries includes the following:

- Evidence for liquid water ocean under the surface of Europa, one of the moons of Jupiter.
- Discovery of a satellite (Dactyl) circling the asteroid Ida.
- Discovery of an intense interplanetary dust storm (the most intense ever observed).
- Discovery of an intense new radiation belt approximately 31,000 miles above Jupiter's cloud tops.
- Detection of Jovian wind speeds in excess of 400 miles per hour.
- Detection of far less water in Jupiter's atmosphere than was previously estimated on the basis of earlier Voyager observations and models of the Comet Shoemaker-Levy 9 impact.
- Detection of far less lightning activity (about 10 percent of that found in an equal area on Earth) than anticipated. Individual lightning events, however, are about 10 times stronger on Jupiter than on Earth.
- Discovery of a nearly equal amount of helium in Jupiter as is found in the Sun (24 percent compared to 25 percent).
- Extensive resurfacing of Io's surface due to continuing volcanic activity since the Voyagers flew by in 1979.
- Preliminary data supporting the tentative identification of intrinsic magnetic fields for two of Jupiter's moons, Io and Ganymede.

The flight team for Galileo ceased operations in February 2003 after a final playback of scientific data from the robotic explorer's tape recorder. The team prepared commands for the spacecraft's on-board computer to manage the remainder of its life. Galileo coasted for the next seven months before taking a 21 September 2003 plunge into Jupiter's atmosphere as a means of ensuring

---

94. "NASA Clears Atlantis for Liftoff Tuesday," *Washington Post*, 15 October 1989, p. A18; "NASA Deploys Guards As Shuttle Launch Nears," *Washington Post*, 16 October 1989, p. A20; Kathy Sawyer, "Galileo Launch Nears; Court Appeal Rejected," *Washington Post*, 17 October 1989, p. A3; and Launch to Landing," *Galileo Messenger*, November 1989, pp. 1–2.

that its nuclear propellant did not cause any mischief in the future, thereby ending what turned out to be a remarkably successful mission.[95]

The Cassini space probe—the largest interplanetary probe ever launched, weighing 6.3 tons and extending 22 feet in length—was a joint NASA, European Space Agency (ESA), and Italian Space Agency (ASI) mission to study Saturn and its rings, moons, and magnetic environment. Launched on 17 October 1997, atop a Titan IV rocket from Cape Canaveral, Florida, Cassini required three RTGs with 72 pounds of plutonium-238 to power a wide array of scientific instruments at Saturn. Like Galileo, although without the Space Shuttle as the Earth-launch vehicle, Cassini would require gravity-assist to reach Saturn in 6.7 years. It followed a Venus–Venus–Earth–Jupiter Gravity Assist (VVEJGA) trajectory that energized the antinuclear community as had nothing since the Galileo launch. Cassini's three General Purpose Heat Source (GPHS) RTGs and 117 lightweight radioisotope heater units (RHUs) provided the necessary electrical power to operate its 19 instruments and maintain the temperatures of critical components as well as the Huygens probe that was destined for deployment by parachute onto the surface of Titan, Saturn's largest moon (see figures 5.7, 5.8, and 5.9). Those three RTGs provided 888 watts of electrical power at mission beginning but would still generate 596 watts after 16 years of operation. As always, Cassini's RTGs were tested extensively to ensure that they could withstand any conceivable destructive force associated with the flight. Also, as had been the practice for many years, independent safety analyses by General Electric, Lockheed Martin, and other technical organizations assessed possible risks from prelaunch fires and explosions, launch accidents, and spacecraft crashes and uncontrolled reentry. Three major reports resulted from those efforts, with the final one prepared a year in advance of the projected launch.[96]

This material, along with additional studies by the Department of Energy and NASA, went to an independent Interagency Nuclear Safety Review Panel responsible for judging whether to recommend a decision in favor of

---

95. Several books on the Galileo mission have been published: Michael Meltzer, *Mission to Jupiter: A History of the Galileo Project* (Washington, DC: NASA SP-2007-4231, 2007); David M. Harland, *Jupiter Odyssey: The Story of NASA's Galileo Mission* (Chichester, UK: Springer-Praxis, 2000); Daniel Fischer, *Mission Jupiter: The Spectacular Journey of the Galileo Spacecraft* (New York: Copernicus Books, 2001); and Cesare Barbieri, Jürgen H. Rahe, Torrence V. Johnson, and Anita M. Sohus, eds., *The Three Galileos: The Man, the Spacecraft, the Telescope* (Dordrecht, The Netherlands: Kluwer Academic Publishers, 1997).

96. Michael V. Frank, "Analysis of Inadvertent Reentry of Cassini RTG's," American Institute of Physics 1-56396-846-0/99, 1999; Chris Barret, "So, What's an RTG and Are They Safe?," unpublished paper in possession of author; and NASA Fact Sheet, "Spacecraft Power for Cassini," July 1999, NASA Historical Reference Collection.

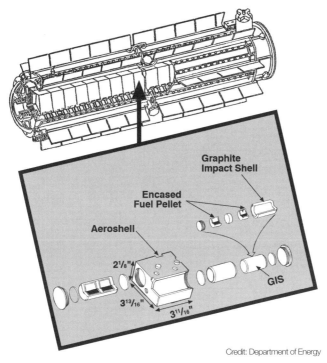

Credit: Department of Energy

FIGURE 5.7. The General Purpose Heat Source (GPHS) has been a mainstay design of the planetary science program since the 1970s. Each of the RTG's 18 modules contained four plutonium-238 fuel pellets, the graphite shell, and the aeroshell.

launch to the President of the United States. As a GAO audit of the Cassini mission documented:

> The processes used by NASA to assess the safety and environmental risks associated with the Cassini mission reflected the extensive analysis and evaluation requirements established in federal laws, regulations, and Executive branch policies. For example, DOE designed and tested the RTGs to withstand likely accidents while preventing or minimizing the release of the RTG's plutonium dioxide fuel, and a DOE administrative order required the agency to estimate the safety risks associated with the RTGs used for the Cassini mission. Also, federal regulations implementing the National Environmental Policy Act of 1969 required NASA to assess the environmental and public health impacts of potential accidents during the Cassini mission that could cause plutonium dioxide to be released from the spacecraft's RTGs or heater units. In addition, a directive issued by the Executive Office of the President requires an ad hoc interagency Nuclear Safety Review Panel. This panel is supported by technical experts from NASA, other federal agencies, national laboratories, and

Credit: NASA

FIGURE 5.8. Lockheed Martin Missile and Space Co. employees Joe Collingwood, at right, and Ken Dickinson retract pins in the storage base to release an RTG in preparation for hoisting operations on 19 July 1997. This RTG and two others were installed on the Cassini spacecraft for mechanical and electrical verification testing in the Payload Hazardous Servicing Facility.

academia to review the nuclear safety analyses prepared for the Cassini mission. After completion of the interagency review process, NASA requested and was given nuclear launch safety approval by the Office of Science and Technology Policy, within the Office of the President, to launch the Cassini spacecraft.[97]

---

97. GAO Report, "Space Exploration," pp. 3–4.

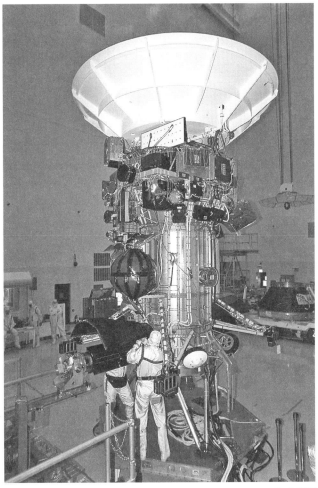

Credit: NASA

FIGURE 5.9. Jet Propulsion Laboratory workers Dan Maynard and John Shuping prepare to install an RTG on the Cassini spacecraft in the Payload Hazardous Servicing Facility (PHSF) on 18 July 1997. The three RTGs that provide electrical power to Cassini on its mission to the Saturnian system underwent mechanical and electrical verification testing in the PHSF.

This detailed and involved process led to the conclusion that, although risk could not be eliminated entirely, the chances of any breech of the plutonium-238 container were exceptionally low. The estimated health effect of an accident was that over a 50-year period not one more person would die of cancer caused by radiation exposure than if there were no accident. These analyses also found that during Cassini's Earth encounter the chance that the

FIGURE 5.10. Environmental health specialist Jamie A. Keeley, of EG&G Florida, Inc., uses an ion chamber dose rate meter to measure radiation levels in one of three RTGs that will provide electrical power to the Cassini spacecraft on its mission to explore the Saturnian system prior to its launch in 1997. The three RTGs and one spare are being tested and monitored in the Radioisotope Thermoelectric Generator Storage Building in KSC's Industrial Area. The RTGs on Cassini are of the same design as those flown on the Galileo and Ulysses spacecraft.

vehicle would accidentally reenter Earth's atmosphere was less than one in a million (see figure 5.10).[98]

None of this review convinced the antinuclear community, however, and it mobilized to prohibit the Cassini launch. The well-organized STOP CASSINI! campaign rested its opposition on a set of charges different from those leveled by the earlier Galileo protesters; it claimed that NASA's technical

---

98. Lisa Herrera, "U.S. Department of Energy Advanced RTG Program," IECEC-98-246, paper presented at the 1998 IECEC conference, Albuquerque, NM; Robert J. Noble, "Radioisotope Electric Propulsion," *Nuclear News*, November 1999; C. Edward Kelly and Paul M. Klee, "Cassini RTG Acceptance Test Results and RTG Performance on Galileo and Ulysses," and NASA, "Cassini Launch Press Kit," October 1997, NASA Historical Reference Collection, pp. 65–69.

risk assessment omitted, neglected, or underestimated the welfare of the public as a whole. Accepting, as the Galileo opponents had not, that NASA had fulfilled the letter of the law, this campaign asserted that the government as a whole had to be redirected away from the use of nuclear power or weaponry in any form whatsoever. Sociologist Jürgen Habermas has suggested that when the "instrumental rationality" of the bureaucratic state intrudes too precipitously into the "lifeworld" of its citizenry, they rise up in some form to correct the state's course or to cast it off altogether. The "lifeworld" is evident in the ways in which language creates the contexts of interpretations of everyday circumstances, decisions, and actions. Habermas argues that the "lifeworld" is "represented by a culturally transmitted and linguistically organized stock of interpretive patterns."[99]

The STOP CASSINI! campaign represented an effort to exile nuclear material from the "lifeworld" of modern America, as protesters expressions of discontent demonstrated, and they could obtain no resolution from the "instrumental rationality" residing in the state. They took direct action and justified it, without a tinge of conscience, as necessary for the greater good. Opponents of Cassini organized a rally of about 1,500 participants at Cape Canaveral in May 1997, with several prominent disarmament leaders speaking. They received publicity from CNN and the NBC local affiliate as well as from print journalists and radio stations. They argued for greater involvement in choosing the technologies used on spacecraft, specifically nuclear power sources. They tried to sensitize the public to dangers from the use of nuclear power for space exploration and addressed not only environmental risks but also the motives behind the reason for using nuclear power.

One protester commented:

> The military has made an unholy alliance with NASA in its quest for space domination. Now people-power and a commitment to compassion and conscience must be brought into an area where it is not wanted and where it is lacking. There must be resistance to the U.S. push to weaponize and nuclearize space ... a renegade government spending massive amounts of money to weaponize and nuclearize space, and at the same time saying that no money is available for schools and other social needs. This issue is not about losing our democracy—we have lost it.[100]

---

99. Jürgen Habermas, *The Theory of Communicative Action*, Vol. 2, *Lifeworld and System: A Critique of Functionalist Reason* (Boston: Beacon Press, 1987), p. 124.
100. The protester is quoted in Victoria Pidgeon Friedensen, "Protest Space: A Study of Technology Choice, Perception of Risk, and Space Exploration," M.S. thesis, Virginia Polytechnic Institute and State University, 1998, pp. 49, 90–94.

The STOP CASSINI! campaign received coverage from many of the major U.S. news outlets, and the Internet buzzed with discussion of its efforts to end the Cassini mission. It deserved credit for gaining the attention of several members of Congress, who demanded additional analysis from NASA and the DOE. Courtroom proceedings, in comparison to those for Galileo, were virtually nonexistent.

When Cassini launched safely on 14 October 1997, the press gave credit to the protesters for forcing NASA to reconsider its use of nuclear power in space and to undertake more extensive testing and verification of systems. A vigil outside the main gate of Kennedy Space Center by the STOP CASSINI! campaign was peaceful. It had raised important questions about this technology and its meaning for society. As one scholar noted, NASA responded poorly to this challenge in terms of public communication. The Agency believed that more information would resolve the crisis, but there is little reason to suppose that this would be the case, as the protest had more to do with ideology and values than with assessments of objective knowledge.[101]

In the end the Cassini mission has been conducted with stunning success. Cassini is the first spacecraft to orbit Saturn, beginning 1 July 2004, and to send a probe (Huygens), launched on 15 January 2005, to the surface of Saturn's moon Titan. But even before its Saturnian encounter, the Cassini mission advanced science by finding individual storm cells of upwelling bright-white clouds in dark "belts" in Jupiter's atmosphere and by conducting a radio signal experiment on 10 October 2003 that validated Einstein's theory of general relativity. At Saturn, Cassini has discovered three new moons (Methone, Pallene, and Polydeuces), observed water-ice geysers erupting from the south pole of the moon Enceladus, obtained images appearing to show lakes of liquid hydrocarbon (such as methane and ethane) in Titan's northern latitudes, and discovered a storm at the south pole of Saturn with a distinct eye wall. Cassini, like Galileo at Jupiter, has demonstrated that icy moons orbiting gas giant planets are potential refuges of life and attractive destinations for a new phase of robotic planetary exploration.[102]

---

101. Ibid., p. 93.
102. Roger Guillemette, "New and Improved Titan Launch Clears Path for Cassini," *Quest: The Magazine of Spaceflight* 5, no. 4 (1996): 38; David M. Harland, *Mission to Saturn: Cassini and the Huygens Probe* (Chichester, UK: Springer-Praxis, 2003); Laura Lovett, Joan Horvath, and Jeff Cuzzi, *Saturn: A New View* (New York: Harry N. Abrams, 2006); Robert T. Mitchell, "The Cassini Mission at Saturn," *Acta Astronautica* 61, no. 1 (2007): 37–43; and Robert T. Mitchell, "The Cassini Mission Exploring Saturn," *Acta Astronautica* 63, nos. 1–4 (2008): 61–67.

8. Conclusion

The use of radioisotope thermoelectric generators to power spacecraft to the outer planets has proven a boon to the space program in the first 50 years of its history. Yet use of this technology invites opposition because of the danger inherent in any launch or reentry to Earth's atmosphere. There have been failures in the past, duly taken notice of by the public, and in each instance refinements and additional requirements to ensure future safety have resulted. This is as it should be. The issue had receded so far from public consciousness by 2006—in no small measure because of the success of the safety program, the efforts to ensure public understanding of how mission surety was undertaken, and the rarity of the use of RTGs—that the New Horizons spacecraft launched to Pluto and the Kuiper Belt in the outer solar system did not receive much public opposition (see figure 5.11). From a societal perspective, however, the protests to past launches raised important questions that remain at the center of the debate. As the scholar Victoria Friedensen has commented:

1. NASA provided inadequate assessment that did not include multi-failure mode testing and had inadequate explanations of the risk.

2. The risk to the citizens and visitors to the region around the Kennedy Space Center, while low in probability, had very high consequence. Nor was the liability to be borne by NASA alone—the public would bear the costs of the consequences. NASA was not perceived as trustworthy enough to prevent accidents.

3. The risk to the global population was untenable for a scientific project. The protestors did not feel that the United States was truly responsible for the lives and well-being of all humans. No one asked global consent before increasing global risk.

4. The potential destructive capabilities that humans have created on Earth must not be carried into the future or onto other planets. The protestors based their opposition on strictly moral, strictly future terms and objected to NASA's counter proposition. An incommensurability of worldviews fueled the controversy.[103]

Such concerns are entirely appropriate and require additional consideration in the future as humanity seeks to extend its presence throughout the solar system. The story of RTGs and space nuclear power is thus one of technological advance and concern from certain segments of society that the

---

103. Friedensen, "Protest Space," pp. 90–91.

Credit: NASA

FIGURE 5.11. New Horizons is the most recent U.S. spacecraft to be powered by RTGs, depicted here in an artist's conception.

consequences of that technological advance might be too great a burden for the public to bear.

Physician and policy analyst Daniel Sarewitz points up the problem in the larger framework of U.S. government science and technology decision-making:

> At present, most citizens have only two options for involving themselves in decision-making about science and technology—the diffuse mechanisms of voting, and the direct but often unmediated local action that is commonly associated with not-in-my-backyard sentiments. A middle ground that enhances opportunities for public participation, while also providing mechanisms for technical input and open dialogue between scientists and the laity remains to be defined.... It does depend on the creation of avenues by which the public judgment can be brought to bear on important issues of science and technology policy, and on granting the public a stake in decision-making processes. The policy goal is not to substitute "common sense" for technical knowledge but to allow democratic dialogue to play its appropriate role in

decision-making that is inevitably dominated not by authoritative data but by subjective criteria and social norms.[104]

Advancing the future of nuclear power for space exploration remains a task not without difficulties. In the aftermath of the controversies over Galileo and Cassini, without significant social input and conscientious efforts to involve a broad constituency, more aggressive efforts will probably have considerable difficulty getting past the stage of paper studies. As it stands at present, the continued use of nuclear power for spacecraft must remain, to adopt a phrase offered by President Bill Clinton concerning the legality of abortion, "safe, legal, and rare."

---

104. Daniel Sarewitz, *Frontiers of Illusion: Science, Technology, and the Politics of Progress* (Philadelphia: Temple University Press, 1996), pp. 182–183.

Chapter 6

NASA AND THE ENVIRONMENT: AN EVOLVING RELATIONSHIP

W. Henry Lambright

The advent of the Space Age has paralleled the rise of the environmental movement. NASA was born in 1958, a year after Sputnik. Although the environmental movement can be traced back many years before the Space Age, it began its modern incarnation in 1962 with Rachel Carson's *Silent Spring*. That book took environmentalism well beyond its traditional conservation thrust and linked it with pollution and the negative impacts of industrial civilization. NASA, as a scientific and technological organization, has intersected with the environmental movement in at least three ways since its formation. First, it had direct impacts through the images of satellites viewing Earth. These environmental satellites have pioneered remote sensing technology. This mission *to* Earth has aimed to inform, improve, and protect the global and regional environment. Protection encompassed early warnings of environmental insults caused by humans or disasters from Mother Nature. Second, indirectly, through its mission *from* Earth, NASA has provided a new understanding of the home planet by the study of other celestial bodies. Comparative planetology is a new field that has come into existence. Third, some environmentalists perceive NASA as having potentially negative impacts on the environment of Earth or other bodies in the solar system and universe.

These three points of intersection do not exhaust all the possible ways NASA and the environment connect, but they are central. They are key aspects of a very large, complex, and diffuse subject. Space technology has clearly had revolutionary impacts. Revolutionary technology—rockets and the spacecraft they propel—have had implications that have cascaded through society over the years. The environment is only one of the impacts of space technology, and NASA only one of the agencies responsible for space technology. The National Oceanic and Atmospheric Administration (NOAA) and the Department of Defense (DOD) have space programs. But NASA has been especially critical owing to its role as a civilian research and development agency responsible for new technology. From its very outset as an agency, when NASA launched meteorological satellites, it actively entered the environmental field at its technological frontier, although it did not articulate that fact at the time. Two decades later, its environmental role had advanced to the point where it publicly proclaimed its "Mission to Planet Earth." In the early twenty-first century, NASA Administrator Sean O'Keefe, in listing NASA's prime purposes, spoke of its responsibility to assist "the home planet."[1] In viewing Earth as a whole, NASA has helped create a new "Earth system science" that integrates the Earth sciences with space technology to enhance understanding and prediction of major geophysical events, from hurricanes to global warming.

But NASA's Earth-oriented mission is only one way the space agency has influenced the environment. The "environmental movement," as it is called, is more than a set of interest groups, individuals, and institutions. It is a set of values, an attitude, and a way of thinking. In understanding NASA's environmental role, it is essential to see how its technology has enabled scientists studying other planets to better understand this one. Space technology opened up the solar system for comparison. Long before "global warming" on Earth became a household word, scientists like the late Carl Sagan were pointing to a "runaway greenhouse" effect on Venus. Astronauts on the Moon compared its desolation to the plenitude of life on Earth. Comparison made many observers realize the apparent uniqueness of the home planet and its vulnerability.

The environmental movement, from Rachel Carson on, has called attention to the negative features of technology, and this feature of the movement has not omitted NASA. NASA is a source of environmental threat in the minds of some because of its space mission. By sending probes to Mars and other planets, some fear contaminating them with Earthly organisms. By returning them (and people) to Earth, some in the environmental movement fear bringing an "Andromeda Strain" back to Earth that will wreck the home planet and wipe out its denizens. Also, some environmentalists note the need of NASA to power spacecraft with nuclear batteries—and perhaps someday with nuclear reactors. There have been acrimonious protests on environmental grounds aimed at stopping certain spaceflights. Perhaps the most serious threat comes from the pollution of the near-Earth space by debris. NASA is one of many space agencies in the world. The space debris problem is real and

---

1. See O'Keefe's address at Syracuse University's Maxwell School of Citizenship and Public Affairs, 12 April 2002, available from *http://www.hq.nasa.gov/office/codez/plans/Pioneer.pdf* (accessed 8 August 2013). The phrase "To understand and protect our home planet" became part of NASA's mission statement that year.

will only grow with time. Such debris threatens space assets in orbit, including the International Space Station (ISS). Some of this debris occasionally falls down to Earth. Space debris has not attracted much attention outside specialist circles, but it may be regarded as an emergent environmental issue.

Like any revolutionary technology, spaceflight has positive and negative impacts. Most of the impacts thus far have been strongly positive. However, NASA's Earth-oriented mission is by no means noncontroversial. Knowledge has impacts on policy. Environmental policy affects institutions and people, sometimes requiring a change in behavior. When policy is controversial, the science behind it can also be an object of political conflict. This has been increasingly the case recently with global warming, perhaps the most contested subject in environmental policy. Ironically, global warming has made NASA controversial with conservatives who oppose emissions controls based on scientific discovery. But it has likewise made NASA controversial with environmentalists who see the administration using research to delay policy when what is needed is regulatory "action."

This chapter seeks not to praise or condemn NASA but to track the history of NASA's intersection with the environmental movement since its earliest days to the present. As noted, there are three dominant themes in this history: mission to Earth, comparative planetology, and possible negative impacts of space technology. The mission to Earth via environmental satellites is by far the most coherent and sustained story. It is a centerpiece of this chapter. However, the other two components are also critical to the intersection. The three themes overlap and influence one another over time. This chapter emphasizes how a space agency evolves an environmental mission and role and contributes positively or negatively to a larger environmental consciousness. Also, because NASA is a research and development (R&D) agency, it has to find ways to relate to operating agencies that use its environmental information—and that relation can be problematic. These issues notwithstanding, the space agency has been a catalyst for environmental thinking and policy over the years. It is an important role likely to continue and could even grow in the future.

The Apollo Years: Beginnings

The Advent of Environmental Satellites

In its early years, especially when it went to the Moon, NASA did not have an overt "environmental" mission. It did have an "applications program." NASA's initial environmental thrust began with its applications program. In 1960, NASA launched the Television Infrared Observation Satellite, or

TIROS, a weather satellite. For the first time, meteorologists saw weather patterns from space, patterns on a huge scale, early information, and the journeys of giant storms over vast distances. They immediately saw how such images would enable improved weather forecasts. In 1961, a TIROS satellite helped track an extremely dangerous hurricane, Carla, which was approaching the Gulf Coast. Early warning led to the evacuation of more than 350,000 people. NASA and the Weather Bureau subsequently worked out a relationship whereby NASA developed new satellites and the Weather Bureau took control of the satellites once they became operational. In 1963, NASA Administrator James Webb reorganized NASA. He created a new Office of Space Science and Applications (OSSA). In his view, the linkage of space science and applications was critical. As NASA did more in space, it would spin off useful applications to Earth. The applications could in turn show politicians and the public why space science was worthy of support.[2]

NASA in the mid-1960s developed Nimbus, a next-generation weather satellite. However, the Weather Bureau found Nimbus more expensive than it could afford. This refusal to accept Nimbus generated a controversy between NASA and the Weather Bureau, embroiling the Defense Department, another user of weather information. The White House came in as an arbitrator. The outcome was an agreement by NASA to upgrade TIROS for the Weather Bureau and retain Nimbus for research purposes. The more advanced Nimbus class of weather satellites was of more interest to scientists than to Weather Bureau operators. Scientists were a different kind of user group, and they saw potential in space satellite observations to learn more about the upper atmosphere in new ways. Atmospheric research became a major focus for NASA as a result, and NASA gradually became involved in understanding climate change. At this point, climate change did not stand for global warming through pollution. The public policy issue was largely in the future, but space technology was beginning to make a difference in this complex area. Thus, weather satellite applications could be for practical and research objectives. They could have dramatic impact in hurricane warning and showed their potential a number of times. For example, in 1969, the extremely disastrous Hurricane Camille was spotted and tracked by weather satellites. Scores of people evacuated who might otherwise have lost their lives.

Also during the late 1960s, NASA developed the first satellites to study land resources. Possible user agencies (the Department of the Interior, the Department of Agriculture) saw potential for them for application to their

---

2.   Homer Newell, *Beyond the Atmosphere: Early Years of Space Science* (Washington, DC: NASA SP-4211, 1980), p. 321. This section on the 1960s is based on Newell's chapter 19, "Space Science and Practical Applications."

missions. They exerted a pull on NASA for further technology development and application. NASA stated that such Earth resource satellites were still experimental and would be ready for operational use in the early 1970s.

Space Science

Space science, however, soon showed how it could also have broad implications for the environment along with applications satellites. In 1962, the first spacecraft successfully to view another planet, Mariner, found Venus incredibly hot. Carl Sagan said this condition was the result of a "runaway greenhouse" effect. Subsequent probes in the decade confirmed this view. The hostility of Venus to anything resembling life moved NASA, OSSA, and the emerging planetary science community to focus ever more on what was always their top priority—Mars. The fabled "Red Planet," the subject of so much speculation for so many years, was now within reach of robotic space probes.[3]

Throughout the remainder of the decade, successive Mariner missions were sent to view Mars from a distance. An atmosphere was detected, but it was extremely thin and would be hostile to humanlike creatures. However, astronomers and other scientists were elated with their view of the other planets. A new field—comparative planetology—evolved, one that could compare and contrast atmospheres of different planets. Understanding planets required more than astronomers, and old disciplines—geology, meteorology, chemistry, and physics—were mutually useful in interpreting other planets. NASA Administrator Webb encouraged the building of academic "space science" as an interdisciplinary enterprise. He insisted that the largest NASA university grants, which sometimes included new "space" buildings, house interdisciplinary teams. He even wanted social scientists and humanists involved in considering the societal implications of space.

Human Spaceflight

Far and away, the top priority of NASA in the 1960s was human spaceflight. Apollo defined NASA, but even Apollo had an environmental dimension. Scientists associated with Apollo worried about the threat of bacterial contamination. Although there were those who were concerned about "forward contamination"—the bringing of bacterial life to the Moon—the major concern of Apollo scientists was bringing back to Earth a humanity-destroying "Andromeda Strain." NASA began sterilizing robotic surveyors sent to the

---

3. Spencer Weart, *The Discovery of Global Warming* (Cambridge, MA: Harvard University Press, 2003), pp. 87–89.

Moon to look for landing sites. There was a debate among technical people about whether and how much sterilization was enough to prevent forward contamination. Decisions were made to have contamination facilities created for Moon samples (rocks, soil, and so on) brought back by astronauts, and even for the astronauts themselves.

A field called exobiology emerged along with comparative planetology.[4] Supported by OSSA, this field was mainly concerned with discovery of extra-terrestrial life, and exobiologists especially worried about the contamination issue. Scientists were fully conscious that the Soviet Union was a potential polluter, as was the United States. They lobbied for international action. In 1967, when the Outer Space Treaty was promulgated, article IX included this statement: "Parties to the Treaty shall pursue studies of outer space including the Moon and other celestial bodies, and conduct exploration of them so as to avoid their harmful contamination and also adverse changes in the environment of the Earth resulting from the introduction of extraterrestrial matter and, where necessary, shall adopt appropriate measures for this purpose."[5] The Outer Space Treaty was signed by many nations, including the United States and the USSR.

The climax of the 1960s came in 1969 with Apollo 11, when men actually landed on the Moon and were returned safely to Earth. However, there was a remarkable prelude to Apollo 11 in Apollo 8, in 1968. This flight in particular underlined the connection of NASA to the environmental movement, which was coming on strong at the same time.

The Inspiration of Apollo's Images

Apollo 8 was the first piloted flight to and around the Moon. Coming soon after the Apollo fire that had killed three astronauts in 1967, Apollo 8 was an astounding feat of skill, technology, and courage. It was also inspirational because of what the astronauts did. On Christmas Eve, the three Apollo 8 astronauts came around the Moon and turned on their television cameras so millions could join them in viewing the Moon below. They read from the book of Genesis and wished the people of Earth a merry Christmas. One of the astronauts, Jim Lovell, drew the Moon-Earth comparison graphically: "The vast loneliness of the Moon up here is awe-inspiring. It makes you realize

---

4. See Steven Dick and James Strick, *The Living Universe: NASA and the Development of Astrobiology* (New Brunswick, NJ: Rutgers, 2004), for the origins of planetary protection, exobiology, and so on, especially chapter 3.

5. "Treaty on Principles Governing the Activities of States in the Exploration and Use of Outer Space, Including the Moon and Other Celestial Bodies," signed at Washington, DC; London; and Moscow, 27 January 1967.

just what you have back there on Earth."[6] For the first time ever, human beings were orbiting a celestial body far away—250,000 miles—from their home planet. They saw Earth as a whole. Another of the Apollo 8 astronauts, Bill Anders, took hundreds of photographs of Earth from the Apollo spacecraft. These photos revealed the sheer beauty of the blue Earth in the desolate blackness of space. Then came Apollo 9, 10, and—finally—Apollo 11. On 20 July 1969, Neil Armstrong climbed down the ladder of his landing craft, placed his foot on the Moon, and took "one small step for [a] man, one giant leap for mankind." It was an epochal moment in human history, and a large portion of the planet's population watched in rapt silence. As with Apollo 8, the astronauts took pictures of Earth. These pictures, especially "Earthrise" from the Moon's horizon, soon became icons of the environmental movement.[7]

Many environmental leaders used the view from space to galvanize public interest for their cause. Stuart Brand, founder and publisher of the *Whole Earth Catalog*, was one. His catalog, begun in 1968, emphasized environmental values, "tools" for environmentally friendly living, and an "Earth consciousness."[8] Another advocate of the "Whole Earth" view was James Lovelock. Working in the late 1960s with NASA's Caltech-based Jet Propulsion Laboratory (JPL), Lovelock began conceptualizing Earth as an evolving, self-regulating system. He called this notion his "Gaia Hypothesis," which held that Earth was a living system, with all parts interacting in a mutually supportive way. Gaia was the name of the ancient Greek Earth goddess.[9]

In late 1972, an Apollo 17 astronaut snapped a picture of the entire Earth. This photo quickly replaced "Earthrise" as the chief image of Earth from space for the environmental movement. According to geographer Neil Mahar, this second image was visual evidence of a need to think of space not as a frontier but as a place within which Earth had its own special identity and human beings everywhere on the planet were interdependent with one another and with nature. This image fit the "Whole Earth" and "living Earth" mood of the times. Environmentalists used the image to bring home the view of Earth as a vulnerable, fragile planet. "Think globally, act locally" became the call to arms.[10] It was ironic that a prime spinoff of Apollo was a new appreciation of Earth.

---

6. Lovell as quoted in Marina Benjamin, *Rocket Dreams* (New York: Free Press, 2003), p. 47.

7. Ibid., p. 49.

8. Benjamin, *Rocket Dreams*, p. 129; Neil Maher, "Gallery: Shooting the Moon," *Environmental History* (July 2004): 527–531.

9. Dick and Strick, *Living Universe*, p. 49.

10. Maher, "Gallery."

The 1970s: NASA in an Environmental Decade

As Apollo journeys to the Moon ended with Apollo 17, the new decade brought the environment to the fore as a national policy priority. The decade began with the creation of the Environmental Protection Agency (EPA) in 1970—an agency embodying the environmental ethic. At the same time, the National Oceanic and Atmospheric Administration (NOAA) was also created. While encompassing the former Weather Bureau, NOAA promised also to give priority to larger ocean and atmospheric issues. Also in 1970, the first Earth Day took place. One of the photographs taken by Apollo 8's crew was used as the flag of the celebration. Earth Day gave way to a major international conference in Stockholm, Sweden. This in turn led to the formation of a United Nations Environmental Program (UNEP). UNEP was intended to provide a global forum for environmental concerns that cut across national boundaries. International concerns included desertification, deforestation, ocean pollution, and transboundary air pollution.

In 1973–74, environmental issues took a new turn as the Arab oil boycott caused energy emergencies to become an overriding challenge. As the United States and other nations coped with energy emergencies, certain environment-friendly technologies, such as wind power, solar energy, and conservation, came to the fore. Space seemed far removed from the era's pressing challenges to the nation. NASA's relevance was in question. Initially, NASA's priorities and those of the nation were at variance. The man who had succeeded Webb as NASA Administrator, and to whom fell the task of selling to President Richard Nixon a post–Apollo program, was Tom Paine. A visionary, Paine wanted to go beyond the Moon to Mars. He wanted a space station and Moon bases. He wanted to build on—not retreat from—the bold program of the 1960s. He used the term "swashbuckling" to describe his view of NASA's spacefaring role. In stressing large-scale exploration, Paine was out of sync with Nixon and congressional sentiment, however. Frustrated, he left and was succeeded by Administrator James Fletcher in April 1971.[11]

The Environmental Administrator

Fletcher recognized there was no hope in selling the kind of space program Paine wanted. The only piloted program Nixon would support was the minimal program that would keep human spaceflight viable. In 1972, Fletcher persuaded Nixon to adopt the Space Shuttle as NASA's flagship for the 1970s. The Shuttle maintained the NASA human spaceflight program as Apollo

---

11. Newell, *Beyond the Atmosphere*, pp. 288–289.

phased down and out.[12] In 1973, Skylab, a mini space station, was launched using mainly Apollo parts and a Saturn V rocket. Skylab astronauts looked outward at the universe and also down at Earth. The Skylab missions continued briefly and then ended.

What else would NASA do? Fletcher had to chart the Agency's course in a way that fit the times. Homer Newell, who worked for Fletcher, described his leader's style as conveying "moderation and cost consciousness." Fletcher sought to project "an image of applying space knowledge and capabilities to problems of concern to the man on the ground, and to do it economically." The Space Shuttle fit this approach in that it was intended to replace the expensive Saturn and other expendable rockets with a cost-effective, reusable vehicle. Fletcher sold the Shuttle as a workhorse that could put large satellites into orbit for "users"—including the Defense Department. The original purpose—transit to and from the space station—was dropped, since the space station was not part of the Nixon decision.[13]

Fletcher especially saw in NASA's applications satellites a way to relate NASA to the new federal priorities, especially the environment. The space historian Roger Launius has written that this practical and societally conscious attitude came easily to Fletcher. According to Launius, Fletcher's interest in the environment was genuine and derived from his Mormon roots and its "stewardship" principles. He explicitly connected NASA with preservation and restoration of the environment. In congressional testimony in 1973, Fletcher declared: "NASA is called the Space Agency, but in a broader sense, we could be called an Environmental Agency…. Everything we do … helps in some practical way to improve the environment of our planet and helps us understand the forces that affect it. Perhaps that is our essential task, to study and understand the Earth and its environment."[14]

Right from the beginning of his tenure, Fletcher emphasized the environmental role of NASA. In 1972, the Earth resources technology satellite was launched as Landsat. It joined the weather and atmospheric science-oriented satellites as showing NASA's importance to environmental values. A series of demonstration programs were launched to persuade user agencies and industry that Landsat imagery could aid them in their decisions. In 1978, the Carter administration touted Landsat as capable of helping forecasts of world food harvests. Sometimes the potential users took a strong interest, and other

---

12. Ibid., p. 289.
13. Ibid., p. 397.
14. Launius, "A Western Mormon in Washington, DC: James C. Fletcher, NASA, and the Final Frontier," *Pacific Historical Review* (1995): 236.

times they thought space-based remote sensing satellites might be a solution looking for a problem.

The Stratospheric Ozone Issue

Throughout his tour as NASA's leader, Fletcher persisted to extol NASA as a friend of the environment. He proactively sought environmental missions for which he thought NASA was relevant. The one that ultimately proved most significant as a Fletcher legacy pertained to stratospheric ozone. In 1973, NASA performed research that indicated the Shuttle might produce chlorine, a highly reactive chemical theorized to deplete ozone in the atmosphere. Was the Space Shuttle an environmental threat? Not long before, environmentalists had helped get the supersonic transport (SST) killed. This concern about ozone depletion was reinforced the next year, when Mario Molina and Sherwood Rowland published a paper in *Nature* arguing that the real danger for ozone depletion was not airplanes or other exotic technology (for example, a Shuttle) but rather a mundane and common family of industrially produced compounds known as chlorofluorocarbons (CFCs). These chemicals were used in spray cans, refrigerators, air conditioners, and the like. At ground level, CFCs were not a threat. High in the stratosphere, however, they were a peril, depleting ozone and thus enhancing skin cancer. A controversy followed, in which scientists, industry, environmentalists, and others contested the "truth."

It was not immediately clear which agency—NASA, NOAA, or the National Science Foundation (NSF)—should take responsibility for determining if ozone depletion was a real or bogus threat, whether caused by CFCs, Shuttles, or airplanes. In any event, given Fletcher's environmental orientation and NASA's need to be relevant to changing national priorities, the space agency went after this mission. In June 1975, Congress passed legislation directing NASA "to conduct a comprehensive program of research, technology, and monitoring of the phenomena of the upper atmosphere." In 1977, the year Fletcher stepped down as NASA Administrator, Congress required NASA to issue biennial reports to Congress on the status of ozone depletion and what was known. The 1975 and 1977 legislation solidified NASA as the lead agency in stratospheric ozone research *and* linked NASA with policy. Fletcher had achieved his aim of moving NASA in an environmental direction.[15]

---

15. W. Henry Lambright, *NASA and the Environment: The Case of Ozone Depletion* (Washington, DC: NASA SP-2005-4538, 2005), pp. 7–8.

Comparative Planetology

Landsat and ozone depletion represented NASA's most direct foray into the environmental field in the 1970s. At the same time, indirectly, the comparative planetology area also advanced NASA's environmental claims. Mariner 9, which went up in 1971, found a huge dust storm on Mars that had altered the planet's climate. Carl Sagan argued that this discovery pointed to issues for climate change on Earth. In 1975, NASA's largest and most ambitious planetary mission up to this time, Viking, was launched. Viking's goal was to land equipment on Mars and determine if life might exist there. The results were largely inconclusive and, for many, negative on the issue of life. Nevertheless, Viking generated enormous interest in Mars and its comparison with Earth. Viking also stimulated interest in NASA's planetary protection requirements. Sterilization techniques were developed and applied to the Viking spacecraft to prevent forward contamination.[16]

Lovelock's Gaia hypothesis was getting better known as Lovelock went to conferences and spoke out. In 1974, he published his thesis "that all living things on Earth, along with the lithosphere, oceans, and atmosphere, act as a unified, synergistic system … analogous to the body of a single organism, which homeostatically controls environmental conditions in the oceans, the atmosphere, and so on, so that they remain within the range needed to support life."[17] Lovelock continued to publish and defend his thesis throughout the decade. In 1979, Gaia got top billing in a "Conference on Life in the Universe." Lovelock's studies of Earth as a system and planetary scientists' work on Mars, Venus, and other planets seemed to be converging at least in ideas, if not yet in government programs.

The environmental movement of the 1970s was an amalgam of science and emotion about Earth. Lovelock's "notion of the Earth as a living being struck a chord with a wide range of nature lovers, from indigenous cultures who thought of the Earth as a sacred spirit to mystics who sought the 'oneness' in nature, and environmentalists who were all too aware of how delicate was the balance of equilibrium in a complex system: cut down a rain forest in one part of the globe and you create a desert in another."[18] There was also a sense of utopia—seeking a better Earth, a purer Earth, a cleaner Earth.

---

16. See Dick and Strick, *Living Universe*, chapter 4, "Vikings to Mars."
17. Ibid., p. 49.
18. Benjamin, *Rocket Dreams*, p. 51.

O'Neill's Appeal

As energy shortages joined environmental interests, the notion of "limits to growth" gained prominence, aided by a book of the same name. One of the more interesting ways in which space enthusiasts, energy-conserving advocates, and environmentalists came together in the 1970s was within the concept of communes in space. Gerard O'Neill, a noted Princeton physicist, published a visionary article in the journal *Physics Today* in 1974. There, he outlined a scheme for human migration beyond planet Earth. As O'Neill continued to write and speak, his ideas attracted a considerable following. What captured the imagination of many people was his concept of colonies in space. Such colonies would be eco-friendly, relying on solar energy and recycling. Stuart Brand, of *Whole Earth Catalog* fame, was a follower, as were others who wanted a better and "greener" life. O'Neill was a serious scientist. He received NASA support to develop his ideas. In 1976, he testified at a standing-room-only congressional hearing. O'Neill showed how environmental values, energy technology, and space exploration could be harmonized. He appealed even to those for whom "growth" was a desirable value. O'Neill showed a way to escape Earth's limits by extending humankind's presence into space.[19] O'Neill's colonies were to be situated at Lagrangian points in space, where gravity between Earth and the Sun would hold the required structures in check and provide stability. His ideas spawned a society of followers, called the L5 society. The momentum behind many of his ideas faded in the 1980s, and the charismatic O'Neill died in 1992.

Another energy-related notion that had currency in space and environmental circles in the 1970s was that of a solar-power satellite (SPS). Studied under NASA and Department of Energy funding, the SPS required an array of solar cells equivalent to Manhattan scale in space. A number of these giant satellite systems would be situated in space at a point where a given array would have a fixed spot above Earth. An SPS would capture solar energy and microwave it to a receiving station on Earth. From there, it could be inserted into an electrical grid. Enough SPS systems and the whole world could be powered by solar energy and the need for electricity-related fossil fuels obviated.[20] Although the idea of SPS solved some environmental issues, it created others. The SPS remained in the "study state" subsequent to the 1970s, never getting any real traction in policy.

---

19. Ibid., pp. 121–138.
20. David Goodstein, *Out of Gas: The End of the Age of Oil* (New York: Norton, 2004), pp. 111–112.

Orbital Debris

Some of these ideas about people living in space or depending on solar-power satellites may have stimulated more attention, at least within NASA, to environmental threats in space. The debris issue continued to grow slowly and quietly. In 1979, NASA established an Orbital Debris Program Office at Johnson Space Center (JSC) to begin tracking the space junk orbiting Earth. Debris, like comparative planetology and space colonies, was one component of NASA's environmental role. NASA gave most attention to the component of putting environmental monitoring satellites into space. It was in this satellite field that NASA's environmental role was most salient and direct. However, debris could strike environmental satellites along with other space assets. At the end of the 1970s, NASA launched two important new Earth observation satellites. One was Nimbus 7, a satellite specially equipped to study problems of the upper atmosphere—namely, ozone depletion. The second was Seasat, a satellite instrumented to test a variety of oceanographic sensors.

Seasat only operated from 28 June 1978 to 10 October 1978, but it showed oceanographers how the view from space could advance their field. Seasat measured sea surface temperatures as well as wind speed, temperature, and other features.[21] Along with atmospheric and land satellites, Seasat pointed up the potential the comparative planetologists had noted when they speculated about doing for Earth what Viking scientists had done for Mars. Like Lovelock, they envisioned that the ability to view Earth as a total system provided new insights and was made possible by space technology.

Turning Concepts into a Program: The 1980s

James Beggs was appointed NASA Administrator in 1981 by President Ronald Reagan. At the time he came into office, the Shuttle program at long last was reaching a point of testing. As Beggs saw it, the time was near when NASA needed another major developmental mission of flagship status to keep the Agency viable. Beggs saw a space station as "the next logical step." It would be a laboratory in space and also a "base camp" for piloted missions back to the Moon and eventually to Mars. The problem he had was to get a presidential decision from Reagan to initiate the program. There were those around Reagan—his science advisor, who thought a space station a bad idea; his budget director, who thought it too expensive—who stood in the way. It would take a while to persuade the President. Reagan was giving priority to

---

21. NASA, "Seasat 1978 Overview," 10 February 1998, *http://southport.jpl.nasa.gov/scienceapps/ seasat.html.*

building up U.S. defense capability, nuclear weapons included. Some space scientists, including Sagan, drawing on work in comparative planetology, argued that nuclear war could have an effect of triggering "nuclear winter." [22]

Environment as a Backup Mission

While he waited for the right moment to approach Reagan on the space station decision, Beggs looked for other initiatives. What was possible? Beggs's Associate Administrator for OSSA, Burton Edelson, was also looking for new missions from his vantage point. Edelson recalled the Viking experience: "Viking's intended landing spot was found to be unsuitable.... [This caused NASA] to call in the nation's leading Mars specialists for an intensive three-week meeting covering all aspects of the planet. The meeting led to the selection of a safe landing area. At the end, one of the scientists involved, Michael McElroy of Harvard University, repeated something he had begun to tell other scientists: 'You know, we've never done anything like this for the Earth.'" [23]

Edelson believed the moment had arrived to take a comprehensive look at planet Earth. He fastened on the concept of "global monitoring" by satellites. As an engineer, Edelson envisioned the launch of very large platforms in space to monitor all the environmental ills of the planet through multiple sensors. Land, atmosphere, ocean, ice—everything would be combined holistically. The program had to be big—the platforms big—because the problems of monitoring global change were large. Also, intuitively, Edelson knew NASA needed big programs to survive, much less prosper. He took his concept to Beggs. Beggs did not want to put OSSA's idea ahead of the space station, but he could see how a global-monitoring mission could help sustain NASA as it moved from the Shuttle to the space station era. He told Edelson to move ahead. [24]

While Edelson, the OSSA Administrator, was developing his program plans, one of the scientists under OSSA was moving much faster. This was James Hansen, who headed the Institute for Space Studies in New York City, a division of the Goddard Space Flight Center, which in turn received most of its funds from OSSA. Hansen was a man who marched to his own drummer. Trained in physics and astronomy under James Van Allen, the discoverer of radiation belts named after him, Hansen started out studying Venus at the

---

22. Dick and Strick, *Living Universe*, pp. 122–125.

23. Burton Edelson, "Mission to Planet Earth," *Science* 227, no. 4185 (25 January 1985): 6.

24. The origins of NASA's "Mission to Planet Earth" are recounted in W. Henry Lambright, "Entrepreneurship and Space Technology: The Ups and Downs of 'Mission to Planet Earth,'" *Public Administration Review* (March–April 1994): 97–104.

University of Iowa. He contemplated how the greenhouse effect had rendered Venus inhospitable to life. Hansen subsequently began working on simulations of Earth's climate for the purpose of understanding the human impact on climate change on his own planet. What he discovered made him very worried. Rather than wait quietly, he decided to sound the alarm. He sent Walter Sullivan, the veteran science writer of the *New York Times*, a report he was about to publish, announcing that Earth was getting noticeably warmer. For the first time, the greenhouse effect made page one of the *New York Times*. The warming, it was stated, could cause a disastrous sea level rise.[25]

Hansen's *New York Times* gambit no doubt raised eyebrows in the Reagan White House, which was blunting the edge of many of the environmental and energy initiatives it had inherited from President Jimmy Carter. Within NASA, Hansen was seen as an asset but also as a maverick. In any event, NASA wanted to mount a program that would determine if Hansen was right. Beggs decided to act. In 1982, NASA convened a scientific group headed by Richard Goody of Harvard to study the concept of linking space to global habitability, and it gave the idea its blessing. Beggs now went to a United Nations (UN) Conference on the Peaceful Uses of Outer Space, where he called for "an international cooperative project to use space technology to address natural and manmade changes affecting habitability of Earth." The reaction was overwhelmingly negative—from other agencies, scientists, and the White House. Beggs had moved too fast. Critics saw NASA as engaged in bureaucratic empire building. "It came across like NASA was trying to take over the world," Edelson recollected.[26] Beggs had broached a technological vision without a political coalition to undergird it.

Beggs subsequently told Edelson to keep "working" on the Earth-monitoring system, but to do so quietly and spend time to build support for the program. Beggs turned to persuading the President to launch the space station program and in 1984 got Reagan to agree. Edelson, meanwhile, mended fences and built support. From 1983 to 1986, NASA engaged in a broad-gauged planning process. A committee—the Earth System Science Committee—involving Earth scientists from many disciplines was established. The "Earth System" concept was key. It conveyed the "holistic view." NASA also invited the NSF, responsible for basic research in Earth science fields, and NOAA, with authority in operational weather and oceans programs, to attend the meetings. At the same time, Edelson set up a new unit

---

25. Weart, *Discovery of Global Warming*, p. 143.
26. Edward Edelson, "Laying the Foundation," *Mosaic* 19, no. 314 (fall–winter): 6–7.

within OSSA (the Earth Sciences and Applications Division) to recruit and strengthen the status of Earth scientists within the space agency.[27]

Ozone Depletion

While planning this comprehensive Earth-monitoring initiative, NASA was becoming increasingly involved in projecting a visible leadership role dealing with the problem of ozone depletion.[28] The wisdom of Fletcher's moves in the 1970s to position NASA in this area of policy became clear in the mid-1980s, as ozone depletion rose on the national and international agenda. In 1985, British scientists using ground-based studies discovered extraordinary and unexpected ozone depletion over Antarctica. NASA followed up with satellite observations confirming what became known as the "ozone hole." The media provided eerie and graphic images of the Antarctic ozone hole and spoke of a global environmental threat, especially skin cancer from unfiltered sunlight, if the hole spread. Environmentalists, the media, and politicians throughout the world demanded action.

Edelson decided the time had come to "surface" again. In 1985, he wrote an editorial for *Science* in which he proclaimed the need for a new "Mission to Planet Earth." Meanwhile, parallel to NASA's planning effort, scientists under the auspices of the National Academy of Sciences were working with colleagues in other countries developing what came to be called a "global change" research agenda. The difference between the two planning exercises was NASA's emphasis on the role of space satellites, whereas this other effort emphasized ground- or ocean-based research. As for satellites, NASA got approval at this time from the White House and Congress to develop a new satellite dedicated to the ozone issue called the Upper Atmosphere Research Satellite (UARS).

In 1986, NASA's Earth System Science Committee produced a report that proposed a very extensive R&D program extending well into the twenty-first century. Its most significant recommendation from a NASA perspective was for the development of an Earth Observation System (EOS). The EOS would constitute the large platforms for Earth monitoring that Edelson had envisioned in 1981. The EOS would consist, at least initially, of two 13-ton platforms carrying a large range of sensors. These platform-sensor systems would provide comprehensiveness and simultaneity of atmospheric, oceanic, and land monitoring. Other platforms could be supplied by other nations. The biggest difference conceptually between the EOS of 1986 and the Edelson

---

27. Lambright, "Entrepreneurship and Space Technology," pp. 99–100.
28. Lambright, *NASA and the Environment.*

vision of 1981 was the linkage of EOS with the space station. It would be physically connected in some way rather than be freestanding in a separate orbit. As OSSA had to build an external coalition, so it had to build an internal coalition. The space station was the 800-pound gorilla within NASA. NASA leadership was anxious to link the station and scientific community. OSSA saw it better to cooperate than compete with the station.[29]

Everything seemed set for NASA to go to the White House and Congress in 1986 to get the new environmental mission and the EOS formally adopted as a major program. NASA's work in ozone depletion was going well. It had played a "lead agency" role in sponsoring expeditions to Antarctica to prove or disprove the CFC theory, and the evidence was mounting that CFCs were indeed the cause of depletion. NASA was linked to policy-makers, via assessments of research, who were moving in the direction of a treaty to deal with CFCs. NASA was an ally in the view of environmentalists in and out of government. Unfortunately, in January 1986, the Space Shuttle Challenger exploded shortly after takeoff, taking seven lives, including Christa McAuliffe, the first teacher in space. For a time, chaos reigned at NASA, and virtually everything was put on hold. NASA was hurt immensely by the fact that Beggs had left NASA in late 1985 to fight what was proved to be a false criminal charge stemming from his pre–NASA work in the private sector. William Graham, Beggs's deputy, sought to run NASA in the immediate aftermath of the Challenger incident, as acting Administrator, and was overwhelmed. Reagan appointed Graham his science advisor and brought James Fletcher back to help NASA in 1986 to help the Agency recover. It would be 1989 before NASA returned the Shuttle to flight.

Fletcher, meanwhile, sought to get his Agency—under attack on all fronts—to think positively about its future. He asked astronaut Sally Ride, first U.S. woman in space, to produce a report on NASA priorities. She did so in 1987. Her report used the phrase "Mission to Planet Earth" (MTPE), listing it prominently among four "leadership initiatives" for the Agency's future. Fletcher embraced the Ride report and the new Earth mission. With the Shuttle grounded and the space station mired in cost overruns and delays, NASA's environmental program could be a source of pride and support. In 1987, the Montreal Protocol was concluded, with NASA pivotal in the effort to provide a base of technical information for the negotiations. The protocol called for emissions control and continuing refinements in the policy to eliminate ozone-destroying chemicals through research.

---

29. Lambright, "Entrepreneurship and Space Technology," p. 100.

Edelson left OSSA in this year and was succeeded by Len Fisk. Fisk inherited a program long in the planning and with a strong constituency of support. It had momentum. Ozone depletion was the vanguard of the global-monitoring effort. "Mission to Planet Earth" was indeed a growing activity of NASA now, with the EOS projected to cost $30 billion over a 15-year period of development and use. All that was needed was adoption by the President and Congress to make the program official.

The Rise of Climate Change

The following year, 1988, NASA continued its ozone-depletion work, waiting for a national policy decision. As it did so, NASA found itself increasingly involved with the climate change issue. This issue was rising fast on the federal agenda, helped by a very hot summer—and NASA scientist James Hansen's willingness to testify before Congress about the threat of global warming. Hansen declared that he could state "with 99 percent confidence" that there was a long-term warming trend under way, and he strongly suspected that the greenhouse effect was to blame. Talking with reporters afterward, he was more direct, saying it was time to "stop waffling, and say that the evidence is pretty strong that the greenhouse effect is here." These statements—and their use by politicians, environmentalists, and media—helped trigger further action by other groups.[30] An international conference of scientists and policymakers in Toronto called for a 20 percent reduction of global carbon emissions by 2005. Two UN organizations—the World Meteorological Organization (WMO) and the United Nations Environmental Program—created a separate body of scientists, the Intergovernmental Panel on Climate Change (IPCC). The IPCC was a large, diffuse, international group of scientists, many government-connected, whose task was to periodically assess the state of knowledge about global warming as an aid to policymakers.[31]

Seeing momentum building on climate change, Reagan's science advisor, Graham, convened an interagency panel, the Committee on Earth Sciences, to get ahead on the issue. Representatives on the committee from NASA, NOAA, and NSF proposed a coordinated interagency program called the U.S. Global Change Research Program (USGCRP), of which NASA's Mission to Planet Earth/EOS would be a significant part. The space and nonspace facets of the program would be integrated. They asked that these agencies' budgets relevant to global change would be considered as a whole. The Office of Management and Budget (OMB), which had been promoting

---

30. Weart, *Discovery of Global Warming*, p. 155.
31. Ibid., p. 158.

interagency coordination, liked what it heard and encouraged the agencies to plan together. In 1988, they readied an initiative to send to the President—USGCRP. NASA's MTPE was thus a component—the largest dimension by far in expense—of this interagency program.[32]

The environment was an issue in the presidential election of 1988. George H. W. Bush said that if elected, he would be the "environmental president." Further, he declared that Americans need not worry about the greenhouse effect because under him the "White House Effect" would take care of the problem. NASA was meanwhile working two fronts at once in global environmental research. It was expanding its ozone-depletion program from Antarctica to include the Arctic. At the same time, it was positioning itself to be active in global warming in a large-scale way via the Earth Observation System. Other environmental issues of concern to NASA—planetary protection and space debris—remained of continuing interest, but they were below the radar screen compared with ozone depletion and climate change.

In 1989, NASA returned to flight with the Shuttle, and Fletcher retired from NASA for the second time. He clearly had encouraged NASA's environmental role—both times as Administrator. George H. W. Bush was now President. NASA, its sister agencies in the USGCRP, and the environmental community, among others, waited to see if he would follow through on his campaign pledge to be the environmental president.

The George H. W. Bush Years

Global Change as a Presidential Priority

Bush appointed Richard Truly, a retired admiral and former astronaut, NASA Administrator in 1989. Truly's priorities lay with piloted spaceflight and especially the Shuttle. His interest in other programs was lukewarm, but he did nothing to constrain NASA's drive for a Mission to Planet Earth. It looked as though NASA were on a growth curve in overall budget now that the Shuttle was flying again, and there would be funding for a number of major programs. Bush appointed Allan Bromley, a Yale chemist, as his science advisor. He told Bromley to propose a few initiatives in science and technology that he could back. Plans for the USGCRP initiative were literally sitting on Bromley's desk, inherited from his predecessor. Bush had economic advisors who were skeptical of the global-warming thesis. They worried that CO_2 emission reduction of the kind environmentalists wanted would wreck the economy. They urged Bush to support research to narrow the uncertainties before taking action to

---

32. Lambright, "Entrepreneurship and Space Technology," p. 100.

regulate emissions, and Bush agreed. Environmentalists saw research as desirable but argued it should not be used as an excuse to delay regulatory policy. In the ozone case, NASA had tied research to policy because policy-makers wanted to act and felt a sense of urgency because of the ozone hole. In global change (that is, climate change), policy-makers felt no such urgency. NASA was given the go-ahead for a major new research program but had no official mandate to connect it to policy.

In July 1989, in a speech commemorating the Apollo Moon landing, Bush proclaimed that NASA would go back to the Moon, this time to stay, and then on to Mars. He also endorsed NASA's Mission to Planet Earth. Congress, controlled by Democrats, dismissed the Moon-Mars decision as empty rhetoric. The MTPE decision was another matter. Al Gore, a leader in the Senate on both environmental and space policy, said MTPE was NASA's real priority.[33] NASA was talking about at least a 15-year program costing $30 billion. Sustained observations were deemed essential to understanding trends in global change. A $30 billion program was a long way from the relatively modest "applications" programs of earlier years. MTPE would not only be a new mission; it was projected to be the biggest robotic program in NASA history. NASA would supply two platforms. The European Space Agency might provide a third, and Japan a fourth. The EOS was therefore conceived as international in participation, a global program in more than one way. It was the Earth scientists' version of Apollo.

In 1990, the President made the USGCRP (and thus MTPE) his first presidential research priority. Congress appropriated $191 million to get the EOS, the centerpiece of MTPE, under way. NASA's new mission was now fully adopted. As if to symbolize the fact that the EOS could stand on its own feet, NASA Administrator Truly decoupled the EOS from the space station in terms of both planned configuration and budget. In surviving to this point, it was helpful to the EOS to be part of the space station. But there was no technical reason why they had to be coupled. Many scientists believed the EOS should be in a very different orbit from the space station. And environmental advocates of MTPE/EOS believed the program was more worthwhile than the space station.

In this year, a prestigious outside group, headed by Norman Augustine, the CEO of Martin Marietta, convened to examine the NASA program at the behest of the White House. There was considerable concern about NASA due to a series of problems, the most notable of which was the Hubble Telescope's blurred vision. The Augustine panel endorsed NASA's Mission to Planet Earth. Like the Ride report, the Augustine report made NASA's

---

33. Ibid., pp. 100–101.

environmental mission virtually coequal with the human and space science (Mission from Earth) programs. Augustine's report recommended substantial raises in the NASA budget to accommodate the various missions.[34]

Senator Gore, who embodied the environmental movement's views, supported MTPE/EOS but advocated that NASA find ways to get scientific knowledge about climate change out sooner than the EOS portended. He noted it would take almost a decade to develop and deploy the two complex sensor-platform systems. To NASA's consternation, its own Hansen suggested a smaller $200 million satellite dedicated to climate ("Climsat"), that could be launched much sooner.[35] Hansen was widely perceived as a close advisor to Gore on climate change. Some observers speculated that if Gore ever became President, he would make Hansen his science advisor. Gore's concern was the pace of the program. Even as NASA's MTPE/EOS program won presidential and congressional support to get adopted, it was sharply criticized by other lawmakers. Not because of pace or purpose, but its cost. The larger policy issue haunting NASA was money to pay for all the programs on its plate. Space, environment, and many other policy initiatives of Bush were giving way to an overriding national priority to restrain budget growth and reduce the deficit accumulated since Reagan took office. There was bipartisan consensus about deficit reduction, not about spending on the global environment.

To enforce budget discipline, Bush and Congress in 1990 agreed to a spending cap. That cap affected most agencies, including NASA. Within NASA, the EOS was projected to ramp up in cost in succeeding years. Unless checked, it could rival the space station, also ramping up at the same time. The administration and Congress grew concerned. Could MTPE achieve its purposes through some less costly technical approach? NASA said its approach was best. Bromley, Bush's science advisor, sought external advice from the National Academy of Sciences (NAS). NAS found that the two-platform concept NASA had proposed was more elaborate than necessary. One platform, not two, would be sufficient for the simultaneous observations NASA desired. The instruments on the second platform might be better arrayed on a number of smaller, more specialized satellites.

Hansen, meanwhile, continued to question his own agency's strategy, publishing an article arguing for his Climsat. He said he was for the EOS, but global warming could not wait for the EOS to be fully deployed. There were various scientific skeptics where global warming was concerned. The scientific body established to assess global warming for the UN, the IPCC, issued its first report in 1990. It confirmed that global warming was taking place but

---

34. Ibid., p. 101.
35. Ibid.

stopped short of saying humans were the cause.[36] Everyone seemed to agree on the need to narrow uncertainty by research. The issue was how much to spend, how long to wait—before policy action.

Full Implementation Thwarted

MTPE, and with it the EOS, was officially adopted in 1990. (MTPE was the overall program in OSSA. The EOS was the principal satellite development project.) But before implementation could begin in earnest, influential critics forced NASA to revise its plans. Bush had created a National Space Council (NSC), headed by Vice President Dan Quayle, to deal with cross-agency space policy issues. The NSC and the OMB both saw a train wreck ahead for NASA because of a static budget and growing programs—the space station and the EOS. They created a blue-ribbon task force, the Engineering Review Board, in 1991 specifically to address the question of EOS design and cost saving. Meanwhile, the media began painting the EOS in the language of "big science," and, in the budget context of 1991, big science was suddenly bad.

Speaking to the Maryland Space Business Round Table, OSSA's Associate Administrator, Lennard Fisk, expressed frustration with those who criticized the EOS for being too big and expensive. He rejected Hansen's interim satellite as well. Fisk argued: "It is simple reality that to determine how the Earth will respond to increased concentrations of greenhouse gases will require a detailed, comprehensive, and complete understanding of how the Earth works. There is no early version of this. No version in which you do some concentrated research in some limited area, and then all truth is revealed. The Earth is too complicated, and its workings too interrelated. If the comprehensive nature of this research effort makes it by definition big science, then so be it. It's a big Earth. And there are big consequences for getting the wrong answers." He explained that the EOS was not a "breakthrough mission." Rather, it was an "evolutionary" program that would provide ever-greater understanding of the impacts of greenhouse gases over time. There would be "simpler missions of smaller spacecraft" along the way, leading to "the more detailed and complete measurements that EOS was to make. EOS would provide simultaneity and comprehensiveness in data about Earth. EOS was the route "to build the definitive story of how the Earth works, and what will be the future of our planet home."[37]

---

36. Weart, *Discovery of Global Warming*, p. 162.
37. Lennard Fisk, "Mission to Planet Earth" (talk to Maryland Space Business Round Table, 26 February 1991).

Unfortunately for Fisk, opponents of the space station used the EOS as leverage against the space station. There was a serious attempt to kill the station in Congress in 1991. The station survived, but the *Washington Post* called the vote "a political victory, not a budget victory." To support the space station, other programs would likely suffer, and that meant the EOS—the other megaprogram. In the summer of 1991, the Engineering Review Board recommended EOS's redesign, with the twin 13-ton platforms split into a fleet of smaller satellites. At the same time, Congress, while appropriating $271 million for the EOS ($65 million less than requested), wrote into the legislation a cap on the program through fiscal year 2000 at $11 billion—$6 billion less than NASA had projected as required for the two-platform design.[38]

Ironically, in September, as NASA contemplated how to comply with the Engineering Review Panel recommendations and congressional strictures, it began implementing its Mission to Planet Earth program. The first satellite under MTPE was the Upper Atmosphere Research Satellite, the $740 million ozone-depletion-monitoring satellite. Approved earlier, UARS was folded into the MTPE program as the initial satellite project. As Fisk had noted, there would be an evolution of spacecraft leading up to the EOS. UARS was the first. It was a fitting start given the impetus ozone depletion had played in launching NASA's expanded environmental mission. By the end of the year, NASA had reluctantly abandoned its two-platform design. It split the two platforms into a fleet of six satellites. These six satellites would carry fewer instruments than the two comprehensive platforms would have done. Simultaneity and comprehensiveness—the twin values OSSA planners wanted—would be sacrificed. The first of the six satellites was scheduled to go up in 1998, about the same time NASA planners had once hoped the EOS as a whole would go up. But money constraints meant stretching the program.

Goldin's "Faster, Better, Cheaper" Mantra

In early 1992, President Bush forced NASA Administrator Truly to resign due to differences in priorities between the White House and the NASA leader. The differences pertained more to human spaceflight than to the EOS, especially the Moon-Mars program Bush had announced. This program had gone nowhere, in part because Truly had not pushed it, but more because Congress did not take it seriously and never funded it. Bush appointed Daniel Goldin, an aerospace executive from TRW, as NASA Administrator. Goldin had crossed swords with NASA previously over the relative merits of the large platforms versus the smaller satellites. Fisk had allegedly threatened

---

38. Lambright, "Entrepreneurship and Space Technology," p. 102.

to withhold contracts from TRW if Goldin continued to push his "small-sat" ideas. Goldin was appointed in part because of his belief that NASA could accomplish its missions more quickly with smaller systems employing more advanced technology. This approach—called "faster, better, cheaper" (FBC)—became Goldin's mantra.[39]

Constrained politically on changing the space station, Goldin particularly emphasized savings in the unmanned part of NASA. He imposed a 30 percent cut in the "run-out budget" (expenditures to 2000) on the EOS. This meant it would fall further, from $11 billion to $8 billion. To enforce his views, he abolished OSSA; moved Fisk to Chief Scientist; and, in place of OSSA, created three smaller (and less powerful) entities. One of these offices was Mission to Planet Earth. Fisk soon left NASA. Without the strong-minded Fisk as a counter, Goldin was able to rework the EOS in his FBC model.[40]

The Arctic Ozone Hole

While these events were transpiring in NASA's organizational context, other important events took place that weakened NASA's credibility as an environmental agency and its effort to have a bipartisan base of political support. NASA's research in the Arctic had expanded. At the beginning of February 1992, project leaders detected signals they believed portended serious depletion problems in the Northern Hemisphere. They debated whether to sound an alarm or wait two months until the expedition's end and make sure concerns about ozone depletion in the Arctic were valid before saying anything in public. An ozone hole over the Arctic could have dire consequences, especially for people who lived in the far Northern Hemisphere. NASA's general rule was, "If we are arguing about it, it doesn't belong in a press release or congressional testimony." Violating that rule, NASA decided to issue a warning. On 3 February, at NASA Headquarters, expedition leaders held a news conference. They said that an ozone hole over the Northern Hemisphere "was increasingly likely" and had to be taken seriously. "We're not concerned with just remote regions now," said Michael Kurylo of NASA. "What we're dealing with extends to very populated regions." With ozone loss increasing, a 30 percent loss by March was possible. "Everyone should be alarmed about this," warned Kurylo.[41]

---

39. Stephanie Roy, "The Origin of the Smaller, Faster, Cheaper Approach in NASA's Solar System Exploration Program," *Space Policy* 14 (1998): 166.

40. Roy, "Origin of the Smaller, Faster, Cheaper Approach," pp. 165–167; Lambright, "Entrepreneurship and Space Technology," p. 103.

41. Lambright, *NASA and the Environment*, pp. 30–31.

The reaction from environmentalists was immediate: "It's frightening," said Liz Cook of Friends of Earth. "If the phenomenon ever occurs on a broader scale, it could be the final curtain for all life in the planet," said Karen Lohr of Greenpeace. The media were also alarmed, and influential media called on President Bush to accelerate the phaseout of ozone-depleting chemicals. A *New York Times* editorial, "The Ozone Hole over Mr. Bush's Head," termed ozone depletion an issue of global importance and noted that "the life-protecting ozone layer may now be thinning above President Bush's summer home in Kennebunkport, Maine." In an editorial called "The Vanishing Ozone Layer," the *Washington Post* declared: "Once again, it turns out that the protective ozone layer in the sky is being destroyed faster than even the pessimists had expected." *Time* magazine's cover headline was "Vanishing Ozone: The Danger Moves Closer to Home."[42]

Feeling pressure to act, Bush signaled a willingness to consider a faster timetable if necessary. A spokesman for the CFC industry said leading firms would find substitutes faster if required. Senator Gore said Bush had had a wake-up call thanks to the "ozone hole…pointed to and predicted about Kennebunkport." It was about time for Bush "to think seriously about doing something," Gore demanded. Gore took the floor of the U.S. Senate to introduce a bill to halt CFC production by 1995. He termed the information in the NASA news conference "an immediate, acute, emergency threat." Following the debate, the U.S. Senate called for a halt as soon as possible, not specifying a date, but voting 96–0 in favor of speed-up. On 11 February 1992, Bush announced that he was ordering American manufacturers to end, by 31 December 1995, virtually all production of chemicals that destroyed ozone. He had that authority under the Clean Air Act, if he found the Montreal Protocol deadline of 2000 inadequate.[43]

The media and environmentalist barrage continued, as did Gore's use of the "crisis" for his own purposes. But in early March, the dreaded ozone hole over the Northern Hemisphere failed to materialize as predicted. Satellite monitoring confirmed there was no emergency. The NASA scientist in charge of monitoring said there was no ozone hole over the Northern Hemisphere. As for Senator Gore's point about Kennebunkport, he said: "I can tell you categorically there is no ozone hole over Kennebunkport. There never has been an ozone hole over Kennebunkport, and I don't really expect one."[44] On 30 April 1992, NASA ended its Arctic project and admitted error in sounding an alarm prematurely. Ozone loss had occurred, but there was no

---

42. Ibid., pp. 31–32.
43. Ibid., pp. 31–32.
44. Ibid., p. 33.

hole. NASA subsequently was strongly criticized by the conservative media, such as the *Washington Times* and the *Wall Street Journal*. NASA went from being a hero to a goat in ozone science and policy, at least in some quarters. NASA's work in ozone depletion would continue in the future, particularly in monitoring ozone loss and thus compliance with the Montreal Protocol.[45]

After 1992, the political spotlight turned increasingly away from ozone and to the global warming issue. But one result of the ozone controversy was that the connection between NASA and Gore was now etched strongly in the minds of many of Gore's political enemies. Gore had published a book in 1992, *Earth in the Balance*, claiming the environmental crisis was grave and needed immediate governmental intervention. He was especially worried about climate change. The ozone-depletion affair made some conservatives believe that Gore and NASA were in alliance, with NASA supplying evidence for Gore's cause. Especially for conservative Republican critics, NASA's credibility was hurt doubly—by being wrong on the Arctic ozone hole and being a supposed tool of Gore and his environmentalist friends.

Bush meanwhile had backed off from his campaign pledge to emphasize environmental values as he saw the economy in trouble. Worried about the economy and his reelection potential, he resisted going beyond research in climate policy. Many European nations wanted targets, deadlines, and regulations. In June 1992, an Earth Summit in Rio de Janeiro took place, and Bush reluctantly attended. It culminated in a Framework Convention on Climate Change, signed by 150 governments. The convention called for emission reduction to 1990 levels by 2000, but there was nothing binding on those who signed.[46] In the presidential campaign of 1992, Bush criticized Gore, the vice-presidential candidate, as "the ozone man." Gore hit back. The election of Bill Clinton as President and Vice President Gore seemed to augur well for NASA's Mission to Planet Earth and the environmental movement.

Clinton-Gore and NASA's Environmental Role

In January 1993, Bill Clinton moved into the White House as President. With Gore as Vice President, everyone expected global warming to get higher priority. In organizing Clinton's White House Office of Science and Technology, Jack Gibbons, Clinton's science advisor, elevated Robert Watson to the new post of associate director of the Office of Science and Technology

---

45. The subsequent ozone history is found in Lambright, *NASA and the Environment*, pp. 33–57.

46. Lamont Hempel, "Climate Policy on the Installment Plan," in *Environmental Policy*, 6th ed., ed. Norman Vig and Michael Kraft (Washington, DC: CQ Press, 2006), pp. 294–297.

Policy (OSTP) for the environment. Watson came from NASA, where he had won acclaim for the way he led the Agency's ozone-depletion program in investigating the Antarctic Ozone Hole and advising policy-makers at the time of the Montreal Protocol. The Clinton-Gore administration gave general support to the existing U.S. Global Change Research Program, NASA's Mission to Planet Earth, and the EOS. But Clinton and Gore were more interested in policy for emissions reduction than the research side of global warming. They already believed global warming was a problem. They wanted "action." Ironically, Mission to Planet Earth and the USGCRP—large-scale research efforts—were more a clear focal point of presidential attention under Bush than they were under Clinton and Gore.[47] Goldin, meanwhile, was retained and given great leeway in running NASA. Clinton abolished the National Space Council. Goldin reported nominally to Gibbons but could see Gore when he needed to do so and established a relatively good relationship with the Vice President.

Goldin brought a new scientist-administrator from outside NASA to lead Mission to Planet Earth, Charles Kennell, an astrophysicist from the University of California, Los Angeles (UCLA). Kennell replaced Shelby Tilford, who had been in charge of the program, and found he had to deal with a number of problems. Under Goldin's pressure, the EOS had been downsized to approximately $7.25 billion for the period to 2000. Moreover, the program had significant technical problems with its data-handling system, known as the Earth Observing System Data and Information System (EOSDIS). Kennell had Goldin's support, now that Goldin had restructured the EOS along FBC lines. Also, Goldin knew that Mission to Planet Earth was one way he could connect NASA to Gore's priorities.

The centerpiece of Mission to Planet Earth remained the EOS. However, other environmental satellite programs with more specific missions were folded into this program. These included Landsat, a program NASA had initiated in the 1970s and then spun off to NOAA in the 1980s. NOAA, in turn, had tried to privatize the program, deemed operational, but to no avail. Landsat images proved useful during the Persian Gulf War, and in 1992 legislation had placed the program under DOD and NASA. Under Clinton, in 1993 it was made a NASA program again, although its operational status made it an uncomfortable fit for NASA. The dilemma was that no other federal agency really wanted to adopt the orphan effort.

Goldin's heart was with "mission *from* Earth," particularly the Mars exploration program, and most of his mental energy and actual time necessarily

---

47. W. Henry Lambright, *The Challenge of Coordinating "Big Science"* (Washington, DC: IBM, 2003), pp. 10–13.

went to the space station effort, by far NASA's most controversial and complex program. However, events external to NASA forced Goldin to think about Mission to Planet Earth. The congressional elections of 1994 brought the Republicans to power in both houses of Congress for the first time in decades. Led by House Speaker Newt Gingrich, they brandished a "Contract with America." That contract called for drastic cuts in government programs, including those that might have anything to do with regulatory/environmental controls affecting the economy.

Saving Mission to Planet Earth

Clinton, stung by the elections and seeing the Republican Congress out to dramatically cut programs and balance the federal budget, sought to retake the initiative in early 1995. He declared that he himself would cut the federal budget, provide middle-class tax relief, and end "big government." The cuts affected most agencies, and NASA was asked to reduce its budget $5 billion below its anticipated spending over the ensuing five years. Goldin was shocked because he had been a "poster boy" for government reinvention and downsizing since the outset of the Clinton-Gore years. He had thought he had taken his lumps, done his duty, and could promise his agency stability. That was not to be. He went along with the White House policy and soon waxed enthusiastic about the cuts and how to use them to reinvent NASA even more than he had already done.

As Goldin was deciding how to adapt NASA to the Clinton reductions, he was hit by additional demands from the House Republicans that the EOS be cut another $2.5 billion below Clinton's parings. Why the EOS? The answer was that the House Republicans saw NASA's Mission to Planet Earth as Gore's baby. Moreover, the EOS was aimed at global warming, and global warming was "political" science. In a press briefing, Representative Dana Rohrabacher (R-CA), chair of the House Science Committee's Energy and Environmental Subcommittee, explained the rationale. He derided the USGCRP and the EOS as "scientific nonsense." Global warming, he said, is at best "unproven, and at worst…liberal claptrap." The chair of the House Science Committee, Robert Walker (R-PA), likewise attacked NASA's Earth Observation System, now calling for a cut of $2.7 billion in it over the ensuing seven years. NASA Administrator Goldin said such a cut would not only hurt NASA but "would dismantle the national approach to U.S. global change research."[48]

---

48. As quoted in W. Henry Lambright, "The Rise and Fall of Interagency Cooperation: The U.S. Global Change Research Program," *Public Administration Review* (January–February 1997): 41.

At Walker's instigation, the National Academy of Sciences evaluated the state of MTPE/EOS. The report provided views that helped both friends and critics of MTPE/EOS make their case. Representative George Brown (D-CA), senior Democrat on the House Science Committee and an EOS advocate, found NAS had "endorsed the scientific underpinnings and research direction of the Global Change Research Program." He noted that the report offered cold comfort to critics of the USGCRP and the EOS. Moreover, the report argued against further cuts, delays, and design changes in the EOS. Noting that Republican leaders in the House and Senate had asked for the report, he declared: "Now we will see if they have the wisdom to heed it." Walker had a different view, noting that the report did indicate room for cost savings in the EOS and a "lack of coordination among agencies involved in USGCRP."[49]

The struggle between the Congress and President over budget cuts was fierce and went way beyond the EOS, involving a host of social, regulatory, and other environmental efforts. The battle caused two shutdowns of government in 1995. In the end, however, the public blamed the Republicans more than Democrats for the shutdowns, and Gingrich backed down. The EOS survived, but additional cuts were almost inevitable given NASA's overall budget, which was contracting under pressure from both Clinton and Congress. The year 1996 saw a continuation of the congressional-presidential struggle, but the issues were different, and the EOS was no longer the major target it had been.

NASA sent various probes up to look at components of the global climate change problem, but not Hansen's Climsat. It struggled to keep the EOS as comprehensive as possible, incorporating up-to-date technology—the "better" in Goldin's faster, better, cheaper mantra. Restructured yet again, the EOS moved in the direction of three moderate-sized satellites, rather than the six marked by the initial restructuring. One satellite would emphasize land, another water, and a third air. The 1990–2000 budget for the EOS, once $17 billion, now fell to $6.8 billion, and Kennell made it clear that "further reductions will translate into knowledge reductions."[50] NASA spoke of a program that had been "restructured," "rescoped," "rebaselined," and "reshaped" since its beginning.

Ironically, as the EOS had contracted, the global warming issue had grown in significance. The IPCC produced its periodic assessment of the science and, in 1996, stated for the first time that "discernible human influence" on

---

49. Ibid., pp. 41–42.

50. "Senators Rally Around NASA, Mission to Planet Earth," *American Institute of Physics Bulletin of Science Policy News* (17 May 1996).

climate systems was now evident.[51] Hansen continued to speak out, comments that no doubt resonated with Gore and the environmental community, grated conservative Republicans in Congress, and caused consternation in NASA's executive suite. Hansen was quoted in *Newsweek* as saying: "As you get more global warming, you should see an increase in the extremes of the hydrological cycle—droughts and floods and heavy precipitation." Writing in the June 1996 issue of *Geophysical Research Letters*, he called 1995 the hottest year for the planet on record.[52] There were still scientific skeptics, and Hansen stood out in his stridency, but a scientific consensus in favor of human impacts on global warming was solidifying.

If anything, the rancorous dispute between Clinton and Congress that shut down the government helped the President get reelected. Again, Clinton retained Goldin as NASA Administrator. Kennell departed and was eventually succeeded by Ghassem Asrar, chief scientist for the EOS. The program had a measure of stability now and was finally building hardware and working out the glitches in the EOSDIS data system. The schedule for the EOS satellites was slipping, but the goal was to launch them one at a time, as they became ready, around the turn of the century. Meanwhile, the program kept active through the smaller Earth probes. In 1997, for example, a satellite to measure tropical rainfall, called the Tropical Rainfall Measuring Mission (TRMM), went up. Although more engaged in emissions policy than climate R&D, the Clinton-Gore White House influenced MTPE/EOS to some degree, stressing a need to go beyond the "Earth system science" paradigm to also study regional impacts. This regional level was where the major risks and vulnerabilities from global warming would show up. NASA adapted, adding regional thrusts and also giving more attention to practical applications of remote sensing in general.[53]

El Niño

One of the application areas where NASA scored positively lay with a particular climate change-related disaster, El Niño. NASA's MTPE included a specialized satellite for measuring sea surface temperatures, called TOPEX/Poseidon, developed with the French. This particular satellite proved especially valuable in detecting El Niño. El Niño was a periodic warming of the Pacific that caused long-range climate effects responsible for droughts in

---

51. Hempel, "Climate Policy on the Installment Plan," p. 294.
52. Sharon Begley, "He's Not Full of Hot Air," *Newsweek* (22 January 1996): 24–27; Randolph Schmid, "NASA: 1995 Hottest Year on Record," *Huntsville Times* (14 July 1996): A4.
53. Lambright, *Challenge of Coordinating "Big Science,"* p. 13.

one country and floods in another. There had been a major, devastating El Niño in 1982–83 that scientists had not detected until it was well under way. NASA, NOAA, and the scientific community labored in the subsequent years to develop the satellite- and ocean-based technology that would enable early warning of the onset. TOPEX/Poseidon, developed in the late 1980s and early 1990s, was to El Niño what UARS was to ozone depletion.

In March 1997, TOPEX/Poseidon "caught red-handed the big rise in sea level that was spreading across the [Pacific] ocean" from the mid-Pacific to the South American coast. This discovery was followed in April by the rise of sea surface temperature that began off the coast of Ecuador and Peru and then spread north and south. Further studies confirmed the El Niño's beginning. This early detection was a demonstration of the power of space technology to advance knowledge of large-scale changes on Earth.[54] El Niño was related to short-term climate policy. The long-term climate policy of global warming remained conflictual.

Global Warming

The year 1997 was quite important in terms of global warming policy. Clinton in June addressed a special session of the UN, calling for "realistic and binding limits" on emissions but offering no specific targets for the United States. A major international conference dealing with emissions was coming up in December in Kyoto. The Senate in July passed a resolution 95–0 instructing the Clinton administration to refrain from signing any forthcoming climate protocol that did not include measures to be undertaken by developing countries. Clinton said in October that the United States would commit to reducing emissions to 1990 levels by 2012 and then pass further restrictions.[55] This statement seemed to be one more thrust in the ongoing battle between Clinton and Congress where climate change was concerned. For NASA, it meant the Agency had to walk a careful line—pursuing "good science" and avoiding the acrimonious tug-of-war on global warming policy between the two branches.

Indeed, NASA now faced a problem new to it—serious opposition from some environmentalists. Most of the time, the environmental movement was mildly supportive of NASA or indifferent. It spent its political capital on issues of immediate consequence. NASA's R&D agenda for climate change was of secondary interest, at best. The environmental groups' position was that R&D was necessary but not sufficient and should not substitute for regulation and

---

54. Madeline Nash, *El Niño* (New York: Warner Books, 2002), pp. 91, 114.

55. Hempel, "Climate Policy on the Installment Plan," p. 294.

other steps to stop global warming. Like Clinton and Gore, environmental groups wanted emissions controls. However, the Cassini launch to Saturn was what really brought NASA into the realm of environmental controversy in a negative way, at least in regard to one aspect of its work.

Nuclear Power in Space

The issue was nuclear power in space. What was new was the level of controversy, not the application. NASA had been using nuclear batteries (batteries powered by the decay of certain nuclear materials) since the 1960s. They were deemed necessary for space vehicles for which solar power and conventional chemical fuels were impractical. As the matter of space debris grew as a concern, the potential threat of nuclear materials rose with it. Even more, the 1980s debate over the Reagan initiative known as Star Wars, which seemed to some critics to presage the militarization of space, made various observers more sensitive to NASA's nuclear connection. The question came to a head for NASA in 1997, when the Cassini mission to Saturn was scheduled to launch.

Cassini used nuclear batteries to power it on its long journey to Saturn. What made Cassini so controversial was that it used far more nuclear materials than most probes and employed a "swing-by" technique to gain momentum—that is, it launched into a particular orbit that swung it back around Earth and then toward Saturn. An opposition group concerned about radioactive debris, the Florida Coalition for Peace and Justice, mobilized grassroots and national opposition. Groups opposed to space-based weapons also became involved as allies. In September, two congressional lawmakers asked for the launch's delay. NASA found itself on the defensive in a public relations and media war. Cassini did go up in October, but NASA was burned by the controversy and made aware that if it was to use nuclear materials in future spacecraft, as it thought desirable and even essential for some missions, it might face opposition from antinuclear forces.[56]

Debris and Global Warming

The nuclear issue exacerbated the space pollution or debris issue. As debris had proliferated over the years, efforts to mitigate it enlarged. In 1995, NASA and DOD issued guidelines regulating the design and operation of spacecraft

---

56. Victoria Friedensen, "Protest Space: A Study of Technology Choice, Perception of Risk, and Space Exploration," 11 October 1999, available at *http://scholar.lib.vt.edu/theses/available/etd-120899-134345/* (accessed 30 July 2014).

so as to guard against the growth of orbital debris.[57] However, NASA was criticized by the National Research Council in 1997 for underestimating the risk of space debris, not only to the Shuttle but to the prospective International Space Station. Nuclear propulsion and near-space pollution were areas where NASA had some measure of control. It could do little about the global warming debate, in which it was enmeshed. No matter what it did or did not do, it was criticized. In December, Al Gore, defying Congress, went to the Kyoto Summit and agreed to binding emissions targets for the United States. The United States, he said, would reduce its emissions 7 percent below 1990 levels by 2008–12. A number of other developed nations, such as Canada, Japan, and many European countries, agreed to emissions reductions also—but not many developing nations, including China and India, did so. While backing his Vice President and making the United States officially party to the convention, Clinton said he would not submit the Kyoto Protocol to the Senate for ratification. The United States would still attempt to meet its obligations under the protocol, Clinton insisted.[58] Meanwhile, scientific study would continue, with NASA the primary funder.

Triana

These larger policy events associated with global warming made the political setting for MTPE/EOS continually unsettled. In 1998, NASA followed the advice of supporters in the White House and Congress and sought to depoliticize its environmental mission somewhat by changing its name from Mission to Planet Earth to the more neutral "Earth Science." The problem in depoliticization was that Vice President Gore would not distance himself. In March 1998, Gore challenged NASA to build a new satellite to provide live images of Earth from outer space. "This new satellite, called Triana, will allow people around the globe to gaze at our planet as it travels in its orbit around the Sun for the first time in history," said Gore. The satellite would act as a mirror for the planet, capturing the motions of changing clouds, the movement of weather systems, and the destructive paths of large fires.[59]

---

57. NASA Office of Safety and Mission Assurance, "NASA Safety Standard: Guidelines and Assessment Procedures for Limiting Orbital Debris," August 1995.

58. Norman Vig, "Presidential Leadership and the Environment," in *Environmental Policy*, 5th ed., ed. Norman Vig and Michael Kraft, pp. 110–111; Hempel, "Climate Policy on the Installment Plan," pp. 299–300.

59. Office of the Vice President, "Vice President Gore Challenges NASA To Build a New Satellite To Provide Live Images of Earth from Outer Space," White House press release, 13 March 1998.

Gore leaned on NASA's Administrator, Goldin, to spend $20 million to put up this satellite. Gore wanted it to provide a 24-hour Earth channel and help create the planetary consciousness that had existed at the time of the first Apollo pictures and subsequent Earth Day. Triana would carry a small telescope or camera and make its observations from 1 million miles away at "L1," a Lagrangian point between Earth and the Sun where gravity forces were balanced. Gore hoped especially to reach young people. The Triana name came from Rodrigo de Triana, the sailor who had first spotted the New World on his journey with Columbus in 1492. Gore's association with the satellite helped and also hurt in getting support, however. It obviously helped with the OMB and Goldin, but it hurt with the Republican Congress. To satisfy Congress, NASA added scientific purposes along with the educational and inspirational features of most interest to Gore. Gore lobbied Congress, but congressional views were decidedly mixed. Some derisively dubbed it "Goresat."[60]

The Triana debate continued into 1999. As NASA added scientific requirements making it more useful (for example, studying impacts of the Sun on Earth's climate), Triana's cost went up. Gore's price of $20 million went up to $50 million and then rose to $220 million, according to some reports. What Gore was really after was to use space for inspiration. He knew that Apollo pictures of Earth from the Moon had helped animate the environmental movement in its early days. Gore believed that continuing to monitor the planet from afar, with pictures easily available to all, would inspire another generation. Goldin spoke up for Triana, but it was subject to partisan criticism because of its origin with a man who wanted to be President.

NASA had Triana reviewed by the National Academy of Sciences. The NAS panel found merit in the program. The panel said that the scientific objectives of the project "are consonant with published science strategies and priorities for collection of climate data sets and the need for development of new technologies." Neal Lane, the President's science advisor, echoed the panel view that the mission would "enhance our understanding of the Earth's energy balance and how it affects our climate systems." The panel also noted that the cost—now put at $75 million—was "reasonable."[61] Congress enacted legislation allowing Triana to be developed but barring its placement in orbit before 1 January 2001. Triana's fate depended on the election.

---

60. "Scripps Institution, GSFC Picked To Put Goresat at L-1 Point," *Aerospace Daily*, article 117803 (29 October 1998).

61. "Statement of Neal Lane, Assistant to the President for Science and Technology on NASA's Triana Mission," *SpaceRef.com*, 8 March 2000, available at *http://www.spaceref. com/news/viewpr.html?pid=1079* (accessed 30 July 2014).

The Space Station Goes Up

In 1999, the first elements of the International Space Station went into orbit. These were U.S.-Russian elements, with many more to come. In 2000, an all-important service module, developed by the Russians, went up; shortly thereafter, "permanent" human habitation began, although assembly had far to go. As before, the debris issue was raised in connection with human safety on the ISS. NASA indicated that further assembly would include protective shields. Meanwhile, astronauts would exercise collision-avoidance procedures as necessary.

The EOS Begins

While the political spotlight was on Triana and the Space Station, NASA's priority in Earth Science was to get the EOS into orbit. The first of the three EOS satellites was Terra. Originally scheduled for a 1998 launch, it was reset for late 1999. On 19 December, the $1.3 billion satellite went up carrying five separate instrument packages, three from NASA and one each from Canada and Japan. Its primary aim was to monitor how solar radiation interacted with Earth's land masses, and oceans. Asrar, Associate Administrator for Earth Science, emphasized the satellite's potential contribution to understanding climate change.[62]

Following adjustments in orbit, Terra produced its first images, released publicly in April. They showed "haze shrouding cities in India, heat radiating from the Sahara, and pollution over the Appalachians." Terra would be in orbit for years and was expected to produce huge volumes of data. NASA said it would evaluate how Terra and the next two EOS satellites performed and how the data were used before proposing major follow-on work. There was concern on the part of Earth scientists and their political supporters that NASA did not have firm plans for Earth monitoring beyond the EOS. Asrar issued a statement in which he promised "the next decade...to be an exciting one." The goal was to move beyond "characterizing the Earth system to genuinely understanding how it works, so that we can begin to predict future change." He indicated that the necessary technological systems would be developed to make this possible.[63] Maybe so, but Senator Barbara Mikulski (D-MD), a senior member of the Senate appropriations committee overseeing NASA, pointed out in April 2000 that there would be a change in administrations

---

62. Michael Mecham, "Terra Launch Puts EOS Program on Track," *Aviation Week and Space Technology* (1 January 2000): 38.

63. Ghassem Asrar, "Earth Science in the New Decade," available at *http://ipp.nasa.gov/innovation/Innovation_83/Wel-EarthScience.html* (accessed 30 July 2014).

and that without firm plans, the program could suffer a hiatus.[64] This would be especially true if the Republicans recaptured the White House.

George W. Bush Takes Over

In January 2001, George W. Bush became President. His victory over Gore, much contested, nevertheless sealed Triana's fate once and for all. He also put much of Earth Science research and development on hold while determining future programs and budgets. Bush retained Goldin while he looked for a successor. During the campaign, he had pledged to curb greenhouse gases. But in March, he announced that the United States would withdraw from the Kyoto Treaty. When his decision was met with outrage by environmentalists and frustration by European allies, Bush asked the National Academy of Sciences for its view of the global warming question. In June, the panel told Bush that not only was global warming a problem, it was getting worse. This finding did not get Bush to change his mind on Kyoto, however. He met with massive protests the next month, when he went to Europe to attend a European Council meeting. Not long afterward, 178 countries, including all European allies, signed a document in Bonn to proceed with the Kyoto agreement without the United States.[65]

Bush countered by saying he would support further research on climate change and policy action as necessary. Meanwhile, the IPCC released its third assessment, providing new and compelling evidence of climate risks, humankind's role, and possible consequences.[66] Bush created a Climate Change Research Initiative (CCRI) that subsumed the USGCRP and the EOS. He put James Mahoney, Assistant Secretary of the Department of Commerce, in charge of the CCRI. Mahoney said he would develop a strategic plan for a government-wide research program to deal with climate change. While the new administration evolved its strategic plan, it also cut NASA's Earth Science budget pending a determination of how the Earth Science program fit into the administration's revised climate change strategy under CCRI. Meanwhile, Terra continued to perform well. It was found extremely helpful in regional problems. Designed to detect pollution, deforestation, and urban growth, Terra plotted smoke, identified hot spots, and helped in fighting forest fires.[67]

---

64. "NASA's Plans for Earth Sciences Reevaluation Draw Mikulski Fire," *Aerospace Daily* (14 April 2000): 73.

65. Vig, "Presidential Leadership and the Environment," p. 119.

66. Hempel, "Climate Policy on the Installment Plan," p. 306.

67. Maggie Fox, "USA: New NASA Satellite Gives Insight into Climate," *Reuters* (19 April 2000); Reuters, "Pollution Revealed by NASA Satellite," *Washington Times* (20 April 2000): A9.

O'Keefe as Administrator

At the beginning of January, Sean O'Keefe joined NASA as Administrator. He retained Asrar as head of Earth Science. O'Keefe was formerly Deputy Director of the OMB and was appointed mainly to address a $4.8 billion overrun on the International Space Station that Goldin left. He had no strong background in space policy. Nor was he particularly attuned to environmental policy. However, as a former Navy secretary and son of a Navy nuclear submariner, O'Keefe was quite interested in nuclear applications to space. Ed Weiler, Associate Administrator for Space Science, believed NASA needed to use nuclear batteries and eventually fission reactors to maximize science payoffs from deep space probes requiring lengthy stays in orbits where solar energy and chemical propulsion were impractical.

One of O'Keefe's first initiatives was to revive the nuclear propulsion activity at NASA, relatively moribund since Cassini. O'Keefe said that nuclear propulsion should be pursued as a priority. In launching what would eventually be called Project Prometheus, O'Keefe took a very different position from NASA's traditional nuclear strategy—stealth. He said NASA had to be open and proactive about nuclear propulsion. Environmentalists and others needed to understand the risks and benefits, and NASA should engage the potential opposition at the outset. In staffing the initiative, O'Keefe gave attention to hiring people for public outreach as well as developing the science and technology. O'Keefe fully intended Prometheus to go beyond nuclear batteries to nuclear fission reactors, seen as needed for human Mars missions as well as distant robotic expeditions. He did not propose specific destinations, however. Instead, he pushed for developing technologies that could enable the reaching of any destinations. Where nuclear propulsion was concerned, O'Keefe knew he had to engage the environmentalists to get acceptance.[68]

Columbia and the Issue of Debris

On 1 February 2003, the Space Shuttle Columbia disintegrated, an event that dealt a severe blow to NASA. The Earth Science mission of NASA was less affected than human spaceflight, to be sure, but still was hurt as the whole Agency reeled. Columbia had other impacts for NASA's environmental role. For example, the fact that debris from Columbia was strewn over East Texas and Louisiana had dramatically raised awareness of the debris issue generally, and that of nuclear debris in particular. As more spacefaring nations became

---

68. W. Henry Lambright with Agnes Schaefer and Jessica Widay, "Federal Agency Strategies for Incorporating the Public in Decision-Making Processes: Case Studies for NASA," report to NASA, 18 April 2005, pp. 46–62.

active, the orbital debris problem would grow, and there was a possibility of debris falling from orbit to an inhabited part of the planet. NASA certainly had already been conscious of the debris-from-orbit problem, and the issue had come up in the Cassini dispute. It had also become a mild issue in considering Space Station risks. In 2001, the Station had to dodge a large tool that an astronaut had previously lost during some external activity. It had to move to a slightly higher orbit. This event had not gotten much publicity, but Columbia made debris a media highlight. Now many people were aware of the debris issue, and NASA was fortunate that Columbia debris did not cause a disaster on the ground. Also, the debris question had to be considered in Prometheus planning.

A more positive issue became apparent in 2003, as NASA-supported researchers reported that the rate of ozone depletion was decreasing. This meant that the Montreal Protocol was working as intended. Evaluating how well this treaty was working by monitoring ozone depletion had become accepted as part of NASA's environmental role, building on NASA's previous work in ozone depletion.[69] This fact also raised a problem for NASA and the nation. NASA's R&D mission required it to advance technology, not simply monitor ozone treaty compliance routinely. What were the boundaries in NASA's environmental mission: Could not NOAA or some other agency take over NASA's ozone-monitoring functions and pay the bill? NASA needed money to tackle new missions, especially in climate change, and the missions had to be defined and approved by the administration and Congress. And there was the perennial problem of Landsat, which had gone through various generations and now, in its seventh, seemed destined to die. Whose responsibility was it? NASA had it by default and wished to disengage.

Aqua and Future Planning

On 4 May 2002, NASA launched the second of its EOS series: Aqua. As its name implied, Aqua's mission was to study the global water cycle. Costing approximately $1 billion, Aqua, like Terra, had a long-term (15-year) mission to build a comprehensive database for detecting environmental changes. As Terra emphasized land surface connections with climate change, so Aqua would concentrate on water dynamics, such as precipitation and evaporation.[70] Meanwhile, EOSDIS, the $2.5 billion computer system, was grappling with the massive data sorting and disseminating task the EOS required

---

69. Lambright, *NASA and the Environment*, pp. 44–45.
70. Brian Berger, "NASA Aqua Mission To Study Global Water Cycle," *Space News* (29 April 2002): 18.

to make the satellites useful. Aqua sent back pictures of sea surface temperature and brightness. Like Terra, Aqua involved other nations (Japan and Brazil). Enlisting international partners had become one way Earth Science managers at NASA had coped with cutbacks in the 1990s. It was a strategy to share costs to retain as much of the science as possible.

In early December 2002, a three-day meeting concerned with planning the new U.S. Climate Change Research Initiative took place. All indications were that Bush, like his father, would support a broad interagency research program. This was good news for NASA and promised to keep the Agency active in global environmental monitoring. The question was what, specifically, it would do. NASA needed decisions. Whatever it did would presumably be linked with an international initiative being planned—a "system of systems."

A System of Systems

The U.S. State Department held an international Earth Observation Summit in the summer of 2003. Many nations attended. The summit was in part a response to the "public diplomacy" problem the President had created with his Kyoto decision. It was an effort for the United States to project leadership in climate change rather than appear a negative force. It was also a sensible idea technically. NASA hoped to build on its remote sensing capability. Other nations had environmental monitoring satellites. Why not link these capabilities into a "system of systems?" the State Department conference asked. Following the conference, NASA announced that it would develop, on an accelerated pace, a new climate change-oriented satellite called Glory. It would incorporate an advanced sensor. Glory would go up in 2007, two years earlier than previously planned. It had been intended that the new sensor would go on a multiagency, operations-oriented technical system called the National Polar-Orbiting Operational Environmental Satellite System (NPOESS) to be launched at the end of the decade. Glory was intended to demonstrate "commitment to studying the causes and consequences of climate change."[71]

The EOS Is Complete: What Next?

On 14 January 2004, President Bush unveiled a new vision for space exploration at NASA: "to the Moon, Mars, and Beyond." In line with the new priority, Administrator O'Keefe subsequently reorganized the Agency, creating a new Exploration Systems Mission Directorate and streamlining other units. He merged NASA's separate Earth Science and Space Science programs into a

---

71. Brian Berger, "Reversing Course," *Space News* (25 July 2005): 10.

single organization, the Science Mission Directorate. In July, NASA launched Aura, the third and final leg of the EOS. Another billion-dollar satellite, Aura would make comprehensive measurements of the atmosphere and also take over for the aging UARS in monitoring ozone trends. It would take the closest look ever at the smoke, aerosols, and other pollutants affecting air quality around the world. The EOS was now complete—land, water, and air.[72] Pulling the data together and determining its meaning was where EOSDIS and the cadre of Earth Science users around the world came in.

At the same time, State Department–initiated planning went forward for linking EOS and post-EOS satellites to the Global Earth Observation System of Systems (GEOSS), as it was called. Fifty nations were now involved in planning this global network. Also in 2004, Mahoney built on 2003 meetings and drafts and produced the finalized Bush Climate Change Science Program strategy. It gave a new legitimacy, under Bush, for larger-scale R&D on climate change and its impacts and role for NASA. The issue critics raised in connection with the report was whether the administration would back the words with money.[73] The financial impact of the President's Moon-Mars vision and NASA reorganization so far did not appear good for Earth science. A House Science Committee analysis found that of NASA's fiscal year 2006 budget request of $5.47 billion for the Science Mission Directorate, only $1.36 billion would be spent on Earth science, a drop of 8 percent below the 2005 level and 12 percent less than the 2004 level. Projections showed a further decline in fiscal year 2007.[74]

Organizations representing the Earth science community complained, as did Congressman Sherwood Boehlert (R-NY), Chair of House Science Committee. Boehlert argued that although he agreed with the exploration priority, it did not have "a blank check." He wanted greater "balance" in NASA programs. A man with environmental credentials, Boehlert made it clear that he wanted to protect Earth Sciences.[75] O'Keefe had worked hard to get the exploration initiative off to a good start. NASA was trying to do something large and new at a time when the nation was engaged in war and facing huge deficits. He used the Exploration Vision to provide a vehicle for deciding

---

72. Tariq Malik, "Last EOS Satellite To Study Air Quality at Earth's Surface," *Space News* (14 June 2004): 8.

73. Richard Kerr, "Climate Change Gets a Qualified Go-Ahead," *Science* (27 February 2004): 1269–1271.

74. Brian Berger, "NASA's Exploration Focus Blamed for Earth Science Cuts," *Space News* (2 May 2005): 4.

75. Ibid.

priorities. The closer to this Exploration mission, the higher the priority programs had. Earth Science seemed distant from the Exploration Vision.

In its effort to save money in Earth Science, NASA decided to cancel the Glory mission and put the spacecraft's featured sensor, a greenhouse gas–measuring instrument, back on NPOESS as originally intended. That was a decision made as part of the fiscal year 2006 budget process under O'Keefe. O'Keefe left NASA in February, and his successor, Mike Griffin, who joined NASA in April, had to deal with criticism of the budget cuts to Earth Science in general and the cancellation of Glory in particular when he became Administrator. The National Academy of Sciences and the National Research Council (NRC) criticized the Earth Science cuts. It produced a report, "Earth Science and Applications from Space: Urgent Needs and Opportunities to Serve the Nation," which questioned the technical feasibility of NASA's relying on NPOESS to host sensors from canceled or scaled-back missions. Glory was but the most notable example of a disturbing trend, the NAS-NRC said, for NASA to depend on NPOESS for its Earth science future. It said that Glory's termination should be reconsidered. Five legislators, including Boehlert, protested the Glory cut. In July, NASA announced that it was reversing course and reinstating the Glory mission. It was now scheduled for 2008 and would incorporate a second instrument, making it even more consequential in climate change research.[76]

The issue of Glory pointed to a conflict between environmental policy goals seen in the State Department initiative and interagency climate change research strategy on the one hand, and the President's space vision on the other. The former favored NASA's Earth Science program. The latter (the President's Exploration policy) put it at risk in NASA priority-setting. A devastating tsunami that hit nations bordering the Indian Ocean in late December 2004 gave Earth Science funding proponents ammunition, however. In mid-February, more than 60 governments approved the GEOSS project, established a secretariat to run it, and approved a 10-year implementation plan. This interagency, international effort required a NASA role.[77]

Griffin as Administrator

The reversal of the Glory decision was but one of a number of changes Griffin was making in decisions inherited from O'Keefe. Another important decision relevant to the environment was to scale back and reorient Prometheus,

---

76. Berger, "Reversing Course."
77. Missy Frederick, "Government Industry Officials Tout Benefits of GEOSS," *Space News* (1 August 2005): 20.

O'Keefe's nuclear initiative. Griffin had to find money somewhere to aug-
ment the Moon-Mars effort. In cutting back, he looked to find funds in
programs whose objectives were very distant or, in his view, not well thought
out to help pay for more immediate priorities, including an augmentation of
Earth Science funds. Griffin clearly had his work cut out for him, implementing the Bush vision while seeking "balance" among NASA programs at a time
of budget constraint. With the EOS now at last a reality, it was obvious in
2005 that NASA had evolved over the years a significant environmental mission, and that mission had enough momentum and support to keep it going.

Issues had to do with future direction, content, and funding. As to direction, climate change and global warming provided a long-run rationale for
the EOS and certain specialized satellites such as Glory. The problem was
how to make room for satellites that performed special or gap-filling roles,
such as the Ice, Cloud, and land Elevation Satellite (ICESat), launched in
2003, did. ICESat took images of the melting Arctic ice caps, the early warning signals of global warming. The old satellites had to give way, in terms
of funding, for newer priorities, like ICESat. But phasing out satellites was
difficult, as NASA found not only with Landsat, but with TRMM. This
rainfall-measuring satellite was slated for termination by NASA's senior program manager. But users and their legislative allies argued that it still functioned and should be maintained. When Hurricane Katrina devastated the
Gulf Coast in late August, early September 2005 satellites of various kinds
from different providers were shown to be extremely useful throughout the
disaster cycle—from early warning through event to postdisaster planning.
The political atmosphere of this time made killing TRMM difficult indeed
for Griffin.

NASA, thanks to Bush's January 2004 decision, had a strategy for human
spaceflight—a vision for the future that served as a basis for prioritizing. It
needed a similar vision and long-term strategy for its role in the environment.
The questions of what would be next beyond the EOS as a centerpiece and
why remained to be answered. Until NASA had a clear strategy, its decisions
would seem ad hoc and fragmented, and its Mission to Planet Earth would
be vulnerable to critics.

Conclusion

NASA has had a major role relative to the environmental movement from
its beginning. As NASA got under way in the in the 1960s, so too did the
modern environmental movement. NASA influenced environmental values
most dramatically in the early days of space through images of Earth taken by
Apollo astronauts. Earth Day and the environmentalism of the 1970s owed

much to ideas about Earth as a living organism in which humanity and the physical environment had to coexist. The mission-from-Earth simultaneously led to a better understanding of Earth based on comparative planetology. Along the way, other themes emerged of environmental interest, such as forward and backward contamination, space debris, and energy.

NASA's most direct, important, and sustained contribution to the environmental movement was through its environmental monitoring satellite programs. These were especially encouraged in the 1970s by Administrator James Fletcher. In the 1980s, NASA built on the policy and technical advances of prior years and played a truly critical role in environmental protection as the lead science agency in ozone depletion. The ozone experience, successful not only in science but in connecting science to policy (as with the Montreal Protocol), became a template for an enlarged effort in climate change. It also provided NASA with an early taste of its need to be careful in issuing warnings, given the controversial nature of environmental issues.

The EOS, planned in the 1980s, became the main priority of NASA's environmental program in the 1990s. In turning EOS plans into hardware reality, NASA faced a twofold challenge. First, the EOS was a massive environmental program in an agency whose space priorities lay more with missions *from* Earth. Second, EOS's greatest strength in terms of connection with environmental values—namely, providing knowledge about climate change—could also be an issue controversial with conservatives who saw climate change (that is, global warming) as "political" science. With the shift of congressional control to Republicans in 1995, NASA's political setting altered dramatically. NASA's change of program name, from Mission to Planet Earth to Earth Sciences, was an effort in part to take EOS out of the line of partisan fire. The EOS survived, albeit in a form vastly shrunken from what it had been planned to be. Funding constraints, related in part to partisan political conflict, downsized the program throughout the 1990s. The structure of the EOS, as it emerged in the early twenty-first century, was commendable but did not provide the simultaneous and comprehensive observations originally foreseen.

What's next for NASA in terms of environmental satellites in the twenty-first century? The answer is not obvious. Since this chapter was written, the Obama administration came to power. It terminated the NPOESS program. However, it gave priority to the issue of climate change. A political issue that hurt NASA environmental satellite funding under Bush became an asset under Barack Obama. The problem was that as Congress moved into Republican control in the Obama years, NASA environmental satellites related to climate change again became a target for cutbacks.

Meanwhile, NASA connects with the environment in ways other than through environmental satellites. These include debris and planetary protection from forward and backward contamination. Finally, the more NASA moves outward to "the Moon, Mars, and Beyond" and learns about other worlds, the more that comparative planetology knowledge will contribute to understanding the home planet. NASA will continue to be a space agency with an environmental mission. That mission is also likely to be controversial. While a national need, the environment, especially global warming, takes NASA into a political thicket. NASA is often caught between environmentalists and their supporters who want it to do more and conservatives who want it to do less, if "more" leads to regulations they abhor. Although NASA clearly has a future in the environmental field, the form that mission will take remains a decision in the making.

Chapter 7

Societal Impacts of Applications Satellites

David J. Whalen

First, I believe that this nation should commit itself to achieving the goal, before this decade is out, of landing a man on the Moon and returning him safely to the Earth....

Secondly, an additional $23 million together with $7 million already available, will accelerate development of the Rover nuclear rocket....

Third, an additional $50 million will make the most of our present leadership, by accelerating the use of space satellites for worldwide communications.

Fourth, an additional $75 million—of which $53 million is for the Weather Bureau—will help give us at the earliest possible time a satellite system for worldwide weather observation.

—John F. Kennedy, "Special Message to the Congress
on Urgent National Needs," 25 May 1961

Part I: Beginnings

Defining what is or is not an applications satellite is occasionally difficult. President Kennedy's two choices, weather and communications satellites, easily fit the bill. Perhaps navigation satellites are also clearly applications satellites. Spy satellites also seem to fit, but I would maintain that Landsat—and possibly commercial land remote-sensing satellites—do not fit. The dictionary has several meanings for "application" that suggest the difference between applications satellites and others—"an act of putting to use: capacity for practical use." This suggests that practical use may be a helpful differentiator. I would add several more constraints: continuous use, and not used exclusively for science or R&D. One more constraint that I would like to add is the requirement that someone be willing to pay for the application satellite—someone other than NASA.

There are three types of applications satellites that have been successfully introduced since the launch of Sputnik in 1957: remote sensing (including reconnaissance and weather), communications, and navigation. Remote sensing initially was the most important. The United States was faced with an enemy whose country, whose entire society, was completely closed. Reconnaissance satellites were among the first space projects pursued—even before Sputnik—in the hope that they would provide greater knowledge about the threat posed by the Soviet Union. Navigation satellites—especially those for use by Polaris submarines—were also seen as having an important military application. In the past decade, navigation satellites have become the most pervasive "dual-use" (military and civilian) satellites.

Principal civilian applications have been communications and weather forecasting. It should be no surprise that Kennedy included these two in his famous 25 May 1961 speech.[1] By the time Kennedy made this speech, Hughes Aircraft Company, AT&T, ITT, and other corporations had invested almost $100 million in communications satellite R&D. These companies saw satellite communications as a profitable commercial enterprise. Others saw satellite communications as something the government should do. Weather satellites did not have a powerful champion at first. Academia was initially more interested in "science" than in applications, more interested in radiation budget experiments than in cloud pictures. But the television infrared observation satellite (TIROS), a product of the reconnaissance satellite program, soon drew interest from academia and the Weather Bureau. Polar-orbiting satellite observations have been continuous since 1960. It would be another decade before geosynchronous meteorological satellites—the kind used to bring weather pictures to our televisions today—would be a reality.

All of these applications satellites have had major impacts on society in the second half of the twentieth century. Communications satellites have had the greatest economic impact. They will soon be a $200 billion industry. Reconnaissance satellites had the greatest geopolitical impact. They probably made the world a safer place, especially in the 1960s and 1970s. Weather satellites have saved thousands of lives and billions of dollars. Civilian use of navigation satellites and specifically the Global Positioning System (GPS), which was originally a military application, has become widespread. Aircraft, ships, boats, cars, and hikers have all become dependent on what is now the most obvious dual-use system. Remote sensing includes reconnaissance, weather, and land-use applications. All three of these have had successes, but land-use observation satellites (civil remote sensing) have never found a champion.

---

1.   "President John F. Kennedy's Challenge to the Nation," NASA, *http://www.nasa.gov/topics/history/features/john_f_kennedy.html* (accessed 24 October 2014).

Started by NASA, transferred to NOAA, and finally spun off to industry, civil and commercial remote sensing have found only one major paying customer: the government—especially the intelligence agencies. This chapter discusses the development of these applications satellites and their societal impacts, which have changed over time. The twentieth-century impacts are still only sketchily known. The twenty-first-century impacts of these and other applications satellites remain to be seen.

1. Before *Sputnik*

As early as the nineteenth century, there were always some who speculated about space. Science-fiction writers were the earliest, but they were quickly followed by rocket pioneers, science fact writers, and the military. Between the end of World War II and the actual launch of Sputnik, interest in space increased dramatically, especially interest in applications satellites.

Science Fiction

Possibly the first science-fiction story about applications satellites was 1869's "The Brick Moon" by Edward Everett Hale (1822–1909).[2] Hale's Moon was designed to act as a navigational beacon for sailors. Jules Verne (1828–1905) is the best-remembered science-fiction writer of the nineteenth century, but he never seems to have written about applications satellites. H. G. Wells (1866–1946) began writing science fiction in the 1890s, but also seems to have neglected applications satellites. George O. Smith (1911–1981) published a short story in 1942, "QRM-Interplanetary," which may have put the idea of communications satellites in the heads of Arthur C. Clarke and John R. Pierce. All of these writers contributed to the excitement and glamour of space but don't seem to have been excited about communications satellites.

Rocket Pioneers

Konstantin Eduardovitch Tsiolkovsky (1857–1935) was born in Russia to a Polish father on 17 September 1857. His seminal article "Exploration of the Universe with Reaction Machines" was first published in the monthly magazine *The Science Review* in 1903. Like many other early writers on space, Tsiolkovsky emphasized the conquest of space rather than space applications. His writings on rocketry were probably the first to explore the science and

---

2.   Edward Everett Hale, "The Brick Moon," *Atlantic Monthly* 14, nos. 10–12 (1869): 451–460, 603–611, and 679–688. Serialized in *The Atlantic, 1869–1870.*

technology of rocket engines. His contributions were eventually recognized by the Soviet Union.

Another pioneer was the American Robert Hutchins Goddard (1882–1945). Like Tsiolkovsky's, Goddard's writings were primarily concerned with the development of rocket technology. Unlike Tsiolkovsky, however, Goddard did not wax lyrical about the wonders of spaceflight—at least in part because some of his early work on rockets was ridiculed. He does not seem to have written much, if anything, about applications satellites. Hermann Julius Oberth (1894–1989) was born in Transylvania, but was ethnically German. His doctoral dissertation on rocketry was rejected by the University of Munich in 1922. In 1923, Oberth published "Die Rakete zu den Planetenräumen" (The Rocket into Interplanetary Space)—a 92-page version of his dissertation that was expanded to 429 pages in 1929. He was later associated with Wernher von Braun's group.

Science Fact

Herman Potočnik (1892–1929), who wrote as Hermann Noordung, was a Slovene rocket engineer and pioneer of cosmonautics (astronautics). He is chiefly remembered for his work addressing the long-term habitation of space. At the end of 1928, Potočnik published his sole book, *Das Problem der Befahrung des Weltraums: Der Raketen-Motor (The Problem of Space Travel: The Rocket Motor)*, in Berlin.[3] He conceived a space station in detail and calculated its geostationary orbit. This is apparently the earliest reference to the use of twenty-four-hour orbits. Arthur C. Clarke (1917–2008), who had apparently never heard of Potočnik, published an article in the October 1945 issue of *Wireless World* entitled "Extra-terrestrial Relays."[4] In this article, Clarke discusses the advantages of twenty-four-hour geostationary orbits that would allow a satellite to maintain position over the same portion of the equator indefinitely. Clarke foresaw the use of space stations at this altitude for radio and television broadcasts. In the following decade (1954), Clarke wrote to Dr. Harry Wexler, then chief of the Scientific Services Division of the U.S. Weather Bureau, about satellite applications for weather forecasting.

RAND, a Douglas Aircraft Company R&D unit at the time, published its famous "Preliminary Design of an Experimental World-Circling Spaceship" on 2 May 1946.[5] Chapter 2 of this document, drafted by Louis Ridenour, was

---

3. Hermann Noordung, *Das Problem der Befahrung des Weltraums: Der Raketen-Motor* (Berlin: Richard Carl Schmidt & Co, 1929).
4. Arthur C. Clarke, "Extra-terrestrial Relays," *Wireless World* 51, no. 10 (1945): 305–308.
5. RAND, "Preliminary Design of an Experimental World-Circling Spaceship," SM-11827, 2 May 1946.

titled "Significance of a Satellite Vehicle." Greatest significance is given to the use of rockets (satellite vehicles) as bombardment vehicles, but next in importance is the observation capability of a satellite over enemy territory. This observation capability would allow accurate bomb damage assessment after raids and weather observation before the raids. The document also discusses the advantages of satellites as communications relay stations. The simplicity of operations if satellites are in geostationary orbits is addressed in passing. It appears that the RAND authors had not read Clarke's article. The value of then-current communications through the ionosphere is given as $10 billion.

In 1947, RAND published the first of many follow-ups to the 1946 report. These reports, prepared under the direction of James E. Lipp, covered a variety of topics. The 1947 report, "Communication and Observation Problems of a Satellite," continued the discussion of satellite communications and brought up the issue of a "spy satellite" for the first time.[6] RAND continued its studies of reconnaissance and weather satellites. In 1951, RAND published a report titled "Inquiry into the Feasibility of Weather Reconnaissance from a Satellite Vehicle," by William Kellogg and Stanley Greenfield, and another on the "Utility of a Satellite Vehicle for Reconnaissance," by James E. Lipp, Stanley M. Greenfield, and R. S. Wehner.[7] Perhaps more important for the space race was an earlier RAND report, "The Satellite Rocket Vehicle: Political and Psychological Problems."[8] This document was considered by Walter McDougall as "the birth certificate of American space policy."[9]

John R. Pierce (1910–2002) was, like Clarke, a science-fiction writer, but he was also an engineering manager at Bell Telephone Laboratories (BTL). In "Don't Write: Telegraph," published in *Astounding Science Fiction* in 1952, Pierce discussed some possibilities regarding communications satellites.[10] In 1954, he was asked to give a space talk to the Princeton section of the IRE (the Institute of Radio Engineers, now the Institute of Electrical and

---

6. D. K. Bailey and A. S. Mengel, "Communication and Observation Problems of a Satellite," RA-15028, RAND, 1 February 1947.

7. William Kellogg and Stanley Greenfield, "Inquiry into the Feasibility of Weather Reconnaissance from a Satellite Vehicle," RAND R-218, RAND, April 1951; James E. Lipp, Stanley M. Greenfield, and R. S. Wehner, "Utility of a Satellite Vehicle for Reconnaissance," RAND R-217, RAND, April 1951.

8. Paul Kecskemeti, "The Satellite Rocket Vehicle: Political and Psychological Problems," RAND RM-567, RAND, 4 October 1950.

9. Walter A. McDougall, … *the Heavens and the Earth: A Political History of the Space Age* (New York: Basic Books, 1985), p. 108.

10. J. J. Coupling [John R. Pierce], "Don't Write: Telegraph," *Astounding Science Fiction* 49 (March 1952). Pierce wrote at least 20 articles for *Astounding Science Fiction* under his pen name, J. J. Coupling, and at least one under his real name.

Electronics Engineers [IEEE]).[11] According to Pierce: "The idea of communication satellites came to me. I didn't think of this as my idea, it was just in the air. Somehow, I had missed Arthur Clarke's paper on the use of manned synchronous satellites for communication." In 1958, Pierce and his colleague Rudolf Kompfner prepared a presentation on satellite communications for a conference. This presentation was later published in the *Proceedings of the IRE*, in March 1959.[12]

The Military

The Navy, the Army, and the Air Force all pursued space applications. The Navy may have been first with its Committee for Evaluating the Feasibility of Space Rocketry (CEFSR), which may have spurred the (Army) Air Force to sponsor the RAND study. The Navy followed up by sponsoring many of the scientific payloads that were launched from White Sands on German V-2s. When the V-2s were exhausted, the Naval Research Laboratory (NRL) funded development of the Viking sounding rocket built by the Glenn P. Martin Company. The Navy Viking and Aerobee sounding rockets would later be the precursors to the Vanguard (improved Viking, Aerobee, and solid) launch vehicle—and eventually the Thor-Delta (Thor, Aerobee, and solid) rocket. The Army V-2 "Rocket Team" under von Braun would eventually become part of the Army Ballistic Missile Agency (ABMA) at Huntsville, Alabama. The (Army) Air Force Intercontinental Ballistic Missile (ICBM) program was canceled just as the Air Force became a separate service in 1947. The Korean War and the Soviet nuclear test slowly brought the Air Force ICBM program back to life and encouraged the Air Force (spurred by RAND) to look at the possibility of developing reconnaissance and other applications satellites.

2. The Pioneers: After Sputnik

In the immediate aftermath of Sputnik, many programs that had been sitting on the shelf or suffering from low priority suddenly became high-priority programs. The military program emphasized reconnaissance, and the NASA program by 25 May 1961 emphasized putting a man on the Moon; but in the short term, applications satellites looked to be productive, cheap, and doable.

---

11. This talk was later published as John R. Pierce, "Orbital Radio Relays," *Jet Propulsion* 25 (April 1955): 153–157.
12. John Robinson Pierce and Rudolf Kompfner, "Transoceanic Communication by Means of Satellites," *Proceedings of the IRE* 47 (March 1959): 372–380.

Reconnaissance

RAND continued to be a major proponent of satellite reconnaissance, but RAND was a think tank, not a manufacturer or an operator. In the mid-1950s, a series of RAND studies—produced by the von Neumann committee (Teapot) and the Killian committee (TCP)—created a major change in U.S. strategic thinking. The von Neumann committee recommended an upgrading of ICBM program design and priority (specifically the Atlas) to take into account the lighter, more powerful thermonuclear weapons then available. The Killian committee found that the United States was vulnerable to a surprise attack and recommended that a more capable reconnaissance satellite be designed and fielded. All of the early RAND recommendations had been for a "direct readout" satellite—one that transmitted pictures to the ground electronically. Many of the studies assumed a standard television camera. Weapon System 117L (WS-117L) was originally an Air Force direct readout satellite.

By 1957, members of the Killian committee were dissatisfied with the Air Force program; they wanted a "film return" satellite, and they wanted the program to be managed by the CIA. The success of the U-2 seemed to indicate that the CIA was better at bringing new technology into operation in a short period of time. Within a few months, the WS-117L program had been reoriented to include CORONA (film return), MIDAS (early warning), and SAMOS (direct readout—later to include film return). The first Discoverer (CORONA) launch was on 28 February 1959; it was a failure—as were most launches over the next two years. The first successes were in August 1960, when space reentry vehicles (SRVs) were recovered from the ocean and in midair. From 1959 to 1972, almost 150 CORONA (KH-1 through KH-4B) satellites were launched on Thor-Agena vehicles. After August 1960, most were successful.[13]

Navigation

In the days immediately following the launch of Sputnik in October 1957, scientists and engineers worked to analyze the spacecraft's signal and its orbit. The Minitrack system designed for Vanguard provided data from which orbits could be calculated, as did optical telescopes and radio telescopes. Bill Guier

---

13. Much of the information on CORONA comes from Merton E. Davies and William R. Harris, "RAND's Role in the Evolution of Balloon and Satellite Observation Systems and Related U.S. Space Technology," RAND R-3692-RC, September 1988; and Dwayne A. Day, John M. Logsdon, and Brian Latell, *Eye in the Sky: The Story of the Corona Spy Satellites* (Washington, DC: Smithsonian Institution Press, 1998).

and George Weiffenbach of the Johns Hopkins University Applied Physics Laboratory (APL) had none of these instruments, but they could listen to the satellite's signal and monitor the change in its frequency due to the Doppler effect. They used this Doppler shift to compute an orbit for the Soviet satellite. Frank McLure (1916–1973) realized that if the orbit were known, the Doppler information could be used to determine the position of the radio receiver on the ground.

In 1958, McLure described the potential for developing a space-based navigation system to Ralph E. Gibson, APL's director. Within a few weeks, McLure and Gibson proposed a navigation system to the Navy. All space projects were transferred to the Advanced Research Projects Agency (ARPA) in 1958, but in 1959 responsibility for Transit—more formally the Navy Navigation Satellite System (NNSS)—was assigned to the Navy. In 1961, the Kennedy administration made the Air Force responsible for all space programs. Success came in spite of "musical chairs" management. The earliest Transits were launched from Cape Canaveral on the Thor-Able and Thor Able-Star. Transits 1A through 3B were spherical. Later Transits were drum-shaped to provide more real estate for solar cells. The last two experimental Transit satellites demonstrated that precise navigation was possible using two frequency beacons broadcasting the satellite ephemerides (orbits). This system was so robust that it was capable of determining the harmonics of Earth's gravitational field and the effects of propagation through the ionosphere. The last satellites were also able to demonstrate the availability of the satellites when in a near-circular orbit at about 1,000 kilometers and inclined about 66 degrees. Transits 4A and 4B also demonstrated the use of nuclear power rather than solar cells. The Transit Research and Attitude Control (TRAC) satellite was launched with Transit 4B to demonstrate the benefits of gravity-gradient stabilization.

The prototype operational satellites, the Transit 5 series, were meant to provide an operational capability while looking at some final design tradeoffs. The most important of these was probably the comparison of nuclear power and solar cells. An additional complication was the decision to launch the operational satellites on the Scout launch vehicle rather than the Thor Able-Star. The nominal reason was lower cost, but the Thor Able-Star had been replaced by the higher performance Thor-Delta, which was in many ways an improved Thor Able-Star. Unfortunately, the Delta was not yet operational from Vandenberg, where polar launches were performed. All experimental launches had been made from Cape Canaveral, but the operational satellite would be in a polar orbit that could be achieved only from Vandenberg. Scout had a much lower payload capacity than either Thor variant. The Transit 5A series used solar power, the 5B series used nuclear power, and the 5E series measured the space environment. The A series was launched on Scout. The B and E series were launched as a dual payload on Thor Able-Star. Transit 5C1 was an improved A series satellite. After the last 5B series satellite, the decision was made to use solar power because of price and politics.

The operational satellites were called Oscars and were initially built by the Naval Avionics Facility Indianapolis (NAFI). NAFI-built Oscars were launched on Thor but lasted only a few weeks. Oscars 1, 2, 3, 5, and 7 were assembled at NAFI and launched, but Oscars 4, 6, 8, 9, and 10 were reworked and assembled by APL before launch. The APL-refurbished satellites worked for about a year. It was eventually discovered that a thermal problem caused the satellites to fail at about the one-year mark. Oscars 11–17 were built at APL. A contract was signed with RCA to build Oscars 18–32. Starting with Oscar 12, Transit satellites tended to last a decade or more. This caused a problem: there were too many satellites on hand. Eventually, they were placed in storage. By the late 1980s, the Scout launch vehicle could launch two Oscars at once. This procedure was known as SOOS (stacked Oscars on Scout). The TIP (Transit Improvement Program) and Nova satellites continued the Transit line, but by the 1980s a new Air Force program was beginning to replace Transit: the Navstar or GPS (Global Positioning System) satellites.

APL built the original Transit satellites for updating the position information on the inertial navigation systems of Polaris submarines. The last group of transit satellites, Oscars 18–32 and the Novas, were built by RCA. The satellites broadcast ephemeris information continuously at 150 and 400 megahertz. The final constellation consisted of six satellites in a polar orbit with a nominal 600-nautical-mile altitude; others were "stored-in-orbit" spares. The first Transit launch was in 1960; the last Transit launch was in August 1988. The system was operational in 1963 and opened to commercial use in 1967. The Transit program terminated navigation service in 1996. Predictable positioning accuracy was 500 meters for a single-frequency receiver and 25 meters for a dual-frequency receiver. Coverage was worldwide but not continuous because of the relatively low altitude of the Transit satellites and the precession of satellite orbits.[14]

Weather

Weather observation in the United States goes back to early colonial times, but systematic recording of weather observation started in 1849 with a cooperative agreement between the Smithsonian Institution and the telegraph

---

14. Much of the material on navigation satellites comes from the Transit History issue of *Johns Hopkins APL Technical Digest* 19, no. 1 (1998).

companies. A National Weather Service was authorized in 1870 and implemented as part of the Army Signal Service. The service was transferred to the Department of Agriculture in 1891 and named the U.S. Weather Bureau. In 1940, the Bureau was transferred to the Department of Commerce.[15] By the 1950s, the idea of weather satellites was beginning to surface. In 1951, RAND published "Inquiry into the Feasibility of Weather Reconnaissance from a Satellite Vehicle," and Arthur C. Clarke depicted polar and geosynchronous "metsats" in the endpapers of *The Exploration of Space*.[16] In 1954, a tropical storm was discovered accidentally when pictures taken from an Aerobee sounding rocket were analyzed. Also in 1954, Dr. Harry Wexler, the Weather Bureau's chief scientist, presented a paper titled "Observing the Weather from a Satellite Vehicle" at the Third Symposium on Space Travel.[17]

In 1955, when the decision was made to launch a satellite during the upcoming International Geophysical Year (IGY), weather observation and radiation balance payloads were considered and eventually flown on Vanguard and Explorer satellites. Harry Wexler (1911–1962) worked at the U.S. Weather Bureau from 1934 until his death in 1962—with some time out from 1942 to 1946, when he served with the Army in World War II. He received his undergraduate education at Harvard and his Ph.D. in meteorology from MIT. Wexler was a proponent of weather observation in space for many years. His 1954 paper, advice on the IGY satellite payloads, and support of television infrared observation satellite (TIROS) were critical to the successes of satellite meteorology. He was a supporter of Verner Suomi's experiments on Explorer.

Verner Suomi (1915–1995) taught high school science in Minnesota until World War II, when he joined the Civil Air Patrol, where he first acquired knowledge of—and a passion for—meteorology. The war brought him to the University of Chicago, where he trained air cadets in basic meteorology. Suomi received his Ph.D. from the University of Chicago in 1953. His doctoral thesis involved measuring the heat budget of a cornfield. Measuring the difference between the amount of energy absorbed and the amount of energy lost in a cornfield led him to think about Earth's heat budget—especially Earth's albedo. The obvious way to measure albedo was from space. Suomi's radiometer was flown on Explorer VII in 1959.

15. Much of the material in this and subsequent paragraphs is from J. Gordon Vaeth, *Weather Eyes in the Sky: America's Meteorological Satellites* (New York: Ronald Press Company, 1965).
16. Arthur C. Clarke, *The Exploration of Space* (New York: Harper, 1951).
17. Harry Wexler, "Observing the Weather from a Satellite Vehicle" (paper presented at the Third Symposium on Space Travel, held at the Hayden Planetarium, New York City, 4 May 1954).

In spite of the influence of scientists like Wexler and Suomi, the first weather satellite was a product of the military. TIROS was RCA's losing entry in the Air Force WS-117L competition that was won by Lockheed in 1956. The Army was persuaded to support the development of TIROS as a polar-orbiting weather satellite. The project was transferred to ARPA and eventually to NASA in 1958. The satellite had two television cameras: one wide-angle and one narrow-angle (high-resolution) on TIROS-1 and -2, both wide-angle on succeeding TIROS satellites. TIROS-8 pioneered the Automatic Picture Transmission (APT) camera system. TIROS satellites had the cameras mounted on the base of the satellite, aligned with the spin axis. This meant that the cameras were Earth-pointing for only a small fraction of their orbits. TIROS-9 pioneered the "cartwheel" configuration, wherein the cameras were mounted on the sides of the spacecraft, the spacecraft spin axis was aligned with orbit-normal, and pictures were taken continuously. All launches were from Cape Canaveral into high-inclination (48-degree) orbits until TIROS-9 and -10 were launched into sun-synchronous polar orbits. Sun-synchronous orbits (SSO) allowed pictures to be taken at the same local time every day (usually early morning).

The Pentagon recognized the disadvantages of the TIROS baseplate-mounted cameras and the advantages of sun-synchronous orbits. Joseph V. Charyk, director of the National Reconnaissance Office (NRO), concluded that NASA development of a better weather satellite would be delayed and expensive. He also was uncomfortable with the international commitments NASA had made to share TIROS weather pictures. Weather information was critical to NRO—too many pictures showed nothing but clouds. In 1961, Charyk sponsored what was to become the Defense Meteorological Satellite Program (DMSP). The program envisioned an improved RCA TIROS launched on a Scout launch vehicle from Vandenberg. The satellite was much lighter than TIROS and carried a single television camera that would "snap" pictures of Earth when the horizon sensors indicated that the camera was pointed in the appropriate direction. Scout launch-vehicle failures led to a decision to use a Thor with a new second stage (Burner). Unfortunately, that launch vehicle would not be available for more than a year. In order to maintain an "intelligence" weather-observation capability, the expensive Thor-Agena combination would be used. This allowed two DMSP/P-35 satellites to be launched together on one rocket. The Thor-Agena variant was known as block 3A (presumably TIROS was block 1A and the Scout-launched DMSP was block 2A). The first Thor-Burner launch was a failure (a second-stage heat shield did not deploy properly), but the rest were very successful. The block 4A, Thor-Burner–launched satellites were heavier than the earlier DMSP satellites, presumably because they now carried two cameras like the civilian

TIROS. At about this time (around 1963), it was decided that the next-generation civilian weather satellite, TOS (TIROS Operational System, also known as ESSA) would be a copy of DOD's DMSP block 4A rather than the NASA Nimbus, which became a research vehicle and later the model for the Earth Resources Technology Satellite (ERTS/Landsat). The block 5 satellites were three-axis-stabilized rather than spin-stabilized. A variant of block 5 became the civilian Improved TOS (ITOS, also known as NOAA). The DMSP program remained classified until late 1972, when DMSP data were routinely delivered to the Weather Bureau.[18]

Communications

While several pre–World War II mentions of satellite communications have been found, the first well-known discussion was Arthur C. Clarke's 1945 article in *Wireless World*.[19] Perhaps of greater importance were later articles by John R. Pierce in *Jet Propulsion* (1955) and *Proceedings of the IRE* (1959).[20] Clarke was a member of the British Interplanetary Society and a budding science-fiction author. Pierce was also a science fiction author, but, more important, he was the Director of Communications Research at AT&T's Bell Telephone Laboratories (BTL). In early March 1958, Pierce and Rudolf Kompfner of AT&T, independent inventors of the traveling-wave tube, saw a picture of the shiny 100-foot sphere that William J. O'Sullivan of the National Advisory Committee for Aeronautics (NACA) Langley Aeronautical Laboratory was proposing to launch into space for atmospheric research.[21] It reminded Pierce of the 100-foot communications reflector he had envisioned in 1954. The recent invention of the maser amplifier made reception of reflected communications signals more practical than it had been in 1954. He visited the NACA at Langley to confirm his understanding of the sphere, and by the end of the month he was discussing the project with Hugh Dryden in Washington.

Later that summer (July), Pierce and Kompfner participated in an Air Force–sponsored meeting on communications at Woods Hole, Massachusetts.

---

18. Most of the material on DMSP is taken from R. Cargill Hall (an NRO historian), "Chapter Three: Weather Reconnaissance," n.d. (1988?), originally classified TOP SECRET/TALENT KEYHOLE. Other sources include the *http://earth.nasa.gov/history* and *http://www.astronautix.com* Web sites.
19. Clarke, "Extra-Terrestrial Relays," pp. 303–308.
20. John R. Pierce, "Orbital Radio Relays," *Jet Propulsion* 25, no. 4 (1955): 153–157; Pierce and Kompfner, "Transoceanic Communication by Means of Satellites," pp. 372–380.
21. Donald C. Elder, *Out from Behind the Eight-Ball: A History of Project Echo* (San Diego, CA: American Astronautical Society, 1995), p. 25.

They were unimpressed with the plans of the Air Force, which to them seemed unrealistic. While there, Pierce met William H. Pickering of JPL, who had received his Ph.D. from Caltech the year before Pierce. The three engineers discussed among themselves the possibility of launching a sphere such as O'Sullivan's for communications experiments. Pickering volunteered the support of JPL (which eventually resulted in use of the JPL Goldstone station as the West Coast station for Echo). To support this plan, Kompfner and Pierce wrote a paper[22] that they presented at an Institute of Radio Engineers (IRE) conference on "extended range communications" at the Lisner Auditorium of George Washington University in Washington, DC, on 6–7 October 1958.[23] In April 1960, AT&T began to prepare a passive follow-on experiment to Echo. Huge, 3,600-square-foot horn antennas would be built on either side of the Atlantic. Forty-kilowatt transmitters would beam television signals up to a duplicate of Echo in a higher orbit—2,000 miles. Unfortunately, the studies showed that passive satellite TV transmission would be of marginal signal quality. In a 13 May 1960 letter to Leonard Jaffe at NASA Headquarters, Kompfner described the current AT&T/BTL research program as shifting to *active* satellites. In this letter, Kompfner reviews the active satellite component/subsystem studies that had been under way since late 1959.[24] Echo 1 was launched into a 1,000-mile circular orbit on 12 August 1960. During the first orbit of the 100-foot sphere, a recording of President Dwight Eisenhower speaking was transmitted from JPL's Goldstone, California, Earth station to AT&T's Holmdel, New Jersey, Earth station. Later experiments included telephone, Teletype, and facsimile transmissions. On 23 August, the first transatlantic voice transmission was executed from Holmdel to Jodrell Bank, England. In spite of Echo's success, it was clear that active, rather than passive, satellites were the technology to develop.[25] By 1960, Pierce had convinced AT&T management to build and launch a medium-Earth-orbit (MEO) satellite system. Pierce had also energized two young engineers—Harold Rosen and Donald D. Williams—at Hughes Aircraft Company to prove wrong his 1959 argument that geosynchronous-Earth-orbit (GEO) satellites were beyond the state of the art.

AT&T's plan to launch a satellite system was put on hold when NASA refused to provide launch services. NASA argued that launch vehicles were

22. Pierce and Kompfner, "Transoceanic Communication by Means of Satellites."

23. J. R. Pierce, *The Beginnings of Satellite Communications* (San Francisco: San Francisco Press, 1968), pp. 9–12.

24. A. C. Dickieson, "TELSTAR, The Management Story," unpublished monograph, Bell Telephone Laboratories, 1970, pp. 34–39.

25. NASA, *Fourth Semi-Annual Report to Congress* (Washington, DC: GPO, 1961), pp. 10–17.

in short supply and must be rationed. The rationing mechanism would be a competition to design an MEO communications satellite. Proposals were submitted to NASA by seven companies—including Hughes (Syncom) and AT&T (Telstar). RCA won the competition with Relay in May 1961, but AT&T was allowed to purchase launch services; by the end of that summer, the Hughes satellite was jointly funded by NASA and DOD.

Less than a year after the 27 July 1961 AT&T-NASA agreement, AT&T was ready to launch its Telstar 1 satellite. The Andover station—with its huge $10 million, 60-foot by 60-foot horn antenna and maser-amplifier receiver—had been ready since the beginning of the year and was officially operational in April.[26] The Pleumeur-Bodou station in France, essentially a copy of the huge AT&T horn, was started in February and finished on 7 July 1962. Three days of tracking and radio frequency calibration ensued between that date and the launch. The British station at Goonhilly was also ready for launch. On 10 July 1962, at 0825 universal time (UT) (4:25 a.m. EST), the Delta carrying Telstar 1 lifted off from its pad at Cape Canaveral. At 1836 UT (2:36 p.m. EST), Pleumeur-Bodou technicians reported they had been tracking the 136-megahertz VHF telemetry beacon since 1802 UT. At 2045 UT (6:02 p.m. EST), Andover was able to track the 136-megahertz beacon for several minutes, allowing AT&T engineers to examine telemetry data, which indicated that all was well.[27] At 2318 UT (11:18 p.m. EST), the first pass with mutual visibility (both stations could see the satellite) between Andover and Europe began. After telemetry indicated that all was well, the command to turn on the communications repeater was sent. At 2325 UT (11:25 p.m. EST), the Andover transmitter was turned on. At 2347 UT (11:47 p.m. EST), the Pleumeur-Bodou station reported receiving an excellent video signal from Andover. Goonhilly was unable to receive a good signal due to ground station problems (a misunderstanding of the polarization of the signal). AT&T had placed a commercially funded communications satellite in orbit before the government-funded projects, but the Communications Satellite Act would be passed in less than two months. AT&T, after spending more than $100 million (in 1960 U.S. dollars), was out of the satellite manufacturing business for good. The Relay and Syncom satellites would be launched over the next two years.

---

26. The Andover station was also referred to as the Rumford, Maine, station, and as Space Hill.

27. Pierce, *Beginnings*, pp. 161–162.

The Communications Satellite Act of 1962

Just after NASA's announcement of the Relay communications satellite program award to RCA, President John F. Kennedy delivered a speech to Congress on "urgent national needs." In this famous 25 May 1961 speech, Kennedy promised to land a man on the Moon and also asked the Congress to provide the funds that "will make the most of our present leadership, by accelerating the use of space satellites for worldwide communications." This speech has been characterized as being driven by the events of April 1961—Yuri Gagarin's orbital flight and the Bay of Pigs—but his comments echo the Wiesner Committee's "Report to the President-Elect of the Ad Hoc Committee on Space," delivered to Kennedy on 10 January 1961.[28] They are also consistent with his State of the Union message of 30 January 1961:

> Finally, this Administration intends to explore promptly all possible areas of cooperation with the Soviet Union and other nations "to invoke the wonders of science instead of its terrors." Specifically, I now invite all nations—including the Soviet Union—to join with us in developing a weather prediction program, in a new communications satellite program and in preparation for probing the distant planets of Mars and Venus, probes which may someday unlock the deepest secrets of the universe.[29]

In any case, politics—Cold War politics—would be a driver in deployment of communications satellites.

NASA Administrator James Webb believed—as did Jerome Wiesner (President's Science Advisory Committee, or PSAC), Lee Loevinger (Department of Justice), Philip J. Farley (Department of State), and Edward Welsh (Space Council)—that satellite communications was an inherently governmental function. Eisenhower and his NASA Administrator, T. Keith Glennan, believed that telecommunications was a private function in the United States. The Federal Communications Commission (FCC) believed that satellite communications should be managed by the international telecommunications companies—much as they had managed the transoceanic telephone cables. NASA believed that the Agency should be in charge of satellite communications. The Space Council—especially the staff as directed by Welsh—worked very hard to address all the issues. Monopoly was of great

---

28. Wiesner Committee, "Report to the President-Elect of the Ad Hoc Committee on Space," 10 January 1961, *http://www.hq.nasa.gov/office/pao/History/report61.html.*

29. John F. Kennedy, "State of the Union," 30 January 1961, *http://www.infoplease.com/t/hist/state-of-the-union/174.html.*

concern. Many were concerned that, while avoiding the Scilla of AT&T, they were creating the Charybdis represented by what later became COMSAT.

The Space Council drafted an Administration Bill in November 1961. It was delivered to Kennedy in December. The Administration Bill provided for a public-private corporation directly regulated by the President. Before it was submitted, Senator Robert Kerr (D-OK) submitted a similar bill that gave more control to the international communications carriers—as the FCC recommended. Kerr negotiated powerfully with the administration, but after several of his amendments were accepted, he supported the Administration Bill (of which he was a cosponsor) and fought a bill introduced by Senator Estes Kefauver (D-TN) that advocated government ownership. The Administration Bill passed the House by a vote of 354 to 9 and, after a liberal filibuster, passed the Senate by a vote of 66 to 11. On 31 August 1962, President Kennedy signed the bill into law.

Congress had first looked at satellite communications as early as 1959 and had spent much of 1961 and half of 1962 in hearings and debates on the subject. The Kennedy administration had put satellite communications on its agenda from the start—much of the Space Council's efforts in 1961 were devoted to satellite communications. All the departments and agencies of government had been heard from, but what was intended or expected wasn't really clear. The State Department had been a constant presence in the Space Council meetings and the congressional hearings. The State Department felt, as did President Kennedy, that the space race was just an extension of the Cold War and that satellite communications was part of the space race. State had wanted a government monopoly on satellite communications. They were afraid that the profit motive would bypass service to the Third World countries that might fall into the communist camp. Although other countries were invited to participate, the assumption had been that the U.S. entity would own the satellites.[30]

The Interim Agreements

COMSAT had been advised by the common carriers—especially AT&T—that bilateral arrangements between COMSAT and each of the foreign post, telegraph, and telephone (PTT) organizations were preferable. AT&T had made bilateral arrangements for all of the submarine telephone cables and for Telstar, but NASA had stepped in and insisted that only the U.S. government

---

30. David J. Whalen, *Origins of Satellite Communications, 1945–1965* (Washington, DC: Smithsonian Institution Press, 2002), pp. 70–100. Edward Welsh interviewed by Thomas Safely, 19 July 1984, COMSAT History Project (CHP) Oral Histories, archived at the Johns Hopkins University.

could negotiate international agreements. As foreign ministries became involved, rather than PTTs, negotiations took a new direction. In fact, even as bilateral negotiations were being considered and before the incorporators had met, a U.K.–Canada–United States (Foreign Ministry/State Department) conference on satellite communications took place in Washington. William Gilbert Carter (of the State Department) met with the Conference of European PTTs (CEPT) and representatives of several European countries in late 1962. This meeting resulted in a query to Dean Rusk asking what was going on. By May of 1963, CEPT formed a new organization, Conference Européenne des Télécommunications par Satellites (CETS), to negotiate as a bloc with the United States. In October of 1963, the International Telecommunications Union (ITU) held an Extraordinary Administrative Radio Conference (EARC) in Geneva to discuss frequencies for satellite communications. The American delegation was led by Ambassador Joseph McConnell, but most of the American negotiations were conducted by Joe Charyk, assisted by Sieg Reiger. Somewhat to the negotiators' surprise, COMSAT got almost everything it wanted out of the conference.[31] One question (of many) that brought about the liberal filibuster against the Communications Satellite Act was whether the act provided for sufficient oversight by the U.S. President of the international negotiations that COMSAT was to undertake. Within the State Department, this was the question of whether satellite communications was "to be seen as a major foreign policy activity of the United States with a strong technical private enterprise component; or was it going to be seen as primarily a technical enterprise component with some foreign policy adjunct to it."[32] State argued that this was a race with the Russians and therefore political. Satellite communications was also global in a way that cables never had been. Finally, demonstrating peaceful space applications that benefited the Third World was of extreme importance. This even led to State's preference for geosynchronous orbit, as it allowed less-developed companies to participate with a single, cheaper Earth station than could be used with the Telstar-type MEO polar orbiter. After NASA diplomacy in 1961 and 1962, and State diplomacy in 1962 through 1964, it was hard to put the genie back in the bottle: satellite communications discussions would be with foreign offices first and PTTs second.[33]

COMSAT, and later Intelsat, had a major problem: Were these organizations "commercial" entities, in the limited sense that government-owned

---

31. Joseph McConnell, interviewed by Frederick Durant III, 18 July 1985, CHP. William Gilbert Carter, interviewed by Nina Seavey, 15 July 1985, CHP.
32. Carter interview, 15 July 1985.
33. Joseph V. Charyk, interviewed by Nina Seavey, 1 April 1986, CHP.

PTTs were "commercial," or were they instruments of foreign policy? If they were "commercial" entities, then their purpose was to earn a profit for their owners by providing global satellite communications. If they were profit-oriented, then decisions should be based on costs and profits. For a long period, the purchase of American satellites by COMSAT and Intelsat was based on the cost-benefit analysis that showed which satellites would provide the best service—and hence greatest profits—at the lowest cost. If Intelsat and COMSAT were instruments of foreign policy, then profits were irrelevant. If these organizations were instruments of technological advance, then each country should obtain "work" (manufacturing contracts) in proportion to their contribution of funding. This later became the European Space Agency's (ESA) principle of *juste retour*.[34] In early 1964, the United States (State Department and COMSAT) met with CETS in Rome. It was clear at this meeting that the Europeans would insist on some amount of control over satellite communications. They did not want to be supplicants asking for permission to use an American national asset. The next meeting was in London with additional participation. Before this meeting, Leo Welch, the COMSAT chairman of the board, insisted that John A. Johnson, who had joined COMSAT from NASA the previous December, be chairman of the U.S. delegation. After some discussion, Johnson was made vice chairman, with Abram Chayes from the State Department serving as nominal chair. At this meeting, it became obvious that there would be two agreements: a government-to-government agreement and a PTT-to-PTT agreement, with COMSAT as the American PTT. The final version of the interim agreements was presented to the world on 20 August 1964 in Vatican City, where 14 countries immediately signed. It is interesting to note that most of the negotiations were between the Europeans (CETS/CEPT) and the United States. It is even more interesting to note that during this negotiation process, COMSAT contracted for the geosynchronous Early Bird and raised $200 million in an initial public offering (IPO).[35] The three most important consequences of this interim agreement were 1) COMSAT would not go it alone, but it would manage the interim system under an Interim Communications Satellite Committee (ICSC); 2) the organization would have both foreign office and PTT representation; and 3) a new definitive agreement would be negotiated in five years.[36]

---

34. Charyk interview, 1 April 1986.

35. Ibid.

36. A dry but fairly complete discussion of both the interim and permanent (definitive) agreements can be found in Marcellus S. Snow, *The International Telecommunications Satellite Organization* (Neu-Isenburg: Momos Verlagsgesellschaft, 1987).

Early Bird and Intelsat II

Early Bird was launched in April 1965 and entered service in June. A few months later, the interim organization adopted the name Intelsat. Four Intelsat II series satellites were launched in 1966–1967; three were successful. The Intelsat II series was launched to support NASA's Apollo program. Early Bird covered only the Northern Hemisphere over the Atlantic Ocean Region (AOR). The Intelsat II series spacecraft covered the globe and were located over both the AOR and the Pacific Ocean Region (POR).

Part II: Maturity

All four major applications—reconnaissance, weather, communications, and navigation—had been "proved" in the decade after Sputnik. These were not R&D projects—they were operational!

1. COMSAT and Intelsat to 1979

Eight Intelsat III series satellites were launched between 1968 and 1970; six were successful. In orbit, these six satellites suffered from thermal problems that caused the despun antenna to stick. The first two series were built by Hughes with no international content. The third series was built by TRW and had some minimal international content. The fourth series was another Hughes satellite, but by this time, the interim arrangements were coming unstuck. By 1969, Intelsat membership had grown to 68. More than 15 countries were operating more than two dozen Earth stations. The hope of bringing modern communications to the Third World was beginning to be realized, but the industrialized nations were still dominant.

The Intelsat Definitive Agreements

When it came time to meet to discuss the definitive arrangements in February 1969, the old disagreements were still present. The ICSC, representing the Intelsat consortium, had been dominated by COMSAT. John A. Johnson had been elected as Chairman of the ICSC—only reasonable considering COMSAT's more-than-60-percent "ownership." Johnson has been described by Brenda Maddox as "the archetypical all-American boy grown up to be bank president. Tall, brusque, with fierce blue eyes, and what seemed to be more than the ordinary number of teeth."[37] Maddox refers to all the COMSAT

---

37.   Brenda Maddox, *Beyond Babel* (New York: Simon and Schuster, 1972), p. 92.

negotiators as "tall, abrasive men, inexperienced in diplomacy."[38] Dr. Reinhold Steiner, who represented Switzerland, Austria, and Liechtenstein, referred to Johnson as "the most xenophobic fellow I've ever met…. [H]e hates foreigners."[39] Many of the Europeans—especially the French—were annoyed at the limited amount of data they were getting. Some of this was blamed on COMSAT and some on the State Department. It was true that the United States wished to limit the export of aerospace knowledge. The Munitions Control Board worried about weapons use; the State Department and the administration worried about diluting the "single global system": Intelsat.[40]

The first Plenipotentiary Conference was held from 24 February to 21 March 1969 in Washington, DC. All but one of Intelsat's 68 member states sent a delegation. As was well understood by then, there was no basis for agreement among the 68 member states. The conference did make clear where there was agreement, where there might be agreement, and where there would never be agreement. Some indication of where things might be headed was the reception that Katherine Johnsen of *Aviation Week & Space Technology* got when she tried to interview the members of the ICSC in 1967: 17 agreed to be interviewed; only John A. Johnson refused. Similarly, at an ICSC meeting in December 1968, a vote was taken as to whether COMSAT should remain as manager: the result was 17 to 1 against.[41] The 1969 conference established a Preparatory Committee, which, while not allowed to negotiate, would prepare draft agreements for consideration by the full conference in 1970. The Preparatory Committee looked at the same issues that the conference had considered: structure, legal, finances, and operations. Three sessions were held in June–July, September, and November–December of 1969. Between 37 and 42 member nations attended, as did observers from non-members, the ITU, and the UN. Progress was made in all areas, and several reports were prepared for the 1970 conference.[42]

---

38. Ibid., p. 86.
39. Quoted in Michael Kinsley, *Outer Space and Inner Sanctums* (New York: Wiley, 1976), pp. 115–116.
40. Many faulted Lewis Meyer at COMSAT for not providing information to the ICSC. The United States had also issued National Security Action Memorandum 338 (NSAM 338) in this timeframe (1966–1967), which prohibited the launch of non-Intelsat communications satellites. This was also the period during which France withdrew from NATO.
41. Katherine Johnsen, "France Backs UN Intelsat Control," *Aviation Week & Space Technology*, 13 February 1967. See also Maddox, *Beyond Babel*, pp. 103–104.
42. Richard R. Colino, "The INTELSAT Definitive Arrangements: Ushering in a New Era in Satellite Telecommunications," European Broadcasting Union, 1973, p. 16.

The Resumed [Second] Plenipotentiary Conference took place between 16 February and 20 March 1970. Sixty-seven of the 75 members of Intelsat participated, and a host of nonmembers observed. The Preparatory Committee reports were read and discussed, but a proposal from Australia and Japan (Document 93) appeared to resolve most of the outstanding issues. Not all issues were resolved—and time did not allow elaborating Document 93 into a formal legal document. An Intersessional Working Group (IWG) was established to develop a draft set of agreements based on Document 93 for presentation at the third conference, scheduled for 1971.[43] In contrast to the Preparatory Committee, the IWG was authorized to negotiate. Three sessions were held in May–June, September–October, and November–December 1970. Between 47 and 48 members attended the IWG sessions—significantly more than had attended the Preparatory Committee sessions. On 31 December 1970, the IWG submitted drafts of the Intelsat Agreement and Operating Agreement to the final conference.[44] The third and final conference began on 14 April 1971. Voting required a two-thirds majority to change bracketed (unresolved) text in the two agreements. On Friday, 21 May 1971, 73 of 78 members attending voted in favor of the final texts of the two agreements—France, the Malagasy Republic, Monaco, and Mexico abstained,[45] and one member was absent. On 20 August 1971, the agreements were opened for signature, and by 14 December 1972, two-thirds of the members had signed—60 days later, on 12 February 1973, the interim agreements were terminated and the new agreement entered into force.[46]

The strange public-private, commercial-political nature of COMSAT was also reflected in the structure of Intelsat. Intelsat had two "governing bodies": nations signed the Intelsat Agreement (also referred to as the Intelsat Treaty), but telecommunications entities (Signatories) signed the Intelsat Operating Agreement. The Intelsat Assembly of Parties consisted of the sovereign governments that signed the Intelsat Agreement. Voting in the assembly was by country: one nation, one vote. Its powers were limited. The Meeting of Signatories consisted of all the telecommunications entities that signed the Intelsat Operating Agreement. Voting in the meeting was on the basis of shares, and the shares were allocated (and paid for) on the basis of usage.

---

43. Australia, Japan, and Canada had a strong influence on Intelsat deliberations as a third party in disputes between the United States and the Europeans. Much later, the Third World countries would favor the United States as more interested in providing them with the best and cheapest technology.

44. Colino, "The INTELSAT Definitive Arrangements," pp. 16–19.

45. Note that three of the four abstainers are Francophone.

46. Colino, "The INTELSAT Definitive Arrangements," pp. 19–21.

This system has been referred to as "one telephone call, one vote." A Board of Governors had functions similar to the ICSC, or to the functions of a commercial board of directors. The board consisted of about 20 members, each having a minimum specified investment, and individual representatives of member-groups whose total investment met the minimum specified (about 2 percent). Finally, there was a manager—COMSAT for six years after the agreements entered into force (terminating 12 February 1979)—reporting to a secretary general until 31 December 1976 and to a director general thereafter. The details of the agreements and the compromises that led to them are detailed in a book by Richard R. Colino of COMSAT. Colino participated in the negotiations and was later Intelsat director general himself. The major antagonists, the United States and the Franco-Europeans, each compromised in some way, but the result was both semicommercial and semipolitical. It could be argued that the State Department got what it wanted because the Third World countries seemed to have been guaranteed international communications at reasonable rates and with some national control—at least control of their own Earth stations. The Europeans continued to complain that satellite contracts went exclusively to the United States (until the 1990s), but the Third World countries preferred cheaper, higher-quality American satellites and launch vehicles.

COMSAT's Intelsat III and IV spacecraft had not officially been slated for geosynchronous orbit when the Intelsat III contract was put out for bids. Hughes decided not to bid an MEO option, which allowed TRW to sneak in a winning bid. There was some attempt at international "sharing" of the contract, but it was minimal (about 6 percent). The Intelsat III series was designed for multiaccess (more than one signal in a transponder). Effective Isotropic Radiated Power (EIRP) increased to 27 decibel-watts (dBW) from Intelsat II's 15 dBW and the 10 dBW of Early Bird. Unfortunately, three of the eight Intelsat III satellites had bad Delta launches at a time when Delta reliability was better than 90 percent. These satellites were the first to provide coverage of the Indian Ocean Region (IOR).

Almost in parallel to the Intelsat III program was the Intelsat IV program. The first three generations of Intelsat had relatively limited capacity. Intelsat IV would be a significant increase in power, number of transponders, mass, and coverage options. The first three generations had Earth coverage only. Intelsat III was considered a big advance because it had a despun antenna that always pointed at Earth, dramatically increasing EIRP. Intelsat IV had two narrow-beam antennas covering the Eastern and Western hemispheres. The Intelsat IV payload design is still the basis for modern communications satellite payload design. The bus design had been proven on TACSAT, and the stabilization technique would be used by Hughes into the twenty-first century. Because of its size, this series would use the Atlas-Centaur launch vehicle instead of the Delta. About 20 percent of the content was provided by international manufacturers.

It was no surprise when Intelsat discovered that North Atlantic traffic was greater than that in the Pacific Ocean Region or the Indian Ocean Region. Two Intelsat IV satellites had insufficient capacity to satisfy traffic demands. A tradeoff was made between waiting for the Intelsat V series and upgrading the Intelsat IV series. The result, Intelsat IVA, had 20 rather than 12 transponders.

Intelsat-IVA-F6 was the last satellite launched by COMSAT as Intelsat "manager." The definitive agreements left the company without a major role in satellite development. COMSAT still monitored construction under contract to Intelsat, but executive decisions were made by Intelsat.

2. COMSAT and Intelsat after 1979

COMSAT was looking for a mission after 1979. The company tried domestic satellites (COMSTAR with AT&T, Satellite Business Systems [SBS] with IBM), broadcast satellites (STC), software, ground systems, and especially Earth stations (RSI). None of them worked. By the mid-1980s, the company's profits were bouncing up and down. It seemed that every other year the company lost tens or hundreds of millions of dollars. Intelsat did well in these years. It was still the largest and most profitable satellite company. It had few business barriers because its owners were the national PTTs. In the late 1980s, competition did affect Intelsat. More dangerous, the U.S. government began seeking to destroy its monopoly status.

Intelsat V

The launch of the three-axis stabilized RCA Satcom F1 in 1975 changed the standard communications satellite design. Future satellites—at least for Intelsat and most of the U.S. domestic communications satellite (DOMSAT) operators—would be three-axis stabilized, with large high-power solar arrays, and with frequency reuse through cross-polarization. In addition, within a few years, the Ku-band frequencies would dominate U.S. and European satellite service. Intelsat V may not have started these trends, but it certainly took advantage of them. The Intelsat V series was the first to be launched on multiple vehicles: Atlas-Centaur (AC) and Ariane (Ar). During the Intelsat V series development, the need for a maritime communications system for lease to Inmarsat resulted in L-band payloads on Intelsat 505–509.

Competition and Other Problems

The FCC was not a particular friend to COMSAT, but neither were they enemies. Because it was a monopoly, COMSAT was often treated by the FCC as if it were AT&T. The Earth station ownership, authorized user, and rate decisions hampered COMSAT's abilities to find new work. As the 1980s began, a new threat appeared. The Nixon administration had opened domestic satellite communications to all "qualified" applicants; now the Reagan administration opened up the international satellite communications business to "separate systems"—separate, that is, from COMSAT and Intelsat. There had long been concerns that COMSAT/Intelsat was a monopoly, and monopolies are "bad." The precipitating event may have been the 11 March 1983 filing by Orion Satellite Corporation for authority to develop a private international communications satellite system. While the FCC was considering the application, the National Telecommunications and Information Administration (NTIA, part of the Commerce Department) and the State Department began feuding over policy.

Less than a year later, on 13 February 1984, Pan American Satellite Corporation (PanAmSat) also filed for authority to launch an international satellite communications system. Several other companies filed between these two, but Orion and PanAmSat would eventually launch satellites. On 28 November 1984, Reagan announced that separate systems were required in the national interest. On the basis of the President's decision, the FCC began granting conditional licenses. On 25 July 1985, the FCC issued its Separate Systems Report and Order. On 1 June 1988, PanAmSat's PAS-1 was launched and eventually drifted to a longitude of 45 degrees west to provide transatlantic service. That same year, the fiber-optic TAT-8 cable began to provide service across the Atlantic—service that was cheaper than Intelsat's. The transatlantic telephone cables had always competed with satellite transmission across the Atlantic, but the savings, if any, were minimal. Fiber-optic cables were cheaper and provided higher-quality transmissions and much higher data rates.

The Rest of the Intelsat Satellites

Hughes had pioneered geosynchronous orbit with Syncom 2. The corporation had been awarded contracts for Early Bird, the Intelsat II series, and the Intelsat IV/IVA series. In 1982, Hughes was awarded the contract for the Intelsat VI series. It would be the last satellite Hughes would build for Intelsat and the last large spin-stabilized satellite. It would also be the last large Intelsat satellite (recognizing that the definition of *large* changes over time). The second launch (603) was on a Titan rocket. The Titan failed, and the satellite was subsequently boosted into orbit by a Space Shuttle "rescue" mission. For a little over $1 billion, this $100 million (probably $150 million, including launch) satellite was reused.

Intelsat VII was the second series built by Ford Aerospace (later Space Systems Loral). If Hughes had built all the even series (II, IV, VI), Ford was starting to win all the odd series (V, VII, and later IX). The Intelsat V series was considered by Santiago Astrain (the first director general of Intelsat) to be the most successful series. Intelsat VI had been designed with the heavy traffic of the North Atlantic in mind. The Intelsat VII series was intended to replace Intelsat V satellites over the Pacific and Indian oceans (the POR and AOR).

Intelsat wanted to ensure that it always had two qualified manufacturers. It did not want to be tied too closely to any one manufacturer. After Intelsat VI, Hughes was no longer on Intelsat's list of preferred manufacturers (a mutual decision). RCA (later GE, then Martin-Marietta, now Lockheed Martin) had built the first commercial operational three-axis satellite. The RCA satellites had dominated the American market—bought by RCA Americom, Southern Pacific (Spacenet), and GTE (GSTAR). It was chosen to build the Intelsat VIII series.

The Intelsat VIII series was less successful than expected. Intelsat returned to Ford/Loral for its next series. This was the first spacecraft of seven Intelsat 9-series satellites ordered from Space Systems/Loral. Intelsat 901 was ordered in December 1996. The 4,723-kilogram mass spacecraft carries 42 C-band and 14 Ku-band operating transponders, using the LS-1300 satellite bus. The satellite's solar arrays generate between 10 kilowatts of power (beginning of life) and 8.6 kilowatts (end of life). The satellite was designed for a 13-year lifespan.

COMSAT and Lockheed

How did it all begin? In 1995, Lockheed and Martin Marietta merged. The earlier combination of RCA Astro and GE Space that had been purchased by Martin Marietta became Lockheed Martin Commercial Space Systems (LMCSS). Martin Marietta had been looking at getting into communications satellite operations rather than (or in addition to) manufacturing. Profit margins in manufacturing looked slim compared with operations. Martin Marietta may have been already looking "across the street" at COMSAT in 1992. Was this a Norm Augustine (of Martin Marietta) and Betty Alewine (COMSAT) deal? What did the Russian deals and Brian Dailey have to do with anything? Russ McFall (of MM, then of LMT) was looking to buy a real satellite operator. Did no one understand the implications of "direct access"? COMSAT and Intelsat at one time had offices facing each other on

L'Enfant Plaza in downtown Washington, DC. Intelsat moved to offices well up Connecticut Avenue near the suburbs. COMSAT was looking for new office space that would be cheaper than L'Enfant Plaza. Rock Spring Plaza in Bethesda, Maryland, was the choice for the new office. Lockheed Martin would eventually be a neighbor.[47] In 1996, C. J. "Pete" Silas was named chairman of COMSAT after the retirement of Melvin Laird.[48] Shortly thereafter, Bruce Crockett left COMSAT to "pursue other business interests." "By removing Crockett, the board essentially repudiated his vision of making COMSAT a communications and entertainment conglomerate." Crockett had changed the name from Communications Satellite Corporation (CSC) to COMSAT. Crockett had been taken in by the glamour of entertainment. Perhaps most annoying to COMSAT executives and board members, Crockett had spent all of his time on entertainment, attending most Denver Nugget NBA and Colorado Avalanche NHL games. Also in 1996, Caleb B. Hurt was elected to the Board of Directors. He was a former president and COO of Martin Marietta, which had just merged with Lockheed. A formal announcement was made on 24 March 1997 that COMSAT was restructuring: this included putting COMSAT Radiation Systems, Inc. (CRSI, another Crockett acquisition), and Ascent up for sale. There were no buyers for Ascent—eventually the stock would be spun off to COMSAT shareholders. CRSI would be easier to sell.[49] During this period, Edwin I. Colodny replaced C. J. Silas as chairman of the board.[50] Colodny was ill at the time. The year 1997 proved to be a difficult one for COMSAT: shareholders were unhappy with the low earnings and slow revenue growth, and at the June shareholders meeting there was a proxy battle. COMSAT sued its ex-CEO, Crockett, accusing him of "conspiring" to replace the company's board.[51] Many old problems persisted and got worse—fiber-optic cable competition; competition from PanAmSat; difficulties selling Ascent and RSI; and the failed merger with CONTEL, Skypix, and Satellite Television Corporation (STC). It was clear that the decades-old task of finding a new focus for COMSAT was a failure. Neil Bauer of Orion,

47. David S. Hilzenrath, "COMSAT Tentatively Agrees To Move Offices to Bethesda," *Washington Post* (15 November 1990): E1–E2; Daniel S. Hilzenrath, "COMSAT Completes Agreement To Move to Bethesda," *Washington Post* (12 July 1991): C1.

48. *S&P Daily News* (23 April 1996) (LN).

49. Mike Mills, "Ailing COMSAT To Sell Unit Making Satellite Dishes," *Washington Post* (24 March 1997): A9; Doug Abrahms, "Comsat Doesn't Want To Play Anymore," *Washington Times* (25 March 1997): B6, B10.

50. *S&P Daily News* (30 April 1997) (LN).

51. "COMSAT Files Suit Against Ex-CEO," *Dow Jones News* (24 April 1997).

one of the "separate systems," could ask: "What does COMSAT really own?"[52] Betty Alewine, who had replaced Crockett at COMSAT World Systems, now replaced him as president of COMSAT. Alewine saw satellite service and network services as the corporation's future.[53] Alewine sold off COMSAT's Philcomsat shares and most of its ICO shares, spun off Ascent, and prepared to sell RSI—all of this in the midst of stockholder revolts and potshots from competitors and Congress. Robert S. Koppel, vice president of international regulatory affairs for WorldCom, spoke for many when he said, "COMSAT's role as middleman is an anachronism." In 1996, one-fifth of COMSAT revenues came from two customers: the U.S. government (14 percent) and AT&T (7 percent); these two customers had represented more than a fourth of its revenues (15 percent and 12 percent, respectively) in 1994. International satellite services (Intelsat and Inmarsat) were profitable; network services were not. In September 1997, COMSAT sold the Clarksburg COMSAT Labs building for $45 million with a 10-year leaseback at $5 million a year. By early 1998, it seemed clear that COMSAT was on the market. Early in the year, COMSAT denied that it was being acquired by Loral.[54] A few months later, the FCC approved the Loral/Orion merger.[55] Retired COMSAT vice president Sidney Metzger claimed COMSAT benefited all consumers. The original AT&T Telstar concept was too expensive.[56] In July, fallout from the Cox report caused aerospace stocks to tumble: COMSAT shares fell to $28.75 from $42 over a "couple months."[57] COMSAT claimed it was in the process of "unlocking the value of investments in Intelsat and Inmarsat." On 20 September, Lockheed announced plans to purchase COMSAT after failing to buy Northrup-Grumman for $8.3 billion. The Lockheed offer was for 49 percent of the COMSAT stock at $45.50 per share, with the last 51 percent to be purchased with one share of Lockheed for two shares of COMSAT. The total value of the purchase was about $2.7 billion.

John Sponyoe, CEO of the newly formed Lockheed Martin Global Telecommunications subsidiary (LMGT), said the deal had been in the

---

52. Quoted in Doug Abrahms, "COMSAT's Dominance of the Skies Under Fire," *Washington Business Times* (23 June 1997): D12–D14.

53. Mike Mills, "A Signal Change at COMSAT," *Washington Business* (18 August 1997): 12–14.

54. "COMSAT Denies Acquisition Report," *Washington Business Journal* (18 February 1998).

55. "FCC Approves Loral/Orion Merger," *Satellite News* (9 March 1998): 8.

56. Sidney Metzger, "COMSAT Success Story," *Space News* (6–12 July 1998).

57. Jerry Knight, "A Probe of US-China Policy Pummels Satellite Stocks," *Washington Business* (13 July 1998): 7. The Cox report was a congressional report alleging Chinese spying on U.S. ICBM, computer, and nuclear weapons designs, aided and abetted by profit-seeking American companies.

works for 18 months (since March 1997). The Lockheed offering of $45.50 was about a third higher than the market price of $34⅛. COMSAT was apparently vulnerable to takeover due to its small size. Lockheed also had $1.7 billion worth of stock in Loral, Iridium, and L-3.[58] Lockheed shares went down $3.50 to $100. Analysts saw the need for vertical consolidation emphasizing services.[59] Lockheed shares fell another $6 to $94, while COMSAT shares climbed. The Bliley bill passed the House, allowing greater than 49 percent ownership but also direct access.[60] Direct access allowed customers to deal directly with Intelsat, bypassing COMSAT. Bear, Sterns & Co. was advising Lockheed. LMGT also owned all or part of LMI, Americom Asia-Pacific (AAP), Astrolink, ACeS, and other communications systems.[61] Lockheed's purchase of COMSAT fit into the pattern of the previous two decades. Hughes had picked up Western Union's Westar in the 1980s and PanAmSat in 1997. RCA Americom had picked up GTE (and Southern Pacific) before being acquired by GE. Loral had bought AT&T's Telstar in 1997. And now Lockheed was buying COMSAT. Lockheed was trying to become "commercial," and this seemed to be the way—especially with the boom in telecommunications generally and "dot-coms" specifically. This would be a really good deal for COMSAT shareholders but not necessarily for Lockheed. Mickey Alpert, a former COMSAT STC executive, did not "think COMSAT [had] a sustainable, ongoing business."[62] But Brian Daley, COO at LMGT, argued that "this initiative is intended to leverage our core skills in satellite communications and take advantage of the higher-margin services business."[63] By 1999, LMGT had other problems. At the 19 May fourth session of the LMI Board of Directors, LMI-2 and LMI-3 were authorized at 83W and 3W.[64] LMI-1 would be launched in September, and it was 75 percent sold. Unfortunately, those transponders didn't stay sold and were leased at prices well below cost. Furthermore, the contract between Lockheed

58. Tim Smart, "Lockheed Seeks To Buy COMSAT for $2.7 billion," *Washington Post* (21 September 1998): A01.

59. "Space Services Could Be Next Fertile Consolidation Ground," *Aerospace Daily* (29 September 1998); "Lockheed Martin-COMSAT To Combine," Lockheed Martin, press release, 20 September 1998; Leslie Wayne, "Lockheed To Buy COMSAT, Satellite Access Seller, for $2.7 Billion," *New York Times* (21 September 1998): A11.

60. "Lockheed Buying COMSAT," *CNNmoney.com* (21 September 1998).

61. "LockMart Swallows COMSAT," *Space Daily* (21 September 1998).

62. "Lockheed Martin's Planned Buyout of COMSAT Will Build Satellite Services," *Satellite News* (28 September 1998): 5.

63. Sam Silverstein, "Lockheed, COMSAT Designing Powerful Union," *Space News* (28 September–4 October 1998): 6.

64. "The Fourth Session of the LMI Board of Directors," Intersputnik International Organization of Space Communications (Intersputnik), press release, 19 May 1999.

and Intersputnik had expensive flaws. Nonetheless, John Sponyoe promised to launch three more spacecraft.[65] On 16 June, Lockheed extended its tender offer for COMSAT to 31 August.[66] On 20 August 1999, COMSAT share-holders voted to accept the Lockheed Martin (LMT) offer (99 percent of votes, 74 percent of shares). The plunge in value of Lockheed shares reduced the value of the deal from $2.7 billion to $2.2 billion.[67] By 8 September, COMSAT shares were down 10 percent, to $30½. Lockheed was still unable to shed its 16 percent of Loral—a condition for its purchase of COMSAT. The merger had to be completed by 18 September.[68] On 15 September, the FCC authorized Lockheed to purchase 49 percent of COMSAT. Lockheed was also authorized to buy COMSAT Government Services, Inc. (CGSI).[69] In addi-tion to buying 49 percent of the shares at $45.50 per share, the remaining 51 percent would be a share-for-share deal (Lockheed had split). Lockheed would also assume $455 million in COMSAT debt.[70] On 16 September, the Department of Justice authorized the Lockheed-COMSAT merger.[71] On 20 September, Lockheed formally bought 26 million shares of COMSAT for $45.50 each; 48 million shares were offered.[72] On 22 September, PanAmSat appealed the FCC's approval of the Lockheed purchase.[73] On 15 October, Lawrence Eagleburger resigned as a COMSAT director; he was replaced by John Sponyoe, the CEO of LMGT.[74] On 22 November, Lockheed attempted to add an amendment to an appropriations bill allowing the purchase of the remaining 51 percent; the amendment failed. Lockheed had to wait for the Orbit Act of 2000 (S 376) to pass.[75]

65. "Lockheed Martin Intersputnik To Begin Communications Service in September," Intersputnik, press release, 7 June 1999.

66. "Lockheed Martin Extends Tender Offer for COMSAT Until August 31," *Aerospace Daily* (16 June 1999).

67. Tim Dobbyn, "COMSAT Shareholders Back Lockheed Sale," *Space.com* (20 August 1999).

68. "COMSAT Shares Fall as Lockheed Bid Falters," *Space.com* (September 1999).

69. "FCC Authorizes Lockheed Martin To Purchase up to 49 Percent of COMSAT," *FCC News* (15 September 1999).

70. Tim Smart, "FCC OKs Lockheed-COMSAT Deal," *WashingtonPost.com* (15 September 1999).

71. "Receives Department of Justice Clearance for COMSAT Corp. Merger," *S&P Daily News* (5 October 1999).

72. "Lockheed Martin Completes COMSAT Tender Offer," *Aerospace Daily* (21 September 1999).

73. "PanAmSat Appeals FCC Approval of Lockheed Martin–COMSAT Deal," *Aerospace Daily* (22 September 1999).

74. "Announces Change to Board of Directors," *S&P Daily News* (6 January 2000).

75. "Lockheed Martin Fails To Get COMSAT Provision Through House," *Aerospace Daily* (22 November 1999).

The Orbit Act of 2000

In 1996, the GAO made a report to Congress, requested by Thomas J. Bliley, Jr. (R-VA), on restructuring.[76] In 1997, Bliley and Edward J. Markey (D-MA) submitted a bill (HR 1872) to privatize Intelsat and Inmarsat.[77] Senator Conrad Burns (R-MT) and his Communications Subcommittee held hearings on 30 July 1997 on "Satellites and the Telecommunications Act." The FCC, the NTIA, and the State Department testified, as did Intelsat, PanAmSat, and Comsat.[78] The claim was made that COMSAT's markup on Intelsat pricing was as much as 86 percent.[79] On 25 March 1998, the House committee passed the Bliley bill (HR 1872, including a variant of "fresh look," an opportunity to renegotiate all COMSAT contracts with Intelsat). Intelsat's Tony Trujillo commented that the bill was fatally flawed.[80] HR 1872 passed the entire House on 6 May 1998. The House bill provided for both "direct access" and a variant of "fresh look." Senator Burns introduced a different bill (S 2365) on 29 July that was seen as more favorable to COMSAT.[81] PanAmSat claimed that S 376 did not go far enough. On 21 January, Tom Bliley asked the FCC to reject any COMSAT ownership greater than 10 percent until after the passage of reform legislation. On 5 May 1999, the full Senate Committee approved Orbit (S 376).[82] On 1 July, S 376 (Orbit) was passed unanimously by the Senate. "Fresh look" was not allowed, but "direct access" was. Trujillo of Intelsat described the bill as "the heavy hand of the U.S. Congress."[83] PanAmSat and many other firms argued that it was unfair to allow Lockheed to buy COMSAT with privileges unchanged. Senator Lott insisted that Burns allow "direct access" by 1 July 2002. The Bliley House bill passed in 1998 had removed almost all COMSAT privileges.[84] In early 2000, Burns tried to cut a deal with Bliley (who had also sponsored the

76. "Telecommunications: Competitive Impact of Restructuring the International Satellite Organizations," GAO/RCED-96-204, GAO Report, July 1996.

77. Doug Abrahms, "Bearing Weight of 141 Nations, Intelsat Tries To Change Itself," *Washington Business Times* (23 June 1997): D13–D14.

78. U.S. Senate Commerce, Science and Transportation Committee, Subcommittee on Communications, Hearing on Satellites and the Telecommunications Act, 30 July 1997.

79. "Ending a High-Flying Monopoly," *Washington Post* (20 March 1998): A10.

80. Quoted in "House Commerce Committee Passes Bliley Bill, but Hurdles Remain," *Satellite News* (30 March 1998): 1.

81. Sam Silverstein, "COMSAT Prefers Senate Bill," *Space News* (3–9 August 1998): 3.

82. "International Satellite Reform Bill Approved by Committee," press release, U.S. Senate Committee on Commerce, Science, and Transportation Committee, 5 May 1999.

83. Quoted in Daniel Sood, "Senate Unanimously Passes Satellite Reform," *Space.com* (1 July 1999).

84. "Senate Satellite Bill Faces Tough Fight in the House," *Aerospace Daily* (6 July 1999).

Telecommunications Act of 1996) and began a partnership with Billy Tauzin (R-LA) and John Dingell (D-MI).[85] On 17 February, the House and Senate conference committee reached a compromise. Direct access was allowed, and the Intelsat IPO was deferred to 1 January 2003.[86] On 4 April, an FCC Public Notice was published to the effect that LMT had applied to transfer control of COMSAT to LMT/CGS.[87] On 31 July, the FCC authorized Lockheed to merge with COMSAT in accordance with the provisions of the 17 March 2000 Orbit Act.[88]

The End of COMSAT

Lockheed Martin Global Telecommunications did not last long. On 7 December 2001 (the 60th anniversary of Pearl Harbor), LMGT announced that it was shutting down its operations. Some units would be absorbed by other Lockheed divisions; some units would be sold off; and some units would simply disappear. Earlier that year, much of COMSAT Laboratories was sold to ViaSat. In January 2002, the sale of COMSAT Mobile Communications to Telenor was finalized. In March 2002, Lockheed sold its remaining COMSAT World Systems facilities to Intelsat. Finally, in 2004, COMSAT General's remaining facilities were also sold to Intelsat by Lockheed. The public-private experiment was over.

The New Intelsat

Passage of the Orbit Act on 17 March 2000 forced Intelsat to "privatize." The two years between the passage of the House bill and the Senate bill allowed plenty of time for Intelsat to plan its future. In any case, none of this was a surprise. Bliley had begun his march toward "privatization" at least as early as 1996. The direction U.S. policy would take was probably clear from the early 1980s, when "separate systems" became U.S. policy. Competition from fiber-optic cables and domestic satellites grown to regional satellites also made it clear that Intelsat must change. At the Twenty-Fourth Intelsat Assembly of Parties in 1999, Intelsat resolved to create a new "private" organization—still owned by the Signatories—and a residual intergovernmental

85. Brody Mullins, "A Satellite Bill: A Career in ORBIT?" *National Journal* (5 February 2000): 405–406.
86. Greg Schneider, "Accord Reached on Bid for COMSAT," *Washington Post* (18 February 2000): E01; Mary Motta, "Lockheed Gets Good News on COMSAT Deal," *Space.com* (19 February 2000).
87. FCC Public Notice, Report No. SAT-00040, 4 April 2000.
88. "Commission Authorizes Lockheed Martin To Take Control of COMSAT," *FCC News* (31 July 2000).

organization (IGO) to protect disadvantaged users. Intelsat LLC was formed and on 18 January 2000 applied for transfer of the Intelsat licenses to the new private organization. On 8 August 2000, the FCC issued its Intelsat LLC Licensing Order. This order allowed Intelsat to transfer its licenses to Intelsat LLC upon "privatization."

On 18 July 2001, Intelsat, the intergovernmental treaty organization, became Intelsat Ltd., a private holding company based in Bermuda with a wholly owned subsidiary, Intelsat LLC, which would hold the U.S. licenses. With this "privatization," Intelsat Ltd. lost all of its immunities as an international treaty organization and became a U.S. telecommunications carrier with no special privileges. On 28 January 2005, the signatories sold their shares in Intelsat Ltd. to a private equity organization, thus completing the privatization.[89] The Intelsat 10 series went to Astrium, a European conglomerate. Intelsat signed two launch services contracts, one with Boeing Launch Services for a Sea Launch Zenit-3SL vehicle and the second with International Launch Services (ILS) for a Proton M/Breeze M vehicle for the launch of the two Intelsat X series satellites (IS-10-01 and IS-10-02). These spacecraft have a design lifespan of 13 years each and are the first Intelsat satellites to use plasma propulsion for in-orbit stationkeeping. EIRP levels are 37 dBW to 42 dBW for zone beams, 37 dBW to 41 dBW for hemi beams (IS-10-02), and up to 54.9 dBW for spot beams (Ku-band). Both satellites were to be launched in 2003, but only 10-02 was launched.[90]

Merger with PanAmSat

After the dot-com crash and general telecom meltdown of the early twenty-first century, market share and profitability became critical. On 28 August 2005, Intelsat and PanAmSat agreed to merge. The merger of the second and fourth largest FSS companies produced a giant that owns between a quarter and a third of all FSS satellites. What makes this merger particularly strange is that PanAmSat was formed as the anti-Intelsat by René Anselmo in 1984. The PanAmSat motto was "Truth and technology will triumph over bullshit and bureaucracy." Anselmo despised the COMSAT-Intelsat monopoly. The PanAmSat mascot was the dog Spot—usually seen urinating on the leg of a representative of "bullshit and bureaucracy." Intelsat agreed to buy PanAmSat Holding Corporation for $3.2 billion in cash. The merged companies would form the world's largest satellite company and give the companies a more

---

89. Details of the "privatization" and the residual IGO (ITSO) can be found in Kenneth D. Katkin, "Communication Breakdown? The Future of Global Connectivity After the Privatization of Intelsat," *http://law.bepress.com/expresso/eps/508*.

90. Intelsat, press release, Hamilton, Bermuda, 15 January 2002.

diversified set of businesses. The new company would own 53 satellites spanning the globe and generate annual revenues of more than $1.9 billion.[91]

On 7 July 2006, the $6.4 billion purchase of PanAmSat was completed, creating a merged company carrying one-quarter of the world's commercial satellite-delivered television programming. The acquisition left PanAmSat a wholly owned subsidiary of Intelsat. The combined company would initially lose money. PanAmSat earned $72.7 million in 2005, but Intelsat lost about $325 million. Intelsat chief executive David McGlade told the *Washington Post* that, given the level of debt and interest payments, the company did not expect to become profitable in the foreseeable future. He said the company's investors had been pleased with Intelsat's positive cash flow and its heavy backlog of orders. The traditional core of Intelsat's business had been telephony, a difficult market in recent years. The combined company would be more diverse with the addition of PanAmSat's television customers.[92]

3. Commercial Satellite Communications

Intelsat was an example of what would later be called a nongovernmental organization (NGO) or a treaty organization. With the exception of COMSAT, most of the Intelsat Signatories were governmental organizations. They were not truly "commercial." They were—again with the exception of COMSAT—monopolistic national telephone companies. They were not broadcasters and were amazingly shortsighted about the benefits of television via satellite, except as a stunt.

Anik and Open Skies

Neither the Communications Satellite Act of 1962 nor the Interim Intelsat Agreements precluded domestic communications satellites (DOMSATs). It was assumed—and later made explicit—that DOMSATs should not interfere with Intelsat. In the United States, both Hughes and COMSAT had discussed providing domestic satellite communications as early as 1964. On 21 September 1965—less than six months after the launch of Early Bird—ABC requested authorization from the FCC to launch a television satellite to link television networks with their local affiliates. The FCC returned the filing, and six months later, on 2 March 1966—less than a year after the launch of Early Bird—the FCC issued a Notice of Inquiry (Docket 16495)

---

91. Yuki Noguchi, "Intelsat Agrees to Merger Deal with PanAmSat," *Washington Post* (29 August 2005): A7.

92. "Intelsat Completes Merger with PanAmSat," 7 July 2006, *http://broadcastengineering. com/beyond_the_headlines/intelsat_mergers_panamsat/?r=1.*

into the matter of "Establishment of Domestic Non-Common Carrier Communications Satellite Facilities by Non-Governmental Entities."[93] It is hard to understand from a half-century later why the FCC dithered— but dither it did. Since the "Above 890 [MHz]" decision of 1959, the FCC had been trying to inject competition into the telecommunications industry. Congress was of a similar mindset—although their goal seemed to emphasize the AT&T monopoly. It was two years from the submission of AT&T's petition for authorization to launch a satellite to the passage of the Communications Satellite Act of 1962. It would take more than seven years to authorize DOMSATs. The FCC was confused about the law; national policy; and the effect on Intelsat, markets, and technology. Like Intelsat and COMSAT, the FCC didn't seem to be sure that there was a viable market for DOMSATs. The Ford Foundation responded with a proposal to create a Broadcasters Non-Profit Service Corporation (BNS). The BNS would provide a total of 44 transponders: six commercial and five noncommercial in each of the four continental U.S. time zones. On 20 October 1966, the FCC issued a "Supplemental Notice of Inquiry." The Ford Foundation and COMSAT suggested that COMSAT launch a pilot program to see if there was a commercial market for the service. The Johnson administration had put together a Task Force on Communications Policy, which reported to Johnson in December 1968 recommending COMSAT's pilot program.[94] The chairman of the task force was Eugene V. Rostow (hence the Rostow report), and the vice chairman was James D. O'Connell. The Nixon administration had a different idea. They rejected the Rostow report and favored "open entry"— that is, any organization with the money to launch a satellite system should be allowed to do so. A small working group was put together in 1969 to formulate Nixon administration policy. Among the members of the group was Clay T. Whitehead. Whitehead's boss, Peter Flanigan, sent a memo to Dean Burch at the FCC recommending open entry (or "open skies") on 23 January 1970. The FCC issued its first report on Docket 16495 a few weeks later. In February 1970, Nixon proposed that an Office of Telecommunications Policy be formed within the White House and headed by Whitehead. Congress had been complaining for some time that the FCC was delaying a decision. The FCC was clearly under the gun. Thirteen entities had filed for authorization to launch DOMSATs. In March 1972, the FCC released a proposed second report and order on DOMSATs, requesting that the filers consolidate their filings. Nobody liked this. The actual second report and order were released on

---

93. Much of the discussion of the early days of DOMSATs is from Robert S. Magnant, *Domestic Satellite: An FCC Giant Step* (Boulder, CO: Westview Press, 1977).

94. President's Task Force on Communications Policy, Final Report (Washington, DC: U.S. GPO, 1969).

16 June 1972 after a 4-to-3 vote by the commissioners. The dissenters objected to the restrictions on AT&T and COMSAT. A final report on DOMSAT was issued on 22 December 1972 that modified (but retained) these restrictions.

Meanwhile, Canada had quickly decided to launch a Canadian satellite to service the Far North. In 1967, the Chapman report recommended that a satellite system be developed. In 1969, Telesat Canada was established. On 9 November 1972, Anik A1—a Hughes HS-333—was launched on an American Delta launch vehicle. The first six commercial communications satellites would all be Hughes HS-333s launched on Delta rockets.

RCA Global Communications began service to Alaska on Anik. RCA may have been first into service, but the telegraph company Western Union was the first U.S. company to launch its own satellite.

RCA built its own satellite using the services of RCA Astro-Electronics in East Windsor, New Jersey, and RCA Canada in Montreal. The RCA satellites had twice the number of transponders and twice the power of the HS-333. They were the first operational (as opposed to the experimental ATS-6 and Symphonie) three-axis stabilized communications satellites.

AT&T had built its own experimental Telstar satellites but opted to buy Hughes satellites for its operational program. More accurately, it leased satellites from COMSAT. AT&T was constrained to provide only point-to-point services—that is, it was prohibited from offering television distribution services.

Indonesia was the third nation to launch a commercial geosynchronous communications satellite business. The Palapa series, like Anik and Westar, was based on the HS-333. Within a few years of the first launch, the tens of thousands of Indonesian islands were connected via satellite.

The Television Revolution

The original U.S. filing for a domestic communications satellite had been made by ABC with encouragement from Hughes. COMSAT and Intelsat had never been much interested in television; some at COMSAT had argued that only four television transponders were necessary—one for each network (ABC, CBS, NBC, and educational television). Some at RCA (the owners of NBC) had argued that at least 20 transponders were needed—one for each of the four networks in each of the four time zones plus one for each National Football League (NFL) game. While a few had seen the future, the explosion in television (especially nonnetwork cable television) was a shock. Within a few years, there were a dozen satellites carrying more than 200 transponders. Two-thirds of the traffic was television—a ratio that persists to this day. By the 1990s in the United States—earlier in Europe—dedicated direct-to-home broadcast satellites had revenues in the tens of billions of dollars.

The Explosion and Consolidation

The 1970s had seen several new satellite communications entities: Telesat Canada, Western Union, RCA, AT&T/COMSAT, Indonesia, and India. The 1980s would see an explosion, followed by ruthless consolidation in the 1990s and 2000s.

Satellite Business Systems

Satellite Business Systems (SBS) was COMSAT's *real* attempt to enter into the domestic satellite communications business. COMSAT had entered the domestic satellite communications field with AT&T in the 1970s. COMSAT provided COMSTAR satellites, which AT&T leased. The first COMSTAR was launched in 1976. Before the COMSTAR launch, COMSAT had taken over an MCI-Lockheed venture and had looked for partners. Eventually, COMSAT teamed with IBM (and Aetna) on SBS, the first serious attempt to use Ku-band frequencies. COMSTAR was profitable but led nowhere. SBS was a disaster. Satellite Television Corporation (STC), COMSAT's venture into direct-to-home television, was also a disaster. Since the late 1960s and early 1970s, when it became obvious that COMSAT would no longer be Intelsat Manager—nor would it be the domestic monopoly operator—COMSAT had been looking for a mission. COMSAT never found that mission. Its success in bringing global satellite communications into being brought COMSAT no continuing purpose.

The SBS satellites had reasonably good performance, but the market IBM had envisioned simply wasn't there. The first five satellites were HS-376 spinners—probably the most successful design ever launched. This success was interesting; the satellites had performance well below comparably priced three-axis satellites built by Ford Aerospace and RCA Astro-Electronics. COMSAT sold its one-third share to IBM and Aetna in 1984. In 1985, SBS was sold to MCI, Aetna receiving cash while IBM received MCI stock and SBS-4 through -6. IBM later sold its Satellite Transponder Leasing Corporation to Hughes Communications, Inc. (HCI), in 1989.[95]

Hughes Communications, Inc.

Hughes had been considering becoming a satellite manufacturer since 1959, when Harold Rosen and Donald Williams came up with the design for what became Syncom. They had prospered as a satellite manufacturer for

---

95. Donald H. Martin, *Communications Satellites, 1958–1995* (Los Angeles: Aerospace Press, 1996), pp. 228–232.

two decades, but they saw opportunities in the marketplace that were not being addressed—or were being addressed poorly. Their first satellites were HS-376s, like SBS, but with conventional C-band payloads.

HCI acquired Western Union's Westar satellites in 1988. In 1989, HCI acquired the SBS satellites from IBM. HCI merged with PanAmSat in 1997—after which Hughes owned two-thirds of the PanAmSat stock. In 2004, after Rupert Murdoch bought Hughes Electronics to control DirecTV, he sold PanAmSat to a private equity firm: KKR. In that same year, Intelsat also was sold to a private equity firm. In 2006, the two companies merged. It is not clear that they will ever be profitable after the private owners raided their treasuries and acquired huge debts, but the merged company controls almost one-third of all FSS transponders.[96]

Pan American Satellite Corporation

Pan American Satellite Corporation (PanAmSat) was the brainchild of René Anselmo. Anselmo founded PanAmSat in 1982—before "separate systems" had been finally approved. His enemy was the COMSAT/Intelsat monopoly. A World War II Marine, a native of Boston—and *not* a native Spanish speaker—Anselmo was able to manage Televisa and create the Spanish International Network. In the period immediately after the Challenger disaster and the failures of several other launch vehicles, Anselmo managed to get a "deal" on both satellite (ex-American Satellite) and launch vehicles. He had difficulty negotiating landing rights but eventually ordered three more satellites. He was ranked #1 in *Space News*'s 2004 (the 15th anniversary of *Space News*) "Top 100."

Anselmo emphasized marketing and lobbying. PanAmSat had no in-house engineering or operations support for years. When PanAmSat "merged" with HCI, the technocratic culture at Hughes was very different from PanAmSat. With the Hughes merger in 1997, PanAmSat now owned the satellite assets of Western Union, Hughes, and SBS.

Societé Européenne des Satellites

Luxembourg has a long history of providing "pirate" radio and television to its European neighbors with "official" government-controlled electronic media. Jan Stenbeck was the first financial backer of Clay Whitehead's Coronet satellite project. This was later taken up by Societé Européenne des Satellites (SES) in 1985. The SES Astra television satellites were first based on RCA/

---

96. Steve Pearlstein, "Sweet Deals Buried Intelsat in Debt," *Washington Post* (18 August 2006): D1, D5. Also Martin, *Communications Satellites*, pp. 232–237.

GE designs—in fact, Astra-1B was originally RCA/GE Satcom K-3. Astra's first customer was Rupert Murdoch's U.K.-based Sky TV. By the end of 1990, all of Astra-1A's transponders were leased.

As the Astra-1 satellites at 19.2E were leased out, it became obvious that a second orbital location would be necessary. The Astra-2 satellites were launched into 28.2E. The money earned from the Astra direct-to-home television satellites allowed SES to buy GE Americom (ex-RCA Americom) in 2001 and to buy shares of AsiaSat in 1999 and shares of Sirius in 2000. In 2006, SES bought New Skies Satellite (NSS), the Intelsat spinoff.

Eutelsat

Eutelsat was brought forth by ESA with the grudging consent of the PTTs. In May 1977, Eutelsat was formed (nominally by the PTTs) and in July awarded a contract for a satellite. The experimental satellites, OTSs (orbital test satellites), were in the process of being launched, and Olympus (also known as H-Sat), the DBS experiment, was being considered. The definitive Eutelsat agreements were agreed upon in 1982. Formal operations began in 1985, two years after the first Eutelsat satellite was launched. The European Communications Satellites (ECS-1 through -5) were designed by ESA and built by the MESH consortium led by Hawker Siddley (British Aerospace).[97]

Eutelsat I (ECS) was succeeded by the Eutelsat II series. A television direct-to-home series, Hot Bird, was launched beginning in 1995. Transatlantic service was begun with the Atlantic Bird series in 2001. In 2006, Eutelsat became the third-largest communications satellite operator after Intelsat/PanAmSat and SES Global.

4. Military Satellite Communications

From NOTUS and Advent to the Initial Defense Communications Satellite Program

Roy W. Johnson presented ARPA's communication satellite program with 1) SCORE, the Atlas package flown in 1958; 2) Courier, another store-and-forward payload, but this time on a real satellite; 3) a polar-orbit real-time repeater; and 4) a 24-hour equatorial orbit (that is, geosynchronous) real-time repeater. Although the terminology was not used in the open sessions,

---

97. Arturo Russo, "The Third Phase of the Telecommunications Programme: ECS, MARECS, and Olympus," in *A History of the European Space Agency, 1958–1987*, vol. 2, *The Story of ESA, 1973–1987*, ed. J. Krige, A. Russo, and L. Sebesta (Noordwijk: European Space Agency, 2000), pp. 229–282. Additional data from Martin, *Communications Satellites*, pp. 282–290.

ARPA was describing the Notus Program: Steer, Tackle, and Decree. Steer is especially interesting because it was a UHF communication satellite for the Strategic Air Command (SAC). In the event of war, SAC bombers would head out over the North Pole en route to the Soviet Union. Steer, a system of polar satellites, would provide decision-makers with the ability to communicate with the bombers. Tackle was an intermediate step to Decree, the 24-hour orbit satellite. Tackle would be a similar satellite in a more easily achievable 10,000-kilometer orbit. The ARPA communications program budget for 1959 was $15 million but would rise to $60 million in 1960 and $100 million per year thereafter. This did not include spending for Centaur, the high-energy upper stage, or Saturn. The problems of the 24-hour orbit satellite might require the payload capability of this much larger launch vehicle.[98] After canceling Advent in the summer of 1962, DOD had decided to develop a smaller, lower-orbit satellite (MACS) for its Initial Defense Communications Satellite Program (IDCSP). In April 1963, Lieutenant General Alfred Dodd Starbird, Director of the Defense Communications Agency (DCA), described the program to Congress, and in May, study proposals were solicited from industry. Two study teams were funded: one led by General Electric, the Advent contractor, and another led by Philco, the Courier contractor.

DSCS-I to DSCS-III

In October 1964, Philco-Ford Corporation (later Ford Aerospace, today Space Systems Loral) was awarded the IDCSP (DSCS I: Defense Satellite Communications System I) contract after an attempt by COMSAT to insist that it had a monopoly on U.S. satellite communications—including military satellite communications. Eugene Fubini, then Director of Defense Research and Engineering (DDRE), had lost all belief in the possible success of geosynchronous-orbit satellites, ensuring that the Initial Defense Communications Satellite Program would be a medium-Earth-orbit (MEO) system. The launch of the 45-kilogram satellites was eventually assigned to Titan. The enormous lift capability of Titan meant that 8 satellites could be launched at a time into near-synchronous orbit. IDCSP/DSCS would use X-band frequencies (8/7 GHz) rather than the commercial C-band

---

98. U.S. Congress, House, 87th Congress, 2nd Session, Committee on Science and Astronautics, *Hearings, Project Advent—Military Communications Satellite Program* (Washington, DC: GPO, 1962), pp. 90–91.

(6/4 GHz). The satellite had one 3-watt, 20-MHz transponder providing an EIRP of 7 dBW. Design lifetime was 18 months, but most lasted far longer.[99]

DSCS II, the follow-on to IDCSP (DSCS I) was a vastly more sophisticated satellite: more than 10 times the mass and power of IDCSP. DSCS II satellites had two narrow-beam and one Earth-coverage antennas with 20-watt (later 40-watt) 50/125-MHz (Earth-coverage) and 50/185-MHz (narrow-beam) transmitters. The antennas were de-spun—EIRP was increased to 28 dBW Earth-coverage and 40 dBW through each narrow-beam antenna (43 dBW if transmitted through only one narrow-beam antenna). These satellites were a scaled-up version of Intelsat III, also built by TRW. They were similar in size and DC power to the contemporaneous Intelsat IV satellites. Two satellites were launched at a time on a Titan IIIC (T34D for last two).

DSCS III was as much a step up in capability as DSCS II had been. In this case, the satellite had beam forming, anti-jam, and nuclear detonation survivability well beyond anything launched heretofore. The satellite had six X-band transponders (two 40 watt and four 10 watt) and a 70-watt UHF transponder. As with DSCS II, DSCS III was a generation behind commercial satellites in its payload and fell further behind with its extremely slow deployment. Early launches were two-on-a-T34D, followed by Atlas II and IIA launches. The last two satellites were launched on Delta 4M rockets, part of the Air Force's Evolved Expendable Launch Vehicle (EELV) program. DSCS III was the first military-communications satellite to be N-S stationkept.

FltSatCom, UFO, and Milstar

If the Defense Satellite Communications System program was to replace the Advent satellite in the Tackle strategic communications portion of NOTUS, there still had to be a tactical satellite to provide UHF communications in the Steer tactical portion of NOTUS. TACSAT had a 30-watt X-band payload and a 230-watt UHF payload. It had almost as much power as DSCS III and pioneered a unique spin-stabilization system that was later used on the Intelsat IV series. The single TACSAT was launched on 9 February 1969. TACSAT and the Lincoln Experimental Satellite (LES) series had demonstrated UHF communications between satellites and mobile tactical users, but an operational system was needed—especially after TACSAT failed in 1972. FltSatCom was to be that operational system, but it would be several years before it was launched. In the meantime, COMSAT provided the Marisat satellites—a variant of the dependable Hughes HS-333 with UHF and L-band capability. Three of these satellites were launched in 1976—one

---

99. Much of the military satellite information is from Martin, *Communications Satellites.*

has just been retired after almost 30 years of service. FltSatCom was built by TRW. The design was reused for the NASA TDRS satellite later. The satellite had a large UHF parabolic receive antenna. In addition to UHF transponders, the satellite had some X-band and Ka-band (30/20 GHz) capability. All were launched by Atlas-Centaur.

The Navy had initially planned a four-satellite constellation similar to DSCS III. This was later doubled using LeaseSat. The UHF follow-on program (UFO) was the first military-communications satellite to model its satellite (HS-601) and procurement (firm-fixed price) on commercial models. At 1,300-kilogram and 3 kilowatts, it was also the biggest military-communications satellite when first launched.

Milstar was intended to replace UFO and implement a new Air Force Strategic Satellite System. Block 1 were low data rate (LDR) satellites while Block 2 were medium data rate satellites. Milstar had extreme robustness and survivability, although Block 2 sacrificed some of this robustness for added payload. Milstar was extremely expensive. It provided less than 5 percent of the communications capacity of a commercial communications satellite for about five times the price. This satellite used some of the higher frequency Ka-band and V-band spectra.

The Wideband Gapfiller Satellite, Advanced Extremely High Frequency, and Transformational Communications Satellite

DSCS III was an aging program in the 1980s. The Service Life Enhancement Program (SLEP) added more capability, but this was still 1960s and 1970s technology. The last DSCS III launch was in 2003. Its wideband replacements were Wideband Gapfiller Satellites (WGS), with WGS-1 launching in 2007, WGS-2 and -3 in 2009, and WGS-4 in 2012. The contract for WGS was awarded to Boeing Satellite Systems (BSS, formerly Hughes) in 2001. The satellite bus is based on the Boeing 702. The UFO was the first military-communications satellite to be firmly based on a commercial satellite bus (with the partial exceptions of DSCS II and Marisat). The last launch was in 1999. The constellation needs to be replaced immediately. Its narrowband replacement is the MUOS (Multi-User Objective System). The MUOS contract was awarded in 2004 to Lockheed Martin, and the first launch was in 2012. The last Milstar was launched in 2003. Although DSCS III has significant protection (anti-jam, nuclear survival), Milstar was the official provider of "protected" service. Its replacement, the AEHF (Advanced Extremely High Frequency) satellite, was launched in 2010. In 2000 the AEHF satellite contract was awarded to a "national team" led by Lockheed Martin. The satellite was based on the Lockheed A2100 satellite bus. The pièce de résistance in the military satellite communications universe was to be TSAT, the transformational communications satellite, but it was canceled in 2009. An Advanced Polar System will be part of TSAT. All of these satellites are late. Each has a price tag in the billion-dollar range (per satellite).

DOD Use of Commercial Communications Satellites

The 1990–1991 Gulf War saw a scramble for commercial satellite bandwidth in which DOD was pitted against commercial news organizations. Military communications satellite capability was not enough! In retrospect, DOD's need for commercial satellite communications during that war was relatively small—about 100 megabits per second (Mbps). In the current action in Iraq and SWA (Southwest Asia), about 3 gigabits per second (Gbps) of commercial satellite communications is in use—a 30-fold increase. As shown in table 7.1, this increase came despite the vastly reduced size of the forces involved.

TABLE 7.1. Increasing Demand for SATCOM Since 1990[100]

	Operations Desert Shield/Storm	Operation Noble Anvil	Operation Enduring Freedom	Operation Iraqi Freedom
Total SATCOM Used (Mbps)	100	250	750	2,400
Total Force Engaged	500,000	51,000	55,000	235,000
Number of 5,000 Military Member Force Increments	100	10.2	11	47
SATCOM Used per 5,000 Military Members (Mbps)	1	24.5	68.2	51.1

Increasing Demand for SATCOM Since 1990

Congress, after hearing the complaints about DOD's competing with news organizations for bandwidth, directed DOD (in the fiscal year 1992 Defense Appropriations Act) to analyze its needs for commercial-satellite communications and discuss these needs with commercial providers. DOD's response was the Commercial Satellite Communications Initiative (CSCI). A result of this initiative was DOD's conclusion that commercial providers of both fixed satellite services (FSS, typical C-band and Ku-band) and mobile satellite services (MSS, typically L-band and S-band) could satisfy many of DOD's needs. DOD specified that new small terminals should have commercial

---

100. Much of the material in this section was obtained from National Research Council, *The Global Positioning System* (Washington, DC: National Academy Press, 1995), pp. 145–276, and *http://www.astronautix.com*.

(C- and Ku-band) as well as military (X-band) capabilities. They also implemented the managed transponder contract (MTC), which allowed DOD to purchase full transponders for at least a year. The MTC contract was awarded to COMSAT in 1995. Not surprisingly, most task orders went to Intelsat—in which COMSAT had a stake—although a few went to PanAmSat or other providers when Intelsat coverage was inadequate. After a few years, DOD end users were complaining that the MTC contract was too constraining. Users wanting a few MHz for less than a year were unable to use the MTC. A new contract, DSTS-G, allowed users to obtain bandwidth of any amount, anywhere, for any amount of time. To avoid what were perceived as high costs on the managed transponder contract, three small businesses received indefinite delivery indefinite quantity (IDIQ) contracts in 2001 that allowed them to compete for DOD's commercial satellite communications business. Because the three vendors were not tied to a single satellite-communications provider, they offered bandwidth from many different providers. The providers competed with each other for the vendors' business, and the vendors competed for DOD's business. DOD established a set of terms and conditions for the DSTS-G program that were incompatible with commercial terms and conditions offered by the bandwidth providers. The DSTS-G contractors signed on to these terms and conditions with very limited ability to "pass through" the DOD requirements. In addition to providing a host of systems engineering services before and after the task order award, the DSTS-G vendors monitored the radio frequency performance of all the links they supplied. After many complaints, a June 2006 DOD Spend Analysis showed that DSTS-G prices were 25 percent below market. But what of the future? High-definition TV and the recovery from the telecom meltdown of 2000 will fill the commercial satellites. It may never be possible again to provide 100 transponders for warfighters in a specific place.

5. Government Meteorological Satellites

The military and the intelligence community were the initial sponsors of weather satellites. NASA took over part of this remit, and by the mid-1960s the Weather Bureau (the Environmental Science Services Administration, or ESSA, then the National Oceanic and Atmospheric Administration, or NOAA) took responsibility for polar weather satellites. Geostationary weather satellites seem to have been championed by NASA and transferred to the Weather Bureau almost immediately after their launch in the 1970s. The military and the intelligence community retained a separate polar weather satellite program into the twenty-first century, but a single National Polar Orbit Environmental Satellite System (NPOESS) was in the works, amid much disarray. TIROS Operational System (TOS) was eventually named ESSA for its sponsor. The TOS had NASA and DOD design participation as well as the inputs of ESSA itself. While the original name implied that the program had gone from "experimental/developmental" to "operational," the Weather Bureau had been using TIROS data for some time—and NASA never relinquished its design and "support" role. ESSA-1, -2, and -9 were launched from Cape Canaveral, and the rest were launched from Vandenberg. A significant Delta-V (velocity) advantage is obtained by launching south from Vandenberg, rather than launching east from the Cape, which requires a "dogleg" maneuver to place the satellite in a polar orbit.

Improved TOS (ITOS)

The Improved TOS (ITOS) was a fairly major departure. Instead of spin-stabilization, three-axis stabilization was used. The satellite was torqued using Earth's magnetic field to maintain one face pointed at Earth at all times. New sensors were carried beyond the TV cameras used until then. The payload of the first two satellites consisted of four TV cameras (automatic picture transmission and wide-angle) similar to those used on the TOS/ESSA satellites. NOAA-2 through NOAA-5 had radiometers instead of television cameras. Satellites subsequent to ITOS-1 were called NOAA-# if they were successful. Since the advent of TOS, ESSA had become the National Oceanic and Atmospheric Administration (NOAA).

TIROS-N

By the late 1970s, the success of both the polar and geosynchronous weather satellites was assured. There had been some concern in the 1950s and 1960s that "cloud pictures" did not fit into the very sophisticated numerical weather models that were being developed to run on advanced computers. A "picture" did not provide the vertical "sounding" of temperature, pressure, and winds that modern meteorology needed. As early as 1954, the value of satellite pictures for major storm (for example, hurricane) observation was recognized. By the time the ITOS series was launched, vertical sounding instruments had been placed on polar weather satellites. The TIROS-N satellite provided more instruments than had ever been flown before. It was almost a research satellite as well as an operational satellite. From the first launch of TIROS-1 in 1960, the civilian weather satellites were launched on Delta launch vehicles— probably the most reliable rocket until the commercial Atlas. The earliest satellites were launched into high-inclination orbits (48°)—and eventually into Sun-synchronous polar orbits (~100°). Altitude varied somewhat but was usually about 900 kilometers. The first two TOS satellites—and the last

TOS—were launched from Cape Canaveral, but all the rest were launched from Vandenberg Air Force Base. TIROS-N and later polar weather satellites would still be launched from Vandenberg, but they had grown from about 100 kilograms to about 1,000 kilograms. Delta lift capacity had also been growing, but not as fast. Future launches would be by Atlas. NOAA-15 through -16 were launched on Titan II, but NOAA-18 saw a return to Delta.

Geosynchronous Weather Observation

In spite of Arthur C. Clarke's work, NASA first looked at geosynchronous orbit (GEO) as a place for weather satellites. A consequence of the stationary orbit over the equator was that a GEO weather satellite could take continuous pictures of about one-third of Earth's surface. The polar weather satellites only took one picture (two if we include night-time infrared pictures) of a given location each day. Verner Suomi joined the National Science Foundation (NSF) in 1962 to serve as Associate Program Director for their Atmospheric Sciences Division. In 1964, he joined the U.S. Weather Bureau, where he served as Chief Scientist for one year. In 1965, Suomi and Robert Parent, a professor in electrical engineering, started the Space Science and Engineering Center (SSEC) at the University of Wisconsin–Madison with funding from NASA and the NSF. While at SSEC, Suomi developed the spin-scan camera. Suomi and Parent's spin-scan camera was launched on ATS-1 in 1966. The camera scanned an east-west strip of Earth with each rotation of the spinning satellite. By the slight tilting of a mirror in the camera with each rotation, a multistrip image of Earth could be created in less than 30 minutes.

Applications Technology Satellites (ATS)

The applications technology satellites had originally been conceived as advanced Syncom satellites. The creation of COMSAT following the Communications Satellite Act of 1962 led to suggestions that communications satellite R&D by NASA was inappropriate as COMSAT was a private entity. NASA was more than willing to add a meteorological payload, and DOD asked that gravity-gradient (GG) stabilization and medium-Earth orbit (MEO) experiments also be conducted.

The first five ATS satellites were all built by Hughes. None of the gravity-gradient experiments worked. All of the cameras and all the C-band transponders worked. ATS-1 and ATS-3 were complete successes, taking the first black-and-white (ATS-1) and first color (ATS-3) pictures of Earth from geosynchronous orbit. The first three ATS launches were on Atlas-Agenas (A-A), the second two on Atlas-Centaurs (A-C), and the sixth on a Titan. ATS-1 and ATS-3 were deactivated in 1978. ATS-5—a nominal failure—provided communications services for many years.

Synchronous Meteorological Satellites (SMS)

The experimental/operational synchronous meteorological satellites (SMS) were built by Ford Aerospace. Their lighter weight allowed the use of the cheaper, more reliable Delta launch vehicle. All carried a Visible Infrared Spin Scan Radiometer (VISSR) built by Hughes Santa Barbara Research Center. SMS-1 was placed over the Atlantic, and SMS-2 was placed over the Eastern Pacific.

Geostationary Operational Environmental Satellites (GOES)

The geostationary operational environmental satellites—GOES-1 through GOES-3—were identical to the SMS-1 and SMS-2 satellites. GOES-4 through GOES-7 were built by Hughes. The more advanced GOES-8 through GOES-12 have an imager much like the advanced TIROS-N polar-orbiting satellites. Attitude control is three-axis, and detailed position and pointing are obtained using the Very High Resolution Radiometer imager. (In 1998, the Advanced Very High Resolution Radiometer imager came online.) As with the polar satellites, GOES now combines operational capabilities and research capabilities. One satellite is usually positioned over the Atlantic at a longitude of 75 degrees west (GOES-East) and another at 135 degrees west (GOES-West). Early in the program, satellites were placed over the Indian Ocean to provide "global" coverage for the Global Atmospheric Research Program (GARP). European, Indian, and Russian satellites now provide Indian Ocean coverage while Japan provides coverage of the Western Pacific. Any "extra" GOES satellites are stored at a longitude of 105 degrees west— ready to replace GOES-East or GOES-West.

International Cooperation

Meteorological data have always been shared with other countries. In 1977, both Europe and Japan launched geosynchronous-orbit weather satellites. When the GOES-NEXT program was delayed, the Europeans loaned NOAA a Meteosat. When the Japanese MTSAT was delayed, NOAA loaned the Japan Meteorological Agency (JMA) a GOES. Starting from the prime meridian, the European Organisation for the Exploitation of Meteorological Satellites (EUMETSAT) covers the Eastern Atlantic from a longitude of 0 degrees east and the Indian Ocean from a longitude of 62 degrees east. The Japan Meteorological Agency covers the Western Pacific from a longitude of

140 degrees east (and 155 degrees east). NOAA covers the Eastern Pacific from a longitude of 135 degrees west and the Western Atlantic from a longitude of 75 degrees west. The five satellites of these three agencies monitor Earth's weather—except for polar latitudes—continuously. These countries also cooperate by sharing polar weather data. Russia also supports this activity.

6. Dual-Use Navigation Satellites

After the poor reliability of the Transit satellites built by the Naval Avionics Facility Indianapolis (NAFI), RCA built the rest.

Developing Navstar GPS

It was always clear that Transit had significant limitations. The accuracy was good enough for nuclear weapons (<1 kilometer) but not good enough for conventional weapons. The fix took some time to obtain—making Transit almost useless for moving objects. The Navy continued research—especially at the Naval Research Laboratory—on improvements. The Air Force started a new program, Project 621B, in 1964. By 1972, DOD wanted just one program: a system that could be used to navigate fast-moving aircraft and even to deliver conventional weapons. In 1973, the Navstar Global Positioning System (GPS) program was approved. Over the next few years, arguments and tradeoffs between the Navy and the Air Force were adjudicated—mostly on their merits—and the GPS Block I launches began in 1978. Transit was kept in operation until 1996.[101]

The Satellites

There is very little difference between the various GPS blocks. In all cases, the satellites are in 12-hour circular orbits inclined 55 degrees to the equator. There are six orbital planes, each containing four satellites. Two L-band frequencies are broadcast (L1: 1,575.42 MHz and L2: 1,227.60 MHz), containing the time (Universal Time Coordinated and a pseudo–random noise code) and satellite position. Differences between the satellite time and the vehicle time provide range measurements—three range measurements allow a position to be determined to within 10 meters to 100 meters. The civilian (SPS) signal has a conditional access (CA) code that degrades accuracy. Military users can get position to within a few meters. A contract for eight Block I

---

101. Much of the material in this section was obtained from National Research Council, *The Global Positioning System* (Washington, DC: National Academy Press, 1995), pp. 145–276, and *http://www.astronautix.com*.

satellites was awarded to North American (Rockwell) in 1974. A contract modification for an additional four Block I satellites was awarded in 1981. Navstar-12 was produced as the Block II qualification model. There was only 1 failure out of 11 launches.

Block II/IIA added nuclear detonation detectors and many improvements. The satellite was still manufactured by Rockwell, but the launch vehicle was now a Delta 6925 for Block II and a Delta 7925 for Block IIA. There was only 1 failure in 33 launches.

After building 44 GPS satellites, Rockwell lost the "replacement" contract to General Electric Astro Space (formerly RCA Astro Electronics, currently Lockheed Martin Commercial Space Systems). The Block IIR satellites were based on the Astro Space series 4000 geosynchronous-communications satellite. The 1989 contract was for 21 satellites. Many improvements in cost, lifetime, autonomy, and precision were made on the IIR series.

Block IIF and Block III

In 1997, the Air Force awarded a contract for six GPS satellites and 27 options to Rockwell (now Boeing). In 2000, the decision was made to rebid the contract. A series of studies for a "generation after next" system, Block III, was begun in 2000. This was revised in 2003 and again in 2005. Lockheed Martin won the Block III competition in 2008, but launch has been delayed until at least 2017.

Civil and Commercial GPS

The decision to allow Transit use by commercial ships seems to have been made at an early date. By 1961, NASA had assumed responsibility for "commercial" maritime satellite navigation applications. By 1962, aircraft navigation was added to the mix. By 1964, the idea of a "traffic management" satellite had been developed. NASA added these tasks to its applications technology satellites program and began thinking about an aeronautical communications satellite. The maritime side of this was given extra urgency when the Torrey Canyon ran aground on the Cornish coast in March 1967, spilling 120,000 tons of oil.[102] By 1995, civilian use of GPS exceeded military use. Ten years later, GPS was the established navigation system—an "international

---

102. Abraham Hyatt, memo to Deputy Administrator and Associate Administrator, "Informal Discussions Regarding Navigation Satellites," 7 September 1961, NHO; Alton B. Moody, "Navigation Satellite Progress," National Electronics Congress, 9 October 1962, NHO; "NAV/TRAF SAT program," *Space Daily* (21 April 1964): 118; Walter Sullivan, "How To Navigate with Satellites," *New York Times* (2 April 1967): E7.

utility." Most other systems were in the process of shutting down. But GPS remained a military system: use could be denied during a military emergency. The Federal Aviation Administration (FAA) started a program in 1995 called the Wide Area Augmentation System (WAAS) that would facilitate the use of GPS for instrument landings. This would obviate the need to build the Microwave Landing System (MLS) scheduled to replace the old Instrument Landing System (ILS). Most MLS systems in the United States have been turned off and replaced by GPS.[103]

Global Orbiting Navigation Satellite System and Galileo

As GPS replaced all previous navigation (and instrument landing) systems, many foreign countries became quite concerned that transportation safety was dependent on an American military system. In part to assuage these fears, and to increase the precision of GPS, selective availability (SA)—a system that ensured lower civilian accuracy—was turned off on 1 May 2000. Somewhat earlier, the Soviet Union had launched its own navigation system, the Global Orbiting Navigation Satellite System (GLONASS), in 1982. The system was fully operational in 1995. Unfortunately, the collapse of the Soviet economy left Russia unable to maintain the system for several years. A planned replenishment was scheduled for completion by 2011. Both GPS and GLONASS are military systems that allow civilian use. The European Galileo system will be completely civilian-run by a private consortium. The four Galileo In-Orbit Validation Experiment (GIOVE) satellites were to be launched by 2008. The 30-satellite operational system will be complete by 2019. Galileo will provide greater accuracy (about 1 meter) and will work in buildings and under trees. Galileo and GPS will be compatible.[104]

7. Mobile Satellite Communications

Almost from its inception, COMSAT was involved with mobile satellites. The interest came from two sources: the maritime industry and the airline industry.

---

103. David Field, "U.S. To Let Airliners Navigate by Its Satellites," *Washington Times* (28 March 1995): B7; Warren E. Leary, "Civilian Uses Are Proposed for Satellites," *New York Times* (1 June 1995): A23.
104. Ibid.

First Is Aerosat

Robert Kester of the FAA's National Aviation Facilities Experimental Center noticed in 1964 that Syncom 2 had Telemetry, Tracking and Command (TT&C) frequencies close to aviation frequencies. He persuaded NASA and Hughes to modify the TT&C equipment slightly. Kester teamed with Roland Boucher at Hughes. Their first experiment involved teletype signals received from the satellite. Somewhat later, the Airline Transport Association arranged for a Pan American World Airways (Pan Am) flight to carry a modified radio and teletype rig across the Pacific. Roland Boucher of Hughes used the VHF CMD/TLM channel of Early Bird to communicate with a Pan American 707 in 1965. ATS-1 and ATS-3 had VHF transponders to experiment with satellite-to-aircraft communications; ATS-5 had an L-band transponder.[105] The FAA was also interested in aircraft-to-satellite communications. Both NASA and the FAA turned to the European Space Research Organisation (ESRO, later the ESA) to garner European support for a transatlantic aeronautical satellite service. In 1965, COMSAT proposed a VHF "Aerosat" to the FAA. The Hughes HS-303A (Intelsat II) would be the basis for an HS-303B with two VHF transponders (later four VHF transponders).

COMSAT also considered "cross-strapped" C/VHF transponders in a 1967 proposal to the FAA. The Europeans feared U.S. North Atlantic hegemony and preferred L-band frequencies—as did FAA research (but not Air Traffic Control) staff. COMSAT proposed VHF and L-band over the Pacific, but this caused concern at NASA with respect to their arrangements with ESRO. Both the Europeans and NASA were opposed to the COMSAT proposal. ARINC[106] requested that COMSAT return to an all-VHF satellite, resulting in a 1968 COMSAT proposal to the FAA for a four-channel VHF satellite. In anticipation of Aerosat, all early 747s had a hump behind the cockpit with a VHF slot-dipole satellite antenna. In 1968, at the European Space Conference (ESC) meeting at Bad Godesberg, there was a discussion of air traffic control satellites. The ESC asked the European Space Research Organization and ELDO to provide research.[107] The International Civil

---

105. Lorenza Sebesta, "The Aeronautical Satellite System: An Example of International Bargaining," in *History of the European Space Agency, 1958–1987*, ed. Krige, Russo, and Sebesta, vol. 2, *Story of ESA*, p. 361.

106. ARINC, as it is known today, was incorporated in 1929 as Aeronautical Radio, Incorporated. It was chartered by the Federal Radio Commission (later the Federal Communications Commission) to serve as the airline industry's single licensee and coordinator of radio communication.

107. John Krige, Arturo Russo, and Lorenza Sebesta, "The Development of ESA," in *History of the European Space Agency, 1958–1987*, ed. Krige, Russo, and Sebesta, vol. 2, *Story of ESA*, Aerosat section, pp. 44–46; Sebesta, "Aeronautical Satellite System," pp. 357–386.

Aviation Organization (ICAO) entered the discussion, forming a panel with members from Australia, Canada, France, the Federal Republic of Germany, Japan, the United Kingdom, the United States, the International Air Transport Association (IATA), the ITU, and the WMO. The IATA and United States favored VHF—all others favored L-band.[108] In 1969, Herman Bondi, the ESRO Director General, approached Thomas Paine, the new NASA Administrator, to coordinate efforts. The first meetings were held in June 1969 at NASA Headquarters. NETCOS, the NASA/ESRO Air Traffic Control System committee, developed an outline of the joint NASA/ESRO project.[109] A NASA/ESRO mission specifications draft was submitted to ICAO. Clay Whitehead's Office of Telecommunications Policy (OTP) stated that the FAA, not NASA, would be responsible for aeronautical satellites. The OTP also decided that L-band, not VHF, would be used. While the Office of Telecommunications Policy favored a purely commercial solution, the Aerosat team would consist of the FAA, the Government of Canada, and ESRO—the FAA to be replaced by a commercial entity; ESRO could choose a U.S. entity. A complicated bidding system was set up to determine that U.S. entity. The candidates were effectively limited to RCA and COMSAT. COMSAT won with a very brief proposal. At about this time, the ITU World Administrative Radio Conference for Space Telecommunications (Geneva, 1971), or WARC 71, allocated two 7.5 MHz bands at 1.6 GHz for maritime satellite service and two 15 MHz bands for aeronautical satellite service at 1.6 GHz (L-band). An intergovernmental organization to operate Aerosat was established in August 1974 by United States, the ESA, and Canada. Australia and Japan could also participate. The Aerosat Council was chaired by David R. Israel, Deputy Associate Administrator of the FAA. In December 1974, three teams were ready to bid on Aerosat: COSMOS (GE, Aerospatiale, MBB, Marconi, Selenia, Siemens, and SAT), MESH (TRW, Matra, Saab, Hawker-Siddeley, Fiat, and ERNO), and STAR (RCA, CSF, BAC, Dornier, Fiat, Montedel, Ericcson, and RCA Ltd.). Hughes did not put in a bid because of the requirement for extensive international content.[110] The Aerosat Space Segment Board, on 22 January 1975, announced the schedule for the request for proposal (RFP) (due on 1 March), proposals (due on 15 June), and award (granted on 15 November). There would be two satellites over the Atlantic separated by 25 degrees. The first launch would be by

---

108. Sebesta, "Aeronautical Satellite System," p. 361.

109. Ibid., p. 363.

110. "Aerosat RFPs To Be Issued in July, Three Bidders Seen," *Defense/Space Daily* (6 December 1974): 192–193.

the end of 1979, the second eight months later.[111] An RFP was issued, and a winning proposal was submitted by GE in September 1976. Three good technical proposals were submitted—GE was cheaper.[112] There was no notice to proceed, however. The airlines revolted—they had not been consulted. By September 1977, the Aerosat Program was over, but the Aerosat Council hung on until 1982.

Next Is Marisat

In the 1960s, there were studies by the Inter-Governmental Maritime Consultative Organization (IMCO), later the International Maritime Organization (IMO), on the use of satellites for maritime communications. In 1971, ITU WARC 71 allocated "two 7.5 MHz bands at 1.6 GHZ for maritime communications and two at 15 MHz for aeronautical communications. Separate from the commercial maritime need, in 1972 the U.S. Navy was looking for a replacement—a "gapfiller"—for Tacsat (and LES-6) until the Fleet Satellite Communications System (FltSatCom) could be launched. COMSAT, teamed with Hughes, proposed a hybrid UHF/L-band satellite with C-band feeder links. The international record carriers (IRCs) complained, resulting in sharing of the venture: RCA 8 percent, WUI 3.41 percent, and ITT 2.3 percent. In February 1973, COMSAT announced a new program at a meeting in London. There was to be no European participation. In May 1973, COMSAT signed a contract with Hughes for $40 million for three satellites. By early 1976, Atlantic & Pacific UHF ground stations were ready in Southbury, Connecticut, and Santa Paula, California. In 1978, an Indian Ocean ground station was in place: the Yamaguchi station of KDD.[113]

The Marisat satellites had relatively limited capability, but it was enough for the U.S. Navy—and the satellites even obtained some commercial business. By 1978, the Navy was worried that FltSatCom would be operational much later and began discussing a Marisat II program.

Inmarsat

At some point, it was decided that INTELSAT could not provide maritime communications. At least part of the rationale was that the USSR wanted to be part of the maritime communications effort but did not want to join INTELSAT. In the 1975–1976 timeframe, IMCO arranged a Convention on

111. ESA, press release, 23 January 1976; COMSAT General, press release, 23 January 1976.
112. ESA, press release, 10 September 1976.
113. Much of the material on Marisat—and some on Aerosat and Inmarsat—was obtained in an interview with Ed Martin of COMSAT.

the International Maritime Satellite Organization–Inmarsat. An Operating Agreement on the International Maritime Satellite Organization–Inmarsat was developed, and the document was opened for signature in 1976. At a meeting in Norway in 1978, the Inmarsat "constitutive" agreements were reached. The United States did not sign at this time. Eventually, Congress enacted the Maritime Satellite Act of 1978 and chose COMSAT as the American "signatory" to the Inmarsat agreements. The Inmarsat agreements entered into force on 16 July 1979. There was substantial controversy over which satellites to use for the new maritime system: Marisats, Intelsat Vs with MCS, or Marecs—an ESA/Eutelsat system based on the ECS satellite.

Part of the Europe–United States political dispute ended with the location of Inmarsat headquarters in London and a bias toward European satellite contracts. When Inmarsat began operations, it leased three Marisats (at longitudes of 15 degrees west, 72.5 degrees east, and 176.5 degrees east), four Intelsat-Vs with MCS (at longitudes of 18.5 degrees west, 63 degrees east, 66 degrees east, and 180 degrees east), and two Marecs (at longitudes of 26 degrees west and 177.5 degrees east). The second generation would be owned by Inmarsat. The contract was awarded to the MESH consortium led by British Aerospace, but with an American (Hughes) payload.

The timing of the third generation was determined by the rapid uptake of Inmarsat services—not the Inmarsat II lifetimes. The new satellites had spot beams and frequency reuse capability. In a reversal of the second-generation satellite contract, the bus was American (the old RCA Astro, then GE, now Lockheed Martin).

LEOs and MEOs

In the 1990s, several new satellite constellations in low-Earth orbit (LEO) and medium-Earth orbit (MEO) were considered and built. All ended up in bankruptcy. Almost all have emerged from bankruptcy and are attempting to build a place for themselves in the market. None has yet to show that this is possible. All were looking at some form of cell phone–like communications.

8. Military Reconnaissance Satellites

The Air Force, the CIA, and CORONA

The Air Force (aided by its RAND think tank) began the development of a reconnaissance satellite on 16 March 1955. The program—initially called Advanced Reconnaissance Satellite (ARS), then SENTRY, and finally SAMOS—was slow to mature. On 7 February 1958, President Eisenhower authorized the Central Intelligence Agency (CIA) to proceed with CORONA.

From 1959 to 1971, CORONA was the principal U.S. reconnaissance satellite (along with a few ARGON and LANYARD special-purpose satellites). SAMOS was eventually canceled. In 1960, a joint program office was formed and designated the National Reconnaissance Office (NRO). The NRO was staffed by the CIA and the Air Force. The existence of the NRO was "revealed" on 18 September 1992. The entire CORONA program was declassified on 24 February 1995. Later programs are still classified, making accurate descriptions difficult.

GAMBIT

The KH-7 and KH-8 GAMBIT satellites provided increased resolution (about 0.5 meter) over the CORONA satellites (about 3 meters). The CORONA satellites had grown in size from 800 kilograms to about 2,000 kilograms (two SRVs), but they were all launched by Thor-Agena launch vehicles. The KH-7 satellites were launched on Atlas-Agenas. The heavier (3,000-kilogram) KH-8 satellites were launched on Titans. About 100 GAMBITs were launched between 1963 and 1984, with about a 95 percent success rate. Early GAMBITs had lifetimes of days, but over time, their lifetimes grew to weeks.

Manned Orbiting Laboratory

The Air Force had always wanted to put humans in space. The Manned Orbiting Laboratory (MOL) was their great opportunity to do so. Although the laboratory had many goals, its primary purpose was the KH-10 (DORIAN) reconnaissance system. The vehicle would have weighed about 15,000 kilograms. First authorized in 1962–1963, the laboratory would eventually be canceled in 1969 after an expenditure of billions of dollars.

The Anti-Ballistic Missile Treaty

Starting in the late 1960s, the United States and the USSR began discussing arms limitation. The Soviet Union had established—or was establishing—rough parity in nuclear weapons and ICBMs. Both sides were developing anti-ballistic missile (ABM) systems. In the process of negotiating the Strategic Arms Limitation Talks (SALT I, the ABM treaty), it was agreed that "national technical means" would be used to verify compliance and that no interference with these means would be allowed. Spy satellites were legal!

Big Bird

It has been argued that the KH-9 was developed as a backup to the Manned Orbiting Laboratory. The vehicle weighed over 11,000 kilograms—almost four times the weight of a KH-8 GAMBIT and not much less than the MOL—but the laboratory mirror was not used until the KH-11 Kennan satellite in 1984. Big Bird carried a television camera as well as film cameras and four SRVs. It was launched by Titan 3D rockets. Big Birds increased satellite lifetimes to months. Of 20 KH-9 launches, only 1 failed, the last launch in 1986. Declassification was progressing until the fall of 1997. There were even plans to place a KH-9 in the then-new Steven F. Udvar-Hazy Center, a Smithsonian annex, near Dulles Airport. According to Dwayne Day, a military space historian who has written about CORONA and other spy satellites, Big Bird probably "gathers dust in a classified warehouse … only a few yards down from the Lost Ark of the Covenant."[114]

The Rest of the Spy Satellites

Perhaps the biggest improvement in spy satellites was the all-electronic, direct-readout KH-11 Kennan/Crystal and its successor, the KH-12/KH-11B Improved Crystal. These satellites finally provided the capability that SENTRY/SAMOS had hoped for: real-time direct readout, facilitated by communications relay satellites including Satellite Data System (SDS), the Tracking and Data Relay Satellite System (TDRSS), and Milstar. About two dozen of these satellites have been launched. The Improved Crystal weighs almost 20,000 kilograms and can only be launched by the Shuttle or Titan 4. Lifetimes are now measured in years. The KH-12 carries about 7,000 kilograms of fuel—its lifetime is over 10 years. In addition to the visible and infrared capabilities of the KH satellites, at least a half-dozen Lacrosse radar satellites have been launched. Image intelligence (IMINT) and human (spy) intelligence (HUMINT) have been supported by various forms of signal intelligence (SIGINT), including satellite SIGINT. These included Navy systems from the 1960s (Grab, Dyno, Poppy), ferrets launched with KH-9 satellites, Air Force systems (Canyon, Vortex, Mercury), and the CIA's Aquacade. Many of these satellites are now in geosynchronous or Molniya orbits—and are all but invisible.

9. Civil Land Remote Sensing Satellites

The Earth Resources Technology Satellite (ERTS, later Landsat) was seen as another task (land remote sensing) NASA should perform. It met with a number of objections over the early years—primarily from DOD (which was worried about compromising spy satellite technology) and the Office of

---

114. Dwayne A. Day, "The Invisible Big Bird: Why There Is No KH-9 Spy Satellite in the Smithsonian," *The Space Review* (8 November 2004), available at *http://www.thespacereview.com/article/263/1* (accessed 28 August 2014).

Management and Budget (which worried about expenses). Initial support came from the Interior (Geological Survey) and Agriculture Departments. NASA tried to push applications by funding Interior and Agriculture studies, but it remained the case that no one wanted to pay for Landsat data, although they were more than happy to accept free data. The first generation (Landsat-1 to -3) was met with little interest, but by the time the second generation was launched, NASA had developed a constituency for Landsat. Unfortunately, there were still no institutions willing to fund Landsat.[115]

The 15 years between the launch of Landsat-5 and the next successful Landsat launch were frustrating for all concerned. In 1979, President Carter had decided to move land remote sensing to NOAA—and eventually to private industry. The Reagan administration tried to accelerate the transfer to private industry, but there were no takers. COMSAT had volunteered to consider taking over the Landsat program if the geosynchronous weather satellites were included. In 1983, Congress made it clear that weather satellites would remain under government control. In 1984, the Department of Commerce released an RFP for privatization and eventually signed a contract with the Earth Observation Satellite Company (EOSAT), consisting of RCA, Hughes, and others, to take over Landsat operations. At about this time, the National Research Council published the report *Remote Sensing of the Earth from Space: A Program in Crisis*, and Congress passed the Land Remote Commercialization Act of 1984. The act gave NOAA responsibility for "licensing" commercial remote sensing satellites.[116]

10. Commercial Remote Sensing Satellites

The French Système Pour l'Observation de la Terre (SPOT) remote sensing satellite had been launched in 1986 and offered "commercial" services. Many countries jumped on the bandwagon, but true commercial programs took a while to move forward. Eventually, the Land Remote Sensing Act of 1992 provided enough incentives for companies to come forward. In 1993, DigitalGlobe/WorldView/EarthWatch (Ball) and Space Imaging (Lockheed) were issued licenses for their first satellites. Orbimage (Orbital Sciences) followed in 1994. About two dozen satellites have been launched since 1997, but commercial markets do not exist to maintain these relatively small and

115. Pamela E. Mack, *Viewing the Earth: The Social Construction of the Landsat Satellite System* (Cambridge: MIT Press, 1990).

116. Space Applications Board, *Remote Sensing of the Earth from Space: A Program in Crisis* (Washington, DC: National Academy Press, 1985).

inexpensive satellites. The largest market is still the intelligence agencies, especially the National Geospatial-Intelligence Agency (NGA).

Part III: Impact and Future

1. Evaluating the 20th-Century Societal Impact of Applications Satellites

From the twentieth century, Americans looked back at the fifteenth century and saw Columbus's discovery of the New World as the most important event. From the twenty-fifth century, Americans may look back at the twentieth century and see the Moon landings as the most important event. It may be more realistic to say that applications satellites had more of an impact than human exploration or the amazing discoveries made in planetary science and astrophysics. There have been commercial, military, safety, and globalization effects that outweigh the lunar landing—at least in the opinion of some.

Commercial

Revenues for commercial satellite applications have been dominated by satellite communications.[117] Revenues approach $200 billion, primarily in services, but also in network and consumer ground equipment. Somewhat surprisingly, sales of commercial GPS receivers have made navigation satellites the second most profitable application. Weather and reconnaissance satellites could be made profitable, but the perception of these as military or safety applications makes this unlikely. Land remote sensing has remained a military/intelligence application. None of the "commercial" remote sensing companies can survive without the intelligence agencies—especially the National Geospatial-Intelligence Agency—as a customer.

Communications Services

The SIA/Tauri "State of the Satellite Industry Report" covering 2014 breaks satellite communications into three areas: consumer services (BSS/DTH), fixed services (FSS), and mobile services (MSS). Revenues from all combined added up to $122.9 billion in 2014. Consumer services dominated with revenues of $100.9 billion. About $25 billion in ground system revenues should also be added. Manufacture of communications satellites generated revenues of about $6 billion, as did launch services.

---

117. Much of the material for this section comes from the Satellite Industry Association (SIA) and Futron. See SIA/Futron, "State of the Satellite Industry Report," June 2006.

Ground Equipment

The SIA/Tauri report states that ground equipment revenues were about $58 billion in 2014. About half of this was satellite communications equipment, but the other half was revenue from the sale of GPS receivers. Unlike satellite services (BSS, FSS, and MSS), revenues from ground equipment are spread across a large number of providers.

Weather Satellites

The commercial value of weather satellites was indicated in the mid-1980s, when the Reagan administration was trying to privatize Landsat. COMSAT offered to take over Landsat only if it was also given the weather satellites. NOAA has provided a compendium of economic statistics in which the costs of weather and climate events are summarized along with some estimates of the benefits of weather forecasting. Severe weather causes damages well in excess of $10 billion every year. Benefits are estimated at over $10 billion per year. Benefits to agriculture, construction, and transportation would presumably increase this total.[118]

Land Remote Sensing Satellites

Various U.S. laws and policy statements eventually created a commercial land remote sensing satellite industry—sparked in part by the apparent success of the French SPOT satellite. Although revenues are fast approaching the billion-dollar range, the "commercial" aspect is minimal.

Satellite Manufacturing

The SIA/Futron figure for satellite manufacturing ($7.8 billion) is dominated by the "commercial" sale of satellites to governments, including the sale of many nonapplications satellites. The total due to applications satellites is probably about $4 billion. Communications satellites represent about half of this total. Most of these manufacturers are subsidiaries of larger companies, making it harder to separate revenues from applications satellites. Almost all of these companies/subsidiaries were near bankruptcy during the telecom/dot-com meltdown at the turn of the century.

---

118. NOAA, "Economic Statistics for NOAA," U.S. Department of Commerce, May 2005, pp. 10, 38.

Launch Industry

Launching applications satellites generates billions of dollars in revenues every year Like satellite manufacturing, the launch industry took a big hit during the telecom/dot-com meltdown at the turn of the century. In the United States at the time, all but Shuttle launches were "commercial." It is noteworthy that the Atlas had a reliability record that was stagnant at about 85 percent when commercial launches became a fact of life. Since then, the Atlas has had a 100 percent reliability record.

U.S. Share

The United States dominated the applications satellite field for many years. This dominance has eroded over time. The United States still dominates satellite navigation and is the most powerful player in the satellite communications and reconnaissance sectors, but it is no longer dominant over the entire applications satellite field. The European Galileo and Russian GLONASS may end U.S. dominance of satellite navigation also. While the United States pioneered weather satellites, those quickly became internationalized. This was why DOD/NRO launched its own weather satellites.

Military Security

> *I wouldn't want to be quoted on this, but we've spent thirty-five or forty billion dollars on the space program. And if nothing else had come out of it except the knowledge we've obtained from space photography, it would be worth ten times what the program has cost. Because tonight we know how many missiles the enemy has and, it turned out, our guesses were way off. We were doing things we didn't need to do. We were building things we didn't need to build. We were harboring fears we didn't need to harbor.*

> —Lyndon B. Johnson, Nashville, March 1967

Without reconnaissance satellites, the United States probably would have deployed 10,000 Minuteman ICBMs—not 1,000. Without reconnaissance satellites, the ABM, SALT, START, and other disarmament treaties and agreements would not be possible. The closest the world came to World War III was the Cuban Missile Crisis of October 1962. Current reconnaissance capabilities would have detected the missiles (and warheads) before their presence became a *crisis*. In 1962, strategic reconnaissance using CORONA satellites was limited by the amount of film on board. Tactical (real-time) reconnaissance was done by aircraft—especially the U-2. The missiles were detected—but too late to avoid a crisis. Since the 1990s, and probably earlier, U.S. military might has included a strong space component: reconnaissance, weather, navigation, and communications. What has been added is a space-enhanced ability to wage war. Many munitions are delivered with GPS-guidance systems. Unmanned aerial vehicles are commanded and their pictures recovered via a communications satellite—typically a *commercial* communications satellite. The United States has waged two wars in the Persian Gulf and another in Afghanistan in which space applications were defining.

Civilian Safety

Applications satellites have changed the risks U.S. citizens face. The hurricanes of 1900 and 1938 came from nowhere and killed people on the shoreline who had no idea a major storm was coming. Galveston on 8 September 1900 had a population of about 36,000; by nightfall, one in six would be dead. The 1938 New England hurricane completely wiped out several vacation areas and flooded sea-level Providence, Rhode Island, and interior Hartford, Connecticut. The Galveston hurricane of 1900 may have been the deadliest natural disaster in the United States, but it doesn't even appear on any list of storms sorted by damage cost. In contrast, in September 2005, Galveston was evacuated over the single bridge linking it to the mainland before Hurricane Rita hit that month. Evacuation was probably easier to enforce after the Hurricane Katrina disaster a month earlier. Katrina was among the most costly hurricanes to hit the United States, but the death toll—in spite of poor evacuation plans—was much lower than it might have been. The inflation-adjusted cost of Katrina damage was 100 times the cost of the 1900 Galveston hurricane damage, but the death toll was one-third. NOAA predicted landfall at New Orleans more than two days in advance. The day before landfall, the local NOAA office recommended immediate evacuation.[119] Weather satellites don't just provide cloud pictures and warnings of hurricanes. They also detect forest fires, volcanoes, and severe storms; and they provide measures of rainfall and winds. Weather satellites allow us to see hurricanes developing and plot their tracks; communications satellites can relay that information anywhere in the world. Any modern explorer can outfit herself with a GPS receiver, a satellite telephone, and a satellite DARS radio for much less than $1,000 (mostly for the telephone). This will enable her to know her position to within meters, keep in touch with world events (including weather), and call for help if necessary.

---

119. From Wikipedia, *http://en.wikipedia.org/wiki/Hurricane_Katrina*, *http://www2.sunysuffolk.edu/mandias/38hurricane*, and *http://en.wikipedia.org/wiki/Hurricane_Rita*.

One World

Submarine telephone and telegraph cables allowed elites to communicate as early as 1858, but satellite communications opened up communications links to the middle classes and even to the "poor" of the industrialized world. Television via satellite has made global news instantaneous. Marshall McLuhan's Global Village has arrived. Unfortunately, it is a very combative and relatively ignorant village with vendettas everywhere, but communication is education as well as propaganda. The benefits of weather satellites are global; weather is global. Nowhere is beyond the reach of a determined explorer with a GPS receiver. Although satellites didn't predict the Indian Ocean tsunami in December 2004 that devastated parts of Indonesia, Thailand, Bangladesh, and other Indian Ocean coastlines, pictures of the devastation were broadcast worldwide in almost real time. Wars are now brought into the living rooms of the middle class around the world. Famine and civil war in Darfur were topics of conversation everywhere. This "global village" is still more a series of tribal villages, but humanity is becoming more and more aware of its connectedness.

2. Looking Toward the Future

As the great American Yogi Berra once said: "It's tough to make predictions—especially if they involve the future." He also said: "The future ain't what it used to be."

Reconnaissance Satellites

The great success of reconnaissance satellites was in allowing the United States visibility into the Soviet Union. That visibility allowed the United States to reduce expenditures on strategic weapons and eventually to reduce strategic weapons. It also allowed the United States to target the Soviet Union more effectively. This "tactical" rather than "strategic" use of spy satellites was obvious in both Gulf Wars. The need for real-time, high-resolution tactical information seems to be satisfied by unmanned aerial vehicles. Global Hawk can loiter for 24 hours while relaying visible, infrared, and radar imagery over a battlefield—or a country. It may be that future reconnaissance satellites will emphasize broad coverage rather than high-resolution "tactical" imagery.

Communications Satellites

It seems likely that satellite communications will grow very little in the immediate future. The C-band and Ku-band sky is full. There is little room for more satellites. More compelling, these satellites are only about 60 percent full. The growth of direct-to-home services—from television to Internet services—should continue, but this may just fill existing satellites. Exploitation of other bands—especially Ka-band—may be more distant than was expected in the 1990s. On the other hand, the transition to high-definition television (HDTV) will use three (or more) times the bandwidth of current television programs. HDTV alone may use up all the available commercial communications satellite bandwidth.[120] During the Second Gulf War, 80 percent of military use of satellite communications was provided via commercial satellites. The military has still not figured out how to fulfill future satellite communications needs.

Weather Satellites

Geosynchronous weather satellites have grown from weather monitors to scientific instruments for weather prediction and climate change forecasts. Capabilities will continue to grow, but it seems unlikely that more satellites will be launched—other than replacements. It may be that polar weather satellites will only exist as scientific satellites, if at all. Global cooperation has been a hallmark of weather satellites. Weather is a global phenomenon. It seems highly likely that a global warning system for all hazards will soon be available.

Navigation Satellites

GPS and its rivals—GLONASS and Galileo—will continue to serve dual uses. This is a real problem. GPS was designed as a weapon. The United States must always have the capacity to turn it off. On the other hand, turning GPS off will surely result in the loss of lives.

Nonmilitary Land Remote Sensing Satellites

The existing "commercial" systems have one "anchor tenant"—the intelligence community. While there are many uses and "needs" for land remote sensing, only the intelligence community is willing to pay. As weather pictures become multispectral and gain higher resolution, the dependence of the intelligence community will only increase.

---

120. "Satellite Services Demand—The Future in High Def," *Futron* (9 June 2006).

3. Conclusion: Technology, Society, and Technology Development

The first space application to be "operationalized" was remote sensing, with the launch of the first Discoverer on 28 February 1959. One could argue that "success" wasn't achieved until August 1960, but in any case, this was the first application where significant funds were expended. These funds came from DOD (the Air Force) and the intelligence community (the CIA). NASA was not involved, nor were commercial firms involved, except as manufacturers or other contractors. In 1972, NASA launched the first ERTS/Landsat satellite, but it is not clear that remote sensing has ever been truly "commercialized," although one can argue that by the 21st century it was possible to buy fairly high-resolution imaging on the open market.

The second space application to be "operationalized" was navigation, with the launch of the first Transit on 17 September 1959. Transit funding came from DOD (the Navy), as did funding for GPS/Navstar (the Air Force) later. In 1967, Transit use by the civilian maritime industry was allowed. While funding for the satellites has come exclusively from DOD, this application has definitely been "commercialized," as evidenced by the billions of dollars expended every year for GPS receivers.

The third space application to be "operationalized" was weather, with the launch of the first TIROS on 1 April 1960. TIROS was based on the RCA proposal for a reconnaissance satellite. Initial funding came from the Army, but the project was transferred to NASA. NASA funded TIROS and many of its upgrades, although many of these upgrades were initially funded by DOD on the Defense Meteorological Satellite Program (DMSP). The Weather Bureau (ESSA) eventually began funding the operation of the satellites and, somewhat later, satellite procurement. NASA seems to have taken the lead on geosynchronous weather satellites—launching the first Synchronous Meteorological Satellite (SMS-1) in 1974. "Commercialization" of this application probably started with the launch of TIROS-1, but transfer to the Weather Bureau didn't formally occur until TOS (the TIROS Operational System) in 1966.

The fourth space application to be "operationalized" was communications, with the first launch of Telstar on 10 July 1962. Earlier dates (Courier and Echo in 1960) and later dates (Syncom 2 in 1963) can be proposed, but it is fascinating to observe that the most "commercial" of all space applications was the last to be actually launched. Not surprisingly, it was the first to be "commercialized"—in every sense of that word—when Early Bird was launched on 6 April 1965. Funding for the earliest communications satellites is complicated. By far, the largest investor was AT&T, but much of that investment was for manufacturing capability. NASA was the second largest investor—funding most of Syncom and all of Relay. Hughes was the third largest investor and may have gained the greatest profit by building proto-Syncom with its own funds. Communications satellites showed the most interesting behaviors—possibly because they are so commercial. The failures—Aerosat, SBS, STC, the LEOs, and the MEOs—all seemed to have misread the market for their offerings. The international projects (Inmarsat and Intelsat) seemed to have generated geopolitical hassles.

It is interesting to note that the earliest space applications are the ones developed by or for the Department of Defense (reconnaissance and navigation). It should be no surprise that the most commercial of all the applications (communications) shows the greatest commercial funding and the earliest "commercialization." The role of NASA is hard to evaluate. NASA seems to have been more of a facilitator than anything else. NASA had no real part in reconnaissance and navigation but certainly "facilitated" the development of weather satellites. It is also possible to claim that NASA "facilitated" the development of communications satellites. If NASA had not been involved, AT&T would have gone ahead with its MEO Telstar system. This might have made it very difficult for the Hughes "better idea" to make it into a marketplace that was dominated by AT&T.

It would be interesting to examine the effect on NASA priorities of its R&D agency status. Any application developed by NASA would have to be given away. Perhaps one measure of NASA's influence would be to examine what would have happened without NASA. Reconnaissance, navigation, and communications satellites would have been developed by DOD and industry, but the weather story is more complicated. The Department of Defense didn't want *its* weather pictures circulated. DMSP is proof of this. It is not clear that the Weather Bureau would have invested the funds that NASA made available. NASA is *still* supporting the development of weather satellites.

Applications satellites are not as glamorous as Moon landings—or Mars landings—but they have made a huge difference in the world we live in: financially, culturally, and in the areas of safety and security. They have created the global village. It is a feisty, angry, violent village, but there are fewer unknowns and a greater chance for peace and prosperity.

Chapter 8

Impacts of the Apollo Program on NASA, the Space Community, and Society

Eligar Sadeh

Introduction

The civil space program in the United States depends on the widespread conviction that our common experience as a state and global community, now and in the future, will be the better for it.[1] One important dimension of this, which is the focus of this chapter, is that society benefits from a civil space program. The societal benefits are a result of state-directed mobilization of resources and investments in the exploration and development of space. The impacts of civil space programs and projects in the United States span from ones specific to technology development and innovation, as well as advances in science and knowledge, to others that entail political, managerial, economic, and educational ones. This chapter is focused on a critical review and evaluation of these impacts in relation to the Apollo program undertaken by NASA. The investigation involves an assessment of the near-term impacts of the Apollo era, defined herein as 1961–72, and a consideration of long-term consequences.

The argument put forward and discussed is that the societal impacts in the near term and societal consequences in the long term are in general unintended, but in some cases there are *intended* impacts and consequences. The preponderance of unintended influences is evident in the ways in which Apollo shaped the broader contours of societal culture—that is, Apollo inspired; it fostered an "imagination capital." This capital was leveraged for political prestige and leadership; for federal and industrial investments in research and development (R&D); and as a means to generate interest in

---

1. This is a view reflected in space history literature. See, among others, Roger D. Launius, "Historical Dimensions of the Space Age," in *Space Politics and Policy: An Evolutionary Perspective*, ed. Eligar Sadeh (Netherlands: Kluwer Academic Publishers, 2002); Walter A. McDougall, *... the Heavens and the Earth: A Political History of the Space Age* (Baltimore, MD: Johns Hopkins University Press, 1997); and Carl Sagan, *Pale Blue Dot: A Vision of the Human Future in Space* (New York: Random House, 1994).

education related to the science, technology, engineering, and math (STEM) disciplines. Intended impacts and consequences are limited to those within the space program itself and to a few specific cases external to the space arena. Apollo determined how NASA approached the management and planning of space programs and projects, and consequences in these areas were present with the implementation of the U.S. Vision for Space Exploration (also known as VSE or Vision). Outside the space arena, there exist impacts and consequences in the areas of city planning, in the systems-architecting discipline, and in the economic and educational areas.

This chapter is divided into four sections, each of which investigates aspects of Apollo program impacts and consequences. The first part deals with what is characterized in this study as the "Apollo Paradigm." An assessment of the political, technological, and exploration dimensions of this paradigm surveys the links between Apollo and impacts on societal culture. The second part examines management and planning impacts and consequences. This involves the impacts of systems management approaches dealing with the development of these management practices in the U.S. Department of Defense (DOD), NASA's application of these practices with the Apollo program, and the longer-term consequences for the present of NASA's approach to management. In addition to this, several other influences dealing with management and planning are evaluated, including the effects of systems management practices used with Apollo; systems engineering applications for city planning; how the best practices and lessons learned from the Apollo program influenced large-scale systems architecting; and the transfer of systems management know-how, applied to Apollo by NASA, to the European space program. The third part of the chapter deals with economic impacts. Therein a number of impacts and consequences are probed, including economic multiplier and productivity impacts, employment, and technology spinoffs as a result of investments made by the government in space-related R&D. In the fourth part of the chapter, the educational impacts concerning STEM disciplines are scrutinized.

Apollo Paradigm

The Apollo program was a watershed or turning point in history.[2] It was an endeavor that demonstrated both the technological and economic prowess of

---

2.	Although there is not a clear definition of a turning point, most historians, if not all, would agree that the "focusing events" that marked the rise of the Space Age, like Sputnik and Apollo, do indeed represent historical turning points. See Roger D. Launius, "Overview: What Is a Turning Point in History, and What Were They for the Space Age?" in *Societal*

the United States and established technological preeminence for the United States over rival states—namely, the former Soviet Union. Attributable to the Apollo program is a paradigm that instilled a certain belief system. This belief system incorporated an ethos that encompasses politics, technology, and exploration. It is in this belief system that Apollo set a new standard by which to gauge human achievement—if humans can put a human on the Moon, then they can do all else, both technically and socially.[3]

Political Ethos

As a large-scale national project, Apollo itself represented an important political symbol.[4] As one scholar wrote: "The quintessential large-scale national technological project, Apollo, was far removed from political and social controversies of the time, alienated essentially no one, and…was experienced vicariously by the public."[5] Apollo served as a unifying symbol in an otherwise fragmented and pluralistic domestic polity. Internationally, the program was propelled by "prestige," an intrinsic element in the international relations between states.[6] Despite these historical claims, there remains some uncertainty as to how symbolism and prestige affect politics beyond the intangible aspects—that is, in a concrete, tangible way.[7] Given this uncertainty, why then do states pursue large-scale national projects, and more to the point here, why did the United States undertake the Apollo program? One answer to this lies in a rational assessment of risks and benefits associated with the endeavor. The political decision-making process that led to Apollo is characterized by such an assessment.[8] The political benefits related to the Cold War

Impact of Spaceflight, ed. Steven J. Dick and Roger D. Launius (Washington, DC: NASA SP-2007-4801), pp. 19–39.

3.   Andrew Chaikin, "The Impact of Apollo" (presented at the Societal Impact of Spaceflight Conference, NASA History Division and National Air and Space Museum Department of Space History, Washington, DC, 19–21 September 2006); Roger D. Launius, *Apollo: A Retrospective Analysis*, Monographs in Aerospace History, no. 3 (Washington, DC: NASA SP-2004-4503, reprinted in July 2004).

4.   Gerald M. Steinberg, "Large-Scale National Projects as Political Symbols," *Comparative Politics* 19, no. 3 (1987): 331.

5.   Murray Edelman, *The Symbolic Uses of Politics* (Urbana: University of Illinois Press, 1974).

6.   Vernon Van Dyke, *Pride and Power: The Rationale for the Space Program* (Urbana: University of Illinois Press, 1964); Hans Morgenthau, *Politics Among Nations: The Struggle for Power and Peace* (New York: Knopf, 1972).

7.   Van Dyke, *Pride and Power*; Amitai Etzioni, *Moon Doggle: Domestic and International Implications of the Space Race* (New York: Doubleday, 1964).

8.   John M. Logsdon, *The Decision To Go to the Moon: Project Apollo and the National Interest* (Cambridge, MA: Massachusetts Institute of Technology [MIT] Press, 1970).

and the United States' national interests outweighed the transaction costs—the economic and technical risks associated with Apollo.

A second answer concerns the political influence of technocratic groups that govern the implementation of the space program, as well as programs such as Apollo; large-scale, state-directed technology development promotes the scientific, professional, and bureaucratic groups.[9] These groups are rooted to the military-industrial complex and are often influential in extracting governmental resources for their preferred programs and projects. This is evident with congressional appropriations for the Apollo program and the justification of those outlays on the basis of benefits to the aforementioned groups. Apollo was justified or rationalized in a number of ways that sought to benefit these groups: to advance science and technology; to promote scientific and technical education; to support national security needs; to apply the knowledge gained in managing Apollo; and to benefit industry through technology R&D, innovation, spinoffs, models of efficiency, and stimulation of the economy.[10]

A third answer deals with the particular role of Apollo as a political symbol. In this regard, Apollo is associated with impacts on national and foreign policies as well as ideological benefits. The prestige factor of Apollo is an important impact that played a role in the Cold War. The rise of the Space Age transformed the Cold War into a total war where national and international prestige—and the wherewithal of states to force technological progress, innovation, and modernization—became essential political goals. For both the John F. Kennedy and Lyndon B. Johnson U.S. presidential administrations, Apollo met vital political needs related to the Cold War confrontation with the former Soviet Union. Space technology was drafted into the cause of national prestige and was embraced as a political panacea. The international image of the United States, as well as its standing in science and technology, advanced considerably after the successful completion of the Apollo 11 mission; people and states abroad knew that the United States had achieved this endeavor.[11]

Part and parcel of Apollo was the "frontier narrative" attached to the program. This narrative—which is associated with historical ideas rooted in exploring, conquering, exploiting, and closing the frontier and is exemplified in United States history by the westward expansion and ideology of Manifest

9. Don K. Price, *The Scientific Estate* (Cambridge, MA: Harvard University Press, 1965).

10. James L. Kauffman, *Selling Outer Space: Kennedy, the Media, and Funding for Project Apollo, 1961–1963* (Tuscaloosa: University of Alabama Press, 1994).

11. United States Information Agency (USIA) Office of Research and Assessment, "Effects of the Moon Landing on Opinions in Six Countries," 12 September 1969.

Destiny[12]—became a way for the public to understand the space program while reaffirming U.S. values and institutions during the uncertain years and challenges of the Cold War.[13] The launching of Sputnik by the Soviet Union in 1957 highlighted these challenges, and this event represents a turning point or watershed event that led to societal sea change.[14]

Sputnik presented national and international challenges to the United States. Nationally, Sputnik challenged the idea of limited government investments in R&D and questioned the superiority of U.S. institutions and values, such as a democratic system of governance, bureaucracy tempered by public and political accountability, political freedoms, and open inquiry and dissemination of knowledge. Internationally, Sputnik suggested Soviet strategic parity with the United States, questioned the military assumptions upon which the "free world" was based, and undermined U.S. world prestige and leadership. Sputnik signaled that U.S. sociopolitical and socioeconomic systems were anachronistic in a Space Age characterized by explosive technological advance.

"Sputnik posed a great challenge," wrote the scholar Walter McDougall. "As a foreign threat with military overtones, it was clearly the government's business. As a blow to U.S. credibility, it seemed to demand a response in kind. As a technocratic accomplishment, involving the integration of science and engineering under the aegis of the state, it called into question the assumptions behind U.S. military, economic, and educational policy—every means by which the mobilization of brainpower is achieved."[15] These challenges resulted in a number of impacts within the scope of this chapter. First, it fostered changes in the role of government regarding R&D and technology utilization. This is entailed in the economic themes related to technology development, innovation, applications, and utilization. Second, it led to the creation of NASA in 1958 and played a role that led to the support and implementation of the Apollo program. Apollo became an "implementation model" to be emulated. Through adaptation of planning and management methods used by DOD in ballistic missile development to the context of Apollo, NASA forged systems and project management models that impacted the administration of

---

12. There are historians who have critiqued the associations made between Apollo and westward expansion and Manifest Destiny. See Patricia N. Limerick, "Imagined Frontiers: Westward Expansion and the Future of the Space Program," in *Space Policy Alternatives*, ed. Radford Byerly (Boulder, CO: Westview Press, 1992).

13. Kauffman, *Selling Outer Space.*

14. Launius, "Overview: What Is a Turning Point in History."

15. McDougall, *...the Heavens and the Earth*, p. 139.

NASA programs other than Apollo. Third, Sputnik ushered in educational reforms that addressed issues with STEM education.

Technological Ethos

The U.S. civil space program instilled a societal belief in the power of science and technology. This technological ethos is best characterized by the "Space Age America" theme advanced by the leadership at NASA, most principally NASA Administrator James E. Webb, and supported by U.S. Presidents and Congresses of the Apollo era through the Johnson administration. This America is one where science and technology are harnessed for peaceful purposes and for solving social issues. It is an America where space promotes education in the sciences and engineering. Space Age America is an America with unlimited promise, potential, and hope that humanity can shape a better future for society. "Here was limitless space, limitless opportunity, limitless challenge…," wrote McDougall. "The activist state fulfilled the individual through education, welfare, incentives, new technology…. Apollo would open up new realms for the individual in stimulation of the economy and elevation of the human spirit. What was more, the space program… seemed a model for society without limits, an ebullient and liberal technocracy… Space Age America."[16]

Space Age America was about how to undertake large-scale endeavors of public value through technocratic governmental agencies and large budgetary outlays. This model for society is based on the idea of the "Moon-Ghetto" metaphor put forward in the Apollo era: if we can go the Moon, then we can use the same know-how in organizing human affairs to solve societal problems and to advance societal goals. The argument was that the Apollo program instilled an ethos in harnessing the power of science and technology for solving social problems, for fostering education in STEM disciplines, and for advancing economic prosperity. It is these ideas that served as some of the philosophical underpinnings of President Johnson's "Great Society" agenda and programs.

One of NASA's missions was to use science and technology emanating from the space program to strengthen the economic and educational interests of the United States. The technocracy and bureaucracy needed to undertake Apollo was viewed by Webb as needing to fulfill societal ends, like stimulation of the economy, education, and new technology harnessed to solve

---

16. Ibid., p. 362.

societal problems.[17] Through Apollo, space became linked to the organizational vitality of the state and to modernization, especially in terms of state-sponsored and state-directed R&D. Webb contended that Apollo was more a management exercise than anything else and that the technological challenge, while sophisticated and impressive, was largely within grasp.[18] More difficult than this was ensuring that those technological skills were properly managed and used, and in this use there are applications in new thought processes and in information and knowledge that serve as a powerful engine of progress relevant to other social goals.

Another impact of the technological ethos are the influences on the level of public confidence in the ability of government to perform; the Apollo program, through the planning and management skills applied therein with successful results, helped to create a culture of competence engendering high levels of public confidence in the U.S. federal government.[19] Trust in government among the public was more than 70 percent with the start of the Apollo program in 1961, and within the 55 percent to 60 percent range during the piloted Apollo missions.[20]

The level of public confidence in NASA as to what the government can do competently is sustained as a longer-term consequence. From the flight of the first Space Shuttle in 1981 to 1994, between 60 and 80 percent of the public approved of the civilian space effort.[21] In 1997, one survey of public attitudes toward the federal government found that 85 percent viewed the government as very successful in working toward the goal of space exploration.[22] This was the highest favorable rating of all the categories considered, including national defense, economic growth, environment, health and safety issues, civil rights, education, crime, poverty, moral values, illegal immigration, and the reduction of drug abuse. The demonstration of competence surrounding Apollo proved that the United States possessed the skill, technology, and

17. James E. Webb, *Space Age Management: The Large-Scale Approach* (New York: McGraw-Hill Book Company, 1969).

18. Launius, *Apollo: A Retrospective Analysis.*

19. Howard E. McCurdy and Roger D. Launius, "If We Can Go to the Moon...Political Power and Public Confidence," undated and unpublished manuscript.

20. Statistical data are graphed in McCurdy and Launius, "If We Can Go to the Moon"; also see Herbert E. Krugman, "Public Attitudes Toward the Apollo Space Program, 1965–1975," *Journal of Communication* (autumn 1977): 87–93.

21. "Public Support for the United States Space Program: Results from a National Tracking Study of Registered Voters," prepared for Rockwell International, Yankelovich Partners, Inc., 1994; "Public Support for the U.S. Space Program," prepared for Rockwell International, Yankelovich Partners, Inc., 1993.

22. Council for Excellence in Government, "Findings from a Research Project About Attitudes Toward Government," 21 March 1997, Washington, DC.

wealth to complete voyages to space; it is this sense of accomplishment that Apollo and NASA symbolized. This helped maintain support for a human spaceflight program when Apollo ended, even as national leaders debated and questioned the goals of the programs—that is, the Space Shuttle and International Space Station (ISS) programs.[23]

Exploration Ethos

There is the argument put forward by those involved in the civil space program, among them the late Carl Sagan, that tangible impacts and consequences are inadequate to sustain political and popular support for space exploration. Rather, an intangible exploration ethos is needed. For Sagan and other space historians, the primary justification of space exploration lies in the imperatives of human nature.[24] Sagan wrote: "We are the kind of species that needs a frontier—for fundamental biological reasons. Every time humanity stretches itself and turns a new corner, it receives a jolt of productive vitality that can carry it for centuries. There is a new world next door. And, we know how to get there."[25]

The exploration ethos of the Apollo era encapsulates this intangible factor. One significant impact of this ethos is how Apollo forced the people of the world to view planet Earth in a new way.[26] One of the Apollo 8 astronauts who circumnavigated the Moon, the first humans exposed to images of the "Earthrise" over the lunar horizon, said that "we came all this way to explore the Moon, and the most important thing is that we discovered the Earth."[27] The "Earthrise" images have had implications that go well beyond the space area—a vision of the planet Earth as a holistic natural and social system.[28]

The "Earthrise" images offer an environmental perspective that played a role in spawning the modern environmental movement and Earth system

---

23. Howard McCurdy, *Space and the American Imagination* (Washington, DC: Smithsonian Institute Press, 1997).

24. One theme among space historians that is reflected in the literature deals with the "human imperative in space exploration." See Roger D. Launius, "Historical Dimensions of the Space Age," in *Space Politics and Policy*, ed. Sadeh, pp. 3–25.

25. Sagan, *Pale Blue Dot*, p. 285.

26. Launius, *Apollo: A Retrospective Analysis*.

27. "William F. Anders," International Space Hall of Fame, New Mexico Museum of Space History, *http://www.nmspacemuseum.org/halloffame/detail.php?id=71* (accessed 25 March 2012).

28. There is some skepticism among historians regarding the purported influence of the "Earthrise" images. This was an issue that was discussed recently at the Societal Impact of Spaceflight Conference, NASA History Division and National Air and Space Museum Department of Space History, 19–21 September 2006, Washington, DC.

sciences. The environmental movement was galvanized in part by this new perception of the planet and the need to protect it and the life that it supports. "Earthrise" as harbinger of Earth observations enabled scientists to study Earth's environmental system in a systemic, holistic fashion. As a social system, "Earthrise" provides humanity with a new perspective, with implications for states and international relations. Apollo set into place images that reflect the globalization that exists today. Sagan stated that the gift of Apollo to humanity, justified by the Cold War and the nuclear arms race, is the stunning transnational vision of "Earthrise" and that global cooperation is the key to humanity's survival.[29]

Associated as well with "Earthrise" are the social and spiritual impacts on the space explorers themselves.[30] The Apollo astronauts represent one set of these space explorers. One work in this area compiled and assessed the views of a number of space explorers, astronauts, and cosmonauts and found that their space experiences are represented by an "overview effect."[31] The views of the space explorers as they related to the "overview effect" cover the following themes: an abiding concern and passion for the well-being of Earth relating to the themes of globalization, transnationalism, and global cooperation; a recognized need for a stewardship perspective and a global participatory management of the planet that is addressed within environmentalism; and an awareness that everything is interconnected, which is the basis for holistic and systemic views and thinking.

Related to the exploration ethos, and particularly to the Apollo astronauts, is Apollo as an iconographic symbol. The societal impact of this is best exemplified by MTV's use of an Apollo image. This image suggests that the mythology of the astronaut in American culture established a representation of the "best" that the United States has to offer the world. This reflects back to the prestige and competence factors discussed earlier in the chapter. Historians have further made the point that the Apollo astronauts served as surrogates for the society that they represented, impacting the way in which humanity views its future. As Roger Launius has said: "The astronauts represented a powerful generational theme, the young, powerful warrior guided by an older, prescient, and often mystical leader or leaders who envision a wonderful future for the nation. In this context, the astronaut is making safe

---

29. Sagan, *Pale Blue Dot.*

30. Denis Cosgrove, "Contested Global Visions: One-World, Whole-Earth, and the Apollo Space Photographs," *Annals of the Association of American Geographers* 84, no. 2 (1994): 270–294. Also see Andre Chaikin, *A Man on the Moon* (London: Viking Penguin, 1994); Chaikin discusses the social aspects of "Earthrise" images.

31. Frank White, *The Overview Effect: Space Exploration and Human Evolution*, 2nd. ed. (Reston, VA: American Institute of Aeronautics and Astronautics [AIAA], 1998).

the way for the civilization to go forward, to progress toward a utopian future elsewhere in the cosmos."[32]

Longer-Term Consequences of the Apollo Paradigm

The conditions of the political system in the 1960s supported the political, technological, and exploration worldviews of the Apollo paradigm. Interestingly, the beginnings of the demise of the paradigm are rooted in the management difficulties faced by Webb after 1965, which culminated in the 1967 Apollo launch pad fire that killed three Apollo astronauts, followed by Webb's resignation as NASA's Administrator in 1968. By the end of the 1960s, the paradigm was no longer valid in changed societal circumstances.[33] A number of factors precipitated the demise of the Apollo paradigm: the counterculture movement in the United States of the 1960s, the development and rise of the environmental movement, the energy crisis of the 1970s, the economic malaise in the United States exemplified by high inflation in the 1970s, a conservative reaction against big government that Space Age America represented, sustained use of satellite systems for Earth observations and robotic probes for planetary and cosmological exploration, the advent of virtual reality systems, and the privatization and downsizing of government activities.

The post-Apollo era was marked by a decline in support for human space exploration as measured by appropriated dollars from the federal government. Fulfilling the challenge of placing humans on the Moon and implementing a foreign policy of détente, which ended the space race and relaxed Cold War tensions between the United States and the former Soviet Union, led to an emphasis on the building of a human spaceflight infrastructure. To this end, economics and enabling technologies were critical supporting variables. Human spaceflight was wedded to space utilization and a "mission to infrastructure" in low-Earth orbit (LEO). This course of action is exemplified by both the United States and Soviet/Russian space programs, and it involved programs and projects like the Salyut and Mir space stations, Apollo-Soyuz, Skylab, the Space Shuttle, Shuttle-Mir, and the International Space Station.

This implied that Apollo's exploration belief system gave way to a "post-Apollo" utilitarian belief in which other social and political concerns dominated space policy in the United States. In short, U.S. space policy became

---

32. Roger Launius, "Heroes in a Vacuum: The Apollo Astronaut as Cultural Icon" (43rd AIAA Aerospace Sciences Meeting, AIAA 2005-702, Reno, NV, 10–13 January 2005).

33. Eligar Sadeh, "Human Mission from Earth: Finding Rationales for Exploration of the Moon and Mars," *Space Policy* 17, no. 3 (2001): 205–212.

ancillary policy.[34] At the time, science and technology became increasingly viewed as "autonomous" forces that could be not be controlled or guided to the benefit of society as thought by the advocates of Space Age America. This was compounded by the fact that the application of technology did not necessarily solve social ills, was very often found to be destructive to the environment, and was used for military purposes, such as the war in Vietnam.

From a utilitarian outlook, space offered a platform for dealing with earthly priorities. Rather than advance prestige and leadership through human space exploration achievements, the United States sought to lead in practical scientific and technological capabilities with tangible economic returns. Even though the rhetoric and metaphors in support of "Apollo-like" political and exploration beliefs resurfaced during the 1980s, concrete political support in terms of funding was absent. Concomitantly, the theme of space utilization was advanced—at the expense of exploration—through government support for the commercialization of space activities. As a result, presidential and congressional politics were incongruous with sustaining the human space exploration efforts begun with Apollo. NASA also encountered organizational changes in its cultural makeup that led to planning problems and errors of judgment as exemplified by the decision to launch the Space Shuttle Challenger in January 1986. By way of illustration, NASA went from an R&D culture during Apollo to an operational one afterward; from a frontier mentality and the propensity to assume risk to a utilitarian (that is, applications and operations) outlook and the propensity to avoid risk; and from an engineering culture to a more bureaucratic, managerial one.[35]

Since the end of the Apollo era, a fundamental concern of the space community is the search for justifications that entail impacts or benefits to support human space exploration missions. This spawned a number of studies and reports in the United States. NASA's post-Apollo plans called for resources to implement the development of a space shuttle, an orbital space station, a nuclear space tug, a human-tended lunar base, and human

---

34. Ancillary policy is a policy of continuation and incrementalism. This is the norm for public policy-making in the United States Congress. Ancillary policy does not solve an identified national problem and is more apt to represent a continuing government commitment, even though the bureaucracy it maintains may have been set up for that purpose long ago. Ancillary policy has low agenda status; it receives only limited public attention, public funds, and efforts of public officials. See Roger D. Launius and Howard E. McCurdy, eds., *Spaceflight and the Myth of Presidential Leadership* (Chicago: University of Illinois Press, 1997).

35. Howard E. McCurdy, *Inside NASA: High Technology and Organizational Change in the U.S. Space Program* (Baltimore, MD: Johns Hopkins University Press, 1993).

expeditions to Mars.[36] In the 1980s and 1990s, a series of reports and initiatives for human space exploration missions were proposed.[37] These reports justified future space program scenarios on the basis of national benefits, such as prestige, leadership, technological development and innovation, and economic growth.

For example, the Space Exploration Initiative proposed by President George H. W. Bush in 1989 was justified with a number of factors that encompassed increasing national prestige, advancing science education, developing technologies, commercializing space, and strengthening the economy.[38] The Ride Report (1987) provided a systematic analysis of the civilian space program to show how the United States had lost its leadership position in space relative to the Soviet Union, principally as it related to maintaining a permanent human presence in LEO.[39] On this basis, Sally Ride developed a space strategic development plan for the twenty-first century based on restoring American leadership status. This requires that the United States possess capabilities that enable it to act independently and impressively in the space environment when and where it chooses.

NASA's strategic-planning process focused on developing its enterprises to meet the goals of various governmental and domestic public constituencies with the benefactors being policy-makers, science communities, aeronautics and aerospace industries, other governmental agencies, the public sector, and academic communities within the United States.[40] A number of different

---

36. Space Task Group, NASA, "The Post-Apollo Space Program: Directions for the Future," September 1969.

37. *Pioneering the Space Frontier*, report of the National Commission on Space (New York: Bantam Books, 1986); Sally K. Ride, *Leadership and America's Future in Space*, report to the NASA Administrator (Washington, DC: Government Printing Office, August 1987); "Report of the 90-Day Study on Human Exploration of the Moon and Mars," NASA report prepared for the NASA Administrator (Washington, DC: NASA, November 1989); Norm R. Augustine, *Report of the Advisory Committee on the Future of the U.S. Space Program* (Washington, DC: Government Printing Office, 1990); Thomas P. Stafford, *America at the Threshold: America's Space Exploration Initiative* (Washington, DC: Government Printing Office, 1991).

38. Stafford, *America at the Threshold*.

39. Ride, *Leadership and America's Future in Space*.

40. NASA Administrator Daniel S. Goldin put into place a strategic-planning process at NASA during his tenure as Administrator from 1992 to 2001 as mandated by the United States Congress—that is, the Congressional Government Performance and Results Act. The common planning process led to a realignment of NASA programs and projects to fit the goals and objectives of that plan. See Charles Pellerin, former NASA Associate Administrator for Strategic Planning, NASA, interview by author on 27 November 1995; and Alan M. Ladwig and Gary A. Steinberg, "Strategic Planning and Strategic Management Within NASA," unpublished report, 1996.

strategic plans were formulated beginning in 1994.[41] For example, the strategic plan from 2006, put into place a "one NASA Vision" that emphasized R&D and an exploration ethos reminiscent of the Apollo era. The 2006 plan was tailored to specify how NASA would implement the goals of the Vision. In the area of human spaceflight, this entails near-term goals of the Space Shuttle's return to flight and the completion of the ISS, as well as longer-term goals of a lunar return program to enable lunar base development and human missions to Mars. Of note in relation to the Apollo-impacts theme of this chapter is that the strategic-planning process is indicative of centralized control on the planning and development of NASA programs and projects, akin to systems management practices used by NASA with Apollo.[42]

On the one hand, there are negative impacts surrounding the demise of the Apollo paradigm in that the premises of the political, technological, and exploration views were not sustained. Yet the paradigm sustained an impact on the civil space program in the United States. The ideas rooted in the paradigm led to consequences in how the space program is rationalized and justified, very often on the basis of societal impacts as the aforementioned examples suggest, and on the planning and management approaches and practices applied by NASA.

Management and Planning Impacts and Consequences

The historical claim concerning the management and planning impacts and consequences of Apollo is that one of the most valuable influences of the Apollo program was human rather than technological. This implies better knowledge on how to plan, manage, and implement great social undertakings that involve the development and application of large-scale technological systems.[43] It is this claim that served as the basis for the Space Age America theme and the technological ethos of the Apollo paradigm. This part of the chapter first explores how NASA adapted management and planning practices used by DOD and then assesses how NASA's use of these practices affected the civil space program in the United States.

---

41. NASA, *The New Age of Exploration: NASA's Direction for 2005 and Beyond* (Washington, DC: NASA, February 2005), available at *http://www.nasa.gov/pdf/107490main_FY06_Direction.pdf* (accessed 25 March 2012); NASA, *2006 NASA Strategic Plan* (Washington, DC: NASA, 2006), available at *http://www.nasa.gov/pdf/142302main_2006_NASA_Strategic_Plan.pdf* (accessed 25 March 2012).

42. Eligar Sadeh, "Management Dynamics of NASA's Human Spaceflight Programs," *Space Policy* 22, no. 4 (2006): 235–248.

43. Dael Wolfle, "The Administration of NASA," *Science* (15 November 1968): 753.

U.S. Department of Defense as a Model

The U.S. space program created an unprecedented demand for managers with both technical and administrative competence in industry and government. In meeting this demand, an enhanced understanding of the application of management to the technology development process was realized. This impact entails matrix-type communications; an environment of managing with high levels of reliability, performance, and accountability; the involvement of top managers in the technology development process; systems management approaches; and new uses of contracting methods—for example, incentive contracting and total package procurement. It is acknowledged and documented in the literature that many of the management models used for civil space were developed by the U.S. military, particularly in the development of ballistic missile programs and space launch vehicles. Of note is that systems management methods were incorporated into NASA from DOD.

Innovation on NASA's part is evident in how management at NASA incorporated DOD "best practices" into a civilian program. NASA was able to integrate effective management controls in the Apollo program. Phased planning and configuration management techniques, used successfully by the U.S. military for ballistic missile development, were integrated into the management of Apollo. DOD's development and application of phased planning for the Titan III program—which entailed defining the project's objectives, costs, and schedules in a preliminary design phase—became a DOD standard by the mid-1960s, one that NASA adopted in 1967.[44] By 1961, the U.S. Air Force (USAF) Ballistic Missile Division developed configuration management, which was also independently created by the Jet Propulsion Laboratory (JPL).[45] Configuration management, further explained in the later section on "Apollo Management as a Model for the Vision," is a managerial technique to control design and technical changes and to link that to cost predictions and controls. The technique emerged as a primary contractor control process, enhanced the reliability of systems, and emerged as a standard process throughout the aerospace industry.[46] Systems management approaches were the "secrets of success" in enabling NASA to meet its lunar goal of placing

---

44. Stephen B. Johnson, "From Concurrency to Phased Planning: An Episode in the History of Systems Management," in *Systems, Experts, and Computers: The Systems Approach in Management and Engineering, World War II and After*, ed. Thomas P. Hughes and Agatha C. Hughes (Cambridge, MA: MIT Press, 2000), pp. 93–112.

45. Stephen B. Johnson, "Space Business," in *Space Politics and Policy*, ed. Sadeh, pp. 241–280.

46. Stephen B. Johnson, *The United States Air Force and the Culture of Innovation, 1945–1965* (Washington, DC: United States Air Force, 2001).

humans on the Moon and returning them to Earth safely during the decade of the 1960s and early 1970s.[47]

The Program Evaluation and Review Technique (PERT) represented another case of NASA incorporating a DOD system management model. PERT is a model for project management that was invented by the U.S. Navy Special Projects Office in 1958 as part of the Polaris mobile submarine-launched ballistic missile project, which was developed as a direct response to the international challenges posed by Sputnik.[48] PERT is a method for analyzing the tasks involved in completing a given project, principally the time needed to complete each task and to complete the total project. The method was applied to simplify the planning and scheduling of large-scale, complex technical projects. NASA incorporated PERT and applied it as a system management practice in dealing with the development of the Saturn space launch vehicle.[49] During the early phases of the Saturn program, NASA's Marshall Space Flight Center management regarded PERT as a very successful effort and as the best source of information available on the status of hardware programs. Notwithstanding this, the PERT network was phased out due to cost considerations by the first Saturn launch in 1967.

This suggests that the overall impacts of PERT are open to question. In many cases, PERT was introduced too late to make much of an impact on funding and schedules. The value of PERT was seen more as a preliminary planning tool and coincidental to managing the ongoing complexity within the Apollo program.[50] As more complexity emerged within the systems used for Apollo, PERT was difficult and costly to use; lagged in real-time usefulness; and was subject to manipulation to avoid exposure of cost, schedule, and technical problems.

---

47. Stephen B. Johnson, *The Secret of Apollo: Systems Management in the American and European Space Programs* (Baltimore, MD: Johns Hopkins University Press, 2002).

48. "PERT did not build the Polaris, but it was extremely useful for those who did build the weapon system to have many people believe that it did ... the program's innovativeness in management methods was ... as effective technically as rain dancing ... it mattered not that management innovations contributed little directly to the technical effort; it was enough that those outside the program were willing to believe that management innovation had a vital role in the technical achievements of the Polaris." See Harvey Sapolsky, *The Polaris Systems Development* (Cambridge: Harvard University Press, 1972), p. 125.

49. Johnson, *Secret of Apollo.*

50. "The only way you ever got PERT really implemented was to go around and ask the guy who was supposed to be doing it where he stood on his PERT program, and you could usually find that he had his own program in his desk drawer. PERT was the thing he was talking to you about, but whether it actually meshed with what was going on ... was in some instances coincidental." George Mueller, interview by Martin Collins, National Air and Space Museum, 2 May 1988.

Apollo Management as a Model for the Vision

NASA leaders acquired and organized unprecedented resources to accomplish the tasks of Apollo. In many ways, Apollo was just as great a management feat as a technical one. The management models and methods developed and used met the enormously difficult engineering, technological, and organizational integration requirements of the Apollo program.[51] Public management of Apollo provided better knowledge on how to plan, manage, and implement great social undertakings that involve the development and application of large-scale technological systems. NASA employed program management concepts that centralized authority and emphasized systems engineering. Systems management approaches were critical to Apollo's success. Understanding the management of complex, technical projects for the successful completion of a heterogeneous task was a critical outgrowth of the Apollo effort.[52]

A comprehensive assessment of Apollo program management identified a number of dynamic and evolutionary management structures and process within an environment of program controls.[53] This encompasses the development of cohesive and flexible patterns of management in NASA and industry; management visibility based on detailed monitoring and auditing systems that allowed for the flow of information both vertically and horizontally; the successful correlation and definition of multiple program interfaces in both NASA and industry; the establishment of real-time, flexible management reporting systems that balanced freedom of innovation with control discipline for the accomplishment of program objectives; and the development of a balance between NASA's in-house and industrial capabilities.

One important theme that emerges is that program management of Apollo combined centralized planning and a hierarchical organization with decentralized and flexible technology development processes. Centralized bureaucratic processes overlaid technical accountability systems characterized by project

51. Webb, *Space Age Management*; Arnold Levine, *Managing NASA in the Apollo Era* (Washington, DC: NASA SP-4102, 1982); W. Henry Lambright, *Powering Apollo: James E. Webb of NASA* (Baltimore, MD: Johns Hopkins University Press, 1995); Sylvia K. Kraemer, "Organizing for Exploration," in *Exploring the Unknown: Selected Documents in the History of the U.S. Civil Space Program*, ed. John Logsdon, vol. 1, *Organizing for Exploration* (Washington, DC: NASA SP-4407, 1995); Robert C. Seamans, Jr., *Project Apollo: The Tough Decisions*, Monographs in Aerospace History, no. 37 (Washington, DC: NASA SP-2005-4537, 2005).

52. Launius, *Apollo: A Retrospective Analysis*.

53. *Apollo Program Management*, Staff Study for the Subcommittee on NASA Oversight, Committee on Science and Astronautics, United States House of Representatives, 91st Cong. (Washington, DC: Government Printing Office, 1969).

management and systems engineering methods. This allowed for organizational accountability. NASA integrated the relatively autonomous technical cultures within its Centers through a centralized management structure that applied the formal controls of systems and configuration management.

A specific way that control was put into place was by imposing an organizational structure on the technical work teams. In relation to the technical engineering teams working on Apollo, engineers initially coordinated changes among themselves in committees. With the integration of systems management into NASA, managers inserted themselves into the engineering teams to understand what was happening and soon required the engineers to give cost and schedule estimates for design and hardware changes.[54] An important method to control the development of technology, in light of the rapid technological innovation and change, was that of configuration management. This method provided an essential link between engineering coordination and centralized organizational control. Even though program controls used for Apollo permitted NASA to have centralized management at Headquarters, the information received there was then distributed to the NASA Centers; managers at Headquarters availed themselves of the technical competence and knowledge at the Centers, and the project mangers at the Centers were kept current on Headquarters activities.[55]

Organizational management practices during Apollo represented a continuing process of adjustment and adaptation to the dynamics of change internal and external to NASA. Flexible management processes were essential to success.[56] NASA's organizational scheme was one of simultaneous centralization and decentralization, a "desired disequilibrium."[57] Organizational flexibility was an essential part NASA's managerial ethos. Webb realized that NASA could not be governed solely by classical principles of "scientific management" that sought to institutionalize stability and order with centralized and hierarchical organizational structures. In order to manage large-scale technological systems and allow for technological innovation, Webb recognized that organizations needed to retain flexible, decentralized management patterns and processes; Webb balanced scientific management based on control with a decentralized technical culture at NASA.

54. Johnson, *Secret of Apollo*.

55. Seamans, *Project Apollo: The Tough Decisions*.

56. Erasmus H. Kloman, "NASA Organization and Management from 1961 to 1965: The Vision and the Reality," in *Issues in NASA Program and Project Management*, ed. Francis T. Hoban (Washington, DC: NASA SP-6101-02, 1989); Phillip K. Tompkins, *Apollo, Challenger, Columbia: The Decline of the Space Program, A Study in Organizational Communication* (Los Angeles: Roxbury Publishing Company, 2005).

57. Webb, *Space Age Management*.

Simultaneous centralization and decentralization were advanced by the "triad" decision-making structure that Webb had established. Webb shared top-level decision-making at NASA with two Associate Administrators, Hugh L. Dryden and Robert C. Seamans, Jr.[58] Webb met the pressures of political accountability being responsive to the concerns of the President and Congress; Dryden ensured technical authority; and Seamans functioned as a bureaucratic manager through the application of systems management approaches at the Agency. The triad successfully navigated among political pressures of accountability and the drive for technical excellence in terms of high performance and high reliability.

The consequences of the Apollo management model concern the direct application of the systems management approaches to the Space Shuttle and ISS programs. This application, however, was a negative one. With the end of Apollo, systems approaches were less effective. This is due to the fact that Apollo was characterized by a "closed systems program" in the sense that the program was largely shielded from external changes.[59] Many of the management problems attributable to the Space Shuttle and the ISS are a result of how these programs are continuously managed with political accountability in mind and within an environment of political change. The operational view of these technological systems further constrained the direct utility of systems approaches used with Apollo. The systems approaches used for Apollo were optimal for the experimental and developmental nature of the technology for that program. These systems approaches were not readily adaptable to management processes with the Space Shuttle and the ISS that often emphasized operational and economic cost-control imperatives. To add to these issues, NASA management departed from the Apollo organizational model in the abandonment of strong Headquarters control and in a weakened emphasis on systems engineering. NASA Centers gained more power, even shunting Headquarters aside in many respects, and NASA emerged more as a bureaucratic agency with a diminished capacity for technical competence and engineering excellence.[60]

---

58. Dr. Hugh L. Dryden was Director of the National Advisory Committee for Aeronautics from 1947 until the creation of NASA. He was named Deputy Administrator of NASA in 1958 and served in that capacity until his death in 1965. In 1960, Robert C. Seamans, Jr. joined NASA as an Associate Administrator. In 1965, after the death of Dryden, Seamans became Deputy Administrator. During his years at NASA, he worked closely with the U.S. Department of Defense in research and engineering programs. Seamans advised NASA on the developments in the military space program that were of relevance for public management at the Agency, such as systems management and systems engineering.

59. P. W. G. Morris, "Science, Objective Knowledge, and the Theory of Project Management," *Proceedings of the ICE—Civil Engineering* 150, no. 2 (1 May 2002): 82–90.

60. McCurdy, *Inside NASA*.

Longer-term consequences of the Apollo-era management models are evident in how NASA implemented the VSE. With this Vision, NASA's Administrator used the phrase that the exploration systems architecture[61] directed at the development of lunar transportation systems: that is, "Apollo on steroids."[62] The idea is that NASA is looking at Apollo as a technical model on how to get back to the Moon. There is a similar analogy in the management area. In making this analogy, the management idea of "desired disequilibrium" that NASA Administrator Webb put forward to describe a need for healthy tension between centralized aspects of management (such as control over cost and schedule) and decentralized aspects of management (such as ensuring that authority over technical competence and engineering is at the NASA Centers) is what NASA is copying.

Griffin stated in 2006: "Webb once characterized his role during the Apollo program in the following way: 'The process of management became that of fusing at many levels a large number of forces, some countervailing, into a cohesive, but essentially unstable whole, and keeping it in the desired direction.' This is it, exactly, and that perspective serves me well today."[63] Rex Geveden, Associate Administrator at the time, added: "In fact, we are looking for an appropriate level of tension, an appropriate level of constructive disagreement, or that desired disequilibrium that Webb referred to, that unfortunately after Apollo was subordinated to program management authority. We want to go to a meeting and to have the engineering director upset with the project manager for not following one of his recommendations. We have not had enough healthy tension in the Agency. That tension should exist all the way to the top of the management chain."[64]

The establishment of the Associate Administrator position at NASA, along with the Deputy Administrator position, put into place a leadership at NASA

61. *Exploration Systems Architecture Study, Final Report* (Washington, DC: NASA TM-2005-21406, November 2005), available online at *http://www.nasa.gov/pdf/140649main_ESAS_full.pdf* (accessed 25 March 2012).

62. NASA Administrator Michael Griffin described the new spacecraft intended to fulfill U.S. President George W. Bush's national vision for a piloted return to the Moon as "Apollo on steroids." "NASA Administrator Griffin and Congress: NASA's Exploration Architecture," *American Institute of Physics Bulletin of Science Policy News* no. 138 (22 September 2005), available online at *http://www.aip.org/fyi/2005/138.html* (accessed 25 March 2012).

63. Michael D. Griffin, "Remarks at the Mars Society Convention" (Washington, DC, 3 August 2006), available online at *http://www.spaceref.com/news/viewsr.html?pid=21597* (accessed 25 March 2012).

64. Eligar Sadeh, "Public Management Dynamics of NASA: Interview with NASA Associate Administrator Rex Geveden," *Astropolitics: International Journal of Space Politics and Policy* 4, no. 1 (2006): 109.

that was based on the triad model that Webb had used during the Apollo era. NASA determined that Webb's management model was NASA's most successful one and decided to emulate that model. In this emulation, it is the same basic type of model in terms of a balance of power between political, institutional or organizational, and technical aspects of management. There are three organizational changes at NASA today that reflect this balance of power. These changes represent the means by which NASA is integrating systems management controls that were applied with the Apollo program. The key changes include the separation of institutional and programmatic managerial authority, independent technical authority, and integrated financial management—NASA's Integrated Enterprise Management Program.[65] Related to these changes are the issues of culture and organizational change at NASA and the impacts of Apollo in this regard.

Culture frames the context for public management in terms of norms of behavior—how organizations do things and task-related behavior. The primary cultures of relevance to NASA are competency and control. The competency culture is decentralized and is characterized by a number of traits: decentralized and informal, redundant patterns of communication and authority based on independent engineering and automatic responsibility for critical review and oversight of technical issues; an exploration ethos and emphasis on R&D directed at technical excellence; risk-taking aimed at avoiding the error of launching an unreliable spacecraft; and "culture of the engineer" and associated values relating to in-house technical capabilities for systems integration as well as contractor oversight and monitoring.[66] Competency is practiced through an emphasis on technical accountability, project- and team-based management approaches, and systems engineering. The "original technical culture" that NASA inherited from its predecessor organizations, the NASA Centers today, is one of competency.

These organizations and their associated cultures are composed of the U.S. National Advisory Committee for Aeronautics and that organization's experience with engineering by technical committee and peer-review processes; the U.S. Army Ballistic Missile Agency and in-house technical development; the U.S. Navy's Research and Ordinance Labs, with a focus on in-house engineering and R&D; and project management methods developed by the U.S.

---

65. An extensive review of these organizational changes is assessed in Sadeh, "Management Dynamics of NASA's Human Spaceflight Programs," and in Sadeh, "Public Management Dynamics of NASA."

66. Gary D. Brewer, "Perfect Places: NASA as an Idealized Institution," in *Space Policy Reconsidered*, ed. Radford Byerly (Boulder, CO: Westview Press, 1989); McCurdy, *Inside NASA*; and Tompkins, *Apollo, Challenger, Columbia*.

Army for missile and rocket development. During the Apollo era, 80 percent of NASA's technical workforce had corporate memory of these organizations, and the original technical culture largely set the context for how the Centers and NASA as an Agency worked.[67]

Control is a culture that permeated NASA from its ties to the U.S. Air Force and program and project management systems in industry. The control culture is a centralized one characterized by the following: hierarchical patterns of communication based on centralized bureaucratic processes and procedures for program and project control through documentation and standard operating procedures; an operational and utilization ethos rooted in notions of efficiency, as well as applications and benefits of space technology; aversion to the risk of launching an unreliable spacecraft; and a "culture of the bureaucratic" that values contracting out and the model of corporate power and systems control.[68] Control is practiced through systems approaches to management and related practices of configuration management.[69]

The cultural traits related to competency and control shifted in NASA's history. In the Apollo era, the original technical culture was predominant. As Apollo moved to fruition, the original technical culture changed as cultural aspects of control took hold. This dynamic of cultural change is largely due to two factors. First are the budgetary contraction and the associated political pressures that NASA faced after Apollo. This resulted in managing to economic considerations and notions of efficiency. These notions led to workforce changes at NASA that emphasized a management culture with reductions in engineers and less in-house technical expertise.[70] Three critical organizational changes that NASA recently implemented—separation of institutional and programmatic authority, independent technical authority, and integrated financial management—represent an attempt to put into place cultural and organizational management changes at NASA that are more reflective of the cultural traits and management practices that existed during the Apollo era.

---

67. Kraemer, "Organizing for Exploration."

68. The model of corporate power indicates that NASA's industrial contractors in the aerospace-defense sector assumed more control and associated responsibilities relative to NASA. During the Apollo era, contractor penetration was practiced by NASA, and NASA possessed the in-house technical capacity for contractor oversight and monitoring as well as more of a role in systems integration. As Apollo came to fruition in 1968–69, the balance of power began to shift as the contractor assumed the lead role in systems integration, and NASA became more of an oversight bureaucracy with diminished capacity for monitoring and penetrating the contractor.

69. Johnson, *Secret of Apollo*.

70. McCurdy, *Inside NASA*.

Systems Approaches to City Planning

The development of systems management approaches in the space program enables systemic design, development, and implementation of large-scale, complex systems. Systems management approaches were viewed to have applications for socioeconomic problems dealing with urban-and city-planning and administration.[71] The evidence shows that impacts are at the level of ideas and potential applications, and not in terms of direct impacts that can be attributed to systems management practices applied to the Apollo program.

A systems approach facilitates a number of aspects that apply to these areas. These range from definition and detailed description of system boundaries to functional descriptions of the system in terms of component subsystems and their operational interactions; determination of objectives and criteria for optimal system performance; examination of alternative configurations of system elements that approximate optimal system performance; the determination of the consequences of each configuration as to feasibility, adaptability, and cost effectiveness; and objective presentation of alternatives to support decision-making. These aspects of a systems approach can allow for the analysis of urban city planning problems in an integrated fashion.[72] In the 1960s, NASA and those in the aerospace community put forward this argument and tried to bring the benefits of management as applied to Apollo to city administration. The basic idea is that both NASA and city institutions require appropriate organizational architectures for successful problem-solving within complex environments that entail organized, disciplined, and highly structured human activities oriented to numerically stable goals.[73] Examples of this in the city setting deal with communication, power, transportation services, crime, pollution controls, and waste management.[74]

---

71. Frank B. Coker, "How To Streamline the Translation of Aerospace Techniques to Non-Aerospace Applications" (paper presented at the AIAA Third Annual Meeting, Boston, MA, 29 November–2 December 1968); Vernal M. Tyler and Carl F. Asiala, "The Aerospace Role in Planning Cities of the Future" (paper presented at the AIAA Third Annual Meeting, Boston, MA, 29 November–2 December 1968); Thomas O. Paine, "Space Age Management and City Administration," *Public Administration Review* 29 (1969): 654–658.
72. Tyler and Asiala, "Aerospace Role in Planning Cities of the Future."
73. Paine, "Space Age Management and City Administration."
74. Harold D. Watkins, "Systems Engineering Aids Social Problems," *Aviation Week and Space Technology* (31 January 1966).

Managerial Heuristics and Systems Architecting

Systems architecting is related to systems management approaches of the 1950s that were formulated to help with the development of ballistic missile programs in the United States. The first standard for systems architecting was developed by the USAF. As discussed earlier, NASA's incorporation and adoption of systems management practices, pioneered in part by the USAF, played a critical role in the managerial success of Apollo. Systems architecting is the art and science of creating and building large-scale, complex systems and then developing system-level solutions.[75] System architects concentrate on initial system definition and design in making use of systems engineering specialties to develop satisfactory and feasible system concepts. This architectural approach is needed most as systems become more complex and heterogeneous.

The influence of best practices and lessons learned from the management and planning of the Apollo program can be thought of as managerial heuristics. Heuristics are simple and efficient rules of thumb proposed to explain how people make decisions typically when facing complex problems or incomplete information. The managerial heuristics derived from the Apollo program impacted the practice of systems architecting.

The Apollo program generated a number of important heuristic perspectives about complex, large-scale civil programs and represents an exemplar case study for the application and formulation of systems architecting in terms of both what to do and what *not* to do. To illustrate, a heuristic that grew out of Apollo is that a system is successful when the natural intersection of technology, politics, and economics is found. Apollo was a successful program because of the significant support across these elements. Purpose orientation is another key element in modern systems architecting; a clear and useful purpose is vital for a successful system. Apollo's purpose and prioritization to put humans on the Moon by the end of the 1960s to demonstrate technological and political superiority over the former Soviet Union represents purpose orientation.

Systems architecting begins with and is responsible for maintaining the integrity of the system's purpose. A system will develop and evolve much more efficiently if there are stable intermediate forms. As the purpose of a system evolves, stable intermediate forms allow the system's functionality to be altered. When purpose changes, the whole program does not need to be terminated, but rather can just fall back to the last stable form and refocus. As

---

75. Mark W. Maier and Eberhardt Rechtin, *The Art of Systems Architecting*, 2nd ed. (London: Taylor and Francis, CRC Press, 2000).

this relates to Apollo, the decision to bypass an orbiting Earth infrastructure, such as a space station element, for supporting exploration of the Moon, and choosing instead a direct lunar mission design—that is, lunar orbiter rendezvous—drove infrastructure and technical requirements that were less reusable when NASA's post-Apollo mission changed.[76] In connection to this consequence is the realization that the best engineering solutions are not necessarily the best political solutions. There were engineers who wanted to approach technology development more incrementally, like Wernher von Braun's incremental approach with the Saturn V space launch vehicle, but the political desire to demonstrate technological and political superiority over the Soviet Union derailed intermediate infrastructure forms. The political pressures to be successful with Apollo contributed to the consequence of failed long-term planning for human space exploration at NASA after Apollo.

The case of post-Apollo planning demonstrates a negative impact of Apollo—how not to develop and sustain a long-term strategic program of human space exploration. This was highlighted earlier with the fact that, due to political pressures to achieve the Apollo goal, the technical system was not designed with stable, intermediate forms to allow use of the Apollo system in ways that were practical politically and economically for other functionality, like developing an infrastructure in LEO. Although NASA leaders understood this problem, the political priority of Apollo thwarted the implementation of any solutions.[77]

---

76. For a historical account of the decision at NASA to go with lunar orbital rendezvous, see James R. Hansen, *Enchanted Rendezvous: John C. Houbolt and the Genesis of the Lunar-Orbit Rendezvous Concept*, Monographs in Aerospace History, no. 4 (Washington, DC: NASA, 1999).

77. Wernher von Braun mapped out a broad and strategic post-Apollo plan with multiple intermediate and stable forms, and Webb argued with President Kennedy in 1962 for a more balanced space program. Webb urged the President to view Apollo as one of NASA's priorities in addition to robotic scientific missions and application satellites.

    There was a tape recording of a White House meeting that took place on 21 November 1962, during which President Kennedy made clear his administration's priority that the United States land on the Moon before the Soviet Union. The tape is particularly noteworthy for the window it provides into presidential decision-making. Faced with the option of directing federal funds more generally across the entire space program, President Kennedy argued with NASA Administrator James E. Webb for a more focused approach toward the lunar landing. Having such a goal, the President argued, would carry the country's entire space effort forward and have the same outcome NASA was seeking. See "Presidential Meeting in the Cabinet Room of the White House," 21 November 1962, Presidential Recordings Collection, tape #63, available online at *http://www.jfklibrary.org* (accessed 12 September 2006).

Transfer of Systems Management to the European Space Program

The transfer of systems management know-how from NASA to the European space effort resulted in European autonomy in space.[78] This facilitated European success with large-scale space projects—namely the European Spacelab and the Ariane space launch vehicle. The United States initially promoted international space cooperation with Europe as part of a strategy to recover the loss of prestige linked to the 1957 Sputnik crisis.[79] This strategy involved the demonstration of political leadership among its European allies by engaging them in cooperative space ventures. Space leadership implied that institutional and resource asymmetries in NASA's favor allowed it to insist upon its preferences for space cooperation—"clearly defined and distinct managerial interfaces," "no exchange of funds," "distinct technical responsibilities," and "protection of sensitive technology"—as preconditions for United States-European cooperation.[80] Europe was willing to accept these preferences, very often as a dependent and junior partner, to realize its specific functional preferences aimed at fostering space science programs, acquiring large-scale systems management and administrative know-how, and developing applied space technology capabilities.[81]

The initial years of cooperation took the form of bilateral arrangements involving launch services provided by the United States in exchange for some form of payload sharing on European scientific satellites. Agreements were reached between NASA and the United Kingdom, Italian, French, and German national space programs. With the institutionalization of a unified European effort in space sciences in 1964, represented by the European Space Research Organisation (ESRO),[82] a series of Memoranda of Understanding (MOUs) were reached with NASA. These MOUs facilitated NASA's launch services for a series of ESRO satellites in exchange for scientific results obtained from these missions. These satellite missions involved ESRO and High Eccentric Orbiting Satellite scientific satellite programs. In both of

78. Johnson, *Secret of Apollo*.
79. Lorenza Sebesta, *United States–European Cooperation in Space During the Sixties*, European Space Agency (ESA) HSR-14 (Noordwijk, Netherlands: ESA Publications Division, July 1994).
80. Eligar Sadeh, "International Space Cooperation," in *Space Politics and Policy*, ed. Sadeh.
81. Johnson, *Secret of Apollo*.
82. The European Space Agency was formed in 1975 on the basis of the European Space and Research Organization.

these programs, the Europeans extensively borrowed from NASA's systems management models used with Apollo.[83]

In addition to this, a policy of technology transfer, which was endorsed by U.S. President Johnson in 1966, was directed at the development of a European-based expendable launch vehicle named Europa.[84] The European Community began these efforts in 1962 with the creation of the European Space Vehicle Launcher Development Organisation (ELDO). The willingness of the United States to allow for some technology transfer, such as in-flight hardware and technical information, was driven by foreign policy preferences. These preferences were to narrow the "technology gap" between the United States and Europe—a gap that was primarily in the managerial and organizational areas related to large-scale systems management capabilities.[85] Narrowing the gap was important to the United States to stimulate economic and industrial growth in Europe and to enhance strategic alliances with Europe vis-à-vis the Soviet Union. Despite the efforts, ELDO failed in its attempts to develop Europa. The Europa program failed due to the inability of ELDO to acquire and adapt to the model of large-scale systems and engineering management. From its inception in 1962, ELDO was organized for failure, and it was disbanded in 1972.

After U.S. presidential and congressional approval of NASA's Space Shuttle Program in 1972, Europe pursued cooperation with NASA on Spacelab. Europe sought cooperation on Spacelab because of a lack of confidence in its own capabilities, especially in large-scale systems management know-how and the belief that its technological and managerial capabilities could only be improved through cooperation with NASA.[86] Cooperation

---

83. Arturo Russo, *The Definition of Scientific Policy: ESRO's Satellite Programme in 1969–1973*, ESA HSR-6 (Noordwijk, Netherlands: ESA Publications Division, March 1993); Arturo Russo, *Choosing ESRO's First Scientific Satellites*, ESA HSR-3 (Noordwijk, Netherlands: ESA Publications Division, November 1992); Arturo Russo, *ESRO's First Scientific Satellite Programme, 1961–1966*, ESA HSR-2 (Noordwijk, Netherlands: ESA Publications Division, October 1992).

84. The restraints on technology transfer were still extensive. Restraints dealt with every aspect of technology critical to the development of communication satellite capabilities and production techniques as well as equipment and manufacturing processes pertaining to satellites and launch vehicles or components thereof. See Lorenza Sebesta, *The Availability of American Launchers and Europe's Decision "To Go It Alone,"* ESA HSR-18 (Noordwijk, Netherlands: ESA Publications Division, September 1996).

85. For an overview of this "technology gap" and U.S. policy on technology transfer to Europe from 1964 to 1972, see Sebesta, *Availability of American Launchers*. Also see Lorenza Sebesta, "The Politics of Technological Cooperation in Space: U.S.-European Negotiations on the Post-Apollo Program," *History and Technology* 11 (1994): 319–326.

86. John M. Logsdon, "U.S.-European Cooperation in Space Science: A 25-Year Perspective," *Science* 223, no. 4631 (1984): 11–16.

on Spacelab proved the European view correct, and Spacelab engendered an "Americanization" of the European space effort in terms of large-scale systems management and organizational techniques. This played an important role in the technical success of Spacelab, the successful development of the Ariane space launch vehicle, and Europe's enhanced space capabilities across the board from the development of space science, telecommunications, and Earth-observing satellite programs in the 1970s and 1980s.

The transfer of system management know-how used with Apollo to European space efforts translated into more of an equal, cooperative partnership with the U.S. civil space program. Such a partnership indicates symmetry in European technological capabilities, interdependent cooperation outcomes in terms of contributions to critical-path technologies and infrastructural components, participation in systems and technical management, and project leadership roles.[87] By the late 1980s, Europe's capabilities in expendable launch vehicle technology, as well as space science, telecommunications, and remote sensing satellites, were not only comparable to those of NASA and the United States but, from a commercial standpoint, were competitive and more successful in capturing market share in the 1980s. In its relations with Europe, the United States is faced with both cooperation and economic competition.

The impacts shifted the balance of power between the U.S. and European space programs. The transfer of systems management triggered a diminished European dependence on the U.S. space program; allowed for Europe to emerge as a genuine, more equal partner with the United States in civil space as exemplified by ESA's involvement with the ISS; and enabled Europe to achieve autonomous space capabilities.

Economics

Examined in this section are the impacts and consequences of Apollo on the U.S. economy. The key question often asked is whether the civil space program is beneficial for the national economy. How NASA affects the U.S. economy consumes a large part of any debate about the Agency's programs and projects.[88] The problems involved in assessing the direct benefits that NASA provides resulted in political advocacy directed at continued increases in spending for the Agency with the perspective that the indirect impacts of NASA's program on the economy are sufficient to justify its cost. In the

---

87. Sadeh, "International Space Cooperation."
88. W. D. Kay, *Defining NASA: The Historical Debate over the Agency's Mission* (Albany: State University of New York Press, 2005).

1970s and 1980s, NASA commissioned a number of studies that undertook a comprehensive economic analysis of the impact of the civil space program on the national economy. NASA officials hoped that the results of the studies would show very robust impacts on the economy, legitimizing the benefits from investments in space.[89]

The common theme is that NASA expenditures, and space activities more generally, affect the economy as a source of job creation and employment, through productivity gains, and through the development of new technologies that are spinoffs from space technologies, creating an economic multiplier effect manifested in a return on investment (ROI).[90] Albeit there is evidence to support these economic impacts to a degree, it is noteworthy that the impacts on the U.S. economy as a whole are not as large as claimed by NASA or by political advocates of the space program. NASA spent approximately $40 billion on R&D from 1961 to 1974. This represented 12 percent of total federal R&D spending in the United States.[91] Even though civil space R&D was a large function of NASA and federal government spending on R&D in space is a sizable share of overall federal R&D spending, the actual ROI of R&D spending was calculated to be at 14 percent. This figure correlated with other types of R&D spending; spending on NASA did bring with it a favorable ROI, but it did not produce dramatic economic benefits that could not be achieved by nonspace-related R&D.

By the 1980s, these impacts lessened. Space-related R&D funding, which reached a peak of more than 20 percent of U.S. R&D in 1965, declined to a low of 3 percent of U.S. R&D in the mid-1980s.[92] Furthermore, it was determined that the relationships between aggregate U.S. technology changes and developments related to R&D spending on NASA are largely speculative.[93]

---

89. Henry R. Hertzfeld, "Space as an Investment in Economic Growth," in *Exploring the Unknown*, ed. Logsdon.

90. Molly K. Macauley, "Economics of Space," in *Space Politics and Policy*, ed. Sadeh; "Final Report: The Economic Impact of NASA R&D Spending," prepared for NASA, contract NASW-2741, Chase Econometric Associates, April 1976; "Short and Long-Term Impacts of NASA R&D Spending on the Economy," unpublished manuscript, 9 February 1976; "Final Report: The Economic Impact of NASA R&D Spending," prepared for NASA, contract NASW-2741, Chase Econometric Associates, April 1975; "Economic Impact of Stimulated Technological Activity," prepared for NASA, contract NASW-2030, Midwest Research Institute, 22 November 1971.

91. Report of the Comptroller General of the United States, "NASA Report May Overstate the Economic Benefits of Research and Development Spending," 18 October 1977.

92. National Science Foundation, *Science and Engineering Indicators* (Washington, DC: United States Government Printing Office, 2000); Hertzfeld, "Space as an Investment in Economic Growth."

93. "The Economic Impact of NASA R&D Spending," prepared for NASA, contract NASW-3346, Chase Econometric Associates, 15 April 1980.

Economists are unable to show a strong positive correlation between R&D spending and overall economic growth.[94] There are two primary reasons for this finding. The first one is that space-related economic data involving data quality and collection are inadequate for economic impact analysis.[95] The second is that most Government-supported R&D is directed to the production of public goods, whose primary social value is not measured in real economic terms.

Public Goods

A host of activities take place in space for the general benefit of society, including the use of space for national defense, environmental monitoring, and the collection of science data and information. According to the legislation that established NASA in 1958, the U.S. space program is to expand knowledge about Earth's atmosphere and about outer space, develop and operate space vehicles, preserve the leadership of the United States in inventing and applying aeronautics and space technology, and cooperate with other nations in space projects. A special characteristic of these activities is that many people can benefit from them simultaneously without reducing their availability to others or adding to the costs of these activities. For instance, the benefits of R&D are available to everyone, and augmenting the number of citizens who benefit does not increase the costs of the activities. Activities with this type of attribute are known as public goods.

A gap in space economics research exists in the measure of intangible impacts associated with space exploration, such as education, national prestige and geopolitical influence, cultural influences, and a greater understanding of space science.[96] To ignore these intangible values leads to underestimating the public-good benefits of space activities. This chapter addresses in part some of these intangible aspects through an assessment of Apollo program impacts dealing with foreign policy, cultural, and educational influences.[97]

---

94. Office of Technology Assessment, *Research Funding as an Investment: Can We Measure the Returns? A Technical Memorandum* (Washington, DC: United States Congress OTA-TM-SET-36, April 1986).

95. Henry R. Hertzfeld, Office of Space Commercialization, *Space Economic Data* (Washington, DC: U.S. Department of Commerce, December 2002).

96. Macauley, "Economics of Space."

97. Economic returns emanating from NASA's space science programs have been studied. Even though this specific impact is beyond the scope of this chapter, it is indicative of impacts as they relate to public goods. See Henry R. Hertzfeld, "Measuring the Economic Returns from Successful NASA Life Sciences Technology Transfers," *Journal of Technology Transfer* 27, no. 4 (2002): 311–320.

Employment

Space activities are often judged as being good for the economy on the basis of direct job-creating potential, such as the number of jobs in the aerospace sector. Space-related jobs are also a cost, not a benefit, to the taxpayers who are not employed in the federal space program. As economists agree, wages belong on the cost side, not the benefit side, of the accounting ledger; for this reason, jobs are not properly the basis of measuring the benefits of space activities.[98] The cost of carrying out any activity—the labor, facilities, and operations—is an expense whether carried out by the government or a private-sector company.

Even if one wanted to make the case for space as a source of aerospace jobs, given that the bulk of space-related jobs are in aerospace, the macroeconomic impact on the United States is relatively small. Aerospace jobs account for less than 0.5 percent of total employment in the U.S. economy. Even at the peak of spending on Apollo in fiscal year 1966, the civil space program employed 400,000, while the total U.S. civilian employment stood at 74 million.[99] A further argument against the job-creation impact is that many of the new technologies developed through space R&D can be considered labor-saving and productivity-type gains. This allows producers of goods and services to employ fewer people and maintain or even increase production levels. Important examples are robotic techniques and automated instrumentation.

Multiplier

Another prevalent view of economic impacts is that space activities lead to multiplier effects on the economy. The multiplier describes a relationship among activities in which one set of economic activities causes a host of other activities to take place, thus cascading the effects throughout the economy.[100] The multiplier theme relates to NASA's contributions to macroeconomic growth in productivity as a result of R&D investments.[101] Macroeconomic studies that assessed productivity impacts, which can be attributed in large part to R&D

---

98. Macauley, "Economics of Space."

99. M. A. Holman, *The Political Economy of the Space Program* (Palo Alto, CA: Pacific Books, 1974).

100. Economic multipliers can be controversial because the initial public investment must originate somewhere, usually from the private sector. Thus the public project, even with superb multipliers, may or may not be better than a private investment based on competitive market and economic factors. David Livingston, "Winning the Public's Support for Space Development Programs and Funding" (Space Technology and Applications International Forum [STAIF] 2005, Albuquerque, NM).

101. *Research Funding as an Investment.*

expenditures with NASA's Apollo program, concluded that there is anywhere from a 7:1 to a 14:1 cost-benefit ratio.[102] Longer-term assessments placed the benefit at 9:1 over a 20-year period (1974–94).[103] These studies tended to indicate significant impacts on economic productivity as a result of civil space R&D. Concomitantly, these studies did contain some major liabilities as to the assumptions made, and subsequent studies refuted the favorable cost-benefit ratio. For example, in a replication of one study that showed the 14:1 return, it was discovered that productivity changes from NASA R&D spending were not statistically different from zero.[104] In 1990, a NASA study concluded that, "because of the small size of NASA spending for R&D, and because of difficulties inherent in quantifying either the costs or benefits of R&D, single number claims … of the economic payoff of NASA R&D can be easily assailed."[105]

The conclusion reached by NASA is that econometric modeling, which underlies macroeconomic studies, deals with an excessive number of variables in the economic equations used and that the economic projections as to multiplier effects are dependent on these variables that do change. The problems with the multiplier approach are so acute that the U.S. Office of Management and Budget (OMB), confronted with frequent use of the multiplier by many federal government agencies, issued guidelines for evaluating the benefits and costs of federal programs. OMB stated, with regard to multiplier effects: "Employment or output multipliers that purport to measure the secondary effects of government expenditures on employment and output should not be included in measured social benefits."[106]

Nevertheless, there was an inspirational value of Apollo highlighted with the Apollo paradigm earlier in this chapter. Though it can be argued that

102. "Final Report: The Economic Impact of NASA R&D Spending," prepared for NASA, contract NASW-2741, Chase Econometric Associates, April 1976; "Final Report: The Economic Impact of NASA R&D Spending," prepared for NASA, contract NASW-2741, Chase Econometric Associates, April 1975.

103. In 1988, the Midwest Research Institute, under contract to the National Academy of Public Administration, performed an analysis that replicated a 1971 Midwest Research Institute study that showed a 9-to-1 cost-benefit ratio for NASA R&D programs. This finding held up to sensitivity analysis; however, NASA did not release the study as it was subject to many technical economic qualifications. See Hertzfeld, "Space as an Investment in Economic Growth."

104. "The Economic Impact of NASA R&D Spending," prepared for NASA, contract NASW-3346, Chase Econometric Associates, 15 April 1980.

105. NASA Office of Special Studies, "Measuring the Impact of NASA on the Nation's Economy," unpublished report, September 1990. Also see Jerome E. Schnee, "Space Program Impacts Revisited," *California Management Review* 20, no. 1 (1977): 62–73.

106. U.S. Office of Management and Budget, "Guidelines and Discount Rates for Benefit-Cost Analysis of Federal Programs," Circular No. A-94, 29 October 1992.

from a strict macroeconomic view, NASA spending affects the economy no differently than other types of federal spending for goods and services, the inspirational value of Apollo did foster a wealth-building process. Wealth building in this context refers to the combined use of engineering, technology, and human skills to maximize the creation, production, and delivery of goods and services that are needed to raise the standard of living, increase employment, spur education, and grow the national economy.[107] Aspects of all these wealth-building processes are evident to an extent in the impacts of Apollo in the areas of public goods, employment, and multiplier impacts, as well as in the technical spinoffs and educational impacts discussed next.

The Apollo program did lead to further development and innovation in some industries, such as in helping to develop telecommunications systems for commercial use and more opportunities for space business in piloted spaceflight, as well as in information technologies.[108] The links between NASA, Apollo, and advances in satellite communications systems are examined in the spinoffs section in this chapter. The Apollo program positioned NASA as a "new" source of government contracts. NASA informally used the "10 percent rule" for contracting: NASA kept 10 percent of funds in-house to train its own engineers and gain experience, and the remaining 90 percent went to industry. This allowed some aircraft companies that were not involved with the space and missile business to enter the field, such as Grumman, which won a contract for the Apollo Lunar Module. Table 8.1 highlights some of the major space projects linked to Apollo that helped to form the development of the space industry in the 1960s. Impacts of Apollo also extended beyond the space industry. One noteworthy example of an industry impact took place in the area of information technologies. North American, the contractor for the Apollo spacecraft shown in table 8.1, established a partnership with International Business Machines (IBM) for an automated system to manage large bills of material for the construction of

---

107. See *Legacy of Apollo: Enduring Gifts to Humanity.* This documentary was produced by the Connell Whittaker Group, LLC. The documentary takes a new look at how the Apollo program radically reinvaded the global community, a new economy, and an environmental consciousness. Also see Livingston, "Winning the Public's Support for Space Development Programs and Funding." David Livingston is the creator of radio interview show called *The Space Show.* He has interviewed thousands of personalities in the space community. One common theme among these individuals is that Apollo was what inspired them to pursue space-related careers, including those aimed at space commercial development. See *http://www.thespaceshow.com* (accessed 13 September 2006).

108. NASA inherited many of its programs from the Department of Defense and inherited the same set of industrial contractors that supported the space and missile business that began in the 1950s. See Johnson, "Space Business."

TABLE 8.1. Major Space Projects of the 1960s

Project Name	Contractor
Mercury Capsule	McDonnell
Gemini Capsule	McDonnell
Saturn IB, Saturn IC, Saturn S-II, Saturn S-IV	Chrysler, Boeing, North American, Douglas
Apollo Command and Service Module	North American
Apollo Lunar Excursion Module	Grumman
Surveyor	Hughes
Lunar Orbiter	Boeing

Information is taken from Johnson, "Space Business."

the spacecraft. This led to the design and development of the Information Control System and Data Language/Interface. In 1969, this development became Information Management System/360, and it was applied to the information technology sector.

Spinoffs

NASA continually makes the case that concrete gains in social or economic value are a result of particular NASA-stimulated products or processes. The claim is that the return benefits of these products and processes (spinoffs) represent a significant dividend to the taxpayer and investment in aerospace-related R&D.[109] This concerns direct and indirect benefits from inventions and innovations as a result of NASA R&D programs, as well as patents and licenses resulting from R&D programs as a measure of the transfer of technology to the private sector.[110]

In the early 1960s, NASA established a technology utilization program with the objective to develop a means of transferring aerospace technology into useful applications by nonaerospace industries. In terms of Apollo-derived inventions, however, the economic impact is minimal. One study showed that none of the identified NASA-derived inventions are "major."[111] Another study that assessed the 1959–79 period documented that of the

109. NASA Office of Special Studies, "Measuring the Impact of NASA on the Nation's Economy," September 1990.
110. *Research Funding as an Investment.* For a history of patents as they apply to NASA, see Sylvia K. Kraemer, "Federal Intellectual Property Policy and the History of Technology: The Case of NASA Patents," *History and Technology* 17 (2001): 183–216.
111. Holman, *Political Economy of the Space Program.*

197 NASA patents licensed to industry, 54 were commercialized.[112] This does represent an impact, yet in the same period, NASA owned more than 3,500 patents.[113]

On the basis of this measure, NASA's technology utilization program was not successful. One reason for this outcome is a factor of how space systems and technologies are optimized for very specialized and complex functions that are unlikely to be adaptable to other needs.[114] A 1972 report regarding Apollo R&D spending found that more than 50 percent of the technology spinoffs were employed within aerospace and defense sectors and that the spinoffs only had a moderate economic impact and relatively low scientific and social impacts.[115] Notwithstanding all this, there is value in information dissemination and publications describing advances in technology that could "in theory" be applied to help solve specific social problems. Most of the NASA documentation on spinoffs with commercial potential does focus on such an approach, which is characterized by cutting-edge research that is under way or recently completed at NASA with plausible commercial realization.

Important impacts of spinoffs are in secondary benefits through adaptation of advances in space technology to commercial development and use. During the Apollo era, NASA R&D played a role in the technology innovation process that established the infrastructure in the development of new industries, such as those based on communications satellites. Firms used NASA contracts to put them into a position to manufacture commercial space systems. This is exemplified by industrial R&D investments; industrial R&D peaked in 1965–66, and this paralleled the growth in federal funding for Apollo in the 1960s.[116]

Despite these developments, the Cold War politics of the Apollo era impacted efforts in industry to capitalize on the technical infrastructure to commercialize space.[117] This is due to a number of factors, encompassing

---

112. Henry R. Hertzfeld, "The Economic Impact of Civilian Space Research and Development Expenditures" (National Academy of Sciences, "National Research Council Colloquium: The Role of Federal R&D," 21 November 1985).

113. Ibid.

114. Richard S. Rosenbloom, "The Transfer of Space Technology," report submitted to the Committee on Space, American Academy of Arts and Sciences, NASA Grant NsG-253-62, March 1965.

115. Martin D. Robbins et al., "Mission-Oriented R&D and the Advancement of Technology: The Impact of NASA Contributions," Denver Research Institute, May 1972.

116. National Science Foundation, *Science and Engineering Indicators* (Washington, DC: U.S. Government Printing Office, 2006).

117. Joan Lisa Bromberg, *NASA and Space Industry* (Baltimore, MD: Johns Hopkins University Press, 1999); Ann Markusen and Joel Yudken, *Dismantling the Cold War Economy* (New York: Basic Books, 1992).

how national security concerns and superpower confrontation issues of the Cold War hindered the ability to form alliances with foreign companies and sell abroad; how private-sector R&D was skewed toward satisfying NASA's technical agenda, which is not particularly congruent with commercial needs; how NASA was given responsibility for commercial policy in the space sector even though its culture was inimical to commercialization; and how companies, through favorable cost-plus contracting and government subsidies, became dependent on the government and tended to follow whatever direction government funding marked out. This overall dependency on government was not necessarily negative for commercial prospects. As discussed earlier, the Apollo era facilitated the development of the technical know-how and infrastructure that were necessary for commercialization. In addition to this, national interests did exist to promote space commerce. The notable case is that of space-based telecommunications systems.

During the Apollo era, the United States pursued a foreign policy strategy of "space diplomacy" based on preeminence and leadership in all space sectors. This strategy viewed the development of international telecommunications systems favorably. Two important developments followed. First, the U.S. Congress passed legislation that created a public, federally funded corporation, Communications Satellite Corporation (COMSAT), with the goal to set up an international communications satellite system as soon as possible. This system was established through the creation of the International Telecommunications Satellite Consortium (INTELSAT), of which COMSAT was the leading entity, as it managed and possessed a majority interest in INTELSAT during the Apollo era. U.S. space policy goals "evinced the same spirit that informed Apollo: do something great in space, do it before the Soviets, and aim it in part at the Third World."[118] These goals focused on national prestige and strengthening relations with developing states while realizing regional and global telecommunications systems with the resulting economic impacts to society.

NASA made contractual investments in developing early (1960–65) telecommunications systems. The first system, a passive communications satellite named Echo, was a NASA-funded project.[119] The second system, an active system known as Telstar, was funded by AT&T in cooperation with NASA

---

118. McDougall, *… the Heavens and the Earth*, p. 359.

119. Donald C. Elder, "Something of Value: Echo and the Beginnings of Satellite Communications," in *Beyond the Ionosphere: Fifty Years of Satellite Communications*, ed. Andrew J. Butrica (Washington, DC: Government Printing Office, 1997).

for space launch.[120] Additional investments included NASA's Relay contract to RCA; COMSAT's selection of Hughes's geosynchronous orbiting satellites for its experimental test and for its initial constellation of four satellites supported by a contract from NASA for the Apollo program; and a NASA contract to TRW in December 1965 for an advanced system of six spacecraft.[121] The U.S. government, in part due to impacts from the Apollo program, influenced advances in telecommunications satellite technology and the development of the satellite communications sector.[122]

Over the course of the Apollo era, there were also a number of documented cases of technology utilization in the areas of electrical machinery, communications equipment, and instruments. A general surge of technological innovation is traceable to technology transfer from Apollo, not as inventions derived from the Apollo program but as improvements and wider application of devices or materials already in existence, including improvements in production processes.[123] The impact was in causing technology advances to occur at an earlier time than would have likely taken place without NASA funding and support.

Within the context of this theme, there are several examples in relation to spinoffs: requirements for computing capability and need for electrical component miniaturization in the Apollo spacecraft design; launch and guidance developments related to Apollo that triggered R&D in microelectronics, computer design, and software; application of digital imaging processing techniques, originally developed for analysis of the Moon photographs, to the enhancement of computerized axial tomography (CAT) scan and magnetic resolution imaging medical data; requirements with Apollo to monitor astronaut body functions, which stimulated progress in medical telemetry; and the demand for metalized films for temperature control on Apollo hardware

---

120. David J. Whalen, "Billion Dollar Technology: A Short Historical Overview of the Origins of Communications Satellite Technology, 1945–1965," in *Beyond the Ionosphere*, ed. Butrica.

121. Johnson, "History of Space Business."

122. Advances in telecommunications satellite technology are due to a number of leading commercial developers, as well as contributors that include not only NASA but also the Department of Defense, the National Science Foundation, universities, research laboratories, and the Jet Propulsion Laboratory. Leading developers included Ball Aerospace, Fairchild, Hughes Aircraft, Lockheed-Martin, TRW, and Ford Aerospace (now Space Systems Loral). See Joseph N. Pelton, "The History of Satellite Communications," in *Exploring the Unknown*, ed. Logsdon.

123. Paul D. Lowman, "T Plus Twenty-Five Years: A Defense of the Apollo Program," *Journal of British Interplanetary Sciences* 49 (1996): 71–79; F. Douglas Johnson et al., "NASA Tech Brief Program: A Cost Benefit Evaluation," Denver Research Institute, May 1977.

and spacecraft, which led to the development and application of such films to food packages, tents, space blankets for accident victims, and flame suits.[124]

One particular area of impact, often claimed by those in the space community and verified in this chapter, deals with the spinoff from the development of the Apollo Guidance Computer (AGC) and the use therein of integrated circuit (IC) technology. The decision in 1962 to design the AGC using IC logic devices was critical to Apollo's computer success.[125] At the same time, IC technology helped create the computer industry by providing users with more speed and functionality.

The first computer to use ICs was the Block I version of AGC. In designing AGC, engineers saw the IC as a way to reduce size and weight. To suit the needs of the Apollo mission, the Massachusetts Institute of Technology (MIT), which was contracted with designing AGC, specified its own IC "logic gate" chips and developed a flight computer that incorporated thousands of these chips.[126] The use of ICs for the AGC impacted the computer industry at an important stage in its development. This helped other industries find applications for ICs, and use of the technology proliferated rapidly.

Technical applications, as a result of spinoffs, are also evident in the longer term relative to Apollo. Since 1976, NASA has published a recurring report, *Spinoffs*, highlighting technical applications as a result of NASA investments. In addition to this, presidential and congressional policies encouraged NASA to move beyond undertaking fundamentally space-based activities to a broader role in providing new technologies for commercial markets on Earth.[127]

Concomitantly, economists urge caution about the use of spinoffs as an appropriate measure of the benefits of space activities. Spinoffs as a measure of technical innovation involve upstream development dealing with R&D investments and then downstream processes of turning these investments into economic value. As this section discussed, the upstream links to Apollo exist, but the downstream processes are limited. Although spinoffs can occur from upstream development, and this is evidence of a potential economic impact, there is an issue when the spinoff technologies versus the economic value become the basis for justifying space expenditures. The argument offered by

---

124. See Office of the Chief Technologist, NASA Spinoff, Apollo Spinoffs, *http://spinoff.nasa. gov/Spinoff2012/pdf/Spinoff2012.pdf* (accessed 25 March 2012).

125. See Eldon Hall, *Journey to the Moon: The History of the Apollo Guidance Computer* (American Institute of Aeronautics and Astronautics, 1996); NASA Office of Logic Design, "A Scientific Study of the Problems of Digital Engineering for Space Flight Systems, with a View to Their Practical Solution," *http://klabs.org/richcontent/Misc_ Content/AGC_And_History/AGC_History.htm* (accessed 15 September 2006).

126. Hall, *Journey to the Moon: The History of the Apollo Guidance Computer*.

127. Kay, *Defining NASA.*

critics of the civil space effort is that if consumers wanted, or there is market demand for, these new products, then funding R&D specifically directed toward those products is more effective. Although spinoffs bring social benefits, their cost of development undertaken as a part of a space project is more expensive then the cost of developing them directly since these benefits are at best indirect in relation to the mission objectives of any space project.[128]

Other Economic Impacts

There are examples where Apollo fostered other economic impacts not discussed in this chapter. These impacts are in two areas. One is the local economic impact as a result of tourists going to visit NASA Centers and view space launches—for example, the impact on restaurants and hotels near the Kennedy Space Center benefiting from expenditures by visitors going to the Center to view launches. Apollo 11 led to local economic impacts on the county in Florida where the Kennedy Space Center is located.[129] These impacts concerned increased spending and tax monies to the county. In this case, costs to the county were minimal in comparison to the economic benefits, and the intangible benefits received from the Apollo 11 launch were believed to be immeasurable, like visibility for the county.

The problem with this impact is that it counts as benefits what are in fact transfers of income from some consumers and producers to other consumers and producers.[130] It also overlooks less desirable transfers of income burdens like traffic congestion and higher prices for residents near NASA Centers. In addition, if the impacts are added to the primary activity, then the multiplier can be large for any activity that involves the public. Hence the impacts related to Apollo in this example are not unique, as there is no discriminating among the economic value of different activities.

The second area of impact deals with economic development as a result of the location of NASA Centers and their contractual partnerships with industry to enable the large-scale technological development needed for Apollo. These partnerships provided jobs and skills to regional areas throughout the United

---

128. Singling out the effect of government influence on the products' markets can also be difficult. For example, Tang, Velcro, and Teflon are all frequently cited as spinoffs from the Apollo program, but they were actually developed *before* the Apollo program began. However, the program might have refined these products or brought them to broader attention and thus expanded commercial markets. Whatever the case, these issues further confound the effort to use spinoffs as a justification for investing in space. See Macauley, "Economics of Space."

129. Brevard County Planning Department, "The Impact of Apollo 11 on Brevard County, Florida," undated document.

130. Macauley, "Economics of Space."

States. The policy of "clustering" around a NASA Center, whereby a core of industries forms around a central NASA hub, led to economic growth.[131] To illustrate, U.S. southern states, due to the location of NASA Centers there, benefited economically, especially given the lower income levels and lesser amounts of industrialization and urbanization that exist there relative to many other parts of the United States. High-income, industrialized, and urbanized states that did in fact receive the larger share of NASA R&D allocations during Apollo and afterwards also benefited economically.[132] The Houston area with the Johnson Space Center, the Orlando area with the Kennedy Space Center, and Sunnyvale in California with the Ames Research Center are illustrative of economic benefits as a result of clustering around a NASA Center.

Research undertaken on the economic impacts of federal R&D allocations to states found that scientific and technical innovations can be expected to bring benefits to the geographic regions in which they are located.[133] Innovations represent long-term impacts, as they deal with a process of technical diffusion incorporating refinements in technology development over many years.[134] There are also benefits to a region on the basis of the technical workforce that space activities generate and require, as this workforce is a versatile resource that is productive and transferable to nonspace sectors. If

---

131. NASA Center histories document and discuss regional impacts based on the policy of "clustering." The NASA Marshall Space Flight Center history by Andrew J. Dunar and Stephen P. Waring, *Power To Explore: A History of Marshall Space Flight Center, 1960–1990* (Washington, DC: NASA SP-4313, 1999), has a chapter on regional economic and social impacts in the Huntsville, AL, area. This and other NASA Center histories make it quite clear that the economic and social impacts on the region in which a NASA Center is located is significant. See Elizabeth A. Muenger, *Searching the Horizon: A History of Ames Research Center, 1940–1976* (Washington, DC: NASA SP-4304, 1985); Henry C. Dethloff, *Suddenly Tomorrow Came…: A History of the Johnson Space Center, 1957–1990* (Washington, DC: NASA SP-4307, 1993); and Mack R. Herring, *Way Station to Space: A History of the John C. Stennis Space Center* (Washington, DC: NASA SP-4310, 1997).

132. Willard I. Zangwill, "Top Management and the Selection of Major Contractors at NASA," *California Management Review* 12, no. 1 (1969): 43–52.

133. During the Cold War, aerospace and defense spending became a major determinant of regional economic prosperity or decay. The West, Gulf, and East Coast regions of the United States reaped enormous gains from the distribution of federal R&D monies. See Ann Markusen, Peter Hall, Scott Campbell, and Sabina Deitrick, *The Rise of the Gunbelt: The Military Remapping of America* (Oxford: Oxford University Press, 1991).

134. One specific area of impact related to innovations took place in advanced manufacturing techniques. John A. Alic, Lewis Branscomb, Harvey Brooks, Ashton B. Carter, and Gerald L. Epstein, *Beyond Spinoff: Military and Commercial Technologies in a Changing World* (Cambridge, MA: Harvard Business School Press, 1992).

many of the aerospace jobs are concentrated in a geographical region, then these jobs can become a benefit to people living in the region.[135]

There is also literature that questions these regional economic impacts. One author on the subject that deals more broadly with government R&D in aerospace and defense argues a "depletion" hypothesis.[136] The argument is that aerospace and defense spending is dictated by the needs of government; fails to generate economic growth; diverts intellectual, financial, and material resources away from civilian and commercial industries; militarizes society; retards R&D; and preempts a significant share of U.S. capital stock.

Education

The quality of STEM education in the United States is an ongoing concern of decision-makers. Following World War II, scientists, engineers, and mathematicians expressed concerns about the quality of precollege instruction in their fields and about the number of students who went on to college and studied STEM subjects. The curriculum was out of date and difficult for teachers to master in order to develop an understanding of the key concepts and ideas in STEM fields.[137] This crisis in education was exacerbated by the beginnings of the Space Age with the launch of Sputnik in 1957.

One of the primary forces shaping the science reforms of the 1950s and 1960s was the National Science Foundation (NSF). The NSF's education effort prior to Sputnik was confined to promoting science fairs and clubs and funding summer institutes for teachers. Following Sputnik, however, the NSF increased its support for curriculum development at a rapid pace. By 1960, the programs of the Education Directorate at NSF represented 42 percent of the NSF annual budget.[138] This science reform movement was sustained through the Apollo program and ended with the Apollo 11 lunar landing.[139]

NASA plays a role in inspiring youth and fostering impacts on STEM education. Investments made by NASA, beginning with the Apollo era and

---

135. Loyd S. Swenson, "The Fertile Crescent: The South's Role in the National Space Program," *Southwestern Historical Quarterly* 71, no. 3 (1968): 377–392.

136. Seymour Melman, *Our Depleted Society* (Austin, TX: Holt, Rinehart, and Winston, 1965). Also see Seymour Melman, *Pentagon Capitalism: The Political Economy of War* (New York: McGraw-Hill, 1970); and Seymour Melman, *The Permanent War Economy: American Capitalism in Decline* (New York: Simon and Schuster, 1974).

137. P. Dow, "Sputnik Revisited: Historical Perspectives on Science Reform" (prepared for the symposium "Reflecting on Sputnik: Linking the Past, Present, and Future of Educational Reform," Washington, DC, 4 October 1997).

138. National Science Foundation, *Science and Engineering Indicators*, 2000.

139. Dow, "Sputnik Revisited."

sustained to the present, encompass curriculum and teaching enhancement activities for primary and secondary schools; supplemental training in STEM subjects for college teachers; cooperative education and work-study programs; and university and college grants and assistantships.[140] During Apollo, there was a dramatic increase in the number of U.S.-citizen students pursuing advanced degrees in STEM disciplines. As the Apollo program was terminated and NASA's funding cut, the number of students going into STEM fields correlated with the downward trend in NASA's budget, especially with regard to graduate studies at the Ph.D. level.[141]

Given NASA's share of the entire federal R&D budget, particularly in the Apollo era, investment in higher education in STEM fields was relatively large as compared to other educational investments. In 1965, for instance, NASA allocations to its university-related programs and research amounted to more than 85 percent of all non-NASA federal appropriations to universities.[142] Following Apollo 11, R&D obligations in the industrial sector declined and did not experience another surge until over a decade later, when Cold War investments in military technology resulted in another period of growth. This decline resulted in a negative impact on STEM education.

These downward trends highlight more current times, and by some accounts, there is a major workforce crisis in the aerospace and defense industrial sector.[143] In physics and advanced mathematics, for example, U.S. high-school seniors score significantly below the international average on performance tests.[144] The trend continues at the undergraduate levels. There is a downward trend in the United States relative to foreign states in science and engineering university degrees granted per capita.[145] At the graduate level, the problem persists for the United States. At U.S. universities, 25 percent of graduate students in the sciences and 40 percent of graduate students in technology, engineering, and math disciplines are foreign nationals.[146]

A contributing factor to these tends is a general disinterest in STEM fields. The argument that money put into the space program is better spent

---

140. These are the principal means by which NASA seeks to promote the continuing replenishment of the nation's STEM workforce.

141. Office of Management and Budget, "Space Activities of the U.S. Government," 2002.

142. Space Policy Institute, "The Apollo Education Initiative: Origins, Activities, and Results," (George Washington University, Washington, DC, June 1990).

143. "Final Report of the Commission on the Future of the United States Aerospace Industry," November 2000.

144. National Science Foundation, *Science and Engineering Indicators*, 2002.

145. Ibid.

146. "Final Report of the Commission on the Future of the United States Aerospace Industry," November 2000.

by putting it directly into the educational system to encourage students to pursue STEM areas is a misconception, as the United States is already one of the top spenders per student in the world.[147] The bottom line is that students need something to inspire their efforts. Thus, the positive impact of space exploration on STEM education is without precedent, as is evident with aspects of the Apollo paradigm and the inspirational value of Apollo as explained earlier in this chapter.

University Programs

During the Apollo era, NASA established a Sustaining University Program (SUP) that envisioned the university as a repository of knowledge to meet public goals and general societal problem-solving.[148] NASA established SUP in 1962. SUP was Webb's primary vehicle for relating NASA to societal purposes.[149] However, evaluations of SUP highlighted the problematic nature of relations between government and universities instead of demonstrating positive societal impacts.[150] The goal of SUP was to further university and college interest in the integration and synthesis of knowledge, but the impacts of the program were limited to the aeronautics and space goals as established by NASA.[151] SUP was not able to institutionalize the educational innovations that it sought. There is no direct evidence that the long-range objectives of the SUP program were met.[152] These objectives involved the development of a university capable of responding as an institution to societal problems and issues, capable of multidisciplinary and interdisciplinary research and teaching, and able to accelerate knowledge transfer from the university or college to society.

---

147. National Science Foundation, *Science and Engineering Indicators*, 2000.

148. W. Henry Lambright, "Using Universities: The NASA Experience," *Public Policy* (1972): 61–82; W. H. Kohl, "The Potential Influence of NASA's Long-Range Technological Needs on the Restructuring of University Engineering Curricula," Advisory Council Meeting, 2–3 October 1969.

149. Webb, *Space Age Management*.

150. The university feels its essential character to be threatened when the government attempts too firmly to direct it along any given path. In addition, science at the university and college levels was more oriented to goals defined by society, like social, urban, and environmental problems.

151. Lambright, "Using Universities"; Task Force To Assess NASA University Programs, Office of Technology Utilization, *A Study of the NASA Sustained University Program* (Washington, DC: NASA SP-185, 1968).

152. *Study of the NASA Sustained University Program*; Thomas W. Adams and Thomas P. Murphy, "NASA's University Research Programs: Dilemmas and Problems on the Government-Academic Interface," *Public Administration Review* 27, no. 1 (1967): 10–17.

Conclusions

The societal benefits that are linked to the Apollo program are both human-centered and technical. Intended societal impacts of Apollo are limited and more or less confined to the Apollo era. Such impacts were found to exist in a number of areas, including city planning, systems architecting, economics, and education. One common theme regarding all these areas is that impacts clearly exist in theory, but are generally limited in practice to a few technical cases, like PERT, Information Management System/360, ICs, and some of the economic and educational benefits that are linked to federal investments in space-related R&D. In the economic and education areas, which are most often associated with impacts on society, the claims were found to be greater than the actual evidence suggests. It was discovered that the intended economic benefits were no different from those that could result from other types of federal R&D investments. And with education, the impacts of Apollo failed to engender any longer-term institutional changes.

The unintended societal impacts and consequences traced to Apollo are much broader and are sustained to the present. Of significance is how the political formulation of Apollo put into place a paradigm that served as the conceptual basis for many impacts. The Apollo paradigm was based on an ethos that the technology and know-how acquired with Apollo could be applied to the space program and elsewhere to establish what was called Space Age America and Space Age management. The European space program benefited from the transfer of managerial know-how used with Apollo. Also, it is this notion of Space Age management that led NASA to emulate the technical and management models used with Apollo to implement more current space exploration strategies and plans.

Unintended impacts and consequences are also shown with the inspirational value of Apollo. This value did more for building wealth than probably any other civil pursuit. This "wealth factor" was evident in a variety of ways. These include the role of Apollo in engendering prestige, confidence, and an air of competence for the U.S. government; the role of Apollo as a "grand laboratory" for the developments, innovations, and applications of technologies for societal benefit; the role that Apollo played in the development of the space industry; and the motivation that Apollo provided for students to pursue STEM disciplines and education. From a broader historical perspective, one cannot predict what space exploration will bring. Undoubtedly, it will bring with it, as Apollo did, changes in how humanity views planet Earth and in how humans and institutions interact with society. Space exploration informs humanity's grandiose search for and interaction with the future. The space odyssey is perpetuated by explorers, discoverers, and seekers. Explorers venture into the unknown cosmos; discoverers pursue cumulative knowledge about Earth and space; and seekers search for underlying models and causal factors to explain cosmological phenomena. Apollo set humanity on a trajectory to interact with the future: to, in the end, become a spacefaring species and society.

Chapter 9

AN ASTROSOCIOLOGICAL PERSPECTIVE ON THE SOCIETAL IMPACT OF SPACEFLIGHT[1]

Jim Pass

Introduction

Astrosociology was introduced in 2004 to begin the process of studying the relationship between space exploration and societies in a systematic manner. This chapter seeks to make the case for two broadly defined arguments: 1) space exploration has already transformed societies to an extensive degree—that is, it has already affected every aspect of social life; and 2) space exploration will continue to transform societies and other aspects of social life to an increasing extent into the future. A third related argument focuses on the idea that we know too little about the impact of space exploration on society and culture, and we need to understand it better in order to take advantage of the knowledge gained. We should look everywhere for potential knowledge and wisdom, which can come from a scientific discipline or a literary source such as science fiction.[2] The major approach of this chapter is to discuss these issues in the context of the need to develop astrosociology.

Historians have probably conducted the greatest amount of research in this area, which is extremely important, but now we have passed the point for social scientists to organize and join them under the banner of astrosociology to construct a single, easily identifiable body of knowledge and related literature. The social and behavioral sciences in addition to the humanities (hereafter referred to as the "social sciences" for brevity) must organize themselves into a single field and work with the space community through formal collaborative channels regarding space issues relevant to both the natural and social sciences. Historically, this has not occurred. Recent events have demonstrated that the time is right for this outdated historical pattern of separation

---

1. I wish to thank former NASA Chief Historian Dr. Steven J. Dick for providing me with this opportunity to introduce astrosociology and its implications to NASA and the space community. In addition, I wish to thank our late colleague, Albert A. Harrison.

2. Simone Caroti, "Defining Astrosociology from a Science Fiction Perspective," *Astropolitics* 9, no. 1 (2011): pp. 39–49.

between the natural sciences and social sciences to end in the area of space theory and research.

Astrosociology fills a void that spans nearly the entire Space Age. Previously, the absence of a social science field that could potentially contribute unique insights focusing on the societal impact of space exploration has slowed the pace of accumulating knowledge in this area of study. In essence, we have settled for less than the full potential of the rich heritage of an entire branch of science.[3] The seriousness of this situation has not continued without notice, however. For example, several sociologists pointed out the indifference of sociology (my particular discipline) during the final two or three decades of the twentieth century. Their recommendations went largely unheeded, however:

> B.J. Bluth,[4] for example, advocated the study of space issues from a sociological perspective long ago. [William Sims] Bainbridge[5] made an important observation about sociology's indifference. Part of his argument involved the recognition that in the face of a substantial interest in space on a societal scale (among citizens and space scientists), sociologists are less well prepared to deal with it compared to the scientists in the so-called "hard sciences." [Alvin] Rudoff,[6] in considering the importance of astrosocial issues, asked a simple though very revealing question: "And where is sociology?" A simple conclusion thus presents itself. Bluth, Bainbridge, Rudoff, along with [Allen] Tough[7] and many others, have long recognized the potential value of the "sociology of space" to the discipline and to society. Proponents of astrosociology continue to marvel at its absence in the face of this untapped potential.[8]

---

3.   Jim Pass, "Astrosociology and Space Exploration: Taking Advantage of the *Other* Branch of Science," *Space Technology and Applications International Forum (STAIF) Conference Proceedings* 969, no. 1 (2008): 879–887, *http://www.astrosociology.org/Library/PDF/STAIF2008_OtherBranch.pdf* (accessed 16 March 2012).

4.   B.J. Bluth, "Sociology and Space Development," in T. Stephen Cheston (Principal Investigator), *Space Social Science, http://er.jsc.nasa.gov/seh/sociology.html* (accessed 16 March 2012).

5.   William Sims Bainbridge, *Goals in Space: American Values and the Future of Technology* (New York: State University of New York Press, 1991).

6.   Alvin Rudoff, *Societies in Space* (American University Studies, Series XI, Anthropology and Sociology, vol. 69), (New York: Peter Lang Publishing, 1996), p. 75.

7.   Allen Tough, "Positive Consequences of SETI Before Detection," *Acta Astronautica* 42, no. 10–12 (1998): 745–748.

8.   Jim Pass, "Inaugural Essay: The Definition and Relevance of Astrosociology in the Twenty-First Century, Part Two: Relevance of Astrosociology As a New Subfield of Sociology," (2004), 6–7, *http://www.astrosociology.org/Library/Iessay/iessay_p2.pdf* (accessed 16 March 2012).

Many social scientists from other disciplines made similar observations and called for the formal establishment of a new field dedicated to the study of "space and society." Moreover, many independent social scientists (such as Albert A. Harrison, Steven Dick, and Ben Finney) have dedicated their careers to the study of space issues. Indeed, Harrison has supported astrosociology from the beginning. As discussed in the next section, the time had finally come in 2004 for social scientists to organize themselves around a single field known as astrosociology.

An astrosociological perspective is important to understanding space exploration because it makes the connection between space and the human cultural drive to explore the unknown. Space exploration represents an imperative based on a number of historical forces.[9] Social and cultural phenomena include the relationship between space and society and thereby reflect a society's involvement in space. They demonstrate a long-term pattern of migration and investigation of new frontiers.[10] Space exploration represents a logical continuation of the expansion of the human presence from its starting point. While humanity has expanded from its common place of origin and explored the surface of Earth rather extensively, space represents a boundless new ecology for humanity to investigate.

As the founder of the field of astrosociology, I would argue that NASA and the entire space community desperately require more of the type of input only possible from social scientists. The era of ignoring them has reached its limit for a number of reasons made evident in this chapter. The slight breakdown of the so-called "great divide" between the space and social science communities, the subject of the first discussion to follow, only recently began. However, this new shift indicates a growing recognition by an increasing number of members within the space community that conceptual, theoretical, and empirical traditions within the social science disciplines possess important value for the future of spaceflight and space exploration.

NASA and other space agencies need to fully understand these shifting conditions in order to take advantage of newly emerging social and cultural patterns by embracing the new field of astrosociology. They require an organized field among the social sciences for the sake of smoother collaboration. Moreover, the space community has yet to tap the theoretical and empirical data pertinent to its future development. Much of this data is readily applicable to space missions and other pertinent aspects of the space community's understanding of space exploration in space and on Earth.

---

9. Ibid.

10. For an excellent discussion, see Ben Finney and Eric M. Jones, *Interstellar Migration and the Human Experience* (Berkeley: University of California Press, 1985).

Recently, the space community began its path toward a comprehensive appreciation for the potential contributions of the social sciences to space-flight, space settlement, utilization and exploitation of space resources, and space exploration. While psychologists have participated with NASA and the space community for decades now, this new appreciation and recognition involves sociology and the other social sciences (i.e., those social sciences traditionally barred or simply ignored by space scientists, engineers, and even space enthusiasts). What accounts for this drastic change? For one thing, space exploration is on the verge of expanding beyond the scope of the traditional space agencies in terms of social control and participation. This expansion will create social forces that affect societies on a much greater and more fundamental basis. This chapter attempts to glimpse into the future based on present social and cultural conditions to make predictions about how spaceflight and space exploration will contribute significantly to social change affecting social systems around the globe, including changes from their traditional orientations and priorities.

Bridging the Great Divide

For nearly the entire course of the Space Age, the space community and social sciences refused to interact with one another to any appreciable extent. Marilyn Dudley-Rowley[11] has characterized this as the great divide between aerospace and sociology (though all the social sciences followed a similar pattern during the past 50 years or so). Until now, engineers could solve the major problems of spaceflight with minimal help from psychologists, including those specializing in human factors research. Soon, this will not prove possible. The growing implications of this fact do not exist without notice.

Recently, space agencies and professional organizations around the world began to take a renewed look at the social and cultural consequences of past events and future implications of their work. A NASA-sponsored conference in September 2006 focused on the societal implications of spaceflight, with the following objective: "The purpose of this conference is to undertake a broad overview of the societal impact of space exploration, especially as illuminated by historical research. The purpose is not to conduct an exercise in public affairs or a debate over public policy, but to examine with rigorous research what the impact has been, both nationally and internationally. This

---

11. Marilyn Dudley-Rowley (aka Marilyn Dudley-Flores), "The Great Divide: Sociology and Aerospace" (2004), *http://astrosociology.org/Library/PDF/submissions/The%20Great%20Divide_ CSA2004.pdf* (accessed 16 March 2012).

is an enormous topic, so we cannot be comprehensive, but we can be broadly representative of the major areas of impact."[12]

The very existence of this conference potentially indicates that we are finally moving in a new and productive direction. The emphasis on the societal impact of the space program in the past provides me with optimism for the future. The participants addressed the historical implications of space exploration quite well at this conference, and this certainly bears on the emphasis in this chapter. However, the unfolding history of space exploration and our ability to take advantage of it for future projects and missions must receive attention. Through a proper understanding of the societal impact of space exploration during the preceding years of the Space Age, and even much earlier in our history, we can become more productive and more appreciative of its effects and, more important, it potential.

I hope that NASA is beginning to take this general area of study seriously enough to provide ongoing conferences about the societal (and even the astrosociological) issues—beyond the good work confined to its History Program Office—and consistent with this, to appropriate funds to conduct relevant research for the benefit of NASA and the space community in general. The implications, if taken seriously, will unquestionably provide benefits for the Agency in all of its operational and research divisions. As a major suggestion, NASA should invite more social scientists to conduct social-scientific research in collaboration with the Agency. After all, social scientists comprise the experts in this area and can adapt a great many concepts and empirical findings from terrestrial social groups to those existing in space. Astrosociologists will add a wealth of theoretical and empirical knowledge gained over hundreds of years to the space community, thereby adding a missing component long overdue while working well with those conducting human factors research.

What has not been clear until recently is that the physical sciences and the social sciences represent complementary approaches, while neither can provide all of the answers to questions becoming more relevant in the future—or perhaps it was never appreciated nearly enough. To use a common metaphor, these two branches of science represent the two sides of the single coin known as space exploration. In the future, the natural and physical scientists within the space community cannot continue to ignore the social science side because humanity will expand into the cosmos; and no human activity, not even space exploration, occurs in a *social* vacuum.

---

12. Steven J. Dick and Roger D. Launius, eds., *Societal Impact of Spaceflight* (Washington, DC: NASA SP-2007-4801, 2007).

The European Space Agency (ESA) also recognizes the astrosociological implications of space exploration. Its Web site discusses the benefits for Europe including 1) competitiveness in high technology, 2) creation of new jobs, 3) quality of life, 4) research and education, and 5) "cultural benefits."[13] Moreover, ESA held an event, Second Space and Society Conference: Space Options for the 21st Century, following the first stand-alone conference sponsored by the International Academy of Astronautics (IAA).[14] Clearly, ESA recognizes the cultural and social benefits of its mission—a very astrosociological approach! Another example is the third Asian Space Conference (ASC), which occurred in March 2007. It catered to the traditional members of the space community, including space scientists and engineers, but it also emphasized sociology, the other social sciences, and the humanities.

These types of official declarations and the corresponding changes in professional conferences around the world represent a historically shifting stance from a dismissal or skepticism of the relevance of the social sciences to one that embraces their contributions that seemed to occur most visibly with the turn of the twenty-first century. From my perspective, they all focus on astrosociological issues and always have, although a formal field was missing before 2004. The next step involves convincing the leadership of space agencies and professional associations within the space community to recognize astrosociology as the scientific approach to study these cultural and social forces in a formal manner. One major way to do this involves recognizing the merits of astrosociology as they did those of astrobiology. In fact, the *astrobiology model* serves as a good example of how to proceed.[15] The two fields are similar in a number of ways, and complementary as well. One of the similar challenges involves convincing a diverse group of scientists and other professionals to both agree with the merits of this new field and then to actually participate in its development. Astrobiology accomplished this feat. In contrast, astrosociology finds itself largely in the midst of the acceptance stage among many in the mainstream, accelerated in 2015 by a new *Journal of Astrosociology*.

Budding astrosociologists now encourage these trends, and take advantage of them when they present themselves, in various ways. As an example, a group from the social science and space communities recently formed

---

13. European Space Agency (ESA), "Home Page," *http://www.esa.int/ESA* (accessed 16 March 2012).

14. European Space Agency (ESA), "Space and Society Conference: Space Options for the 21st Century," *http://www.esa.int/SPECIALS/TTP2/SEM8OVRMTWE_2.html* (accessed 19 March 2012).

15. Albert A. Harrison, "Overcoming the Image of Little Green Men: Astrosociology and SETI," (2005), *http://www.astrosociology.com/library/pdf/submissions/overcoming%20lgm_harrison.pdf*, (accessed 12 March 2012).

the Astrosociology Working Group (AWG) within the American Institute of Aeronautics and Astronautics (AIAA), now under the sponsorship of the Society and Aerospace Technology Technical Committee, and recently upgraded to the Astrosociology Subcommittee. Inroads such as this within the space community will result in a permanent presence of the field of astrosociology and should result in its full acceptance by space agencies and professional organizations over time. This would certainly represent a boon to this developing field for both the social science and space communities.

In May 2008, the Astrosociology Research Institute (ARI) was inaugurated.[16] ARI represents the next step in fulfilling our mission of developing astrosociology in ways that only a formally structured organization can accomplish. The assistance of two particular categories of supporters serves as a strong driver of ARI programs. Specifically, ARI seeks to provide assistance, resources, and opportunities for 1) students who wish to pursue astrosociology in their academic careers, and 2) faculty and independent scholars who wish to accommodate students interested in astrosociology and/or for those who wish to pursue the field themselves as professional astrosociologists.

Understanding the various elements of the societal impact of space exploration represents the core objective of the new field of astrosociology. In this instance, the fact that NASA sponsored a conference centered on the societal impact of spaceflight is remarkable in itself because the Agency is essentially acknowledging the relevance of the social sciences to the space program—an admission that was not forthcoming in official channels for most of its history. Spaceflight and exploration have always affected societies on Earth. The aerospace industry affects the workforce, its accomplishments impact the larger culture, and the first landing on the Moon captivated people around the world, for example, yet NASA never elicited the assistance of the social science community in a big way. On the other hand, the social science community never strongly pushed for entrance into the space community's exclusive club, even while social and cultural forces related to space reshaped the social structures and cultures (and subcultures) of the United States and other social systems around the world. Social scientists must take much of the blame for

---

16. ARI is a nonprofit public benefit corporation with a 501(c)(3) status, for which I serve as the Chief Executive Officer. Joining me as officers are Christopher M. Hearsey (Deputy Executive Officer) and Simone Caroti (Secretary/Treasurer), both of whom replaced Thomas Gangale and Marilyn Dudley-Flores, who were the original officers, early advocates of the field, and early collaborators following the onset of the astrosociology movement. The official Web site for the Astrosociology Research Institute is *http://www.astrosociology.org*. The Web site *http://www.astrosociology.com* continues to exist to accommodate the Virtual Library as an archive to match older references. (The new Virtual Library refers to *http://www.astrosociology.org* and duplicates older references.)

neither addressing the impact of these societal forces nor forcefully demonstrating the importance of social scientific research to the space community.

Two facts seem incontrovertible. First, the general absence of the social sciences in the study of "space and society" issues over the course of the Space Age for so long has severely limited our knowledge about the impact of exploration on society (despite the good work by a limited number of historians and social scientists). Second, and relatedly, if we remedied the unheeded need to bring in astrosociologists and independent social scientists, it would greatly assist the space community in catching up with the past as well as understanding the present and future consequences of space exploration on society. We are indeed late in addressing this fundamental area of inquiry from an organized social-scientific perspective, but it seems that we have finally begun to do so.

For example, the discipline of sociology should address the impact of human spaceflight and exploration and, in fact, all social and cultural phenomena related to space, due to its very orientation. Philosophers and others created sociology over 200 years ago in order to study *all* aspects of social life. Space-related phenomena are no less important aspects of social life than acts of deviance or the social forces related to social movements, both of which receive an inordinate amount of sociological scrutiny.

> The promise of sociology resides in its potential capacity to recognize the connections between individuals and both the social structures and the cultural communities comprising their society.[17] This fundamental application of the sociological imagination seems obvious when considering the historical development of the discipline, and its attention to "normal" social phenomena. However, the ongoing failure to apply the sociological imagination to an understanding of [social phenomena related to space] demands special consideration of astrosociological issues. Modern human activities...related to space, characterizing the *space age*, have been taking place since [before the] 1950s, yet their impact on society over the years is largely unknown due to a significant level of sociological indifference and perhaps even a certain level of contempt.[18]

Similar arguments apply to the other social science disciplines, such as psychology, economics, history, political science, and anthropology, and indeed astrosociology has transformed itself into a multidisciplinary field. The contributions of other social scientists to the work of historians would provide an

---

17. C. Wright Mills, *The Sociological Imagination* (New York: Oxford University Press, 1959).
18. Pass, "Inaugural Essay, Part Two," 5–6.

additional dimension to our understanding of the impact of space exploration. We have much to learn about the impact of all space-related patterns of
behavior on the social structures and cultures of societies around the world.
Astrosociology can organize this effort due to its very concentration on these
types of issues.

This chapter attempts to address some of the major areas of specialization
within the new field of astrosociology that are pertinent to the purpose and
spirit of the NASA conference cited earlier. In fact, the very focus of astrosociology centers on the interface between space and society from a social science perspective. As will become evident, although most of the space-related
behavior occurs on Earth at the present time, an increasingly larger proportion will begin to occur over the next few decades in space and on bodies
other than Earth. Social and cultural forces will affect migrating populations
isolated from Earth. It will serve these populations much better to account
for these effects and plan for positive social conditions rather than reacting
to negative social patterns that develop and prove difficult to alter once they
become established.

I did not take an outlandish position when I argued that the serious development of a social science field such as astrosociology was long overdue in
2004. The suddenness of the collaboration between the space and social science communities seemed too easily accomplished unless some underlying
set of like-minded principles existing on unofficial channels of communication. The timing just happened to coincide with the development of astrosociology; it seems that the proponents of astrosociology ended up in the right
place at the right time. While astrosociology is not yet deeply entrenched
in the space community, it definitely has become a recognizable presence
at space and social science conferences since 2004. As a new field, astrosociology will find acceptance only if it proves itself as an important scientific
field—a reality that seems assured by the timing of events currently set into
motion. With the acceptance of the members of the space community and
a growing number of social scientists gravitating toward this new field, it
seems as though it is finally time for the social sciences to enter the Space Age
in a formal manner. We can hope, at least, that sociology and the social sciences enter the Space Age now that we have celebrated the 50th anniversary
of spaceflight.

Definition, Scope, and Relevance of Astrosociology

Although many social scientists had conducted astrosociological research for
years before the establishment of astrosociology, the field is bound to become
more successful when the social scientists interested in studying space are

organized rather than isolated, without adequate support from their disciplines or adequate collaboration between one another. Lack of organization also carries with it the reduced likelihood that students in social science programs will select space exploration as their career focus. In contrast to previous calls for a new field, astrosociology has demonstrated surprising growth in both the social science and space communities. Within the social science community, astrosociology is now developing on a two-prong basis as 1) a sociological subfield, and 2) most importantly, as a general social science field. Within the space community, astrosociology represents the field comprised of the social sciences with which it can easily identify and collaborate in a way that builds constructive knowledge in innovative ways not possible before.

The basic definition of *astrosociology* has evolved to the study of *astrosocial phenomena* (that is, the social, cultural, and behavioral patterns related to outer space) with the addition of the "behavioral" component in 2009.[19] Astrosocial phenomena, logically enough, comprise a subset of all social, cultural, and behavioral phenomena. I coined the term "astrosocial" to refer specifically to phenomena characterized by a relationship between human behavior and space phenomena. For example, an undiscovered asteroid hurtling toward Earth is a space phenomenon. In contrast, the discovery of that asteroid and subsequent actions to cope with it as a potential threat represent astrosocial phenomena. This particular category of social phenomena has traditionally received very little attention from the social science and space communities. Yet the Societal Impact of Spaceflight conference that took place in September 2006 did, in fact, focus on astrosocial phenomena! Such phenomena date back to the time that the first human being looked up at the heavens and attempted to comprehend the meaning of space and the planets and stars within it. Human spaceflight represents a much shorter historical period, although it nevertheless affected society for more than 50 years in a number of different ways. While the level of understanding in this area is currently limited due to the small number of contributions from most of the social science perspectives, the significance of these effects finally began to garner serious attention in recent years. After a long, unproductive period before 2003, astrosociology now possesses the great potential to bridge the great divide.

While astrosocial phenomena do indeed exist, identifying and understanding them requires the rigor of science. According to Howard E. McCurdy, and

---

19. Jim Pass, "Pioneers on the Astrosociological Frontier: Introduction to the First Symposium on Astrosociology," *Space, Propulsion & Energy Sciences Forum: SPESIF-2009*, 1103, no. 1 (2009): 375–383, doi: 10.1063/1.3115541, (accessed 12 March 2012).

others, "Events in spaceflight have social, cultural, and ideological effects."[20] He also warns, however, that we must be careful when attributing events as causes that trigger various effects because what appears to be correlation and even causation is often just a coincidental set of circumstances.[21] Because astrosociology is a scientific field, astrosocial phenomena are only defined as such when there is a verifiable connection between society (including social, cultural, and behavioral effects) and outer space. Otherwise, they are phenomena of other varieties. The astrosociologist must be careful to make the correct distinction between the two. Furthermore, astrosocial phenomena often exist as subtle or disguised patterns that are difficult to identify. Scientific investigation is the only course of action as space advocacy is no substitute for the scientific method based on the two-way interaction between theory and observation.

It is important to remember that the focus of space exploration remains on space, of course, but astrosocial phenomena currently affect terrestrial societies most strongly.

> Astrosociology seeks to involve sociologists in the study of space-centered human activities whether these occur on Earth or in space itself.[22] Perhaps the most obvious opportunities are in human space exploration. Here, social scientists might study social policy and political support, space advocacy groups, private sector businesses and public agencies that have vested interests in space, and groups of people who live on the high frontier.[23]

The significance of an astrosociological perspective relates more to its future than to its past, although present and past impacts require study as well. Indeed, as a species, we have only engaged in spaceflight for just over 50 years. "Recognizing that contemporary human efforts in space are best viewed as the tip of an iceberg and as possible precursors of grander future efforts," writes Harrison, "astrosociology proposes to move sociology into the

---

20. Howard E. McCurdy, "Has Spaceflight Had an Impact on Society? An Interpretative Framework," in *Societal Impact of Spaceflight*, Steven J. Dick and Roger D. Launius, eds. (Washington, DC: NASA SP-2007-4801, 2007), p. 8.

21. Ibid., 4–16.

22. Harrison, "Overcoming the Image of Little Green Men," p. 9; Jim Pass, "Inaugural Essay: The Definition and Relevance of Astrosociology in the Twenty-First Century, Part One: Definition, Theory, and Scope" (2004), *http://www.astrosociology.com/Library/Iessay/iessay_p1.pdf* (accessed 12 March 2012); see Jim Pass, "Invitation to Astrosociology: Why the Sociologist–Space Enthusiast Should Consider It" (2005), *http://astrosociology.org/Library/PDF/submissions/Invitation%20to%20Astrosociology.pdf* (accessed 16 January 2012).

23. Harrison, "Overcoming the Image of Little Green Men," 9.

space age.... Astrosociology deals with the broad, societal contexts of activity pertaining to space, as well as actual space exploration including human space exploration and the search for extraterrestrial life."[24] Harrison's comments concerning sociology pertain to all of the social sciences, and thus they all have the opportunity to contribute to the development of astrosociology.

In initiating the development of astrosociology, I envisioned three major goals: 1) to organize social scientists interested in space, 2) to develop a coherent body of knowledge and associated literature, and 3) to form a stable, formalized collaborative structure between the social sciences and the space community. Each of these goals receives ongoing attention and thus improvement on a slow and steady basis. The growing group of pioneering astrosociologists has made important strides within the space community and among individuals within the social science community. Ironically, sociology and the other social sciences continue to ignore it at the macro level even as more individual scholars and scientists, mostly young professionals and students, begin to gain a greater sense of excitement as a result of their recognition of the value of astrosociology. It seems that established professional social scientists not already interested in space will stay away from it, and even criticize it, while the current new generation of students includes individuals much more open to the idea of pursuing astrosociology as a career specialization. The status quo usually proves difficult to change in any discipline or organization. This represents a major challenge for contemporary proponents.

Based on the foregoing discussion, it remains important to remember that one does not need to be a sociologist to become an astrosociologist! The perfect analogy is astrobiology. Those who flocked to the pursuit of astrobiology included biologists, to be sure, but they also included planetary scientists, geologists, paleogeologists, astronomers, chemists, computer scientists, roboticists, artists, and those from a myriad of different disciplines in addition to biologists. Anyone interested in the relationship between space and society (including human behavior in all of its aspects) is a potential astrosociologist. Despite difficulties in academic programs as they now exist, those who pursue astrosociology will find themselves working on the cutting edge of a new field of study—they will be part of the cadre of astrosociology's pioneers who shape the development of a new field.[25]

---

24. Ibid., 14.
25. Pass, "Pioneers on the Astrosociological Frontier."

Applied Astrosociology

Incontrovertibly, *applied astrosociology*—defined as the purposeful application of astrosociological knowledge to solve practical problems and other issues—will become more relevant in the future as humans begin to create a permanent presence in space. As I wrote in 2006, "*Applied astrosociology* is defined as the application of astrosociological (or astrosocial) knowledge to address practical problems or concerns related to astrosocial phenomena. In addition to solving space-related problems, then, applied astrosociologists employ the same knowledge gained for space research toward solving real issues related to astrosocial phenomena that exist in space or on the Earth."[26] This practical dimension of astrosociology becomes important because the knowledge gained through research possesses applications for both terrestrial and extraterrestrial issues.

The need to take into account the social and cultural ramifications of space exploration undoubtedly awaits humanity in the future. In fact, even now the need to understand the effects of space exploration on terrestrial societies has begun to receive an increased level of attention. Moreover, the planning of space settlements and other human space missions has begun to include astrosociological input. While the possibilities exceed our current imaginations, applied astrosociology will demonstrate its applicability to the areas listed here as a starting point.

Some of the relevant areas of applied astrosociology include long-duration space missions, social problems and their solutions, planetary defense, spacefaring societies, space settlements (i.e., "space societies"), implications of astrobiology and the Search for Extraterrestrial Intelligence (SETI), space policy and space law, astrosocial change, medical astrosociology, and interplanetary relations. Most of these areas receive attention later in this chapter. The field of astrosociology actually covers a much larger scope than listed here because it includes any type of social or cultural pattern that involves human interaction.[27] The application of astrosocial knowledge to solving a problem or meeting a goal makes this new field even more important due to the challenges humanity faces in the near future. Space has affected societies tremendously thus far in the history of human societies, but much more comprehensive, transformative effects lurk in the future. The Space Age has only begun.

---

26. Jim Pass, "Applied Astrosociology: The New Imperative to Protect Earth and Human Societies," (AIAA 2006-7511), (2006), 5, *http://arc.aiaa.org/doi/pdf/10.2514/6.2006-7511* (accessed 10 March 2012).

27. For a more comprehensive discussion of the scope of astrosociology and astrosociological specializations, see Jim Pass, "Viewpoint: Astrosociology As the Missing Perspective," *Astropolitics* 4, no. 1 (2006): 91.

Even this partial listing of topics related to applied astrosociology demonstrates that 1) a great many topics do in fact exist, 2) astrosociology will only become more relevant as space exploration advances, and 3) a single field is necessary to bring organization to all these seemingly disparate topics that normally remain isolated from one another within their social science disciplines. Without something like astrosociology to provide recognition and order to the relationship of space exploration and societal consequences, the importance of astrosocial phenomena to societies around the world will remain unclear to most and studied on an individual basis to a large extent. This long-standing status quo has served us poorly in the past. Ignorance is far from bliss when it comes to the effects of astrosocial phenomena. They affect us on a daily basis so we should understand them. Part of the development of astrosociology can occur by conducting research, presenting papers at conferences, and seeking publication of works. For the most part, its development has depended on these activities. However, these activities do have limitations. Astrosociology faces challenges within the relevant programs of educational institutions. Inroads in this general area remain untapped yet indispensable for long-term success.

Expanding Space into Social Science Classrooms

Space undoubtedly brings excitement into the classroom. Students participate more intensely in their schoolwork and become more likely to study longer during their time away from school. However, not every student is drawn to the natural sciences, math, or engineering. Many students become attracted to the social sciences and humanities instead.

> We must open ourselves up to more inclusive possibilities for the future. Rather than utilizing space exploration to stir the imaginations of only potential space scientists and engineers among our youth, we should also do so to motivate all potential scientists. As an additional step, we must utilize space exploration to encourage potential *astrosociologists* to follow a different, though related, path. In order to ensure the greatest, most comprehensive understanding of humanity's destiny in space, we must encourage students in the physical sciences and engineering disciplines, who serve as the usual targets. Moreover, we need to encourage those in the social and behavioral sciences, and the humanities, to become involved in the study of astrosocial phenomena in an astrobiological context.[28]

---

28. Jim Pass, "Astrosociological Implications of Astrobiology (Revisited)" (2010), 415, doi: 10.1063/1.3326269 (accessed 12 March 2012).

With the growing importance of studying astrosocial phenomena, it now seems like a good time to expand the scope of space in the classroom, to broaden the treatment of space phenomena so as to emphasize their social and cultural implications as well as their practical applications.

To prepare ourselves for the gargantuan task of conducting research on the impact of astrosocial phenomena, which remains inadequately studied, we must incorporate space into the social science classrooms of our schools at all levels. NASA's current approach on education focuses on the various STEM (science, technology, engineering, and mathematics) disciplines of natural and physical sciences, technology, engineering, and mathematics. However, it almost totally disregards the social sciences, specifically, astrosociology. We need to bring the social sciences into the current effort and reference the new orientation by changing the acronym to STEMA—the *A* refers to the field of astrosociology and a focus on astrosocial phenomena—to reflect a more inclusive approach to education involving outer space.[29] The *A* represents all of the social sciences and humanities. Members of the public generally support the dream of space exploration,[30] and the student population probably supports it even a bit more strongly. Many social science students will become more enthusiastic about their studies, as I found in an informal survey in three of my sociology classes in 2005–2006.[31] Even students who go into the social sciences outside of astrosociology will find space an interesting topic and become better-informed members of the public.

The social sciences have made innumerable discoveries regarding human behavior and social life within terrestrial societies, most of which have nothing to do with space. However, many research findings are universal enough to be applicable to space research and may serve as a complementary collection of insights outside the scope of the space community's everyday concerns. The means in which social structures function in the form of social groups of all sizes—from couples to organizations and from institutions to societies—provides invaluable data for groups going into space. In addition, research on isolated groups with military and other forms of command structures provides invaluable lessons for future groups in space. More general research findings involving social interaction in a variety of social settings provide untapped information for the space community.

---

29. Pass, "Enhancing Space Exploration by Adding Astrosociology to the STEM Model."
30. Howard E. McCurdy, *Space and the American Imagination* (Washington, DC: Smithsonian Institution Press, 1997).
31. Pass, "Enhancing Space Exploration by Adding Astrosociology to the STEM Model," 10–14.

With the space community increasingly cognizant of the need to include the social sciences even in the near term, it seems wise for NASA and other space agencies to bring the social sciences (that is, astrosociology) into the classroom in an active manner alongside the traditional STEM fields *or* at least advocate openly for their inclusion in the classroom. High-ranking NASA officials from the Education Division have acknowledged the need, and they have claimed that NASA does not discourage astrosociology-related subjects in the classroom. At the same time, NASA only promotes the STEM fields when, in reality, we need to train space scientists, engineers, technicians, mathematicians, and *astrosociologists* for space exploration in the near future. The discussions in the next section should make this statement abundantly clear. We must emphasize STEMA, rather than STEM, to meet the needs of the future generations of space explorers. Many current students at various academic levels would find inspiration in the introduction of space into their social science classes, just like those who gravitate toward the natural science and engineering disciplines. Many of them would feel uninspired to pursue academic careers without it, potentially contributing to the unacceptable dropout rate at many schools.

Only a formal collaboration between the space and social science communities can ensure a comprehensive understanding of the issues involving humanity's movement into space.[32] We should encourage all scientists, scholars, and the various technicians interested in space to study it regardless of their perspective or background.[33] The future demands a viable multidisciplinary field.[34] The status quo will not serve us as well into the future as it did during the initial half century of the Space Age. Admittedly, this proposition serves to encourage the development of the new field that I founded, but there are additional benefits at the societal, group, and individual levels of analysis that require application of the scientific method.

The justifications of the need to develop the field of astrosociology will become all the more unmistakable in the discussions that follow. Next, some of the important specializations of astrosociology receive brief attention. The purpose of the following section is to present the reader with an appreciation of the astrosociological approach as well as to demonstrate the need to

---

32. Jim Pass, "Astrosociology and the Space Community: Forging Collaboration for Better Understanding and Planning," *The Space Review*, 8 August 2005, *http://www.thespacereview.com/article/424/1* (accessed 18 March 2012).

33. Jim Pass, "Invitation to Astrosociology: A New Alternative for the Social Scientist-Space Enthusiast," in *Living in Space: Cultural and Social Dynamics, Opportunities, and Challenges in Permanent Space Habitats*, eds. Sherry Bell and Langdon Morris (ATWG [self-published], 2009), 71–89.

34. Pass, "Astrosociology and the Space Community."

develop this new field. In the process, scientists and scholars from both the social science and space communities will hopefully recognize the importance of astrosociology to their disciplines and the benefits of collaboration.

The Future of Exploration, Society, and Astrosociology

As humanity moves further into the twenty-first century, space exploration and its related activities will increasingly alter societies in both predictable and unfathomable ways. Its significance will grow in all facets of astroso-cial phenomena (for example, scientific inquiry and discovery, technological innovation, travel, tourism, settlement, and exploitation). Unless postindus-trial societies plummet into dystopias or otherwise experience compromised growth, astrosocial phenomena will become ubiquitous within societ-ies around the world. The large scope of potential topics falling under the purview of astrosociology dictates that few of them can receive attention in this chapter.

Nevertheless, contemplating the future with a focus on astrosocial phe-nomena, however brief and limited the treatment offered here, easily dem-onstrates the need to develop astrosociology by both the social science community and the space community. In fact, it seems incredible that it has taken this long to establish such a field. Astrosocial phenomena will only become more forceful in reshaping societies, and therefore we must begin to address these social forces and even attempt to take advantage of them. The latter is impossible, of course, without a proper understanding of the phenomena involved. And the rich tradition of understanding within the realm of the social sciences will serve to bolster our overall appreciation and understanding of astrosocial phenomena. The topics that follow include both theoretical and applied astrosociology, which normally possess an interactive relationship to one another.

Spaceflight Altered: Phasing Out the Traditional Crew Model

As Thomas Gangale[35] has pointed out quite clearly, the ability to continue with the crew model of spaceflight will prove impossible to continue as social groups and larger populations move farther and farther from Earth. Impracticability makes impossible the current protocol in which crewmem-bers follow along with items on a list with mission control, step-by-step, to fit the maximum number of activities into the mission. "For nearly half a

---

35. Thomas Gangale, "Practical Problems in Astrosociology," *Space* (AIAA 2006-7474) (2006), *http://arc.aiaa.org/doi/abs/10.2514/6.2006-7474* (accessed 10 March 2012).

century, nearly every minute of every flight has been scripted and directed by Mission Control," Gangale writes. "Although there has been some loosening up during long-duration, near-Earth orbit missions, sometimes forced by a revolt of the crew, this culture of control remains largely intact and well entrenched. Human spaceflight is expensive, and the culture of control is inevitably driven to squeeze every last drop of value from every minute of human labor in space."[36]

When distance along with social, cultural, and psychological impediments render instantaneous communications impossible, and as the delay of replying becomes increasingly burdensome, those living aboard spacecraft and in other space locales will find themselves on their own to a greater extent. They will select their own pace of work and social life in other areas. Earth-based authority will lessen as dictates from the home planet evolve from "mission control" to more of a "help-desk" or mission-assistance relationship.[37] Members of space missions and populations living in space "societies" (that is, settlements) will need to become autonomous over time as they face local problems and developments. Shorter-duration flights (though longer by contemporary standards) and two-way missions will involve these problems to a lesser extent.

What effect will this shift of authority have on the societal impact of spaceflight? The simple answer to this question is that the nature of the impact of space exploration will change.

Isolation due to an increased distance and duration of the mission contributes to unintended circumstances as past experiences on space stations have hinted. The effects of isolation on the crew behavior require understanding and planning. Because greater distance results in increased isolation, long-duration missions far from Earth will pose significant social-cultural and psychological challenges. The nearly full social control of space crews becomes unsustainable once their proximity to Mission Control makes instantaneous communication impossible. Crews become more autonomous under such circumstances. Controllers on Earth lose their authority as certain social conditions come into play. In addition, the duration of the mission alone produces important effects, even when the distance is not great, such as when orbiting Earth. The longer the mission, the more isolation results in a greater social cohesion of the group in space and more problems developing between the crew and the ground.

New norms (social rules of conduct) and values (ideas about what is important) develop. As time goes by, these new norms and values increasingly

---

36. Ibid., 3.
37. Ibid., 2.

favor group needs and desires over those of mission controllers. These social and cultural trends intensify if a long duration accompanies a great distance. One-way missions to settlements, for example, will almost certainly involve an intensification of this process of separation. Increasingly, crewmembers will make decisions for themselves and "inform" Earth-based officials of their choices. At some point, when members of space societies finally view themselves as autonomous on social, economic, political, and other dimensions, there is the potential that they will cease to inform their original controllers and seek to set up interplanetary relations with them. Although this last development seems like a far distant reality, now is the time to think about preparations for this outcome. Increasing isolation and duration of spaceflight will begin soon, even with today's level of technology, so we must think about the progression toward these "far-off" developments as points along a continuum rather than eventualities so distant that we do not need to worry about them. We should collect astrosociological data from the current point onward.

An important astrosocial force comes from the isolation of populations away from Earth. Social groups on Earth have a long history of becoming ethnocentric when forming their in-group identity and contrasting their membership from the out-group (that is, outsiders who are not members). Ethnocentrism (allegiance to one's group) develops and leads members of the isolated group to thinking of themselves as an independent entity to some extent independent of terrestrial authorities. Members begin to feel that their own group is superior to others, in this case, groups that reside on Earth. The original relationship with Earth will become more confrontational to the extent that ethnocentricity increases.

Applied astrosociologists can assist in the creation of social systems and their cultures for space travel and settlement before the mission begins. We can encourage social cohesion and ethnocentrism from the very beginning of training. In this way, we can make the survival of the crew or settlement socially sustainable. And realistically, we can construct institutions and groups to deal with the inevitable independence of the space-bound population. This step will probably prove difficult to accept because the affiliated group on the Earth will need to relinquish a great amount of social control to make it possible. Next, let us consider the example of a space settlement or "space society" that cannot operate utilizing the traditional crew model because of a new level of social-cultural complexity.

Phasing out of the traditional crew model will unquestionably become a necessity due to the practical limitations of communications and the other forces. The consequences of isolation will necessitate the planning and instituting of new organizational models. We must prepare for such occurrences from the initial planning stages. As Harrison has stated quite clearly, we need to expand human factors beyond the study of "biomedical adaptation and the human-machine interface, and to include personality, social relationships, and the broad political and organizational contexts for human activity in space. Increasing crew size, increasing diversity of crew composition, increasing mission duration, and burgeoning technology prompt us to rethink the human side of spaceflight."[38] In other words, we need to expand our astrosociological imaginations when it comes to spaceflight as well as other issues involving humans and space.

Constructing Space Societies

Many future problems will require the contributions of social scientists in collaboration with scientists, architects, and engineers within the space community. A good example involves the placement of human populations into enclosed, isolated habitats far from Earth. Although engineering solutions can construct a physical environment that can sustain the population on a biological basis, this capability cannot ensure the success of the settlement due to its inattention to the critical issues related to the social environment. Moreover, the interior of the habitat must encourage a healthy social life for the residents of the settlement. In general, then, as the complexity of space exploration increases in the future, the need for astrosociology will only increase as social-scientific issues become increasingly pertinent and recognizable, contributing both to problems and solutions to those problems.

Even today, NASA and other space agencies, as well as researchers in corporations and educational institutions, make plans for space settlements they target for the near future (between 2020 and 2040). Unfortunately, they usually do so without much regard for social-scientific aspects of their missions. The contributions by astrosociologists, while becoming more recognized, remain best characterized almost as afterthoughts. Special accommodation efforts on the part of space community scientists, engineers, space architects, and others must occur in order to allow astrosociologists to participate in the planning stages of space settlements (or space societies) and other projects.

We must view permanent space settlements as the standard—a goal to strive for even at present. To a noteworthy extent, at some point in the future human exploration will require moving into the solar system in a methodical and calculated manner, establishing a permanent presence along the way. Using this methodology, the explorers can remain in a particular part of the solar system and even move farther away from Earth, rather than coming

---

38. Albert A. Harrison, *Spacefaring: The Human Dimension* (Berkeley: University of California Press, 2001), p. 278.

all the way back to terrestrial soil and starting over again from scratch. Populations will exist in space either on a permanent or long-term basis, and thus the traditional crew model as described earlier cannot work. This will save time and resources, although it will also provide a great number of challenges that are biological, psychological, and social-cultural in nature.

The definition of space society remains a vital concept for the future of space travel, exploration, and settlement. This term implies that any social group sent into space for an extended period must be capable of sustaining itself on a variety of different dimensions that we take for granted within our own societies here on Earth. As I wrote in 2006, "For present purposes, *space society* is defined as a space colony/settlement in which members of the population 1) share a common culture, 2) live within a closed physical environment, and 3) cooperate with one another, social groups, and institutions in order to meet the social needs of all its citizens."[39] The social environment is just as vital for long-term survival as the physical environment.[40] Traditionally, planners concentrated on the physical structure and life-support systems of a settlement. Because the social sciences did not participate in the planning, the internal social environment was most commonly overlooked. Practically, however, we are sending human beings to live in these physical environments. We must consider what happens *after* they arrive at these physical environments, even if they allow for the biological survival of humans. Survival is not the standard we should use. We should think in terms of livability.

When thinking about space societies, one must think about the survival of the population as a social structure, whatever its size may be. Even a small group represents a microsociety because that social structure must provide for the needs of each of its members. The larger the size of the population, the more complex becomes the overall social structure and the more impersonal become the relationships. Social structures require construction lest we leave the members of these isolated groups of people to fend for themselves in the midst of inevitable chaos. Social structures (groups, organizations, institutions) provide regulation to social life and deal with social problems once they arise. Conflict can arise to an extent that people harm one another even while their physical environment keeps them safe on a biological level.

Thus the larger culture and its subcultures also require planning, just as the physical elements of the settlement. The larger culture serves to regulate social life so that the members of the settlement understand what is expected

---

39. Jim Pass, "The Astrosociology of Space Colonies: Or the Social Construction of Societies in Space," *AIP Conference Proceedings*, 813, no. 1 (2006), 1153–1161, doi: 10.1063/1.2169297, (accessed 10 March 2012).

40. Ibid., 7.

of them. The formal rules or expectations, or norms, must exist from the very beginning. Also, the informal rules—at least the important ones—require familiarity by settlers, including terms of their enforcement. These types of ideas must receive a great deal of attention in the selection of the population and the training of those chosen to live in the space society. On Earth, the socialization process normally requires a period of many years to acclimate the new citizen into the social milieu. In this situation it must occur over a very short period. In addition to learning the norms or rules of behavior, the new citizen of a space society must learn the important values. Social life becomes more organized and predictable when ideas and things valued by society receive widespread agreement. These elements of the culture serve to bind the citizens together and regulate social life.

Space societies represent one future element of space exploration and spaceflight. These types of manifestations of humanity's expansion into our solar system will also affect those of us living on Earth. Space societies are unlikely to be self-sufficient, at least the early ones, so economic trade will become necessary. Again, to the extent that these slivers of human civilization can survive away from the home planet, a new sort of planetary relations (think "foreign relations" between nations on Earth) will become necessary. The separation of human groups from terrestrial societies will also result in a high degree of ethnocentrism in which the space society citizens view themselves as independent.

At this point in our history, however, most current space missions and space science discoveries directly affect terrestrial societies as they have since the inception of the Space Age. We are orbiting Earth or utilizing scientific knowledge and technological innovation for our benefit here on our home planet. We collect scientific knowledge for its own sake and for application to terrestrial concerns. The technologies developed for the space program may find applications in other areas of social life such as medicine, law enforcement, public safety, or aviation. How does space exploration affect societies today? It does so in ways we tend to overlook. At the same time, it could do much more if social scientists worked together with space scientists, engineers, and architects to solve some of the most serious problems facing humanity on Earth. The consideration of social problems by the space community is extremely relevant today—the problems are serious issues and they affect even space scientists—and space can offer unique insights and solutions that may otherwise never receive consideration due to a lack of social-scientific input.

Utilizing Space to Solve Social Problems

Many critics of the space program point to the idea that terrestrial social problems should receive attention and even resolution before we "waste" federal dollars on a "useless" program such as one dedicated to space exploration. Contrary to this opinion, evidence that confirms the benefits of space exploration is readily available. Depending on the study, the U.S. space program returns from 7 to 10 times on its investment. Not many other federal programs can claim the same thing. This is not the point, however. From a social science perspective, the space program exists, it affects society, and thus it demands study! I refer to the discipline of sociology here, but the other social science disciplines regard space in much the same way. As a community, sociologists tend to fall into this category despite their mandate to study *all* social phenomena.[41]

As a subset of all existing phenomena, astrosocial phenomena never did receive their due attention even while the most exciting events witnessed by human beings, such as the Apollo landings on the Moon and the spectacular photographs sent by the Voyager probes, were taking place. One must ask a simple question in this regard: Why? In 2004, I offered one possible explanation for this confounding situation: "Why do astrosocial phenomena continue to exist for nearly 50 years since the advent of the Space Age without significant recognition or attention by sociologists? Based on personal communications with critics within the sociological community, one reason looms large as a vital contributor to this problem. Quite simply, human behavior related to space lacks legitimacy as a substantive area of scholarly treatment."[42] This situation is not reasonable, obviously, because astrosocial phenomena exist in the daily lives of ordinary citizens. [43] Most of these patterns of interaction may take place on Earth, but they still relate to space and more will take place in space as this century advances.

One reason that so many sociologists and social scientists in general seem reluctant to recognize astrosociology as a legitimate field relates to their mistaken assumption that it somehow relates to the pseudosciences due to its focus on outer space. "These critics argue that astrosociology must focus on

41. Peter L. Berger, *Invitation to Sociology: A Humanist Perspective* (New York: Anchor Books, Doubleday, 1963).

42. Jim Pass, "Space: Sociology's Forsaken Frontier" (2004), *http://astrosociology.org/Library/PDF/submissions/Space_Sociology%27s%20Forsaken%20Frontier.pdf* (accessed 12 March 2012).

43. Jim Pass, "The Potential of Sociology in the Space Age: Developing Astrosociology To Fill an Extraordinary Void," (2006), 3–4, *http://astrosociology.org/Library/PDF/submissions/Potential%20of%20Astrosociology.pdf* (accessed 11 March 2012).

the several pseudosciences that relate to space in some way including astrology as well as paranormal topics, alien detections on or near Earth, alien abductions, UFOs, crop circles, and cattle mutilations. Such assumptions are erroneous. These topics do not fall under the purview of the space sciences. Therefore, the controversies related to the perceived illegitimacy of these topics are not under review here given that such topics are not relevant to astrosociology."[44] Astrosociology faces a similar challenge that the Search for Extraterrestrial Intelligence (SETI) overcame in this general area in which critics disparage the legitimacy of the field based on an incorrect characterization of it.[45] In time, this unfounded view of astrosociology will fade into a positive, objective analysis of the field. Important reasons exist for this change of view, including practical reasons related to applied astrosociologists seeking to find long-term solutions to enduring social problems.

Astrosocial phenomena represent an important subject for reasons that involve much more than the resolution of problems related exclusively to space exploration. Their study can result in benefits that contribute to solving social problems on Earth. Besides assisting the developing nations that face staggering socioeconomic conditions, the problems plaguing human societies in general on a global basis require serious attention. Space-capable nations already cooperate with one another to conduct space missions. They could likewise use those same relationships to tackle the global social problems that threaten all societies.

Examples of social problems include explosive population growth, energy crises, global warming, pollution and other forms of environmental stress, destruction of habitats, global catastrophes (such as volcanic eruptions, tsunamis, and earthquakes), and impact by a cosmic body.[46] These represent social problems because they are harmful and they are either societally induced or affect the well-being of societies, or both. A pertinent subclassification of social problems is *astrosocial problems*, which includes those that are amenable to solution by space-based resources.[47] Thus, in the context of this discussion, I really refer to *astrosocial problems* when referring to *social problems*. Realistically, however, most social problems can be mitigated by the establishment of new astrosocial phenomena, as we have seen the unrealistic scenario in which space prisons can ameliorate crime on Earth in science-fiction literature and other forms of entertainment. In the distant future, space societies can assist by moving large numbers of people off the surface

---

44. Pass, "Space: Sociology's Forsaken Frontier," 22.
45. Harrison, "Overcoming the Image of Little Green Men."
46. Pass, "The Potential of Sociology in the Space Age."
47. Ibid.

of Earth for other reasons, such as the overpopulation on Earth. This migration pattern will also protect our species in case large-scale natural disasters or cosmic impacts result in the loss of large numbers of people.

Social problems exist within the social lives of individuals and as larger, stable patterns of social interaction. Likewise, the solutions to these problems will come from all aspects of social life including those related to astrosocial phenomena. Space-based solutions to terrestrial problems represent nothing new in the course of the Space Age. Technology transfers and spinoffs from spaceborne research find application in other societal institutions, such as medicine and other economic industries. The difference that astrosociology brings involves the insights brought to bear on these problems within their social and cultural contexts. Solutions will require a strong collaboration between astrosociologists and members of the space community to determine current conditions and social forces that produce harm, as well as finding socially/culturally viable (and acceptable) solutions.

Moreover, social problems (e.g., deviance) will follow human populations into space. We need to apply all methods at our disposal to solve them, whether they manifest themselves in space or in terrestrial environments. There is no doubt that space scientists and engineers can assist us on Earth to solve our problems. Conversely, astrosociologists can minimize these problems in space habitats to a much greater extent than would be the case without their involvement through their recognition and application of preventative solutions. Social scientists have studied these problems for hundreds of years within terrestrial societies. We would do well to put that acquired knowledge into practical use in outer space. Many of the problems that arise will involve relationships between people, rather than people with their machinery, so we need to deal with that eventuality.

Perhaps the greatest "social problem" that currently limits the participation of social scientists in the solution of the traditional problems traces back to their assessment of space as an unimportant area of concern and thus unworthy of funding for research. This perspective is ludicrous if one simply considers the potential of space-based solutions to such problems as understanding global warming and the ozone hole, mapping various resources using space-based radar, studying the weather, and accessing alternative resources for energy use and other needs. The successful development of astrosociology focuses on the connections between space and society, demonstrating the significance of astrosocial phenomena and their potential for solving some of the seemingly intractable problems of societies around the world. Individuals from social science disciplines increasingly defy the prevalent "wisdom" of their leaders by deciding to support astrosociology. The need exists for them to bring astrosociology into the classroom. To put it mildly, that is a good thing for the social science community and for the space community.

Astrosociological Contributions to Planetary Defense

There are social problems and then there are social problems. Planetary defense may be the ultimate social problem because the impact of an asteroid or comet fragment may place entire societies and even entire civilizations in grave peril. The human species and individuals are in danger, of course, but so are societies and their cultures. In this case, a space phenomenon becomes relevant to societies due to the very real threat of the survival of human beings along with their social systems and cultures. The field of astrosociology becomes relevant in this regard in terms of identifying this problem in a new context for those within the space community. In addition, it becomes relevant because of its contributions for reducing the potential consequences of this problem in collaboration with space-community experts. The problems posed involve astrosocial phenomena both in the sense of human beings recognizing the problem and societies (not just people) trying to devise solutions to protect themselves. Solutions will require collaboration between experts in the social science and space communities that include making tradeoffs between optimal protection schemes and practical implementations based on complex socioeconomic variables.

Although the general likelihood of an asteroid striking Earth that is capable of destroying societies and threatening the human species is not high at any given moment, the likelihood of a strike by a society-killer or even a civilization-killer is approximately 1 in 20,000 (based on an estimate by David Morrison).[48] This is similar to the odds that an individual has of dying in a plane crash—more likely than dying in a tornado or from a bite or sting from an insect, but less likely than dying in a car crash or from electrocution.[49] If a nation were devastated by a cosmic impact during the next two generations, it would certainly represent a stroke of bad luck. However, that it will occur at some point in the future is a fact.

If the threat is indeed real and inevitable, even if we cannot predict when it will occur, then we have an obligation to address the problem. As I wrote in 2006:

---

48. David Morrison as cited in Jim Pass, "Applied Astrosociology: The New Imperative to Protect Earth and Human Societies," *American Institute of Aeronautics and Astronautics Space Conference & Exposition Proceedings* (AIAA 2006-7511) (2006), 3, doi: 10.2514/6.2006-7511 (accessed 14 March 2012).

49. Ibid.

The proposal offered here will change societies around the world in ways unseen in the past. This new strategy represents an *ideal type* (i.e., the extreme form of implementation). The ideal type represents the best system humanity can produce to detect, defend, and survive any threat by any object regardless of reaction time or cost to implement. Meeting the requirements of ideal type would be impossibly expensive and require an outrageously large infrastructure, so we must decide as individual societies and as a species how close we wish to approximate this model of the "perfect" planetary defense system.[50]

The perfect plan—one that would protect all of a society's social and cultural assets, including vital ideas and physical structures—could never become reality. Its exorbitant cost and cutting-edge technology to implement it would prove prohibitive. Anything built represents some sort of compromise from the ideal type, something that will fall far short of the perfect system.

I proposed a three-stage plan of protection that includes 1) detection; 2) defense, and, if all else fails; 3) survival.[51] For the most part, the space community addresses the first two components of the plan while the third component rarely receives its due attention. We tend to strive for a perfect detection system coupled with a perfect defensive system, although detection schemes receive most of the attention. When the experts realize that anything rivaling perfection is impossible, they rarely take the next step of planning for the survival of the social and cultural structures that shape our sense of social reality. The astrosociological approach will hopefully change this mainstream approach in line with a greater concentration on the survival component of the overall defense system. This includes the reactions of populations to the announced threat of a cosmic impact.[52] Government agencies must prepare themselves.

The comprehensive, three-prong strategy that I propose combines the space community and the astrosociology community in a way that becomes mutually beneficial for each one as well as for the overall project. If we committed to a global strategy, ethical concerns (a specialty of social scientists) would inevitably arise. For example, what types of contributions would developing nations make if they could not provide financing? Perhaps they could provide land, labor, or other resources if pieces of the system needed placing within

---

50. Pass, "Applied Astrosociology," 7.
51. Ibid., 8.
52. Albert A. Harrison, "Psychological Factors Influencing Responses to Major Near Earth Object Impacts," *2004 Planetary Defense Conference: Protecting Earth from Asteroids* (AIAA 2004-1462) doi: 10.2514/6.2004-1462.

their borders.[53] If a developing nation were the target of a rogue space object whose impact had no global repercussions, would developed nations spend the resources to attempt an interception? Astrosociologists can bring insights and research methods that enhance the planning of a planetary defense system in an additive way to the normal concerns of space professionals.

We must come to terms with the fact that we are a fragile species living on a fragile planet. This fragility extends to our societies and their cultures. With this in mind, it becomes rather obvious that we need to protect our species and our planet; but we must also protect our societies and our cultures to the best of our abilities. The expertise in the areas of social systems, cultures, and human behavior—as related to space—falls under the purview of astrosociology. Acknowledgment of the value of the social science perspective to this field of study does indeed occur from time to time.[54] The problem lies in the formal cooperation with social scientists on an ongoing basis. Astrosociologists become valuable partners when uniting the survival component with the other two. Together with experts in the space community, astrosociologists specializing in this area can better understand the problem and thereby develop practical solutions in a cooperative manner. Passing on the problem to future generations is arguably irresponsible, considering we possess the science, knowledge, and technology to provide at least partial solutions today.

Transformation into Spacefaring Societies[55]

Social change is probably the most fundamental law of both sociology and astrosociology. We cannot always predict the course of the change, but we can state emphatically that change will occur as time passes. We can try to make predictions and adjustments, however, based on what happened in the past compared with how things operate today. One prediction that makes sense in the midst of accelerating social change based largely on innovations in science and technology concerns the role of space exploration. Are we moving toward a spacefaring society in the United States or something else? Failed social systems have shown us that although progress is common, it is never assured.[56]

---

53. Pass, "Applied Astrosociology."

54. Ibid.

55. For a general discussion, see Jim Pass and Albert A. Harrison, "Shifting from Airports to Spaceports: An Astrosociological Model of Social Change toward Spacefaring Societies," *American Institute of Aeronautics and Astronautics Space Conference & Exposition Proceedings* (AIAA 2007-6067) (2007), doi: 10.2514/6.2007-6067 (accessed 15 March 2012).

56. Jim Pass, "An Astrosociological Perspective on Space-Capable vs. Spacefaring Societies," *Physics Procedia* 20 (2011): 369–384, doi: 10.1016/j.phpro.2011.08.033.

Nevertheless, one could easily argue that the movement toward a spacefaring society is exceptionally plausible given the trends of the twentieth century. It is an argument that finds support here.

The United States and the few other nations capable of spaceflight on their own accord do not exemplify spacefaring societies as many laypersons, dictionaries, and space professionals commonly define it. This characterization only implies that they are examples of what I have termed *space-capable societies*, those that possess at least the capacity to reach low-Earth orbit.[57] The mere ability to reach space successfully represents a minimum requirement— merely a starting point: "All social scientists [i.e., astrosociologists] should reserve the term *spacefaring society* for properly characterizing the transformation of an entire social system that reaches a threshold in which a specific set of social and cultural characteristics exist."[58] In other words, a spacefaring society would represent a new set of socioeconomic conditions in which contemporary developed nations change from postindustrial organization to one significantly and closely intertwined with space.

Thus astrosocial phenomena affect a spacefaring society in all aspects of its functioning. That is, all social institutions incorporate space into their daily operations. Even foreign policy involves space-based cooperative and conflict negotiations. Space policy comes to the forefront:

> A unique set of social conditions typifies a spacefaring society. Every major institution is highly involved in some way with carrying out space policy as a high priority, and thus space law is well developed. A space-based economy flourishes, for example. Astrosocial phenomena are highly pervasive and vital for the society's survival. Space issues are intertwined in a multitude of ways into the everyday social interactions taking place in subcultures, social groups, organizations, and institutions. The larger culture reflects the importance of astrosocial phenomena through their incorporation as highly important values, strong norms protecting them, and their omnipresence in a space-dominated material culture.[59]

Again, while there is no guarantee for the long-term transformation into a spacefaring society, it is arguably a likely scenario unless dystopian social conditions intervene. One should be careful not to view spacefaring societies as utopias, either. While social conditions will improve, significant social problems will likely remain with solutions continuing as works in progress.

---

57. Pass, "Inaugural Essay, Part Two," 2.
58. Ibid., 15.
59. Pass, "Inaugural Essay, Part Two," 17–18.

One social pattern that seems rather obvious (and referenced earlier in this chapter) involves a new reliance on space for successful commerce. The entire economic system would involve space-based activities closely related to terrestrial activities. Astrosocial phenomena would be ubiquitous in everyday social life. The mining of asteroids for use on Earth (but also for space societies) is just one example of this.[60] Early on, a space tourism industry could conceivably develop into a large industry that helps establish the path toward a spacefaring future. The privatization of space will be critical, as will the guidance of government regulations that attempt to create a stable regulatory environment and a fair "marketplace" that serve at least some of the interests of the nation. Private industry and government (including NASA) will need to improve their working relationships even more strongly, as each possesses different, though complementary, characteristics and capabilities. Space advocacy groups will need to recognize this and develop more mainstream personas within the population of any particular society moving along the path toward a spacefaring future.

Societies in the Space Age continue to change partly due to astrosocial forces. From the time that the Soviet Union employed societal resources to achieve sending Sputnik into space in 1957, the scientific and technological innovations that developed were adapted to solving problems on Earth, including the improvement of the standard of living for space-capable nations. Efforts to send mammals and then a human into space in 1961 increased the transfer of space-based innovations to terrestrial uses. The history of the Space Age is replete with examples of technology transfers and spinoffs resulting in products and processes that improved the social conditions of societies around the world. The very movement toward a spacefaring society promises to continue and probably accelerate this socioeconomic pattern.

Astrosociologists represent the proper scientists to study these types of emerging patterns. Space scientists and engineers focus on issues that have traditionally ignored astrosocial phenomena, as they should, given the purpose of their work and the history of separation from the social science community. Even so, this does not imply that astrosociological issues are irrelevant to the space community, and certainly not to the societies in which their work occurs. This fact demonstrates only that different scientists concentrate on their own traditional areas of study, and relationships between the social and natural sciences tend to remain separated from one another.

Aerospace workers and NASA employees conduct activities that fall into the category of astrosocial phenomena, another demonstration of the

---

60. For a full discussion, see John S. Lewis, *Mining the Sky: Untold Riches from the Asteroids, Comets, and Planets* (New York: Basic Books, 1996).

relationship between members of the space community and astrosociology. As such, both greater understanding and planning result from a formal collaborative structure between space researchers from the social sciences and those from the natural sciences—that is, between those within the astrosociology community and the space community as we currently recognize them. It is indisputable that space exploration affects societies on Earth. It is also indisputable that social scientists possess the expertise in the area of human behavior. Space scientists, engineers, and architects possess knowledge exclusive to the space community. Thus, each community provides insights unique to its distinctive perspective. By bringing the two together, we achieve insights never attainable within one community or the other. The development of astrosociology within the social sciences and as a collaborative field with the space community can finally provide realization of this potential.

Although this wisdom is not new, it deserves mention in this context. We must expand to explore the unknown as a society, even as a civilization and ultimately as a species, or we shall meet the same fate as past societies that failed to do so. We will perish from Earth and make no human impact on the universe beyond our home planet. Students of history know this, and sociologists and other social scientists know this, yet the bulk of our population may not take it to heart; and most politicians—who control the resources and set the tone through the regulatory environment they create—may lack the long-term vision necessary to move in a direction consistent with the development of a spacefaring society. Astrosociologists can provide policymakers with historical data about past failures and informed advice about how to proceed in the future as space exploration becomes more and more entrenched in everyday social life.

Astrosociological Ties to Astrobiology and SETI

The contemporary search for extraterrestrial (nonhuman) life reflects the greatest level of sophistication yet of an investigation into the one question that has generated great curiosity and speculation within human minds throughout history: Are we alone in the universe? Never before have human beings delved into an organized search for life elsewhere in our galaxy with the current unprecedented utilization of science and technology. On an objective basis, either a positive or negative answer is monumental, although it seems that ruling out extraterrestrial life is an impossible task. The actual discovery of extraterrestrial life, on the other hand, would hit humanity over the head like an asteroid plummeting to Earth.

What does astrosociology have to do with astrobiology and SETI? Probably the main aspect of this relationship deals with the fact that the search for nonhuman extraterrestrial life in the cosmos, whether successful or not, possesses social and cultural implications—and that societal effects are too rarely investigated.[61] Although social scientists have made important strides in this area, the nature of the exploration of this relationship will become more rigorous with the development of astrosociology. Assuming that astrosociology attains even a glimmer of the popularity of astrobiology, a greater understanding of this relationship should develop. The outcomes of these studies that combine the two fields will present us with new benefits. A brief discussion of some key issues should suffice to demonstrate several of the interconnections between astrobiology and astrosociology. The social-scientific dimension has much to contribute; much more than it was able to offer during the first 50 years of the Space Age—largely because humanity is posed to increase its presence in space in the twenty-first century.

Starting in the twentieth century, hope has increased regarding the likelihood of the existence of extraterrestrial life.[62] And within the past decade or so, the prospect for detection of extraterrestrial life has grown even more dramatically. In Earth's most inhospitable environments, we recently began to find a surprising characteristic of terrestrial life that challenges previous assumptions. Life on Earth seems much more determined to survive than previously assumed, as we find it thriving in environments once thought to be inhospitable to living organisms: "Terrestrial *extremophiles* or organisms adapted to live in extreme conditions demonstrate that life is far more tenacious than previously believed. They include tubeworms that represent life not dependent on the Sun's energy, discovered deep on the ocean floor," and others that live in caves and other extreme terrestrial biosphere niches.[63] Investigators have also discovered life in other environments. We now know that hearty organisms exist in harsh places that include the rods of nuclear power plants, acidic lakes, and boiling hot springs.[64] These extremophiles provide researchers with a greater expectation that extraterrestrial organisms could demonstrate the same tenacity. Is our galaxy devoid of any type of life? Many astrobiologists believe it unlikely with the discovery of extremophiles as

61. Pass, "Astrosociological Implications of Astrobiology (Revisited)"; Steven J. Dick, "Cultural Aspects of Astrobiology," in *When SETI Succeeds: The Impact of High-Information Contact*, ed. Allen Tough,(Bellevue, WA: Foundation for the Future, 2000), pp. 145–152; Albert A. Harrison, *After Contact: The Human Response to Extraterrestrial Life* (New York: Perseus Publishing, 1997).

62. See Steven J. Dick, *The Biological Universe: The Twentieth Century Extraterrestrial Life Debate and the Limits of Science* (Cambridge: Cambridge University Press, 1996); Pass, "Astrosociological Implications of Astrobiology (Revisited)."

63. Pass, "Astrosociological Implications of Astrobiology (Revisited)."

64. Pass, "Astrosociological Implications of Astrobiology (Revisited)."

well as amino acids and various organic compounds floating in nebulae and other environments. But the absence of direct proof still prevails.

SETI, now considered part of astrobiology, concentrates on the loftier goal of detecting *intelligent* extraterrestrial life. This brings up related questions to the one posited earlier: Are we the only intelligent beings in the Milky Way galaxy? Even more profoundly, are we the only intelligent beings in the entire universe? The growing number of extrasolar planets, most notably "super Earths," discovered by astronomers and cosmologists, provides greater hope for the existence of beings who can build civilizations that produce a significant level of science and technology capable of creating radio signals, including television and radio broadcasts. If such civilizations exist, the likelihood is that they will possess social, scientific, and technological characteristics far advanced of us.

The search itself affects human beings, societies, science, and technology in positive ways. Allen Tough,[65] who coined the modern usage of the term *astrosociology* and inspired me to begin development of the new field, wrote about the positive social consequences of SETI in 1995. Among other things, he pointed out that SETI causes humanity to contemplate a host of significant questions:

> For some people both inside and outside the SETI community, the concrete search activities trigger thought and discussion of the following questions. Who is out there? What are they like, what are their fundamental values and priorities, where are they heading, what do they know about us, and what sorts of detectable communications might they be using? What role will they play in our long-term future? Might some extraterrestrials be so alien, so deeply weird, that we cannot even imagine their thought patterns, communications, and behavior? Where is cosmic evolution heading, and where is human civilization heading?[66]

How humanity deals with questions concerning extraterrestrial life reflects back on humanity. Therefore, this is a serious area of scientific investigation and research (from both an astrobiological and astrosociological perspective) and it is unrelated to a search for "little green men."[67] It addresses humanity's relationship to the rest of the universe and various other astrosociological

---

65. Allen Tough, "Positive Consequences of SETI Before Detection," preprint of paper IAA-95-IAA.9.2.06 for the 46th International Astronautical Congress, October 1995, Oslo, Norway, *http://www.astrosociology.org/Library/PDF/Positive%20Consequences%20of%20SETI%20Before%20Detection.pdf* (accessed 18 March 2012).

66. Ibid., 2–3.

67. Harrison, "Overcoming the Image of Little Green Men."

issues such as whether a particular society's attention is directed inward or toward the cosmos. An inward outlook results in a greater focus on terrestrial social problems approached with terrestrial solutions, for example, and a reduced focus on space activities. The decreased impact of astrosocial phenomena that occurs under this scenario tends to slow progress toward transforming a space-capable society into a spacefaring society.

The connection between astrobiology and astrosociology relates most fundamentally to the fact that the very exercise of conducting astrobiological and SETI research is fundamentally a category of human behavior, an important part of the astrosociology's purview.

> As soon as we begin to discuss astrobiological issues, including SETI, we inevitably begin to consider social and cultural issues. The search for life of any variety inevitably creates repercussions for societies and their citizens. Space scientists (e.g., astrobiologists, planetary geologists, even aerospace engineers) discuss astrosociological issues when attempting to explain (or justify) their efforts to the public. We can see it in many of the documentaries focusing on alien life … aired on television and in the other media. Thus, astrobiological issues possess fundamental, and thus unavoidable, astrosociological implications. The common questions bring this home. First, are we alone in the universe? Where do we fit into the big cosmological picture? Is it important for us to seek answers in an organized way? What happens to our societies if we discover extraterrestrial intelligence? Alternatively, what happens even if we "only" discover microbial life? The two subdisciplines are intertwined. Cultural and social considerations from a [astro]sociological perspective are imperative additions to considerations currently discussed by those from the [space community]. Again, it is the development of a multidisciplinary approach that provides [the] greatest potential for achieving our greatest level of understanding.[68]

Regarding the last point, along with the other areas of specialization, then, this area of research should involve a strong collaboration between astrobiologists and SETI researchers from within the space community and those astrosociologists who specialize in these areas. Astrobiology and SETI will likely become a major specialization within the field of astrosociology for social scientists, as many already study these issues as individual investigators.

---

68. Jim Pass, "The Sociology of SETI: An Astrosociological Perspective" (2005), 16–17, *http://www.astrosociology.org/Library/PDF/submissions/Sociology%20of%20SETI.pdf* (accessed 18 March 2012).

Detection of extraterrestrial life would alter the human condition as we know it. The human reaction would vary by culture and more finely by sub-culture. The leaders of societies in their various capacities must be careful not to underestimate the impact of positive detection:

> There are important questions about the days, weeks, and months after detection, then the intervening years as we get used to the idea that we are not alone, and then the long-term consequences, which will mirror in their significance the discoveries of Galileo, Newton, and Darwin. These questions embrace most fields of human endeavor. They are of obvious import for science. However, the point of this meeting is to examine the broader societal issues, and so focus on human behavior. Key areas are anthropology, sociology, and individual, group, and social psychology. Reactions will vary according to the social attributes of individuals, and the social, economic, and political contexts within which the discovery has occurred. Other important questions are the history of analogous events in our past; political, institutional, international, governmental and legal affairs; the effects on different organized and diffuse religions; the media; and education. There are broad cultural and ethical issues.[69]

The implications quoted here demonstrate that many social scientists actually conducted what amounts to astrosociological research before the establishment of the field—and the same is true for the other specializations of astrosociology. Furthermore, it points to at least a certain level of interest by NASA in astrosociological issues, as it sponsored this conference held at Ames Research Center back in 1999.

Authorities will need to prepare themselves for outbursts of hysteria, joy, violence, and a host of different emotional reactions. Religious groups, as well as other types of groups, will vary in terms of how they react and whether they even accept the findings that indicate extraterrestrial life.[70] Detection may even accelerate the growth of space programs of various societies. It may unite humanity a bit more closely than the past. Conversely, it may have the opposite effect if some nations reject the findings and separate themselves

---

69. Albert A. Harrison and Kathleen Connell, eds., *Workshop on the Societal Implications of Astrobiology: Final Report* (Mountain View, CA: Ames Research Center, NASA Technical Memorandum, 1999), 47, *http://astrobiology.arc.nasa.gov/workshops/societal/societal_report.pdf* (accessed 18 March 2012).

70. See Douglas A. Vakoch, "Roman Catholic Views of Extraterrestrial Intelligence: Anticipating the Future by Examining the Past," in *When SETI Succeeds*, ed. Allen Tough (Bellevue, WA: Foundation for the Future, 2000, pp. 165–174; Douglas A. Vakoch and Y. S. Lee, "Reactions to Receipt of a Message from Extraterrestrial Intelligence: A Cross-Cultural Empirical Study," *Acta Astronautica* 46, no. 10–12 (2000): 737–744.

from nations that accept the existence of extraterrestrial life. We can expect a heightened level of social conflict and other reactions that require study. "Astrosociology can ... help us understand how various agencies and organizations (alone or in combination) are likely to react to discovery."[71] Another area of interest relates to how we respond to a message from an extraterrestrial intelligence. What type of responding message(s) do we send back? Do we respond at all? What happens if some entities (including nations) want to respond and others do not? Each of these issues requires continuing theorization and investigation by social scientists—that is, by astrosociologists and collaborating organizations and individual researchers. As with the other areas of specialization, collaboration with space scientists and engineers will undoubtedly result in innovative ideas and approaches to the search.

The relationship between astrobiology and astrosociology exists because the social interactions carried out by humans to accomplish the search for extraterrestrial life are based on a set of organized social structures coupled with a subculture focused on seeking out life in the universe. Even without an announcement of success forthcoming in the near future, and even without consideration of the implications if such an announcement became a reality, the very attempt to seek out life in an organized manner merits the attention of astrosociologists from a number of disciplines, including sociology, psychology, anthropology, and history. If this is the case, astrosociology must investigate this behavior along with the implications of long-term failure and, of course, success. The social and cultural implications of this work make it too important to ignore. In fact, it is imperative that astrosociologists participate alongside their space-community counterparts to attain comprehensive knowledge; both for its own sake and for practical application should some type of reaction prove necessary.

Conclusions

The astrosociological perspective brings the social sciences into the Space Age by fostering the creation and development of a field dedicated to the study of the impact of space exploration. That is, astrosociology precisely focuses on the social, cultural, and behavioral effects of space exploration (that is, astrosocial phenomena). Created in 2004, it continues today to investigate the issues related to forms of astrosocial phenomena neglected in the past. Although we are far behind where we should be in understanding these issues, we have nonetheless begun to address them in a formal and organized

---

71. Albert A. Harrison, "The Search for Extraterrestrial Intelligence: Astrosociology and Cultural Aspects," *Astropolitics* (Special Issue: Astrosociology) 9, no. 1 (2011), 74.

manner—an approach that is more than 50 years overdue. And while no one advocates that all social scientists and humanities scholars who focus on space issues should become astrosociologists, those who do take this new path will benefit by the coordination of effort and recognition of their work, not to mention the satisfaction of conducting pioneering work.

The seemingly coincidental events of 1) bridging the great divide (collaboration), and 2) developing astrosociology demonstrate that the "societal impact of spaceflight and space exploration" represents an extremely timely topic for scientists interested in outer space from within both the space and social science communities. Applied astrosociology will prove itself invaluable as space missions begin to involve large crews, long-duration spaceflights, and isolated space environments (e.g., space societies). Social problems on Earth and in space will experience at least partial amelioration as well. Theoretical astrosociology will posit the roots of social change produced by astrosocial forces related to areas of social life such as space law[72] and ideas in culture. (Regarding the latter, a good example involves to the connections between the Apollo Moon landings and folk beliefs that were common in a number of different subcultures around the world.)[73] Because tangible benefits, or simply the knowledge gained, exist on Earth as well as in space, astrosociology is relevant to humans in both past and contemporary societies. Before there was spaceflight, there was astronomy that produced astrosocial phenomena, which in turn impacted on ancient cultures.

Although the great divide between the social sciences and the space community spans nearly the entire length of the Space Age, this has changed in recent years. The permanent development of astrosociology creates a recognizable and relevant field for space scientists, engineers, academic researchers, space advocates, and the public. Astrosociology can serve to bind the social science and space communities together by providing a new field that involves the fundamental elements of both camps. A real need exists for the creation of a permanent collaborative structure between the two so that space scientists, engineers, and other space professionals can take advantage of astrosociological knowledge and astrosociologists can take advantage of space-related knowledge. Collaboration allows each side of the divide to increase its level of knowledge about astrosocial phenomena—the interrelationship between space phenomena and social/cultural phenomena.

---

72. Christopher M. Hearsey, "The Nexus Between Law and Astrosociology," *Astropolitics* (Special Issue: Astrosociology) 9, no. 1 (2011): 28–38.

73. For example, see Virgiliu Pop, "Space Exploration and Folk Beliefs on Climate Change," *Astropolitics* (Special Issue: Astrosociology) 9, no. 1 (2011): 50–62.

An added benefit for the space community relates to the movement of astrosociologists into the space community as planners alongside their space-community counterparts so that social-scientific knowledge becomes available for practical application to missions and projects. With the space agencies and professional associations within the space community finally acknowledging the importance of astrosociological issues for themselves as well as for societies around the world, the time to develop astrosociology within both the space and social science communities now makes sense for everyone. Astrosociology can bridge the two communities by emphasizing the integration of space science and social science in the study and planning of future missions.[74] Astrosociology can serve as a uniting force resulting in a formal collaborative structure between the natural sciences and social sciences in the particular area of space research.

The societal impact of spaceflight and space exploration has increased as the Space Age has continued, with a surge of influences during the Apollo era and the two preparatory programs preceding it. Before that, humans contemplated their place in the universe and the possibility of life beyond Earth. These ideas have affected cultural development in past societies, and remnants still exist in contemporary cultures around the world. Today, current programs such as the International Space Station (ISS) and the Cassini mission continue to influence social change. Having celebrated the 50th anniversary of the Space Age, the influences of space exploration will only increase if we continue to follow a similar pattern into the distant future. A radical change in the future involves adding the social sciences to mainstream space research, which represent the other side of the coin known as "space exploration." Future endeavors will receive advantages from the greater level of comprehensiveness gained through the cooperation between the two major scientific branches.

Unless we plan to send only robots to Mars for the unforeseeable future, we must assume that human exploration will occur, including the need for human beings to live there. That is, assuming that private spaceflight increases and NASA eventually intends to send human beings beyond low-Earth orbit to the Moon, Mars, and beyond, then we must initiate a scheme looking seriously at astrosociological issues in space, not to mention doing something about our continuing ignorance of how space exploration affects societies existing on Earth. Social science research requires the participation of trained social scientists (including astrosociologists) rather than continuing to rely on

---

74. For more details, see Jim Pass, "Developing Astrosociology for the Space Sciences"(2006), *http://astrosociology.org/Library/PDF/submissions/Developing%20Astrosociology.pdf* (accessed 18 March 2012).

speculative declarations unsupported by proper empirical investigation. To address this need, we need to train astrosociologists.

Astrosociology exists precisely to study the societal impact of space exploration and spaceflight. No doubt should exist about the relevance of this new multidisciplinary field despite arguments to the contrary. What critics usually fail to appreciate is the fact that the societal impact of space exploration and spaceflight currently affects terrestrial societies most strongly. It is true that the ISS houses an ongoing crew of six or fewer for a sustained period, and the average stay of any particular crew member lasts only six months. However, the experience and data acquired onboard the ISS have produced invaluable astrosocial knowledge for human societies on Earth in addition to that for future space missions. Moreover, the space program in the United States has contributed in innumerable ways to the character of social life, including the high standard of living and quality of life. It has contributed to the accelerated pace of social change and will most probably continue to do so on an even more pronounced basis. Those involved in projects for the space program work on the planet Earth and contribute significantly to the economy and other societal dimensions as posited in this chapter.

Even 50 years represents only a tiny blip in the existence of humanity. Nevertheless, we find ourselves close to the point in which the procedures that made past missions possible, including the Moon landing, can no longer do so for missions beyond our nearest neighbor. We will need to adapt, and we should employ astrosociologists to assist in the planning and evaluations of mission details. All the while, our societies will likely continue to change in new and unexpected ways that bring us closer to a spacefaring societal structure. How do we know when that point in our history arrives? What are its important characteristics? What does this mean for humanity? Answers to such questions will come from astrosociologists.

When space exploration increases to the point at which human beings live in space on a more sustained or permanent basis, terrestrial societies will experience the effects—but so will human space societies. Space travel and exploration will become normal aspects of social life carried out by Earthlings living in multiple locations. Earth societies will take on increasingly spacefaring characteristics. Moreover, an additional element will develop. Space societies on planetary bodies and within space-based habitats will denote the spread of humanity beyond Earth. These human beings may regard themselves as a type of human being, although not necessarily Earthlings. They will probably move along new evolutionary paths due to differences in gravity, radiation, other conditions, and even genetic manipulation.[75] Humans may come to regard themselves as Martians, for example, and thus evaluate themselves as something quite different.

As implied throughout this overall discussion, the societal impacts of spaceflight and space exploration will most likely increase as the twenty-first century advances. Human populations will find themselves increasingly distant, and thus isolated, from the bulk of humanity living on Earth. Logically, it seems that we should prepare for this eventuality by learning the greatest number of lessons possible before venturing too far into the cosmos.[76] Critics of spaceflight often argue that these types of ventures are unsustainable and too dangerous for humans. We should determine early on if this is correct or discover the limitations of human biology if living in space is possible. In doing so, we can apply those lessons to new social environments beyond Earth. More important for humanity, these lessons will also find application in social life in terrestrial societies. As we move toward spacefaring societies and a potential single spacefaring human civilization, the knowledge accumulated in this area will prove invaluable.

In large measure, the message of this chapter relates to the proposition that spaceflight and space exploration already impact positively on terrestrial human societies, and furthermore that these sorts of impacts, along with those of new categories, will impact humanity throughout the solar system in the years to come. This list provides a partial summary of the benefits of space exploration from an astrosociological perspective. Through space exploration, we investigate (among other things):

1. the structure of the universe and other matters via the space sciences;
2. ways of improving technologies to study space phenomena, leading to social change that improves social conditions on Earth;
3. better methods of understanding Earth and thereby formulating better solutions for protecting it into the future;
4. new ways to ameliorate seemingly intractable social problems of all types;
5. how to utilize space technologies for societal applications (i.e., technology transfers and spinoffs);

---

75. For example, see Dawn L. Strongin and E. K. Reese, "Earthlings on Mars: The Physiological Psychology of Cultural Change, in *Living in Space: Cultural Dynamics, Opportunities, and Challenges in Permanent Space Habitats*, eds. Sherry Bell and Langdon Morris (ATWG: [self-published], 2009).

76. Jim Pass, "Moon Bases As Initial 'Space Society' Trials: Utilizing Astrosociology to Make Space Settlements Livable," *Space Technology and Applications International Forum (STAIF) Conference Proceedings* 880, no. 1 (2007): 806–813, *http://astrosociology.org/Library/PDF/ STAIF2007_Moon%20Base.pdf* (accessed 19 March 2012).

6. how human beings can reshape space environments through the construction of both physical environments and social environments that characterize space societies;

7. the future of our terrestrial societies (including movement toward a spacefaring future);

8. the possibility of extraterrestrial life (both simple and intelligent) beyond Earth;

9. and, ultimately, how well humans function biologically and socially beyond Earth.

The lessons we learn in these areas of investigation, as the last item implies, teach us in the end about who we are as members of particular societies and as members of the humanity. Are we destined to remain on Earth without exploring space? What if the conditions on our planet deteriorate and finally fail to sustain us?

So why should contemporary societies allocate serious resources to explore outer space? Perhaps the most important reason from an astrosociological perspective relates to the idea that the exploration of space in all of its manifestations produces benefits for societies that invest in such ventures. In contrast, societies that have forsaken expansion in some form of exploration eventually find themselves on the decline and possibly on their way to extinction. Space exploration shapes space-capable societies more strongly because of its impact on their cultures, including national pride and excitement generated by the exploration itself. But the impact of space exploration also positively affects humanity as a whole. Even citizens of developing countries ultimately benefit in numerous ways. Space exploration affects all of humanity due to a multitude of different forces it creates, including technological innovations, scientific discoveries, inspirational mission successes, and the potential to bring together nations in cooperation despite disagreements in other areas of social life. It provides developing states with the opportunity to join the world of nations in a way that would prove impossible on their own. The very existence of astrosocial phenomena means that social scientists need to study them. Beyond gaining new knowledge, there are also practical benefits. With the assistance of astrosociologists, it becomes possible to identify problems, devise solutions, and implement them more readily, and to share them among all of humanity's nation-states more quickly and justly.

Humanity has only begun to experience the positive effects of space exploration. Even so, our understanding of astrosocial phenomena continues to lag far behind the impact of their contemporary and historical influences due to the historical absence of the field now known as astrosociology. While we must therefore continue to pursue space exploration for all the reasons provided in this chapter and elsewhere; we must also expand our study of its effects from an astrosociological perspective in an earnest manner. After all, human spaceflight is ultimately about humans—a fact we must continue to bear in mind as the effects of space impact more forcefully on humanity at all levels of social reality as the twenty-first century unfolds, whether individuals live in space or on Earth.

About the Authors

WILLIAM SIMS BAINBRIDGE earned his doctorate from Harvard University with a dissertation on the space program, published as *The Spaceflight Revolution*. He is the author of 21 books, 4 textbook-software packages, and about 250 shorter publications in the social science of technology, information science, and culture. *Goals in Space* was a questionnaire study of motivations for space exploration, *Dimensions of Science Fiction* explored popular conceptions of the future in space, and *The Virtual Future* (2011) considered some computer-enabled alternatives if human spaceflight proves too costly. Among his recent projects are editing *The Encyclopedia of Human Computer Interaction* (2004) and coediting *Nanotechnology: Societal Implications—Improving Benefits for Humanity* (2006), *Managing Nano-Bio-Info-Cogno Innovations: Converging Technologies in Society* (2006), *Progress in Convergence* (2006), and *Leadership in Science and Technology* (2012). Bainbridge represented the social sciences on five advanced technology initiatives at the National Science Foundation: High Performance Computing and Communications, Knowledge and Distributed Intelligence, Digital Libraries, Information Technology Research, and Nanotechnology. He also represented computer and information science on the Nanotechnology and Human and Social Dynamics initiatives, among others.

ANDREW BUTRICA earned a doctorate in the history of technology and science from Iowa State University in 1986. He was a Chercheur Associé at the Center for Research in the History of Science and Technology (Centre de Recherches en Histoire des Sciences et Techniques) at the Cité des Sciences et de l'Industrie in Paris before embarking on a career as a research historian. Among other books, he has written *To See the Unseen* (1996), which won the Leopold Prize of the Organization of American Historians, and *Single Stage to Orbit: Politics, Space Technology, and the Quest for Reusable Rocketry* (2003), a history of the Reagan Revolution and the DC-X experimental rocket, which won the 2005 Michael C. Robinson Prize of the National Council on

Public History. More recently, Butrica has written monographs on NASA's role in the manufacture of integrated circuits during the Apollo era and the Agency's contributions to the early history of microelectromechanical systems (MEMS). Currently, he is writing a history of NASA's deep space navigation.

STEVEN J. DICK served as NASA Chief Historian and Director of the NASA History Office from 2003–2009, as the 2011–2012 Charles A. Lindbergh Chair in Aerospace History at the National Air and Space Museum, and as the 2013–2014 Baruch S. Blumberg NASA–Library of Congress Chair in Astrobiology. He obtained his bachelor's degree in astrophysics (1971) and his master's and doctoral degrees (1977) in history and the philosophy of science from Indiana University. He worked as an astronomer and historian of science at the U.S. Naval Observatory in Washington, DC, for 24 years, including 3 years on a mountaintop in New Zealand, before joining NASA in 2003. Among his books are *Plurality of Worlds: The Origins of the Extraterrestrial Life Debate from Democritus to Kant* (1982) (translated into French), *The Biological Universe: The Twentieth Century Extraterrestrial Life Debate and the Limits of Science* (1996), and *Life on Other Worlds* (1998), the latter translated into Chinese, Czech, Greek, Italian, and Polish. Dick's most recent books are (with James Strick) *The Living Universe: NASA and the Development of Astrobiology* (2004) and a comprehensive history of the U.S. Naval Observatory, *Sky and Ocean Joined: The U.S. Naval Observatory, 1830–2000* (2003). The latter received the Pendleton Prize of the Society for History in the Federal Government.

Dick is also editor of *Many Worlds: The New Universe, Extraterrestrial Life, and the Theological Implications* (2000); (with Keith Cowing) *Risk and Exploration: Earth, Sea, and Stars* (2005); (with Roger Launius) *Critical Issues in the History of Spaceflight* (2006) and *Societal Impact of Spaceflight* (2007); *Remembering the Space Age* (2008); (with Mark Lupisella) *Cosmos and Culture: Cultural Evolution in a Cosmic Context* (2009); and *NASA's First 50 Years: Historical Perspectives* (2010). His book *Discovery and Classification in Astronomy: Controversy and Consensus* was published by Cambridge University Press in 2013. He is the recipient of the Navy Meritorious Civilian Service Medal, the NASA Exceptional Service Medal, the NASA Group Achievement Award for his role in NASA's multidisciplinary program in astrobiology, and the 2006 LeRoy E. Doggett Prize for Historical Astronomy from the American Astronomical Society. He has served as chairman of the Historical Astronomy Division of the American Astronomical Society, as president of the History of Astronomy Commission of the International Astronomical Union, and as president of the Philosophical Society of Washington. He was elected a Fellow of the American Association for the Advancement of Science in 2011 and is a corresponding member of the International Academy of Astronautics. Minor planet 6544 Stevendick is named in his honor.

W. HENRY (HARRY) LAMBRIGHT is a professor of public administration and international affairs and also political science at the Maxwell School of Syracuse University. He is the author or editor of seven books and more than 275 articles, papers, and reports. His books include *Powering Apollo: James E. Webb of NASA* (1995) and *Space Policy in the 21st Century* (2003). A long-standing student of leadership and change in government, he recently received support from IBM for monographs on other NASA Administrators: *Transforming NASA: Dan Goldin and the Remaking of NASA* (2001), *Executive Response to Changing Fortune: Sean O'Keefe As NASA Administrator* (2005), and *Launching a New Mission: Michael Griffin and NASA's Return to the Moon* (2009). Lambright is also currently researching a book under NASA sponsorship on Mars exploration.

He has served as a guest scholar at the Brookings Institution, as the director of the Science and Technology Policy Center at the Syracuse Research Corporation, and as the director of the Center for Environmental Policy and Administration at the Maxwell School of Syracuse University. He has also served as an adjunct professor in the Graduate Program of Environmental Science in the College of Environmental Science and Forestry at the State University of New York. Lambright's interest in NASA and space policy goes back to his years as a graduate student. Early in his career, he served as a special assistant at NASA, working in the then Office of University Affairs and writing speeches for NASA Administrator Tom Paine. The recipient of a range of grants from federal and private organizations for his research in science and technology policy, Lambright is frequently cited in the media. He teaches courses in science, technology, and public policy as well as energy and environmental policy. He has a bachelor's degree from Johns Hopkins University and master's and doctoral degrees from Columbia University.

ROGER D. LAUNIUS is a senior curator in the Division of Space History at the Smithsonian Institution's National Air and Space Museum in Washington, DC. Between 1990 and 2002 he served as NASA Chief Historian. A graduate of Graceland College in Lamoni, Iowa, he received his doctorate from Louisiana State University–Baton Rouge in 1982. He has written or edited more than 30 books on aerospace history, including *Globalizing Polar Science: Reconsidering the International Polar and Geophysical Years* (2010); *Smithsonian Atlas of Space Exploration* (2009); *Robots in Space: Technology, Evolution, and Interplanetary Travel* (2008); *Societal Impact of Spaceflight* (2007); *Critical Issues in the History of Spaceflight* (2006); *Space Stations: Base Camps to the Stars* (2003; 2nd edition 2009), which received the history manuscript prize of the

American Institute of Aeronautics and Astronautics (AIAA); *Reconsidering a Century of Flight* (2003); *To Reach the High Frontier: A History of U.S. Launch Vehicles* (2002); *Imagining Space: Achievements, Possibilities, Projections, 1950–2050* (2001); *Reconsidering Sputnik: Forty Years Since the Soviet Satellite* (2000); *Innovation and the Development of Flight* (1999); *Frontiers of Space Exploration* (1998; revised edition 2004); *Spaceflight and the Myth of Presidential Leadership* (1997); and *NASA: A History of the U.S. Civil Space Program* (1994, revised edition 2001). Launius is a Fellow of the American Association for the Advancement of Science, the International Academy of Astronautics, and the American Astronautical Society and is an Associate Fellow of the AIAA. He also served as a consultant to the Columbia Accident Investigation Board in 2003 and presented the Harmon Memorial Lecture on the history of national security space policy at the United States Air Force Academy in 2006. He is frequently consulted by the media for his views on space issues and has been a guest commentator on National Public Radio and major television networks.

JIM PASS received his doctorate in sociology at the University of Southern California in 1991. He created the field of astrosociology in 2003 to begin the process of studying the relationship between space exploration and societies in a systematic manner. The intention from the beginning was to organize individual social scientists already conducting research into space issues and to attract new ones to join the effort with the goal of building a recognizable body of astrosociological literature. As an avid space enthusiast, he longed to combine his interest in space exploration with sociology. In December 2002, he came across an online article by Allen Tough about the social implications of the Search for Extraterrestrial Intelligence (SETI) that mentioned the importance of establishing a new field that focused on the social ramifications of space exploration. "Astrosociology" was mentioned as one of two possible labels for such a new field. Upon seeing this term, Pass purchased the domain *Astrosociology.com* and set out to define this new field. Several months later, in July 2003, he uploaded the site's first content. While astrosociology started out as a sociological subfield, it became a multidisciplinary field that includes the social and behavioral sciences, the humanities, and the arts. Pass has taught sociology at Long Beach State and Long Beach City College in California for more than 15 years, although today he devotes most of his time working to advance the development of astrosociology along with his colleagues and a growing number of supporters. He serves as the chief executive officer of the Astrosociology Research Institute (ARI). The Web site for ARI is now *Astrosociology.org*. The goal of ARI is to develop astrosociology as an academic field with a strong emphasis on attracting students to assist in its development. He was instrumental in establishing the Astrosociology Subcommittee within the American Institute of Aeronautics and Astronautics. Pass has written more than 30 conference papers and publications that address the reasoning for developing astrosociology and provide introductions of several of its subfields. Most of Pass's papers are available in the ARI virtual library at *Astrosociology.org* along with those of other authors.

ELIGAR SADEH is the president of Astroconsulting International, which empowers space and defense programs with the critical skills to optimize outcomes. He serves as a research professor with the Center for Space Studies at the University of Colorado, Colorado Springs; as part of the adjunct faculty at the International Space University; and as editor of *Astropolitics*, an academic journal focused on space policy and technology. Previously, Sadeh held professorships in space and defense studies in the department of space studies at the College of Aerospace Sciences of the University of North Dakota and at the United States Air Force Academy. He also worked for Lockheed Martin as a systems engineer. Sadeh served as a research associate with the Eisenhower Center for Space and Defense Studies at the United States Air Force Academy and with the Space Policy Institute of George Washington University. Sadeh instructs professional development courses for the University of Colorado, the NASA Academy of Program/Project and Engineering Leadership, and the American Institute of Aeronautics and Astronautics. He has written a number of publications, including *Space Strategy in the 21st Century: Theory and Policy*, *The Politics of Space: A Survey*, and *Space Politics and Policy: An Evolutionary Perspective*, as well as articles in *Astropolitics*, *Space Policy*, and *Acta Astronautica*. He earned a doctorate that focused on science and technology studies from Colorado State University, a master's degree in International Studies from the Hebrew University, and a bachelor's degree in aerospace engineering sciences from the University of Colorado.

DAVID J. WHALEN is an associate professor of space studies at the University of North Dakota. He was chair of the space studies department from 2007 to 2010. He has been an engineer and engineering manager in the communications satellite industry for more than 30 years. He has also worked on weather satellites (INSAT, GOES-NEXT), Earth-observing satellites (Landsat), and science satellites (GRO, Hubble). He is an Associate Fellow of the American Institute of Aeronautics and Astronautics. He is a member of the AIAA History Technical Committee and was a member of the AIAA Communications Systems Technical Committee from 2007 to 2010. Over the past 20 years, Whalen has written about space history and space policy in addition to his engineering work. He holds a bachelor's degree in astronomy from Boston University; a master's degree in astronomy from the University of Massachusetts; a master's degree in business administration from the

College of William and Mary; and a doctorate in science, technology, and public policy from George Washington University. He has taught university and industrial courses in orbit determination and maneuver planning, satellite communications, space policy, and the history of technology. His book *Origins of Satellite Communications, 1945–1965* (2002) was published by Smithsonian Institution Press. He is currently finishing a book about the Comsat corporation. He has made many presentations at AIAA, the Pacific Telecommunications Conference, the Society for the History of Technology Conference, and NASA conferences. He has published articles and book reviews on space history, space policy, and space technology in a variety of publications, including *IEEE Technology and Society* and *Technology and Culture*.

The NASA History Series

Reference Works, NASA SP-4000:

Grimwood, James M. *Project Mercury: A Chronology*. NASA SP-4001, 1963.

Grimwood, James M., and Barton C. Hacker, with Peter J. Vorzimmer. *Project Gemini Technology and Operations: A Chronology*. NASA SP-4002, 1969.

Link, Mae Mills. *Space Medicine in Project Mercury*. NASA SP-4003, 1965.

Astronautics and Aeronautics, 1963: Chronology of Science, Technology, and Policy. NASA SP-4004, 1964.

Astronautics and Aeronautics, 1964: Chronology of Science, Technology, and Policy. NASA SP-4005, 1965.

Astronautics and Aeronautics, 1965: Chronology of Science, Technology, and Policy. NASA SP-4006, 1966.

Astronautics and Aeronautics, 1966: Chronology of Science, Technology, and Policy. NASA SP-4007, 1967.

Astronautics and Aeronautics, 1967: Chronology of Science, Technology, and Policy. NASA SP-4008, 1968.

Ertel, Ivan D., and Mary Louise Morse. *The Apollo Spacecraft: A Chronology, Volume I, Through November 7, 1962*. NASA SP-4009, 1969.

Morse, Mary Louise, and Jean Kernahan Bays. *The Apollo Spacecraft: A Chronology, Volume II, November 8, 1962–September 30, 1964*. NASA SP-4009, 1973.

Brooks, Courtney G., and Ivan D. Ertel. *The Apollo Spacecraft: A Chronology, Volume III, October 1, 1964–January 20, 1966*. NASA SP-4009, 1973.

Ertel, Ivan D., and Roland W. Newkirk, with Courtney G. Brooks. *The Apollo Spacecraft: A Chronology, Volume IV, January 21, 1966–July 13, 1974*. NASA SP-4009, 1978.

Astronautics and Aeronautics, 1968: Chronology of Science, Technology, and Policy. NASA SP-4010, 1969.

Newkirk, Roland W., and Ivan D. Ertel, with Courtney G. Brooks. *Skylab: A Chronology*. NASA SP-4011, 1977.

Van Nimmen, Jane, and Leonard C. Bruno, with Robert L. Rosholt. *NASA Historical Data Book, Volume I: NASA Resources, 1958–1968*. NASA SP-4012, 1976; rep. ed. 1988.

Ezell, Linda Neuman. *NASA Historical Data Book, Volume II: Programs and Projects, 1958–1968*. NASA SP-4012, 1988.

Ezell, Linda Neuman. *NASA Historical Data Book, Volume III: Programs and Projects, 1969–1978*. NASA SP-4012, 1988.

Gawdiak, Ihor, with Helen Fedor. *NASA Historical Data Book, Volume IV: NASA Resources, 1969–1978*. NASA SP-4012, 1994.

Rumerman, Judy A. *NASA Historical Data Book, Volume V: NASA Launch Systems, Space Transportation, Human Spaceflight, and Space Science, 1979–1988*. NASA SP-4012, 1999.

Rumerman, Judy A. *NASA Historical Data Book, Volume VI: NASA Space Applications, Aeronautics and Space Research and Technology, Tracking and Data Acquisition/Support Operations, Commercial Programs, and Resources, 1979–1988*. NASA SP-4012, 1999.

Rumerman, Judy A. *NASA Historical Data Book, Volume VII: NASA Launch Systems, Space Transportation, Human Spaceflight, and Space Science, 1989–1998*. NASA SP-2009-4012, 2009.

Rumerman, Judy A. *NASA Historical Data Book, Volume VIII: NASA Earth Science and Space Applications, Aeronautics, Technology, and Exploration, Tracking and Data Acquisition/Space Operations, Facilities and Resources, 1989–1998*. NASA SP-2012-4012, 2012.

No SP-4013.

Astronautics and Aeronautics, 1969: Chronology of Science, Technology, and Policy. NASA SP-4014, 1970.

Astronautics and Aeronautics, 1970: Chronology of Science, Technology, and Policy. NASA SP-4015, 1972.

Astronautics and Aeronautics, 1971: Chronology of Science, Technology, and Policy. NASA SP-4016, 1972.

Astronautics and Aeronautics, 1972: Chronology of Science, Technology, and Policy. NASA SP-4017, 1974.

Astronautics and Aeronautics, 1973: Chronology of Science, Technology, and Policy. NASA SP-4018, 1975.

Astronautics and Aeronautics, 1974: Chronology of Science, Technology, and Policy. NASA SP-4019, 1977.

Astronautics and Aeronautics, 1975: Chronology of Science, Technology, and Policy. NASA SP-4020, 1979.

Astronautics and Aeronautics, 1976: Chronology of Science, Technology, and Policy. NASA SP-4021, 1984.

Astronautics and Aeronautics, 1977: Chronology of Science, Technology, and Policy. NASA SP-4022, 1986.

Astronautics and Aeronautics, 1978: Chronology of Science, Technology, and Policy. NASA SP-4023, 1986.

Astronautics and Aeronautics, 1979–1984: Chronology of Science, Technology, and Policy. NASA SP-4024, 1988.

Astronautics and Aeronautics, 1985: Chronology of Science, Technology, and Policy. NASA SP-4025, 1990.

Noordung, Hermann. *The Problem of Space Travel: The Rocket Motor*. Edited by Ernst Stuhlinger and J. D. Hunley, with Jennifer Garland. NASA SP-4026, 1995.

Gawdiak, Ihor Y., Ramon J. Miro, and Sam Stueland. *Astronautics and Aeronautics, 1986–1990: A Chronology*. NASA SP-4027, 1997.

Gawdiak, Ihor Y., and Charles Shetland. *Astronautics and Aeronautics, 1991–1995: A Chronology*. NASA SP-2000-4028, 2000.

Orloff, Richard W. *Apollo by the Numbers: A Statistical Reference*. NASA SP-2000-4029, 2000.

Lewis, Marieke, and Ryan Swanson. *Astronautics and Aeronautics: A Chronology, 1996–2000*. NASA SP-2009-4030, 2009.

Ivey, William Noel, and Marieke Lewis. *Astronautics and Aeronautics: A Chronology, 2001–2005*. NASA SP-2010-4031, 2010.

Buchalter, Alice R., and William Noel Ivey. *Astronautics and Aeronautics: A Chronology, 2006*. NASA SP-2011-4032, 2010.

Lewis, Marieke. *Astronautics and Aeronautics: A Chronology, 2007*. NASA SP-2011-4033, 2011.

Lewis, Marieke. *Astronautics and Aeronautics: A Chronology, 2008*. NASA SP-2012-4034, 2012.

Lewis, Marieke. *Astronautics and Aeronautics: A Chronology, 2009*. NASA SP-2012-4035, 2012.

Flattery, Meaghan. *Astronautics and Aeronautics: A Chronology, 2010*. NASA SP-2013-4037, 2014.

Management Histories, NASA SP-4100:

Rosholt, Robert L. *An Administrative History of NASA, 1958–1963*. NASA SP-4101, 1966.

Levine, Arnold S. *Managing NASA in the Apollo Era*. NASA SP-4102, 1982.

Roland, Alex. *Model Research: The National Advisory Committee for Aeronautics, 1915–1958*. NASA SP-4103, 1985.

Fries, Sylvia D. *NASA Engineers and the Age of Apollo*. NASA SP-4104, 1992.

Glennan, T. Keith. *The Birth of NASA: The Diary of T. Keith Glennan*. Edited by J. D. Hunley. NASA SP-4105, 1993.

Seamans, Robert C. *Aiming at Targets: The Autobiography of Robert C. Seamans*. NASA SP-4106, 1996.

Garber, Stephen J., ed. *Looking Backward, Looking Forward: Forty Years of Human Spaceflight Symposium*. NASA SP-2002-4107, 2002.

Mallick, Donald L., with Peter W. Merlin. *The Smell of Kerosene: A Test Pilot's Odyssey*. NASA SP-4108, 2003.

Iliff, Kenneth W., and Curtis L. Peebles. *From Runway to Orbit: Reflections of a NASA Engineer*. NASA SP-2004-4109, 2004.

Chertok, Boris. *Rockets and People, Volume I*. NASA SP-2005-4110, 2005.

Chertok, Boris. *Rockets and People: Creating a Rocket Industry, Volume II*. NASA SP-2006-4110, 2006.

Chertok, Boris. *Rockets and People: Hot Days of the Cold War, Volume III*. NASA SP-2009-4110, 2009.

Chertok, Boris. *Rockets and People: The Moon Race, Volume IV*. NASA SP-2011-4110, 2011.

Laufer, Alexander, Todd Post, and Edward Hoffman. *Shared Voyage: Learning and Unlearning from Remarkable Projects*. NASA SP-2005-4111, 2005.

Dawson, Virginia P., and Mark D. Bowles. *Realizing the Dream of Flight: Biographical Essays in Honor of the Centennial of Flight, 1903–2003*. NASA SP-2005-4112, 2005.

Mudgway, Douglas J. *William H. Pickering: America's Deep Space Pioneer.* NASA SP-2008-4113, 2008.

Wright, Rebecca, Sandra Johnson, and Steven J. Dick. *NASA at 50: Interviews with NASA's Senior Leadership.* NASA SP-2012-4114, 2012.

Project Histories, NASA SP-4200:

Swenson, Loyd S., Jr., James M. Grimwood, and Charles C. Alexander. *This New Ocean: A History of Project Mercury.* NASA SP-4201, 1966; rep. ed. 1999.

Green, Constance McLaughlin, and Milton Lomask. *Vanguard: A History.* NASA SP-4202, 1970; rep. ed. Smithsonian Institution Press, 1971.

Hacker, Barton C., and James M. Grimwood. *On the Shoulders of Titans: A History of Project Gemini.* NASA SP-4203, 1977; rep. ed. 2002.

Benson, Charles D., and William Barnaby Faherty. *Moonport: A History of Apollo Launch Facilities and Operations.* NASA SP-4204, 1978.

Brooks, Courtney G., James M. Grimwood, and Loyd S. Swenson, Jr. *Chariots for Apollo: A History of Manned Lunar Spacecraft.* NASA SP-4205, 1979.

Bilstein, Roger E. *Stages to Saturn: A Technological History of the Apollo/Saturn Launch Vehicles.* NASA SP-4206, 1980 and 1996.

No SP-4207.

Compton, W. David, and Charles D. Benson. *Living and Working in Space: A History of Skylab.* NASA SP-4208, 1983.

Ezell, Edward Clinton, and Linda Neuman Ezell. *The Partnership: A History of the Apollo-Soyuz Test Project.* NASA SP-4209, 1978.

Hall, R. Cargill. *Lunar Impact: A History of Project Ranger.* NASA SP-4210, 1977.

Newell, Homer E. *Beyond the Atmosphere: Early Years of Space Science.* NASA SP-4211, 1980.

Ezell, Edward Clinton, and Linda Neuman Ezell. *On Mars: Exploration of the Red Planet, 1958–1978.* NASA SP-4212, 1984.

Pitts, John A. *The Human Factor: Biomedicine in the Manned Space Program to 1980.* NASA SP-4213, 1985.

Compton, W. David. *Where No Man Has Gone Before: A History of Apollo Lunar Exploration Missions.* NASA SP-4214, 1989.

Naugle, John E. *First Among Equals: The Selection of NASA Space Science Experiments.* NASA SP-4215, 1991.

Wallace, Lane E. *Airborne Trailblazer: Two Decades with NASA Langley's 737 Flying Laboratory.* NASA SP-4216, 1994.

Butrica, Andrew J., ed. *Beyond the Ionosphere: Fifty Years of Satellite Communications.* NASA SP-4217, 1997.

Butrica, Andrew J. *To See the Unseen: A History of Planetary Radar Astronomy.* NASA SP-4218, 1996.

Mack, Pamela E., ed. *From Engineering Science to Big Science: The NACA and NASA Collier Trophy Research Project Winners.* NASA SP-4219, 1998.

Reed, R. Dale. *Wingless Flight: The Lifting Body Story.* NASA SP-4220, 1998.

Heppenheimer, T. A. *The Space Shuttle Decision: NASA's Search for a Reusable Space Vehicle.* NASA SP-4221, 1999.

Hunley, J. D., ed. *Toward Mach 2: The Douglas D-558 Program.* NASA SP-4222, 1999.

Swanson, Glen E., ed. *"Before This Decade Is Out..." Personal Reflections on the Apollo Program.* NASA SP-4223, 1999.

Tomayko, James E. *Computers Take Flight: A History of NASA's Pioneering Digital Fly-By-Wire Project.* NASA SP-4224, 2000.

Morgan, Clay. *Shuttle-Mir: The United States and Russia Share History's Highest Stage.* NASA SP-2001-4225, 2001.

Leary, William M. *"We Freeze to Please": A History of NASA's Icing Research Tunnel and the Quest for Safety.* NASA SP-2002-4226, 2002.

Mudgway, Douglas J. *Uplink-Downlink: A History of the Deep Space Network, 1957–1997.* NASA SP-2001-4227, 2001.

No SP-4228 or SP-4229.

Dawson, Virginia P., and Mark D. Bowles. *Taming Liquid Hydrogen: The Centaur Upper Stage Rocket, 1958–2002.* NASA SP-2004-4230, 2004.

Meltzer, Michael. *Mission to Jupiter: A History of the Galileo Project.* NASA SP-2007-4231, 2007.

Heppenheimer, T. A. *Facing the Heat Barrier: A History of Hypersonics.* NASA SP-2007-4232, 2007.

Tsiao, Sunny. *"Read You Loud and Clear!" The Story of NASA's Spaceflight Tracking and Data Network.* NASA SP-2007-4233, 2007.

Meltzer, Michael. *When Biospheres Collide: A History of NASA's Planetary Protection Programs.* NASA SP-2011-4234, 2011.

Center Histories, NASA SP-4300:

Rosenthal, Alfred. *Venture into Space: Early Years of Goddard Space Flight Center.* NASA SP-4301, 1985.

Hartman, Edwin P. *Adventures in Research: A History of Ames Research Center, 1940–1965.* NASA SP-4302, 1970.

Hallion, Richard P. *On the Frontier: Flight Research at Dryden, 1946–1981.* NASA SP-4303, 1984.

Muenger, Elizabeth A. *Searching the Horizon: A History of Ames Research Center, 1940–1976.* NASA SP-4304, 1985.

Hansen, James R. *Engineer in Charge: A History of the Langley Aeronautical Laboratory, 1917–1958.* NASA SP-4305, 1987.

Dawson, Virginia P. *Engines and Innovation: Lewis Laboratory and American Propulsion Technology.* NASA SP-4306, 1991.

Dethloff, Henry C. *"Suddenly Tomorrow Came…": A History of the Johnson Space Center, 1957–1990.* NASA SP-4307, 1993.

Hansen, James R. *Spaceflight Revolution: NASA Langley Research Center from Sputnik to Apollo.* NASA SP-4308, 1995.

Wallace, Lane E. *Flights of Discovery: An Illustrated History of the Dryden Flight Research Center.* NASA SP-4309, 1996.

Herring, Mack R. *Way Station to Space: A History of the John C. Stennis Space Center.* NASA SP-4310, 1997.

Wallace, Harold D., Jr. *Wallops Station and the Creation of an American Space Program.* NASA SP-4311, 1997.

Wallace, Lane E. *Dreams, Hopes, Realities. NASA's Goddard Space Flight Center: The First Forty Years.* NASA SP-4312, 1999.

Dunar, Andrew J., and Stephen P. Waring. *Power to Explore: A History of Marshall Space Flight Center, 1960–1990.* NASA SP-4313, 1999.

Bugos, Glenn E. *Atmosphere of Freedom: Sixty Years at the NASA Ames Research Center.* NASA SP-2000-4314, 2000.

Bugos, Glenn E. *Atmosphere of Freedom: Seventy Years at the NASA Ames Research Center.* NASA SP-2010-4314, 2010. Revised version of NASA SP-2000-4314.

Bugos, Glenn E. *Atmosphere of Freedom: Seventy Five Years at the NASA Ames Research Center*. NASA SP-2014-4314, 2014. Revised version of NASA SP-2000-4314.

No SP-4315.

Schultz, James. *Crafting Flight: Aircraft Pioneers and the Contributions of the Men and Women of NASA Langley Research Center*. NASA SP-2003-4316, 2003.

Bowles, Mark D. *Science in Flux: NASA's Nuclear Program at Plum Brook Station, 1955–2005*. NASA SP-2006-4317, 2006.

Wallace, Lane E. *Flights of Discovery: An Illustrated History of the Dryden Flight Research Center*. NASA SP-2007-4318, 2007. Revised version of NASA SP-4309.

Arrighi, Robert S. *Revolutionary Atmosphere: The Story of the Altitude Wind Tunnel and the Space Power Chambers*. NASA SP-2010-4319, 2010.

General Histories, NASA SP-4400:

Corliss, William R. *NASA Sounding Rockets, 1958–1968: A Historical Summary*. NASA SP-4401, 1971.

Wells, Helen T., Susan H. Whiteley, and Carrie Karegeannes. *Origins of NASA Names*. NASA SP-4402, 1976.

Anderson, Frank W., Jr. *Orders of Magnitude: A History of NACA and NASA, 1915–1980*. NASA SP-4403, 1981.

Sloop, John L. *Liquid Hydrogen as a Propulsion Fuel, 1945–1959*. NASA SP-4404, 1978.

Roland, Alex. *A Spacefaring People: Perspectives on Early Spaceflight*. NASA SP-4405, 1985.

Bilstein, Roger E. *Orders of Magnitude: A History of the NACA and NASA, 1915–1990*. NASA SP-4406, 1989.

Logsdon, John M., ed., with Linda J. Lear, Jannelle Warren Findley, Ray A. Williamson, and Dwayne A. Day. *Exploring the Unknown: Selected Documents in the History of the U.S. Civil Space Program, Volume I: Organizing for Exploration*. NASA SP-4407, 1995.

Logsdon, John M., ed., with Dwayne A. Day and Roger D. Launius. *Exploring the Unknown: Selected Documents in the History of the U.S. Civil Space Program, Volume II: External Relationships*. NASA SP-4407, 1996.

Logsdon, John M., ed., with Roger D. Launius, David H. Onkst, and Stephen J. Garber. *Exploring the Unknown: Selected Documents in the History of the U.S. Civil Space Program, Volume III: Using Space.* NASA SP-4407, 1998.

Logsdon, John M., ed., with Ray A. Williamson, Roger D. Launius, Russell J. Acker, Stephen J. Garber, and Jonathan L. Friedman. *Exploring the Unknown: Selected Documents in the History of the U.S. Civil Space Program, Volume IV: Accessing Space.* NASA SP-4407, 1999.

Logsdon, John M., ed., with Amy Paige Snyder, Roger D. Launius, Stephen J. Garber, and Regan Anne Newport. *Exploring the Unknown: Selected Documents in the History of the U.S. Civil Space Program, Volume V: Exploring the Cosmos.* NASA SP-2001-4407, 2001.

Logsdon, John M., ed., with Stephen J. Garber, Roger D. Launius, and Ray A. Williamson. *Exploring the Unknown: Selected Documents in the History of the U.S. Civil Space Program, Volume VI: Space and Earth Science.* NASA SP-2004-4407, 2004.

Logsdon, John M., ed., with Roger D. Launius. *Exploring the Unknown: Selected Documents in the History of the U.S. Civil Space Program, Volume VII: Human Spaceflight: Projects Mercury, Gemini, and Apollo.* NASA SP-2008-4407, 2008.

Siddiqi, Asif A., *Challenge to Apollo: The Soviet Union and the Space Race, 1945–1974.* NASA SP-2000-4408, 2000.

Hansen, James R., ed. *The Wind and Beyond: Journey into the History of Aerodynamics in America, Volume 1: The Ascent of the Airplane.* NASA SP-2003-4409, 2003.

Hansen, James R., ed. *The Wind and Beyond: Journey into the History of Aerodynamics in America, Volume 2: Reinventing the Airplane.* NASA SP-2007-4409, 2007.

Hogan, Thor. *Mars Wars: The Rise and Fall of the Space Exploration Initiative.* NASA SP-2007-4410, 2007.

Vakoch, Douglas A., ed. *Psychology of Space Exploration: Contemporary Research in Historical Perspective.* NASA SP-2011-4411, 2011.

Ferguson, Robert G., *NASA's First A: Aeronautics from 1958 to 2008.* NASA SP-2012-4412, 2013.

Vakoch, Douglas A., ed. *Archaeology, Anthropology, and Interstellar Communication.* NASA SP-2013-4413, 2014.

Monographs in Aerospace History, NASA SP-4500:

Launius, Roger D., and Aaron K. Gillette, comps. *Toward a History of the Space Shuttle: An Annotated Bibliography.* Monographs in Aerospace History, No. 1, 1992.

Launius, Roger D., and J. D. Hunley, comps. *An Annotated Bibliography of the Apollo Program.* Monographs in Aerospace History, No. 2, 1994.

Launius, Roger D. *Apollo: A Retrospective Analysis.* Monographs in Aerospace History, No. 3, 1994.

Hansen, James R. *Enchanted Rendezvous: John C. Houbolt and the Genesis of the Lunar-Orbit Rendezvous Concept.* Monographs in Aerospace History, No. 4, 1995.

Gorn, Michael H. *Hugh L. Dryden's Career in Aviation and Space.* Monographs in Aerospace History, No. 5, 1996.

Powers, Sheryll Goecke. *Women in Flight Research at NASA Dryden Flight Research Center from 1946 to 1995.* Monographs in Aerospace History, No. 6, 1997.

Portree, David S. F., and Robert C. Trevino. *Walking to Olympus: An EVA Chronology.* Monographs in Aerospace History, No. 7, 1997.

Logsdon, John M., moderator. *Legislative Origins of the National Aeronautics and Space Act of 1958: Proceedings of an Oral History Workshop.* Monographs in Aerospace History, No. 8, 1998.

Rumerman, Judy A., comp. *U.S. Human Spaceflight: A Record of Achievement, 1961–1998.* Monographs in Aerospace History, No. 9, 1998.

Portree, David S. F. *NASA's Origins and the Dawn of the Space Age.* Monographs in Aerospace History, No. 10, 1998.

Logsdon, John M. *Together in Orbit: The Origins of International Cooperation in the Space Station.* Monographs in Aerospace History, No. 11, 1998.

Phillips, W. Hewitt. *Journey in Aeronautical Research: A Career at NASA Langley Research Center.* Monographs in Aerospace History, No. 12, 1998.

Braslow, Albert L. *A History of Suction-Type Laminar-Flow Control with Emphasis on Flight Research.* Monographs in Aerospace History, No. 13, 1999.

Logsdon, John M., moderator. *Managing the Moon Program: Lessons Learned from Apollo.* Monographs in Aerospace History, No. 14, 1999.

Perminov, V. G. *The Difficult Road to Mars: A Brief History of Mars Exploration in the Soviet Union.* Monographs in Aerospace History, No. 15, 1999.

Tucker, Tom. *Touchdown: The Development of Propulsion Controlled Aircraft at NASA Dryden.* Monographs in Aerospace History, No. 16, 1999.

Maisel, Martin, Demo J. Giulanetti, and Daniel C. Dugan. *The History of the XV-15 Tilt Rotor Research Aircraft: From Concept to Flight.* Monographs in Aerospace History, No. 17, 2000. NASA SP-2000-4517.

Jenkins, Dennis R. *Hypersonics Before the Shuttle: A Concise History of the X-15 Research Airplane.* Monographs in Aerospace History, No. 18, 2000. NASA SP-2000-4518.

Chambers, Joseph R. *Partners in Freedom: Contributions of the Langley Research Center to U.S. Military Aircraft of the 1990s.* Monographs in Aerospace History, No. 19, 2000. NASA SP-2000-4519.

Waltman, Gene L. *Black Magic and Gremlins: Analog Flight Simulations at NASA's Flight Research Center.* Monographs in Aerospace History, No. 20, 2000. NASA SP-2000-4520.

Portree, David S. F. *Humans to Mars: Fifty Years of Mission Planning, 1950–2000.* Monographs in Aerospace History, No. 21, 2001. NASA SP-2001-4521.

Thompson, Milton O., with J. D. Hunley. *Flight Research: Problems Encountered and What They Should Teach Us.* Monographs in Aerospace History, No. 22, 2001. NASA SP-2001-4522.

Tucker, Tom. *The Eclipse Project.* Monographs in Aerospace History, No. 23, 2001. NASA SP-2001-4523.

Siddiqi, Asif A. *Deep Space Chronicle: A Chronology of Deep Space and Planetary Probes, 1958–2000.* Monographs in Aerospace History, No. 24, 2002. NASA SP-2002-4524.

Merlin, Peter W. *Mach 3+: NASA/USAF YF-12 Flight Research, 1969–1979.* Monographs in Aerospace History, No. 25, 2001. NASA SP-2001-4525.

Anderson, Seth B. *Memoirs of an Aeronautical Engineer: Flight Tests at Ames Research Center: 1940–1970.* Monographs in Aerospace History, No. 26, 2002. NASA SP-2002-4526.

Renstrom, Arthur G. *Wilbur and Orville Wright: A Bibliography Commemorating the One-Hundredth Anniversary of the First Powered Flight on December 17, 1903.* Monographs in Aerospace History, No. 27, 2002. NASA SP-2002-4527.

No monograph 28.

Chambers, Joseph R. *Concept to Reality: Contributions of the NASA Langley Research Center to U.S. Civil Aircraft of the 1990s.* Monographs in Aerospace History, No. 29, 2003. NASA SP-2003-4529.

Peebles, Curtis, ed. *The Spoken Word: Recollections of Dryden History, The Early Years.* Monographs in Aerospace History, No. 30, 2003. NASA SP-2003-4530.

Jenkins, Dennis R., Tony Landis, and Jay Miller. *American X-Vehicles: An Inventory—X-1 to X-50.* Monographs in Aerospace History, No. 31, 2003. NASA SP-2003-4531.

Renstrom, Arthur G. *Wilbur and Orville Wright: A Chronology Commemorating the One-Hundredth Anniversary of the First Powered Flight on December 17, 1903.* Monographs in Aerospace History, No. 32, 2003. NASA SP-2003-4532.

Bowles, Mark D., and Robert S. Arrighi. *NASA's Nuclear Frontier: The Plum Brook Research Reactor.* Monographs in Aerospace History, No. 33, 2004. NASA SP-2004-4533.

Wallace, Lane, and Christian Gelzer. *Nose Up: High Angle-of-Attack and Thrust Vectoring Research at NASA Dryden, 1979–2001.* Monographs in Aerospace History, No. 34, 2009. NASA SP-2009-4534.

Matranga, Gene J., C. Wayne Ottinger, Calvin R. Jarvis, and D. Christian Gelzer. *Unconventional, Contrary, and Ugly: The Lunar Landing Research Vehicle.* Monographs in Aerospace History, No. 35, 2006. NASA SP-2004-4535.

McCurdy, Howard E. *Low-Cost Innovation in Spaceflight: The History of the Near Earth Asteroid Rendezvous (NEAR) Mission.* Monographs in Aerospace History, No. 36, 2005. NASA SP-2005-4536.

Seamans, Robert C., Jr. *Project Apollo: The Tough Decisions.* Monographs in Aerospace History, No. 37, 2005. NASA SP-2005-4537.

Lambright, W. Henry. *NASA and the Environment: The Case of Ozone Depletion.* Monographs in Aerospace History, No. 38, 2005. NASA SP-2005-4538.

Chambers, Joseph R. *Innovation in Flight: Research of the NASA Langley Research Center on Revolutionary Advanced Concepts for Aeronautics.* Monographs in Aerospace History, No. 39, 2005. NASA SP-2005-4539.

Phillips, W. Hewitt. *Journey into Space Research: Continuation of a Career at NASA Langley Research Center.* Monographs in Aerospace History, No. 40, 2005. NASA SP-2005-4540.

Rumerman, Judy A., Chris Gamble, and Gabriel Okolski, comps. *U.S. Human Spaceflight: A Record of Achievement, 1961–2006.* Monographs in Aerospace History, No. 41, 2007. NASA SP-2007-4541.

Peebles, Curtis. *The Spoken Word: Recollections of Dryden History Beyond the Sky.* Monographs in Aerospace History, No. 42, 2011. NASA SP-2011-4542.

Dick, Steven J., Stephen J. Garber, and Jane H. Odom. *Research in NASA History.* Monographs in Aerospace History, No. 43, 2009. NASA SP-2009-4543.

Merlin, Peter W. *Ikhana: Unmanned Aircraft System Western States Fire Missions.* Monographs in Aerospace History, No. 44, 2009. NASA SP-2009-4544.

Fisher, Steven C., and Shamim A. Rahman. *Remembering the Giants: Apollo Rocket Propulsion Development.* Monographs in Aerospace History, No. 45, 2009. NASA SP-2009-4545.

Gelzer, Christian. *Fairing Well: From Shoebox to Bat Truck and Beyond, Aerodynamic Truck Research at NASA's Dryden Flight Research Center.* Monographs in Aerospace History, No. 46, 2011. NASA SP-2011-4546.

Arrighi, Robert. *Pursuit of Power: NASA's Propulsion Systems Laboratory No. 1 and 2.* Monographs in Aerospace History, No. 48, 2012. NASA SP-2012-4548.

Goodrich, Malinda K., Alice R. Buchalter, and Patrick M. Miller, comps. *Toward a History of the Space Shuttle: An Annotated Bibliography, Part 2 (1992–2011).* Monographs in Aerospace History, No. 49, 2012. NASA SP-2012-4549.

Gelzer, Christian. *The Spoken Word III: Recollections of Dryden History; The Shuttle Years.* Monographs in Aerospace History, No. 52, 2013. NASA SP-2013-4552.

Ross, James C. *NASA Photo One.* Monographs in Aerospace History, No. 53, 2013. NASA SP-2013-4553.

Launius, Roger D. *Historical Analogs for the Stimulation of Space Commerce.* Monographs in Aerospace History, No 54, 2014. NASA SP-2014-4554.

Buchalter, Alice R., and Patrick M. Miller, comps. *The National Advisory Committee for Aeronautics: An Annotated Bibliography*. Monographs in Aerospace History, No. 55, 2014. NASA SP-2014-4555.

Chambers, Joseph R., and Mark A. Chambers. *Emblems of Exploration: Logos of the NACA and NASA*. Monographs in Aerospace History, No. 56, 2015. NASA SP-2015-4556.

Electronic Media, NASA SP-4600:

Remembering Apollo 11: The 30th Anniversary Data Archive CD-ROM. NASA SP-4601, 1999.

Remembering Apollo 11: The 35th Anniversary Data Archive CD-ROM. NASA SP-2004-4601, 2004. This is an update of the 1999 edition.

The Mission Transcript Collection: U.S. Human Spaceflight Missions from Mercury Redstone 3 to Apollo 17. NASA SP-2000-4602, 2001.

Shuttle-Mir: The United States and Russia Share History's Highest Stage. NASA SP-2001-4603, 2002.

U.S. Centennial of Flight Commission Presents Born of Dreams—Inspired by Freedom. NASA SP-2004-4604, 2004.

Of Ashes and Atoms: A Documentary on the NASA Plum Brook Reactor Facility. NASA SP-2005-4605, 2005.

Taming Liquid Hydrogen: The Centaur Upper Stage Rocket Interactive CD-ROM. NASA SP-2004-4606, 2004.

Fueling Space Exploration: The History of NASA's Rocket Engine Test Facility DVD. NASA SP-2005-4607, 2005.

Altitude Wind Tunnel at NASA Glenn Research Center: An Interactive History CD-ROM. NASA SP-2008-4608, 2008.

A Tunnel Through Time: The History of NASA's Altitude Wind Tunnel. NASA SP-2010-4609, 2010.

Conference Proceedings, NASA SP-4700:

Dick, Steven J., and Keith Cowing, eds. *Risk and Exploration: Earth, Sea and the Stars*. NASA SP-2005-4701, 2005.

Dick, Steven J., and Roger D. Launius. *Critical Issues in the History of Spaceflight*. NASA SP-2006-4702, 2006.

Dick, Steven J., ed. *Remembering the Space Age: Proceedings of the 50th Anniversary Conference*. NASA SP-2008-4703, 2008.

Dick, Steven J., ed. *NASA's First 50 Years: Historical Perspectives.* NASA SP-2010-4704, 2010.

Societal Impact, NASA SP-4800:

Dick, Steven J., and Roger D. Launius. *Societal Impact of Spaceflight.* NASA SP-2007-4801, 2007.

Dick, Steven J., and Mark L. Lupisella. *Cosmos and Culture: Cultural Evolution in a Cosmic Context.* NASA SP-2009-4802, 2009.

Index

Page numbers in **bold** indicate tables and illustrations.

A

ABC
   polling activity by, 1, 70–71
   television satellite, launch of, 459–60, 461
ABM (anti-ballistic missile) systems, 480
ABM treaty (Strategic Arms Limitation
   Talks), 480
abortion, 381
accelerometer
   airbag system MEMS accelerometer, 261
   commercial applications for, 322
   concept and function of, 308–11, 320
   cost of fabrication, 321–22
   development of, NASA role in, 320–23
   electromechanical accelerometer, 261
   impact and importance of, 320–23
   MEMS accelerometer, NASA role in
      development of, 251, 252, 314
   MEMS accelerometer research, 308–14
   MEMS accelerometer research, funding
      for, 313–14
   patents for and related to, 251, 320–23,
      321n274, 323n285
   technical problems with, 321–22
ACF Industries, 169
AcQtrac System, 111
Adams, Carsbie, 81–83, 94
Adin, Louis E., 136
Admiral, 167
Adolphsen, John W., 222
Advanced Bionics, 137
Advanced Extremely High Frequency
   (AEHF) satellite, 467–68
Advanced Reconnaissance Satellite (ARS),
   479
Advanced Research and Technology, Office
   of (OART), 183, 195, 264

Advanced Research Projects Agency
   (ARPA), 434, 437, 464–65
Advanced Trauma Life Support (ATLS),
   American College of Surgeons, 96–97
Advani, Gulu T., 208
Advent, 465, 466
AEHF (Advanced Extremely High
   Frequency) satellite, 467–68
Aerobee sounding rocket, 432, 436
Aerofit, 127
aeronautical research and development,
   spinoffs from, 80
Aerosat, 476–78
Aerosat program, 490
aerospace
   aerospace medicine technologies, 85
   divide between sociology and, 537,
      538–43, 571–72
   poll questions about funding for, 68
Aerospatiale, 477
Aetna, 462
Afghanistan, 486
Aftergood, Steven, 334–35
age and support for space program, 35–36,
   37–41
Agriculture, Department of, 386–87, 436,
   482
airbags, automobile, 251, 261
aircraft, transistor purchases for, 166
Air Force, U.S.
   Advanced Minuteman Missile program
      and integrated circuits development
      and use, 152–54, 155, 176–77
   applications satellites, interest in, 432, 489
   Ballistic Missile Division, 504
   Cambridge Research Laboratory, 194
   configuration management practices, 504

data bank of electronics parts, use of by, 193
Electronics Systems Division, 194
Evolved Expendable Launch Vehicle (EELV) program, 466
fly-by-wire systems and haptic information displays, 118–19
ICBM program, 432, 433
integrated-circuit research, funding for, 183n98
integrated circuits development and Air Force electronics miniaturization program, 171–72
Mead triode funding from, 208
meeting about communications at Woods Hole, 438–39
Molecular Electronics (Molectronics) Program, 171–72
nanotechnology program of, 172
navigation satellite program, 473, 489
Project 621B program, 473
reconnaissance satellite development by, 337–38, 479–81
resonant gate transistor, development of under contract with, 258–59n24
Rome Air Development Command, 193, 196
Stanford Electronics Laboratories, funding for, 288–89
Strategic Satellite System, 467
systems management practices in, 511, 513
tactile communication, support for research on, 120
transistor use by, 166
Air Force Weapons Laboratory, 219
Airline Transport Association, 476
air pollution and funding for anti–air pollution programs, poll question about, 16, 54
air-traffic control telemetry transmitter, 212
Alaska
    Anik satellite service in, 461
    telemedicine programs in, 112–13
albedo, 436
Albert Einstein Institute, 267
Aldebaran, 361
Alewine, Betty, 451, 453
Alexandria Digital Library, 117
Alferov, Zhores I., 174n65
Allison, David, 202
alloy-junction transistors, 159

Almond, Gabriel, 9–10
Alpert, Mickey, 454
Alpha Microelectronics Corporation, 191
Altair, 116
Alto, 116
aluminum
    honeycomb material for seat cushions made from, 131
    integrated circuits, aluminum leads on, 200, 203n162, 220, 223
    metallurgical incompatibility of gold and, 202–3, 203n162
    triode fabrication with, 208
aluminum oxide, 208
Amelco, 176, 207n177
American College of Surgeons, Advanced Trauma Life Support (ATLS), 96–97
American Foundation for the Blind, 120
American Institute for Electrical Engineering (AIEE), 290
American Institute of Aeronautics and Astronautics (AIAA), Astrosociology Working Group (AWG), 541
American Institute of Physics, 262
American Nuclear Society, 118
American Physical Society, 59–60, 254–55
American Society of Biomechanics, 88
American Society of Mechanical Engineers (ASME), 262
American Telephone & Telegraph. *See* AT&T (American Telephone & Telegraph)
Americom Asia-Pacific (AAP), 454
Ames Research Center/Ames Research Laboratory
    accelerometer development, role in, 252, 314
    aerodynamics and wind tunnel research at, 265, 270, 271, 278
    anti-shock garment research, 93–95
    biomedical instrumentation research at, 264–85
    biomedical instrument design research at, 270–71n71, 270–74
    biomedical pressure sensor, funding for, 299, 302–3, 314–15, **326**
    Biomedical Research Branch, **326**
    Biomedical Research Division, 280, 281
    biomedical research funding from, 267–70, 268n58, 324
    biosatellite program, 263, 269
    Biotechnology Division, 264, 272–73, 281

bone density analyzer, research to support development of, 86, 87

collaboration and friction between scientists and Instrument Division at, 265, 270

data bank of electronics parts, use of by, 193

electronics research and development at, 183–85

Electronics Research Branch/Electronic Instrument Development Branch, 271, 271n74, 272–73, 274, 279, 284, **326**

Environmental Biology Division, 264–65, 271, 280

Exobiology Division, 264

gas chromatograph research, support and funding for, 303–4, 306–8, 314, 315, 316–20, **326**

Human Performance Branch, 272–73

Instrumentation Division, 266, 270–74, 278–79, 281, 284

Instrument Research Division, 278

integrated circuit research at, 282–83

integrated-circuit technology use in biomedical instrumentation, rejection of, 272, 274

Life Detection Systems Branch, 303–4

Life Sciences Directorate, 279–80

Life Sciences Division, 87, 285

Life Sciences Office, 264

location of, 253, 323

Measurement Sciences Branch, 275, 279, 326, **326**

Measurements Research Branch, 270, 271n74, 283–84

MEMS research at, 252, 274–85, 278n110

MEMS research funding from, 275

Moffett Field laboratories at, 271

name change of, 271

Physiology Branch, 271, 280

Pioneer Project, acknowledgment for assistance in receiving radio signals from probes, 361

protective seat system research and temper foam development role, 131, 133

regional impact of, 529

Stanford biomedical research, role in, 264–69, 283–85, 284n136, 323–26, **326**

Stanford MEMS research, role in, 283–85, 284n136, 323–26, **326**

tactile communication, support for research on, 120, 121, 124

amplifiers

amplifier microcircuits, testing of for nitrogen, helium, and hydrogen, 205–6

biomedical use of, 100

Centaur-Mariner H loss and failure of, 244–45

maser amplifier, 438

miniaturization of, 170, 171

solid-state amplifiers, invention of, 158

transistor use in, 166

ancillary policy, 500–501, 501n34

Anders, Bill, 389

Anderson, G. M., 343

Andover station (Rumford, Maine, station/ Space Hill), 440, 440n26

Andreasen, George B., 128

Andromeda Strain, 384, 387–88

Angell, James B.

accelerometer research, 309–10, 313–14

background and experience of, 286

biomedical pressure sensor research, role in, 278, 299, 314–15

Electrical Engineering Department role of, 294

gas chromatograph research, report on funding for, 307–8

Integrated Circuits Laboratory role of, 293

NIH research, role in, 292

Solid-State Electronics Laboratory role of, 286

Wise's research under, 295

Angell, William W., 292

Anik, 461

animals

avian biorhythms research, 271–72

biomedical instrumentation research to monitor, 263, 282

biomedical pressure sensor research with, 300, 314–15

capacitance transducer testing on dogs, 270

implantable telemetry system for monitoring blood flow in, 293

MEMS and monitoring of, 251

monitoring during test flights, 100, 147

monkeys, start of work with, 282

size of instruments for use in, 282

trials in laboratories on, 263

anisotropic etching, 257, 300–302, 301n190,
    304, 305–6, 311–12, 315, 323, 327–28
Anliker, Max, 269
anodic bonding (electrostatic bonding,
    Mallory process), 258, 306, 311, 312
Anselmo, René, 458, 463
Anstead, Robert J., 222
Antarctica, 398, 399, 401, 409
anti-ballistic missile (ABM) systems, 480,
    485
antimony, 215–16
anti-shock garments, 93–97, 143
apnea monitor, 267
Apollo guidance and navigation computers
    integrated circuits development and use in,
        149–59, 176–77, 225–28, 242, 527
    MIT design and development of, 527
    number of integrated circuits purchased
        for, 150–51, 150nn4–5, 155,
        226–27, 242
    price of integrated circuits for, 154
    suppliers of integrated circuits for, 151,
        176, 226
    testing integrated circuits for, 220
Apollo Paradigm, 492–533, 503
    demise of, 500
    exploration ethos, 498–500, 498n24,
        498n28, 533–34
    longer-term consequences of, 500–503,
        501n34, 502n40
    political ethos, 493–96, 495n12
    technological ethos, 496–98, 503
Apollo program
    Apollo 1 mission and fire, 15, 500
    Apollo 8 mission, 15, 16, 388–89, 390,
        498
    Apollo 9 mission, 389
    Apollo 10 mission, 389
    Apollo 11 mission, 16, 17, **334, 341,
        350,** 353–54, 356, 388, 389, 528
    Apollo 12 mission, **350,** 353–54, **353,**
        356, 357
    Apollo 13 mission, 33–34, **350,** 358
    Apollo 14 mission, **350,** 353–54, 356
    Apollo 15 mission, **350,** 353–56
    Apollo 16 mission, **350,** 353–56
    Apollo 17 mission, 17, **341, 350,**
        353–56, 389, 390
    Apollo Lunar Surface Experiments
        Package (ALSEP), **334, 350,**
        352–58, **353**
    Apollo-Soyuz mission, 17, 500

appreciation of Earth as spinoff of, 389,
    533–34
budget and funding for, 34, 494
Command and Service Module, contrac-
    tor for, **523**
command module recovery system, 131
confidence and trust in government
    during, 497–98
contamination concerns and, 387–88
economic impact of, 492, 517–30,
    519n97, 522n107, 529n131,
    529nn133–134, 533
educational impact of, 491, 530–32,
    531n140, 533
electronics for, 178
emotional bonds with astronauts, poll
    questions about, 33–34
end of, 17
faked Moon landing, opinions about, 17
history of space program, communication
    of, 147
hybrid circuit use in, 200n153
as iconographic symbol, 499–500
images from Apollo mission and apprecia-
    tion for Earth, 388–89, 390, 416,
    424, 498–99, 498n28
importance of landing a man on Moon,
    opinions about, 17
inspirational value of, 388–89, 416,
    521–22, 533–34
integrated circuit manufacturing and, viii
Lunar Excursion Module, contractor for,
    523
management and planning impacts and
    consequences, 492, 495–96, 497,
    503–17, 533
management structure for, 351–52
pacemaker technology development and,
    103
prestige from, 352, 491–92, 493–96, 533
public opinions and attitudes toward,
    12–17, 19–20
reliability program for, 201
RTG use in spacecraft for, **334, 341,** 349,
    350, 351
scientific objectives of, 352–58
societal impact of, viii, 491–92, 533–34
transistor failure analysis research for
    lunar equipment package, 244
as turning point in history, 492–93,
    492–93n2
Apollo Program Prize, 92

Apple II, 116
applications compared to spinoffs, 78
applications satellites
    civilian safety and, 486
    definition of, 427–28
    direct-readout satellites, 433
    early concepts and programs, 385–87,
        432–45, 489–90
    environmental mission of NASA and,
        391–92
    film return satellites, 433
    future of, 487–88
    geostationary orbit, 430–31, 469, 472,
        482
    geosynchronous-Earth-orbit (GEO)
        satellites, 428, 439, 450, 465, 481,
        488, 526
    interest in before Sputnik, 429–32
    launch of, revenue from, 485
    manufacturing and sales of, 484
    NASA role in development of, 489–90,
        525–26, 526n122
    revenues from commercial satellites, 483
    rocket pioneers and, 429–30
    science-fiction stories about, 429, 438
    societal impact of, viii, 428–29, 483–87,
        489–90
    space technology and development of, 78
    types of, 428
    United States domination of, 485
    See also communications satellites; envi-
        ronmental/Earth resources satellites;
        navigation satellites; reconnaissance
        satellites; weather/meteorology
        satellites
Applications Technology Satellites
    ATS-1, 112, 113, 115, 471–72, 476
    ATS-3, 471–72, 476
    ATS-5, 472, 476
    ATS-6, 112–13, 115, 461
    development and purpose of, 471–72
    launch vehicles for, 471
Applied Physics Laboratory, Johns Hopkins
    University, 98, 104–6, 342–43, 345,
        434, 435
Aqua, 420–21
Aquacade system, 481
Aramark Corporation, 95
Arctic, 401, 406–8
Arctic Council, 113
Arecibo radio telescope, 100
Argonne National Laboratory, 118

ARGON satellites, 480
Ariane launch vehicle, 449, 515
Ariel 1, 240
ARINC (Aeronautical Radio, Incorporated),
    191, 476, 476n106
Armstrong, Neil, 281, 389
Armstrong Medical, 97
Army, U.S.
    applications satellites, interest in, 432
    Ballistic Missile Agency, 432, 510–11
    Combat Casualty Care Guidelines, 96–97
    Electronics Command, 191
    integrated-circuit research, funding for,
        183
    rocket development by and V-2 Rocket
        Team, 432
    Signal Corps Micromodule program,
        170–71, 170n52
    Signal Corps Research and Development
        Laboratories, 121, 289
    Signal Corps use of DTL circuits, 228
    Signal Corps use of transistors, 165
    Signal Service and weather observations,
        436
    weather satellite development, 437
Arnaud, Sara, 87, 88–89
ARPA (Advanced Research Projects
    Agency), 434, 437, 464–65
ascendancy, 64
Ascent, 452, 453
Asian Space Conference, 540
AsiaSat, 464
Asrar, Ghassem, 412, 417, 419
Associated Press polls, 22, 31, 71
Association of South Eastern Asian Nations,
    316
asteroids, mining of, 564
Astrain, Santiago, 451
Astra television satellites, 463–64
Astrionics Laboratory, 186, 222
Astrium, 458
astrobiology
    background of astrobiologists, 546
    popularity of, 566
    recognition of discipline, 540
    relationship to astrosociology, 540, 546,
        565–70
    space exploration and, viii
astrochemistry, viii
astrogeology, viii
Astrolink, 454
astrology, 44, **45**

astronauts/humans
    biomedical instrumentation research to
        monitor, 263, 264, 265, 282, 526
    human body in free fall, research on
        dynamics of, 269
    as iconographic symbol, 499–500
    implantable telemetry system for moni-
        toring blood flow in, 293
    MEMS and monitoring of, 251
    monitoring heart of, 107–8, 109
    telemetry and monitoring condition of,
        111, 526
astrosociology
    academic programs for, 546, 548–51
    applied astrosociology, 547–48, 571
    background of astrosociologists, 546
    benefits of perspective, 570–76
    benefits of space exploration from per-
        spective of, 574–76
    coining of term, 567
    crew model of spaceflight, 551–54, 571
    definition and scope of, 543–51
    development of, viii–ix, 535–38, 543,
        570–71
    divide between aerospace and sociology,
        537, 538–43, 571–72
    formal establishment of discipline, call
        for, 536–37
    founding, 535, 537, 570
    future of exploration, society, and,
        551–70
    goals of, 546
    justifications of the need to develop,
        550–51
    pioneers in development of, 546
    planetary defense, contributions to,
        560–62
    recognition of discipline, 540–41, 557–58
    relationship to astrobiology, 540, 546,
        565–70
    relevance and significance of, 545–46, 548
    social problems, using space to solve,
        557–60, 571, 574
    social science classrooms, expanding space
        into, 548–51
    societal impact of spaceflight, conferences
        on, 538–39, 540, 541, 544
    societal impact of spaceflight, study of,
        572–76
    spacefaring societies, transformation into,
        562–65, 573–75

    space societies, creation of, 554–56, 571,
        573–75
Astrosociology Research Institute (ARI),
    541, 541n16
Astrosociology Working Group (AWG),
    American Institute of Aeronautics and
    Astronautics (AIAA), 541
astrotheology, viii–ix
AT&T (American Telephone & Telegraph)
    antitrust suit against, 158–59, 159n29
    communications satellite development by,
        428, 438–40, 489–90
    domestic communications satellites, 449,
        462
    FCC DOMSAT report and, 461
    leasing of satellites by, 461
    monopoly status of, 460
    satellite manufacturing by, end of, 440
    Telstar, 440, 442, 453, 454, 461, 490,
        525–26
    transistor use by, 165, 166
Atalla, M. M. "John," 239
Atlantic Bird, 464
Atlas rockets
    Atlas-Agena launch vehicles, 471, 480
    Atlas-Centaur launch vehicles, 359, 449,
        467, 471
    development of, 7
    propellant for, 82
    reliability of, 470, 485
    satellite launches with, 466, 470, 471
atmosphere
    comparative planetology and research
        on, 387
    satellites and research on, 386
    See also ozone depletion
Atomic Energy Commission (AEC). See
    Energy, Department of/Atomic Energy
    Commission (AEC)
atoms, Feynman challenge to manipulate
    material on atomic level, 254–56,
    255–56n13
Atoms for Peace initiative, 338–40
atom-smasher research tool, 139. See also
    cyclotron
Augustine, Norman "Norm," 402–3, 451
Aura, 422
Australia, 447, 447n43, 477
Austria
    Intelsat representation, 446
    Sputnik, public knowledge of launch of, 9

automobiles
    airbags, MEMS use in, 251, 261
    engine control sensors, 261
    MEMS use in, 254, 254n8, 260–61
    pollution requirements and manifold abso-
        lute pressure (MAP) sensors, 260
    skid control and MEMS, 261
    tire inflation level and MEMS, 261
avian biorhythms research, 271–72
aviation
    airplane accidents, research to improve
        survivability of, 131, 133, 143
    haptics and, 118–19
    mobile satellite communications, 475–78

B
BAC, 477
Back, R. E., 203
Bahan-Szucs, June C., 140–41
Bainbridge, William Sims, 536
Bakalar, David, 176n70
Bakalar, Leo, 176n70
Bakken, Earl, 99–100, 106
balance
    age and, 93
    flooring surfaces and, 93
    free-fall sensation and weightlessness, 90
    impact of spaceflight on, 90, 93
    patents related to evaluation of, 91–92
    weightlessness and sense of balance, 90
balance evaluation systems
    Balance Master, 90–93
    EquiTest, 90–93
    importance of, 143–44
    NASA contributions to development of,
        92–93
    patents related to, 91–92
    studies about, 110
    success of, 143–44
Ball Aerospace, 526n122
Bandung High Tech Valley, 316
Bandung Institute of Technology, 316
Banting Institute, 96
Banting Research Foundation Hall of Fame,
    96
barcodes, 146
Bardeen, John, 158
Barth, Phillip, 313
Bassous, Ernest, 260
Battelle Memorial Institute, 128–29, 196,
    207, 207n175, 207n177

batteries
    pacemaker batteries, **62**, 98, 99, 100,
        101, 103, 104, 105–6, 146
    rechargeable pacemakers, 62, 98, 103–4,
        105, 146
    satellites, charging batteries for, 106
    solar power for charging, 106, 343
    spacecraft, provision of electrical power
        with, 332
    telemetry and charging of, 106
    useful life for providing power for space-
        craft, 343
Bauer, Neil, 452–53
Bauer, Raymond A., 12, 59, 79–80
Beadle, George, 268
Beam, Benjamin "Ben," 277, 278–79
Bean, Alan, **353**, 357
Beaudoin, Jacques, 289–90
Beckman, Arnold, 164
Beckman Instruments, Inc., 136
Beggs, James, 395–96, 397, 399
Bein, Michael, 349
Belgium, 9
Bell, Alexander Graham, 166
Bell, David, 228
Bellamy, Ronald, 96–97
Bell Telephone Laboratories
    anisotropic etching invention at, 301n190
    communications satellite development by,
        438–40
    computer-aided design of circuits, semi-
        nar on, 219
    Digital Device Integration Department,
        295
    epitaxial deposition process and transistor
        manufacturing, 161–62
    ion implantation process, 213–14
    mesa technique and transistor manufac-
        turing, 160–61
    MOS transistor development at, 239
    patents for transistors, 158–59, 159n29
    planar process and transistor manufactur-
        ing, 161n34
    semiconductor research at, 199
    transistor development by, 158–59
Bendix Aerospace Systems Division, 356
Bendix Radio Division, 219
Bendix Research Laboratories, 198
Benefits Briefing Notebook (Denver Research
    Institute), 233–37, 233n263
Bennett, Gary L., 342

Berdahl, Bonnie J., 318
Berlin, Leslie, 152, 154–55, 156, 225–26, 242
Beth Israel Hospital, 98
bidirectional telemetry, 103
Big Bird, 480–81
Bijlefeld, Marjolijn, 78, 96
Billingham, John, 281
Binswanger (Bying), Gustav, 275n94
Bioastronautics Roadmap, 85
Bio-Impedance Technology, 110
biological clock research, 269, 272
Biomedical Monitoring Systems, Inc., 284
biomedical research
accelerometer development, 251, 308–14
Ames research, 264–85
Ames-Stanford biomedical research link, 264–69, 283–85, 284n136, 323–26, **326**
astronauts/humans, instrumentation research to monitor, 263, 264, 265, 282, 526
avian biorhythms research, 271–72
biomedical sensors, research on, 113–14, 257
brain probe, 294–99, 298n180
Congress and funding for, 232–33
funding for, 267–70, 268n58, 315, 324
growth of field and research programs, 290–94
instrument design research at Ames, 270–71n71, 270–74
NASA contributions to, 145–46
NASA interest in, 263–64, 526
NIH support for Stanford biomedical research, 290–93
piezoresistive pressure sensor research, 299–303, 310, 314–15, **326**
pressure sensor development, 257
sensors and transmitters, implantation of, 265
Stanford Electrical Engineering Department and School of Medicine, collaboration between, 291–92, 304
technology utilization, technology transfer, and dissemination of information, 265–69
Bionic Ear Institute, 138
biopotential data/electrocardiogram (ECG) signals, 273
biosatellite program, 263, 269

Biostim, 137
BioZ impedance cardiographs, 110–11
bird biorhythms research, 271–72
Birden, John, 336–37
Black, Joan, 42
black plague, 203n162
Blacks, poll questions about improving conditions of, 46–47, **46**
Bliley, Thomas J. "Tom," Jr., 454, 456–57
blind and visually disabled people
assistive technologies for, stimulation of innovation in, 123–24
haptics and assistive devices for, 117–20, 124
OCR devices, 120
patents for devices to assist, 120, 123–24, 125
research on applications to assist, 119, 291, 293
tactile reader for. *See* Optacon (Optical TActile CONverter) device
Bliss, James C., 117, 119, 120–21, 122–24, 125, 292
blood pressure
capacitance transducer on cardiac catheter for measuring, 270
disposable blood pressure sensors, 262
remote blood pressure monitoring transducer, 267
Blue, James, 140–41
Blume, Stuart S., 137–38
Bluth, B. J., 536
Boehlert, Sherwood, 422, 423
Boeing Corporation, 219, **523**
Boeing Launch Services, 458
Boeing Satellite Systems (BSS), 467
Bohrer, Philip J., 323n285
Bondi, Herman, 477
bone density
bone loss during long-duration spaceflights, 86
bone loss in weightlessness, 86, 88–89
Gait Scan system, 87
Mechanical Response Tissue Analyzer (MRTA), 86–90, 110, 143–44
Boucher, Roland, 476
Bowman-Gray School of Medicine Department of Biomedical Engineering, Wake Forest College, 267
brain probe, 294–99, 298n180
Brand, Stuart, 389, 394
Brattain, Walter H., 158

Brazil, 9, 421
"The Brick Moon" (Hale), 429
British Aerospace (Hawker Siddley), 464, 477, 479
Broadcasters Non-Profit Service Corporation (BNS), 460
Bromley, Allan, 401, 403
Brown, George E., Jr., 364n80, 411
budgets and funding
  accelerometer research, funding for, 313–14
  Apollo program, 34, 494
  balancing the budget, poll questions about, 69
  biomedical pressure sensor, NASA funding for, 299, 302–3, 314–15, **326**
  biomedical research, funding for, 267–70, 268n58, 315, 324
  Contract with American and budget cuts, 410–11
  deficit reduction and spending cap under Bush administration, 403
  Earth Observation System funding, 400, 403, 404, 405, 406, 409, 410–11, 425
  electronics costs as portion of NASA missions, 178
  electronics research funding, 183–84, 183n98
  environmental satellites, funding for, 425
  gas chromatograph research, funding for, 279–82, 285, 303–4n200, 306–8, 307nn215–16, 314, 315, 316–18
  government activities, poll questions about funding, 45–59, **46, 57, 58**
  Integrated Circuits Laboratory, funding for, 287–88, 289, 292–93, 323, 324, **326**
  International Space Station funding, 419
  MEMS research funding, 262, 275, 323–26, **326**
  Mission to Planet Earth/Earth Science funding, 400, 401, 402, 403, 404, 410–11, 418, 422–23
  Moon program, poll questions about value of, 12–16, 34
  NASA budget and funding, 403, 404, 410–11, 422, 518–19
  national debt, poll questions about, 5
  policy decisions, results of polls, and decisions about, 73–74
  poll questions about, 5–6, 18–20, **18, 20,** 28, 30, 31, 38–39, 43–44, **45,** 53–54, 67–74
  probe mission priorities over human spaceflight programs, 364, 364n80
  scientific research, funding of, 56–58, **58**
  space race between United States and Soviets, role in, 8
  Stanford Electronics Laboratories, funding for, 288–89, 323, 324
  Stanford University, NASA funding for research at, 287, 289, 323–26, **326**
  Triana funding, 416
Buehler, William J., 126–29
Bugos, Glenn, 265, 270, 277
Burch, Dean, 450
Burke, Robert, 78
Burns, Conrad, 456–57
Burroughs, 154–55, 226
Burshell, Alan L., 88, 96
Bush administration and George H. W. Bush
  deficit reduction efforts under, 403
  election of, 25
  environmental and energy policies under, 401–8, 409
  environmental values campaign pledge, 401, 408
  Space Exploration Initiative proposal, 502
Bush administration and George W. Bush
  election of, 418
  environmental policy and space vision, conflict between, 423
  environmental policy under, 418–24, 425
  Kyoto Protocol withdrawal by, 418, 421
  Mars, plans for exploration of and manned missions to, 67, 68–69, 71
  Moon, Mars, and Beyond space vision, 421–24, 426
  space program changes under, 67
Bying (Binswanger), Gustav, 275n94

C
cadmium, 215, 216
calculators, hand-held, **62,** 146
California Institute of Technology (CalTech)
  American Physical Society meeting at, 254–55
  medical technology research at, 85
  patent waiver requests, 247, **247, 248**
Callaghan, Christopher, 87
Cambridge Instruments, 221, 222

cameras
  spin-scan camera, 471
  TIROS satellite cameras, 437–38
Camille, Hurricane, 386
Canada
  Aerosat program, 477
  communications satellite launch by, 461
  Cosmos 954 incident and radioactive
    debris in, 365–66
  Intelsat agreements, 447n43
  Kyoto Summit and Protocol, 415
  Sputnik, public knowledge of launch of, 9
  telemedicine project, cooperative role in,
    113
  Terra satellite instrument package, 417
cancer
  cure for, opinions about, 4, 49–50
  detection of, biomedical sensor research
    for, 113–14
  fast-neutron cancer tumor treatment,
    138–41
  telemedicine and care for children with,
    112
Canon, 123
cantilever beams, 259–60, 300, 310–12, 322
Canyon system, 481
capacitive pressure sensors/transducers, 258,
  258n23, 270, 270–71n71
capacitors
  integrated circuit use of, 157, 168, 173,
    174, 175
  thin-film technology and formation of,
    168
  Tinkertoy project and silk-screen printing
    of, 169
Cape Canaveral
  Cassini launch from, 372
  Cassini protests at, 376–78
  RTG delivery to, 345
  satellites launched from, 343, 434, 437,
    440, 470, 471
Carderock Division, Naval Surface Warfare
    Center, 126
cardiac-output monitor and impedance
    cardiography, 107–11, 143–44
CardioDynamics, 110–11
cardiovascular system. *See* heart and cardio-
    vascular system
Carla, Hurricane, 386
Carle, Glenn C., 303–4, 307, 318, 319–20,
    326
Caroti, Simone, 541n16

Carpini, Michael Delli, 42
Carson, Rachel, 383, 384
Carter, William Gilbert, 443
Carter administration and Jimmy Carter
  environmental and energy initiatives
    under, 397
  funding for nuclear technology, 55
  land remote sensing, privatization of, 482
  Landsat, capabilities of, 391–92
  nuclear power for spaceflight, morato-
    rium on, 366
  presidential performance polls about, 55
Case Western Reserve University/Case
    Institute of Technology
  merger to create Case Western Reserve
    University, 256n16
  nanotechnology research at, 255–56n13
  pressure sensor development by, 257
Cassini
  characteristics and size of, 372
  economic impact statement for, 347–48
  launch of, 378, 414
  opposition to and protests against launch
    of, 367, 376–78, 381, 414
  RTG use on, **334**, 340, **350**, 367,
    372–78, **373**, **374**, **375**, **376**
  safety analyses and reports, 372–77
  Saturn mission of, 372, 378, 414
  social change and, 572
  success of, 378
  Venus–Venus–Earth–Jupiter Gravity
    Assist for, 372, 414
C-band frequencies, 451, 463, 465–66,
    468–69, 471, 478, 487–88
CBS (Columbia Broadcasting System)
  polling activity by, 1, 24, 68
  television satellite, launch of, 461
  transistor manufacturing by, 165
CBS Laboratories
  patent waiver requests, 247, **247**, **249**
  semiconductor research at, 214
Centaur launch vehicles
  ARPA communications satellite program,
    465
  Atlas-Centaur launch vehicles, 359, 449,
    471
  Centaur-Mariner H, 244–45
  Galileo and Ulysses launch vehicles,
    367–68
Center for Integrated Systems, 294
Center for Science, Mathematics, and
    Engineering Education, 11

Center for Telemedicine, Texas Tech
     University, 113
Central Intelligence Agency (CIA)
     applications satellites, interest in, 489
     as defense consumer of innovations, 233
     integrated circuits use by, 153
     Meindl proposal for research for, 290
     reconnaissance satellites, 479–81
     satellite program management by, 433
centralized processes, decentralized processes,
     and control, 503, 506–11, 511n68
Central Research Laboratories, 118
ceramic dielectric material, 276, 276n97
Certified Production Line (CPL) concept,
     234–36, 235–36n272
Ceruzzi, Paul E., 153, 225
CFCs (chlorofluorocarbons), 392, 399, 407
Chaffee, Roger, 15
Chaisson, Eric, 44
Challenger, 21, 24–28, 33, 367, 369, 399
Chardack, William C., 99–100
Charyk, Joseph V., "Joe," 437, 443
Chayes, Abram, 444
China
     Kyoto Summit and Protocol, 415
     Moon mission, plans for, 72
     spying by, alleged reports of, 453, 453n57
chip, 155. *See also* integrated circuits
chlorofluorocarbons (CFCs), 392, 399, 407
chromatograph, gas
     concept and function of, 303–5
     fabrication process, 305–6, 310, 317
     funding for, 279–82, 285, 303–4n200,
          306–8, 307nn315–316, 314, 315,
          316–18, **326**
     impact and importance of, 279, 316–20
     objective of research on, 280–81
     patents for, 316–17, 317n254
     research and development of, 303–8
Chrysler, **523**
CIGNA HealthCare, 111
circadian rhythm research, 272, 273, 280
cities, poll question about funding for aid
     to, 16, 46–47, **46**, 54, 55
city planning, systems management
     approach to, 492, 512, 533
Clark, Graeme, 137, 138
Clarke, Arthur C., 429, 430, 431, 432, 436,
     438, 471
Clatterbridge Neutron Cancer Therapy
     Cyclotron, 139
Clean Air Act, 407

Cleveland Clinic
     fast-neutron cancer tumor treatment
          through, 138, 139, 140–41
     founder of, 95
     pressure suit use during surgery at, 94
climate change/global warming
     atmospheric research and understanding
          of, 386
     causes of, 404, 411–12
     concern about by scientists, 384, 411–12
     confirmation of, 403–4
     controversies over policy and regulatory
          action, 385
     dust storms on Mars and, 393
     economic concerns about environmental
          initiatives, 401–2
     effects of, 412–13
     emissions control and Montreal Protocol,
          399, 407, 408, 409, 420, 425
     emissions policy under Clinton, 409,
          412, 413–14, 415
     global carbon emissions, call for reduc-
          tion of, 400
     global changes research, 398, 400–402,
          409
     global warming, pollution, and concept
          of climate change, 386
     greenhouse effect and, 384, 387, 396–97,
          400
     human impact on, 397
     increase in, 418
     increase in concern about, 400–401, 408,
          411–12
     Kyoto Summit and Protocol, 413, 418,
          421
     NASA plans for Earth monitoring
          beyond EOS, 417–18, 424, 425–26
     pollution and, 386
     regional impacts, study of, 412
     research on and policy development to
          address, 401–8, 410–12
     skepticism about, 401–2, 403–4, 410–12
     space technology use to address natural
          and manmade changes, 397, 412–13
     See also ozone depletion
Climate Change Research Institute (CCRI),
     418
Climate Change Science Program, 422
Climsat, 403, 404, 411
Cline, James E., 222
Clinton administration and Bill Clinton
     abortion policy of, 381

election of, 408
environmental policy under, 408–18
reelection of, 412
clocks, quartz, 146
closed systems program, 508
CNN polls
budget and funding questions, 5
Mars, questions about exploration of, 69
start of polling activities, 1
cobalt, 203
Cochlear, 137
cochlear implants
concept and function of, 134–35
importance of, 143
invention of, 135–38
NASA contributions to development of, 135, 138
patents for, 135–36, 137, 138
CODSIA (Council of Defense and Space Industry Associations), 190, 190n120
cold, opinions about cure for, 49–50
Cold War
communications satellites and, 441, 442
electronics technology and, 178, 181
prestige from Apollo program and, 352, 493–94, 533
public opinions and attitudes and, 8
space race and, 442
technology transfer and politics of, 524–25
Colino, Richard R., 448
college students
Harvard students, poll of about space goals, 59, **60**, 63–67
opinions and attitudes of, 49–50
Collingwood, Joe, **374**
Colodny, Edwin I., 452–53
Colorado State University, 141
Columbia, 30–32, 34–35, 67, 419–20
Columbia Accident Investigation Board, 188
Columbia Broadcasting System. *See* CBS (Columbia Broadcasting System)
Columbia University
Cartwright Prize, 95
electronics research laboratory at, 288
Combat Casualty Care Guidelines, 96–97
Commerce, Department of
Business and Defense Services Administration, 153, 233
Weather Bureau transfer to, 436
Commercial Satellite Communications Initiative (CSCI), 468–69

Committee for Evaluating the Feasibility of Space Rocketry (CEFSR), 432
"Communication and Observation Problems of Satellites" (RAND), 431
Communications Satellite Act (1962), 440–42, 443, 459, 460, 471
Communications Satellite Corporation (CSC), 452. *See also* COMSAT
communications satellites
active satellites, 439, 525–26
as applications satellites, 427–28
Cold War politics and, 441, 442
commercial communications satellites, use of by DOD, 468–69, 488
commercial development and launch of, 428, 441–42, 459–64, 489–90
design of, 448–49, 450–51, 462, 466
direct access, 454, 456–57
domestic communications satellites (DOMSATs), 449, 459–63
early concepts and programs, 431–32, 438–45, 489–90, 525–26, 526n122
economic impact of, 428
future of, 487–88
government role in development and launch of, 428, 441–42, 525–26, 526n122
growth and consolidation in, 462
increasing demand for, 468–69, **468**
launch vehicles for, shortage of, 439–40
maritime communication system, 449
military satellite communications, 464–69
mobile satellite communications, 475–79
monopoly over, concerns about, 441–42, 449
monopoly over, end to, 450
passive satellites, 439, 525
profits from, 483
public attitudes toward space exploration and benefits of, 65–66
public understanding of operation of, 41
satellite communications as foreign policy activity, 443–44, 525–26
Shuttle flights for launching, 21
societal impact of, 428, 487
space technology and development of, **62**, 78, 146
telecommunication companies as managers of, 441–42
telemedicine projects and, 111–17
See also COMSAT
Compac, 169

comparative planetology
    atmospheric research through, 387
    environmental mission of NASA and, 393
    objective and purpose of, 383, 387
    uniqueness and vulnerability of Earth, realization of through, 384, 425
complementary transistor logic (CTL) circuits, 204
Components Division/Electronic Components Research Division, 198
Component Technology Laboratory/ Electronic Materials Laboratory, 198
Computer-assisted Survey Methods (CSM) Program, University of California, Berkeley, 7
computers
    computer-aided design of circuits, 218–19, 243
    personal computers, development of, 116
    See also Apollo guidance and navigation computers
Computers in Spaceflight (Tomayko), viii
COMSAT
    Aerosat program, 476–77
    competition for, 450, 452
    contracts in proportion to contribution of funding, 444
    costs and profits, 443–44, 449
    creation of, 441–42, 525
    data from, limits on, 446, 446n40
    definitive agreements, 445–49
    DOD contract for transponders, 469
    domestic communications satellites, 449, 462
    end of, 457
    FCC DOMSAT report and, 461
    last satellite launched by, 449
    leasing of satellites from, 461
    Lockheed merger, 451–57
    markup on pricing by, 456
    mission of, 449
    monopoly status of, 441–42, 450
    negotiations and interim agreements, 442–44
    office of, 451–52
    pilot program for DOMSATs, 460
    purchase of by Lockheed, 453–55, 456–57
    restructuring of, 452–53
    Satellite Business Systems (SBS), 462

COMSAT Government Services, Inc. (CGSI), 455
COMSAT Mobile Communications, 457
COMSAT Radiation Systems, Inc. (CRSI), 452
COMSAT World Systems, 457
COMSTAR satellites, 462
Conference Européenne des Télécommunications par Satellites (CETS), 443
Conference of European PTTs (CEPT), 443, 444
Conference on the Peaceful Uses of Outer Space, United Nations, 397
configuration management practices, 504, 507, 511
Congress, U.S.
    ancillary policy, 501n34
    Clean Air Act, 407
    Communications Satellite Act, 440–42, 443, 459, 460, 471
    Contract with American and budget cuts, 410–11
    government shutdowns by, 411, 412
    Land Remote Sensing Act, 482
    Maritime Satellite Act, 479
    Orbit Act, 454, 455–57
    space race between United States and Soviets, role in, 8
    technology utilization, politics, and, 231–37, 232n256, 233n263, 235–36n272
    Triana debate in, 416
Conrad, Pete, 357
conservation, energy emergencies and, 390, 394, 500
contamination and forward contamination concerns, 384, 387–88, 393, 425, 426
CONTEL, 452
Continental Devices, 180
control, centralized processes, and decentralized processes, 503, 506–11, 511n68
Cook, Liz, 407
cool suit treatment for hypohidrotic ectodermal dysplasia, 94–95
Coon, Grant W., 270–71, 270–71n71
Cooper, Irving, 83
copper, 203
Corbin-Farnsworth, Inc., 267
Corliss, William R., 337

CORONA satellites, 433, 479–80, 481, 485–86
Coronet satellite project, 463
corporate power, model of, 511, 511n68
Cosmos 954 incident, 365–66
Cosmos and Culture, ix
COSMOS consortium, 477
Cotton, John, 88
Council of Defense and Space Industry Associations (CODSIA), 190, 190n120
Coupling, J. J. (John R. Pierce), 429, 431–32, 431n10, 438–39
Courier, 464, 465–66, 489
Cox report, 453, 453n57
crew model of spaceflight, 551–54, 571
Crile, George Washington, 95–96
crime, poll questions about spending to curb, 5, 16, 46–47, **46**, 54
Cristofori, Bartolomeo, 89
Crockett, Bruce, 452
Cryofit couplings, 127
cryogenic gases/liquids
    medical use of, 82–83
    propellant use of, 82
    space program use of and general development of industry, 83
cryosurgery
    cryoprobe for, 83
    experiments with, 83
    liquid nitrogen use for, 82–83
CSCI (Commercial Satellite Communications Initiative), 468–69
CSF, 477
CTS Microelectronics, Inc., 199
Cuban Missile Crisis, 485–86
Cull, Ronald C., 342
culture and cultural traits of NASA, 506–11
curium-242, 349
cyclotron, 138–41

D

Dactyl, 371
Dailey, Brian, 451
d'Alembert, Jean le Rond, 230
Daley, Brian, 454
Danellis, Joan Vernikos, 279–80
Dassoulas, John, 343
data bank of electronics parts, 191, 193–94
David, Moses, 21
David Clark, 97
Deboo, Gordon J., 271–73, 279, 282

debris/space debris
    concern about, 401, 414–15
    falling from orbit to Earth, possibility of, 385, 419–20
    mitigation efforts, 414–15, 426
    pollution of near-Earth space by, 384–85
    radioactive debris, 347–48, 358, 365–66, 414, 419–20
    threat to space assets in orbit from, 385, 395, 415, 417, 420
    tracking space junk, 395
decentralized processes, centralized processes, and control, 503, 506–11, 511n68
Decree, 465
Deep Space 1 probe, 141
deep space missions
    nuclear power systems for, 331, 333–34, **334**, 419
    photovoltaic solar cells, limitation of use for, 332–33
    RTG use during, 351, 358–64, 367–68
Deep Space Network (DSN), 361
Defense, Department of (DOD)
    applications satellites, interest in and funding for, 489, 490
    Certified Production Line (CPL) concept, adoption of, 235, 235–36n272
    commercial communications satellites, use of by DOD, 468–69, 488
    Commercial Satellite Communications Initiative (CSCI), 468–69
    communications satellite development by, 526n122
    communications satellites, increasing demand for, 468–69, **468**
    data bank of electronics parts, use of by, 193
    as defense consumer of innovations, 233
    integrated circuits use by, 153, 177
    Landsat program, 409
    management and planning practices in, 495–96, 503–5
    NASA programs from, 522n108
    Nimbus weather satellite use by, 386
    nuclear power systems for spacecraft, role in development of, 338–39
    sealed container, procedures for locating leaks in, 217, 220
    space program of, 383–84
    space race between United States and Soviets, role in, 8

Stanford Electronics Laboratories, funding for, 289
 transistor use by, 165–66
 Transit satellite, support for launch of, 344–45
Defense Advanced Research Projects Agency (DARPA)
 Defense Sciences Office and Air Force nanotechnology program, 172
 DLI project role, 116
 MEMS research funding from, 262
Defense Meteorological Satellite Program (DMSP), 437–38, 489, 490
Defense Satellite Communications System I (DSCS I), 465–66
Defense Satellite Communications System II (DSCS II), 466
Defense Satellite Communications System III (DSCS III), 466, 467
Defense Sciences Office, Defense Advanced Research Projects Agency (DARPA), 172
defibrillators, implantable, 98, 105
Delta rockets
 reliability of, 448, 470
 satellite launches with, 440, 448, 461, 466, 470, 471, 472, 474
 Thor-Delta rockets, 432, 434–35
Denmark, 113
Denver Research Institute, University of Denver, 233–37, 233n263
depletion hypothesis, 530
Desert Shield/Desert Storm Operations, **468**
deuterium, 205
diaphragm production, process for, 257, 300–302, 315
Dick, Steven, 537
Dickinson, Ken, **374**
Diderot, Denis, 230
diffusion, embodied, 79
diffusion, horizontal and vertical, 79–80
Diffusion of Innovations (Rogers), 79
diffusion process for semiconductor manufacturing, 160, 202–3, 214–15, 214n202
Digisonic programme, 137
DigitalGlobe/WorldView/EarthWatch, 482
Digital Library Initiative (DLI), 116–17
DiMaggio, Paul, 102
Dimeff, John, 273, 277–79, 278n107, 281, 282–85, 284n137
Dimeff Associates, 284
Dingell, John, 457

diodes
 germanium use in, 198
 gold-flake contamination, 180–81
 integrated circuit use of, 157, 168, 175, 181
 ion implantation process, 214
 Ranger program and failure of, 180–81, 182
 sales of compared to sales of transistors, 164
 silicon use in, 159
 solid-state diodes, 157
diode-transistor logic (DTL) circuits, 204, 207n175, 227–28
DirectTV, 463
dirigibles, 157
Discovery
 bone loss in weightlessness research on, 89
 Glenn return to space on, 30
 launch of and Columbia accident, 32
Dividends from Space (Ordway, Adams, and Sharpe), 81–82
Docotte, Behrman, 136
domestic communications satellites (DOMSATs), 449, 459–63
Donaldson, Ralph W., Jr., 279, 303–4n200, 307, **326**
Dong, Eugene, Jr., 292
"Don't Write: Telegraph" (Pierce/Coupling), 431
Doppler effect, 434
DORIAN satellites, 480
Dornier, 477
Douglas, **523**
Doyle, James H., 136
drug addiction, poll questions about spending on, 46–47, **46**
drug-delivery systems, implantable, 104–5, 106
Dryden, Hugh L., 438, 508, 508n58
Dryden Flight Research Center, 185
dryers and washers, balancing, 251
DSCS I (Defense Satellite Communications System I), 465–66
DSCS II (Defense Satellite Communications System II), 466
DSCS III (Defense Satellite Communications System III), 466, 467
DSTS-G, 469
dual-energy x-ray absorptiometry (DXA), 88, 89

Dudley-Rowley, Marilyn (Marilyn Dudley-Flores), 538, 541n16
Dukakis, Michael, 25
Duke University Medical Center, 267
Dummer, Geoffrey W. A., 172–73, 174
Dunlap, W. Crawford, 198
Dyna Med, 94, 95
Dyna Med Anti-Shock Trouser (DMAST), 94–97
Dynamic Systems, 131–32
Dyno system, 481

E
Eagleburger, Lawrence, 455
Early Bird, 444–45, 448, 450, 459, 476, 489
Earth
    appreciation of as spinoff of Apollo, 389, 533–34
    Earth-monitoring system program, 396, 397–98, 400
    Earth system science, creation of, 384
    heat budget of, 436
    images from Apollo mission and appreciation for, 388–89, 390, 416, 424, 498–99, 498n28
    images from Triana and appreciation for, 415–16
    IMP and data collection about atmosphere, 238–39
    living system, Earth as, 389, 393, 395
    magnetosphere, 239
    Moon-Earth comparison, 388–89
    NASA plans for Earth monitoring beyond EOS, 417–18, 424, 425–26
    uniqueness and vulnerability of, realization of, 384, 425
    See also Mission to Planet Earth (MTPE)/ Earth Science
Earth, mission from
    Goldin interest in, 409
    priority of, 409–10, 425
    understanding Earth through, 383, 425
Earth Day, 390, 424–25
Earth in the Balance (Gore), 408
Earth Observation Satellite Company (EOSAT), 482
Earth Observation Summit, 421
Earth Observation System (EOS)
    adoption of, 400, 404
    Aqua satellite, 420–21
    Aura satellite, 422
    budget and funding for, 400, 403, 404, 405, 406, 409, 410–11, 425
    CCRI creation and, 418
    Clinton administration support for, 409
    completion of, 422, 424
    depoliticization of, 425
    design of, 398–99, 402, 403, 404–5, 425
    development and purpose of, 398–99
    environmental research with, 401, 402–3, 410–12, 424, 425
    evolution of spacecraft leading to, 404, 405
    faster, better, cheaper revision of, 406, 409, 411
    GEOSS link to, 422
    international partners in, 398, 402, 421
    MTPE role of, 402, 404, 409
    NASA plans for Earth monitoring beyond, 417–18, 424, 425–26
    regional impacts, study of, 412
    satellite fleet to replace original platform concept, 405
    schedule for satellites, 412
    space station connection to, 399, 402, 405
    support for implementation of, 404–5
    Terra satellite, 417, 418, 420, 421
    USGCRP role of, 400–401
Earth Observation System Data and Information System (EOSDIS), 409, 412, 420–21, 422
Earth resources satellites. *See* environmental/ Earth resources satellites
Earth Resources Technology Satellite (ERTS)/Landsat, 391–92, 409, 420, 424, 438, 481–82, 484, 489
"Earthrise," 389, 498–99, 498n28
Earth Science program. *See* Mission to Planet Earth (MTPE)/Earth Science
Earth Sciences, Committee on, 400
Earth Sciences and Applications Division, 397–98
Earth Summit, 408
Earth System Science Committee, 397, 398
EarthWatch/DigitalGlobe/WorldView, 482
Echo satellite, 439, 489, 525
economics
    Apollo program, impact of, 492, 517–30, 519n97, 522n107, 529n131, 529nn133–134, 533
    Cassini economic impact statement, 347–48
    communications satellites, impact of, 428

contracts and space projects, 522–23, 522n108, **523**

economic concerns about environmental initiatives, 401–2

Moon, economic development of, 100

multiplier, 520–23, 520n100, 521n103

public goods, 519, 519n97, 522

regional impact of NASA Center, 528–29, 529n131

return on investment (ROI), 518

space program, economic benefits of, 61–62, **62**

spinoff technologies, 523–28

technology development and, 78

tourism, impact of, 528, 564

wealth-building processes, 522, 533

Edelson, Burton, 396, 397–400

education

Apollo program impact on, 491, 530–32, 531n140, 533

funding for, poll questions about, 16, 46–47, **46**, 53–54, 55, 69

medical education, Web-based digital library of videos for, 113

NASA educational efforts, 44–45

science education in United States and launch of Sputnik, 10–11, 496, 530

social science classrooms, expanding space into, 548–51

space program as educational tool, 67

support for space program and, 35–36, 41–45, 74

Sustaining University Program (SUP), 532, 532n150

See also science, technology, engineering, and math (STEM) disciplines; science, technology, engineering, math, and astrosociology (STEMA) disciplines

Education, Office of, 287–88, 292, 292–93n166

educational television, 461

Effective Isotropic Radiated Power (EIRP), 448, 458, 466

EG&G Florida, Inc., 376

Einstein, Albert, 378

Eisenhower administration and Dwight Eisenhower

atomic power, support for peaceful uses of, 338, 339–40, **339**

communications satellite development by commercial, private companies, support for, 441

Echo 1 satellite and recording of, 439

reconnaissance satellite development, authorization of, 479

RTG development for spacecraft, support for, 339–40, **339**

space race between United States and Soviets, role in, 8

electricity/electrical power

for flying saucers, 134

ion drive propulsion, 141

methods for providing, 332, **335**

requirement for flying in space, 331–32

research and development activities to advance power technologies, 341–42

See also radioisotope thermal generators (RTGs)

Electric Spacecraft, 134

electrocardiogram (ECG) signals/biopotential data, 273

electrocardiogram (ECG) telemetry unit, 267

electrocardiograph, 107

Electrochemical Society Meeting, 214n202

electrodes, 99, 100, 267

electroencephalogram (EEG), 273

electrometers, 278, 278n107

Electronic Components Research Division/ Components Division, 198

Electronic Industries Association, 164

Electronic Materials Laboratory/Component Technology Laboratory, 198

Electronics and Control Directorate, 180, 183, 183n96, 188–89, 195

electronics and microelectronics

Apollo and Minuteman programs and development of, 153, 225

challenges for NASA, 178–83

conditions encountered in space, resistance to harsh, 181–82, 197, 201, 206–13

costs as portion of NASA missions, 178

electrical circuit design and integrated circuits, 156–57

flight failures and electronic failures, 179–81

microelectronics use by NASA, 181

military concepts for miniaturization of electronics, 169–72, 169n49, 170n52

miniaturization of electronics and integrated circuits, 168, 526

miniaturization of electronics and launch
    of satellites, 179
production quality assurance program,
    234–36, 235–36n272
reliability requirements for Air Force,
    155, 171
reliability requirements for NASA, 155,
    243–45
spacecraft electronics and the space race,
    178–79
standards and specification requirements,
    development of by NASA, 181, 182,
    190–94, 196, 235–36, 235–36n272
symposia on microelectronics, 190–91,
    190n122
temperature and failure of, 223–25
See also integrated circuits; transistors
Electronics Components Laboratory, 198,
    201, 213
Electronics Components Reliability Center
    (ECRC), 196
electronics research by NASA
    failure and reliability research, 196, 197,
        201–6
    funding for, 183–84, 183n98
    integrated-circuits research, 185–86,
        186n107, 186n109
    packaging and interconnection problems,
        196
    personnel and facilities for, 183–86,
        183n96
    programs and resources for, 183–87, 195,
        228, 243–45
    project- and mission-orient research,
        183–87
    reliability offices, programs, and hand-
        books, 187–94, 191n127, 243–45
    standards and specification requirements,
        development of, 181, 182, 190–94,
        196, 235–36, 235–36n272
    testing method standards, 196
    thin-film technology research, 185–86
    university-industry-government complex,
        228
Electronics Research Center (ERC)
    biomedical instrumentation research at,
        264, 265, 275
    closing of, 216n209, 244, 275
    Components Division/Electronic
        Components Research Division, 198
    Components Standards Branch, 193

Component Technology Laboratory/
    Electronic Materials Laboratory, 198
computer-aided design of circuits, semi-
    nar on, 219
contracted research activities for, 196
data bank of electronics parts, 191,
    193–94
Electronic Components Research
    Division, 198
Electronics Components Laboratory, 198,
    201, 213
Failure Mechanisms Branch, 198,
    216–17, 222
formation of, 195, 197
handbooks on electronics parts, 191–93,
    191n127
integrated-circuit fabrication processes
    research, 199–200
integrated circuits research by, 186n105,
    243–44
laboratories at, 197–200
location of, 194
Microelectronics Subcommittee, 198
Microwave Radiation Laboratory, 211
opening of, 194–95
personnel for, 196, 197–99, 199n151
PREDICT facility, 199–201, 243, 282–83
Qualifications and Standards Laboratory,
    189, 194, 198, 199, 201, 213, 220,
    244, 245
scope of activities and research at,
    194–97, 243–44
space-charge-limited triode, development
    of, 209–11, 243
Tech Briefs from, 229–30, 246, **246**
transistor failure analysis research at, 244
Electronics Research Task Group (ERTG),
    194–97, 196n142
    Components and Technology Division,
        196
    Electromagnetic Division, 195
    Instrumentation and Data Processing
        Division, 195–96
electron microscopes, 200, 220–23,
    222n231, 234, 236–37
Electro-Optical Systems, **247**, **248**, 255
electrostatic bonding (anodic bonding ,
    Mallory process), 258, 306, 311, 312
Ellis, Richard T., 105
Ellul, Jacques, 232n256
El Niño, 412–13

embodied diffusion, 79

emitter-coupled logic (ECL) circuits,
    207n175

employment opportunities as benefit of
    space program, 67, 520

Enceladus, 378

*Encyclopédie ou Dictionnaire raisonné des
    sciences, des arts et des métiers par une
    société de gens de lettres* (Diderot and
    d'Alembert), 230

Enduring Freedom, Operation, **468**

energy
    conservation and energy emergencies,
        394, 500
    environmental and energy initiatives, 397
    oil boycott and energy emergencies, 390

Energy, Department of/Atomic Energy
    Commission (AEC)
    as defense consumer of innovations, 233
    integrated circuits use by, 153
    management structure for RTG program,
        351–52
    nuclear power systems for spacecraft, role
        in development of, 338–40, **339**,
        343–46, 351
    radioisotope development by, 336
    research and development activities to
        advance power technologies, 341–42
    RTGs, support for use of, 351
    safety reviews and approvals before launch
        of RTGs, 365, 372–73
    solar-powered satellite (SPS), 394–95
    Transit satellite, support for launch of,
        344–45

Engineering in Medicine and Biology
    Society, 290

engineers/engineering
    applications satellites, interest of engi-
        neers in in, 430–32
    astrosociological issues and, 564–65,
        571–72
    convergence of branches of technology,
        145–46
    credit to engineers for innovation, 106–7

Engler, Richard, 338–39

Enterprise, 19

environment
    Bush (G. H. W.) administration policies,
        401–8, 409
    Bush (G. W.) administration policies,
        418–24, 425
    Clinton administration policies, 408–18

contamination and forward contamina-
    tion concerns, 384, 387–88, 393,
    425, 426

depoliticization of environmental mis-
    sion, 415–16, 425

economic concerns about environmental
    initiatives, 401–2

energy and environmental initiatives, 397

environmental policy, 385

funding to improve and protect, poll
    questions about, 46–47, **46**, 54, 69

images from Apollo mission and apprecia-
    tion for Earth, 388–89, 390, 416,
    424, 498–99, 498n28

images from satellites and, 383

images from Triana and appreciation for
    Earth, 415–16

impact of NASA on, viii, 383–85

insults to, early warnings of, 383

Montreal Protocol and environmental
    policy, 399, 407, 408, 409, 420, 425

NASA environmental mission, 384,
    384n1, 385, 390–92, 395, 398–405,
    415–16, 420, 424–26

NASA plans for Earth monitoring
    beyond EOS, 417–18, 424, 425–26

national policy and, 390

space vision and environmental policy,
    conflict between, 423

threat from technology, 384–85

See also pollution

environmental/Earth resources satellites
    budget and funding for, 425
    civil land remote sensing satellites, 481–82
    Climsat, 403, 404, 411
    development and launch of, 385–87,
        395, 398
    global monitoring satellites, 396
    Glory, 421, 423, 424
    Ice, Cloud, and land Elevation Satellite
        (ICESat), 424
    importance of research with, 425
    Landsat/Earth Resources Technology
        Satellite (ERTS), 391–92, 404, 409,
        420, 424, 438, 481–82, 484, 489
    land-use observation satellites, 428–29
    mission to Earth via, 385
    priority of, 425
    remote sensing technology for, 383,
        391–92
    smaller satellites, support for by Goldin,
        405–6

system of systems by linking satellite
    capabilities, 421, 422
Triana, 415–16, 418
See also Earth Observation System (EOS)
environmental movement
    attitudes, values, and way of thinking of,
        384, 424–25
    beginning of, 383, 424
    opposition to NASA by environmental-
        ists, 413–14, 500
    science-emotion amalgam about Earth
        and, 393
Environmental Protection Agency (EPA),
    390
Environmental Science Services
    Administration (ESSA)
    ESSA satellites/TIROS Operational
        System (TOS), 438, 470–71, 489
    polar weather satellites, responsibility for,
        469–71
    See also National Oceanic and
        Atmospheric Administration
        (NOAA)
epitaxial deposition process, 161–62
EquiTest, 90–93
Ericcson, 477
ERNO, 477
Esaki, Leo, 208
Eshleman, Von, 324
ethylene glycol, 217
Europa, 371, 516
Europe
    cooperative space ventures with United
        States, 515–17, 516n84, 533
    environmental satellite coverage by, 472
    Kyoto Protocol withdrawal by Bush, reac-
        tion to, 418
    Kyoto Summit and Protocol, 415
    meteorological data sharing and interna-
        tional cooperation, 472–73
    space program of, transfer of systems
        management to, 515–17, 533
European Communications Satellites, 464
European Organisation for the Exploitation
    of Meteorological Satellites
    (EUMETSAT), 473
European Space Agency (ESA)
    astrosociological implications of space
        exploration, recognition of, 540
    Cassini space probe role, 372
    EOS role of, 402
    Eutelsat, 464

formation of, 515n82
juste retour principle, 444
Space and Society Conference, 540
systems management practices, transfer
    to, 515–17, 533
European Space Conference (ESC), 476–77
European Space Research Organisation
    (ESRO), 476, 476–77, 515, 515n82
European Space Vehicle Launcher
    Development Organisation (ELDO),
    476–77, 516
Eutelsat, 464, 479
Everhart, Thomas E., 222, 237
Evolved Expendable Launch Vehicle (EELV)
    program, 466
exobiology
    objective and purpose of, 388
    Stanford research on, 324
exploration ethos, 498–500, 498n24,
    498n28, 533–34
Exploration of Space, The (Clarke), 436
"Exploration of the Universe with Reaction
    Machines" (Tsiolkovsky), 429–30
Exploration Systems Mission Directorate,
    421
Explorer IMP series. *See* Interplanetary
    Monitoring Platform (IMP)
Explorer satellites
    Explorer 1 satellite, 10
    Explorer VI satellite, 436
    transistor use in, 167
    weather observation payloads on, 436
Extraordinary Administrative Radio
    Conference (EARC), 443
extraterrestrial life
    detection of, 569–70
    emotional reactions to finding, 569–70
    exobiology and discovery of, 388
    intelligent extraterrestrial life, goal of
        detecting, 567
    messages from, 570
    messages to, 363–64, 570
    search for, 547, 565–70
"Extra-terrestrial Relays" (Clarke), 430
extravehicular activity (EVA), 356

F

F-14 fighter jet hydraulic system, 127, 129
Fagerdala World Foams, 133
Failure Mechanisms Branch, 198
Fairchild, 526n122
Fairchild Camera and Instruments, 164

Fairchild Semiconductor
    commercialization of and commercial
        applications for integrated circuits,
        153–55, 226–27, 226n242
    founding of company, 164–65
    gold doping technique, 202
    integrated circuit invention by, 121,
        174–75
    integrated circuit manufacturing by,
        153–54, 175, 176, 242
    location of, 175
    mesa technique and transistor manufac-
        turing, 160–61
    Micrologic Elements, 153–54, 177
    Moore's law and number of components
        on single circuit, 178
    MOS circuits manufactured by, testing
        for radiation resistance of, 207n177
    MOS technology research at, 239–40
    off-the-shelf integrated-circuit product
        line, development of, 151–52,
        151n8
    planar process and transistor manufactur-
        ing, 161, 161n34
    price of integrated circuits from, 153–55,
        226
    specialized integrated circuit manufactur-
        ing, 227
    supplier of integrated circuits for Apollo
        program, 151, 176, 226–27
    testing integrated circuits manufactured
        by, 204
    transistor prices and sales numbers, 164
Farber, Louis I., 212
Farley, Philip J., 441
farmers, poll question about funding for
    subsidies for, 16
Farnsworth, Philo, 122
fast-neutron cancer tumor treatment,
    138–41
fast-scan infrared microscope, 223–25
Federal Aviation Administration (FAA)
    as defense consumer of innovations, 233
    integrated circuits use by, 153
    National Aviation Facilities Experimental
        Center, 476
    Wide Area Augmentation System
        (WAAS), 475
Federal Communications Commission
    (FCC)
    COMSAT monopoly and treatment of by
        FCC, 450

    domestic communications satellites,
        459–61
    Intelsat licensing order and privatization,
        458
    Lockheed purchase of COMSAT,
        approval of, 455–57
    Orbit Act and direct access to Intelsat, 456
    telecommunication companies as manag-
        ers of communications satellites,
        support for, 441
Feedback, Project, 337–38
ferrite memory technology, 197
Feynman, Richard P., 254–56
FI-100 transistor, 241
Fiat, 477
fiber-optic cables, 450, 452–53, 457
films, development of, 526–27
Finger, Harry, 351, 352
finite element simulation, 88
Finland, 113
Finney, Ben, 537
Fischell, Robert E., 104–6, 107
Fisk, Lennard, "Len," 400, 404–6
fixed satellite services (FSS), 458, 463,
    468–69, 483, 484
Flanigan, Peter, 450
flashbulb memories, 33
Fleet Satellite Communications System
    (FLTSATCOM), 466–67, 478
Fletcher, James C.
    Centaur launch vehicle, decision to not
        use, 368
    environmental interests of, basis for, 391
    environmental mission of NASA under,
        390–92, 398, 399, 425
    leadership style of, 391
    patent for biological impedance electrode
        assembly, 109
    retirement and legacy of, 392, 401
    return to NASA after Challenger acci-
        dent, 399
    as successor to Paine, 390
flight controls, 118–19
flip-flop circuits, 154, 173, 249
Florida Coalition for Peace and Justice, 414
Floyd, Samuel R., 222
FLTSATCOM (Fleet Satellite
    Communications System), 466–67, 478
fluorocarbon liquid, locating leaks in sealed
    containers with, 218
flying saucers, electrostatic propulsion of, 134
FM transmitters, 212, 212n195

Ford Aerospace (Space Systems Loral), 451, 453, 454, 455, 462, 465, 472, 526n122
Ford Foundation, 460
foreign aid, poll questions about spending on, 46–47, **46**
Framework Convention on Climate Change, 408
France
  Aerosat program, 477
  cooperative space ventures with, 515
  data from COMSAT, limits on, 446
  Intelsat agreements, 447, 447n45
  NATO, withdrawal from, 446n40
  scientific development and military strength of United States and Soviets, opinions about, 10, **10**
  Sputnik, public knowledge of launch of, 9
  Système Pour L'Observation de la Terra (SPOT) satellite, 482–83, 484
  TOPEX/Poseidon development, 412
Franks, Wilbur, 96
Friedensen, Victoria, 379
Frobenius, Wolf Dietrich, 309
frontier narrative, 494–95, 495n12
Frosch, Carl, 296
Fryer, Thomas Benton, "Tom," 271–74, 278, 279, 281, 282, 283–84nn132–133, 283–85, **326**
Fubini, Eugene, 465
fuel cell technology, 332
Furash, Edward E., 59
Futron Corporation, 72–73

G
Gagarin, Yuri, 4, 19
Gaia Hypothesis, 389, 393, 395
Gait Scan, 87
Galileo
  discoveries made by, 371
  end of mission and plunge into Jupiter's atmosphere, 371–72
  environmental impact statement on, 368–69, 370
  Jupiter mission of, 367–72, 378
  opposition to and protests against launch of, 367–72, 381
  permission to deploy from Space Shuttle, 368–70
  RTG use on, **334**, 340, **350**, 367–68
  solar array retrofit of, 368
  Venus–Earth–Earth Gravity Assist for, 368, 369–70, 371

Galileo In-Orbit Validation Experiment (GIOVE) satellites, 475, 485, 488
gallium arsenide, 215–16
gallium phosphide, 296, 296n176
Galls, 95
Gallup, George, Jr., 29
Gallup Organization, 7
Gallup polls
  budget and funding questions, 5, 12–14, 18–19
  Challenger accident and questions about manned space program, 26
  Columbia accident and questions about space program, 30–32
  German bombs, opinions about, 3
  goals in space, attitudes toward, 62–63
  man flying to Moon, changing opinions about, 3–4, 4n5
  Mars, questions about exploration of and manned missions to, 16, 71–72
  Mars, questions about life on, 71
  national defense and military capabilities in space, questions about, 24
  space exploration, questions to children about interest in, 44–45
  space race between United States and Soviets, questions about, 8, 11
  start of polling activities, 1
  travel into space, questions about personal interest in, 35–36, 37–38
  Voyager 2 discoveries and questions about manned space program, 28–29
Galveston hurricane, 486
GAMBIT satellites, 480–81
Gangale, Thomas, 541n16, 551–52
Ganymede, 371
Garbarini, Robert F., 189
Gasch, Oliver, 370
gas chromatograph. *See* chromatograph, gas
gases, measurement and reporting on employee exposure to, 308
Gaspra (asteroid 951), 369, 371
GE Americon/RCA Americom, 451, 454, 464
Gedankenexperiment (thought experiment), 244
Gemini program
  contractor for capsule, **523**
  electronics for, 178
  history of space program, communication of, 147
  public opinions and attitudes toward, 51

gender and support for space program,
36–38, **39**
General Dynamics, 247, **247**, **248**
General Electric (GE)
Aerosat program, 477, 478
alloy-junction transistor fabrication by,
159
communications satellite manufacturing
by, 451, 463–64, 465
diodes manufactured by, failure of, 180
germanium diode and transistor research
at, 198
GE Space, 451
pacemaker manufacturing by, 101
RTG manufacturing by, 356
RTG safety analysis by, 372
scanning electron microscope development by, 222
transistor manufacturing by, 165
transistor manufacturing for defense
consumers, 166
Valley Forge Space Technology Center,
196
General Electric Astro Space, 474, 479
General Electric Company, Hirst Research
Laboratories, 275–76, 275n94,
275–76nn96–97
General Instrument, 207n177
General Micro Electronics Company, 241–42
General Motors Corporation, Cadillac
Division, 254n8
General Precision, Inc.
Librascope division, 203–4, **247**, **248**
solid-state laboratory, 276
General Purpose Heat Source (GPHS)
system, 340, 372, **373**
general relativity theory, 378
General Social Survey (GSS), 6, 18–19, **18**,
20–21, **20**, 37, 38–39, 40–41, 43–44,
73
Geophysics Corporation of American
(GCA), 197
Georgia Institute of Technology, 85, 262
GEOSS (Global Earth Observation System
of Systems), 422, 423
geostationary operational environmental
satellites (GOES), 472
geostationary orbit satellites, 430–31, 469
geosynchronous-Earth-orbit (GEO) satellites, 428, 439, 450, 465, 471, 481,
488, 526
Gephart, Landis S., 187, 201

germanium
depositing films with controlled-hole
mobility, development of process
for, 209
ion implantation in, 214
mesa technique and transistor manufacturing, 160–61
miniature devices made from, 253–54
piezoresistivity of, 256–57
planar process and, 161
semiconductor switch use of, 296
transistor radio use of, 167, 167n45
use of in diodes, 198
use of in transistors, 158, 159, 198
Germanium Products, 165
Germany
Aerosat program, 477
cooperative space ventures with, 515
robot bombs, opinions about existence
of, 3
scientific development and military
strength of United States and
Soviets, opinions about in West
Germany, 10, **10**
Sputnik, public knowledge of launch of, 9
Geveden, Rex, 509
Ghandi, Sorab K., 202–3
Gibbons, Jack, 408, 409
Gibson, Ralph E., 342–43, 434
Gilfillan, S. C., 81
Gillespie, Mark, 18
Gingrich, Newt, 410, 411
Giovannetti, Angelo, 279
glacier activity, 254
Glaser, Robert J., 292
Glenn, John, 18, 30
Glennan, T. Keith, 441
Glenn P. Martin Company, 432
Global Atmospheric Research Program
(GARP), 472
Global Earth Observation System of
Systems (GEOSS), 422, 423
Global Hawk, 487
Global Orbiting Navigation Satellite System
(GLONASS), 475, 485, 488
Global Positioning System (GPS)
civil and commercial GPS, 474–75
civilian safety and, 486
development of, 473, 489
future of, 488
impact on society, 428
munitions use of, 486

precursor to, 104
profits from sales of receivers for, 483,
    484
satellite blocks, 473–74
space technology and development of, 78
Transit satellite replacement with, 435
Wide Area Augmentation System
    (WAAS), 475
global village, 487
Glory, 421, 423, 424
glow-charge/glow-discharge system, 298,
    298n180
glycerol, 217, 220
Goddard, Robert Hutchins, 82, 138, 430
Goddard Space Flight Center
    data bank of electronics parts, use of by,
        193
    electronics research and development at,
        183–86, 243–45
    electron microscope research at, 222–23,
        236
    Institute for Space Studies, 396
    integrated circuit failures and projects at,
        245
    integrated circuit research at, 201
    integrated circuit testing methods,
        research on, 222–23
    Interplanetary Monitoring Platform
        (IMP), 238–42, 238n282
    Quality Assurance Division, 222–23
    semiconductor screening technique used
        at, 244
    silicon solid-state devices, research on
        effect of space radiation on, 207
Goertz, Raymond C., 118
GOES (geostationary operational environ-
    mental satellites), 472
gold
    gold doping technique, 202
    gold-flake/gold-ball contamination,
        180–81, 243, 245
    integrated circuit use of, 202–3,
        203n162, 243, 245
    metallurgical incompatibility of alumi-
        num and, 202–3, 203n162
    nickel as replacement for, 203
Goldin, Daniel S.
    administrator role of, 405–6, 409, 412,
        418
    biomedical sensors, research on, 114
    budget and funding reductions and
        NASA program, 410–11

Clinton administration and NASA pro-
    grams under, 409–12
faster, better, cheaper mantra of, 405–6
neck pillow from Tempur-Pedic for,
    133–34
smaller satellites, support for, 405–6
strategic planning process for NASA,
    502n40
Triana, pressure to build and launch,
    415–16
Golovin, Nicholas E., 201
Good Samaritan Hospital, 92
Goody, Richard, 397
Goodyear Aerospace Corporation, 128–29,
    130
Google, 117
Goonhilly station, 440
Gore, Al
    climate change, concerns about, 412
    defeat of, 418
    Earth in the Balance, 408
    election of as vice president, 408
    environmental interests of, 408
    environmental research, concern about
        pace of, 403
    Kyoto Summit atfgtendance and signing
        of Protocol, 415
    MTPE priority for NASA, 402
    ozone depletion, NASA, and, 408
    ozone depletion policy and halt of CFC
        production, support for, 407
    ozone man label for, 408
    Triana, support for, 415–16
Govan, Duncan E., 292
GPS. *See* Global Positioning System (GPS)
Grab system, 481
Graham, William, 399, 400
gravity
    adjustment to following long-duration
        spaceflight missions, 90
    balance sense and, 90
    Grand Tour of solar system and slingshot
        effect of gravity assist process, 359
    gravitational stress on fluid balance and
        blood circulation, study of, 269, 280
    Lagrangian points, 394, 416
    manipulation of material on atomic level
        and, 255
    Venus–Earth–Earth Gravity Assist for
        Galileo mission, 368, 369–70, 371
    Venus–Earth gravity assists for Ulysses
        mission, 368

Venus–Venus–Earth–Jupiter Gravity
Assist for Cassini mission, 372, 414
See also weightlessness/zero gravity
Greatbatch, Wilson, 99–101, 105, 106
Great Britain/United Kingdom
Aerosat program, 477
cooperative space ventures with, 515
Royal Radar Establishment, 172–73
scientific development and military
strength of United States and
Soviets, opinions about, 10, **10**
space race between United States and
Soviets, opinions about, 11–12
Sputnik, public knowledge of launch of, 9
Great Society programs, 496
Greenfield, Stanley, 431
greenhouse effect
Bush policy toward, 401
climate change and, 397, 400
on Earth, 397
sensor for measuring, 423
on Venus, 384, 387, 396–97
Greenland, 113
Griffin, Michael "Mike," 423–24, 509,
509n62
Grissom, Gus, 15
Gruber, William H., 78–79
Grumman, 522, **523**
g-suit (anti-gravity suit)
blackout prevention for pilots, use of for,
95–96
hemophiliac children, treatment of with,
94
invention of, 96
GTE (GSTAR), 451, 454
Guier, William H. "Bill," 342–43, 433–34
Gulf Wars, 409, 468, 486, 487, 488

H
Habermas, Jürgen, 377
Hale, Edward Everett, 429
Hall, Eldon C., 150–51, 226
Hall, Robert N., 198
halogen leak detector, 218
Hamiter, Leon C., Jr., 223, 225, 235
handbooks on electronics parts, 191–93,
191n127
Hansen, Carl L., 321n274
Hansen, James, 396–97, 400, 403, 404,
411, 412
haptics, 117–20, 124
harpsichord, 89

Harris, William B., 176n70
Harrison, Albert A., 537, 545–46, 553–54
Harrison, Donald C., 266, 268–69
Harrison, John D., 127
Harris polls
awareness of space issues and response to
poll questions, 41
benefits of space program, questions
about, 61–62, **62**
budget and funding questions, 5, 13,
14–16
Challenger accident and questions about
manned space program, 26–27, 28
Columbia accident and questions about
space program, 30
emotional impact of major spaceflight
events, questions about, 33–34
goals in space, attitudes toward, 60–61
Kennedy administration, polls about,
47–48
Mars, questions about exploration of and
manned missions to, 16–17, 68–69
Mars exploration, poll questions about, 70
Moon, opinions about scientific exploration station on, 16–17
nuclear power, attitude toward to, 366
scientific research, questions about funding of, 56–57
Shuttle flights, opinions about practical
uses of, 21
travel into space, questions about personal
interest in, 37
Hartman, Edwin P., 271, 278
Harvard Business Review readers, poll of,
59, **60**
Harvard University
electronics research laboratory at, 288
ERC location and, 194
poll of students about space goals, 59, **60**,
63–67
Harvey, Douglas G., 337
Haskins Institute, 123
Hawker Siddley (British Aerospace), 464,
477, 479
Haymaker, Webb E., 264
Hazeltine Research, Inc., 276, 276n97
HDTV (high-definition television), 488
Health, Education, and Welfare,
Department of
Occupational Safety and Health Act
and measurement and reporting on
employee exposure to gases, 308

Office of Education and Optacon development, 124–25
health, poll questions about funding to improve and protect, 46–47, **46**, 53–54, 55
Health and Human Services, Department of, 84
Healthcare Open Systems and Trials (HOST) Consortium, 112
hearing aids, 166–67, 177
Hearsey, Christopher M., 541n16
heart and cardiovascular system
cardiac-output monitor and impedance cardiography, 107–11, 143–44
cardiac prosthesis for heart valve prolapse mitigation, 127
cardiotachometer research, 272–73
cardiovascular physiology, study of biomechanics within field of, 269
cardiovascular system, evaluation of during circulatory stress, 268–69
defibrillators, implantable, 98, 105
electrodes, development of improved, 267
gravitational stress on fluid balance and blood circulation, study of, 269
heart wall accelerations and silicon MEMS accelerometer research, 309–14
implantable telemetry system for monitoring blood flow, 293
renal artery blood flow, research on measurement of, 269
See also pacemaker technology
Heinlein, Robert A., 118, 140
heliopause, 363
helium
gas chromatograph use of, 303, 305
Helium-3, harvesting for use of as fuel, 100
Jupiter and Sun, helium content of, 371
presence of when testing transistors and microcircuits, 206
sealed container, locating leaks in, 220
sealed container, locating leaks in with, 218
Helliwell, Robert A., 324
hemophiliac children, g-suit treatment for, 94
heuristics, 513
Hewlett-Packard Company, 260, 294, 317
Higashi, Robert E., 321n274, 323n285
high-definition television (HDTV), 488
high-Q (high-quality) electromechanical filter, 259, 259n25

highways, poll question about funding for building, 16
Hillenga, Sheri, 94, 95
Hirst, Hugo, 275n94
Hirst Research Laboratories, General Electric Company, 275–76, 275n94, 275–76nn96–97
H. J. Brunnier Associates, 286–87
Hoerni, Jean, 161, 202
Hogrefe, Arthur F., 106
Honeywell, 257, 321, 323n285
horizontal diffusion, 79–80
Hot Bird, 464
House, William, 138
HS-303A satellites, 476
HS-333 satellites, 461, 466–67
HS-376 satellites, 462, 463
H-Sat (Olympus), 464
Hubble Space Telescope, 44–45, 402
Hughes Aircraft Company
Aerosat program, 477
ATS satellites from, 471
Earth Observation Satellite Company role, 482
last satellite built for Intelsat, 450–51
Microelectronics Division, 209–11
PanAmSat, 454
patent waiver requests, **247**, **248**
satellite development and manufacturing by, 428, 439, 445, 448–49, 450–51, 461, 462–63, 489–90, 526, 526n122
Surveyor project, **523**
Syncom/Syncom 2, 440, 450, 462–63, 476, 489, 490
Westar, 454
Hughes Communications, Inc. (HCI), 462–63
Hughes Research Laboratory, 215–16
Hughes Santa Barbara Research Center, 472
humans. *See* astronauts/humans
humidified air for patients, system to supply, 108
Hurt, Calab B., 452
Huygens probe, 372, 378
hybrid circuits, 200, 200n153, 201
hydrazine, 301
hydrogen, liquid, 82
hydrogen contamination and failure of integrated circuits, 205–6, 205nn169–170, 243
hydrogen fluoride, 312

hypohidrotic ectodermal dysplasia, cool suit
    treatment for, 94–95
hypotension and anti-shock garments,
    93–97, 143

I

Iacocca, Lee, 260
IBM (International Business Machines)
    computer-aided design of circuits, semi-
        nar on, 219
    domestic communications satellites, 449,
        462, 463
    Information Management System/360,
        522–23, 533
    information management system develop-
        ment by, 522–23
    ink-jet printer technology development
        by, 260
    integrated circuit manufacturing by, 176
    MOS technology research at, 239–40
    patent waiver requests, 247, **247**, **248**
    research lab of, 119
    specialized integrated circuits for, 227
    thermal-conductivity detectors for
        gas chromatographs, funding for
        research on, 303–4, 307
    Thomas J. Watson Research Center, 260
    transistor use in computers manufactured
        by, 167
    Watson Research Center, 219
Ice, Cloud, and land Elevation Satellite
    (ICESat), 424
Iceland, 113
Ida (asteroid 243), 369, 371
impedance cardiography and cardiac-output
    monitor, 107–11, 143–44
impedance plethysmograph, 109
Improved TOS (ITOS)/NOAA satellites,
    438, 470, 471
India, 110, 415, 462, 472
Indian Ocean tsunami, 423, 487
Indonesia
    high-tech area in, 316
    Palapa communications satellites, 461, 462
Indonesian Academy of Sciences, 316
Indonesian Institute of Sciences, 316
Industrial Development Engineering
    Associates (IDEA), Regency Division,
    167
Industrial Products and Practices (Denver
    Research Institute), 234

industrial revolution, third, 156
Information Dynamics Corporation, 193
Information Management System/360,
    522–23, 533
infrared microscope, fast-scan, 223–25
Initial Defense Communications Satellite
    Program (IDCSP), 465–66
Injury Prevention and Control (Varghese), 97
ink-jet printer technology, 254, 260
Inmarsat (International Maritime Satellite
    Organization), 449, 453, 456–57,
    478–79
"Inquiry into the Feasibility of Weather
    Reconnaissance from a Satellite Vehicle"
    (Kellogg and Greenfield), 431, 436
inspirational aspects of space exploration,
    64, 388–89, 416, 521–22, 533–34
Institute for Space Studies, 396
Institute of Electrical and Electronics
    Engineers (IEEE)
    Engineering in Medicine and Biology
        Society, 290
    formation of, 290
    *IEEE Transactions on Biomedical
        Engineering*, 302
    IEEE Transactions on Electron Devices,
        307, 310, 313
    International Solid-State Circuits
        Conference, 302
    Journal of Microelectromechanical Systems,
        262
    Medal of Honor award from, 290n156
    Professional Group on Bio-Medical
        Electronics, 290
    Solid-State Circuits Conference, 309–10,
        314
    Solid State Device Research Conference,
        214n202
    Workshop on Micro Electro Mechanical
        Systems, 263
Institute of Radio Engineers (IRE)
    communications satellite talk by Pierce
        to, 431–32, 439
    Dummer talk at symposium sponsored
        by, 172
    Professional Group on Medical
        Electronics, 290
Instrument Landing System (ILS), 475
insulin delivery, programmable implantable
    system for, 104–5, 106

integrated-circuit case studies
  component development for resistance
      to radiation and space conditions,
      201, 206–13, 207n175, 207n177,
      212n195, 214–15, 243
  fabrication processes research, 210, 213–
      19, 214n202, 215n205, 216n209,
      218n214, 243
  failure and reliability research, 201–6,
      203n162, 205nn169–170, 243–45
  testing methods research, 201, 219–25,
      222n231, 243
integrated circuits
  acceptance of and psychological impact of
      NASA use of, 151, 226, 226n242
  Air Force electronics miniaturization pro-
      gram and development of, 171–72
  aluminum leads, 200, 203n162, 220, 223
  avian biorhythms research, rejection of
      integrated circuit use in, 272
  Centaur-Mariner H loss and failure of,
      244–45
  Certified Production Line (CPL) concept,
      234–36, 235–36n272
  commercialization of and commercial
      applications for, 149–50, 151–52,
      151n8, 153–55, 156, 177, 226–27,
      242–43
  components, increase in number on
      single circuit, 177–78
  components of, 155–56, 168, 177–78
  computer-aided design of, 218–19, 243
  concept and function of, 155–56, 168
  conditions encountered in space, resis-
      tance to harsh, 181–82, 197, 201,
      206–13, 243
  cost of compared to cost of individual
      components, 154, 226
  cost of integrated circuits for Minuteman
      II program, 152
  cost/price of, 153–54, 176–77, 226,
      242–43, 243
  defense and non-defense consumers of,
      153
  development and use of, Air Force role in,
      152–54, 155, 176–77, 242
  development and use of, NASA role in,
      viii, 104, 149–52, 153–54, 155,
      176–77, 225–28, 242–45, 252, 527
  diffusion process for manufacture of, 160,
      202–3, 214–15, 214n202
  diode use in, 157, 168, 175, 181

  drawbacks of and barriers to using,
      181–83
  electrical circuit design and, 156–57
  fabrication of, NASA role in, viii
  fabrication process and quality of, 182–83
  fabrication processes research, 199–200
  fabrication process improvements, 200,
      201, 213–19, 214n202, 215n205,
      218n214, 243, 276–77, 277n100
  failure and reliability, causes of, 201–6,
      243–45
  first use of in space, 238, 238n281, 243
  gold use in fabrication of, 202–3,
      203n162, 243, 245
  handbook on reliability of silicon inte-
      grated circuits, 192–93
  hybrid circuits, 200, 200n153, 201
  hydrogen contamination and failure of,
      205–6, 205nn169–170, 243
  importance of, 156–57
  industry to manufacture, development of,
      175–78, 175n69
  inspection techniques, 200
  invention of, 121, 121–22, 168, 172–75,
      174n65
  ion implantation process, 213–16,
      214n202, 243
  microwave integrated circuits, 211–13,
      212n195, 243
  miniaturization of electronics and, 168
  monolithic integrated circuits, 168, 175
  Moon mission plans and, 177
  nickel use in fabrication of, 203
  number produced and shipped, 176–77,
      226–27
  number purchased for Apollo program,
      150–51, 150nn4–5, 155, 226–27,
      242
  number purchased for Minuteman II
      program, 152, 155, 226–27, 242
  Optacon device development and,
      121–22
  pacemaker technology development and,
      103, 104
  patents for, 121–22, 174, 175
  photolithographic procedures for manu-
      facture of, 162–64, 174
  planar technique for manufacturing,
      174–75
  production quality assurance program,
      234–36, 235–36n272

research activities of NASA, 185–86,
  186n107, 186n109
resistance to use of, 152
scribing silicon wafers, 217, 218,
  218n214
sealed container, locating leaks in,
  217–18, 220, 243
separation from wafer, processes to avoid
  damage during, 216–19, 243
silicon solid-state devices, effect of space
  radiation on, 206–7, 207n175,
  207n177
silicon use in, 152, 155, 159, 173–75
silicon wafers, size of, 177–78
specialized integrated circuits, 227–28,
  237–42
standardization of by NASA, 182–83, 193
Tech Briefs on, 229–30, 246, **246**
temperature and failure of, 223–25
temperature telemeter, rejection of integrated circuit use in, 274
testing methods, nondestructive, 201,
  219–25, 222n231, 234, 236–37
transistor technology compared to, 156
transistor use in, 157, 168, 172–75, 177,
  181
wafers, separation into individual circuits,
  200
Integrated Electronics for Medical
  Applications Program, 291, 303
Integrated Enterprise Management
  Program, 510
Intel Corporation, 294
intelligence and response to poll questions,
  43–45, **45**
Intelsat (International Telecommunications
  Satellite Consortium)
  competition for, 449, 450, 457
  costs and profits, 443–44
  creation of, 525
  data from COMSAT, limits on, 446,
    446n40
  definitive agreements, 445–49
  DOD contract for transponders, 469
  Intelsat II series satellites, 445, 450, 451,
    476
  Intelsat III series satellites, 445, 448, 466
  Intelsat IV/IVA series satellites, 445,
    448–49, 450, 451, 466
  Intelsat V series satellites, 449, 451, 479
  Intelsat VI series satellites, 450–51
  Intelsat VII series satellites, 451

Intelsat VIII series satellites, 451
Intelsat IX series satellites, 451
Intelsat Agreement/Intelsat Treaty, 447–48
Intelsat Operating Agreement, 447–48
Intelsat X series satellites, 458
Interim Communications Satellite
    Committee and renaming as
    Intelsat, 444–45
  management role of, 445
  maritime communications and, 478
  markup on pricing by COMSAT, 456
  membership in, 445
  monopoly status of, 449, 450
  office of, 451–52
  Orbit Act and direct access, 454, 456–57
  PanAmSat merger, 458–59
  privatization of, 456–58
  profits and success of, 449, 453
  sale of to private equity firm, 463
  structure of, 447–48
  telephony as core business of, 459
Interagency Nuclear Safety Review Panel,
  372–76
Inter-Governmental Maritime Consultative
  Organization (IMCO)/International
  Maritime Organization (IMO), 478–79
Intergovernmental Panel on Climate Change
  (IPCC), 400, 403–4, 411–12, 418
Interim Communications Satellite
  Committee (ICSC), 444
Interior, Department of, 386–87, 482
International Academy of Astronautics
  (IAA), 540
International Air Transport Association
  (IATA), 477
International Atomic Energy Agency,
  United Nations, 302, 303
International Civil Aviation Organization
  (ICAO), 476–77
International Components Symposium, 173
International Conference on Solid State
  Sensors and Actuators, 262–63
International Geophysical Year (IGY), 436
International Launch Services (ILS), 458
International Maritime Organization
  (IMO)/Inter-Governmental Maritime
  Consultative Organization (IMCO),
  478–79
International Meeting on Chemical Sensors,
  263
International Space Station (ISS)
  astrosocial knowledge from, 573

budget and funding for, 419
collision-avoidance procedures and protec-
    tive shields for debris threat to, 417
completion of, 503
crew on, size of and length of stay by, 573
elements of, launch into orbit of, 417
planning for, 409–10
social change and, 572
space debris threat to, 385, 415, 417, 420
space utilization and, 500
systems management practices in, 508
International Telecommunications Union
    (ITU), 443, 446, 477, 478
Internet
    collaborative use of, tools for, 117
    creation of, 116
    DLI projects, 116–17
    World Wide Web growth and, 116
Interplanetary Monitoring Platform (IMP)
    Ariel 1, 240
    Explorer 12, 240
    Explorer 18/IMP 1/IMP A, 238–39,
        238n282, 240–41
    Explorer 33/IMP D, 240–41
    Explorer 34/IMP E, 240–41
    Explorer 35/IMP F, 240
    Explorer series, 238, 238n282
    first use of integrated circuits in space,
        238, 238n281, 243
    specialized integrated circuits for, 237–42
    standardization of design for, 240
    success of missions, 238
    supplier of MOS transistors for, 241–42
Inter-University Center for Microelectronics,
    316
Io, 363, 371
ion drive propulsion, 141
ion implantation, 213–16, 243
Iraq, 468
Iraqi Freedom, Operation, **468**
iron, 203
iron lung, respirator collar for, 109
Islamic Academy of Sciences, 316
isolation on crew behavior, 552–53, 571
isotopic power, 337, 343. *See also* radioiso-
    tope thermal generators (RTGs)
isotropic etching, 257, 301, 305–6
Israel, 110
Israel, David R., 477
Italian Space Agency (ASI), 372
Italy
    cooperative space ventures with, 515

impedance cardiography research, 110
scientific development and military
    strength of United States and
    Soviets, opinions about, 10, **10**
Sputnik, public knowledge of launch of, 9
It Came From Outer Space (Bijlefeld and
    Burke), 78
Itek Corporation, 198, 198n148
ITOS (Improved TOS)/NOAA satellites,
    438, 470, 471
ITT, 204, 428

J
Jaffe, Leonard, 439
Japan
    Aerosat program, 477
    Aqua satellite role of, 421
    EOS role of, 402
    Intelsat agreements, 447, 447n43
    Kyoto Summit and Protocol, 415
    meteorological data sharing and interna-
        tional cooperation, 472–73
    Sputnik, public knowledge of launch of, 9
    Terra satellite instrument package, 417
    transistor radio production by, 167n45
JBliss Imagine Systems, 123–24
JBliss Low Vision Systems, 124
Jeffrey, Kirk, 98, 102, 106
Jennings, Thomas, 96
Jerman, John Hallock "Hal," 304, 306,
    316–17
Jet Propulsion Laboratory
    communications satellite development by,
        526n122
    configuration management practices, 504
    data bank of electronics parts, use of by,
        193
    electronics research and development at,
        183–85, 186n109
    Goldstone station, 439
    integrated circuit failures and projects at,
        245
    Mariner '69 program, 244, 244n295
    Pioneer Project, acknowledgement for
        assistance in receiving radio signals
        from probes, 361
    Ranger program testing at, 180
    RTGs, support for use of, 351
    Voyager 2 discoveries, research on com-
        munication to public about, 143
    Voyager Biological Laboratory, 264
Jodrell Bank radio telescope, 11

Johnsen, Katherine, 446
Johns Hopkins University, Applied Physics
    Laboratory, 98, 104–6, 342–43, 345,
    434, 435
Johnson, John A., 444, 445–46
Johnson, Lyndon B.
    communications policy under, 460
    Europa launch vehicle, authorization for
        development of, 516
    Great Society programs under, 496
    nuclear power systems for spacecraft, sup-
        port for use of, 346
    presidential performance polls about,
        50–51
    prestige from Apollo program and Cold
        War, 494
Johnson, Robert G., 321n274, 323n285
Johnson, Roy W., 464
Johnson Space Center/Manned Spacecraft
    Center
    balance research at, 93
    impedance cardiography research at, 108
    Orbital Debris Program Office, 395
    regional impact of, 529
    transistors research for, 244
Joint Services Electronics Program, 218,
    288, 289, 292, 293, 295, 313–14
Jordan, Kenneth, 336–37
Journal of Microelectromechanical Systems,
    262
*Journal of Micromechanics and
    Microengineering*, 262
junction transistors, 159
Jupiter
    Galileo mission to, 367–72
    knowledge/information about and
        Cassini mission, 378
    knowledge/information about and
        Galileo mission, 371, 378
    knowledge/information about and
        Pioneer 10 mission, 359
    knowledge/information about and
        Pioneer 11 mission, 359
    knowledge/information about and
        Voyager probes, 363
    RTG use for missions to explore, **334**,
        340, **350**, 362–63, 367–68
    scientific interest in, 367
    Voyager probes missions to, 362–63
Jupiter rockets, 82, 137

K
Kahng, Dawon, 239
Kane, Thomas R., 269
Katrina, Hurricane, 424, 486
Katz, Elihu, 79
Kauffman, George, 127
Kaufman, Harold R., 141
Keeley, Jamie A., **376**
Kefauver, Estes, 442
Kelley, Albert J. "Al"
    biomedical sensors and transmitters,
        implantation of, 265
    diode failure and Ranger program, 180–81
    electronics research and development
        programs, survey of, 184, 185, 195
    Electronics Research Center creation, role
        in, 195
    electronics research mission of NASA,
        183, 183n96
    microelectronics use by NASA, 181
Kellogg, William, 431
Kelvin, Lord, 256
Kennedy administration and John F.
    Kennedy
    applications satellite development and use
        under, 427, 428, 441
    Communications Satellite Act, signing of
        by, 442
    Moon mission plans under, 177, 183, 441
    Moon mission priority under, 514n77
    nuclear power systems for spacecraft, sup-
        port for use of, 339
    presidential performance polls about,
        47–49
    prestige from Apollo program and Cold
        War, 494
    satellites using RTGs, support for launch
        of, 345
Kennedy Space Center
    antinuclear protests at, 370–71
    Cassini protests at, 378
    Radioisotope Thermoelectric Generator
        Storage Building, 376
    regional impact of, 529
    rocket program at, 137
    RTG use and risks to citizens in area, 379
    tourism impact at, 528
    Voyager probes launches from, 362
Kennell, Charles, 409, 411, 412
Kentucky residents, poll of about space
    program, 61–62, **62**, 146
Kerr, Robert, 442

Kester, Robert, 476
KH-7 GAMBIT satellites, 480
KH-8 GAMBIT satellites, 480, 481
KH-9 GAMBIT satellites, 480–81
KH-10 DORIAN satellites, 480
KH-11 Kennan/Crystal satellites, 481
KH-12/KH-11B Improved Crystal satellites, 481
Khrushchev, Nikita, 48, 49, 51
Kiefer, Heather Mason, 37–38
Kilby, Jack S., 121, 173–74, 174n65
Killian committee (TCP), 433
kinesthetic sense, 118, 119
Kinnen, Edwin, 109
Kissiah, Adam, 135–38, 143
KKR, 463
Kleiman, Herbert S., 151, 155
Klein, Harold, 264
Knutti, James W., 258
Kompfner, Rudolf, 432, 438–39
Konigsberg, Eph, 282
Konigsberg Instruments, Inc., 282
Koppel, Robert S., 453
Korean War, 8, 432
Kovacs, Gregory T. A., 309, 317
Kraemer, Sylvia K., 19
Kraus, Jerome, 152
Kreuzi, John, 270
Kroemer, Herbert, 174n65
Ku-band frequencies, 449, 451, 458, 462, 468–69, 487–88
Kubicek, William G., 108–9
Kubokawa, Charles, 131, 134
Kuhn, Larry, 260
Kuiper Belt, 379
Kulite Semiconductor Products, 257, 258n23, 308–9
Kurtz, Anthony D., 257
Kurylo, Michael, 406
Kurzweil, Ray, 123
Kurzweil Computer Products, 123
Kurzweil Reading Machine, 122
Kyoto Summit and Protocol, 413, 418, 421

L
L5 society, 394
Lacrosse radar satellites, 481
Lagrangian points, 394, 416
Laird, Melvin, 452
Land Remote Sensing Act, 482

land remote sensing satellites, 383, 391–92, 481–83, 488. *See also* environmental/Earth resources satellites
Landsat/Earth Resources Technology Satellite (ERTS), 391–92, 409, 420, 424, 438, 484, 489
land-use observation satellites, 428–29
Lane, Neal, 416
Langley Research Center
    atmospheric research sphere development at, 438
    electronics research and development at, 183–86, 186n107
    integrated circuit research at, 201
    Nitinol wire research role of, 128–29
    thin-film technology research for, 208–9
LANYARD satellites, 480
laser-scribing silicon wafers, 218, 218n214
latex paint, 62, **62**, 146
launch window, 144
Launius, Roger, 12, 19, 499–500
law enforcement programs, poll question about funding for, 16, 55
Lawrence, Ernest O., 139
Lazarsfeld, Paul F., 79
L-band frequencies, 212, 449, 466, 468–69, 473, 476–77, 478
Leary, Timothy, 21
LeaseSat, 467
Lécuyer, Christophe, 154–55, 225, 226, 242, 288, 289, 293
Lederberg, Joshua, 267–68, 324
LEDs (light-emitting diodes), 296, 296n176
Leonov, Aleksei, 13
Leslie, Stuart W., 288, 289, 323
Lewis Research Center/Glenn Research Center
    cyclotron at and fast-neutron cancer treatment, 138, 140–41
    electronics research and development at, 183–85
    ion drive research at, 141
    LED research, funding for, 296, 296n176
    nuclear reactor power for spacecraft, research on, 140–41
    Radiation Physics Branch, 140
    Ranger program testing at, 180
Librascope division, General Precision, Inc., 203–4, **247**, **248**
Liechtenstein, 446
Life Science Programs, Office of, 263
Life Support Products, 97

lifeworld, 377
light-emitting diodes (LEDs), 296, 296n176
Lilienfeld, Julius E., 157–58
Lillehei, C. Walton, 99–100
Lincoln Experimental Satellite (LES)
    LES 6, 478
    LES 8, **334**, **350**
    LES 9, **334**, **350**
    UHF communications with, 466
Linvill, Candace "Candy," 119–20, 121, 288
Linvill, John G.
    Integrated Circuits Laboratory grant for
        work on Optacon, 287–88
    Integrated Circuits Laboratory role of,
        293
    NIH research, role in, 292
    Optacon, development and manufactur-
        ing of, 117, 119–21, 122, 288n150
    patent for Optacon and NASA role in
        development of, 125
    solid-state electronics program, role in,
        286
Linvill, Marjorie, 119
Lipp, James E., 431
Lipsey, Mark, 20–21
liquid-fuel rockets, 82, 138, 144
Liquid Sunmate, 132
Litant, Irving, 216–18, 216n209
Literary Digest poll, 2
Liu, Jian, 321n274
Lockheed Martin
    AEHF satellite program, 467–68
    communications satellite manufacturing
        by, 451, 526n122
    COMSAT merger, 451–57
    merger to form, 451
    MUOS program contract, 467
    RTG safety analysis by, 372
Lockheed Martin Commercial Space
    Systems (LMCSS), 451, 474, 479
Lockheed Martin Global
    Telecommunications (LMGT),
    453–55, 457
Lockheed Missiles and Space Company,
    247, **247**, **248**
Lockney, Daniel P., 134
Loevinger, Lee, 441
logic circuits, 157
Lohr, Karen, 407
Los Alamos Scientific Laboratory, 219
Los Angles Time polls, 22, 24, 27
Lott, Trent, 456

Louisiana, 419–20
Lovell, Jim, 388–89
Lovelock, James, 389, 393, 395
low-Earth orbit (LEO) spacecraft and satel-
    lites, 479, 490, 500, 502, 514
Lowell, Art, 242
Lozier, David, 361
Luetscher, John A., Jr., 269
Lunar Orbiter, **523**
Luxembourg, 463–64

M
MacIntosh, 116
Maddox, Brenda, 445–46
magnetism
    patents for exploitation of, 105
    satellite control and guidance with, 105
magnetometers, 278, 278n107
magnetosphere, 238, 239
magnets, patents related to permanent, 275,
    275–76n96
Mahajan, Harpreet, 36
Mahar, Neil, 389
Mahle, Stephen, 100–101
Mahoney, James, 418, 422
Malagasy Republic, 447, 447n45
Mallory process (anodic bonding, electro-
    static bonding), 258, 306, 311, 312
Malmivuo, Jaakko, 111
management and planning impacts and con-
    sequences, 492, 495–96, 497, 503–17,
    533
    centralized processes, decentralized
        processes, and control, 503, 506–11,
        511n68
    configuration management practices, 504,
        507, 511
    DOD management and planning prac-
        tices, 495–96, 503–5
    heuristics, 513
    Program Evaluation and Review
        Technique (PERT), 505, 505n48,
        505n50, 533
    systems architecting, 492, 509, 513–14,
        533
    systems management approach to city
        planning, 492, 512, 533
    systems management practices, 492,
        495–96, 504–5, 506–10, 512–13,
        515–17, 533
    triad decision-making structure, 508
Manifest Destiny, 494–95, 495n12

manifold absolute pressure (MAP) sensors
     and pollution requirements, 260
manipulators
     master-slave manipulators, development
          of, 118
     remote manipulation and robot arms, 118
     waldoes, 118
Manned Orbiting Laboratory (MOL),
     480–81
Marconi, 477
Marecs, 479
Mariner missions
     Centaur-Mariner H, 244–45
     Mariner 5 (Mariner Venus 67), 244
     Mariner 9 mission, 393
     Mariner '69 program, 244, 244n295
     Mars missions, 387, 393
     Venus missions, 387
Marisat satellites, 466–67, 478, 479
maritime communication system, 449,
     466–67, 475, 478–79
Maritime Satellite Act, 479
Market Opinion Research, 69–70
Markey, Edward J., 367, 456
Marquis, Donald G., 78–79
Mars
     colonization of, 65, 70
     dust storms on, 393
     exploration of and manned missions to,
          poll questions about, 16–17, 68–72
     failed mission to and public support for
          space program, 18–19
     forward contamination concerns and
          exploration of, 384, 393
     joint missions to, poll questions about,
          69, 70
     knowledge/information about and
          Mariner missions, 387, 393
     landing area for Viking probe, 396
     life on, development of sensors to look
          for, 264, 304, 318–20
     life on, poll questions about, 71
     mission to, plans for, 67, 390, 402, 405,
          409, 503, 572–73
     Moon, Mars, and Beyond space vision
          and missions to, 421–24, 426
     nationalistic motives for exploration of, 70
     Pathfinder probe for exploration of,
          70–71, **334**, 367
     probe missions to explore, 364, 387
     Prometheus project and propulsion tech-
          nology for mission to, 419

public support for Mars program, 6
     RTG use for missions to explore, **334**,
          349, **350**
     Spirit and Opportunity rovers for explo-
          ration of, 31, 69, 71, **334**
     travel to, questions about personal inter-
          est in, 37–38
     Viking probe program, 264, 280–81, 304,
          317–20, 318n260, 393, 395, 396
Marshall Space Flight Center
     Astrionics Laboratory, 186, 222
     Certified Production Line (CPL) concept,
          development of, 235
     data bank of electronics parts, use of by,
          193
     electronics research and development at,
          183–86, 243
     electron microscope research at, 222, 236
     fast-scan infrared microscope, role in
          development of, 223–25
     integrated circuit failures and projects at,
          245
     integrated circuit research at, 201
     integrated circuit testing methods,
          research on, 222
     PERT use at, 505
     Quality and Reliability Assurance
          Laboratory, Parts and
          Microelectronics Technology
          Branch, 192
     Saturn launch vehicles electronics, con-
          version to integrated circuits, 186
     silicon integrated circuits, handbook on
          reliability of, 192–93
Martin, Steve, 364
Martin Marietta
     communications satellite manufacturing
          by, 451
     merger with Lockheed, 451
Martin Nuclear Division, 343, 345
maser amplifier, 438
Mason, Albert E., Jr., 212
Mason, Samuel J., 123
Massachusetts Institute of Technology (MIT)
     Apollo Guidance Computer design and
          development, 527
     Apollo Program Prize, 92
     computer-aided design of circuits,
          218–19
     Electronic Systems Laboratory, 208
     ERC location and, 194
     Instrumentation Laboratory, 194

Lincoln Laboratory, 194, 198–99
  Mead triode research at, 208
  medical technology research at, 85
  MEMS research at, 262
  patent waiver requests, 247, **247**, **249**
  Research Laboratory of Electronics, 286, 288
Matra, 477
Matta, R. K., 222, 237
Maynard, Dan, **375**
Mayo, Isaac, 127
Mazlish, Bruce, 78
MBB, 477
McAuliffe, Christa, 26, 399
McCabe, Frank, 96
McClure, Frank T., 342–43
McConnell, Joseph, 443
McCullough Building, 286–87, 294, 304
McCurdy, Howard E., 544–45
McDonnell, **523**
McDougall, Walter, 431, 495
McElroy, Michael, 396
McFall, Russ, 451
McGlade, David, 459
MCI, 462
McLellan, William H., 255, 255–56n13
McLeod, Jack, 8–9
McLuhan, Marshall, 487
McLure, Frank, 434
McWhorter, Malcolm, 293
Mead, Carver A., 207–8
Mead triode, 207–8
Mechanical Response Tissue Analyzer (MRTA), 86–90
  clinical use of, 89–90
  development of, 86–87
  importance of, 86, 143–44
  patent for, 86–87
  reliability of, 87–89
  studies about, 110
  success of, 89–90, 143–44
MED-EL, 137
Medicaid, poll question about funding for, 16
medical data, digital transmission of, 267
medical education, 113
medical technologies
  aerospace medicine, 85
  biomedical research and, 145
  Harvard students, poll of about space goals and development of, 66
  high-performance requirements in, 129

importance of, 143–44
MEMS use in, 254, 262
NASA contributions to, viii, 85, 141–47
NIH research on, 84, 85
NSF funding for research on, 84–85
space technology and development of, 78
timeliness of medical innovations, 144
medium-Earth-orbit (MEO) satellite, 439–40, 443, 448, 465, 471, 479, 490
Medtronic, 99, 100–101, 103
Mehregany, Mehran, 255–56n13
Meindl, James D., 117, 121, 122, 258, 269, 283, 285, 289–94, 290n156
Meinhard, J. E., 205–6
memory foam. *See* temper foam
MEMS. *See* microelectromechanical systems (MEMS)
Mercury, Project
  contractor for capsule, **523**
  electronics for, 178
  history of space program, communication of, 147
  impedance cardiography research during, 108
  medical research for, 85
Mercury rockets, 137
Mercury system, 481
Merton, Robert K., 81
mesa technique, 160–61, 210
MESH consortium, 464, 477, 479
metal-oxide semiconductor field-effect transistors (MOSFETs), 191, 239–42, 258–59
metal-oxide semiconductor/metal-oxide silicon (MOS) technology, 192, 207, 207n177, 239–42, 248, 258, 276–77
meteorology satellites. *See* weather/meteorology satellites
Methone, 378
Metzger, Sidney, 453
Mexico
  Intelsat agreements, 447
  Sputnik, public knowledge of launch of, 9
Meyer, Lewis, 446n40
Michelson, Robin P., 136, 138
Michigan State University, National Superconducting Cyclotron Laboratory, 139
Microcircuit Line Certification Program, 234–36, 235–36n272
microcircuits, 150n5. *See also* integrated circuits

microelectromechanical systems (MEMS)
　Ames research on, 252, 274–85, 278n110
　Ames-Stanford MEMS research, start of,
　　283–85, 284n136, 323–26, **326**
　beginning of, 254–56
　brain probe research, 294–99, 298n180
　cantilever beam fabrication, 259–60, 300,
　　310–12, 322
　capacitive pressure sensors, 258, 258n23
　commercial applications for, 251,
　　253–54, 254n8, 260–61
　concept and function of, 253–54
　development of, collaboration with
　　Stanford University for, viii, 252
　development of, NASA role in, viii,
　　251–53, 314, 323–26, **326**
　early advances in, 258–61
　fabrication processes, 253
　Feynman challenge to manipulate
　　material on atomic level, 254–56,
　　255–56n13
　first devices, 256–58
　flexible beam fabrication, 309
　funding for research on, 262, 275,
　　323–26, **326**
　gas chromatograph research, 279,
　　280–81, 282, 285, 303–4n200,
　　303–8, 307nn315–316, 310, 314,
　　315, 316–20, **326**
　growth of field and market, 261–63
　history of, 253–63
　impact of research on, 252–53, 314–23,
　　318n260
　journals and conferences on, 262–63
　patents for, 251
　photolithographic procedures, 296,
　　297–98
　piezoresistive pressure sensor research,
　　299–303, 310, 314–15, **326**
　research on, 251–52
　sacrificial layering technique, 259–60
　silicon accelerometer research, 308–14
　silicon crystalline orientation, etching
　　processes, and, 301–2, 327–28
　size of, 253
　surface machining and micromachinging,
　　invention of, 258–59n24, 258–60
　term MEMS, first use of, 252, 252n2, 261
　university research programs on, 262
Microelectronic Device Data Handbook,
　191–93, 191n127

microelectronics. *See* electronics and
　microelectronics
Microelectronics Reliability Program,
　189–94, 234, 245
Microelectronics Subcommittee, 198
Microelectronics Symposia, 190–91,
　190n122
Micrologic Elements, 153–54, 177
micromachining, 253, 257, 297
Micromodule program, 170–71, 170n52
micromotor, 255–56n13
microprocessors, 155, 156. *See also* inte-
　grated circuits
Micro-Scan, 222, 236, 237
microscopes
　electron microscopes, 200, 220–23,
　　222n231, 234, 236–37
　fast-scan infrared microscope, 223–25
　magnification levels for inspections, 220
　sealed container, observing bubbles to
　　locating leaks in, 220
　testing integrated circuits and transistors
　　with, 219–20
Microsensor Technology, Inc. (MTI
　Analytical Instruments), 261, 306,
　316–17
microwave amplifier, 210
microwave integrated circuits, 211–13,
　212n195, 243
Microwave Landing System (MLS), 475
microwave ovens, **62**, 146
MIDAS satellites, 433
Mikulski, Barbara, 417–18
military and defense
　applications satellites, interest in, 432
　commercial communications satellites,
　　use of by DOD, 468–69, 488
　communications satellites, 464–69
　communications satellites, increasing
　　demand for, 468–69, **468**
　defense policies, changing opinions
　　about, 25
　defense spending, poll questions about,
　　46–47, **46**, 53–54, 57, **57**, **58**, 69
　Harvard students, poll of about space
　　goals and defense policies, 66
　Integrated Circuits Laboratory, funding
　　for, 288, 323
　militarization of space, concerns about,
　　414
　military defense and nuclear weapon space
　　stations, poll questions about, 16

miniaturization of electronics concepts,
169–72, 169n49, 170n52
national defense and military capabilities
in space, opinions about, 21–25
reconnaissance satellites and national
defense, 25, 485–86
Stanford Electronics Laboratories, funding for, 288–89, 323, 324
Strategic Defense Initiative (SDI)/Star
Wars, 21–25, 37, 56, 66, 414
transistor use by, 165–66
weather satellite development, 437
military anti-shock trouser (MAST), 96–97
Milky Way galaxy, 359, 360
Milstar, 467, 481
Minitrack system, 433
Minnesota Impedance Cardiograph (MIC),
108–9, 111
Minuteman II program
cost of integrated circuits for, 152
integrated circuits development and use
in, 152–54, 155, 176–77, 225–28,
242
number of integrated circuits purchased
for, 152, 155, 226–27, 242
suppliers of integrated circuits for, 152,
176, 226
transistors and microcircuits used on,
testing of for nitrogen, helium, and
hydrogen, 205–6
Minuteman missiles
miniaturization of electronics and accuracy of, 179
Minuteman I, 82
reconnaissance satellites and deployment
of, 485
testing of, 137
Mir space station, 500
missiles and intercontinental ballistic missiles (ICBMs)
anti-missile defense system, poll question
about, 16
development of, 7, 432
disarmament treaties and agreements,
480, 485
German robot bombs, opinions about
existence of, 3
ICBM program, upgrade recommendation for, 433
liquid fuels for, 82, 144
miniaturization of electronics and range
and accuracy of, 179

race between United States and Soviets to
develop, 7–8
SDI and defense against attacks by,
21–25, 37, 56
solid fuels for, 82, 144
transistor purchases for, 166
transistor use on, 165
See also Minuteman missiles
Mission to Planet Earth (MTPE)/Earth
Science
administrators of, 409
adoption of, 400, 404
Augustine report on, 402–3
budget and funding for, 400, 401, 402,
403, 404, 410–11, 418, 422–23
Clinton administration support for, 409
depoliticization of, 415–16, 425
environmental role of NASA and, 384,
398–405, 421–25
environmental satellites and, 385
EOS role in, 402, 404, 409
implementation of, 405
name change to Earth Science, 415, 425
office creation after elimination of OSSA,
406
origins of, 396
purpose of, 383
regional impacts, study of, 412
USGCRP role of, 400–402
See also ozone depletion
Mitre Corporation, 194
mobile satellite communications, 475–79
mobile satellite service (MSS), 468–69, 483,
484
MOL (Manned Orbiting Laboratory),
480–81
Molecular Electronics (Molectronics)
Program, 171–72
Molecular Electronics for Radar
Applications (MERA) program,
211–13
Molina, Mario, 392
Moll, John L., 296–97
Molniya orbit, 481
Monaco, 447, 447n45
Monroe, Alan, 1
Monsanto Research Corporation, Mound
Laboratory, 336, 338
Montedel, 477
Monte Fiore Hospital, 267
Montreal Protocol, 399, 407, 408, 409,
420, 425

Moon
    Apollo 11 and landing on, 388
    bases on, plans for, 390, 503
    budget and funding questions and value
        of Moon project, 12–16, 34
    Chinese mission to, plans for, 72
    colonization of, 60, 65
    contamination concerns about samples
        from, 388
    desolation of compared to Earth, 384
    economic development of, 100
    faked landing on, opinions about, 17
    forward contamination concerns and
        exploration of, 387–88, 425, 426
    importance of landing a man on, opinions
        about, 17
    knowledge/information about and
        ALSEP, **334**, **350**, 352–58, **353**
    Lunar Orbiter, **523**
    man flying to, changing opinions about,
        3–4, 4n5
    mission to, Kennedy administration plans
        for, 177, 183
    mission to, priority of under Kennedy,
        514n77
    mission to and integrated circuits, 177
    Moon, Mars, and Beyond space vision
        and missions to, 421–24, 426
    Moon-Earth comparison, 388–89
    public support for Moon program, 6
    return to, plans for, 67, 72, 73, 402, 405,
        572–73
    routine landings on and public opinion
        toward space budget, 20
    RTG use for missions to explore, **334**,
        349, **350**, 351
    scientific exploration station on, opinions
        about, 16–17
    spacecraft to return to, 509, 509n62
    space station on, public opinions about, 68
    travel to, questions about personal interest
        in, 35, 37–38, **39**
Moon-Ghetto metaphor, 496
Moore, Betty, 207
Moore, Gordon, 178, 207
Moore, Pamela, 93
Moore's law, 178
Morrell, Frank, 268
Morrell, Lenore K., 268
Morrison, David, 560

MOS (metal-oxide semiconductor/metal-
        oxide silicon) technology, 192, 207,
        207n177, 239–42, 248, 258, 276–77
Mosaic, 116
Motorola
    hybrid circuit development by, 200n153
    integrated circuit manufacturing by,
        175n69, 176
    location of, 175n69
    mesa technique and transistor manufac-
        turing, 160–61
    MOS circuits manufactured by, testing for
        radiation resistance of, 207n177
    testing integrated circuits manufactured
        by, 204
    transistor manufacturing by, 165
    transistor radio production by, 167
Mound Laboratory, Monsanto Research
        Corporation, 336, 338
MTI Analytical Instruments (Microsensor
        Technology, Inc.), 261, 306, 316–17
Mueller, George E., 356–57
Multi-User Objective System (MUOS), 467
Munitions Control Board, 446
MUOS (Multi-User Objective System), 467
Murdoch, Rupert, 463, 464
Murphy, J. L. "Larry," 235

N

nanotechnology
    Air Force program for, 172
    birth of, 254–56
    convergence of branches of technology,
        145–46
    high-performance requirements and, 129
    motor, creation of nanotechnology-scale,
        255–56n13
    negative consequences of, 80
    positive consequences of, 80
    size of, 253
    societal implications of, 80, 145
NASA
    Committee on Earth Sciences role, 400
    creation of and Sputnik, 10, 383
    cultural worldviews, contributions to
        enrichment of, ix
    culture and cultural traits, 506–11
    defense and non-defense consumers of
        innovations from, 153, 233
    DLI project role, 116
    DOD programs transferred to, 522n108
    educational efforts of, 44–45

environmental mission of, 384, 384n1, 385, 390–92, 395, 398–405, 415–16, 420, 424–26
individual achievement at, difficulty in identifying, viii
leadership structure at, 508, 509–10
management structure and power of Centers, 508
mission of, 496–97, 500–503, 502n40
North Eastern Operations Office, 194, 195, 197
ozone depletion research and credibility of, 408
post-Apollo goals and support for, 19–20, 500–503, 514, 514n77
public confidence in and support for, 18–20, 29–30, 497–98
regional impact of NASA Center, 528–29, 529n131
relevance of, questions about, 390
reorganization of, 386, 421–23
science and technology, contributions to, vii–viii, 141–47
societal impact of spaceflight, conference on, 538–39, 541, 544
strategic planning process, 502–3, 502n40
technical culture at Centers, 510–11
universe, contributions to understanding of, ix
NASA Headquarters
Electronics and Control Directorate, 180, 183, 183n96, 188–89, 195
History Program Office, 539
management structure and power of, 508
Office of Advanced Research and Technology (OART), 183, 195, 264
Office of Life Science Programs, 263
Office of Reliability and Quality Assurance, 187–89, 196
Office of Reliability and Systems Analysis, 187
Office of Space Science and Applications, 183, 189
Office of Technology Utilization/Office of Technology Utilization and Policy Planning, 229, 229n253, 233–34, 265–66, 266n48
Parts Steering Committee, 188–89
Reliability and Quality Assurance Meeting, 189–90
NASA Space Act Award, 137

Nashner, Lewis, 91–93
Nathanson, Harvey C., 258–59, 258–59n24
National Academy of Engineering, Russ Prize, 99–100, 106
National Academy of Sciences
Earth Science cuts, criticism of, 423
educational reform, symposium about, 11
global changes research agenda of, 398
global warming review by, 418
MTPE/EOS evaluation and report by, 411
Triana review by, 416
National Academy of Sciences, Institute of Medicine, 137–38
National Aeronautics and Space Act, 230–31
National Aerospace Plane, 32
National Aviation Facilities Experimental Center, 476
National Bureau of Standards, 169
National Cancer Institute, 113–14, 139
National Center for Research Resources, 315
National Commission on Technology, Automation, and Economic Progress, 231–32
national defense. *See* military and defense
National Environmental Protection Act (NEPA), 370
National Football League (NFL), 461
National Geospatial-Intelligence Agency (NGA), 483
National Institute for Occupational Safety and Health (NIOSH), 287, 304, 307–8, 307nn215–216, 316
National Institute of General Medical Sciences, 287, 290–91, 315
National Institute of Neurological Disorders and Stroke, 293, 298
National Institute on Deafness and Other Communication Disorders, 134
National Institutes of Health (NIH)
biomedical engineering, support for, 290–91
biomedical pressure sensor, funding for, 303, 314–15
brain probe research, funding for, 296, 298
federal government department responsible for, 84
gas chromatograph research, funding for, 315

goals and mission of, 84
Integrated Circuits Laboratory, funding for, 287
integrated-circuit technology, investment in, 291–93
Meindl proposal for research for, 290–93
research activities at, 84, 85
Research Career Development Award, 92
Stanford biomedical research, support for, 290–93
Visually impaired people, support for research on applications to assist, 120, 291
National Library of Medicine, 117
National Oceanic and Atmospheric Administration (NOAA)
Committee on Earth Sciences role, 400
creation of, 390
Earth science research by, 397
land remote sensing, transfer to, 482
Landsat program, 409
meteorological data sharing and international cooperation, 472–73
NOAA satellites/Improved TOS (ITOS), 438, 470, 471
ozone-depletion research, role in, 392
ozone monitoring by, 420
polar weather satellites, responsibility for, 469–71
space program of, 383–84
See also Environmental Science Services Administration (ESSA)
National Opinion Research Center (NORC), 20–21
National Polar-Orbiting Operational Environmental Satellite System (NPOESS), 421, 423, 425, 469–70
National Reconnaissance Office (NRO), 437, 480
National Research Corporation, 214–15
National Research Council (NRC), 415, 423, 482
National Science Foundation (NSF)
Committee on Earth Sciences role, 400
communications satellite development by, 526n122
Directorate for Biological Sciences, 84
DLI project role, 116
Earth science research by, 397
governance of, 84

McCullough Building construction, funding for, 287
medical research funding by, 84–85
MEMS research funding from, 262
MEMS term usage and workshops sponsored by, 252n2
ozone-depletion research, role in, 392
public awareness of space issues, commissioning of surveys to measure, 42, 44, 56
research role of, 84
science education in United States and launch of Sputnik, 530
Stanford Electronics Laboratories, funding for, 289
Stanford University, funding for, 287
Visually impaired people, development of assistive technologies for, 123–24
National Semiconductor Corporation, 199, 257–58
National Space Council (NSC), 404, 409
National Superconducting Cyclotron Laboratory, Michigan State University, 139
National Telecommunications and Information Administration (NTIA), 450, 456
National Weather Service, 436
Naval Ammunition Depot/Naval Weapons Support Center, 245
Naval Avionics Facility/Naval Avionics Facility Indianapolis, 214, 435, 473
Naval Research Laboratory, 139, 271, 432, 473, 510–11
Naval Surface Warfare Center, Carderock Division, 126
Naval Weapons Laboratory/Naval Ordnance Laboratory, 125–26, 510–11
navigation satellites
accuracy of, 435
as applications satellites, 427–28
development and launch of, 428, 433–35
dual-use applications, 428
dual-use satellites, 473–75
early concepts and programs, 433–35, 489, 490
future of, 488
importance of, 428

Navy program for, 104–5, 342–43,
434–35
polar orbit of, 434
profits from, 483
RTG use on, **350**
societal impact of, 428
space technology and development
of, 78
See also Global Positioning System
(GPS); Transit satellites
Navstar GPS. *See* Global Positioning
System (GPS)
Navy, U.S.
applications satellites, interest in, 432
Bureau of Weapons, 242
Fleet Satellite Communications
System (FLTSATCOM),
466–67, 478
integrated-circuit research, funding
for, 183
magnetism, patents related to the
exploitation of, 105
medical aviation laboratory, 281
navigation satellite program, 104–5,
342–43, 434–35, 473–75
Nitinol wire development and use by,
126–28
objectives for satellites, 348
Office of Naval Research, 119–20,
124, 288
Office of Naval Research meeting on
electronics reliability, 190n122
reconnaissance satellites, 481
rocket development by, 432
Special Projects Office, 505n50
Tinkertoy project, 169–70, 169n49
UHF follow-on program (UFO), 467
Navy Navigation Satellite System
(NNSS), 434. *See also* Transit
satellites
NBC
polling activity by, 1, 22, 27
television satellite, launch of, 461
Neptune
knowledge/information about and
Voyager probes, 363
RTG use for missions to explore, **334**
slingshot effect of gravity assist process
and length of time for probe to
reach, 359
Voyager 2 discoveries about and opin-
ions about space program, 28–29

Voyager probes missions to, 363
Netherlands, 110
Neurelec, 137
NeuroCom International, 90–93
Neurological Sciences Institute, 92
Newell, Homer, 391
New Horizons, **334**, **350**, 367, 379, **380**
Newman, Tom, 255, 255–56n13
Newport, Frank, 29
news and news stories
results of polls as news, 1
science knowledge and news sources, 9
Voyager 2 discoveries, news stories
about, 143
New Skies Satellite (NSS), 464
Newsweek
poll for about Mars exploration, 70
poll for following Challenger accident
and questions about manned
space program, 26
New York State Department of Health,
Bureau of Emergency Medical
Services, 97
New York Times polls
national defense and military capabilities in space, questions about, 24
space station on Moon, questions
about, 68
start of polling activities, 1
nickel, 203
Nimbus weather satellite
controversy over cost and capabilities
of, 386
development of, 386, 438
Nimbus 7, 395
Nimbus B-1, **350**
Nimbus III, **334**, **350**
research with, 386, 438
semiconductor screening technique
for, 244
Nitinol wire
applications for, 127–30
dental use of, 125, 128, 129
manufacturing of, difficulty and cost
of, 129
memory-metal properties of, 125,
126, 129–30
name of, basis for, 125
NASA contributions to progress in
understanding and using, 125,
128–30

nickel-titanium alloy research and development of, 125–26
patents for, 126–28, 129–30
nitric acid, 312
nitrogen, liquid
coolant use of, 82
medical uses for, 82–83
propellant use of, 82
testing uses for, 82
nitrogen, presence of when testing transistors and microcircuits, 205–6
nitrogen bomb test, 220
Nixon administration and Richard M. Nixon
communications policy under, 450, 460
military defense and nuclear weapon space stations, poll questions about, 16
NASA programs under, 390
presidential performance polls about, 51–53
satellite communications, end to monopoly on, 450
space program funding under, 140
Nobel Prizes
cyclotron invention, 139
integrated circuit invention, 174, 174n65
physiology or medicine research, 267–68
quantum electrodynamics research, 254
transistor invention, 158
tunneling effect, discovery of, 208
Noble Anvil, Operation, **468**
nongovernmental organization (NGO), 459
nonrandom sampling methodology, 2–3
Noordung, Hermann (Herman Potočnik), 430
Norden, 205–6
North American Rockwell/North American Aviation
Apollo program, contract for, 522–23, **523**
Autonetics Division, 205–6, 205n170, **247, 248**
General Atomic facilities, 205–6
GPS satellite contract, 474
hydrogen contamination and failure of integrated circuits, research on, 205–6, 205nn169–170
patent waiver requests, 247, **247, 248**
Systems Dynamic Group, 131
Northeastern University, 203
Northrup-Grumman, 453
Norway, 9, 113
NOTUS program, 464–65, 466
Nova satellites, 435

Noyce, Robert N., 121, 174–75, 226, 242–43
NPOESS (National Polar-Orbiting Operational Environmental Satellite System), 421, 423, 425, 469–70
nuclear power
half-life of fuel, 343, 364–66
Polaris ballistic missile submarines, 342, 428, 435, 505, 505n48
public opinions and attitudes toward, 366–67
safety issues and concerns, 364
submarines powered by, 364
trains and airplanes powered by, opinions about, 4
nuclear power plants
Carter administration and funding for nuclear technology, 55
safety issues and concerns, 364
Three Mile Island incident, 366–67
Nuclear Safety Review Panel, 372–76
nuclear systems for space power
advantages of, 333
approval process for launching satellites powered by, 348
Atoms for Peace initiative, 338–40
comparison to other types of power, **335**
controversies over use of, viii, 331, 414
deep space missions, use as power for, 331, 333–34, **334**, 419
development of, 336–37
environmental concerns about, 384
evolution of systems, 331, 340–42, **341**, 349–64
high tide of space nuclear power, 349–64
military defense and nuclear weapon space stations, poll questions about, 16
moratorium on after Cosmos 954 incident, 366
nuclear reactor power for spacecraft, Lewis Research Center research on, 140–41
nuclear rocket engines, development of, 140, 141
opposition to and protests against use of, 345, 369–71, 376–78, 379, 384, 414
origins of use for spaceflight, 337–42, **341**
Prometheus, Project, 419, 420, 423–24
safety issues and concerns, 335, 344–47, 358, 379–81, 384
types of, 333–34

See also radioisotope thermal generators
(RTGs)
nuclear weapons and defense capabilities,
395–96
Nunn, Timothy A., 315
Nye, Patrick W., 122–23

O

Oates, Ronald W., 131
Obama, Barack, 425
Oberth, Hermann Julius, 430
"Observing the Weather from a Satellite
Vehicle" (Wexler), 436
Occupational Safety and Health Act, 308
ocean monitoring
Aqua satellite, 420–21
El Niño and TOPEX/Poseidon, 412–13
Seasat, 395
O'Connell, James D., 460
Odum Institute for Research in Social
Science, University of North Carolina–
Chapel Hill, 7
oil boycott and energy emergencies, 390,
500
O'Keefe, Sean, 137, 384, 419, 421–24
Oklahoma City bombing, 33
Olympics, 21–22
Olympus (H-Sat), 464
O'Neill, Gerard, 394
online polling, 2–3
open skies/open entry policy, 460–61
opinion research, 2
Opportunity, 31, 69, 71, **334**
Optacon (Optical TAcatile CONverter)
device
benefits of, 123
cessation of manufacture of, 123
concept and function of, 117
development of, 119–25
development of, contribution of government agencies to, 119–20, 124–25
haptics research and, 117–20, 124
importance of, 144
integrated circuits and, 104, 121–22
Integrated Circuits Laboratory grant for
work on, 287–88
manufacture of, 122, 288n150
miniaturization of, funding for, 287–88
NASA contributions to development of,
117, 120, 121, 124–25
patent for, 120, 125
research with, 117–19

stimulation of innovation in assistive
technologies, 123–24
optical character reading (OCR) devices,
120, 123–24
Orbimage, 482
Orbit Act, 454, 455–57
Orbital Debris Program Office, 395
Orbiting Geophysical Observatory, 238
Ordway, Frederick, 81–83, 94
Oregon Health and Sciences University, 92
Orion Satellite Corporation, 450, 452–53
The Orthotic Group, 87
Oscars, 435
O'Sullivan, William J., 438, 439
Outer Space Treaty, 388
overview effect, 499
oxygen, liquid, 82
Oyama, Vance, 264, 304, 318
ozone depletion
Antarctica, ozone research in, 398, 399,
401, 409
Arctic, ozone research in, 401, 406–8
causes of, 392, 399, 407
decrease in, 420
increase in ozone hole, 406–8
Montreal Protocol and policy to address,
399, 407, 408, 409, 420, 425
NASA environmental mission and
research on, 398–400, 425
research on and policy development to
address, 392, 401–2, 406–8
satellites to monitor and study, 395,
398–400

P

pacemaker technology, 98–107
batteries for pacemakers, **62**, 98, 99, 100,
101, 103, 104, 105–6, 146
bidirectional telemetry and, 103
collaboration and development of, 106–7
concept and function of pacemakers, 98,
106
electrodes, 99, 100
history of development of, 98–101
innovation in, 102–3
integrated circuit development and, 103,
104
MEMS and pacemaker regulation, 251
NASA contributions to, 100–107, 143,
146
patents for, 99, 100, 102–3
programmable pacemakers, 101–7

purpose and benefits of pacemakers, 98
rechargeable pacemakers, 62, **62**, 98,
    103–4, 105, 146
telemetry and, 101, 103, 104, 106
Pacesetter Systems, Inc., 98, 101–3, 104
Paine, Thomas "Tom," 390, 477
paint, latex, 62, **62**, 146
Palapa communications satellites, 461, 462
Pallene, 378
Pan American Satellite Corporation
    (PanAmSat), 450, 452, 454, 455, 456,
    458–59, 463, 464, 469
Pan American World Airlines (Pan Am), 476
paraplegics, alternating pressure seats for, 109
Parent, Robert, 471
Parkinson's disease, 83
Parts Steering Committee, 188–89
patents
    balance evaluations system patents, 91–92
    cochlear implant patents, 135–36, 137,
        138
    impact of innovations, estimating
        through patent records, 83
    integrated circuit patents, 121–22, 174,
        175
    licensing of innovations, 102–3, 229, 230
    Nitinol wire patents, 126–28, 129–30
    Optacon device patent, 120, 125
    pacemaker technology patents, 99, 100,
        102–3
    patent pools, 102
    patent rights and patent waivers, 230–31,
        247–49, **247**, **248**, **249**
    patent thickets, 102
    prior art, 321
    progress as series of patentable inventions,
        106–7
    protection of innovations with, 79, 107
Pathfinder probe, 70–71, **334**, 367
Patrick Air Force Base, 345
Patterson, Robert P., 109
Payload Hazardous Servicing Facility, **374**,
    375
Perner, Frederick Andrew, 303–4, 307, 316,
    318n260, 319–20
Pershing rockets, 137
Persian Gulf Wars, 409, 468, 486, 487, 488
Personal Influence (Katz and Lazarsfeld), 79
Pescador, Héctor, 136
Petersen, Kurt, 262, 317, 321
Pew Research Center for the People and the
    Press, 7, 33, 57–58

pharmaceutical products, development of,
    21
Philco-Ford Corporation, 465. *See also* Ford
    Aerospace (Space Systems Loral)
Philco/Philco Corporation
    microelectronics research by, 186
    MOS circuits manufactured by, testing
        for radiation resistance of, 207n177
    patent waiver requests, **247**, **248**
    transistor manufacturing by, 165
Phillips, Mary, 94
phosphorus, 215–16
photolithographic procedures
    integrated circuit manufacturing, 174
    MEMS accelerometer research, 311–12
    MEMS brain probe manufacturing, 296,
        297–98
    MEMS gas chromatograph, 304
    transistor manufacturing, 160, 161,
        162–64
photovoltaic solar cells, 332–33
piano, 89
Pickering, William H., 439
Pierce, John R. (J. J. Coupling), 429,
    431–32, 431n10, 438–39
piezoresistive effect and piezoresistive prop-
    erty of material, 256–57
piezoresistive pressure sensors, 256–58,
    299–303, 314–15, **326**
pilots, pressure suits to prevent blackout,
    95–96
Pion, Georgine, 20–21
Pioneer 5 probe, 11
Pioneer 10 probe
    end of mission, 361
    mission of, 359–62
    radio signals from, length of time to
        receive, 361
    RTG use on, **334**, 349, **350**, 359–62, **360**
Pioneer 11 probe
    end of mission, 360
    mission of, 359–62
    RTG use on, **334**, 349, **350**, 359–62,
        360
Pioneer Venus project, 364n80
Pittendrigh, Colin S., 269, 272
planar process, 161, 161n34, 210
planets/solar system
    colonies on/settlement of, 60, 65, 70, 394,
        538, 547, 554–56, 571, 573–75
    gas chromatograph as planetary probe,
        317–20, 318n260

Grand Tour of solar system and slingshot effect of gravity assist process, 359

Grand Tour of solar system by probes, 349, 359, 362–64, 364nn79–80

icy moons orbiting gas planets, 378

launches of planetary probes to explore, 364, 364n79

life on, development of sensors to look for, 264

Moon, Mars, and Beyond space vision and exploration of, 421–24, 426

planetary defense, astrosociological contributions to, 560–62

planetary protection, concern about, 401

RTG use for missions to explore, **334**, 340, 349, **350**, 358–64

Venus–Earth–Earth Gravity Assist for Galileo mission, 368, 369–70, 371

Venus–Earth gravity assists for Ulysses mission, 368

Venus–Venus–Earth–Jupiter Gravity Assist for Cassini mission, 372, 414

See also comparative planetology

plants, 263

Plessey Company, 173

plethysmography, 107–8, 109

Pleumeur-Bodou station, 440

Plonsey, Robert, 111

Pluto

New Horizons mission to explore, 379

Pioneer 11 mission to and beyond, 360

RTG use for missions to explore, **334**, **350**

plutonium-238, 334, 336, 343–45, 347–48, **350**, 356, 358, 362, 364–65, 372, 375

plutonium-239, 347

point-contact (Type A) transistors, 158–59

Polaris ballistic missile submarines, 342, 428, 435, 505, 505n48

Polaris rockets, 82

polar-orbiting satellites, 428, 434, 437

political ethos, 493–96, 495n12

polls and surveys of public opinion

age and support for space program, 35–36, 37–41

age effects, 39–41

analytic techniques, 1

Apollo, opinions and attitudes toward, 12–17

Apollo, trends in support after, 18–32

astrology, questions about, 44, **45**

atomic power for trains and airplanes, 4

awareness of space issues and response to poll questions, 41–43, **43**

budget and funding questions, 5–6, 12–16, 18–20, **18**, **20**, 28, 30, 31, 38–39, 43–44, **45**, 53–54

budget and funding questions for future programs, 67–74

cancer, cure for, 4

Challenger accident and questions about manned space program, 24–28

cohort effects, 39–41, **41**

college students, opinions and attitudes of, 49–50, 59, **60**, 63–67

Columbia accident and questions about space program, 30–32

education and support for space program, 35–36, 41–45, 74

emotional bonds with astronauts, poll questions about, 33–34

emotional impact of major spaceflight events, questions about, 32–35

functions of surveys, 1

future of spaceflight, attitudes toward, 67–74

gender and support for space program, 36–38, **39**

German robot bombs, opinions about existence of, 3

goals in space, attitudes toward, 59–67

government activities, poll questions about funding, 45–59, **46**, **57**, **58**

intelligence and response to poll questions, 43–45, **45**

Kentucky residents, poll of about space program, 61–62, **62**, 146

knowledge of public and basis for responses to, 3

man flying to Moon, changing opinions about, 3–4, 4n5

Mars, opinions about exploration of and manned missions to, 16–17, 68–72

motivations for space exploration, 45–59

NASA, public confidence in and support for, 18–20, 29–30, 497–98

national defense and military capabilities in space, questions about, 21–25

news and news stories about, 1

personal impact of space exploration, 32–45

policy decisions and results of polls, 1, 73–74

presidential elections and, 2

questionnaire design, 1, 3
questions, framing of, 5–6
reliability and validity of, 2–3
safety of space program, questions about, 26–27, 67–68
sampling methodologies, 2–3
science and technology, public confidence in and support for space program, 20–21, **20**
scientific research, questions about funding of, 56–58, **58**
Shuttle flights, opinions about practical uses of, 21
space exploration, questions to children about interest in, 44–45
space race between United States and Soviets, questions about, 8, 11–15, 19–20, 48–49, 50–51, 59
Space Shuttle, opinions and attitudes toward, 18, 19, 21–32
Sputnik, opinions and attitudes toward, 7–12
statistical significance and sampling methodologies, 2–3
telephone service and, 2
theories, testing with poll data, 1
travel into space, questions about personal interest in, 35–39, **39**
trends, tracking will poll data, 1
value of, 73–74
Voyager 2 discoveries and questions about manned space program, 25–26, 27, 28–29
See also presidential performance polls
pollution
funding for anti-air and anti-water pollution programs, poll question about, 16, 54
global warming and, 386
Harvard students, poll of about space goals and pollution removal, 67
manifold absolute pressure (MAP) sensors and pollution requirements, 260
polonium-210, 338, 343
Polydeuces, 378
Pope, Jack M., 272–73
Poppy system, 481
potassium hydroxide, 301, 311
potentiometer, 278
Potočnik, Herman (Hermann Noordung), 430

poverty and funding for anti-poverty programs, poll question about, 16, 53–54
Powell, Walter, 102
power dissipation, 223, 224
Pratt & Whitney, 197
PREDICT (Process Reliability Evaluation and Determination of Integrated Circuit Techniques) facility, 199–201, 243, 282–83
"Preliminary Design of an Experimental World-Circling Spaceship" (RAND), 430–31
presidential performance polls
Carter administration, 55
Johnson administration, 50–51
Kennedy administration, 47–49
Nixon administration, 51–53
Reagan administration, 55–56
pressure sensor (strain gauge), 256–58, 278–79
pressure suits
anti-shock garments, 93–97
blackout prevention for pilots, use of for, 95–96
cool suit treatment for hypohidrotic ectodermal dysplasia, 94–95
g-suit (anti-gravity suit), invention of, 96
g-suit treatment for hemophiliac children, 94
Preston, Jane, 115
prisons, poll questions about spending on, 5
Pritchard, Robert L., 289
Das Problem der Befahrung des Weltraums: Der Raketen-Motor (The Problem of Space Travel: The Rocket Motor, Potočnik), 430
Process Reliability Evaluation and Determination of Integrated Circuit Techniques (PREDICT) facility, 199–201, 243, 282–83
Professional Group on Medical Electronics, Institute of Radio Engineers (IRE), 290
Programalith, 101
Program Evaluation and Review Technique (PERT), 505, 505n48, 505n50, 533
Prometheus, Project, 419, 420, 423–24
Proton M/Breeze M vehicle, 458
public goods, 519, 519n97, 522
Public Health Service, 293
public opinions and attitudes
awareness of space issues and response to poll questions, 41–43, **43**

changes in, vii
communications satellites, understanding
of operation of, 41
goals in space, attitudes toward, 59–67, **60**
Harvard students, opinions of about
space goals, 59, **60**, 63–67
human spaceflight risks, opinions about,
72
impact of space program on, 6–7
importance of understanding changes
in, vii
intelligence and, 43–45, **45**
interest in people before interest in ideas,
29–30
joint missions and cooperation between
United States and other countries,
public attitudes toward, 51–52, **52**,
54, 60, **61**, 69
knowledge of and basis for, 3
nuclear power, opposition to, 366–67
opinion research, 2
policy decisions and results of polls, 1,
73–74
science and technology, public confidence
in and support for space program,
20–21, **20**, 497–98
science awareness survey and launch of
Sputnik, 8–9
science knowledge and news sources, 9
social-psychological research, 2
space exploration and public interest
in science and technology, 6, 42,
44–45, 56–59, 65–66
space race between United States and
Soviets, 7–15, **10**, 19–20, 48–49,
50–51, 59
spinoff technologies, misconceptions
about, 146
themes that shape, vii
See also polls and surveys of public opinion
Pudgee, 132
purple plague, 203, 204
pyracatechnol, 301

Q
"QRM-Interplanetary" (Smith), 429
Quake, Stephen R., 321n274
Qualifications and Standards Laboratory,
189, 194, 198, 199, 213, 244
Quality Engineering and Evaluation
Laboratory, Naval Ammunition Depot/
Naval Weapons Support Center, 245

Quantitative Computed Tomography, 89
quantitative ultrasound (QUS), 88, 89
quantum electrodynamics research, 254
quartz clocks, 146
Quayle, Dan, 404
Quek, Francis, 124
quota sampling methodology, 2

R
Radford, Wade E., 106
radiation
data collection on space radiation,
238–39, 238n282
radiation-resistant components, development of, 181–82, 197, 201, 206–13,
214–15, 243
radioactive debris, 347–48, 358, 365–66,
414, 419–20
Radiation Incorporated, 207n177
radiflo test, 220
radioisotope heater units (RHUs), 362, 372
radioisotope thermal generators (RTGs)
accident environments, 335
advantages of, 334–35, **335**
characteristics and size of, **360**, 361–63,
372, **373**
comparison to other types of power, **335**
concept and function of, 333–34
controversies over use of, viii
deep space missions, use as power for,
333–34, **334**, 367–68
design of, 335, 347–48
development of, 336–37, 342
difficulties associated with systems, 357
evolution of, 340–42, **341**, 349–64
failure of, injection into higher orbit as
disposal method after, 349
failure of and radioactive debris, 358
high tide of space nuclear power, 349–64
impact-resistant container to protect from
accidental release of plutonium,
362–63, 365
location of on Pioneer 10 and 11 spacecraft, **360**
malfunction of satellite using and release
of debris, 347–48
management structure for program,
351–52
opposition to and protests against use of,
345, 369–71, 376–78, 379
origins of use for spaceflight, 337–42,
339, 341

plutonium-238 use in, 334, 336, 343–45,
347–48, **350**, 356, 358, 362,
364–65, 372, 375
power/wattage provided by, 335–36,
340–41, 348, 361, 372
safety and reliability of, 335–36, 338–40
safety issues and concerns, 335, 344–47,
358, 364–78, 379–81
safety reviews and approvals before launch
of RTGs, 365
satellite and spacecraft use of, **350**
satellite applications, early use for,
342–49, 434
support for use of, 351
testing and safety program for, 348, 379
See also SNAP (Systems for Nuclear
Auxiliary Power)
Radioisotope Thermoelectric Generator
Storage Building, Kennedy Space
Center Industrial Area, 376
Ragent, Boris, 275, 279
Ragone, Jared, 88
"Die Rakete zu den Planetenräumen" ("The
Rocket into Interplanetary Space,"
Oberth), 430
Rambo, William, 292, 293, 324
Ramey, Robert L., 208–9
random sampling methodology, 2
RAND/RAND Corporation
applications satellites, interest in, 430–31
nuclear power, origins of use for space-
flight, 337
reconnaissance satellite development, 479
reconnaissance satellite recommendations
from, 431, 433
weather satellite recommendations from,
431, 436
Ranger program
diode failure and, 180–81
electronics failures, 180, 180n83
Ranger 1, 180n83
Ranger 2, 180n83
Ranger 3, 180n83
Ranger 4, 180n83
Ranger 5, 180
Ranger 6, 180
Ranger 7, 180
Rapp, Roxy, 95
Raychem Corporation, 127–28
Raytheon/Raytheon Corporation
electronics research at, 198, 199

fast-scan infrared microscope, role in
development of, 223–24
Space and Information Systems Division,
220
transistor manufacturing by, 165
transistor manufacturing for defense
consumers, 166
RCA (Radio Corporation of America)
Aerosat program, 477
antitrust suit against, 159, 159n29
Earth Observation Satellite Company
role, 482
Micromodule program, 170–71, 170n52
MOS manufacturing by, 276
patents for transistors, 159n29
Relay satellite, 440, 441, 490, 526
Satcom F1, 449
Satcom K-3, 464
satellite development and manufacturing by, 435, 437, 451, 461, 462,
463–64, 473
scanning electron microscope development by, 222
Semiconductor and Materials Division,
276
supplier of integrated circuits for
Minuteman II program, 152, 226
television transponders, number needed,
461
transistor manufacturing by, 159, 165
transistor manufacturing for defense
consumers, 166
transistor radio production by, 167
RCA Americom/GE Americon, 451, 454,
464
RCA Astro-Electronics/RCA Astro, 451,
461, 462, 474, 479
RCA Canada, 461
RCA Global Communications, 461
RCA Ltd., 477
Reagan administration and Ronald Reagan
anti-missile defense system, poll question
about, 16
attitude of Harvard students toward, 59,
66
Challenger accident and speech to nation,
25–26
defense capabilities and nuclear weapons
priorities under, 395–96
deficit growth under, 403
emotional impact of shooting of, poll
questions about, 33

environmental and energy policies under, 397, 400–401

land remote sensing, privatization of, 482, 484

polling about budget under, 5–6

presidential performance polls about, 55–56

satellite communications, end to monopoly on, 450

SDI, proposal to develop, 21–25, 37, 56, 66, 414

space policies of, opinions about, 28

space station project decision by, 395–96, 397

reconnaissance satellites

as applications satellites, 427–28

development and launch of, 337–38, 428, 432, 433

direct-readout satellites, 433

early concepts and programs, 431, 433, 490

film return satellites, 433

future of, 487

Harvard students, poll of about space goals and defense policies, 66

impact on society, 428

importance of, 428

military satellites, 479–81

military security and, 25, 485–86

national defense and, 25

nuclear power systems for, 337–38

profits from, 483

remote sensing technology use by, 428–29

rectifiers, 164

Redfern, Mark, 93

"Red Gold," 95

Redstone rockets, 82, 137

Regency TR1 transistor radio, 167

Reiger, Sieg, 443

Relay satellite, 440, 441, 490, 526

Reliability and Quality Assurance, Office of, 187–89, 196

Reliability and Quality Assurance Meeting, 189–90

Reliability and Systems Analysis, Office of, 187

Reliability Handbook for Silicon Monolithic Microcircuits, 192–93

Remington Rand Univac computer, 167

remote sensing technology

applications satellites use of, 428–29, 489

civil land remote sensing satellites, 481–82

commercial remote sensing satellites, 482–83

environmental satellites use of, 383, 391–92

future of remote sensing satellites, 488

land remote sensing, privatization of, 482, 484

profits from satellites, 483, 484

renal artery blood flow, research on measurement of, 269

Rensselaer Polytechnic Institute, 202–3

Report to the President on Government Contracting for Research and Development (Bell), 228

Research Triangle Institute, 186n105, 237–38, 267

resistor-capacitor-transistor logic (RCTL) circuits, 207n175

resistors

integrated circuit use of, 157, 168, 173, 174, 175

thin-film technology and formation of, 168

Tinkertoy project and silk-screen printing of, 169

resistor-transistor logic (RTL) circuits, 207n175

resonant gate field effect transistor/resonant gate transistor, 258–59, 258–59n24

resources, Harvard students, poll of about space goals and development of, 66–67

Rice University, 219

Ride, Sally, 37–38, 399, 502

Ridenour, Louis, 430–31

Rita, Hurricane, 486

Rock, Bernard, 352

Rockefeller University, 267

"The Rocket into Interplanetary Space" ("Die Rakete zu den Planetenräumen," Oberth), 430

rockets

cryogenic gases/liquid propellants, 82

development of, 432

ion drive propulsion for, 141

liquid-fuel rockets, 82, 138, 144

nuclear rocket engines, development of, 140, 141

pioneers in development of, 429–30

race between United States and Soviets to develop, 7–8

scientific payloads launched with, 432
societal impact of, 383–84
solid-fuel rockets, 82, 144
space applications for liquid-fuel rocket
    technology, 82, 144
Rockwell International
    poll for about Mars exploration, 69–70
    poll for following Challenger accident,
        27–28
    See also North American Rockwell/North
        American Aviation
Roco, Mihail, 145
Rogers, Everett M., 79
Rogers Commission Report, 27–28
Rohrabacher, Dana, 410
Roper Center, University of Connecticut, 7
Roper polls, 5–6, 23, 26
Rosen, Harold, 439, 462–63
Rostow, Eugene V., 460
Rowland, Sherwood, 392
Royal Air Force, 172
Royal Radar Establishment
    (Telecommunications Research
    Establishment), 172–73
Roylance, Lynn Michelle, 308–14, 320–23,
    321n274
RTGs. *See* radioisotope thermal generators
    (RTGs)
rubber, synthetic, 62, **62**, 146
Rudoff, Alvin, 536
Rumford, Maine, station (Andover station/
    Space Hill), 440, 440n26
Rumsfeld, Donald, 232–33
Rusk, Dean, 443
Russ Prize, National Academy of
    Engineering, 99–100, 106

S
Saab, 477
Saby, John, 159
sacred spirit, Earth as, 393
sacrificial layering technique, 259–60
Sagan, Carl, 370, 384, 387, 393, 498, 499
SALT I (Strategic Arms Limitation Talks),
    480, 485
Salyut space station, 500
Samadikun, Samaun
    acknowledgement of assistance in research
        by, 277, 279, 282, 284, 285
    anisotropic etching, role in development
        of variation of, 300–302, 315, 323

Bandung High Tech Valley, role in cre-
    ation of, 316
piezoresistive pressure sensor research,
    299–303, 310, 315
SAMOS satellites, 433, 479–80, 481
Sander, Craig S., 258
Sandia Corporation, 219
Sandler, Harold "Hal," 280, 281–85, 307,
    326
Sarewitz, Daniel, 380–81
SAT, 477
Satellite and Telecommunications Act,
    456–57
Satellite Business Systems (SBS), 449, 462,
    463, 490
Satellite Data System (SDS), 481
"The Satellite Rocket Vehicle" (RAND), 431
satellites
    approval process for launching nuclear-
        powered satellites, 348
    atmospheric research with, 386
    batteries for, charging of, 106
    bidirectional telemetry with, 103
    cooperative space ventures between
        United States and Europe, 515–17,
        516n84
    environment and images from, 383
    information from, interest in, 428
    launch of from Shuttle, 391
    magnetism and control and guidance of,
        105
    manufacturing and sales of, 484
    miniaturization of electronics and launch
        of, 179
    nuclear power use for, early applications,
        342–49
    nuclear power use for, origins of, 337–42,
        341
    number of satellites in orbit, opinions
        about, 11–12
    power requirements for, 331–32
    public attitudes toward space exploration
        and benefits of, 65–66
    scientists as users of information from, 386
    solar-powered satellite (SPS), 394–95
    solar power use on, 344, 348
    space race between United States and
        Soviets, 7–12
    space technology and development of,
        62, 146
    transistor use in, 167

See also applications satellites; environmental/Earth resources satellites

Satellite Television Corporation (STC), 449, 452, 462, 490

Satellite Transponder Leasing Corporation, 462

Sato, Yasushi, 201

Saturn
  Cassini mission to, 372, 414
  knowledge/information about and Cassini mission, 378
  knowledge/information about and Pioneer 11 mission, 359–60
  knowledge/information about and Voyager probes, 363
  RTG use for missions to explore, **334**, **350**, 362–63, 372–78
  Voyager 2 discoveries about, research on communication to public about, 143
  Voyager probes missions to, 362–63

Saturn rockets
  contractors for, **523**
  electronics on, conversion to integrated circuits, 186
  improved version of, 67
  propellants for Saturn V moon rocket, 82
  satellites launched with, 465
  Saturn V rocket, 391, 514

S band, 212

Scapicchio, Anthony J., 216–17, 216n209

Schmitt, Harrison, 353

Schramm, Wilbur, 9

Schwinger, Julian, 254

science, technology, engineering, and math (STEM) disciplines
  educational impact of Apollo program, 491, 530–32, 531n140, 533
  interest in and promotion of education related to, 491–92, 549, 550
  Sputnik and interest in, 10–11, 496

science, technology, engineering, math, and astrosociology (STEMA) disciplines, 548–51

science and technology
  advances in, consequences of and decision-making about, 379–80
  Cold War and technology transfer, 524–25
  convergence of branches of technology, 145–46
  defense and non-defense consumers of innovations, 153, 233

diffusion of technical knowledge, 229–31, 246, **246**

engineers and credit for innovation, 106–7

environmental threat from, 384–85

individuals responsible for contributions, difficulty in identifying, viii

inventions, simultaneous by different individuals, 81

licensing of innovations, 102–3, 229, 230

migration of technology from original applications, 78–79

NASA contributions to, vii–viii, 141–47

personal credit for contributions to, viii

practical capabilities, development of, 500–503

progress as series of patentable inventions, 106–7

public confidence in and support for space program, 20–21, **20**

public interest in, 6, 29–30

public understanding of, NSF polls on, 42, 44, 56

public understanding of through spinoffs, 80–81

scientific development and military strength of United States and Soviets, opinions about, 10, **10**

scientific research, questions about funding of, 56–58, **58**

societal impact of research programs, 231–37, 232n256, 233n263, 235–36n272

societal impacts of space technology, 383–84

space exploration and public interest in, 6, 42, 44–45, 56–59, 65–66

space program technology and general development of industries or scientific communities, 83–85

spillover, 79

successful technology, delays in adoption of, 89–90

technological ethos, 496–98, 503

technology transfer, 78–85, 101, 141–47, 516n84

technology utilization, politics, and Congress, 231–37, 232n256, 233n263, 235–36n272

technology utilization, technology transfer, and dissemination of information, 229–31, 265–69, 523–28

See also spinoff technologies
Science and Technology Policy, Office of
   (OSTP), 408–9
science fiction and applications satellites,
   429, 438
Science Mission Directorate, 421–22
Scientific and Technical Information
   Program (NASA), 89–90
scientists
   applications satellites, interest in, 430–32
   astrosociological issues and, 564–65,
      571–72
   collaboration and friction between
      scientists and Instrument Division at
      Ames, 265, 270
   credit for innovation by, 106–7
   global warming concerns of, 384
   information from satellites, interest in, 428
   satellite information, use of by, 386
SCORE, 464
Scout launch vehicle, 434–35, 437
SDI (Strategic Defense Initiative)/Star Wars,
   21–25, 37, 56, 66, **341**, 395–96, 414
Seaborg, Glenn, 344–46, 351
Sea Launch Zenit-3SL vehicle, 458
sealed container, locating leaks in, 217–18,
   220, 243
Seamans, Robert C., 187, 195, 508, 508n58
Search for Extraterrestrial Intelligence
   (SETI), 547, 565–70
Seasat, 395
seat cushions
   aluminum honeycomb material for, 131
   characteristics of ideal cushioning mate-
      rial, 131
   survivability of airplane accidents and,
      131, 133, 143
   temper foam applications, 131–32, 133,
      134, 143
Seattle Cancer Care Society, 140
Second-Order Consequences (Bauer), 79–80
Seebeck, Thomas Johann, 336
Selenia, 477
semiconductors
   Commerce Department data on semicon-
      ductor use, 153
   diffusion process for semiconduc-
      tor manufacturing, 160, 202–3,
      214–15, 214n202
   Fairchilden and employees leaving com-
      panies and starting new companies,
      164–65

integrated circuit development, incentive
   for, 150
location of semiconductor industry, 175,
   175n69
metal-oxide semiconductor/metal-oxide
   silicon (MOS) technology, 192,
   207, 207n177, 239–42, 248, 258,
   276–77
semiconductor switch use of silicon
   instead of germanium, 296
transistor invention and development of
   semiconductor industry, 157
Senefelder, Alois, 162
sensors
   aerospace medicine and development of,
      85
   biomedical sensors, research on, 113–14,
      257
   biomedical sensors and transmitters,
      implantation of, 265
   blood pressure sensors, disposable, 262
   capacitive pressure sensors/transducers,
      258, 258n23, 270, 270–71n71
   development of to measure effects of
      space environment on, 263
   journals and conferences on, 262–63
   life on other planets, development of sen-
      sors to look for, 264
   manifold absolute pressure (MAP) sensors
      and pollution requirements, 260
   miniature devices made from silicon and
      germanium, 253–54
   piezoresistive pressure sensors, 256–58,
      299–303, 310, 314–15, **326**
   pressure sensor (strain gauge), 256–58,
      278–79
   silicon-based strain gauge, 257
Sensors and Actuators, 262, 263
SENTRY satellites, 479, 481
Service Life Enhancement Program (SLEP),
   467
Shapiro, Robert, 36
Sharpe, Mitchell, 81–83, 94
Shepard, Alan, 13
shock and anti-shock garments, 93–97, 143
Shockley, William, 158, 164
Shockley Semiconductor Laboratory, 164,
   175, 289–90
Shoemaker-Levy 9 comet, 371
Shostak, Arnold, 119
Shuping, John, **375**
Siemens, 477

Sigelman, Lee, 42
Signetics, 204, 207n177, 227–28
silane, 276–77, 277n100
Silas, C. J. "Pete," 452
Silent Spring (Carson), 383
silicon
    anisotropic etching to micromachine,
        257, 300–302, 301n190, 304,
        305–6, 311–12, 315, 323, 327–28
    anodic bonding (electrostatic bonding,
        Mallory process), 258, 306, 311, 312
    challenges of working with, 159
    conditions encountered in space, resis-
        tance to harsh, 181–82, 197, 201,
        206–13, 243
    crystalline orientation and anisotropic
        etching, 301–2, 327–28
    diaphragm biomedical pressure sensor,
        299–303, 310, 314–15, **326**
    diaphragm production, process for, 257,
        300–302, 315
    diodes, use of in, 159
    epitaxial deposition process and transistor
        manufacturing, 161–62
    gold use in fabrication of devices made
        with, 202–3, 203n162
    growth of individual crystals, develop-
        ment of process for, 159–60
    growth of individual crystals and control
        of purity levels, 161–62
    handbook on reliability of silicon inte-
        grated circuits, 192–93
    hybrid circuits, 200, 200n153, 201
    integrated circuit silicon wafers, size of,
        177–78
    integrated circuits made with, drawbacks
        of and barriers to using in space,
        181–82
    integrated circuit use of, 152, 155, 159,
        173–75
    ion implantation in, 214–15
    isotropic etching to micromachine, 257,
        301, 305–6
    mesa technique and transistor manufac-
        turing, 160–61
    mesa technique and triode fabrication,
        210
    miniature devices made from, 253–54
    piezoresistivity of, 256–57
    planar process and transistor manufactur-
        ing, 161, 161n34
    planar process and triode fabrication, 210

research activities of NASA, 185–86
scribing wafers, 217, 218, 218n214
semiconductor switch use of, 296
separation from wafer, processes to avoid
    damage during, 216–19, 243
shaping crystals to control resistance, 173
space radiation effects on, 206–7,
    207n175, 207n177, 214–15
strain gauge, silicon-based, 257
surface machining and micromachining,
    invention of, 258–59n24, 258–60
transistor use of, 152, 159–60
wafers, separation into individual circuits,
    200
"Silicon as a Structural Material" (Petersen),
    262
silicon carbide, 215–16
silicon dioxide/silicon oxide, 161, 161n34,
    162, 174, 205, 276–77, 297–98,
    305–6, 311–12, 313
silicon nitride, 276–77, 277n100
Silicon Valley
    development of electronics industry in,
        252–53
    Indonesian version of, 316
    NASA role in development of, 252–53,
        323
    Stanford integrated circuit research, sup-
        port from companies in, 294
Simmons, Blair, 138
Simulator Program with Integrated Circuits
    Emphasis (SPICE), 218
Skylab space station, 17, 86, 235, 245, 391,
    500
Skypix, 452
Sky TV, 464
Slovic, Paul, 366–67
Small Business Innovation Research (SBIR)
    grants, 87, 123–24
Smith, Charles S., 256–57
Smith, David B. D., 272–73
Smith, George O., 429
Smithsonian Institution, 435–36, 481
smoke detectors, 146
SNAP (Systems for Nuclear Auxiliary
    Power)
    characteristics and size of, 348
    development of, 337–38
    failure of, injection into higher orbit as
        disposal method after, 349
    high tide of space nuclear power, 349–64
    mercury shield around, 344

numbering of, 338
satellite and spacecraft use of, **350**
SNAP-1, 338
SNAP-2, 338, **353**
SNAP-3, 338, 339–40, **339**
SNAP-3B, 344–46
SNAP-8, 338
SNAP-9A, 347–48
SNAP-10, 338
SNAP-10A reactor, 348–49
SNAP-11, 349
SNAP-19, 361–62
SNAP-27, 356, 357, 362
See also radioisotope thermal generators
     (RTGs)
SNAPSHOT, 348–49
social communication, 79
social problems, using space to solve,
     496–97, 557–60, 571, 574
social-psychological research, 2
social science classrooms, expanding space
     into, 548–51
Societé Européenne des Satellites (SES),
     463–64
Society and Aerospace Technology Technical
     Committee, 541
sociology of space, 536. *See also*
     astrosociology
sodium-potassium alloy, 338
solar power
     comparison to other types of power, **335**
     energy emergencies and advances in, 390
     Harvard students, poll of about space
          goals and development of, 66–67
     ion drive propulsion and solar cells, 141
     photovoltaic solar cells, 332–33
     satellite batteries, charging with, 106, 343
     satellites, use of on, 344, 348
     satellites, use on, 434–35
     solar array retrofit of Galileo, 368
solar-powered satellite (SPS), 394–95
solar system. *See* planets/solar system
solar wind, 238–39
solid-fuel rockets, 82, 144
Sony/Sony Corporation, 167n45, 208
Southern Pacific (Spacenet), 451, 454
South Texas Hospital, Harlingen, 112
South Texas Regional Family Medicine
     Grand Rounds Virtual Video Library,
     113
Southwest Asia, 468

Soviet Union/Russia
     breakup of Soviet Union, 25, 29
     Cosmos 954 incident, 365–66
     environmental satellite coverage by, 472
     joint missions and cooperation with
          United States, public attitudes
          toward, 51–52, **52**, 54, 60, **61**, 69
     liquid-fuel rockets, invention of, 138
     meteorological data sharing and interna-
          tional cooperation, 473
     reconnaissance satellite information
          about, 428
     relationship with United States and SDI,
          21–25
     scientific development and military
          strength of United States and
          Soviets, opinions about, 10, **10**
     SDI and defense against attacks by, 21–25
     spacecraft electronics, 178–79
     space nuclear power development by, **341**
     space race between United States and,
          7–15, **10**, 19–20, 48–49, 50–51,
          59, 495, 514n77
     Sputnik, public knowledge of launch of, 9
     technology competition between United
          States and, 340
     telemedicine project, cooperative role in,
          113
Soyuz 2229 biosatellite, 88–89
Soyuz spacecraft and Apollo-Soyuz mission,
     17, 500
space
     colonies in/settlement of, 60, 65, 70, 394,
          538, 547, 554–56, 571, 573–75
     Lagrangian points, 394, 416
     militarization of, concerns about, 414
     peaceful use of outer space, conference
          on, 397
     societal implications of, 387
Space Age America theme, 496–98, 500,
     501, 503, 533
Space and Society Conference, 540
space charge, 210
space-charge-limited triode, 197, 209–11,
     243
Space Council, 441–42
spacecraft
     contamination and forward contamina-
          tion concerns and sterilization of,
          384, 387–88, 393, 425, 426
     lunar transport system, design of, 509,
          509n62

nuclear power use for, origins of, 337–42, **341**

nuclear reactor power for, Lewis Research Center research on, 140–41

radio signals from, length of time to receive on Earth, 361

societal impact of, 383–84

space debris. *See* debris/space debris

space diplomacy and foreign policy strategy, 525

space exploration

ascendancy and, 64

benefits of from astrosociological perspective, 574–76

benefits of space program, 491, 533–34

benefits of space program, communication of, 142–43, 146–47

colonization goals, 60, 65, 70

contamination concerns and, 387–88, 425, 426

economic benefits of space program, 61–62, **62**

employment opportunities as benefit of space program, 67, 520

environmental policy and space vision, conflict between, 423

excitement of, 64–65

exploration ethos, 498–500, 498n24, 498n28, 533–34

false stereotypes about benefits to society, 143

goals in space, attitudes toward, 59–67, **60**

history of space program, communication of, 146–47

human spaceflight risks, public opinions about, 72

inspirational aspects of, 64, 388–89, 416, 521–22, 533–34

justification of and imperatives of human nature, 498, 498n24

knowledge and understanding from, 535–38

knowledge from, public attitudes toward, 66

large-scale plans under Paine, 390

Moon, Mars, and Beyond space vision, 421–24, 426

new disciplines spawned from, viii–ix

overview effect and, 499

personal impact of, 32–45

privatization of, 573

probe mission priorities over human spaceflight programs, 364, 364n80

public confidence in and support for space program, 18–20, 29–30, 497–98

public interest in science and technology and, 6

public opinions and attitudes toward, 73–74

Reagan space policies, opinions about, 28

safety of space program, poll questions about, 26–27, 67–68

science and technology, public confidence in and support for space program, 20–21, **20**

social problems, using space to solve, 496–97, 557–60, 571, 574

societal impact of, 78–81, 535–43, 572–76

space utilization and, 500–501

spinoff-centric view of value of space, 142

spirit of exploration and human need to explore, 64

support for, decline in, 500

swashbuckling spacefaring role of NASA, 390

transformation from, 535

transistor purchases for, 166

travel into space, questions about personal interest in, 35–39, **39**

unity and, 66

See also budgets and funding

Space Exploration Initiative (SEI), **341**, 342, 502

spacefaring societies, transformation into, 562–65, 573–75

spaceflight/human spaceflight

balance and, 93

benefits of from astrosociological perspective, 574–76

bone loss/calcium loss during long-duration spaceflight missions, 86

contamination concerns and, 387–88, 425, 426

crew model of, 551–54, 571

environmental dimension of, 387–88

gravity, adjustment to following long-duration spaceflight missions, 90

haptics and, 118–19

high-performance requirements in, 129

isolation on crew behavior, 552–53, 571

Moon, Mars, and Beyond space vision,
    421–24, 426
    privatization of, 573
    societal impact of, 572–76
    space utilization and, 500–501
    support for, decline in, 500
    tourism, space, 72–73, 564
Space Foundation, 31
Space Foundation, Hall of Fame, 96
Space Hill (Andover station/Rumford,
    Maine, station), 440, 440n26
Space Imaging, 482
Spacelab, 515, 516–17
Space Nuclear Systems Division, 351
space policy
    ancillary policy, 500–501, 501n34
    birth certificate of, 431
    Reagan space policies, opinions about, 28
space race
    Cold War politics and, 442
    Moon mission priority under Kennedy
        and, 514n77
    public opinions and attitudes toward
        United States–China race, 72
    public opinions and attitudes toward
        United States–Soviet/Russia race,
        7–15, **10**, 19–20, 48–49, 50–51, 59
    science education in United States and
        launch of Sputnik, 10–11, 496, 530
    spacecraft electronics and, 178–79
    Sputnik, satellites, and start of, 7–12, 495
    technological capabilities, opinions about,
        9–10, **10**
    who was winning, opinions about, 11–15
space science, interdisciplinary study of, 387
Space Science and Applications, Office of
    (OSSA)
    creation of, 386
    Earth-monitoring system program, 396,
        397–98, 400
    Earth Sciences and Applications Division,
        397–98
    electronics research funding from budget
        of, 183
    elimination of, 406
    Microelectronics Reliability Program role
        of, 189
    purpose of, 386
SpaceShipOne, 73
Space Shuttle program
    approval for, 390–91
    beginning of, 17

Challenger accident, 21, 24–28, 33, 367,
    369, 399
Columbia accident, 30–32, 34–35, 67,
    419–20
cost-effective, reusable vehicle, value of,
    391
development of successor vehicle, 32
end of, 32, 67
environmental threat from, 392
first Shuttle launch, 19
Galileo deployment from, 368–70
history of space program, communication
    of, 147
launches as commercial, 485
microcircuit line certification for, 235
practical uses of Shuttle flights, opinions
    about, 21
propellants for Shuttles, 82
publicity about Shuttle capabilities, 19
public opinions and attitudes toward, 18,
    19, 21–32
purpose of, 391
return to flight after accidents, 26–32,
    399, 401, 503
satellite launches from, 391
Shuttle-Mir mission, 500
space debris threat to shuttles, 415
space utilization and, 500
STS-8 use of impedance cardiography,
    109
systems management practices in, 508
space sickness, 90
space societies, creation of, 554–56, 571,
    573–75
space stations
    base camp purpose of, 395
    early concepts of, 430
    EOS connection to, 399, 402, 405
    on Moon, opinions about, 68
    opposition to and attempts to cancel
        program for, 405
    plans for, 390
    poll questions about, 16–17
    Reagan administration decision about,
        395–96, 397
    United Nations space station, opinions
        about, 17
    See also Skylab space station
Space Systems Loral (Ford Aerospace),
    451, 453, 454, 455, 462, 465, 472,
    526n122
space utilization, 500–501

spacewalk, 13, 51

Spanish International Network, 463

Sperry Rand Research Center, 276, 277n100

Sperry Semiconductor Division, 199

SPICE (Simulator Program with Integrated Circuits Emphasis), 218

spillover, 79

Spinoff report, vii–viii, viii n2, 77, 85–86, 143, 527

spinoff technologies
  applications compared to spinoffs, 78
  attribution to individuals, viii
  attribution to NASA and accuracy of claims, vii–viii, 141–47, 528n128
  benefits of, 61–62, **62**, 65, 80–81, 523–28
  benefits of to NASA, 101–2
  catalyst function to facilitate innovation, 82–85
  Cold War and technology transfer, 524–25
  communication of history of space program and, 146–47
  definition of spinoff, 77–78, 142–43
  economic impact of, 523–28
  false stereotypes about benefits to society, 143
  high-performance requirements in space-flight and medical care and, 129
  importance of, 143–44
  importance of, debate about, 78
  invention walking on its own legs, 78–79
  misconceptions about, 146
  NASA report on, vii–viii, viii n2, 77, 85–86, 143, 527
  programs responsible for, 77
  public understanding of science and technology through, 80–81
  space program and spinoff-centric view of value of space, 142
  space program technology and general development of industries or scientific communities, 83–85
  spillover and, 79
  studies of potential societal implications of space exploration and, 78–80
  successes compared to spinoffs, 89–90
  technology transfer, 78–85, 101, 141–47
  technology utilization, politics, and Congress, 231–37, 232n256, 233n263, 235–36n272
  technology utilization, technology transfer, and dissemination of information, 229–31, 265–69, 523–28
  tracking back to historical origins, 81–83

spin-scan camera, 471

Spirit, 31, 69, 71, **334**

spirit of exploration, 64

Spirkovska, Lilly, 124

Sponyoe, John, 453–54, 455

SPOT (Système Pour L'Observation de la Terra) satellite, 482–83, 484

Sprague Electric Company, 214–15

Spriggs, James O., 189

springback foam. *See* temper foam

Sprint, 112

Sputnik and Sputnik II
  batteries for electrical power for, 332
  challenge to US by launch of, 495–96
  exhibition of and observation of Soviet spacecraft electronics, 179
  launch of, 7
  public claims of seeing, 8
  public opinions and attitudes toward, 7–12
  science awareness survey and launch of, 8–9
  science education in US and launch of, 10–11, 496, 530
  as turning point in history, 492–93n2

SRI International (Stanford Research Institute), 119, 120, 130

Standard Parts Program, 235

Stanford Center for Radar Astronomy, 323–24

Stanford University
  accelerometer research, 308–14
  antiwar protests at, 290
  Biomedical Instrumentation Laboratory, 268, 324
  biomedical research at, Ames role in, 264–69, 283–85, 284n136, 323–26, **326**
  biomedical research at, NIH support for, 290–93
  biomedical research funding from NASA, 267–70, 268n58, 324
  Biomedical Technology Transfer team, 266
  bone density analyzer, research to develop, 86–87
  brain probe research, 294–99, 298n180

capacitive pressure sensor development at,
258, 258n23
Cardiology Division, School of Medicine,
266–69
Center for Integrated Systems, 294
Center for Materials Research, 287
Computer Systems Laboratory, 294
DLI project role, 117
Electrical Engineering Department, 121,
286, 291–92, 294–96, 298, 299,
324
Electrical Engineering Department and
School of Medicine, collaboration
between, 291–92, 304
electrode development and MEMS
research at, 267
exobiology research at, 324
gas chromatograph research, 279,
280–81, 282, 285, 303–4n200,
303–8, 307nn315–316, 310, 314,
315, 316–20, **326**
Information Systems Laboratory, 294
Integrated Circuit Fabrication Laboratory,
287
Integrated Circuits Laboratory, 252, 258,
285–94
Integrated Circuits Laboratory, funding
for, 287–88, 289, 292–93, 323,
324, **326**
Laboratory for Advanced Materials
Research, 294
McCullough Building, 286–87, 294, 304
MEMS development and collaboration
with, viii, 252
MEMS research at, Ames role in,
283–85, 284n136, 323–26, **326**
MEMS research funding from NASA, 275
NASA funding for research at, 287, 289,
323–26, **326**
Optacon device, role in development of,
117, 119–22
piezoresistive pressure sensor research,
299–303, 314–15, **326**
Solid-State Electronics Laboratory, 286,
289, 292–93n166, 293–94
Space Engineering Building, 324
Stanford Electronics Laboratories, 291
Stanford Electronics Laboratories, fund-
ing for, 288–89
Stanford Electronics Laboratories/
Stanford Electronics Research
Laboratories, 286, 287

Varian Laboratory, 287
Starbird, Alfred Dodd, 465–66
STAR consortium, 477
Starr, Arnold, 292, 295
Star Wars/Strategic Defense Initiative (SDI),
21–25, 37, 56, 66, **341**, 395–96, 414
State, Department of
data from COMSAT, limits on, 446
Earth Observation Summit, 421
environmental policy and space vision,
conflict between, 423
government monopoly over satellite communications, support for, 442
Orbit Act and direct access to Intelsat,
456
satellite communications as foreign policy
activity, 443–44, 525–26
satellite system of systems, 421, 422
space diplomacy and foreign policy
strategy, 525
space race as extension of Cold War, 352
Third World, communication capabilities
for, 448
Transit satellite, resistance to launch of,
344–45, 346
state secrets, 79
steatite, 169, 169n49
Steele, Charles R., 86, 88–89
Steer, 465, 466
Steiner, Reinhold, 446
Stenbeck, Jan, 463
Stencel Aero Engineering, 131
Stephenson, L. M. "Bill," 304
Stereoscan, 221
Stereotoner, 122
Steven F. Udvar-Hazy Center, 481
Stockholm, environmental conference in,
390
strain gauge (pressure sensor), 256–58,
278–79
Strategic Air Command (SAC), 7, 465
Strategic Arms Limitation Talks (SALT I),
480, 485
Strategic Defense Initiative (SDI)/Star Wars,
21–25, 37, 56, 66, **341**, 395–96
Strategic Satellite System, 467
Strickland, Donald, 59–60
Strontium-90, 344
Sullivan, Walter, 397
Sunmate, 132
sun-synchronous orbits, 437
Suomi, Verner, 436–37, 471

supersonic transport (SST), 392
Surveyor, Project, 349, 351, **523**
Surveys of Public Understanding of Science
    and Technology, 42, 44, 56
Sustaining University Program (SUP), 532,
    532n150
Sweden, 113
Swinehart, James, 8–9
Switzerland, 446
Sylvania, 165, 166
Symphonie, 461
synchronous meteorological satellites (SMS),
    472, 489
Synchrony, 101
Syncom/Syncom 2, 440, 450, 462–63, 476,
    489, 490
Système Pour L'Observation de la Terra
    (SPOT) satellite, 482–83, 484
systems architecting, 492, 509, 513–14, 533
Systems for Nuclear Auxiliary Power. *See*
    SNAP (Systems for Nuclear Auxiliary
    Power)
systems management practices, 492,
    495–96, 504–5, 506–10, 512–13,
    515–17, 533

T

Tackle, 465, 466
TACSAT, 448, 466
tactile communication, research on, 120,
    121, 124
tactile reader for the blind. *See* Optacon
    (Optical TAcatile CONverter) device
Tang, 146, 527n128
tantalum pentoxide, 208
Tatum, Edward L., 267–68
Tauzin, Billy, 457
TDRS (Tracking and Data Relay Satellite)/
    TDRSS (Tracking and Data Relay
    Satellite System), 467, 481
Teal, Gordon, 159–60
Teapot (von Neumann committee), 433
Tech Briefs, 229–30, 246, **246**
Technical Data Bank, 191, 193–94
Technion, 96
Technology Utilization/Technology
    Utilization and Policy Planning, Office
    of, 229, 229n253, 233–34, 265–66,
    266n48
Teflon, 62, **62**, 146, 527n128

telecommunication industry
    monopolies and competition in, 450,
        460–61
    open skies/open entry policy, 460–61
Telecommunications Act, 456–57
Telecommunications Research
    Establishment (Royal Radar
    Establishment), 172–73
Teledyne, 176
telehealth, 114, 114n101
Telehealth Center, 113
telemedicine
    advantages and benefits of, 114–15
    challenges and limitations of, 115–16
    concept and definition of, 111–12
    DLI projects and, 116–17
    licensure of medical professionals and,
        115–16
    NASA contributions to programs and
        research on, 112, 113–17
    programs and experiments using, 112–13
    technologies for, 112–17
telemetry
    aerospace medicine and development of,
        85
    astronauts, monitoring condition of with,
        111, 526
    battery charging and, 106
    bidirectional telemetry, 103
    implantable telemetry system for moni-
        toring blood flow, 293
    pacemaker technology and, 101, 103,
        104, 106
Telenor, 457
telephones
    integrated circuit use in, 156
    MEMS use in mobile telephones, 254
    push-button telephones, **62**, 146
    transistors use in telephones, telephone
        lines, and switching equipment,
        158, 159, 166
telephone service and polls, 2
Telesat Canada, 461, 462
telescopes, 82
Telesensory Systems, 122, 123, 288n150
Televisa, 463
television industry
    ABC satellite, launch of, 459–60, 461
    broadcast satellites, 461
    direct-to-home broadcast satellites, 461
    electronic system as basis for, 122

HDTV (high-definition television), 488
television transponders, number needed,
    461
Television Infrared Observation Satellite
    (TIROS)
    cameras on, 437–38
    development and launch of, 385–86,
        437–38, 489
    importance of, 386
    ITOS (Improved TOS)/NOAA satellites,
        438, 470, 471
    sun-synchronous orbits, 437
    support for development of, 436
    TIROS-1, 437, 489
    TIROS-2, 437
    TIROS-8, 437
    TIROS-9, 437
    TIROS-10, 437
    TIROS-N, 470–71, 472
    TIROS Operational System (TOS)/ESSA
        satellites, 438, 470–71, 489
    upgrade of, 386
    weather information from, 386, 428
Telstar, 440, 442, 453, 454, 461, 490,
    525–26
temperature
    digestible temperature transmitter, devel-
        opment of, 273
    temperature extremes in space, compo-
        nent resistance to, 181–82, 197,
        201, 209–11
    temperature telemeter research, 267,
        273–74
temper foam
    importance of, 143
    invention of, 131
    medical applications for, 130, 132, 134
    memory properties of, 130
    NASA contributions to development of,
        131, 132–34
    seat-cushion applications for, 131–32,
        133, 134, 143
    Tempur-Pedic mattress use of, 132–34
Terra, 417, 418, 420, 421
Terry, Stephen C., 282, 303–8, 313,
    316–18, 319, 321
Tesla coil, 134
Texas
    Columbia accident and debris in, 419–20
    telemedicine programs in, 112, 113,
        114–15
Texas A&M University, 139

Texas Department of Health, 112
Texas Instruments
    amplifier microcircuits made by, testing
        of, 205–6
    electron microscope research at, 237
    hearing aid applications for transistors
        manufactured by, 166–67
    integrated circuit invention by, 121,
        173–75
    integrated circuit manufacturing by,
        153–54, 175n69, 176, 242
    integrated-circuit research at, 186
    location of, 175n69
    mesa technique and transistor manufac-
        turing, 160–61
    microwave integrated circuits, 211–13,
        212n195
    Molecular Electronics for Radar
        Applications (MERA) program,
        211–13
    patent waiver requests, 247, **247**, **249**
    price of integrated circuits from, 152–54
    Series 51 computer, 176
    Series 51 Solid Circuits, 153–54
    silicon integrated circuits, handbook on
        reliability of, 192–93
    silicon transistors, development of,
        159–60
    supplier of integrated circuits for
        Minuteman II program, 152, 176,
        226
    transistor manufacturing by, 165, 166–67
    transistor radio production by, 167,
        167n45
Texas Tech University, Center for
    Telemedicine, 113
thermal-conductivity detectors for gas chro-
    matographs, 303–4, 307, 319–20
thermodilution, 108
thick-film technology, 186, 186n107
thin-film technology
    capacitor formation with, 168
    conditions in space and use of, 182, 197,
        243
    germanium films, development of process
        for depositing, 209
    high-temperature thin films, 197
    hybrid circuits, 200, 200n153, 201
    ion implantation and thin-film technique,
        214
    miniaturization of electronics and, 168

radiation-resistant components, development of, 206–11
research activities of NASA, 185–86, 186n107, 200
resistor formation with, 168
space-charge-limited triode, 197, 209–11, 243
transistors, inability to form from, 168, 181
Third World, communication capabilities for, 443–44, 445, 447n43, 448, 487, 525
Thomas, Rosita, 25
Thomas J. Watson Research Center, 260
Thompson Ramo Wooldridge. *See* TRW (Thompson Ramo Wooldridge)
Thor rockets
  propellant for, 82
  Thor-Able rockets, 434
  Thor Able-Star rockets, 346–47, 434, 435
  Thor-Agena launch vehicles, 433, 437, 480
  Thor-Burner launch vehicles, 437
  Thor-Delta rockets, 432, 434–35
thought experiment (*Gedankenexperiment*), 244
3M, Unitek division, 128
Three Mile Island incident, 366–67
Tikniks, 100
Tilford, Shelby, 409
Time polls, 24
Tini Alloy Company, 130
Tinkertoy project, 169–70, 169n49
Tipton, Steve, 21
TIROS Operational System (TOS)/ESSA satellites, 438, 470–71, 489
Titan, 372, 378
Titan rockets
  propellant for, 82
  satellite launches with, 450–51, 465, 466, 471, 480, 481
  Titan II ICBMs, 82
  Titan III program, 504
  Titan IV launch vehicle, 372
Tokyo Tsushin Kogyo (Totsuko), 167n45
Tomayko, James E., viii
Tombs, Nigel C., 274–77, 275–76nn96–97, 277n100, 278, 284, 285, **326**
Tomonaga, Sin-Itiro, 254
Tonga Trench, 358
TOPEX/Poseidon, 412–13
Torrey Canyon, 474

TOS (TIROS Operational System)/ESSA satellites, 438, 470–71, 489
Totsuko (Tokyo Tsushin Kogyo), 167n45
Tough, Allen, 536, 567
tourism, space, 72–73, 564
tourism to NASA Centers, 528
tracheotomy tubes, transmitters to monitor, 273
Tracking and Data Relay Satellite (TDRS)/ Tracking and Data Relay Satellite System (TDRSS), 467, 481
trade secrets, 79
Transensory Devices, Inc., 261
Transformation Satellite Communications System (TSAT), 468
transistor radios, 167, 167n45
transistors
  alloy-junction transistors, 159
  commercial applications for, 166–67, 167n45, 177
  diffusion process for manufacture of, 160
  epitaxial deposition process for manufacture of, 161–62
  expansion of transistor industry and sale of transistors, 164–67
  failure analysis research on, 244
  germanium use in, 158, 159, 198
  gold use in fabrication of, 202–3
  integrated circuit technology compared to, 156
  integrated circuit use of, 157, 168, 172–75, 177, 181
  invention and manufacturing processes for, 157–64
  ion implantation process, 214
  junction transistors, 159
  limitations in manufacturing and reliability of, 167–68
  mesa technique for manufacture of, 160–61, 210
  metal-oxide semiconductor field-effect transistors (MOSFETs), 191, 239–42
  military and defense consumers of, 165–66
  military concepts for miniaturization of electronics, 169–72, 170n52
  NASA demand for, 166
  patents for, 158, 159n29
  photolithographic procedures for manufacture of, 160, 161, 162–64

planar process for manufacture of, 161,
	161n34
point-contact (Type A) transistors,
	158–59
price of, 164, 165, 242–43
reliability of, 160
reliability requirements for NASA and
	military, 166
resonant gate field effect transistor/reso-
	nant gate transistor, 258–59
sealed container, locating leaks in,
	217–18, 220
silicon use in, 152, 159–60
thick-film transistors, 186, 186n107
transistor-transistor logic (T2L) circuits,
	207n175
Transit Improvement Program (TIP), 435
Transit Research and Attitude Control
	(TRAC) satellite, 434
Transitron Electronic Corporation, 176,
	176n70
Transit satellites
	accuracy of, 435
	building and assembly of, 435
	characteristics and shape of, 434
	commercial ships use of, 474
	development of, 104–5, 342–43, 489
	launch of, 434, 435
	limitations of, 473
	Oscars, 435
	precision of navigation with, 434, 473
	reliability of, 473
	responsibility for, 434
	RTG use on, 342–47, **350**, 434
	solar power use on, 434–35
	Transit 1A, 343
	Transit 1B, 343
	Transit 4A, **334**, 344–46, **350**, 434
	Transit 4B, **334**, 344–46, **350**, 434
	Transit 5A series, 434–35
	Transit 5BN-1, **334**, 346–47, **350**
	Transit 5BN-2, **334**, 346–47, **350**
	Transit 5BN-3, 347, **350**, 351
	Transit 5B series, 434–35
	Transit 5C1, 435
	Transit 5E series, 434–35
	Transit 5 series, 434–35
	Transit Triad-01-1X, **334**, **350**
Transportation Systems Center, Department
	of Transportation, 216n209
treaty organization, 459
Trent, Robert L., 199

triad decision-making structure, 508
Triana, 415–16, 418
Trilogy, 101
triodes
	Mead triode, 207–8
	space-charge-limited triode, 197, 209–11,
		243
	thin-film triode, advantages of, 210–11
tritium and tritium tracer, 205–6, 205n169,
	220
Triton, 363
Tropical Rainfall Measuring Mission
	(TRMM), 412, 424
Trujillo, Tony, 456
Truly, Richard, 401, 402, 405
TRW (Thompson Ramo Wooldridge)
	Aerosat program, 477
	communications satellite development
		and manufacturing by, 445, 448,
		466, 526n122
	FLTSATCOM (Fleet Satellite
		Communications System), 467
	Goldin role at, 405–6
	instrument development to search for life
		on Mars, 318, 319, 320
	patent waiver requests, 247, **247**, **249**
TSAT (Transformation Satellite
	Communications System), 468
Tsiolkovsky, Konstantin Eduardovitch, 138,
	429–30
tsunami, 423, 487
Tung-Sol, 165
tunnel emission and tunneling effect, 208
Tyco Laboratories, **247**, **248**

U

U-2 program, 433, 485–86
UARS (Upper Atmosphere Research
	Satellite), 398, 405, 413, 422
UHF communication satellites, 465,
	466–67, 478
UHF follow-on program (UFO), 467
UHF silicon transistors, 212
Ulysses, **334**, 340, **350**, 367–68
United Kingdom. *See* Great Britain/United
	Kingdom
United Nations Conference on the Peaceful
	Uses of Outer Space, 397
United Nations Environmental Program
	(UNEP), 390, 400

United Nations Intergovernmental Panel on
Climate Change (IPCC), 400, 403–4,
411–12, 418
United Nations International Atomic
Energy Agency, 302, 303
United States (U.S.)
cooperative space ventures with Europe,
515–17, 516n84, 533
defense capabilities and nuclear weapons
priorities under Reagan administration, 395–96
frontier narrative of, 494–95, 495n12
government, confidence and trust in
during Apollo program, 497–98
joint missions and cooperation with
Russia, public attitudes toward,
51–52, **52**, 54, 60, **61**, 69
Kyoto Protocol withdrawal by, 418, 421
liquid-fuel rockets, invention of, 138
relationship with Soviets and SDI, 21–25
science education in and launch of
Sputnik, 10–11, 496, 530
scientific development and military
strength of United States and
Soviets, opinions about, 10, **10**
space race between Soviets and, 7–15,
10, 19–20, 48–49, 50–51, 59, 495,
514n77
Sputnik as challenge to, 495–96
superpower status of, 29
technology competition between Soviets
and, 340
United States Climate Change Research
Initiative, 421
United States Global Change Research
Program (USGCRP), 400–402, 409,
410–11, 418
United States Information Agency (USIA)
poll, 11–12
Unitek division, 3M, 128
unity, peace, and space exploration, 66
universe
NASA contributions to understanding
of, ix
other cultures throughout, understanding
potential for, ix
University College of North Wales, 236–37
university-industry-government complex,
228
University of Alaska, 112, 113
University of California, Berkeley
computer-aided design of circuits, 218

Computer-assisted Survey Methods
(CSM) Program, 7
MEMS research at, 262
micromotor creation at, 255–56n13
University of California, Santa Barbara, 117
University of Connecticut, Roper Center, 7
University of Connecticut poll, 31
University of Denver, Denver Research
Institute, 233–37, 233n263
University of Iowa, School of Dentistry,
Department of Orthodontics, 128
University of Leipzig, 157
University of Michigan, 93, 298
University of Minnesota, 108–9, 111
University of North Carolina–Chapel Hill,
Odum Institute for Research in Social
Science, 7
University of North Carolina Medical
School, 267
University of Texas Health Sciences Center
at San Antonio (UTHSCSA), 112, 113
University of Texas Medical Branch, 113
University of Virginia, Research Laboratories
for the Engineering Sciences, 208–9
University of Washington, 112
University of Washington Medical Center,
139–40
University of Wisconsin, 99, 262
Upper Atmosphere Research Satellite
(UARS), 398, 405, 413, 422
uranium 235, 365–66
Uranus
knowledge/information about and
Voyager probes, 363
RTG use for missions to explore, **334**
Voyager 2 discoveries about, research on
communication to public about,
143
Voyager 2 discoveries about and opinions
about space program, 25–26, 27
Voyager probes missions to, 363
USA Today polls
budget and funding questions, 5
Challenger accident and questions about
manned space program, 26
emotional impact of major spaceflight
events, questions about, 34–35
goals in space, attitudes toward, 60
Mars, questions about exploration of, 69
travel into space, questions about personal
interest in, 36, 38
U.S. News and World Report, 26

uterus, measurement of motion of, 310
"Utility of a Satellite Vehicle for
    Reconnaissance" (Lipp, Greenfield, and
    Wehner), 431

V

V-1 buzz bomb missiles, 3
V-2 rockets
    development of, 7
    poll questions about, 1, 3
    propellants for, 82
    scientific payloads launched with, 432
Vacca, Gene A., 264
vacuum tubes, 156, 165, 167, 169
Van Allen, James, 396
Van Allen radiation belts, 209, 238–39, 396
Vandenberg Air Force Base, 348–49, 434,
    437, 470, 471
van der Waals attraction, 255, 255n11
Vanguard launch vehicle, 432
Vanguard satellite, 7, 167, 433, 436
Varghese, Mathew, 97
Varian Laboratory, 287
Väsamed, 111
Velcro, 146, 527n128
Venus
    greenhouse effect on, 384, 387, 396–97
    hostility of environment on, 387, 397
    knowledge/information about and
        Mariner missions, 387
    Venus–Earth–Earth Gravity Assist for
        Galileo mission, 368, 369–70, 371
    Venus–Venus–Earth–Jupiter Gravity
        Assist for Cassini mission, 372, 414
Verne, Jules, 429
Vernikos Danellis, Joan, 279–80
vertical diffusion, 79–80
VHF transponders/satellites, 476–77
ViaSat, 457
Vietnam War
    anti-shock garment use during, 96
    financing of, public opinion about, 15,
        54
    military use of technology during, 501
    public disillusionment during, 20–21
    public opinion about handling of, 50
Viking probe program
    budget and funding for, 364n80
    instruments and sensors for use in
        missions, 264, 280–81, 317–20,
        318n260

Mars missions, 264, 280–81, 304,
    317–20, 318n260, 393, 395, 396
    Oyama role in missions, 304
    RTG use during, **334**, 349, **350**
    Viking 1, **334**, **350**
    Viking 2, **334**, **350**
Viking sounding rocket, 432
Villanova University, 218–19
Villard, Oswald G., Jr., 324
Virginia Tech, 124
Visible Infrared Spin Scan Radiometer
    (VISSR), 472
Vision for Space Exploration (VSE), 492,
    503, 509–10
Visually disabled people. *See* blind and visu-
    ally disabled people
VMH Visual Communications, 95
von Braun, Wernher, 7, 430, 432, 514,
    514n77
von Neumann committee (Teapot), 433
Vortex system, 481
Voskhod 2 spacecraft, 13
Voyager 1 probe
    budget and funding for, 364n80
    mission of and information from, 359,
        362–64
    phonograph records of information about
        life and culture on Earth for extra-
        terrestrial life, 363–64
    RTG use on, **334**, 349, **350**, 362–63
Voyager 2 probe
    budget and funding for, 364n80
    discoveries by and opinions about
        manned space program, 25–26, 27,
        28–29
    discoveries made by, research on commu-
        nication to public about, 143
    mission of and information from, 359,
        362–64
    phonograph records of information about
        life and culture on Earth for extra-
        terrestrial life, 363–64
    RTG use on, **334**, 349, **350**, 362–63
    termination shock encounters of, 363
Voyager Biological Laboratory, 264
VTEL Corporation, 112

W

WAC Corporal rocket, 7
Wade, Serena, 9

Wake Forest College, Bowman-Gray School of Medicine Department of Biomedical Engineering, 267

"Waldo" (Heinlein), 118

Walker, Robert, 410–11

Wall Street Journal, 27

Wang, Frederick E., 126–28

Wanlass, Frank, 241–42

Warschauer, Douglas M., 198–99

washers and dryers, balancing, 251

Watergate scandal, 21

water pollution and funding for anti-water pollution programs, poll question about, 16, 54

Watson, Robert, 408–9

Watt, Chauncey W., 193

Wayne State University Physicians Group, 139

wealth-building processes, 522, 533

Weapons System 117L (WS-117L), 433, 437

weather
  National Weather Service, authorization of, 436
  recording of weather observations, 435–36

Weather Bureau, 386, 390, 436, 469

weather/meteorology satellites, 435
  as applications satellites, 427–28
  cameras on, 471
  civilian safety and, 486
  development and launch of, 385–86, 428
  early concepts and programs, 431, 435–38, 489, 490
  environmental role of, 384
  future of, 488
  geostationary orbit satellites, 469, 472, 482
  geosynchronous satellites, 428, 471, 488
  government satellites, 469–73
  importance of information from, 386, 437
  international cooperation and data sharing, 472–73
  practical and research objectives of, 386
  profits from, 483
  public attitudes toward space exploration and benefits of, 65–66
  remote sensing technology use by, 428–29
  RTG use on, **350**

societal impact of, 428, 483, 484, 486
space technology and development of, 78
synchronous meteorological satellites (SMS), 472, 489

weather prediction methods
  satellites and, 365–86
  space technology and development of, **62**, 146

weather telemetry transmitter, 212

Webb, James E.
  Apollo program and stimulation of economy, education, and technology, 496–97
  balanced space program, support for, 514n77
  centralized processes, decentralized processes, and management style of, 507–8, 509–10
  communications satellites, support for government role in, 441
  electronics, importance of to space exploration, 178
  management difficulties faced by, 500
  Paine as successor of, 390
  reliability office changes under, 187
  reorganization of NASA under, 386
  research activities under, 228
  Space Age America theme under, 496
  space science study, support for, 387
  Sustaining University Program under, 532
  triad decision-making structure established by, 508

Wehner, R. S., 431

Weiffenbach, George C., 342–43, 434

weightlessness/zero gravity
  balance sense and, 90
  biomedical instrument development to determine effects of, 263
  bone loss in, 86, 88–89
  free-fall sensation and, 90

Weiler, Ed, 419

Weissman, Robert H., 298n180

Welch, Leo, 444

welfare and relief, poll question about funding for, 16, 46–47, **46**, 53–54, 55

Wells, H. G., 429

Wells, Oliver C., 222, 237

Welsh, Edward, 441

Westar, 454, 463

Western Electric, 158–59, 159n29, 164, 165, 166, 176

Western Union
  communications satellite launch by, 461,
     462
  Westar, 454, 463
Westinghouse/Westinghouse Electric/
     Westinghouse Research Laboratories
  accelerometer development by, 309
  amplifier microcircuits made by, testing
     of, 205–6
  electron beam and diffusion process,
     214n202
  electron microscope research at, 222, 237
  integrated circuit manufacturing by, 176
  integrated circuit testing by, 222
  Molecular Electronics Division, 171, 191
  Molecular Electronics Program, 171
  MOS circuits manufactured by, testing
     for radiation resistance of, 207n177
  patent waiver requests, 247, **247**, **248**,
     249
  supplier of integrated circuits for
     Minuteman II program, 152, 226
  surface machining, invention of by,
     258–59, 258–59n24
  transistor manufacturing by, 165
Wexler, Harry, 430, 436–37
Wheatstone, Charles, 299
White, Ed, 15
White, James A., 266
Whitehead, Clay T., 450, 463, 477
Whitten, Pamela, 115
Whole Earth Catalog, 389, 394
Wickstrom, Robert A., 258–59, 258–59n24
Wide Area Augmentation System (WAAS),
     475
Wideband Gapfiller Satellites (WGS), 467
Wiesner, Jerome, 441
Wilhelms, Don, 357–58
Williams, Donald D., 439, 462–63
wind power, 390
Winget, Charles M., 271–73, 280
Wise, Kensall D. "Ken"
  anisotropic etching, role in development
     of variation of, 300–302, 315, 323
  biomedical pressure sensor research, 299,
     300–302, 310, 315

  brain probe research, 294–99,
     298–99n183
  gas chromatograph research, 280–81,
     317–18
  MEMS research at Stanford, start of, 274,
     277, 284
Witsoe, David A., 109
Wolfe, Tom, 29–30
Workshop on Solid-State Sensors and
     Actuators, 263
World Administrative Radio Conference for
     Space Telecommunications (WARC
     71), 477, 478
WorldCom, 453
World Meteorological Organization
     (WMO), 400, 477
WorldView/DigitalGlobe/EarthWatch, 482
Wright, G. T., 209
Wright-Patterson Air Force Base
  Aeronautical Research Laboratory, 198
  Air Force Systems Command, 211–12

X

X-33 aircraft, 32
X-band frequencies, 465–66, 467, 469
Xerox, 294
x-ray screening procedures and x-ray specto-
     graph, 220

Y

Yankelovich polls, 5, 25
Yarsky, Christine, 93
Yost, Charles A., 131–32, 133, 134, 143
Young, Laurence R., 92
Young, Richard, 264

Z

Zenith, 167
Zeppelin, Ferdinand von, 157
zinc, 215, 216
Zoex, 94, 95
Zogby polls, 72–73
Zoll, Paul M., 98–99
Zuleeg, Rainer, 210–11